Microbial Diversity and Bioprospecting

Microbial Diversity and Bioprospecting

EDITED BY Alan T. Bull

ASM PRESS   WASHINGTON, D.C.

Copyright © 2004   ASM Press
                        American Society for Microbiology
                        1752 N Street, N.W.
                        Washington, DC 20036-2904

Library of Congress Cataloging-in-Publication Data

Microbial diversity and bioprospecting / edited by Alan T. Bull.
    p.; cm.
Includes bibliographical references and index.
    ISBN 1-55581-267-8 (hardcover)
    1. Microbial diversity. 2. Microbial ecology. 3. Biotechnology. 4. Microbial genetics.
    [DNLM: 1. Ecosystem. 2. Microbiology. 3. Biotechnology. 4. Variation (Genetics) QW 4 M62255 2004] I.
Bull, Alan T.
    QR73.M53 2004
    660.6′2—dc21   2003008182

All Rights Reserved
Printed in the United States of America

10 9 8 7 6 5 4 3 2 1

Address editorial correspondence to: ASM Press, 1752 N St., N.W., Washington, DC 20036-2904, U.S.A.

Send orders to: ASM Press, P.O. Box 605, Herndon, VA 20172, U.S.A.
Phone: 800-546-2416; 703-661-1593
Fax: 703-661-1501
Email: books@asmusa.org
Online: www.asmpress.org

Cover figure: The cover collage symbolizes the principal themes of diversity and prospecting that recur throughout the book and incorporates images kindly provided by Pedro Brandao, Phil Cash, Daphne Stoner, and Stephen Wrigley.

CONTENTS

CONTRIBUTORS

F. Abe
The DEEPSTAR Group, Japan Marine Science and Technology Center (JAMSTEC), 2-15 Natsushima-cho, Yokosuka 237-0061, Japan

G. Antranikian
Technical University of Hamburg-Harburg, Biotechnology/Technical Microbiology, Kasernenstrasse 12, D-21071 Hamburg, Germany

Linda L. Blackall
Advanced Wastewater Management Centre, Department of Microbiology and Parasitology, The University of Queensland, Brisbane, 4072 Queensland, Australia

Torben Vedel Borchert
Novozymes A/S, Krogshøjvej 36, 2880 Bagsværd, Denmark

James Borneman
Department of Plant Pathology, Boyce Hall 3489, University of California, Riverside, Riverside, CA 92521

John A. Breznak
Department of Microbiology and Molecular Genetics, 6190 Biomedical and Physical Sciences, Michigan State University, East Lansing, MI 48824-4320

Alan T. Bull
Research School of Biosciences, University of Kent, Canterbury, Kent CT2 7NJ, United Kingdom

D. K. Button
Institute of Marine Science, University of Alaska, Fairbanks, AK 99775

Phillip Cash
Department of Medical Microbiology, University of Aberdeen, Foresterhill, Aberdeen AB25 2ZD, United Kingdom

Brent C. Christner
Department of Land Resources and Environmental Sciences, 304 Leon Johnson Hall, Montana State University, Bozeman, MT 59717

Arnold L. Demain
Charles A. Dana Research Institute (R.I.S.E.), HS-330 Drew University, Madison, NJ 07940

Holly Doremus
University of California at Davis, School of Law, 400 Mrak Hall Drive, Davis, CA 95616

Genoveva F. Esteban
Center for Ecology and Hydrology, Dorset, Winfrith Technology Centre, Winfrith Newburgh, Dorchester, Dorset DT2 8ZD, United Kingdom

Guy Fauque
UR 101 Extrêmophiles, Institut de Recherche pour le Développement IFR-BAIM, Universités de Provence et de la Méditerranée, ESIL, Case 925, 163 avenue de Luminy, F-13288 Marseille Cedex 09, France

Hans-Peter Fiedler
Mikrobiologisches Institut, Universität Tübingen, Auf der Morgenstelle 28, D-72076 Tübingen, Germany

Bland J. Finlay
Center for Ecology and Hydrology, Dorset, Winfrith Technology Centre, Winfrith Newburgh, Dorchester, Dorset DT2 8ZD, United Kingdom

Christopher L. Finan
Institute of Biological Sciences, University of Wales,
Aberystwyth, Aberystwyth SY23 3DD,
United Kingdom

Stephen S. Fong
Department of Bioengineering, University of
California, San Diego, 9500 Gilman Drive,
La Jolla, CA 920093-0419

John C. Fry
Cardiff School of Biosciences, Main Building,
Cardiff University, Park Place, Cardiff CF10 3TL,
United Kingdom

Rodrigo Gámez
Instituto Nacional de Biodiversidad, 3100
Santo Domingo, Heredia, Costa Rica

Royston Goodacre
Institute of Biological Sciences, University of Wales,
Aberystwyth, Aberystwyth SY23 3DD,
United Kingdom

Michael Goodfellow
School of Biology, University of Newcastle,
Claremont Road, Newcastle upon Tyne NE1 7RU,
United Kingdom

Lorena Guevara
National Institute of Biodiversity (INBio), 3100
Santo Domingo, Heredia, Costa Rica

Jo Handelsman
Department of Plant Pathology, University of
Wisconsin, 1630 Linden Drive, Madison, WI 53706

B. P. Hedlund
Department of Biological Sciences, University of
Nevada, Las Vegas, 4505 Maryland Parkway,
Las Vegas, NV 89154-4004

Russell T. Hill
Center of Marine Biotechnology, University of
Maryland Biotechnology Institute, Columbus Center
Suite 236, 701 East Pratt Street, Baltimore, MD
21202-4031

Carola Holmström
Department of Microbiology and Immunology,
School of Biotechnology and Biomolecular Sciences,
University of New South Wales, Sydney,
NSW 2052, Australia

Koki Horikoshi
The DEEPSTAR Group, Japan Marine Science and
Technology Center (JAMSTEC), Yokosuka
237-0061, Japan

Peter Jeffries
Department of Biosciences, University of Kent,
Canterbury, Kent CT2 7NJ, United Kingdom

Peter Kämpfer
Institut für Angewandte Mikrobiologie, Justus-
Liebig Universität Giessen, Heinrich-Buff-Ring
26-32, D-35392 Giessen, Germany

Arseny S. Kaprelyants
Bakh Institute of Biochemistry, Leninskii Prospekt
33, 117071 Moscow, Russia

Chiaki Kato
Department of Marine Ecosystems Research, Japan
Marine Science and Technology Center (JAMSTEC),
Yokosuka 237-0061, Japan

Flora N. Katz
Fogarty International Center, National Institutes of
Health, 31 Center Drive, Bethesda, MD 20892-2220

Laura A. Katz
Department of Biological Sciences, Smith College,
Northampton, MA 01063

Douglas B. Kell
Department of Chemistry, Faraday Building,
Sackville Street, UMIST, P.O. Box 88, Manchester
M60 1QD, United Kingdom

Staffan Kjelleberg
Department of Microbiology and Immunology,
School of Biotechnology and Biomolecular Sciences,
University of New South Wales, Sydney,
NSW 2052, Australia

Randy Lee
Ecological and Cultural Resources, Idaho National
Engineering and Environmental Laboratory,
Idaho Falls, ID 83415-2213

James M. Lynch
Forest Research, Alice Holt Lodge, Farnham, Surrey
GU10 4LH, United Kingdom

Galya V. Mukamolova
Institute of Biological Sciences, University of Wales, Aberystwyth, Aberystwyth SY23 3DD, United Kingdom

Karen E. Nelson
Institute of Genome Research TIGR, 9712 Medical Center Drive, Rockville, MD 20850

Bernard Ollivier
Directeur UR101 Extrêmophiles, Institut de Recherche pour le Développement, IFR-BAIM, Universités de Provence et de la Méditerranée, ESIL, Case 925, 163 avenue de Luminy, F-13288 Marseille Cedex 09, France

Bernhard Ø. Palsson
Department of Bioengineering, University of California, San Diego, 950 Gilman Drive, La Jolla, CA 92093-0419

R. J. Parkes
Department of Earth Sciences, University of Cardiff, Main Building, P.O. Box 914, Cardiff CF10 3YE, United Kingdom

David W. Pearce
Department of Economics, University College London, Gower Street, London WC1E 6BT, United Kingdom

Fergus G. Priest
School of Life Sciences, John Muir Building, Heriot-Watt University, Riccarton, Edinburgh EH14 4AS, United Kingdom

John C. Priscu
Department of Land Resources and Environmental Sciences, 309 Leon Johnson Hall, Montana State University, Bozeman, MT 59717

Jennifer L. Reed
Department of Bioengineering, University of California, San Diego, 9500 Gilman Drive, La Jolla, CA 92093-0419

Ron Rope
Ecological and Cultural Resources, Idaho National Engineering and Environmental Laboratory, Idaho Falls, ID 83415-2213

Joshua P. Rosenthal
Fogarty International Center, National Institutes of Health, 31 Center Drive, Bethesda, MD 20892-2220

Ramon Rosselló-Mora
Group d'Oceanografia Interdisciplinar, Institut Mediterani d'Etudis Avancats (CSIC-UIB), Miquel Marques 21, E-07190 Esporles (Illes Balears), Spain

Thomas Schäfer
Novozymes A/S, Krogshøjvej 36, 2880 Bagsvaerd, Denmark

Kornelia Smalla
Federal Biological Research Centre for Agriculture and Forestry, Institute for Plant Virology, Microbiology and Biosafety, Messeweg 11-12, D-38104 Branschweig, Germany

James E. M. Stach
Research School of Biosciences, University of Kent, Canterbury, Kent CT2 7NJ, United Kingdom

James T. Staley
Department of Microbiology, University of Washington, Seattle, WA 98195

Peter Steinberg
Centre for Marine Biofouling and Bio-Innovation, School of Biological Science, University of New South Wales, Sydney, NSW 2052, Australia

Daphne L. Stoner
Biotechnology Department, Idaho National Engineering and Environmental Laboratory, P.O. Box 1625, Idaho Falls, ID 83415-2203

William R. Strohl
Department of Biologics Research, Merck Research Laboratories, P.O. Box 2000, Mail drop RY80Y-215, Rahway, N.J. 07065

Hanne Svarstad
NINA, P.O. Box 736 Sentrum, 0105 Oslo, Norway

Giselle Tamayo
National Institute of Biodiversity (INBio), 3100 Santo Domingo, Heredia, Costa Rica

Kerry ten Kate
Insight Investment, 33 Old Broad Street, London
EC2N 1HZ, United Kingdom

Constantinos E. Vorgias
National and Kapodistrian University of Athens,
Faculty of Biology, Department of Biochemistry-
Molecular Biology, Panepistimiopolis-Zographou,
15701 Athens, Greece

Alan C. Ward
School of Biology, University of Newcastle,
Claremont Road, Newcastle upon Tyne NE1 7RU,
United Kingdom

Pete Wellsbury
Department of Earth Sciences, University of Bristol,
Bristol BS8 1RJ, United Kingdom

Luke White
Programmatic Software Development, Idaho
National Engineering and Environmental
Laboratory, Idaho Falls, ID 83415-3419

Stephen K. Wrigley
Cubist Pharmaceuticals (UK) Ltd., 545 Ipswich
Road, Slough, Berkshire SL1 4EQ,
United Kingdom

Christine Yeates
Advanced Wastewater Management Centre,
The University of Queensland, St. Lucia,
Queensland 4072, Australia

Michael Young
Institute of Biological Sciences, University of Wales,
Aberystwyth, Aberystwyth SY23 3DD,
United Kingdom

Hongjuan Zhao
Institute of Biological Science, University of Wales,
Aberystwyth, Aberystwyth SY23 3DD,
United Kingdom

FOREWORD

This is an exciting time for those involved in bioprospecting, especially in the pharmaceutical area; indeed this field is at a crossroads in its development. Whereas ingenuity, innovation, and product introduction have been deaccelerated by the mega mergers of the pharmaceutical industry, the new opportunities now available for the development of new drugs are staggering. Indeed, the high cost of these novel opportunities to "big pharma" has in part contributed to the downgrading of natural product research and development. The loss of interest in this most important aspect of new drug innovation is of course only temporary because its replacement by combinatorial chemistry, genomics, and high-throughput screening has not been productive. This opinion is not a conclusion arrived at by biologists; rather, it has been stated publicly by prominent medicinal chemists. So what is to be done? The answer is a synergistic combination of the traditional and the new, i.e., combining intelligent bioprospecting of nature's diversity with the novel techniques of genomics, proteomics, metabolomics, metagenomics, combinatorial chemistry, combinatorial biosynthesis, high-throughput screening, and bioinformatics. We are fortunate that, during this period of downgrading of natural products by the major drug companies, a number of smaller biotechnology companies have picked up the slack by entering the arena of natural product discovery. They are using some of the newer techniques in their efforts as well as expanding the search to relatively ignored environments such as the ocean.

When one speaks about natural products, included are biopharmaceutical (i.e., erythropoietin) primary metabolites such as amino acids and vitamins, secondary metabolites (penicillin G), products discovered in nature but made by chemical synthesis (thienamycin), and chemical derivatives of natural products (clarithromycin). Successful applications have included antibiotics, antitumor agents, enzyme inhibitors, antiparasitic agents, bioherbicides, algicides, and bioinsecticides. Of great importance for pharmaceutical discovery are new targets such as receptor-ligand binding, reporter genes, adhesion, proteosome action, signal transduction, and cell-to-cell communication. For antitumor agents, recent targets have included protein kinase C, farnesyl protein transferase, P53-related targets, proteosomes, and telomerase.

Many of the new developments in industrial microbiology derive from the birth of molecular biology in the 1950s and of recombinant biotechnology in the 1970s. Of special interest is the area of industrial enzymes that has made major strides because of cloning and the complementary techniques of protein engineering and directed molecular evolution. These enzymes have great use in food processing, detergents, cleaning of contact lenses, biosensors, and molecular biology (DNA polymerase for the polymerase chain reaction, and restriction enzymes). Enzymes and cellular bioconversions have been applied in the preparation of chiral drugs that are currently desired by industry and health authorities. Many industrial enzymes are derived from thermophilic, alkaliphilic, or psychrophilic microbes ("extremophiles") from areas of high biodiversity interest. Environments of interest for bioprospecting are soil, the marine environment including the deep biosphere, the icy biosphere, marine sponges, and insects. Plants harbor many microbial symbionts that are a good source of alkaloids and other drugs, biocontrol agents, plant growth stimulators, agents protecting plants from abiotic stress, and for ecosystem restoration. Other current or potential benefits of exploring microbial diversity include (i) the economical and environmentally important replacement of chemical processes by biological procedures in the manufacture of riboflavin, acrylamide, 7-aminocephalosporanic acid, and 7-aminodeacetoxycephalosporanic acid; (ii) control of agricultural pollution by the use of feed enzymes such as phytase; (iii) replacement of petroleum, plastics, and other materials by bioprocessing of renewable raw materials; (iv) bioremediation and biodegradation of polluting materials; (v) discovery of new plant growth promoting microbes; and (vi) novel antifoulants and antibiofilm agents. In relation to the antifoulants, it was surprising for me to read in the present book that biofouling, i.e., the colonization of surfaces in aqueous environments by living organisms, costs the shipping industry over five billion dollars per year!

There is no doubt that biodiversity is being lost throughout the world. This is unfortunate because we need biodiversity to provide novel microbes and novel products. We are told that only 0.5 to 1% of living bacterial species and 5% of living fungal species have been cultured. Of great interest are new methods that allow isolation of previously uncultured microbes, e.g., low-nutrient media, long incubation times, dilution to extinction, ecosystem mimicry, syntrophy, and cell-to-cell communication. Of use in these efforts have been micromanipulation, optical tweezers, atomic force microscopy, and density gradient centrifugation. The newly cultured species may not do very well in industrial fermentors, but cloning of their production and regulatory genes into industrial bacteria and fungi would allow scale-up and industrial production. The metagenomics approach is a complementary development that allows expression of environmental DNA and mRNA. This very exciting area has already yielded known antibiotic products such as violacein, indirubin, and fatty dienic alcohols and new antibiotics such as acyltyrosines, terragine, and turbomycin. In addition to these secondary metabolites, new enzymes, antiporter and antibiotic resistance determinants, have been isolated from environmental nucleic acids.

It is clear that the large pharmaceutical companies will soon adapt their genomic, combinatorial chemistry, and high-throughput screening efforts to new natural product scaffolds. Such novel structures are sorely needed and will be provided by the proper utilization of biological diversity. It is extremely fortunate for the field that this book has been assembled at this time with contributions from the desks of the world's best minds of microbial diversity and bioprospecting. Of course, microbes are not the only source of remarkable drugs, but inclusion of plants and other life forms would have required much greater time and effort and would have delayed publication of this useful compilation, which is needed Now!

Arnold L. Demain
Madison, New Jersey

PREFACE

ORIGINS

This book is born out of a lifetime's fascination for microorganisms, a fascination that has been nurtured by many people and many experiences and, not the least, by undergraduate courses in systematic biology, an increasingly rare feature of degree curricula. Serendipity, I freely confess, has played a major role in sustaining this fervor for microbiology, and a significant turning point came in the 1970s when I joined the then Panel on Applied Microbiology of the United Nations Environment Programme and the United Nations Educational, Scientific, and Cultural Organization. This U.N. involvement enabled me to work with a group of extraordinarily committed humanitarian and knowledgeable microbiologists from the world over. The effect of working with Martin Alexander, Goran Heden, Roger Porter, David Pramer, Maurits la Riviere, Jacques Senez, and H. Taguchi, and others too numerous to name, was electrifying and permanent in so many ways but especially in revealing how microbiology can be developed for the common good. The panel experience had a variety of consequences for me, among them an entrée into international science and opportunities to see first hand a wide range of microbial technology being exploited in developing countries for both traditional and innovative processes. Years later this amalgam of experience led to two particular opportunities to bring microbial diversity and bioprospecting together in quite dramatic fashion—one in Indonesia that was relatively low tech and the other in Japan that was decidedly high tech.

Alfred Russel Wallace, "one of the neglected giants of the history of science and ideas," has long been one of my heroes, for, as Peter Raby concludes in his splendid biography of Wallace (P. Raby, *Alfred Russel Wallace: A Life,* Princeton University Press, 2001), "There is, finally something heroic about a man who independently constructs a theory of natural selection

... and spends the rest of his life proclaiming the ideals of co-operation and altruism as the way to hasten the perfecting of the human." Consequently the chance to retrace a few of Wallace's steps in the Malay Archipelago intermittently over a period of 15 years, but collecting microorganisms rather than insects or birds of paradise, came as a piece of tremendous good fortune. Results of the ensuing biotechnology training and research program, which included mycorrhizal inoculant technology, bioremediation, biopesticides, and applied microbial taxonomy, are summarized elsewhere (*Indones. J. Biotechnol.,* Special Issue, June 2000), but the enduring memory is of the spectacular biodiversity of that remarkable archipelago. At the westernmost peninsula of Java, facing the Sunda Strait, lies Ujung Kulon, an area that was inundated by 10- to 15-m tidal waves in 1883 following the eruption of Krakatau. Ujung Kulon now is a national park devoid of human inhabitants and notable for containing the remaining (very small) population of Javan rhino. It was here some years ago that I witnessed a glittering display of bioluminescence. Our camp, where evenings were shared with an assortment of macaques, deer, monitor lizards, geckos, and bats, was close to the shore which, on one occasion, was intensely illuminated as teeming populations of microinvertebrates were oxygenated in the breaking tide. However, a few years later even this microbiological display was eclipsed by the coral reef communities off the north coast of North Sulawesi. These reefs are one of the most spectacular in the whole Indo-Pacific region, and the realization that a very high proportion of their invertebrate biomass comprised microorganisms about which so little was known was a forceful reminder of how fragmented and incomplete was our knowledge of microbial diversity. The English naturalist Sidney Hickson, following his observations of these ecosystems, wrote "A coral reef cannot be properly described. It must be seen to be thoroughly appreciated" (Hickson, *A Naturalist in North Celebes,*

Murray, 1889); how right he was! Readers will not be surprised to find marine invertebrates and their microbial symbionts featured in this book. Geothermal and other extreme environmental locations also are common in many parts of the archipelago and have provided further insights to the diversity of the microbial world. However, the opportunity to prospect microorganisms of truly extreme habitats was presented when Koki Horikoshi invited us to collaborate with the DeepStar program of the Japan Marine Science and Technology Agency (http://www.jamstec.go.jp). Our interest has been on the actinomycete diversity in very deep-sea sediments including those below the seafloor, where the extent of taxonomic diversity again is remarkably high. The exploration of these newly discovered biospheres is exciting and promising for bioprospecting, and various aspects of this novel field are contained in this book.

Edward Wilson probably has done more than any other individual to awaken the interest of both scientists and the public in biodiversity and why it should be promoted to a front page issue. His writings are rich in knowledge, challenging questions, and memorable imagery, and provide an especially pervading sense of wonder. Wonder, that emotion excited by what surpasses expectation and the desire to know, is something I very much hope that readers will encounter throughout the course of this book, for, as Francis Bacon declared, "For all knowledge and wonder (which is the seed of knowledge) is an impression of pleasure in itself" (F. Bacon, *Proficience and Advancement of Learning Book I.i.3, 1605*).

REVOLUTION

During the lifetime of my generation there has occurred an unprecedented change in which biological systems—from cell to biome—are viewed and investigated. As I attempt to show in chapter 24, this has been a revolution of genuinely Kuhnian proportions. Within the span of 50 years we have progressed from speculative debates about the organization of DNA in bacteria (see, for example, E.T.C. Spooner and B.A.D. Stocker [eds.], *Bacterial Anatomy*, Cambridge University Press, 1956) to a position in which students can manipulate, with facility and rationality, the DNA within and between species, and even Domains! The introduction and adoption of the techniques of molecular biology have occurred with unbelievable rapidity and ease such that they now permeate the whole spectrum of biological research. Questions can now be posed, and answered, that were inconceivable and/or had minimal expectation

of being resolved prior to the molecular biology era. The impact on our approach to and understanding of phylogeny and evolution, biodiversity and ecology, infection and therapy has been immense; the impact on the ways in which we exploit genetic resources in the context of biotechnology is no less impressive. In short we are of a generation that has seen the emergence of a new and powerful discipline, albeit with ill-defined boundaries, called bioinformatics. A word of caution is necessary here: bioinformatics, powerful though it is, is not the sole technoscientific driver of biotechnology; innovative developments in chemistry, chemical and biochemical engineering, computer science, and nanotechnology, for example, all have engaged with biology in transforming the search for novel drugs, chemicals, materials, and so on. Nevertheless, there has been a demonstrable paradigm shift in the way in which we do, or can do, search and discovery in biotechnology, that is, the shift from traditional biology based on specimen collection, observation, and experimentation to bioinformatics based on data collection, storage, and mining. One of the questions emerging from this paradigm shift is whether bioinformatics, in concert with approaches such as combinatorial chemistry, will displace the traditional biological approach or will exist in synergy with it. Present evidence strongly suggests that a synergy will become established.

The extent to which we are able to integrate diverse technical and analytical capabilities will be critical for the future of microbiology and biotechnology. The multidisciplinary approach has long and widely been appreciated as an essential underpin for successful biotechnology, but a comparable recognition in microbiology and among microbiologists has been slower to emerge. Recently this point was made very clear by Ed DeLong, who concluded that, "The challenge to future microbial biologists is that they must become as conversant in Earth science as nanotechnology, as familiar with systems ecology as genomes, and as well versed in global information systems as bioinformatics" (E. F. DeLong, Towards microbial systems science: integrating microbial perspective, from genomes to biomes. *Environ. Microbiol.* 4:9–10, 2002). Just as integrated technology approaches are seen as increasingly necessary for addressing the big questions in microbiology, there is a growing sense that severe reductionist science, epitomized by molecular biology, has deflected attention away from an understanding of the complexities of biological system properties. I am pleased, therefore, that many of the contributions in this book directly confront or elude to these issues of technology integration and the holistic perspective.

MICROBIAL DIVERSITY AND BIOPROSPECTING

Although this book has had a long gestation, now seems to be the appropriate time to bring together its principal themes and ideas. Equipped with a formidable set of new tools based on molecular biology, chemometrics, computing, statistics, and so on, and firmly fixed in the postgenomics era, we can begin to evaluate the effects that developments made possible by these tools are having on the way we go about exploring microbial diversity and searching for exploitable biology. The scope of this book is broad so that, in addition to those topics that might be anticipated, the reader will find other topics that are rarely considered in the context of microbial search and discovery, among them questions of biogeography, extinction, and the value of biodiversity. We also consider the implications of the Convention on Biological Diversity for microbial prospecting activities.

The book is organized in nine sections that deliberate on biotechnology and the case for natural product discovery; the resource that the microbial world presents for biotechnology; why it is important to take an ecological perspective when engaging in search and discovery; the distribution of microorganisms at the global scale and early attempts at microbial biocartography; the paradigm shift embodied in bioinformatics; some illustrations of microbial prospecting activities and their results in a range of high and not so high-added-value industries; the loss of evolutionary history and the conservation of microbial gene pools; microbiology, biotechnology innovations, and the Convention on Biological Diversity; and, finally, what we perceive as the value of biodiversity and how valuations might be made. Such a book is very unlikely to be exhaustive in its coverage; this one naturally reflects a personal survey of the landscape, and I would be pleased to have readers' thoughts on omissions and amendments that might be made. Readers also will be aware of the problems that often beset multiauthored books. Consequently, with the exception of the first and last sections, I have attempted to set the scene and provide continuity within and between the sections by way of short preambles. A few topics and authors have been lost along the way, but overall it has been possible to keep to the original plan and content.

QUAERENDO INVENIETIS!

If the reader will indulge me briefly, I wish to switch, finally and hopefully to some purpose, from microorganisms to music and in particular to Johann Sebastian Bach. A significant part of Bach's life overlapped with those of van Leeuwenhoek and Linnaeus, but this narrative concerns Frederick the Great rather than either of these notable biologists. Bach enthusiasts will recall that Frederick and Bach eventually met at Potsdam in 1747 and how the king invited Bach to improvise on a "royal theme," the result being one of the grandest and most complex inventions in the history of music—*The Musical Offering*. This composition of two fugues, ten canons, and a trio sonata was inscribed *Regis Jussu Cantio Et Reliqua Canonica Arte Resoluta* (at the king's command, the song and the remainder resolved with canonic art). The acrostic of Bach's dedication reveals "ricercar," the term for an earlier composition in the style of a fugue incorporating the most extreme devices of counterpoint, but also an adroit message in Italian—to seek out, with an implication of effort required in the search. The second ricercar of *The Musical Offering* is an astonishing six-part fugue, astonishing to the extent that Douglas Hofstadter likened it to "the playing of sixty simultaneous blindfold games of chess, and winning them" (D. Hofstadter, *Gödel, Escher, Bach*, Vintage Books, 1980). Moreover, among the canons, which were presented to the king as uncompleted musical puzzles, is one marked *quaerendo invenietis*—by seeking, you will discover. I like to think of the totality of this musical composition as the apt metaphor for biodiversity, biocomplexity, and bioprospecting, the more so because Bach and many of his contemporaries regarded music as a science; indeed Bach became a member of Lorenz Mizler's Society for Musical Sciences at the time of the *Offering's* composition. This exhortation is emphasized by Bill Strohl in the epilogue to his chapter and this, in essence, is the theme of this book.

ACKNOWLEDGMENTS

The immediate precursor of this book was an article invited by Roy Doi for *Microbiology and Molecular Biology Reviews* (A. T. Bull, A. C. Ward, and M. Goodfellow, search and discovery strategies for biotechnology: the paradigm shift, **64**:573–606). My first thanks, therefore, to Roy for his encouragement and to Alan Ward and Mike Goodfellow for many years of stimulating and eclectic discussion and productive collaboration. As mentioned in the preface, I have incubated the contents and ideas contained in these pages for several years, and in consequence many people either wittingly or unwittingly have contributed to my thinking and to keeping fun firmly on our agenda. So many thanks to the generations of graduate students and postdocs and colleagues in various organizations and places who have helped to keep me enthused and enlightened on microorganisms.

I thank Frank Bisby for his kind gift of the *Catalogue of Life* 2002 annual checklist; and Iain Prance and Dan Simberloff, my editors for *Biodiversity and Conservation*, for helping to sustain my commitment to biological diversity in the wider context and for their staunch support of the journal since its foundation. David Wynn-Williams of the British Antarctic Survey, a generous friend and fine microbiologist, was killed during the preparation of this book (see *Extremophiles* **6**:265–266, 2002); I am especially pleased, therefore, that the book contains an excellent account of the "icy biosphere" with which David's career was so closely concerned.

The enthusiastic interactions that I have enjoyed with the contributors to this book have been marvelous, and I am greatly indebted to them for preparing such thoughtful and timely accounts of their specialist subjects—my warmest thanks to you all. It is a pleasure to acknowledge the support for my own research into microbial diversity and biotechnology, some of which is mentioned in this book, from the Biotechnology and Biological Sciences Research Council and the Natural Environment Research Council (U.K.), The British Council, the Department for International Development (U.K.), the International Institute for Biotechnology, and the Japan Marine Science and Technology Center.

Greg Payne has been the ideal editor with whom to work—ever solicitous, patient, and enthusiastic; to him and the production team at the ASM Press I wish to convey very special thanks and appreciation for all their efforts in bringing this book to fruition.

Finally, I want to dedicate this book to my wife Jenny for her steadfast support over the years and not the least during the preparation of this book.

Alan T. Bull
Canterbury, January 2003

I. INTRODUCTION: THE RATIONALE

Microbial Diversity and Bioprospecting
Edited by Alan T. Bull
© 2004 ASM Press, Washington, D.C.

Chapter 1

Biotechnology, the Art of Exploiting Biology

ALAN T. BULL

EXPLOITABLE BIOLOGY

The business of the biotechnology-based industries is founded on the search for exploitable biology, or, in the terms of the Convention on Biological Diversity, genetic resources. Article 2 of the convention defines genetic resources as "genetic material of actual or potential value" and inclusive of whole organisms, cell components and extracts, and nucleic acids. A typical search and discovery program in biotechnology might involve the collection of appropriate genetic resources, screening for a property of interest, and, finally, the development of a commercial product or process. In order to maximize the scope of screening activities, companies may include a range of genetic resources in their search and discovery programs. Microorganisms and plants are prioritized by most companies but, as the survey of Laird and ten Kate (1999) revealed, a wider sweep of biota is made by the biotechnology industries as a whole (see Table 1 with respect to pharmaceutical companies).

The point is made throughout this book that biotechnology impacts a very much broader range of industries than just pharmaceuticals, and, in consequence, the search strategy embraces a correspondingly broader range of biota than the foregoing, as is the case with biocatalysts and biomaterials, for example.

At this point it is important to give meaning to the term exploitable biology. To many readers it will be synonymous with natural products and, in the context of pharmaceuticals, can be subdivided into (i) biologicals, or products obtained directly from genetic resources or via recombinant DNA technology, e.g., L-lysine and interferons; (ii) natural products that are identical to biologicals but are manufactured by semisynthetic or total chemical syntheses, e.g., cyclosporin A; (iii) chemically modified derivatives of natural products, e.g., semisynthetic cephalosporins;

and (iv) chemically synthesized compounds derived from a natural product structural type, e.g., acyclovir. In 1997, 42% of sales of the top 25 drugs worldwide arose from these categories of natural products (Newman and Laird, 1999) and thereby affirmed the continued importance of genetic resources for this industrial sector. Beyond pharmaceuticals, however, exploitable biology has a much more expansive frame of reference that includes raw materials, catalysts, a plethora of novel products, and biomimetics. As such, these resources reflect the major applications of biotechnology, namely, the replacement of fossil fuels and raw materials by renewable raw materials, the replacement of extant nonbiological products and processes by ones based on biology, and the introduction of completely innovative goods and services. I have previously evaluated the broader range of biotechnology-biodiversity synergies that are characterized by a number of mutually beneficial flows of materials, services, and ideas (Bull, 1996) that include biological detection, circumscription, and phylogeny; search and discovery; ecosystem function; bioremediation and biorestoration; and genetically designed organisms and processes.

Although the use of renewable raw materials is largely presaged by the price of fossil fuels, various strategic developments are in place for generating bulk industrial feedstocks and biofuels. These initiatives are based on crop plants that can yield low-cost raw materials such as starch and vegetable oils and their microbiological conversions to valuable feedstocks. Good examples of this approach are polyester fiber and biodegradable polylactide plastics production. Polytrimethylene terephthalate fiber is superior to polyethylene terephthalate, but the chemical route to its manufacture from ethylene oxide is not economically competitive. However, a recent development by DuPont and Genencor International has provided a competitive route to 1,3-propanediol production—the key intermediate

Alan T. Bull • Research School of Biosciences, University of Kent, Canterbury, Kent CT2 7NJ, United Kingdom.

Table 1. Prioritization of genetic resources
for screening purposes[a]

Resource	Priority ranking (no. of companies)				
	1	2	3	4	5
Plants	12	5		1	1
Microorganisms	8	6			
Fungi	3		1		
Marine organisms	2	6	2		
Insects			4		

[a] Data from Laird and ten Kate (1999).

in polytrimethylene terephthalate synthesis—in which glucose is converted in a one-step fermentation process by a recombinant strain of *Escherichia coli* (Bull, 2001). Similarly, Cargill Dow is making chiral lactic acids by fermentation of starch-derived dextrose, from which are produced lactide isomers; the latter are combined to generate a family of novel polymers (Organization for Economic Co-operation and Development [OECD], 2001). Also, in the United States, an alternative feedstock program led by the Department of Energy and Applied Carbo-Chemicals is developing a bioprocess to produce succinic acid from renewable farm crops that will be the starting point for the manufacture of chemical feedstocks such as *N*-methyl pyrrolidone.

The new generation of biotechnology-based processes is being driven strongly by developments in biocatalysis. In the past the deployment of biocatalysts was thought to be compromised by their fragility under industrial process conditions, high costs, and low volumetric productivities. This situation has changed so dramatically that, as a result of molecular and bioprocess engineering and the discovery of novel catalysts such as extremozymes, the customization of enzymes for an ever-widening range of industrial and domestic applications has become a reality. The advantages of developing enzymes as industrial catalysts include their abilities to provide stereo- and regioselective compounds and pure enantiomeric isomers, their operational cleanliness compared with conventional chemical catalysts, and the possibilities that their use affords for truncating traditional chemisynthetic processes (Bull et al., 1999). Referring again to renewable raw materials, the further development of enzymes for converting plant biomass into sugars and chemical feedstocks at competitive costs has major strategic significance. Here the requirements include high-activity and low-cost cellulases that may result from the engineering of known proteins and/or from bioprospecting. The improvements in high-fructose syrup production from starches are encouraging in

this context. Starch is hydrolyzed with α-amylase and then saccharified with glucamylase, and the resulting glucose is converted to fructose by glucose isomerase. A disadvantage of the currently used α-amylases is their requirement for calcium ions; however, alternatives have been discovered that lack this requirement and that also have greatly increased thermostabilities that are produced, for example, by *Pyrococcus furiosus* (Zeikus et al., 1998). Similarly, isomerases with high-temperature optima and enhanced thermostabilities have been discovered in organisms such as *Thermotoga neapolitana* (Zeikus et al., 1998), properties that complement the favored production of fructose at high temperatures. Although this is not the place to detail the impressive penetration and future prospects of biocatalysis in industrial processing, the interested reader will find comprehensive information in Roberts (1998), Marrs et al. (1999), Liese et al. (2000), and Schäfer and Borchert (chapter 33).

The case for basing bioprospecting on microbial genetic resources can be made on several grounds: (i) the record of discovery of bioactive compounds, enzymes, and a profusion of other exploitable properties; (ii) the extraordinarily high taxonomic and genetic diversity of microorganisms; (iii) the unrivaled ability of microorganisms to colonize the Earth's environments; and (iv) the relative facility with which microorganisms can be grown and preserved, and their genotypes and phenotypes manipulated, in the laboratory. There are salutary caveats to the last point, however. Many microorganisms, including pathogens, endosymbionts, and extremophiles, require very exacting conditions for their cultivation ex situ, although only a very small and arguably unrepresentative fraction of microbial diversity has yet been cultivated. Moreover, whereas the concept of exploitable microbiology based on the isolation of organisms, cultivation to elicit maximum gene expression, and screening for desired properties (Bull et al., 1992) remains the practice in many biotechnology companies and industries, major changes in the approach to search and discovery are occurring. Such developments have brought about a paradigm shift in exploitable biology from that based on the above philosophy to one based increasingly on bioinformatics. Thus, we are experiencing the introduction of an unprecedented range of technologies that include gene and genome prospecting; combinatorial biosynthesis; metabolic pathway engineering; and the artificial evolution of genes, proteins, and pathways. Although this paradigm shift is neither absolute nor universal, it is so profound that its effect on search and discovery and implications for biotechnology will pervade much of the discussion in this book.

THE IMPACT OF BIOTECHNOLOGY AND ITS CONTRIBUTION TO SUSTAINABLE INDUSTRY

Modern biotechnology represents a cluster of enabling technologies that have sprung from radical innovations and not simply from incremental advances in science. This radical change is best illustrated by the emergence of industrial biocatalysis and the spectacular development and applications of molecular biology. Biotechnology comprises a considerable spread of industrial activities based on "the application of scientific and engineering principles to the processing of materials by biological agents to provide goods and services" (Bull et al., 1982) and has penetrated a range of manufacturing, service, and agricultural industry sectors (see chapter 29). The range of techniques defined by biotechnology is large and varied and none of them are likely to be applicable across all of the above sectors. However, the versatility of biotechnology is such that many industries that have not previously considered using biological systems are now exploring these options.

The main drivers of biotechnology are economic (market pull), technological practicability, and public demand (and acceptance), each of which reflect the positions of the respective stakeholders, namely, industry, government, and society. Whereas economic considerations have been foremost in determining the adoption of biotechnology, the changing public and government perceptions of the global environment are driving the introduction of clean technology, and in particular clean biotechnology. Therefore, two salient questions need to be asked of biotechnology: (i) can it provide a cheaper option than a convention process and (ii) are economic competitiveness and environmental friendliness compatible objectives? There are several points in the life cycle of a product or process in which opportunities arise for increasing its sustainability, and not the least by introducing biotechnology (Fig. 1). The recent OECD (2001) enquiry into biotechnology for sustainable industrial development demonstrated that "the application of biotechnology invariably led to a more sustainable process, a lowered ecological footprint in the widest sense, by reducing some or all energy use, water use, wastewater or greenhouse gas production." This conclusion was reached by scrutinizing 21 case studies of companies in a very wide range of industrial sectors and countries who had adopted biotechnology, primarily on economic grounds, but who had also

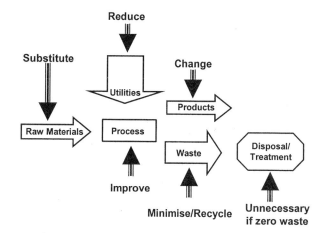

Figure 1. Improvement options for sustainable processing

achieved the type of overall improvements in sustainability featured in Fig. 1.

Examples of raw material substitution were mentioned above. The following selection of cases illustrates how biotechnology can affect process substitution, process improvement, energy saving, waste minimization, and pollution control while reinforcing the benefits of microbial prospecting and biocatalyst development. Full details of these case studies can be found in the OECD report.

Process Replacement: Riboflavin (Hoffmann-LaRoche, Germany)

Riboflavin (vitamin B_2), an important food and feed supplement, formerly was manufactured via chemical routes, but most producers now have installed biotechnological processes. Hoffmann-LaRoche has developed a one-step fermentation process based on a metabolically engineered strain of *Bacillus subtilis* that uses glucose as the feedstock. This case is notable because of the life-cycle analysis made on the chemical and biotechnological processes; this has revealed that the latter had a 75% saving on nonrenewables, 50% and 66% reductions in emissions to air and water, and a 50% reduction in operating costs.

Process Replacement: Acrylamide (Mitsubishi Rayon, Japan)

Acrylamide is a commodity chemical with a wide range of industrial applications, where again an innovative biotechnological route has been developed as an alternative to a traditional chemical manufacturing process. Nitrile hydratase-producing bacteria were isolated that converted acrylonitrile to the amide and were subsequently manipulated to generate high-enzyme-producing and substrate-resistant strains. Acrylamide concentrations up to 50% (wt/vol) and substrate conversions of 100% are now possible with the production strain *Rhodococcus rhodochrous* J-1. "Green" auditing (sensu stricto OECD, 2001) of the chemical and biocatalytic processes show that energy costs are greatly reduced (80%) in the *Rhodococcus* process, and waste arisings are down, although an overall cost benefit and safer operation is achieved.

Product Improvement: (*S*)-Chloropropionic Acid (Avecia, United Kingdom)

The driving force in this case was the increased pressure on the pharmaceutical and agrochemical manufacturers to produce homochiral enantiomers, rather than racemic mixtures, in order to affect cost savings through feedstock use and compliance with regulations. The crux of the biotechnological manufacture of (*S*)-chloropropionic acid was the isolation of a *Pseudomonas putida* strain from soil that produced *R*- and *S*-specific hydrolytic dehalogenases. By inactivating the gene coding for the *S*-specific enzyme, the bacterium is used to dehalogenate the *R*-enantiomer in commodity racemic choropropionic acid (CPA) and produce (*S*)-CPA and (*S*)-lactic acid. The (*S*)-CPA is then used in the manufacture of homochiral herbicides of the Fusilade family.

Process Innovation: Semisynthetic Antibiotics (Dutch State Mines, The Netherlands)

Semisynthetic cephalosporins, such as cephalexin, cefadroxil, and cephadrine, traditionally were manufactured by multistep chemical processes that generated considerable waste. Dutch State Mines has developed a process for producing 7-aminodesacetoxycephalosporanic acid (7-ADCA), a key intermediate in the synthesis, from penicillin G that combines chemical and biocatalytic steps. Even more encouraging from a clean technology point of view is the development of a complete biotechnological route to 7-ADCA, achieved through the construction of a recombinant *Penicillium chrysogenum* strain into which was cloned penicillin G expandase, and a new dicarboxylic acid acylase for side chain hydrolysis. Compared with the earlier chemical process for making 7-ADCA, the new fermentation route produces greater purity of product, greatly increased energy efficiency, and very little requirement for organic solvents. Although life-cycle assessments have not been published for these cephalexin processes, waste generation has been reduced by the factors of 2 and 10,

respectively, for the six-step hybrid technology and the four-step direct fermentation-cum-biocatalysis routes of production.

Agricultural Pollution Control: Feed Enzymes

Feed enzymes for application in high-intensity livestock and poultry rearing is my final example illustrating the contribution that biotechnology can make to industrial sustainability. Phytic acid (*myo*-inositol hexaphosphate) is a major storage form of phosphorus in plants but is indigestible for certain animals. Phytase is an enzyme found in many fungal and plant taxa and, when added to animal feed, hydrolyzes the phytic acid with the release of assimilable phosphorus. This practice obviates the need to incorporate inorganic phosphates into feed and may reduce phosphate excretion up to 30% in pig feces. Isolation of a heat-resistant phytase from *Aspergillus* enables the enzyme to be added during feed formulation without it becoming inactivated (see chapter 24 for a consensus design of highly thermostable fungal phytase). Forsberg and his colleagues (Golovan et al., 2001) subsequently have augmented the natural repetoire of digestive enzymes in monogastric animals with protease-resistant salivary phytase ex *E. coli*. Complete digestion of phytic acid is achieved with transgenic pigs, which now reduces fecal phosphorus excretion up to 75% and has a great sparing effect on the need for inorganic phosphorus supplementation.

BIOPROSPECTING: SCOPE AND ISSUES

Innovative biotechnology is intimately associated with the discovery of novel exploitable biology, be it in terms of organisms per se, their activities, metabolic products, or genes (Fig. 2). In turn, discovery is dependent on the recognition of novel organisms, the definition of biotechnology targets, and the invention of search and screening strategies, issues that are focal points of this book.

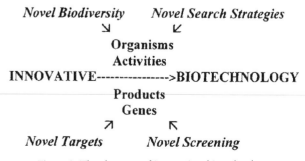

Figure 2. The elements of innovative biotechnology

Following the identification of a biotechnological target, one question that arises is what microorganism, or group of microorganisms, might be the most propitious for focusing the search strategy? Some guides for organism selection include (Bull et al., 2000) (i) playing the percentage game, e.g., actinomycetes for biopharmaceuticals; (ii) interrogating taxon-chemistry and taxon-property databases—if they exist (see chapter 28); (iii) focusing on novel and neglected taxa; (iv) recovering isolates from unusual or little-explored ecosystems; and (v) matching the target with members of previously unscreened but known taxa, e.g., HIV-inactivating protein cyanovirin-N produced by cyanobacteria.

Is Exploitable Microbiology Exhausted?

Whatever the specific approach that is made to organism selection, the strategy is predicated by the extent to which the exploitable microbiology is or is not exhausted. It is impossible to answer this conundrum in any systematic way, but reference to the role of natural products in the discovery of new pharmaceuticals—the most critical and intensely probed sector for genetic resource exploration—may provide some clues. Numerous commentators have opined over the past few years that the role for natural products in "improving the human condition . . . will continue as long as there are unexplored sources" of natural products (Clark, 1996). Nisbet and Moore (1997) go further and conclude that without natural products there would be a significant therapeutic deficit in clinical areas including neurodegenerative and cardiovascular diseases, solid tumors, and immune-inflammatory diseases. One of the most revealing facts is the unabated reporting of novel natural chemotypes having interesting structural and biological activity features (Kuo and Garrity, 2002; Pietra, 2002). Combinatorial chemistry and combinatorial biosynthesis are frequently raised in this context and posited as displacing natural products in the pharmaceutical industry. However, it can be argued that the innovation of combinatorial programs with novel molecular scaffolds of natural origin that are already known to possess significant bioactivity greatly enhances the possibilities for creating serious drug candidates. Demain (1998) is unequivocal in his criticism of companies who are dropping natural product discovery programs to support combinatorial chemistry efforts; rather he sees these approaches as complementary and not mutually competitive.

When posing the question, are actinomycetes exhausted as a source of secondary metabolites?, Bérdy (1995) highlighted several research needs, including (i) selective isolation techniques for discovering new producer strains; (ii) increased emphasis on "neglected, rare, fastidious species"; and (iii) dereplication by all possible means to reduce the rediscovery rate. Bérdy's second point is clearly made by a survey of publications patterns. Galvez et al. (1998) reported that little or nothing was published on 17.5% of formally described bacteria between 1991 and 1997 and that the publication rate on another 56% was very low. It is a reasonable assumption that the position is even more extreme with regard to other groups of microorganisms. The observations of Lazzarini et al. (2000) emphatically underscore this point. Surveys of bioactive products produced by actinomycetes revealed that numbers of so-called rare taxa were increasingly significant sources of new antibiotics. These authors make two well-informed comments: (i) given the use of robust selective isolation methods, such taxa "are not rare at all and can be recovered from many soil samples"; (ii) the high throughput of strains strategy adopted by industry for drug discovery "does not normally allow a complete taxonomic description of all isolates," ipso facto it is important not to neglect taxonomic circumscription and dereplication.

Recent inventories of natural products (www.chemnetbase.com) contain 164,000 entries and, as of 1996, approximately 30,000 bioactive natural products had been described from bacteria (33%), plants (27%), fungi (26%), and animals sources (13%) (Henkel et al., 1999). Returning to the actinomycetes, Watve et al. (2001) have asked the intriguing question of the largest antibiotic-producing taxon: how many antibiotics are produced by the genus *Streptomyces*? In other words, are those taxa that are popularly screened for bioactive compounds sufficiently well explored? Watve and his colleagues constructed a model that predicted that *Streptomyces* organisms are capable of producing as many as 10^5 antibiotics! These authors argue that the fall in discovery represents a decline in screening efforts rather than an exhaustion of compounds. Among the criticisms that can be leveled at this type of analysis is the equivocal taxonomic characterization of organisms.

A crucial question is whether natural products represent a structurally unique pool of compounds that cannot be surplanted by combinatorial chemicals. Here the findings of the Bayer group (Henkel et al., 1999) have major significance: their objective was "to describe statistically the differences between structural properties of natural products and synthetic chemicals so as to derive statements into the complementarity of these pools of structures. The investigation took structural information from five databases covering natural products, chemicals, and drugs and evaluated it in terms of molecular weight, heteroatom distribution, the average number of rings and chiral centers, and pharmacophoric groups per molecule. Even when redundancy of structures in the Bioactive Natural Product Database was accounted for, nearly 12,000 structurally diverse types of molecule were defined. The authors concluded that their statistical evaluation revealed a distinct difference in the structural properties of natural products relative to synthetic compounds and that the potential of new natural products was not exhausted: the proposition that natural products have already been sufficiently examined was definitively rejected.

A complementary analysis has been made by Pietra (2002) by focusing on molecular skeletons, whose diversity was described by a semiquantitative molecular complexity index. For his analysis Pietra used the complexity metric (S) and size metric (H) developed by Whitlock (1998), S being the sum of constants times the number of rings, unsaturations, heteroatoms and chiral centers, and H being the number of bonds, in the molecule. Thus, natural products could be defined in terms of their structural complexity (S) and specific molecular complexity (S/H), the latter being largely independent of molecular size. These indices were computed for seven classes of biogenic molecular skeletons (alkaloids; peptides and proteins; fatty acids; polyketides; shikimates; and carbohydrates), and the complexity of each class was related to its distribution in a range of 7 terrestrial and 10 marine ecosystems and in fossil molecules. Graphical representation of these data enables a ready comparison of natural product diversity and ecosystem distribution. Thus, isopenoids showed the largest skeletal variety in all ecosystems, the maximum being found in the Indo-Pacific Ocean, and reflects the coral biodiversity of that region. Pietra has assembled and analyzed an enormous amount of data to which the interested reader is directed for specific information on biodiversity, natural product, and biogeographic relationships. The marine environment is a remarkable source of novel natural products with over 10,000 compounds having been described (Proksch et al., 2002). To date, the majority of these compounds that have entered clinical trials (principally anticancer drugs) have been isolated from invertebrates, and, apart from the primacy of drug efficacy, a number of other complementary issues arise from this fact: (i) the problem of supply, (ii) the feasibility of chemical synthesis, (iii) the ultimate source of the compounds—invertebrate or microbial symbiont?, and (iv) the culturability of such symbionts. On the other hand there are abundant opportunities for exploiting the natural products of cultivable marine microorganisms as the current investigation of North Sea bacteria testifies (Wagner-Döbler et al., 2002).

It should be acknowledged that not everyone subscribes to the view that extensive biodiversity prospecting is an inherently advantageous means of discovering exploitable biology. The most recent counterargument of this type has been advanced by Tulp and Bohlin (2002) who, in the particular context of drug discovery, opine that "because potential lead compounds are likely to be found in many different species, there are no obvious reasons why success needs to be sought in 'biodiversity prospecting.' We consider that it is sufficient to focus on a limited number of species that preferably belong to the same ecosystem." Tulp and Bohlin postulate that, despite the enormous diversity of natural chemical molecules, "it is inconceivable that the number of different biological functions is near this number"; in other words, important molecular mechanisms are ubiquitous, and the number of molecular targets is comparatively small (e.g., only approximately 600 different targets so far identified in the human genome). This hypothesis warrants rigorous testing but appears to be premature on a number of grounds. Thus, the reductionist approach may be too simplistic even in the context of novel drug discovery. Even the rational selections of ligands that bind target receptors at the highest affinity may not guarantee the highest biological activity (Bull et al., 2000), e.g., thrombopoietin receptors (Livnah et al., 1996; Cwirla et al., 1997). Moreover, many human diseases of interest to the pharmaceutical industry involve multiple gene pathways, environmental interactions, and genetic predisposition rather than simply direct causal effects (Kola, 1999). Thus, effective search and discovery for new drugs will likely require more holistic screening strategies (e.g., cell function assays) that reflect the epigenetic nature of certain disease conditions. Therefore, arguments pertaining to a limited number of species, and ecosystems, are not well founded. The value of prospecting in different ecosystems and biomes is clearly evident in Pietra'a compilations (see above), while it is prudent to sample from a number of environments rather than exhaustively from a smaller number (Kuo and Garrity, 2002). Tulp and Bohlin give no indication of what might constitute the "limited number of species." It is important to reiterate the significance of intraspecific variation in search and discovery activities; *Trichoderma polysporum* may have been identified in many countries since its initial isolation in Norway (see Svarstad, chapter 40), but the genetic diversity of this fungus in relation to secondary metabolite synthesis has not been established. This point raises the generic issue of dereplication, both of organisms and of their products, which is discussed below. Finally, bioprospecting in the full context of biotechnology has a very much wider frame of reference and portfolio of targets than drugs, examples of which occur throughout this book.

The Importance of Dereplication

Dereplication can be defined as the process of recognizing identical species or strains of organisms, and of identical natural products, and thereby preventing redundant and wasteful screening effort (Bull et al., 1992; Kuo and Garrity, 2002; Wagner-Döbler et al., 2002). Strain discrimination at the infraspecific level is essential because the genetic diversity within a species frequently determines the capacity to produce secondary metabolites and enzymes. Chemometric whole-organism fingerprinting methods such as pyrolysis mass spectrometry (PyMS), Fourier-transform and dispersive Raman spectroscopy, and matrix-assisted laser desorption MS and electrospray MS are increasingly deployed for discrimination (Goodfellow et al., 1997; Goodacre et al., 1998; Tindall et al., 2000) and identification (van Baar, 2000). PyMS has been used in our laboratory to circumscribe novel, industrially promising actinomycetes isolated from deep-sea sediments. Moreover, excellent congruence between numerical taxonomy and PyMS has been found in double-blind analyses of representative rhodococcal isolates indicating that pyrogroups are directly ascribable to observed phenotypic variation (Colquhoun et al., 2000) and, in consequence, have real value in screening programs. We have also shown high congruence between PyMS analysis and molecular fingerprinting (PCR-restriction fragment length polymorphism and single-strand conformational polymorphism), thus reinforcing the discriminatory power of PyMS at the infraspecies level (Brandao et al., 2002) and demonstrating that PyMS dereplication can be applied directly to organisms recovered on primary isolation plates.

The dereplication argument also applies to other facets of genetic resource screening. Although the advent of high-throughput DNA sequencing has enabled the rapid and economic production of very large sequence data sets, the need has arisen to dereplicate sequence libraries. One such procedure compares all sequences to each other, groups similar sequences, and identifies a representative sequence from each group, thereby eliminating duplication (Seguritan and Rohwer, 2001). However, the utility of such programs is dependent on the value that is set for percent sequence identity (PSI) and the aim of the particular application; thus, the 97% PSI selected by the authors will not discriminate between strains below species level, or in some cases even between species.

Finally, dereplication methods are equally significant for natural product chemists (VanMiddlesworth and Cannell, 1998). Databases of natural product structures containing high-pressure liquid chromatography, MS, nuclear magnetic resonance, and other information can be used to interrogate purified compounds and microbial fermentation broths. Such methods are used widely in industry for identification (e.g., alkaloid and sesquiterpenes [Bradshaw et al., 2001]), for investigating microbial metabolism (e.g., analysis of *Streptomyces citricolor* [Abel et al., 1999]), and to identify individual components in fermentation broths.

REFERENCES

Abel, C. B. L., J. C. Lindon, D. Noble, B. A. M. Rudd, P. J. Sidebottom, and J. K. Nicholson. 1999. Characterization of metabolites in intact *Streptomyces citricolor* culture supernatants using high-resolution nuclear magnetic resonance and directly coupled high-pressure liquid chromatography-nuclear magnetic resonance spectroscopy. *Anal. Biochem.* 270:220–230.

Bérdy, J. 1995. Are actinomycetes exhausted as a source of secondary metabolites? p. 13–34. *In* V. G. Debabov, Y. V. Dudnik, and V. N. Danilenko (ed.), *Proceedings of the 9th International Symposium on the Biology of Actinomycetes*. All-Russia Scientific Research Institute for Genetics and Selection of Industrial Microorganisms, Moscow, Russia.

Bradshaw, J., D. Butina, A. J. Dunn, R. H. Green, M. Hajek, M. M. Jones, J. C. Lindon, and P. J. Sidebottom. 2001. A rapid and facile method for the dereplication of purified natural products. *J. Nat. Products* 64:1541–1544.

Brandao, P. F. B., M. Torimura, R. Kurane, and A. T. Bull. 2002. Dereplication for biotechnology screening: PyMS analysis and PCR-RFLP-SSCP (PRS) profiling of 16S rRNA genes of marine and terrestrial actinomycetes. *Appl. Microbiol. Biotechnol.* 58:77–83.

Bull, A. T. 1996. Biotechnology for environmental quality: closing the circles. *Biodiver. Conserv.* 5:1–25.

Bull, A. T. 2001. Biotechnology for industrial sustainability. *Korean J. Chem. Eng.* 18:137–148.

Bull, A. T., G. Holt, and M. D. Lilly. 1982. *Biotechnology—International Trends and Perspectives*. OECD Publications, Paris, France.

Bull, A. T., M. Goodfellow, and J. H. Slater. 1992. Biodiversity as a source of innovation in biotechnology. *Annu. Rev. Microbiol.* 42:219–257.

Bull, A. T., A. W. Bunch, and G. K. Robinson. 1999. Biocatalysts for clean industrial products and processes. *Curr. Opin. Microbiol.* 2:246–251.

Bull, A. T., A. C. Ward, and M. Goodfellow. 2000. Search and discovery strategies for biotechnology: the paradigm shift. *Microbiol. Mol. Biol. Rev.* 64:573–606.

Clark, A. M. 1996. Natural products as a resource for new drugs. *Pharm. Res.* 13:1133–1141.

Colquhoun, J. A., J. Zulu, M. Goodfellow, K. Horikoshi, A. C. Ward, and A. T. Bull. 2000. Rapid characterisation of deep-sea actinomycetes for biotechnological screening programmes. *Antonie Leeuwenhoek* 77:359–367.

Cwirla, S. E., P. Balasubramanian, D. J. Duffin, C. R. Wagstrom, C. M. Gates, S. C. Singer, A. M. Davis, R. L. Tansik, L. C. Mattheakis, C. M. Boytos, P. J. Schatz, D. P. Baccanari, N. C. Wrighton, R. W. Barret, and W. J. Dower. 1997. Peptide agonist of the thrombopoietin receptor as potent as the natural cytokine. *Science* 276:1696–1699.

Demain, A. L. 1998. Microbial natural products: alive and well in 1998. *Nat. Biotechnol.* 16:3–4.

Galvez, A., M. Maqueda, M. Martinez-Bueno, and E. Valdivia. 1998. Publication rates reveal trends in microbiological research. *ASM News* 64:269–275.

Golovan, S. P., R. G. Meidinger, A. Ajakaiye, M. Cottrill, M. Z. Wiederkehr, D. J. Barney, C. Plante, J. W. Pollard, M. Z. Fan, M. A. Hayes, J. Laursen, J. P. Hjorth, R. R. Hacker, J. P. Phillips, and C. W. Forsberg. 2001. Pigs expressing salivary phytase produce low-phosphorus manure. *Nat. Biotechnol.* 19:741–745.

Goodacre, R., E. M. Timmis, R. Burton, N. Kaderbhai, A. M. Woodward, D. B. Kell, and P. J. Rooney. 1998. Rapid identification of urinary tract infection bacteria using hyperspectral whole-organism fingerprinting and artificial neural networks. *Microbiology* 144:1157–1170.

Goodfellow, M., R. Freeman, and P. R. Sisson. 1997. Curie-point pyrolysis mass spectrometry as a tool in clinical microbiology. *Zentbl. Bakteriol.* 285:133–156.

Henkel, T., R. M. Brunne, H. Müller, and F. Reichel. 1999. Statistical investigation into the structural complementarity of natural products and synthetic compounds. *Angew. Chem. Int. Ed.* 38:643–647.

Kola, I. 1999. Complex traits, genes and polymorphisms and the drug discovery/development process. *Curr. Opin. Biotechnol.* 10:589–592.

Kuo, A., and G. M. Garrity. 2002. Exploiting microbial diversity, p. 477–520. *In* J. T. Staley and A.-L. Reysenbach (ed.), *Biodiversity of Microbial Life.* Wiley, Chichester, United Kingdom.

Laird, S. A., and K. ten Kate. 1999. Natural products and the pharmaceutical industry, p. 34–77. *In* K. ten Kate and S. A. Laird (ed.), *The Commercial Use of Biodiversity.* Earthscan Publications Ltd., London, United Kingdom.

Lazzarini, A., L. Cavaletti, G. Toppo, and F. Marinelli. 2000. Rare genera of actinomycetes as potential producers of new antibiotics. *Antonie Leeuwenhoek* 78:399–405.

Liese, A., K. Seelbach, and C. Wandrey. 2000. *Industrial Biotransformations.* Wiley-VCH, Weinheim, Germany.

Livnah, O., E. A. Stura, D. L. Johnson, S. A. Middleton, L. S. Mulcahy, N. C. Wrighton, W. J. Dower, L. K. Joliffe, and I. A. Wilson. 1996. Functional mimicry of a protein hormone by a peptide agonist: the EPO receptor complex at 2.8 Ångstrom. *Science* 273:464–471.

Marrs, B., S. Delagrave, and D. Murphy. 1999. Novel approaches to discovering industrial enzymes. *Curr. Opin. Microbiol.* 2:241–245.

Newman, D. J., and S. A. Laird. 1999. The influence of natural products on 1997 pharmaceutical sales figures, p. 333–335. *In* K. ten Kate and S. A. Laird (ed.), *The Commercial Use of Biodiversity.* Earthscan Publications Ltd., London, United Kingdom.

Nisbet, L. J., and M. Moore. 1997. Will natural products remain an important source of drug research for the future? *Curr. Opin. Biotechnol.* 8:708–712.

OECD. 2001. *The Application of Biotechnology to Industrial Sustainability.* OECD Publications, Paris, France.

Pietra, F. 2002. Biodiversity and natural product diversity. Tetrahedron Organic Chemistry Series Volume 21, Pergamon, Amsterdam.

Proksch, P, R. A. Edrada, and R. Ebel. 2002. Drugs from the sea—current status and microbiological implications. *Appl. Microbiol. Biotechnol.* 59:125–134.

Roberts, S. M. 1998. Preparative biotransformations: the employment of enzymes and whole cells in synthetic organic chemistry. *J. Chem. Soc. Perkin Trans.* 1:157–168.

Seguritan, V., and F. Rohwer. 2001. FastGroup: a program to dereplicate libraries of 16S rDNA sequences. *BMC Bioinformatics* 2:9.

Tindall, B. J., E. Brambilla, M. Steffen, R. Neumann, R. Pukall, R. M. Kroppenstedt, and E. Stackebrandt. 2000. Cultivable microbial biodiverity: gnawing at the Gordian knot. *Environ. Microbiol.* 2:310–318.

Tulp, M., and L. Bohlin. 2002. Functional versus chemical diversity: is biodiversity important for drug discovery? *Trends Pharmacol. Sci.* 23:225–231.

van Baar, B. L. M. 2000. Characterisation of bacteria by matrix-assisted laser desorption/ionisation and electrospray mass spectrometry. *FEMS Microbiol. Rev.* 24:193–219.

VanMiddlesworth, F., and R. J. P. Cannell. 1998. Dereplication and partial identification of natural products, p. 279–327. *In* R. J. P. Cannell (ed.), *Methods in Biotechnology,* vol. 4: *Natural Products Isolation.* Humana Press Inc., Totowa, N.J.

Wagner-Döbler, I., W. Beil, S. Lang, M. Marinus, and H. Laatsch. 2002. Integrated approach to explore the potential of marine microorganisms for production of bioactive metabolites. *Adv. Biochem. Eng.* 74:207–238.

Watve, M. G., R. Tickoo, M. M. Jog, and B. D. Bhole. 2001. How many antibiotics are produced by the genus *Streptomyces? Arch. Microbiol.* 176:386–390.

Whitlock, H. W. 1998. On the structure of total synthesis of complex natural products. *J. Org. Chem.* 63:7982–7989.

Zeikus, J. G., C. Vieille, and A. Savchenko. 1998. Thermozymes: biotechnology and structure-function relationships. *Extremophiles* 2:179–183.

II. MICROBIAL DIVERSITY: THE RESOURCE

This section of the book seeks to set the scene and explore the resource that biological diversity, and particularly microbial diversity, presents for biotechnology.

First there is the remarkable compass of the microbial world containing the objects "of potentially endless study and admiration" (Wilson, 1995). At the conclusion of his narrative, *Naturalist*, Edward Wilson reveals that microbial ecology would be his métier if he could do it all over again: "The jaguars, ants and orchids would still occupy distant forests in all their splendour, but now they would be joined by an even stranger and vastly more complex living world virtually without end." The speed and facility with which molecular biological analyses have been applied to exploring this complexity have been truly amazing. Now, and highly relevant to the success of microbial prospecting, attention is focusing on the scale of microbial diversity to be found in the world's profusion of habitats. Microbiologists have been rather slow to adopt and develop statistical tools for making meaningful estimations of diversity, but a number of recent seminal publications (Hughes et al., 2001; Dunbar et al., 2002; Martin, 2002; Curtis et al., 2002) point the way to more sophisticated analyses that can reveal the taxonomic richness (species richness?) and genetic diversity of microbial communities. Thus, the application of statistical estimators not only should assist investigations of spatial, resource, and temporal factors that affect microbial diversity, but also inform bioprospecting strategies. Implicit in this recommendation, of course, is the necessity for microbiologists to interact increasingly with other disciplines—notably mathematics, statistics, and computer science—in order to address new challenges.

Second, and with a sense of the inevitable, even foreboding, we are confronted with the concept of species, and most acutely by the prokaryotic species. Several contributors refer to the *species genome* concept proposed by Lan and Reeves (2000), resulting from comparative intraspecies genomic analyses, and not limited to pathogenic microorganisms. And from the perspective of the species genome, Doolittle (2002) challenges us to appreciate the true enormity of microbial diversity: "How many different genomes are hiding behind a single environmental . . . , phylotype, and how many under the phylotype clusters are commonly found?"

Nevertheless, we are left with the need for operational definitions in microbiology (species, operational taxonomic unit, or whatever) in order to provide the *lingua franca* for identification and classification, and hence effective, unambiguous communication within clinical, industrial, patent, and a host of other professional activities. This point is emphasized by Ramon Rosselló-Mora and Peter Kämpfer (chapter 3), who argue that the species should be a pragmatic unit through which, although not necessarily easy to circumscribe, operational and predictive taxonomy can be constructed. In this context the polyphasic taxonomic approach remains the best option for building robust taxonomic and identification schema and for the detection of rare and novel taxa of biotechnological importance. Polyphasic taxonomy is supported (confused?) by a myriad of technological innovations that Fergus Priest (chapter 5) puts into perspective and that has become practicable with the advent of rapid data acquisition and manipulation systems and associated databases. Consequently it becomes incumbent on microbiologists to adhere to recommended minimal standards when describing new taxa (see, for example, *Staphylococcus* species; Freney et al., 1999). In a recent pithy commentary on the state of taxonomy, Valdecasas and Camacho (in press) repeat the warning that the diminishing pool of taxonomic specialists "and the undervaluation of their work . . . could lead to specialists from other disciplines working with meaningless data."

This section concludes with brief reference to microbial diversity from a taxonomic-cum-phylogenetic perspective. Laura Katz draws attention to the relatively unexplored nature of eukaryotic microorganisms and provides an overview of selected eukaryotic lineages. Regrettably, despite our best endeavors, the complementary chapter on prokaryotic microorganisms failed to materialize; consequently, the reader will find that the following authors provide excellent starting points for exploring bacterial and archaeal diversity and phylogeny: Brown and Koretke (2000), DeLong and Pace (2001), Gupta and Griffiths (2002), and Matte-Teilliez et al. (2002). Although these topics are revisited quite regularly in the review literature, reference here is important because of the speed of new discoveries (cf. the number of Divisions in the Domain *Bacteria*), the progressive unravelling of major lineages (especially within the *Eukarya*), and the specialist (restricted) knowledge imposed on many of us by the downright enormity of the microbial world. The diversity of ciliates illustrates well this last point; a recent monograph on soil ciliates from the Etosha region and Namib Desert of Namibia (Foissner et al., 2002) alone records 1 new order, 3 new families, 34 new genera, and 128 new species. Thus, a comprehensive knowledge of microbial diversity becomes an essential prerequisite in our attempts to understand ecology on the one hand and to devise intelligent bioprospecting strategies on the other.

REFERENCES

Brown, J. R., and K. K Koretke. 2002. Universal trees. Discovering the archaeal and bacterial legacies, p. 19–55. *In* F. G. Priest and M. Goodfellow (ed.), *Applied Microbial Systematics.* Kluwer Academic Publishers, Dordrecht, The Netherlands.

Curtis, T. P., et al. 2002. Estimating prokaryotic diversity and its limits. *Proc. Natl. Acad. Sci. USA* **99:**10494–10499.

DeLong, E. F., and N. R. Pace. 2001. Environmental diversity of bacteria and archaea. *Syst. Biol.* **50:**470–478.

Doolittle, F. 2002. Diversity squared. *Environ. Microbiol.* **4:**10–11.

Dunbar, J., et al. 2002. Empirical and theoretical bacterial diversity in four Arizona soils. *Appl. Environ. Microbiol.* **68:**3035–3045.

Foissner, W., et al. 2002. Soil ciliates (Protozoa, Ciliophora) from Namibia (Southwest Africa), with emphasis on two contrasting environments, the Etosha Region and the Namib Desert. *Denisia,* vol. 5. Oberösterreichischen Landesmuseums, Linz, Austria.

Freney, J., et al. 1999. Recommended minimal standards for description of new staphylococcal species. *Int. J. Syst. Bacteriol.* **49:**489–502.

Gupta, R. S., and E. Griffiths. 2002. Critical issues in bacterial phylogeny. *Theoret. Pop. Biol.* **61:**423–434.

Hughes, J., et al. 2001. Counting the uncountable: statistical approaches to estimating microbial diversity. *Appl. Environ. Microbiol.* **67:**4399–4406.

Lan, R., and P. R. Reeves. 2000. Intraspecies variation in bacterial genomes: the need for species genome concept. *Trends Microbiol.* **8:**396–401.

Martin, A. P. 2002. Phylogenetic approaches for describing and comparing the diversity of microbial communities. *Appl. Environ. Microbiol.* **68:**3673–3682.

Matte-Teilliez, O., C. Brochier, P. Forterre, and H. Philippe. 2002. Archaeal phylogeny based on ribosomal proteins. *Mol. Biol. Evol.* **19:**631–639.

Valdecasas, A. G., and A. I. Camacho. 2003. Conservation to the rescue of taxonomy. *Biodiver. Conserv.* **12:**1113–1117.

Wilson, E. O. 1995. *Naturalist.* Allen Lane. The Penguin Press, London, United Kingdom.

Microbial Diversity and Bioprospecting
Edited by Alan T. Bull
© 2004 ASM Press, Washington, D.C.

Chapter 2

An Overview of Biodiversity—Estimating the Scale

ALAN T. BULL AND JAMES E. M. STACH

GLOBAL BIODIVERSITY

What is Biodiversity?

The term "biological diversity" was introduced by Elliot Norse and colleagues (Harper and Hawksworth, 1995) to define diversity at three levels of complexity: (i) genetic (intraspecies diversity), (ii) species (numbers of species), and (iii) ecological (community diversity), but subsequently the contracted expression "biodiversity" has become the common parlance (Wilson and Peter, 1988). This definition of biodiversity may lack precision, and various revisions have been proposed, for example, organismal diversity in place of species diversity in order to acknowledge diversity at higher taxonomic levels. Indeed, each of the components in the definition of biodiversity has a hierarchical structure spanning biomes to niches (ecological), domains to populations (species and organismal), and populations to nucleotide sequences (genetic). For present purposes here, we use the term species as the basic unit of biodiversity while appreciating that there is not a universally accepted definition of a species (Heywood and Watson, 1995). We return to this issue, in the particular context of microorganisms, later in this section of the book. Nonetheless, any imprecision in defining biodiversity, and the plurality of its perception by different interest groups, can be regarded as a strength in making biodiversity a unifying concept and by focusing different disciplines and interests on common goals, namely, the understanding, conservation, and sustainable use of genetic resources.

The simplest measure of biodiversity is *species richness*, that is, the number of species present in a given area or habitat. Also known as α-diversity, species richness defines the richness of a potentially interactive assemblage of organisms; γ-diversity similarly measures within-area diversity but related to a large area such as a biome or bioregion. β-Diversity is a measure of between-area diversity and can be expressed as a simi-larity index. More ecologically informative measures of biodiversity incorporate *species abundance*; generally speaking, the more equally abundant are the species in the area, the more biodiverse it is regarded. Another component of biodiversity is *functional diversity*, which, ideally, is measured by the range of species traits in a habitat or in a regional or experimental species pool (Tilman and Lehman, 2001).

Biologists, including microbiologists, refer to rare species often in anecdotal rather than scientific terms. *Rarity* can be categorized according to local population size (large or small), habitat specificity (wide or narrow), and geographical range (large or small). On the basis of the eight possible combinations of these states, Rabinowitz et al. (1986) established seven variant forms of rarity and characterized *common* species as having large populations and ranges and wide specificity.

Finally it is pertinent to consider *taxic diversity*. Whereas measures of species richness treat all species equally, taxic diversity is a measure of the taxonomic dispersion of species and counts the higher taxa represented in an area or habitat (Heywood and Watson, 1995). Thus, marine ecosystems have much greater taxic diversity than freshwater or terrestrial ones in terms of their animal phyla: the numbers of the total (and endemic) animal phyla in these ecosystems are, respectively, 28 (13), 14 (0), and 11 (1) (Harper and Hawksworth, 1995). As these authors question, if one were planning to conserve ecosystems on the basis of the greatest diversity of natural products, "what measure of biodiversity would be the best guide?" Phylum or endemic richness might be good candidates, but species richness could be very misleading.

Species Inventories

Basic curiosity begs us to ask the question, how many species are there on Earth? Although exciting insights into the scale of biodiversity have been gained in

Alan T. Bull and James E. M. Stach • Research School of Biosciences, University of Kent, Canterbury, Kent CT2 7NJ, United Kingdom.

recent years, the answer remains elusive. Indeed, the situation is not much better that that stated by May (1988): "we do not know to within an order of magnitude how many species of plants and animals (and other species) we share the globe with. . . . At the theoretical level, things are even worse: we cannot explain from first principles why the global total is of the order of 10^7 rather than 10^4 or 10^{10}." The global numbers of recognized species are reported periodically (e.g., Hammond, 1995a [1.75 million]; Reaka-Kudla, 1997 [1.87 million]) but are only approximations due to various factors, the first of which is an accelerating rate of species collection and description (e.g., angiosperms [Prance, 2001]). Hammond (1995a) noted that the rate of new species description was about 13,000 per year (a rate that had been constant for one or two decades), whereas more recent estimates suggest that the current rate is sixfold greater (Purvis and Hector, 2000). However, the fact that not all of these species become accepted as valid by the appropriate taxonomic authorities cautions that the rate may be significantly lower. The second factor is due to discoveries of completely new higher taxa (e.g., the animal phyla Loricifera and Cycliophora [Kristensen, 1983]) and the algal classes Pelageophyceae and Bolidophyceae [Andersen et al., 1993; Guillou et al., 1999]. Third, there are inadequate taxonomic studies resulting in misidentification and synonymies that in many cases are not systematically recorded (cf. molluscs versus insects [Hammond, 1995a]). And finally, there is a disproportionate taxonomic effort, especially for the more poorly described taxa (e.g., soil microarthropods [André et al., 1994]; nematodes [Coomans, 2000]).

Species 2000 (http://www.sp2000.org/) has the realistic and practical objective of creating a uniform and validated index of all known species on Earth, including microorganisms. Data on approximately 260,000 species and approximately 500,000 names (including synonyms) appear in the "Catalogue of Life" year 2002 annual checklist (Froese and Bisby, 2002), and there are hopes that the tally will reach 500,000 species by 2003. The name service on the website presently covers about three million species names (bacteria, archaea, fungi, plants) and it hoped to extend this by a further four million names from the *Zoological Record*. The Natural History Museum, London, is developing its WORLDMAP project (http://www.nhm.ac.uk/science/projectsworldmap/index.html) that can be used to access biodiversity distributions and feature hot spots and regions of endemicity. Another ambitious agenda has been set by the All Species Foundation (Lawler, 2001), nothing less than to inventory all species on Earth, again inclusive of microorganisms, during an estimated minimum time frame of 20 years. Given that a complete all-taxa biodiversity inventory, inclusive of microorganisms of any given habitat, still remains to be reported (Bull et al., 2000), the latter objective may prove impossibly ambitious: time will tell, but May's (1995) expectation of having "in 50 years . . . a fairly good idea of roughly how many species there currently are on earth" is a more tenable one. These proposals bring us to the critical issue of estimating possible species richness.

Biodiversity Estimates

Various empirical and theoretical attempts have been made to estimate global species richness (see May, 1992; Hammond, 1995a,b), a brief synopsis of which follows.

1. Time-series descriptions of new species. The rates at which new species are reported differ greatly with the category of organism. Thus, half of all bird species were described by 1845, and the curve has now reached a plateau, whereas most known species of arthropods have been recorded in the past two decades and the rate is not declining. During the period 1978 to 1987, the numbers of recorded birds, arachnids, nematodes, and fungi increased by 0.005, 1.8, 2.4, and 2.4%, respectively.

2. Taxon-to-taxon ratios. Considering the ratio of fungi to plants, for example, which in the United Kingdom is 8.4:1 (Hawksworth, 2001), and the known global vascular plant tally of 270,000, an estimate for global fungal diversity of 2.37 million can be derived. Similarly, the ratio of all insect to butterfly species in the United Kingdom is just under 330 and, because the world diversity of butterflies is known quite accurately, a global insect diversity of 5.75 million species is estimated.

3. Size relationships. In crude terms there are systematically more species of small animals than large animals, and for each 10-fold reduction in body size, there is an equivalent 100-fold increase in species numbers. This relationship, which appears to break down for organisms that are less than about 1 mm long, extrapolates to about 10 million species of animals.

4. Intensive sampling and extrapolation. The classic illustration of this approach is Terry Erwin's (1982) insecticidal fogging experiment to recover beetles and all other insects from the canopy of a single tree species in the Panamanian rainforest. A succession of extrapolations based on the number of tree-specific beetles, the diversity of tropical tree species, the ratio of canopy beetles to all canopy insects, and the ratio of canopy to ground insects produced an estimated 30 million tropical arthropods.

5. Species accumulation curves. These are plots of the cumulative number of species (S_n) recorded within a defined area, volume, or habitat as a function (n) of the effort required to find them (typically the number of samples, clones, etc., examined). Such plots, also known as species-area curves, are dependent simply on the presence or absence of data, and S_n approaches S_{max} asymptotically. Both parametric and nonparametric models can be used to estimate species richness from such sampling efforts, and among the commonly deployed models are continuous log-normal distributions (e.g., Preston, 1962) and nonparametric estimators (e.g., Chao, 1984). An excellent exposition of estimating species richness through extrapolation is given by Colwell and Coddington (1995), and examples of its application will appear later in this and in other chapters.

6. Molecular phylogenies. The development of phylogenetic markers, notably small-subunit (ss) rRNA genes, and their use in exploring microbial diversity completely revolutionized our conception of microbial species richness. This innovation has led to the discovery of numerous novel microbial taxa, from species to division status, and is discussed in detail later in this chapter and in section III of this book.

7. Taxonomists. Finally, the opinions of specialist taxonomists can be counseled, and Hammond's (1992, 1995b) estimates of global species richness take note of such views.

Arguably, the most useful estimates remain those of Hammond (1995a), who compiled a working figure total of about 13.6 million species and attempted to put some probable accuracy on the estimates (Table 1). Hammond cautioned that the figures for bacteria, viruses, and algae were frankly speculative, whereas those for fungi and protozoa were very insecurely based. Thus, these figures convey the measure of our ignorance, particularly of microbial diversity.

It is not difficult to find problems with the processes by which estimated diversity numbers have been derived: reliability of data and overreliance on single data sets; undersampling; ratios of one group to another group of organisms varying with geographic location; wide variation in the study of different groups and the "general lack of concordance between patterns for well-known (and species-poor) groups and the hyperdiverse groups of organisms" (Hammond, 1995a); assumptions based on as yet unproven host specificities; and, again, the issue of species definition and circumscription and the inconsistency of terms used to describe the basic units of organism diversity (cf. species, morphospecies, operational, or recognizable taxonomic unit) and so on. However, such endeavors have been crucial in highlighting the importance of biodiversity, erecting testable hypotheses, and establishing research agendas. Estimates of the type shown in Table 1 are continually under review as the cases of fungi and insects demonstrate.

Hawksworth's (1991) persuasive analysis of the magnitude of fungal diversity was based on several independent data sets from which he opined that a conservative estimate of fungal diversity was 1.5 million species. This estimate was considered conservative because of the modest 270,000 figure used for global plant numbers, no account being taken of the fungus:insect ratio, or of endophytic fungi, and the probable geographic variation in fungus:plant ratios. One of the most valuable outcomes of Hawksworth's reasoning was the publication of fungal diversity estimates based on different and complementary ap-

Table 1. Described and estimated species numbers[a]

Group	No. of species[b]			Accuracy
	Described	Estimated (range)	Estimated (working figure)	
Plants	270	300–500	320	Good
Chordates	45	50–55	50	Good
Arthropods	1,065	2,375–101,200	8,900	Moderate
Molluscs	70	100–200	200	Moderate
Nematodes	25	100–1,000	400	Poor
Protozoa	40	60–200	200	Very poor
Algae	40	150–1,000	400	Very poor
Fungi	75	200–9,900	1,500	Moderate
Bacteria	4.9	50–3,000	1,000	Very poor
Viruses	4	50–1,000	400	Very poor
Others	115	200–800	250	Moderate
Totals	1,753	3,635–111,655	13,620	Very poor

[a] Adapted from Hammond (1995a) with amendments from Hawksworth (2001) and Froese and Bisby (2002).
[b] In thousands.

proaches. Such approaches have produced diversity values ranging from 0.5 to 9.9 million, including predictions of ascomycete and endophytic species alone reaching 7 and 1.3 million, respectively. The picture is further confused by the realization that cryptic species of fungi are commonplace, i.e., biological species concealed within described morphospecies and subsequently circumscribed particularly on the basis of molecular typing. Crypticity alone could promote the level of fungal diversity to an astonishing 5.1 million species (Hawksworth, 2001). On revisiting his earlier estimate, Hawksworth accepts criticisms such as fungal diversity is greater in the tropics than in temperate regions and that extrapolations based on host specificity need to be reviewed. Nevertheless, he recommends that the working figure of 1.5 million be retained until new and independent data sets become available.

May (1995) has raised a number of important conceptual issues regarding species number estimates and, among others, focused on Erwin's (1982) assessment of arthropod diversity. The issues attracting May's concern include (i) extrapolations between geographic regions, (ii) unequivocal definition of host specificity, and (iii) validity of group-to-group extrapolations, e.g., to what extent are canopy beetle species representative of other insects? Recently Ødegaard (2000) reviewed the estimates of arthropod diversity and, largely because of uncertain host specificity and a paucity of data sets and using a modified version of Erwin's estimates, revised them down to 4.8 million species. These revisions notwithstanding and accepting the working figures stated in Table 1, the question remains—where are the missing species?

Focusing first on fungi, a number of attempts have been made to suggest where the missing 1.43 million species can be found. Hyde (2001) reviews data that indicate that they may occur within poorly researched countries, hosts, and habitats and are predominantly microfungal species. For example, initial studies of palms resulted in the discovery of a high proportion of new species (up to 50% in some cases), whereas 5 to 7% of beetle species may be hosts for ectoparasitic ascomycetes (*Laboulbeniales*). Based on their data sets from Sulawesi, Weir and Hammond (1997) estimate that the species diversity of *Laboulbeniales* alone is between 10,000 and 50,000. Similarly, fungal endophytes (fungi producing asymptomatic infections in plants) have been recorded from every plant species so far examined. A recent study of fungal endophytes in tropical forest woody angiosperms is noteworthy because it used frequency- and similarity-based indices to analyze patterns of host specificity and spatial heterogeneity (Arnold et al., 2001). Results from this study support the con-

tention that tropical endophytes are a significant component of fungal species richness and, most importantly, recommend research methods that might be deployed for assessing γ-diversity within this group. Finally, Sipman and Aptroot (2001) also point to the tropics and the Southern Hemisphere as regions where novel lichen diversity will be found. They predict that the 4,000 or so "missing" lichens will be discovered primarily in tropical primary forests on leaves and bark in the canopy.

A second exciting opportunity for extending our knowledge of global biodiversity is presented by little-studied habitats, the most expansive of which are marine habitats, and the least explored of which are the deep seas. The Earth's oceans cover more than 70% of its surface, with a high proportion being more than 2,000 m deep; consequently, the logistics of obtaining samples from deep-sea environments (particularly if it is required to maintain them at in situ pressures and temperatures and prevent contamination) are extremely demanding compared with terrestrial sampling. The abyssal and hadal oceans (depths below 2,000 and 6,000 m, respectively) used to be regarded as biological deserts, but over the past decade the analogy of the deep seas as rainforests has become more pertinent. The pioneering research of Grassle and colleagues (see Grassle and Maciolek, 1992) produced estimates of the macrofaunal species richness of the deep sea by extrapolations from large data sets obtained from the continental slope and rise of the eastern United States. They concluded that the diversity could conservatively exceed 10 million species and observed that about 60% of the species they recovered had not been described previously. The bases for such estimates have been criticized by several authors, but as May (1992) concluded, the implication of this work should prompt "more taxonomists to turn their attention to the oceans." At about the same time, Lambshead (1993) was drawing attention to the possible scale of marine meiofauna diversity and suggested that it could be as high as 10^8. Although interstitial meiofauna such as nematodes may be the dominant deep-sea animals, both in terms of species richness and abundance (Hammond, 1995b), such astonishingly high figures may not be realized because of invalid extrapolations from small scale to regional or global scales. That the marine environment is a major source of unrevealed biodiversity is further reinforced if we take account of the high degree of endemism recorded in the deep sea (see "What is Biodiversity?" above). Finally, we should also note the recent discoveries of deep-sea environments such as subseafloor sediments, cold seeps, brine lakes, carbonate mounds, mud volcanoes, hydrocarbon seeps, and gas hydrates; these systems open up entirely new opportunities for biodiversity exploration. Subseafloor sediments, for example, which may extend down several kilometers, are known to have a significant prokaryotic biota (see chapter 12).

Taxonomy Is Not a Luxury

Before leaving this overview of biodiversity, some brief reference to taxonomy is appropriate. Taxonomy is that branch of biological systematics that deals with the circumscription and description of organisms (α-taxonomy) and with the classification of organisms (β-taxonomy) on the basis of phenetics (overall phenotypic similarity) or phylogenetics (evolutionary history). Regrettably, taxonomy, and particularly α-taxonomy, is frequently treated in dismissive and deprecatory terms. Taxonomy, however, emphatically is not a luxury, and to trivialize it merely as stamp collecting is unhelpful and ignorant. Taxonomic inventorying determines the present state of biodiversity and how it can be assessed, and it also becomes an integral part of databases on ecosystem functioning and for search and discovery programs. The competitive image of taxonomy and its weak ability to attract major research funds means that the size of the overall workforce is seriously inadequate for the tasks now being asked of it. Not that this paradoxical situation has passed without serious comment. Edward Wilson, with his customary prescience, made the position abundantly clear in *The Diversity of Life* (Wilson, 1992). With reference to such ecologically important organisms as termites and abundant soil animals such as oribatid mites, Wilson reported at that time that specialist taxonomists dealing with these groups numbered three worldwide and only one in North America, respectively. Considering that the smaller mites such as *Actinedia* are even more neglected and rarely identified (André et al., 1994), the situation affords little confidence in our ability to make an all-inclusive bioinventory. Considerable problems attend many other groups; for example, Lambshead (1993) opined that there were only 20 marine nematode taxonomists in the world describing, on average, 10 new species a year (see above). The diminishing pool of taxonomic specialists "and the undervaluation of their work . . . could lead to specialists from other disciplines working with meaningless data" (Valdecasas and Camacho, 2003). At the same time that Wilson was publishing his exuberant account of biodiversity, Gaston and May (1992) were evaluating the state of the taxonomic workforce in Australia, the United States, and the United Kingdom. The relevance of their conclusions has basically not changed during the past decade; typically a small minority of taxonomists is engaged with microorganisms; and an equally small minority of taxonomists is based in developing countries where, ironically, the greatest proportion of the world's terrestrial biodiversity is located. Gaston and May drew attention to the pressing need to coordinate information that already exists, and in this context the achievements and objectives of *Species 2000* are encouraging. But, overall, Cooman's (2000) comment seems depressingly legitimate: "despite political rhetoric in support of the need for biodiversity research and warnings for the devastating consequences of a lack of qualified taxonomists by leading scientists in the field, very little is done."

MICROBIAL DIVERSITY

Concerning the Unit of Count

Attempts to quantify microbial diversity usually follow those established for macroorganisms, namely, making inventories of species numbers. The figures given in Table 1 all involve, in one way or another, the concept of species—an entity that "reside(s) somewhere within a continuum of genetic diversity extending from the individual to the kingdom. . . . But how can we rely upon them when the unit of count is an undefined quantity?" (Martin and Salamini, 2000). The problem of species definition becomes progressively severe as one moves from animals and plants to microorganisms, and, although this topic is examined in detail in chapter 3, a few remarks are needed to guide the present discussion.

A very large literature exists on species concepts (see, for example, Minelli, 1993 and Quicke, 1993), and operational concepts tend to differ from speciality to speciality in biology. In the second half of the 20th century, microbial systematics was influenced by two major efforts to construct new theoretical and practical approaches: these have developed as numerical taxonomy and phylogenetic systematics. Numerical taxonomy is a phenetic system based on overall phenotypic similarities that establishes operational taxonomic units (OTUs), i.e., organisms that a taxonomist wishes to treat separately and that can include individuals and demes. In contrast, phylogenetic systematics (cladistics) reveals, or attempts to reveal, evolutionary relationships between a group of organisms and uses DNA base composition, DNA homology, genome fingerprinting, and gene sequencing to establish such relationships.

It has been implied earlier that prokaryotes exhibit considerable intraspecies variation both in terms of sequence variation and in the presence or absence of particular genes. Thus, it is important to emphasize that the genome of a single prokaryote strain is very unlikely to represent the genetic capacity of a prokaryotic species. Lan and Reeves (2000) conceptualized the position as follows: "Only if we have the sequence for all the DNA important for a species do we really have the *species genome,* as distinct from an individual genome." The species genome can be viewed as a set of core genes present in at least 95% of all strains and a set of auxilliary genes in 1 to 95% of all strains. Lan and Reeves recognize that these cutout values are arbitrary, but it is now possible to quantify these gene sets, particularly the core genes, by using techniques such as suppressive subtractive hybridization and DNA microarrays (Lan and Reeves, 2000; Boucher et al., 2001). Evidence is accumulating to show that the gene content of strains of the same species can differ by as much as 20%. Complete genome sequencing of *Escherichia coli* O157:H7 has revealed that it has 1,387 genes not found in *E. coli* K-12, and that K-12 has 528 genes not found in O157:H7 (Perna et al., 2001). However, such genome divergences are not restricted to pathogenic taxa but, for example, have been reported in planktonic crenarchaeotes (Béjà et al., 2002). The intraspecies genome divergences arise as a result of horizontal gene transfer and gene loss. Lawrence (2001) has referred to this genome dynamism in terms of genetic headroom—information that can be lost without altering the metabolic capabilities of the organism. Thus, an organism with large genetic headroom "can explore novel ecologies with impunity" and "offers potential metric for the propensity of a lineage to speciate" (see chapter 4).

The deployment of molecular biological methods has had a dramatic effect on the resolution of critical taxa that include cryptic species (not distinguishable on the basis of morphology) and pleomorphic species (for example, fungi having more than one independent form). A cogent illustration of the problem of pleomorphism is provided by entomopathogenic fungi. Thus, *Cordyceps,* a teleomorphic (sexual morph) genus, is connected with at least 17 anamorphic (asexual morph) genera. It follows that the nature of the microbial species is not simply a matter for philosophical debate but has a real practical need. As Goodfellow et al. (1997) emphasized, "It is the implementation of taxonomic concepts and practices which gives rise to identification and typing systems, procedures for quality control and risk assessment, protocols for the analysis and characterization of biodiversity, hypotheses about the evolution of prokaryotes, and improved procedures for the selective isolation and use of microorganisms in biotechnological processes." Diversity estimates of prokaryotes sometimes are regarded as meaningless because of the difficulty in applying the species concept to them, but this position also has changed as a consequence of adopting the polyphasic species approach. This operational species concept, based on good-quality genotypic and phenotypic data (Busse et al., 1996; Bull et al., 2000), can result in well-described species and stable nomenclature and identification schemes. Accordingly, the term species should be used only when characterization has been made on polyphasic grounds, whereas terms such as OTU and phylotype should be used for circumscriptions based on phenetic and gene sequence, respectively.

Numbers and Diversity

The figures presented in Table 1 suggest that the percentages of protozoa, algae, fungi, bacteria, and viruses so far described and authenticated are, at best, 20, 10, 5, 0.5, and 1, respectively. Moreover, a conspicuous omission from the table is the domain *Archaea;* initially viewed as extremophiles, these prokaryotes are now also known, largely as a result of culture independent surveys, to be widespread in nonextreme habitats (DeLong and Pace, 2001). The number of cultured species of archaea was reported as 179 and that of bacteria as 4,422 (Rainey and Ward-Rainey, 2000). The interim conclusions on the most specious groups of bacteria, based on culturability, are the *Actinobacteria,* the low-G+C gram positives, and the *Proteobacteria.* Numbers of a different type have been computed by Whitman et al. (1998) for the total population of prokaryotes on Earth. Reasoning that such an assessment could be made by analyzing a range of representative habitats, it was found that most prokaryotes occurred in soil, sediment, soil subsurfaces, and the oceans, habitats that are discussed in detail in section III of this book. The respective numbers in soil, subsurface, and oceans were estimated to be 3.5, 1.4, and 0.12 $\times$ 10^{30}; the numbers in other habitats, such as animals, were at least 6 orders of magnitude less. Several caveats must be placed on these estimations including large variations resulting from sampling effort and extrapolations and integrations from representative data. Nevertheless, these estimates carry some striking inferences. As the authors remark, the large populations imply that events that are rare in the laboratory may occur frequently in nature, and they point to an enormous potential to accumulate mutations and thereby acquire genetic diversity. Once again this fact emphasizes the importance of interrogating intraspecies diversity in the course of biotechnology screening programs and the need to discriminate (dereplicate) organisms at this level.

Uncertainty of another sort is found in the systematics of eukaryotic microorganisms. The term protist was formerly used to encompass protozoa, algae, and fungi, but it has become clear that it does not define a clade nor do these organisms form a monophyletic taxon; hence, the kingdom *Protista* becomes invalid as a natural group (Andersen, 1998). Patterson (1999) specified protists sensu stricto as a paraphyletic group (comprising a common ancestor and some but not all descendants) of eukaryotes that is a sister group to choanoflagellates (animals), charophytes (land plants), and chytrids (fungi). In these terms, Patterson has defined over 70 monophyletic lineages without identifiable sister taxa and more than 200 additional lineages that have been described but remain unplaced in a phylogenetic scheme.

A point was made above concerning disproportionate taxonomic effort: the situation is no less problematical with respect to microorganisms. A survey of publication patterns made by Galvez et al. (1998) revealed that nothing was published on 97 genera of bacteria between 1991 and 1997, and little was published on a further 310 genera; in contrast, the publication rate for just eight genera was over 500 per year. Not surprisingly those bacteria with the highest publication rates had major clinical, medical, or biotechnological importance. This markedly skewed effort is unlikely to have altered much in the intervening years; moreover, it is reasonable to assume that the situation is just as or even more distorted with regard to other microorganisms. In consequence, not only are we woefully ignorant about the scale of microbial diversity, but also about the nature of a single component of that diversity. It is not difficult to pinpoint neglected areas of microbial diversity research, whether as a result of funding priorities, tunnel vision, or logistical problems. Thus, the case of phototrophic sulfur bacteria is cogently advanced by Overmann (2001): whereas current research is focused on a few model species, "recent studies indicate that we might miss fundamental features of the photosynthetic way of life if we keep on studying only a handful of easily culturable strains and if we disregard those species which actually dominate in nature, using them merely as colourful curiosities, suitable only for textbook illustrations."

Although studies of microbial symbionts have advanced dramatically in recent years (see chapters 18, 19, and 20 of this book), our appreciation of their diversity is in its infancy (Polz et al., 2000). Among the cases of endosymbiotic associations, that of intestinal ciliates and methanogenic bacteria is particularly interesting. The survey made by Hackstein and colleagues (Hackstein, 1997) showed that there are "thousands of potential vertebrate host races and

species, and ten-thousands of potential arthropod host strains," leading him to predict that if a high degree of host specificity could be established, the numbers of ciliate and methanogen species would need to be revised upwards. Evidence from subsequent studies of intestinal, and freshwater, anaerobic ciliates supported the notion of high specificity of the association (van Hoek et al., 2000). Such findings go a long way to uphold earlier suggestions that within the great diversity of arthropods at least one new species of bacteria may be contained in every species of insect. Further evidence for high host specificity comes from Cavanaugh's laboratory (Polz et al., 2000) where the interest is in episymbiotic bacteria of marine animals (nematodes and the hydrothermal shrimp genus *Rimicaris*) and colonial ciliates. The nematodes are considered the most spectacular and diverse of these episymbioses; thus, symbioses involving the nematode subfamily *Stilbonemtinae* are highly host specific and exclusive and are characterized by very regular epibiont growth. Other closely related nematodes support more diverse epibacterial populations. Recall that Lambshead's estimate of marine nematode diversity is 1 million (Lambshead, 1993)!

The Phylogenetic Framework

In a recent, beautifully written essay, Carl Woese (2002) recounts the amazing revolution that occurred in microbiology following the development and application of ribosomal RNA gene sequencing, a revolution in which he was the key figure. As Woese points out, the advent of gene sequencing enabled two longstanding barriers in microbiology to be overcome: "the lack of a comprehensive evolutionary framework and the limitation imposed by the need to culture (microorganisms) to identify them," and, prior to this transformation, "microbiology could not begin to cope with the problem of microbial diversity." The impact of cultivation-independent molecular phylogenetic studies on microbial diversity is attested to by many of the discussions in this book, so only a few broad remarks are made here.

The original circumscription of the domain *Bacteria* (Woese, 1987) identified 12 monophyletic divisions based on rRNA gene sequence data, a number that has now risen to over 40 (DeLong and Pace, 2001). Two examples demonstrate the dramatic rate of discovery. In just 2 years following recognition of the candidate (sensu no cultured members) division *Acidobacterium* (Barnes et al., 1999), over 250 rDNA sequences representing at least eight major subdivisions had been reported and claims made for its presumptive metabolic and genetic diversity to rival those of the *Proteobacteria*. A second candidate division designated WS6 contains among the largest phylogenetic divergence of any known bacterial division and the prospect of interesting metabolic novelty (Dojka et al., 2000). Major taxonomic diversity also occurs in the domain *Archaea,* and culture-independent surveys have revealed novel types in a wide variety of habitats. Recently Karner et al. (2001) reported that one group of archaea, the pelagic crenarchaeota, are extremely abundant in the ocean and estimated that the world oceans contain over 10^{28} archaeal cells. Their estimate of the total prokaryote population in the oceans (4.4×10^{28}) is not so different from that proposed by Whitman et al. (1998).

Molecular sequencing of microbial eukaryotes is in its infancy, but even the small number of initial studies have revealed a large diversity of protists that previously has gone undetected by classical methods (Moreira and Lopez-Garcia, 2002; Massana et al., 2002). To date most studies have been made of marine nano- and picoplankton that contain organisms with photo-, hetero-, and mixotrophic metabolisms; a high proportion of the reported phylotypes correspond to novel taxa. As with most of the newly discovered prokaryotic lineages, many of these novel eukaryotes are very divergent from cultivated members and, in all cases, present an urgent challenge to develop procedures for their cultivation. Although many of the novel eukaryotic phylotypes are affiliated with established protist groups (e.g., alveolates and heterokonts), others appear to be located toward the base of the eukaryotic tree (Lopez-Garcia et al., 2001). Culture-independent studies of eukaryotic diversity in anoxic sediments from the Guaymas Basin, for example, have revealed novel protist diversity, the full extent of which remains to be determined (Edgcomb et al., 2002). A similar analysis of marine and freshwater anoxic sediments identified many previously unrecognized eukaryotes (Dawson and Pace, 2002) and notably members of seven clades that were not specifically related to any known kingdom-level lineages.

Small-subunit (ss) rRNA gene sequencing also is being used to establish phylogenetic relationships among fungi, an analysis that is increasingly necessary because many recently described fungi are not ascribable to the conventional taxonomic groups. In overall terms, ss rRNA gene sequencing has confirmed the separation of higher (ascomycetes, basidiomycetes) and lower (zygomycetes, chytrids) fungi into two monophyletic lineages and the fact that anamorphic fungi (deuteromycetes) do not form a distinct taxon (Hamamoto and Nakase, 2000). However, the application of gene sequence analysis to fungal systematics produces a dilemma that has theoretical and practical repercussions for a group of organisms that traditionally have been classified on the basis of phenotypic characters.

A final remark is necessary about this phylogenetic landscape. Phylogenetic tree construction has become part and parcel of microbial diversity research and has catalyzed a resurgence of interest in taxonomy. However, to what extent do such trees represent phylogenetic reality or natural classifications? Thus, a major challenge to the veracity of "universal" phylogenetic trees based on gene sequencing has been the revelation of substantial horizontal gene transfer and lineage-specific gene loss (Doolittle, 1999). Although operational genes (housekeeping) have been subject to extensive horizontal transfer compared with informational genes, there is some evidence for the transfer of rRNA genes (e.g., Eardly et al., 1996). The availability of complete genome sequences opens up alternative procedures for tree construction that is well demonstrated by the recent analyses of Wolf et al. (2001). The limitations notwithstanding, the sequencing of conserved genes such as those encoding ss rRNA has great practical value in circumscribing microbial taxa. And in cases in which the conservation of rRNA genes is so high as to preclude discrimination at the species level (Fox et al., 1992), the sequences of other less-conserved genes, exemplified by RNA polymerase and DNA gyrase subunit genes, may enable closely related species to be resolved. For example, *gyrB* gene sequence comparisons have proved to be useful for defining intra- and some intergeneric relationships within the *Enterobacteriaceae,* whereas 16S rDNA sequences were useful for revealing more distant phylogenetic relationships in this family (Dauga, 2002).

Estimating and Comparing the Diversity of the Uncountable

"Silent species" is the term Edward Wilson uses to describe bacteria and other microorganisms that do not respond to standard cultivation techniques (Wilson, 1992). Current estimates predict that only 1% of all bacteria are presently in culture (Bakken, 1995). Most microbiologists would agree that purely cultivation-based approaches to estimating bacterial diversity are Sisyphean (Curtis et al., 2002), making molecular methods for the detection of unculturable bacteria routine (see chapter 9). The first insight of the scale of microbial diversity came from low-resolution reassociation kinetic analyses of environmental DNA (Torsvik et al., 1990). These authors found that the number of bacterial genomes present in soil and marine sediment was as great as 6,500 and 11,400, respectively, representing an estimated

20,000 to 37,000 species (Torsvik et al., 2000). This dramatic finding represents the first limitation for those wishing to compare microbial diversity in different environments and samples, i.e., that tens of thousands of samples would be needed to ensure complete coverage (not including the likelihood of catching the same species twice). Dunbar et al. (2002) derived null models to estimate bacterial diversity in Arizona soils. They calculated that the documentation of half of the species in their communities would require surveys of 16,284 to 44,000 individuals; quantitative comparison for half of the community would be at least 10-fold higher for each community. The second limitation is that species are defined inconsistently (see chapter 3); most estimations of bacterial diversity are based on the 16S rDNA gene, with species (or OTUs) being assigned to groups that show arbitrary percentage differences. For instance, Kroes et al. (1999) defined an OTU as a 16S rDNA sequence group in which sequences differed by less than 1%; other values of 3 and 5% have been employed (McCaig et al., 1999; Martin, 2002).

These problems are not the sole preserve of microbiologists, however, and ecologists studying macroorganisms are faced with similar problems and have designed tools enabling estimations that are applicable to microbial communities. Statistical approaches for estimating and comparing microbial communities can be divided into those that estimate species richness, i.e., the number of types, and those that compare genetic diversity between samples. The accuracy of both approaches depends on how well a sample reflects the true diversity of the sample. In any sampling regimen, the number of types increases with effort until all types are observed. This relationship allows accumulation or rank abundance curves to be plotted, the shape of which provide information on how well the community has been sampled relative to its total diversity. Hughes et al. (2001) showed that our ability to sample bacterial diversity in the human mouth may be equal to that of sampling moth diversity in a few hundred square kilometers of tropical forest. Currently, the exhaustive counting of microbes is experimentally insuperable; however, in most cases microbiologists are concerned with comparing diversity between samples; therefore, if both samples are treated equally, good predictions can be made.

A number of statistical models have been used to compare microbial community richness; all rely on a species or OTU definition. As stated above, currently there is no consensus on this definition. Stackebrandt and Gobel (1994) used criteria based on the correlation between 16S rDNA and DNA-DNA pairing studies in cultured bacteria to show that two bacteria that share less than a 97.5% 16S rDNA similarity are unlikely to exhibit more than a 60 to 70% DNA similarity, and hence be related at the species level. Therefore, sequence groups that show less than 2.5% 16S rDNA sequence dissimilarity could be a reasonable definition for a bacterial OTU. It is possible to more accurately define an OTU when investigating diversity within specific groups (e.g., classes, genera); in our laboratory we use a definition of less than 1% 16S rDNA sequence dissimilarity to define an actinomycete OTU, which is based on comprehensive 16S rDNA and DNA-DNA pairing data restricted to this class of organisms (J. E. M. Stach, unpublished data). Once an OTU has been defined, it is possible to compare and estimate species richness. A commonly applied approach is rarefaction, which compares observed richness in environments that have been unequally sampled (Hughes et al., 2001). Rarefaction is limited due to the fact that comparisons are based on samples, not on communities; variation is calculated based on the reordering of subsamples within the collected sample, not on the precision of the observed richness (Hughes et al., 2001). Second, rarefaction analyses do not exclude the possibility that accumulation curves cross at a higher sample size; when curves intersect, the ranking of communities by observed species richness depends on sample size, creating inconsistency in comparisons of diversity (Lande et al., 2000). In contrast to rarefaction, richness estimators estimate the total richness of community from a sample. These estimators are divided into three classes: extrapolation from accumulation curves, parametric estimators, and nonparametric estimators. The theory and definitions of these estimators are fully reviewed by Hughes et al. (2001); that paper is considered essential reading for those wishing to apply estimators, and only a brief reiteration is covered here.

Curve extrapolation methods benefit from the fact that once a species has been counted, it does not need to be counted again; the drawback is that for diverse communities, in which only a small fraction of species is detected, several curves often fit equally well but predict different asymptotes (Soberón and Llorente, 1993). Parametric estimators estimate the number of unobserved species in the community by fitting sample data to models of relative species abundance. Curtis et al. (2002) assume log-normal abundance curves and propose that the total number of individuals in a sample (N_T) is given by the area under the curve, and N_{max} and N_{min} are the numbers of individuals of the most and least abundant species or OTUs. The dispersion of the curve (narrow or broad) is proportional to a constant (the spread factor; a in Curtis et al.'s nomenclature, which is an inverse measure) whereas Preston's (1962) canonical hypothesis states that the peak of the individuals' curves (the product of N and S, the numbers of species at any given abundance) will coincide with N_{max}. Curtis and his colleagues opine that it is feasible to estimate S_T (the total number of species or OTUs) by determining the spread factor either by applying the canonical hypothesis or by assuming a value of 1 for N_{min}. Application of this estimator allowed Curtis et al. (2002) to estimate bacterial diversity on a small scale and speculate about diversity at a larger scale; the authors predict that the diversity of the sea may be unlikely to exceed 2×10^6 whereas a metric ton of soil could contain 4×10^6 different taxa. Parametric estimators are currently hindered by the fact that there is still dispute over which models fit distribution patterns best (Hughes et al., 2001) and that large sample sets will be required to resolve this debate.

The final class of estimator, nonparametric estimators, are very promising for microbial studies. Nonparametric estimators are adapted from mark-release-recapture studies and consider the proportion of species that have been observed before to those that are observed once. The nonparametric estimators ACE (abundance-based coverage estimator) (Chao and Lee, 1992) and Chao1 (Chao, 1987) have recently been applied to microbial data sets (Hughes et al., 2001; Dunbar et al., 2002; Stach and Burns, 2002). Nonparametric estimators were developed to incorporate the fact that equal catchability is unattainable in natural populations, even with randomized capture locations on each sampling occasion (Chao, 1987), and to incorporate the fact that many captured individuals (OTUs) are caught rarely in the sample. The evaluation of estimators is based on their precision and bias; to calculate bias, the true species richness of a sample needs to be known. As this is impractical for most microbial communities, a relative comparison of samples will rely on the precision of the estimator (assuming that bias does not influence the relative order of estimates between samples [Hughes et al., 2001]). Precision describes the variation in the number of species expected to be observed if the community were sampled repeatedly (Hughes et al., 2001). In contrast to rarefaction analysis, richness estimators account for the shape of the accumulation curve, thus predicting an intersection point and thereby better predicting relative species richness. The fact that the richness estimator variance relates to the precision of the estimate allows investigators to extrapolate from the range of variance to calculate how many samples would be needed to observe a significant difference between the samples under investigation.

The richness estimators described above have great utility in describing the relative species richness of different communities; however, the caveat is that these estimators do not describe the genetic diversity between samples. This is principally due to the fact that OTUs (however defined) are counted equally, disregarding any divergence or phylogenetic uniqueness. The problem is highlighted by Martin (2002) (again required reading for those wishing to estimate microbial diversity). Consider two hypothetical communities in which the numbers of species, the richness profiles of species, and the rarefaction profiles are identical but that differ in the magnitude of phylogenetic diversity (i.e., the degree of divergence among the sampled sequences). Application of the above estimators would predict that the communities were equally diverse, missing the fact that one contained a greater genetic diversity. The methods by which the phylogenetic diversity of microbial communities can be compared are reviewed by Martin (2002), and only a brief description is offered here. First, the degree of differentiation between the microbial communities can be compared by examining the genetic diversity within each community to the total genetic diversity of the communities combined: $F_{ST} = (\theta_T - \theta_W)/\theta_T$, where θ_T is the genetic diversity for all communities combined and θ_W is the genetic diversity within each community averaged over all the communities being compared. This comparison is subject to the limitations inherent in calculating θ from sequences libraries generated by PCR, i.e., estimation will be affected by PCR bias, drift, and selection (von Wintzingerode et al., 1997). Diversity can also be compared by considering the phylogeny as a cumulative function of the number of lineages relative to time (Martin, 2002). This method relies on tree drawing such that all branch lengths from all terminal taxa to the common ancestor (root) are equal and subsequent division of the tree into arbitrary units of time. The distribution of divergence times are then summarized as a lineage-per-time plot. Fitting best-fit regression equations to the plots yields information as to whether an excess of highly divergent or closely related species are present. This method has the advantage of being unaffected by gene frequencies and therefore avoids some of the bias introduced by molecular techniques. Singleton et al. (2001) advocate a similar approach; they compare the sampling coverage of a community with the evolutionary distances between species and provide the computer program LIBSHUFF (http://www.arches.uga.edu/~whitman/libshuff.html) that calculates whether the communities under investigation are significantly different from each other.

The statistical testing of genetic dissimilarity is an important requirement; two communities that have no OTUs in common may harbor nearly identical phylogenetic diversity (Martin, 2002). Statistical significance can be calculated for F_{ST} (F_{ST} test) by randomly assigning sequences to populations and running 1,000 permutations. An alternative approach is to test whether the distribution of unique sequences between communities exhibits significant covariation with phylogeny (P test [Martin, 2002]). In this test an objective criterion, such as parsimony, is used to estimate the minimum number of changes to switch from one community to the other; significance of the observed covariation is inferred from determining the expected number of changes under the null hypothesis that the community does not covary with phylogeny. Martin (2002) proposes a combination of the F_{ST} and P tests to reveal information about the differentiation between communities. Significance for both tests indicates clear differentiation between communities; insignificance for both tests indicates that the samples are drawn from the same pool of sequences; significance for the P test coupled with insignificance for the F_{ST} test might reflect high levels of diversity in both communities (relative to combined data) but that the phylogenetic lineages differ in each community; finally, a significant F_{ST} test combined with an insignificant P test indicates that the within-community diversity is significantly less than when the two communities are combined and that each community may contain unique sequences distributed across the phylogeny of all samples. Statistical testing of genetic disparity also offers a means to detect which groups of sequences (genera or species) account for differences observed between two communities (by adding or removing sequences and rerunning P and F_{ST} tests), allowing system function to be linked with species composition that is a prerequisite in deciphering the role of microorgansims in ecological processes (Dunbar et al., 2002). The explanatory power of such statistical comparisons is significant when considering ecosystem function; in most, but certainly not all, cases microbiologists will need to know whether two communities have different genetic diversity, not whether two communities have different numbers of species. Of course the application of both richness estimators and statistical methods for calculating phylogenetic disparity are complementary and provide a wealth of information to aid experimental design and hypothesis formulation. In particular, armed with knowledge about the richness and diversity of microbial communities, microbiologists can begin to investigate the spatial, resource, and temporal parameters that affect microbial diversity, questions that until recently were intractable.

Marine Bacterial Diversity: a Case Study

Recently we have applied the statistical methods described above to bioprospect actinomycete di-versity in a deep-sea sediment core (south edge of Saharan debris flow; 27°02.39′N, 18°29.02′W, 3,814 m deep). The core was divided into three sections, representing distinct geological periods, at 5 to 12 cm, 15 to 18 cm, and 43 to 46 cm deep below the seafloor and are referred to as the top, middle, and bottom, respectively. Actinomycete 16S rDNA clone libraries were generated from DNA extracted from the sediments using primers specific for the class *Actinobacteria* (J. E. M. Stach et al., 2003, http://microbe2.ncl.ac.UK/MMB/AAR.htm) and the recovered 16S rDNA sequences were placed into OTUs based on the definition above. Phylogenetic analysis revealed that all sequences belonged to the actinomycete line of descent; a high proportion of sequences were most closely related to the members of the families *Streptomycineae*, *Frankineae*, and *Corynebacterineae*. However, 91% of the sequences displayed less than 99% 16S rDNA homology and therefore are likely to be novel species or genera (Stackebrandt and Gobel, 1994). Accumulation curves plotted for each sediment section community indicated that each had been sampled with roughly equal intensity in proportion to the overall richness, with only a slight increase in coverage in proportion to increasing depth. The curves matched those expected when each type sampled represents an individual (Hughes et al., 2002). The nonparametric estimator ACE (Chao and Lee, 1992) was applied to the data sets using a log transformation to calculate 95% confidence intervals, as the distribution of estimates is not normal (Chao, 1987). Estimates revealed that diversity was highest in the top community and decreased in proportion with depth; 903 (95% confidence intervals, 892 and 915); 453 (218 and 815); and 226 OTUs (156 and 331), respectively. Extrapolation from the range of confidence intervals made it possible to reject the null hypothesis (i.e., that there was no difference between the species richness in the three communities) at the 0.05 significance level.

Application of the LIBSHUFF program (Singleton et al., 2001) indicated that the top community was significantly different from both the middle and the bottom community (P = 0.001). However, the middle and bottom communities were deemed to be similar (P = 0.855). This method is still based on OTUs and therefore does not indicate the degree of phylogenetic diversity. Consequently, lineage-per-time plots were made to investigate whether an excess of divergent or closely related lineages were present. The plots revealed a slight excess of divergent lineages in the top community when compared with the middle community. However, both top and middle communities closely matched the theoretical predictions of diversification corresponding to a constant birth and death rate (Martin, 2002). The plot for the bottom community was convex, indicating an excess of closely related species. In order to further investigate phylogenetic diversity within the three sediment communities, F_{ST} and P tests were applied (Martin, 2002). Both tests were significant for all community combinations, indicating that there was less genetic diversity within each community than for any two communities combined and that the different communities harbored distinct lineages (Martin, 2002).

Bioprospecting of deep-sea sediments by the combined application of methods described above allows us to conclude that the differing sections of the core profile contained significantly different actinomycete populations, both in terms of species richness and phylogenetic diversity; thus, each section represents a unique environment for biotechnological exploitation. Furthermore, such information enables the bioprospector to evaluate how thoroughly a particular environment has been screened by comparing diversity assessments before and after sampling (i.e., culture-independent versus culture-dependent prospecting).

REFERENCES

Andersen, R. A. 1998. What to do with Protists? *Austr. Syst. Bot.* **11**:185–201.

Andersen, R. A., G. W. Saunders, M. P. Paskind, and J. Sexton. 1993. Ultrastructure and 18S rRNA gene sequence for *Pelagomonas calceolata* gen. and sp. nov. and the description of a new algal class, the Pelageophyceae *classis nov. J. Phycol.* **29**:701–715.

André, H. M., M. I. Noti, and P. Lebrun. 1994. The soil fauna: the other last biotic frontier. *Biodiver. Conserv.* **3**:45–56.

Arnold, A. E., Z. Maynard, and G. S. Gilbert. 2001. Fungal endophytes in dicotyledonous neotropical trees: patterns of abundance and diversity. *Mycol. Res.* **105**:1502–1507.

Bakken, L. R. 1995. Separation and purification of bacteria from soil. *Appl. Environ. Microbiol.* **49**:143–169.

Barnes, S. M., S. L. Takala, and C. R. Kuske. 1999. Wide distribution and diversity of members of the bacterial kingdom *Acidobacterium* in the environment. *Appl. Environ. Microbiol.* **65**:1731–1737.

Béjà, O., E. V. Koonin, L. Aravind, L. T. Taylor, H. Seitz, J. L. Stern, D. C. Bensen, R. A. Feldman, R. V. Swanson, and E. F. DeLong. 2002. Comparative genomic analysis of archaeal genotypic variants in a single population and in two different oceanic provinces. *Appl. Environ. Microbiol.* **68**:335–345.

Boucher, Y., C. L. Nesbø, and W. F. Doolittle. 2001. Microbial genomes: dealing with diversity. *Curr. Opin. Microbiol.* **4**:285–289.

Bull, A. T., A. C. Ward, and M. Goodfellow. 2000. Search and discovery strategies for biotechnology: the paradigm shift. *Microbiol. Mol. Biol. Rev.* **64**:573–606.

Busse, H.-J., E. B. M. Denner, and W. Lubitz. 1996. Classification and identification of bacteria: current approaches to an old problem. Overview of methods used in bacterial systematics. *J. Biotechnol.* **47**:3–38.

Chao, A. 1984. Non-parametric estimation of the number of classes in a population. *Scand. J. Stat.* **11**:265–270.

Chao, A. 1987. Estimating the population size for capture-recapture data with unequal catchability. *Biometrics* **43**:783–791.

Chao, A., and S.-M. Lee. 1992. Estimating the number of classes via sample coverage. *J. Am. Stat. Assoc.* **87**:210–217.

Colwell, R. K., and J. A. Coddington. 1995. Estimating terrestrial biodiversity through extrapolation, p. 101–118. *In* D. L. Hawksworth (ed.), *Biodiversity, Measurement and Estimation*. Chapman & Hall, London, United Kingdom.

Coomans, A. 2000. Nematode systematics: past, present and future. *Nematology* **2**:3–7.

Curtis, T. P., W. T. Sloan, and J. W. Scannell. 2002. Estimating prokaryotic diversity and its limits. *Proc. Natl. Acad. Sci. USA* **99**:10494–10499.

Dauga, C. 2002. Evolution of the gyrB gene and the molecular phylogeny of Enterobacteriaceae: a model molecule for molecular systematic studies. *Int. J. Syst. Evol. Microbiol.* **52**:531–547.

Dawson, S. C., and N. R. Pace. 2002. Novel kingdom-level eukaryotic diversity in anoxic environments. *Proc. Natl. Acad. Sci. USA* **99**:8324–8329.

DeLong, E. F., and N. R. Pace. 2001. Environmental diversity of bacteria and archaea. *Syst. Biol.* **50**:1–9.

Dojka, M. A., J. K. Harris, and N. R. Pace. 2000. Expanding the known diversity and environmental distribution of an uncultured phylogenetic division of bacteria. *Appl. Environ. Microbiol.* **66**:1617–1621.

Doolittle, W. F. 1999. Phylogenetic classification and the universal tree. *Science* **284**:2124–2128.

Dunbar, J., S. M. Barns, L. O. Ticknor, and C. R. Kuske. 2002. Empirical and theoretical bacterial diversity in four Arizona soils. *Appl. Environ. Microbiol.* **68**:3035–3045.

Eardly, B. D., F. S. Wang, and P. vanBerkum. 1996. Corresponding 16S rRNA gene segments in Rhizobiaceae and Aeromonas yield discordant phylogenies. *Plant Soil* **186**:69–74.

Edgcomb, V. P., D. T. Kysela, A. Teske, A. de Vera Gomez, and M. L. Sogin. 2002. Benthic eukaryotic diversity in the Guaymas Basin hydrothermal vent environment. *Proc. Natl. Acad. Sci. USA* **99**:7658–7662.

Erwin, T. L. 1982. Tropical forests: their richness in Coleoptera and other arthropd species. *Coleopt Bull.* **36**:74–75.

Fox, G., J. D. Wisotskey, and P. Jurtshuk. 1992. How close is close: 16S rRNA sequence identity may not be sufficient to guarantee species identity. *Int. J. Syst. Bacteriol.* **42**:166–170.

Froese, R., and F. A. Bisby (ed.). 2002. Species 2000 & ITIS Catalogue of Life: Indexing the World's Known Species. CD-ROM. Species 2000, Los Baños, Philippines.

Galvez, A., M. Maqueda, M. Martinez-Bueno, and E. Valdivia. 1998. Publication rates reveal trends in microbiological research. *ASM News* **64**:269–275.

Gaston, K. J., and R. M. May. 1992. Taxonomy of taxonomists. *Nature* **356**:281–282.

Goodfellow, M., G. P. Manfio, and J. Chun. 1997. Towards a practical species concept for cultivable bacteria, p. 25–59. *In* M. F. Claridge, H. A. Dawah, and M. R. Wilson (ed.), *Species: The Units of Biodiversity*. Chapman & Hall, London, United Kingdom.

Grassle, J. F., and N. J. Maciolek. 1992. Deep-sea species richness: regional and local diversity estimates from quantitative bottom samples. *Am. Nat.* **139**:313–341.

Guillou, L., M. J. Chrétiennot-Dinet, L. K. Medlin, H. Claustre, S. Loisseaux-de Goër, and D. Vaulot. 1999. *Bolidomonas*: a new genus with two species belonging to a new algal class, the Bolidophyceae (Heterokonta). *J. Phycol.* **35**:368–381.

Hackstein, J. H. P. 1997. Eukaryotic molecular biodiversity: systematic approaches for the assessment of symbiotic associations. *Antonie Leeuwenhoek* **72**:63–76.

Hamamoto, M., and T. Nakase. 2000. Phylogenetic relationships among fungi inferred from small subunit ribosomal RNA gene sequences, p. 57–71. *In* F. G. Priest and M. Goodfellow (ed.), *Applied Microbial Systematics*. Kluwer Academic Publishers, Dordrecht, The Netherlands.

Hammond, P. M. 1992. Species inventory, p. 17–39. *In* B. Groombridge (ed.), *Global Biodiversity, Status of the Earth's Living Resources*. Chapman & Hall, London, United Kingdom.

Hammond, P. M. 1995a. Described and estimated species numbers: an objective assessment of current knowledge, p. 29–71. *In* D. Allsopp, R. R. Colwell, and D. L. Hawksworth. (ed.), *Microbial Diversity and Ecosystem Function*. CAB International, Wallingford, United Kingdom.

Hammond, P. M. 1995b. The current magnitude of biodiversity, p. 113–138. *In* V. H. Heywood (ed.), *Global Biodiversity Assessment*. Cambridge University Press, Cambridge, United Kingdom.

Harper, J. L., and D. L. Hawksworth. 1995. Preface, p. 5–12. *In* D. L. Hawksworth (ed.), *Biodiversity, Measurement and Estimation*. Chapman & Hall, London, United Kingdom.

Hawksworth, D. L. 1991. The fungal dimension of biodiversity: magnitude, significance, and conservation. *Mycol. Res.* 95:641–655.

Hawksworth, D. L. 2001. The magnitude of fungal diversity: the 1.5 million species estimate revisited. *Mycol. Res.* 105:1422–1432.

Heywood, V. H., and R. T. Watson (ed.). 1995. *Global Biodiversity Assessment*. Cambridge University Press, Cambridge, United Kingdom.

Hughes, J. B., J. J. Hellmann, T. H. Ricketts, and B. J. M. Bohannan. 2001. Counting the uncountable: statistical approaches to estimating microbial diversity. *Appl. Environ. Microbiol.* 67:4399–4406.

Hyde, K. D. 2001. Where are the missing fungi? Does Hong Kong have any answers? *Mycol. Res.* 105:1514–1518.

Karner, M. B., E. F. DeLong, and D. M. Karl. 2001. Archaeal dominance in the mesopelagic zone of the Pacific Ocean. *Nature* 409:507–509.

Kristensen, R. M. 1983. Loricifera, a new phylum with Aschelminthes characters from the meiobenthos. *Z. Zool. Syst. Evol.* 21:163–180.

Kroes, I., P. W. Lepp, and D. A. Relman. 1999. Bacterial diversity within the human subgingival crevice. *Proc. Natl. Acad. Sci. USA* 96:14547–14552.

Lambshead, P. J. D. 1993. Recent developments in marine benthic research. *Oceanis* 19:5–24.

Lan, R., and P. R. Reeves. 2000. Intraspecies variation in bacterial genomes: the need for species genome concept. *Trends Microbiol.* 8:396–401.

Lande, R., P. J. DeVries, and T. R. Walla. 2000. When species accumulation curves intersect: implications for ranking diversity using small samples. *Oikos* 89:601–605.

Lawler, A. 2001. Up for the count. *Science* 294:769–770.

Lawrence, J. G. 2001. Catalyzing bacterial speciation: correlating lateral transfer with genetic headroom. *Syst. Biol.* 50:470–496.

Lopez-Garcia, P., F. Rodriguez-Valera, C. Pedros-Alio, and D. Moreira. 2001. Unexpected diversity of small eukaryotes in deep-sea Antarctic plankton. *Nature* 409:603–607.

Martin, A. P. 2002. Phylogenetic approaches for describing and comparing the diversity of microbial communities. *Appl. Environ. Microbiol.* 68:3673–3682.

Martin, W., and F. Salamini. 2000. A meeting at the gene. Biodiversity and natural history. *EMBO Rep.* 1:208–210.

Massana, R., L. Guillou, B. Diez, and C. Pedro-Alio. 2002. Unveiling the organisms behind novel eukaryotic ribosomal DNA sequences from the ocean. *Appl. Environ. Microbiol.* 68:4554–4558.

May, R. M. 1988. How many species are there on Earth? *Science* 241:1441–1449.

May, R. M. 1992. How many species inhabit the Earth? *Sci. Am.* 267:42–48.

May, R. M. 1995. Conceptual aspects of the quantification of the extent of biological diversity, p. 13–20. *In* D. L. Hawksworth (ed.), *Biodiversity, Measurement and Estimation*. Chapman & Hall, London, United Kingdom.

McCaig, A. E., L. A. Glover, and J. I. Prosser. 1999. Molecular analysis of bacterial community structure and diversity in unimproved and improved upland grass pastures. *Appl. Environ. Microbiol.* 65:1721–1730.

Minelli, A. 1993. *Biological Systematics. The State of the Art.* Chapman & Hall, London, United Kingdom.

Moreira, D., and P. Lopez-Garcia. 2002. The molecular ecology of microbial eukaryotes unveils a hidden world. *Trends Microbiol.* 10:266–267.

Ødegaard, F. 2000. How many species of arthropods? Erwin's estimate revisited. *Biol. J. Linn. Soc.* 71:583–597.

Overmann, J. 2001. Diversity and ecology of phototrophic sulfur bacteria. *Microbiol. Today* 28:116–118.

Patterson, D. J. 1999. The diversity of eukaryotes. *Am. Nat.* 65:S96–S124.

Perna, N. T., G. Plunkett III, V. Burland, B. Mau, J. D. Glasner, D. J. Rose, G. F. Mayhew, P. S. Evans, J. Gregor, H. A. Kirkpatrick, G. Posfai, J. Hackett, S. Klink, A. Boutin, Y. Shao, L. Miller, E. J. Grofbeck, N. W. Davis, A. Lim, E. T. Dimalanta, K. D. Potamousis, J. Apodaca, T. S. Anantharaman, J. Lin, G. Yen, D. C. Schwartz, R. A. Welch, and F. R. Blattner. 2001. Genome sequence of enterohaemorrhagic *Escherichia coli* O157:H7. *Nature* 409:529–533.

Polz, M. F., J. A. Ott, M. Bright, and C. M. Cavanaugh. 2000. When bacteria hitch a ride. *ASM News* 66:531–539.

Prance, G. T. 2001. Discovering the plant world. *Taxon* 50:345–359.

Preston, F. W. 1962. The canonical distribution of commonness and rarity, Parts I and II. Ecology 43:185–215, 410–432.

Purvis, A., and A. Hector. 2000. Getting the measure of biodiversity. *Nature* 405:212–219.

Quicke, D. L. J. 1993. *Principles and Techniques of Contemporary Taxonomy*. Blackie Academic & Professional, London, United Kingdom.

Rabinowitz, D., S. Cairns, and T. Dillin. 1986. Seven forms of rarity and their frequency in the flora of the British Isles, p. 182–204. *In* M. Soulé (ed.), *Conservation Biology: The Science of Scarcity and Diversity*. Sinauer Associates, Sunderland, Mass.

Rainey, F. A., and N. Ward-Rainey. 2000. Prokaryotic diversity, p. 31–42. *In* J. Seckbach (ed.), *Journey to Diverse Worlds. Adaptation to Exotic Environments*. Kluwer Academic Publishers, Dordrecht, The Netherlands.

Reaka-Kudla, M. L. 1997. The global diversity of coral reefs: a comparison with rain forests, p. 83–108. *In* M. L. Reaka-Kudla, D. E. Wilson, and E. O. Wilson (ed.), *Biodiversity II: Understanding and Protecting Our Biological Resources*. Joseph Henry Press, Washington, D.C.

Singleton, D. R., M. A. Furlong, S. L. Rathbun, and W. B. Whitman. 2001. Quantitative comparisons of 16S rRNA gene sequence libraries from environmental samples. *Appl. Environ. Microbiol.* 67:4374–4376.

Sipman, H. J. M., and A. Aptroot. 2001. Where are the missing lichens? *Mycol. Res.* 105:1433–1439.

Soberón, J., and J. Llorente. 1993. The use of species accumulation functions for the prediction of species richness. *Conserv. Biol.* 7:480–488.

Stach, J. E. M., and R. G. Burns. 2002. Enrichment versus biofilm culture: a functional and phylogenetic comparison of polycyclic aromatic hydrocarbon-degrading microbial communities. *Environ. Microbiol.* 4:169–182.

Stach, J. E. M., L. A. Maldonado, A. C. Ward, M. Goodfellow, and A. T. Bull. New primers for *Actinobacteria*: application to marine and terrestrial environments. *Environ. Microbiol.*, in press.

Stackebrandt, E., and B. M. Göbel. 1994. Taxonomic note: a place for DNA-DNA reassociation and 16S rRNA sequence analysis in the present species definition in bacteriology. *Int. J. Syst. Bacteriol.* 44:846–849.

Tilman, D., and C. Lehman. 2001. Biodiversity, composition, and ecosystem processes: theory and concepts, p. 9–41. *In* A. P. Kinzig, S. W. Pacala, and D. Tilman (ed.), *The Functional Consequences of Biodiversity. Empirical and Theoretical Extensions.* Monographs in Population Biology 33. Princeton University Press, Princeton, N. J.

Torsvik, V., J. Goksoyr, and F. L. Daae. 1990. High diversity in DNA of soil bacteria. *Appl. Environ. Microbiol.* 56:782–787.

Torsvik, V., F. L. Daae, R. A. Sandaa, and L. Øvreas. 2000. Molecular biology and genetic diversity of microorganisms, p. 45–57. *In* J. Seckbach (ed.), *Journey to Diverse Worlds. Adaptation to Exotic Environments.* Kluwer Academic Publishers, Dordrecht, The Netherlands.

Valdecasas, A. G., and A. I. Camacho. 2003. Conservation to the rescue of taxonomy. *Biodiver. Conserv.* 12:1113–1117.

van Hoek, A. H. A. M., T. A. van Alen, V. S. I. Sprakel, J. A. M. Leunissen, T. Brigge, G. D. Vogels, and J. H. Hackstein. 2000. Multiple acquisition of methanogenic archaeal symbionts by anaerobic ciliates. *Mol. Biol. Evol.* 17:251–258.

von Wintzingerode, F., U. B. Gobel, and E. Stackebrandt. 1997. Determination of microbial diversity in environmental samples: pitfalls of PCR-based rRNA analysis. *FEMS Microbiol. Rev.* 21:213–229.

Weir, A., and P. M. Hammond. 1997. Laboulbeniales on beetles: host utilization patterns and species richness of the parasites. *Biodiver. Conserv.* 5:701–719.

Whitman, W. B., D. C. Coleman, and W. J. Wiebe. 1998. Prokaryotes: the unseen majority. *Proc. Natl. Acad Sci.* USA 95:6578–6583.

Wilson, E. O. 1992. *The Diversity of Life.* Allen Lane. The Penguin Press, London, United Kingdom.

Wilson, E. O., and F. M. Peter (ed.). 1988. *Biodiversity.* National Academy Press, Washington, D.C.

Woese, C. R. 1987. Bacterial evolution. *Microbiol. Rev.* 51:221–271.

Woese, C. R. 2002. On the evolution of cells. *Proc. Natl. Acad. Sci.* USA 99:8742–8747.

Wolf, Y. I., I. B. Rogozin, N. V. Grishin, R. L. Tatusov, and E. V. Koonin. 2001. Genome trees constructed using five different approaches suggest new major bacterial clades. *BMC Evol. Biol.* 1:1–38.

Microbial Diversity and Bioprospecting
Edited by Alan T. Bull
© 2004 ASM Press, Washington, D.C.

Chapter 3

Defining Microbial Diversity—the Species Concept for Prokaryotic and Eukaryotic Microorganisms

RAMON ROSSELLÓ-MORA AND PETER KÄMPFER

Have enough words been said and written on the subject of what species are? How many evolutionary biologists sometimes wish that not one more word, in speech or text, be spent on explaining species? How many biologists feel that they have a pretty good understanding of what species are? Among those who do, how many could convince a large, diverse group of scientists that they are correct? (Hey, 2001)

In this way Jody Hey states the feelings of many biologists when they hear the words "species concept." Indeed, having to write about what a species is, and to convince the reader that what is being exposed is the best that we can get, is a very difficult task. The species concept is a recurrent topic of discussion among philosophers and taxonomists of all disciplines and has not yet come to a satisfactory end. There is a considerable disagreement among microbiologists about how the species is conceived, and there are arguments that the current concept is too conservative and leads to an underestimation of the real diversity (Cohan, 2002; Dykhuizen, 1998; Whitman et al., 1998). Additionally, this conservative nature of the concept is considered to be a significant disadvantage in as much as it is not comparable with the concepts designed for higher eukaryotes (Staley, 1997; Ward, 1998). However, from such discussions it seems that we are not always aware of the debates being held among taxonomists of higher eukaryotes and/or philosophers dealing with such concepts. The species concept, which actually originates in the higher eukaryote taxonomies, is being continuously discussed and reformulated without reaching a common agreement (Hey, 2001; Hull, 1997; May, 1990; Mayden, 1997). In our case, specifically for prokaryotes, there are few formulated concepts that can be adopted (Rosselló-Mora and Amann, 2001), which

simplifies the discussion about what we mean when we talk about species.

Species are the basis of the taxonomic scheme. They are the universally accepted lowest taxonomic cluster of living organisms; among many uses, they are considered to be the units for measuring biodiversity (Claridge et al., 1997) and evolution (Hull, 1976). In this chapter we discuss this unit and how we can recognize it. Our discussion is mostly from a "prokaryotologist" point of view, as prokaryotes may be endowed with the widest diversity among the living organisms (Whitman et al., 1998). Additionally, microscopic eukaryotes are circumscribed in species whose conception is dependent on the nature of the group they are assigned to and the analysis techniques that taxonomists use to classify them (Brasier, 1997; John and Maggs, 1997; Perkins, 2000; Purvis, 1997). Indeed, a proper revision of the species concepts devised for each of the different groups of microscopic eukaryotes would require a complete chapter for each of them.

WE (TAXONOMISTS) ARE NOT ALONE . . .

. . . in the world of microbiologists. There is someone out there waiting for the microbial taxonomists to achieve a classification system that is useful for his or her own special purposes. Pluralists would recommend constructing a system for each discipline that is directed to certain purposes. However, most systematists define themselves as monists. This means that one can achieve a single organization system that can be used for all (many) special purposes (Hull, 1997). And this should be the aim of a taxonomist, to generate a framework that is used as a common language among all microbiologists. This

Ramon Rosselló-Mora • Grup d'Oceanografia Interdisciplinar, Institut Mediterrani d'Estudis Avançats (CSIC-UIB), Miquel Marqués 21, E-07190 Esporles, (Illes Balears) Spain. **Peter Kämpfer** • Institut für Angewandte Mikrobiologie, Justus-Liebig Universität Giessen, Heinrich-Buff-Ring 26-32, D-35392 Giessen, Germany.

framework should be operational and predictive. The end users do not want to deal with the theoretical background that has been set up to construct the system, but rather to use it without major problems. This is one side of operationalism; one has to be able to identify new isolates at the species level by using those techniques that are available. One cannot be restrictive regarding the amount of approaches that can be used for identification within a given system. This means that the broader the set of interrelated techniques is, the more operative the system will be. Not being restrictive may at the same time produce a system that is predictive. If taxa are extensively characterized, this will lead to a fine-tuned prediction of putative characters for new isolates that require identication. And this is exactly what nontaxonomists, e.g., medical microbiologists, microbial ecologists, and industrial microbiologists, are expecting from us.

Of course any classification scheme will be artificial. Microbiologists have adopted the monistic background of botanists and zoologists. This also means that we agree that a single way can be used for dividing the world into kinds and organizing these kinds into a single hierarchy of laws (Hull, 1997). In this regard, despite the artificialness of the system, there is a universality in the application of it for all living organisms. All taxa above species are abstract entities and are thus comparable among all taxonomies that have adopted the same system. The circumscription of these taxa, however, is dependent on the criteria used to define species, and there is no agreement among the many biological disciplines that construct taxonomies (Rosselló-Mora and Amann, 2001). Species are practical entities for whom circumscriptions may vary depending on the concept applied (biological, phenetic, phylogenetic, evolutionary [Hull, 1997]). However, we could achieve a universal concept if there was a consensus among taxonomists on a philosophical background to apply for the common concept.

WHAT IS A SPECIES, OR WHAT CAN WE REGARD AS A SPECIES?

Species are categories that are motivated by recurrent observations about the world. Humans are great observers of patterns of repetition and devise categories as a response for such observations (Hey, 2001). In the traditional view, the species category is a class defined in terms of properties that particular species possess (e.g., reproductive isolation); particular species are classes defined in terms of the properties that organisms possess (e.g., pigmentation); and

particular organisms are individuals whose names are not defined at all (Hull, 1976). An organism is a member of its species, and each species is a member of the species category. The term species can be used to refer either to a taxonomic category that corresponds to an abstract concept devoid of any spatiotemporal location or to a concrete collective entity made up of real organisms localized in space and time, i.e., taxon (Van Regenmortel, 1997). Taxa, named as natural kinds of organisms, are the result of two processes: (i) the evolutionary processes that have caused biological diversity and (ii) the human mental apparatus that recognizes and gives names to patterns of recurrence (Hey, 2001).

The duality of meanings of the term species has led to confusion and philosophical debates. Actually, biologists accept that genera and higher taxa are artificial, abstract constructions of the mind, but it is interesting to recognize that many accept that species are real (Van Regenmortel, 1997). Indeed, an argument against viewing species as an Aristotelian class is that they are immutable and timeless. Species categories are seen as units of evolution that change through time. For this reason it has been proposed that species should be regarded as individuals, i.e., practical entities, and not as abstract classes (Hull, 1976). Species are constituted of organisms in the same way as individuals are constituted of cells and organs.

However, the main problem of the species concept is its formulation. It seems very difficult to achieve a satisfactory formulation of a concept that embraces all individuals of any different kind into discrete, recognizable, and naturally occurring related groups of organisms. Indeed, a universal species concept is difficult to achieve (Hull, 1997), and to date at least 22 different concepts have been formulated (Mayden, 1997). One of the important problems when formulating a species concept is that systematists tend to have different goals or, even if they agree in principle on what a species concept should provide, frequently prefer different species concepts (Hull, 1997). It appears that the differences among the concepts devised for eukaryotes versus prokaryotes are especially severe. As May (1988) stated, the basic notions about what constitutes a species would be necessarily different for vertebrates than for bacteria. However, it seems that a more uniform conception is being used for prokaryotes than for eukaryotes for which most of the concepts have been devised (Rosselló-Mora and Amann, 2001).

As Hull (1997) advised, scientific concepts should meet several criteria: universality, applicability, and theoretical significance. The two last criteria tend to be in opposition to each other; the more theoretical the concept, the more difficult it is to apply. However, there is an agreement that the species concepts should be as operational (applicable) as possible, and this means that such units should be defined in the framework of the type of data available. Among the myriad of concepts devised, Hull sorts them into three categories with different theoretical commitment and argues that the universality of a concept can best be achieved with pragmatic and low thereotical load, concepts such as those based on overall similarity. Some biologists prefer a pragmatic concept with a low theory load such as the phenetic or polythetic species concept ("a similarity concept based on statistically covarying characteristics which are not necessarily universal among the members of the taxa" [Hull, 1997; Sokal and Crovello, 1970]). Others, however, prefer the highly theoretical evolutionary species concept ("an entity composed of organisms which maintains its identity from other such entities through time and over space, and which has its own independent evolutionary fate and historical tendencies") as the primary concept to be universally applied (Mayden, 1997). Between these concepts there is a wide range of formulations with different theoretical commitments that are being applied to the different classifications of living organisms. Among them are several that have some significance for the prokaryotic world: (i) the phylogenetic species concept with its two different formulations, the monophyletic (or autapomorphic) species concept that is seen as "the least inclusive monophyletic group definable by at least one autapomorphy" and the diagnostic species concept defined as "the smallest diagnosable cluster of individual organisms within which there is a parental pattern of ancestry and descent"; (ii) the biological species concept (BSC) as "groups of interbreeding natural populations that are reproductively isolated from other such groups"; and (iii) the cohesion species concept as "the most inclusive population of individuals having the potential for phenotypic cohesion through intrinsic cohesion mechanisms (among them are gene flow, natural selection, and ecological developmental and historical constraints)" (Hull, 1997).

The phenetic (or polythetic) species concept is considered to be the most operational and universally applicable among those formulated (Hull, 1997). This is extremely useful for dealing with biological entities endowed with intrinsic variability because it can accommodate individual members that lack one or more characters considered typical of the class (Van Regenmortel, 1997). Here species are defined by a combination of characters, each of which may also occur outside the given class and may be absent in any member of the class. However, the success of cir-

cumscribing species in such terms is directly dependent on the amount of characters being used. In the absence of any scientific theory, the only difference between a natural and an artificial classification is the number of characters used. A natural classification is constructed using a large number of characters, whereas an artificial classification is constructed using only a few (Hull, 1970). However, evolutionists or advocates of other concepts disregard this formulation because it has no temporal dimension as they see dimensionality as a virtue (Hull, 1997).

The species concept adopted for prokaryotes is strongly influenced by phenetics. And among all taxonomies of living organisms, it may be the best example of where classification has been constructed by using species circumscribed in part from large numbers of covariant characters. However, this has been a controversial issue through the history of microbiology, and there is now an ongoing tendency to disregard phenetics. Perhaps viruses are the only biological entities for which a polythetic species concept has been formulated and accepted by a large part of the virology community (Van Regenmortel, 1997). Although prokaryotes are currently underrepresented within the species catalog of living organisms, we are convinced that the concept currently in use is satisfactory enough. This is universally applicable among prokaryotes and succeeds for identification purposes (Rosselló-Mora and Amann, 2001).

THE SPECIES CONCEPT FOR MICROSCOPIC EUKARYOTES

Eukaryote classification, in particular of plants and animals, has a much longer history than that of prokaryotes and thus has been discussed and debated more broadly. Modern classification first began with the work of Linnaeus in the 18th century. Linnaeus' notion of species was characterized by three different attributes: (i) distinct and monotypic, (ii) immutable and created as such, and (iii) breeding true (Claridge et al., 1997). He already implied that a species should be defined in terms of sexuality. During the 18th and 19th centuries, naturalists and museums started to collect specimens from all over the world, and the tradition was reinforced that species, and indeed higher taxa, must be based on morphological characters recognizable in preserved specimens. This is the earliest species concept, a morphological species concept or morphospecies, which is not a true concept but the description of a technique that can be stated as "a community, or a number of related communities, whose distinctive morphological characters are, in the opinion of a competent systematist, sufficiently definite to entitle it, or them, to a specific name" (Claridge et al., 1997). However, during these two centuries two distinct tendencies emerged with naturalists emphasizing breeding criteria and systematists emphasizing morphological differences. With the beginning of the 20th century, and particularly with the publication of *Systematics and the Origin of Species* (Mayr, 1942), the BSC was established unifying genetics, systematics, and evolutionary biology (Claridge et al., 1997). This is the most widely known and most controversial species concept to date. Most nontaxonomists refer to a true breeding concept when trying to define a species. The BSC has been refined over the past 50 years and has been successfully applied to many of the animal lineages for which the concept was originally conceived. However, over time, taxonomists of all different disciplines have shown this concept to be unsuccessful in accommodating the smallest recognizable units that they created. The BSC has been shown to be successful for animals, particularly insects (Claridge et al., 1997), most of the invertebrates (Knowlton and Weigt, 1997; Minelli and Foddaiy, 1997), and vertebrates (Cracraft, 1997; Corbet, 1997). However, such a concept is difficult to apply to plants (Gornall, 1997) and animals who reproduce parthenogenetically (Hunt, 1997; Foottit, 1997).

Lower eukaryotes are still mostly classified by using differential morphological traits observed under the microscope. Taxonomists of algae, fungi, or protozoa mainly adopted the morphological species concept (John and Maggs, 1997; Brasier, 1997) or the similarity species concept (Perkins, 2000). However, there is a growing belief that, based solely on morphology, the classification system remains unstable, that molecular biology should be introduced for taxonomic purposes, and that a phylogenetic species concept be adopted based on the analysis of DNA gene sequences and/or DNA fingerprinting techniques such as randomly applied polymorphic DNA or restriction fragment length polymorphism (John and Maggs, 1997; Perkins, 2000). However, it is recognized that discontinuities in morphological variation will undoubtedly remain the principal practical approach to species-level circumscription of many lower eukaryotes (John and Maggs, 1997). Katz (2002) has pointed out that vertical transmission of heritable material, a cornerstone of the Darwinian theory of evolution, is inadequate to describe the evolution of eukaryotes, particularly microbial eukaryotes, because eukaryotic cells and eukaryotic genomes are chimeric, having evolved through a combination of vertical (parent to offspring) and lateral (transspecies) transmission. Observations on widespread chimerism in eukaryotes have led to new and revised hypotheses for the origin

and diversification of eukaryotes that provide specific predictions on the tempo (early versus continuous transfers) and mode (nature of donor and recipient lineages) of lateral gene transfers. It is essential for further testing of hypotheses on the origin and diversification of eukaryotes to sequence and analyze complete genomes from a sufficient number of diverse eukaryotes and prokaryotes combined with sequences of targeted genes from a broad phylogenetic sample (Katz, 2002).

THE SPECIES CONCEPT FOR PROKARYOTES

The field of prokaryote taxonomy experienced most of its growth during the 20th century. The classification system as well as the Linnaean nomenclature were adopted as an analogy to established systems for eukaryotes, in particular the Botanical Code (Sneath, 1992). Throughout the history of prokaryote taxonomy, much more attention has been paid to the nomenclature of taxa (Buchanan et al., 1948; Cowan, 1965; Sneath, 1992; Trüper, 1996, 1999) than to the practical circumscription of the species concept applied to prokaryotes.

Today's prokaryotic species concept results from empirical improvements of what has been thought to be a unit. The circumscription of the species has been optimized through the development of microbiological methods that reveal both genomic and phenotypic properties of prokaryotes, which cannot be retrieved through simple observation. Recently, the current definition of the prokaryotic species has been heavily criticized by some nontaxonomists as too conservative and ill-defined (Dykhuizen, 1998; Ward, 1998; Whitman et al., 1998). On the other hand, many other microbiologists find the current concept acceptable (Goodfellow et al., 1997; Rosselló-Mora and Amann, 2001; Stahl, 1996; Stackebrandt et al., 2002). We argue in the following that, on the basis of the current state of available techniques and information on microorganisms, this concept has shortcomings, but is the most practicable one for the moment. It also fulfills several important requirements for a concept, e.g., the resulting classification scheme is stable, operational, and predictive.

The Concept and Its Definition

Early definitions of bacterial species were often based on monothetic groups. This means that such groups were recognized by a unique set of features considered to be both sufficient and necessary for the definition of the group, and they were invariable (Goodfellow et al., 1997). The rigidity of this circumscription basis led to identification limitations in as much as strains varying in key characters would not be identified as a member of an existing taxon. Additionally, simultaneous classifications and uneven research efforts (e.g., environmentally relevant organisms were underclassified, and those industrially significant were overclassified) often led to nomenclatural confusions in which a single species could be simultaneously classified under several different names (Goodfellow et al., 1997; Van Niel and Allen, 1952).

In the middle of the past century, the simultaneous discovery of DNA as an information-containing molecule and the development of the numerical taxonomy (Sneath and Sokal, 1973) contributed to the changes of the conception of what a species could be and improved the circumscriptions of such units. Particularly important was the tendency to abandon a monothetic in preference to a phenetic or polythetic approach, where species were circumscribed on the basis of a large set of independent covarying characters, each of which may also occur outside the given class, and thus were not exclusive of the class (Van Regenmortel, 1997). There is no official definition of a species in microbiology; however, a prokaryote species is generally considered to be "a group of strains that show a high degree of overall similarity and differ considerably from the related strain groups with respect to many independent characteristics" or "a collection of strains showing a high degree of overall similarity, compared to other, related groups of strains" (Colwell et al., 1995). In most cases, phenetic analyses have been done on the basis of the results of phenotypic data, most often biochemical properties. However, although not included in the same numerical analyses, bacterial phenotype is often extensively studied by the use of chemotaxonomic markers such as fatty acid profiles, polyamines, and quinones (Goodfellow and Minnikin, 1985). As pointed out by Young (2001), the goal of phenetic classification is to create clustered groups of strains, established as a hierarchy of species and genera on the basis of their overall similarities. At every taxonomic level, taxa share common characters and can be circumscribed by a description that distinguishes their members from others at the same level. Phenotypic data include either single-character data or multiple-character data. Single-character biochemical and physiological reactions (e.g., staining reactions, cell and flagellar morphology, pigment production, nutritional characters) describe a small component of the total bacterial phenotype. Multiple-character features comprise complex patterns (e.g., results of chemotaxonomic studies such as comparisons of cell wall composition, fatty acid and protein profiling, but also the results of DNA fingerprinting approaches, etc.) that can be compared directly, for these data, computer-assisted, numerical analyses have also been devised (Goodfellow et al., 1985).

The inclusion of genomic properties to the set of characters used to circumscribe species has promoted important changes and has influenced the definition of a species (Stackebrandt et al., 2002). Particularly decisive was the inclusion of whole-genome hybridizations (Grimont, 1988). The establishment of the DNA hybridization technique as the standard for the delineation of bacterial species was based on the results of numerous investigations in which good agreement was found between DNA relatedness values and corresponding results based on numerical phenotypic and chemotaxonomic data (Goodfellow et al., 1997). In the recommendation made by a group of renowned taxonomists in the late 1980s (Wayne et al., 1987), some numerical boundaries were proposed and for most microbiologists taken as firm limits of what a species should be. Thus, organisms having DNA-DNA reassociation values of 70% or higher, and/or a ΔT_m of 5°C or less, when accompanied by a phenotypic property, would be recognized as a single species. Experience in hybridization experiments had shown that such values should not be taken as absolute boundaries and that, in some cases, more relaxed limits should be taken as limits of single genomic units (Rosselló-Mora and Amann, 2001; Ursing et al., 1995).

The results of DNA-DNA similarity studies have had a tremendous impact on how prokaryotic species are circumscribed and, since the establishment of the technique, most of the new species characterizations included and/or were based on DNA reassociation experiments. Thus, species were tailored by using such results, and any further improvement or application of any new technique to the species definition would be referenced to whole-genome comparisons as the measuring stick (Lan and Reeves, 2000; Stackebrandt and Goebel, 1994; Vandamme et al., 1996). The primacy of DNA reassociation experiments in defining species has been criticized, especially because of the pitfalls of the techniques used (see below) and the lack of enough published data (Goodfellow et al., 1997; Sneath, 1983; see also Staley, chapter 4). These reasons together with the inability to create a cumulative database have led to a recommendation of its substitution (Stackebrandt et al., 2002), but without disregarding its overall value.

A remarkable breakthrough in microbial systematics occurred when the 16S rRNA gene sequence was used in cladistic analyses to draw genealogical (phylogenetic) trees that represent lines of descent (Woese, 1987). As noted by Young (2001), surprisingly, the term phylogeny in this context is rarely defined precisely. Microbes have left no detailed fossil record, and phylogenetic inferences depend almost entirely on the indirect evidence provided by the analysis of sequence data that are considered to express rates of genetic change (nucleotide base mutation) with time. In this context for microorganisms, it is assumed that changes in molecular sequences can be used to infer reliable historical relationships. This assumption, in connection with the easy use of PCR, to amplify conserved 16S rRNA or 23S rRNA and to sequence it has resulted in an explosion of so-called phylogenetic studies.

Phylogenetic reconstructions based on rRNA gene analysis confirmed the monophyletic nature of a given group of strains. Since then, the recognition of many paraphyletic or polyphyletic badly classified species has been possible and has led to further reclassifications. The 16S rRNA gene sequence lacks resolving power at the prokaryotic species level, as defined by DNA-DNA similarity, but permits the identification of the phylogenetic position of new organisms (Ludwig and Schleifer, 1994; Stackebrandt and Goebel, 1994). Its use in new species characterization has been routinely used although it has never been implicitly necessary. In some cases new classifications could have been based solely on phylogenetic reconstructions, but this has been heavily criticized (e.g., Niklas et al., 2001; Palleroni, 1997; Schachter et al., 2001). However, at this point in time, the use of complete 16S rRNA gene sequences in new species classifications is officially recommended as one of the indispensable parameters to be provided (Stackebrandt et al., 2002). Currently the recommended prokaryotic species definition demands the identification of a single group of strains by a discriminative phenotypic property accompanied by their phylogenetic affiliation showing monophyly of its members and a genomic coherency (Stackebrandt et al., 2002). This could be achieved by the simultaneous application of a set of techniques addressed to the understanding of the diversity of the strains under study, following the so-called polyphasic approach (Vandamme et al., 1996).

The current conception of prokaryotic species results from the empirical improvements of what has been thought to be a natural unit in relation to the development and application of techniques of analysis to systematics (Rosselló-Mora and Amann, 2001). The current basis is supported by the nearly 7,700 prokaryotic species that appear to be correctly classified (Euzéby, 2002, http://www.bacterio.cict.fr/). These efforts and the empirical background must not be underestimated when trying to introduce radical changes into the already established classification system.

Several attempts have been undertaken to reformulate the concept of prokaryotic species (Cohan, 2002; Lan and Reeves, 2000; Ward, 1998). However, radical changes in the conception of species would imply important rearrangements of the current classification scheme, and this would lead to confusion. Indeed, taxonomists of higher eukaryotes are reluctant to such changes unless they are demanded by expediency (Lewin, 2001; Niklas et al., 2001). In the recent review on the species concept for prokaryotes (Rosselló-Mora and Amann, 2001) it was concluded that no existing concept can accommodate the current circumscription of the species, but results from the application of phenetic, phylogenetic, and the genomic cluster definitions provide the most appropriate way forward (Rosselló-Mora and Amann, 2001).

The currently practiced species concept in microbiology corresponds partially to several concepts designed for eukaryotes. It is phenetic in that the basis for the understanding of the taxon's coherency and internal diversity is based on the numerical analysis of independently covarying characters, which are not necessarily universally present in the taxon. It is phylogenetic in that the members of these units have to show a common pattern of ancestry, i.e., they must be monophyletic. Finally, it is based on a genotypic (genomic) cluster definition (Mallet, 1995) in the way that genome comparisons, although indirect, give objective numerical frontiers to the unit circumscription and, as discussed above, guarantee the close genealogical relationship of the strains included in a cluster. We suggest that it be referred to as the phylophenetic species concept, indicating the combination of a phenetic evaluation of the unit with requirements for monophylism of its components. A phylophenetic species is "a monophyletic and genomically coherent cluster of individual organisms that show a high degree of overall similarity with respect to many independent characteristics, and is diagnosable by a discriminative phenotypic property" (Rosselló-Mora and Amann, 2001). Or, as pragmatically reformulated, "a species is a category that circumscribes a (preferably) genomically coherent group of individual isolates/strains sharing a high degree of similarity in (many) independent features, comparatively tested under highly standardized conditions" (Stackebrandt et al., 2002).

This definition includes the following requisites: (i) a species should be a monophyletic group of organisms with a high degree of genomic similarity, (ii) the absolute genomic boundaries for the circumscription of each independent species should be particularly defined after the analysis of their phenotype, (iii) the internal homogeneity or heterogeneity of the group can be understood only after the phenetic analysis of as many characters as possible, and (iv) a prokaryotic species should not be classified unless it can be recognized by several independent identification approaches and given a phenotypic set of determinative properties.

The Polyphasic Approach

Prokaryote systematics has undergone spectacular changes in recent years by taking full advantage of developments in chemistry, molecular biology, and computer science to improve the understanding of the relationships between microorganisms and the underlying genetic mechanisms on which they are based (Goodfellow et al., 1997). A relatively large set of techniques are being used routinely for prokaryote classification. However, it is of primary importance to understand at which level these methods carry information. The kind of information that each technique retrieves is directly related to its resolving power, and the correct use of this information is essential to guarantee the adequate classification of a taxon. Extensive reviews on the application of different techniques to prokaryote taxonomy and their resolving power have been published elsewhere (Rosselló-Mora and Amann, 2001; Vandamme et al., 1996).

Currently, prokaryote taxonomists agree that a reliable classification can be achieved only by the exploration of the internal diversity of taxa by a wide range of techniques in what is generally known as the polyphasic approach in the reinterpretation of Vandamme et al. (1996). This approach implies that the two sources of information, genomic parameters and phenotype, must be simultaneously investigated as extensively as possible. Genomic parameters are gained from all data that can be retrieved from nucleic acids, either directly through sequencing or indirectly through parameters such as G+C mol%, DNA-DNA similarity, 16S rRNA gene sequence analysis, or DNA-based typing methods such as restriction fragment length polymorphism, low-frequency restriction pattern analysis, randomly amplified polymorphic DNA, amplified fragment length polymorphism, repetitive extragenic palindromic-PCR (Rep-PCR), and amplified ribosomal DNA restriction analysis, among others.

Phenotype refers to the way in which the genotype is expressed, the visible or otherwise measurable physical and biochemical characteristics of an organism, a result of the interaction of genotype and environment. Phenotype information is retrieved by the use of classical phenotypic analyses including chemotaxonomic studies. Classical phenotypic analyses embrace most of the techniques used to identify or classify microorganisms used in the majority of the

laboratories, including morphological observations of cells and colonies (shape, endospore, flagella, inclusion bodies) and physiological and biochemical features (temperature growth range, salt concentrations, resistances to antimicrobial agents, enzyme activities, metabolism of compounds) among others. Chemotaxonomic markers refer to those chemical constituents of the cells that are useful to characterize prokaryotes, including components such as cell wall composition, cellular fatty acids, isoprenoid quinones, and polyamines. The phenotype typing methods produce single-strain fingerprints that are useful for establishing relationships within a given taxon, which are generally useful for species but they lack discriminative power in higher taxa. Techniques such as serotyping, electrophoretic profiles (whole-cell protein profiles, lipopolysaccharide profiles, multilocus enzyme electrophoresis), and spectroscopy (Fourier-transform infrared spectroscopy, UV resonance Raman spectroscopy) provide strain-unique patterns that might be useful for identification and discrimination purposes.

In a recent meeting, the current definition of species (i.e., the minimal requirements to circumscribe a species properly) was reevaluated within the framework of the new methods that have recently appeared (Stackebrandt et al., 2002). There was a common agreement that new techniques should be evaluated to replace DNA reassociation to give a more accurate hint about the coherency of a group of strains. Methods like multilocus sequence typing or DNA-based typing methods (DNA profiling) such as amplified fragment length polymorphism, Rep-PCR, and PCR-restriction fragment length polymorphism have been recommended to be evaluated as an alternative to DNA-DNA hybridization studies. However, for now whole-genome hybridizations will remain as a standard for showing genomic coherency among strains. Additionally, the ad hoc committee also highlighted the necessity of establishing standardized methods of reporting phenotypic and genotypic data and the inaccuracy or lack of reproducibility of some of the published data. Altogether, the report encourages microbiologists to make descriptions of the organisms they want to classify by an extensive analysis of the phenotype and the genomic parameters of (if possible) more than a single strain and by the use of comparable methods applied to reference strains of closely related taxa (Stackebrandt et al., 2002).

The Role of Genetic Exchange

The existence of lateral gene transfer for bacterial evolution was observed as long ago as 1928 (Veal et al., 1992). However, the extent of lateral gene transfer within the prokaryotic world is now being demonstrated after whole-genome sequence comparisons (Ochman et al., 2000; see also Nelson, Chapter 25). Strong signs of large lateral gene transfer activities among prokaryotes have been found, and indicate that such genomes may incorporate large numbers of unique sequences, often from very divergent organisms. Especially dramatic are the genome differences among strains belonging to the same species, such that in extreme cases two genomes may even differ as much as 20% in gene composition (Boucher et al., 2001; Lan and Reeves, 2000). As pointed out by Young (2001), transfer of parts of the 16S rDNA molecules had been suggested earlier (Sneath, 1993), and the highly conserved nature of 16S rDNA makes it a candidate as a vector of gene transfer (Strätz et al., 1996). In addition, evidence for its recombination has been published (Eardly et al., 1996; Yap et al., 1999).

Genetic exchange among genomes tends to compromise the standard measures to circumscribe species, especially those that retrieve single phenotypic traits as biochemical tests. Indeed, it seems that the type of biochemical differences traditionally used to distinguish strains and species more often results from the acquisition and loss of genes than from mutations in otherwise conserved sequences (Boucher et al., 2001). However, with the increase of available genomic data it seems that horizontal transfer activities might not hinder the recognition of discrete units that we call species. Of course lateral gene transfer provides a venue for bacterial diversification by the reassortment of existing capabilities, thus speciation. However, it seems that there are constraints in how genomes are constructed (Ochman et al., 2000). These constraints are, among others, responsible for the fact that the strains that can be recognized within a single species share a common backbone of genes (Lan and Reeves, 2000).

A very suggestive idea has been formulated by Lan and Reeves (2000), which is the species genome concept. Genes found in most individuals are called the core set of genes for the species and are the genes that determine those properties characteristic of all members of a given species. It is proposed that the core genes are present in at least 95% of strains, whereas an auxiliary set is found in 1 to 95% of strains. Genes less frequent than 1% should be considered foreign or on the decline. This latter set of genes and its phenotypic expression may be responsible for the identification of a new isolate as a member of an existing species. The broader the group of strains studied, the better that the core set of genes will be delimited.

The current circumscription of the species needs not be troubled by the occurrence of gene exchange.

It is true that acquisition of genomic material through lateral gene transfer can lead to misclassifications or misidentifications because of their influence on the phenotype. However, extensive phenetic analysis in which a large set of characters are used would minimize the influence of gene exchange as well as of homoplasies (Rosselló-Mora and Amann, 2001). Additionally, it seems that the extent of shared genes among strains of a single species do correlate with whole-genome hybridization results despite the pitfalls of the method (Lan and Reeves, 2000).

The Uncultured Organisms

In order to properly circumscribe a prokaryotic species it is necessary to analyze pure cultures of at least one strain. To date, the amount of information that we can retrieve from uncultured microorganisms does not match with the minimal set of information that we need to properly classify a species. However, new molecular techniques allow the retrieval of some information on uncultured organisms and permit the recognition of their uniqueness within the hitherto established classification scheme. In this regard, the International Committee on Systematic Bacteriology implemented the category of *Candidatus* to define putative taxa of prokaryotes. This category is used for "describing prokaryotic entities for which more than a mere sequence is available but for which characteristics required for description according to the International Code of Nomenclature of Bacteria are lacking" (Murray and Schleifer, 1994; Murray and Stackebrandt, 1995). Such descriptions should include not only phylogenetic information, but also information on morphological and ecophysiological features as far as they can be retrieved in situ, together with the natural environment of the organism. It is important to note that *Candidatus* is not a rank but a provisional status, and efforts to isolate and characterize the members of the putative taxa should be made to enable their definitive classification. In this regard, efforts to properly describe yet uncultured organisms and giving them a *Candidatus* status are encouraged (Stackebrandt et al., 2002).

Guidelines to the Recognition of a Prokaryotic Species

Although microbiologists have achieved a rather stable classification system using the current circumscription of prokaryotic species, it is important to make a correct use of the definition in order to avoid confusions and misclassifications. Some recommendations for classifying a new prokaryotic species are as follows:

1. Try to isolate or collect an adequate number of strains of the taxon to be studied and use all of them for comparisons. Avoid, although sometimes this is impossible, the description of a species based on a single strain because this could hamper the correct identification of new isolates.

2. Try to recognize the closest related taxa through 16S rRNA analysis and phenotypic characteristics. Include at least the type strains of these related taxa in the taxonomic analyses. The need of the simultaneous use of type strains in laboratory experiments will be necessary until standardized methods and enough cumulative databases are available to the scientific community.

3. Do not use values of 70% DNA similarity (or 5°C a ΔT_m) as absolute limits for circumscribing a new species. The current concept allows more relaxed DNA-DNA similarity frontiers, and an internal genomic heterogeneity is permitted. A single species can consist of several genomic groups (genomovars) that do not necessarily have to be classified as different species. This will be possible when a phenotypic property is found that identifies each of them. As reassociation experiments might be substituted by alternative techniques, the implementation of such techniques should be in sufficient congruence with the results retrieved by DNA-DNA hybridization experiments (Stackebrandt et al., 2002).

4. Make an effort to characterize the phenotype of the organisms. Although commercially available test kits based on biochemical and physiological characters may be useful (Analytab Products, Biolog), sometimes the information retrieved from the results might be insufficient. A critical view on the usefulness of these test kits, often designed for medically important organisms, is demanded. The phenotype is described not only by metabolism, but in addition there are chemotaxonomic markers (fatty acids, polyamines, quinones), which are in some cases more stable, i.e., independent from cultivation conditions that produce important information on organisms. The more exhaustively the phenotype is described, the better the circumscription.

5. Be generous with time and effort when taxonomically analyzing the strains. The classification of a species is not an easy task and should not be underestimated.

6. Follow the nomenclature rules (Sneath, 1992; Trüper, 1999); this is the best discussed and established side of prokaryotic taxonomy. Avoid using words that are hard to pronounce if you do not want to annoy your colleagues.

Of course, these are recommendations and may sometimes be difficult to follow. It is still necessary to obtain pure cultures of microorganisms to recognize them as a new species. We cannot retrieve enough information from uncultured organisms to achieve a correct and stable classification.

Epilogue

Species should be a pragmatic unit. Pragmatic does not mean that the unit is easy to circumscribe (this in terms of applying methodologies), but that with this unit we will construct a taxonomic schema that is operational and predictive. Microbiologists are, in many cases, one step ahead of other taxonomists. We have achieved a polythetic taxonomy that covers as much as possible the phenotype and genome and have been flexible with regard to character variation. A classification system cannot be based only on genomic characters because the system should be predictive in terms of how the organism is and not only what is it. The end users of the taxonomic schema want to understand potential dangers (clinical microbiologists), potential functions in their environment (ecologists), and potential biotechnological applications (industry and pharmacy). Perhaps we will be able to abandon the phenotyping when genome sequencing is routine and we can assign a function for all putative genes. However, even knowing the gene pool and the putative function of each gene, this does not mean that we understand how the organism functions. For example, we cannot retrieve from the genome information the pattern (amount and composition) of fatty acids that this very genome will exhibit, but only (and only probably) the putative composition. An organism is not only its genotype but the multiple combinations of its expression, and that is what most microbiologists (not taxonomists) expect to understand when they deal with a name.

Acknowledgments. R.R.-M. acknowledges the Ministerio de Ciencia y Tecnología for support of research activities on microbial systematics. Natushcka Lee is acknowledged for helpful and constructive discussions.

REFERENCES

Boucher, Y., C. L. Nesbø, and W. F. Doolitle. 2001. Microbial genomes: dealing with diversity. *Curr. Opin. Microbiol.* 4: 285–289.

Brasier, C. M. 1997. Fungal species in practice: identifying species units in fungi, p. 135–170. *In* M. F. Claridge, H. A. Dawah, and M. R. Wilson (ed.), *Species: The Units of Biodiversity.* Chapman & Hall, London, United Kingdom.

Buchanan, R. E., R. S. John-Brooks, and R. S. Breed. 1948. International bacteriological code of nomenclature. *J. Bacteriol.* 55:287–306.

Claridge, M. F., H. A. Dawah, and M. R. Wilson. 1997. Practical approaches to species concepts for living organisms, p. 1–15. *In* M. F. Claridge, H. A. Dawah, and M. R. Wilson (ed.), *Species:*

The Units of Biodiversity. Chapman & Hall, London, United Kingdom.

Cohan, F. M. 2002. What are bacterial species? *Annu. Rev. Microbiol.* **56**:457–487.

Colwell, R. R., R. A. Clayton, B. A. Ortiz-Conde, D. Jacobs, and E. Russek-Cohen. 1995. The microbial species concept and biodiversity, p. 3–15. *In* D. Allsopp, R. R. Colwell, and D. L. Hawksworth (ed), *Microbial Diversity and Ecosystems Function*. CAB International, Oxon, United Kingdom.

Corbet, G. B. 1997. The species in mammals, p. 341–356. *In* M. F. Claridge, H. A. Dawah, and M. R. Wilson (ed.), *Species: The Units of Biodiversity*. Chapman & Hall, London, United Kingdom.

Cowan, S. T. 1965. Principles and practice of bacterial taxonomy—a forward look. *J. Gen. Microbiol.* **39**:148–159.

Cracraft, J. 1997. Species concepts in systematics and conservation biology—an ornithological viewpoint, p. 325–340. *In* M. F. Claridge, H. A. Dawah, and M. R. Wilson (ed.), *Species: The Units of Biodiversity*. Chapman & Hall, London, United Kingdom.

Dykhuizen, D. E. 1998. Santa Rosalia revisited: why are there so many species of bacteria? *Antonie Leeuwenhoek* **73**:25–33.

Eardly, B. P., F. S. Wang, P. van Berkum. 1996. Corresponding 16S rRNA gene segments in *Rhizobiaceae* and *Aeromonas* yield discordant phylogenies. *Plant Soil* **186**:69–74.

Euzeby, J. 2002. List of bacterial names with standing nomenclature. http://www.bacterio.cict.fr/

Foottit, R. G. 1997. Recognition of parthenogenetic insect species, p. 291–308. *In* M. F. Claridge, H. A. Dawah, and M. R. Wilson (ed.), *Species: The Units of Biodiversity*. Chapman & Hall, London, United Kingdom.

Goodfellow, M., and D. E. Minnikin (ed.). 1985. *Chemical Methods in Bacterial Systematics*. Academic Press, London, United Kingdom.

Goodfellow, M., D. Jones, and F. G. Priest (ed.). 1985. *Computer-Assisted Bacterial Systematics*. Academic Press, London, United Kingdom.

Goodfellow, M., G. P. Manfio, and J. Chun. 1997. Towards a practical species concept for cultivable bacteria, p. 25–60. *In* M. F. Claridge, H. A. Dawah, and M. R. Wilson (ed.), *Species: The Units of Biodiversity*. Chapman & Hall, London, United Kingdom.

Gornall, R. J. 1997. Practical aspects of the species concept in plants, p. 171–190. *In* M. F. Claridge, H. A. Dawah, and M. R. Wilson (ed.), *Species: The Units of Biodiversity*. Chapman & Hall, London, United Kingdom.

Grimont, P. A. D. 1988. Use of DNA reassociation in bacterial classification. *Can. J. Microbiol.* **34**:541–546.

Hey, J. 2001. The mind of the species problem. *Trends Ecol. Evol.* **16**:326–329.

Hull, D. L. 1970. Contemporary systematic philosophies. *Annu. Rev. Ecol. Syst.* **1**:19–54.

Hull, D. L. 1976. Are species really individuals? *Syst. Zool.* **25**:174–191.

Hull, D. L. 1997. The ideal species concept—and why we can't get it, p. 357–380. *In* M. F. Claridge, H. A. Dawah, and M. R. Wilson (ed.), *Species: The Units of Biodiversity*. Chapman & Hall, London, United Kingdom.

Hunt, D. J. 1997. Nematode species: concepts and identification strategies exemplified by the *Longidoridae*, *Steinernematidae* and *Heterorhabditidae*, p. 221–246. *In* M. F. Claridge, H. A. Dawah, and M. R. Wilson (ed.), *Species: The Units of Biodiversity*. Chapman & Hall, London, United Kingdom.

John, D. M., and C. A. Maggs. 1997. Species problems in eukaryotic algae: a modern perspective, p. 83–108. *In* M. F. Claridge, H. A. Dawah, and M. R. Wilson (ed.), *Species: The Units of Biodiversity*. Chapman & Hall, London, United Kingdom.

Katz, L. A. 2002. Lateral gene transfers and the evolution of eukaryotes: theories and data. *Int. J. Syst. Evol. Microbiol.* **52**:1893–1900.

Knowlton, N., and L. A. Weigt. 1997. Species of marine invertebrates: a comparison of the biological and phylogenetic species concept, p. 199–220. *In* M. F. Claridge, H. A. Dawah, and M. R. Wilson (ed.), *Species: The Units of Biodiversity*. Chapman & Hall, London, United Kingdom.

Lan, R., and P. R. Reeves. 2000. Intraspecies variation in bacterial genomes: the need for a species genome concept. *Trends Microbiol.* **8**:396–401.

Lan, R., and P. R. Reeves. 2001. When does a clone deserve a name? A perspective on bacterial species based on population genetics. *Trends Microbiol.* **9**:419–424.

Lewin, R. A. 2001. Why rename things? *Nature* **410**:637.

Ludwig, W., and K. H. Schleifer. 1994. Bacterial phylogeny based on 16S and 23S rRNA sequence analysis. *FEMS Microbiol. Rev.* **15**:155–173.

Mallet, M. 1995. A species definition for the modern synthesis. *Trends Ecol. Evol.* **10**:294–299.

May, R. M. 1988. How many species are there on Earth? *Science* **241**:1441–1449.

May, R. M. 1990. How many species? *Phil. Trans. R. Soc. London Ser. B* **330**:293–304.

Mayden, R. L. 1997. A hierarchy of species concepts: the denouement in the saga of the species problem, p. 381–324. *In* M. F. Claridge, H. A. Dawah, and M. R. Wilson (ed.), *Species: The Units of Biodiversity*. Chapman & Hall, London, United Kingdom.

Mayr, E. 1942. *Systematics and the Origin of Species from the Viewpoint of a Zoologist*. Columbia University Press, New York, N.Y.

Minelli, A., and Foddai, D. 1997. The species in terrestrial non-insect invertebrates (earthworms, arachnids, myriapods, woodlice and snails), p. 309–325. *In* M. F. Claridge, H. A. Dawah, and M. R. Wilson (ed.), *Species: The Units of Biodiversity*. Chapman & Hall, London, United Kingdom.

Murray, R. G. E., and K. H. Schleifer. 1994. Taxonomic notes: a proposal for recording the properties of putative taxa of prokaryotes. *Int. J. Syst. Bacteriol.* **44**:174–176.

Murray, R. G. E., and E. Stackebrandt. 1995. Taxonomic note: implementation of the provisional status *Candidatus* for incompletely described prokaryotes. *Int. J. Syst. Bacteriol.* **45**:186–187.

Niklas, K. J., K. de Queiroz, M. Donoghue, H. Gest, J. Favinger, and D. M. Abbey. 2001. Taxing debate for taxonomists. *Science* **292**:2249–2250.

Ochman, H., J. G. Lawrence, and E. A. Groisman. 2000. Lateral gene transfer and the nature of bacterial innovation. *Nature* **405**:299–304.

Palleroni, N. J. 1997. Prokaryotic diversity and the importance of culturing. *Antonie Leeuwenhoek* **72**:3–19.

Perkins, S. L. 2000. Species concepts and malaria parasites: detecting a cryptic species of *Plasmodium*. *Proc. R. Soc. London* **267**:2345–2350.

Purvis, O. W. 1997. The species concept in lichens, p. 109–134. *In* M. F. Claridge, H. A. Dawah, and M. R. Wilson (ed.), *Species: The Units of Biodiversity*. Chapman & Hall, London, United Kingdom.

Rosselló-Mora, R., and R. Amann. 2001. The species concept for prokaryotes. *FEMS Microbiol. Rev.* **25**:39–67.

Schachter, J., R. S. Stephens, P. Timms, C. Kuo, P. M. Bavoil, S. Birkelund, J. Boman, H. Caldwell, L. A. Campbell, M. Chernesky, G. Christiansen, I. N. Clarke, C. Gaydos, J. T. Grayston, T. Hackstadt, R. Hsia, B. Kaltenboeck, M. Leinonnen, D. Ocjius, G. McClarty, J. Orfila, R. Peeling, M.

Puolakkainen, T. C. Quinn, R. G. Rank, J. Rauslton, G. L. Ridgeway, P. Saikku, W. E. Stamm, D. Taylor-Robinson, S.-P. Wang, and P. B. Wyrick. 2001. Radical changes to chlamydial taxonomy are not necessary just yet. *Int. J. Syst. Evol. Microbiol.* **51:**249.

Sneath, P. H. A. 1983. Distortions of taxonomic structure from incomplete data on a restricted set of reference strains. *J. Gen. Microbiol.* **129:**1045–1073.

Sneath, P. H. A. 1992. *International Code of Nomenclature of Bacteria.* American Society for Microbiology, Washington D.C.

Sneath, P. H. A. 1993. Evidence from *Aeromonas* for genetic crossing-over in ribosomal sequences. *Int. J. Syst. Bacteriol.* **43:**626–629.

Sneath, P. H. A., and R. R. Sokal. 1973. *Numerical Taxonomy.* W. H. Freeman & Co., San Francisco, Calif.

Sokal, R. R., and T. J. Crovello. 1970. The biological species concept: a critical evaluation. *Am. Nat.* **104:**127–153.

Stackebrandt, E., and B. M. Goebel. 1994. Taxonomic note: a place for DNA-DNA reassociation and 16S rRNA sequence analysis in the present species definition in bacteriology. *Int. J. Syst. Bacteriol.* **44:**846–849.

Stackebrandt, E., W. Frederiksen, G. M. Garrity, P. A. D. Grimont, P. Kämpfer, M. C. J. Maiden, X. Nesme, R. Rosselló-Mora, J. Swings, H. G. Trüper, L. Vauterin, A. C. Ward, and W. B. Whitman. 2002. Report of the ad hoc committee for the re-evaluation of the species definition in bacteriology. *Int. J. Syst. Evol. Microbiol.* **52:**1043–1047.

Stahl, D. A. 1996. Molecular approaches for the measurement of density, diversity and phylogeny, p. 102–114. *In* C. J. Hurst, G. R. Knudsen, M. J. McInerney, L. D. Stetzenbach, and M. V. Walter (ed.), *Manual of Environmental Microbiology.* ASM Press, Washington D.C.

Staley, J. T. 1997. Biodiversity: are microbial species threatened? *Curr. Opin. Biotechnol.* **8:**340–345.

Strätz, M., M. Mau, and K. N. Timmis. 1996. System to study horizontal gene exchange among microorganisms without cultivation of recipients. *Mol. Microbiol.* **22:**207–215.

Trüper, H. G. 1996. Help! Latin! How to avoid the most common mistakes while giving Latin names to newly discovered prokaryotes. Microbiologia **12:**473–475.

Trüper, H. G. 1999. How to name a prokaryote? Etymological considerations, proposals and practical advice in prokaryote nomenclature. *FEMS Microbiol. Rev.* **23:**231–249.

Ursing, J. B., R. A. Rosselló-Mora, E. Garcia-Valdés, and J. Lalucat. 1995. Taxonomic note: a pragmatic approach to the nomenclature of phenotypically similar genomic groups. *Int. J. Syst. Bacteriol.* **45:**604.

Vandamme, P., B. Pot, M. Gillis, P. De Vos, K. Kersters, and J. Swings. 1996. Polyphasic taxonomy, a consensus approach to bacterial systematics. *Microbiol. Rev.* **60:**407–438.

Van Niel, C. B., and M. B. Allen. 1952. A note on *Pseudomonas stutzeri. J. Bacteriol.* **64:**413–422.

Van Regenmortel, M. H. V. 1997. Viral species, p. 17–24. *In* M. F. Claridge, H. A. Dawah, and M. R. Wilson (ed.), *Species: The Units of Biodiversity.* Chapman & Hall, London, United Kingdom.

Veal, D. A., H. W. Stokes, and G. Daggard. 1992. Genetic exchange in natural microbial communities, p. 383–430. *In* K. C. Marshall (ed.), *Advances in Microbial Ecology,* vol. 12. Plenum Press, New York, N.Y.

Ward, D. M. 1998. A natural species concept for prokaryotes. *Curr. Opin. Microbiol.* **1:**271–277.

Wayne, L. G., D. J. Brenner, R. R. Colwell, P. A. D. Grimont, O. Kandler, M. I. Krichevsky, L. H. Moore, W. E. C. Moore, R. G. E. Murray, E. Stackebrandt, M. P. Starr, and H. G. Trüper. 1987. Report of the ad hoc committee on reconciliation of approaches to bacterial systematics. *Int. J. Syst. Bacteriol.* **37:**463–464.

Whitman, W. B., D. C. Coleman, and W. J. Wiebe. 1998. Prokaryotes: the unseen majority. *Proc. Natl. Acad. Sci. USA* **95:**6578–6583.

Woese, C. R. 1987. Bacterial evolution. *Microbiol. Rev.* **51:**221–271.

Yap, W. H., Z. Zhang, and Y. Wang. 1999. Distinct types of rRNA operons exist in the genome of the actinomycete *Thermomonospora chromogena* and evidence for horizontal gene transfer of an entire rRNA operon. *J. Bacteriol.* **181:**5201–5209.

Young, J. M. 2001. Implications of alternative classifications and horizontal gene transfer for bacterial taxonomy. *Int. J. Syst. Evol. Microbiol.* **51:**945–953.

Microbial Diversity and Bioprospecting
Edited by Alan T. Bull
© 2004 ASM Press, Washington, D.C.

Chapter 4

Speciation and Bacterial Phylospecies

JAMES T. STALEY

Speciation is the evolutionary process whereby one species separates into two. The speciation event that occurs is intrinsic to the organism that is evolving. However, the speciation process occurs in the environment in which the species lives. So, both the organism and the extrinsic environmental factors play a role in speciation. The extrinsic factors of the environment that affect speciation may include physical, geographical, chemical, and biological factors. These ecological factors not only provide the setting in which speciation occurs, but they comprise the selective force for the speciation process. However, it is the response of the organism to the environmental factors that is the hallmark of speciation. The organism's response takes the form of genetic variation in which mutation and horizontal gene transfer play roles, along with natural selection. In this chapter I begin by discussing animal and plant speciation followed by bacterial speciation.

SPECIATION IN ANIMALS AND PLANTS

The animal and plant concepts of species and speciation serve as an excellent reference point to begin our discussion. First, the species concept is important in understanding the speciation process in that it provides the constraint necessary for evaluating the fulfillment of the speciation event. A variety of species concepts have been proposed for microorganisms (Rosselló-Mora and Amann, 2001; Rosselló-Mora and Kämpfer, Chap. 3); however, three major species concepts are currently favored (Freeman and Herron, 1998). One is termed the biological species concept (BSC). First proposed by Ernst Mayr (1942), this is used as the actual legal definition of a species for the Endangered Species Act in the United States. According to the BSC, speciation in animals and plants occurs through reproductive isolation within subpopulations of a species. When reproductive isolation occurs, the separated populations can no longer exchange genetic material. Over time, mutation, genetic drift, and natural selection give rise to two distinct populations that become separate species. In the classic scenario, these two separate species cannot successfully interbreed with one another even if they are subsequently brought back together. Thus, according to the BSC, plant or animal species successfully breed to produce fertile progeny with other members of their species. It is important to recognize that the BSC is not applicable to organisms that reproduce asexually.

The morphospecies concept is based on morphological differences among organisms. This concept has been particularly useful in paleontology because there is little other evidence to describe a species apart from its appearance in fossilized material. Because of its simplicity and wide applicability for macroscopic organisms, the morphospecies concept has been favorably received by many plant and animal taxonomists. Also, it is commonly applied to eukaryotic microorganisms by taxonomists who use microscopic evidence of morphological traits. However, it has several limitations. For example, the morphological criteria used by one investigator may not agree with those used by another. Most importantly, from our standpoint, it is not useful for *Bacteria* and *Archaea* because most of these organisms are too simple morphologically.

The third concept is the phylogenetic species concept (PSC) that is based on the evolutionary relatedness among organisms (Rosen, 1978; Nelson and Platnick, 1981). According to the PSC, species are considered as an "irreducible cluster of organisms, diagnosably different from other such clusters and within which there is a parental pattern of ancestry and descent" (Craycraft, 1989). More recently, with the availability of molecular sequencing approaches, species can be considered as the fundamental monophyletic group based on traits, in par-

---

James T. Staley • Department of Microbiology, University of Washington, Seattle, WA 98195.

ticular, appropriate proteins and genes whose sequences can be subjected to phylogenetic analysis. Two special advantages of this approach are that the results can be tested and, especially important for our consideration, it applies to all organisms including *Bacteria* and *Archaea*.

Three different mechanisms of speciation are known in animals and plants: allopatric speciation, parapatric speciation, and sympatric speciation.

Allopatric speciation is brought about by the physical separation of a population into two geographic areas. This process is thought to be the most common speciation process for plants and animals. Isolated islands in the Pacific Ocean, such as the Hawaiian Islands, provide numerous examples of allopatric speciation. The Hawaiian Islands arose in the mid-Pacific about 70 million years ago, thousands of kilometers from other land masses and therefore separated from extant terrestrial plant and animal life. Rare colonization events, due to air and water dispersal, introduced a few seeds, birds, and small animals to the newly formed islands. A few of these species successfully colonized the islands, beginning when the first islands formed to the northwest of the 2,000-km-long archipelago. Some of the species that had been transported to the islands from other places became established as founder species. Over time these species evolved by mutation and selection to occupy niches in the new habitat and became separate species from the original founder populations. New species also arose through further allopatric speciation events within the Hawaiian Islands as new islands were subsequently formed farther south as the Pacific plate shifted to the northwest over the hot spot that generated all of the islands.

Parapatric speciation occurs within a species in the same location. This process occurs in organisms in which natural selection dominates over genetic exchange. So, either decreased genetic exchange or increased pressures of natural selection can account for splitting an existing species into two. Examples of this are found in habitats that have been disrupted. For example, in mining areas, toxic mine tailings containing heavy metals such as zinc and lead may be left exposed. Over several centuries, new species of plants have evolved that are resistant to the heavy metals. They do not interbreed with the original species because they flower at different times.

Like parapatric speciation, sympatric speciation occurs within a species in the same locale. This process usually involves polyploidy, an increase in the number of chromosomes, which can occur in plants, leading to genetic isolation of the polyploid from the parent species.

Even in plant and animal life, the speciation process is poorly understood. Darwinian views of speciation, termed phyletic gradualism, assumed that speciation was a gradual process in which small variations occurred over many millenia before a distinct new species was formed. However, another model, that of punctuated equilibrium, has been proposed more recently to explain what appear to be abrupt changes in species composition over relatively short intervals in the fossil record (Eldredge and Gould, 1972). The most abrupt changes in the fossil record have been correlated with sudden environmental changes caused by cosmic events such as meteor impacts, some of which are known to have caused the extinction of up to 90% of animal species. These extinction events led to relatively rapid evolution of new species to fill vacant and novel niches.

An important aspect of speciation in plants and animals is that it does not necessarily involve dramatic changes in the genome. As discussed below, the genetic differences between humans and chimpanzees, our most closely related species, are minor. The human has increased brain capacity and the ability to carry out verbal communication. Although the genetic differences have resulted in major changes between the roles and activities of humans versus chimpanzees, they are derived from relatively minor changes in the genome. Indeed, it has been estimated that less than 1% of the genome has seen dramatic change. Likewise, small variations of the *hox* genes of animals can result in major morphological changes quickly leading to the creation of novel species.

BACTERIAL SPECIES CONCEPT

The concept of the species is of paramount importance when considering the process of speciation as illustrated for plants and animals. Of the three species concepts used for macrobes, only the PSC is of use for *Bacteria* and *Archaea*. Like plants and animals, a bacterial species has a monophyletic ancestry. However, bacteria are haploid organisms throughout their life cycles and they also differ from the standpoint of sexuality and gene exchange.

Despite the lack of a clear concept of the bacterial species as an evolving entity, a functional species definition exists. The definition is based on the degree of DNA/DNA reassociation (Brenner et al., 2000) between strains that are being compared. Two organisms are considered to comprise the same species if their purified DNA exhibits greater than 70% hybridization (Wayne et al., 1987). In contrast, the degree of DNA/DNA hybridization between humans and our closest relative, chimpanzees, is 98.6% (Sibley and Ahlquist, 1987).

Table 1. Comparison of *E. coli* with its primate host species[a]

Organism	% DNA/DNA reassociation	Mol% G+C	16S-18S rRNA variability	Genome size range
E. coli	>70	48–52	>15 bases	ca. 25%
H. sapiens	98.6[b]	42	?	?
Primate order	>70[c]	42[d]	<16 bases[e]	?

[a] Adapted from Staley (1999).
[b] Comparison between *H. sapiens* and chimpanzee.
[c] Comparison between *H. sapiens* and lemurs.
[d] Value for all primates.
[e] Mouse 18S rRNA differs from human rRNA by about 16 bases.

What about other molecular properties such as DNA base composition (mol% G+C), the range in 16S and 18S rDNA sequence values, and genome size? In this regard it is interesting to compare *Escherichia coli* with one of its host mammalian species *Homo sapiens* (Staley, 1997). Whereas the mol% G+C of humans is about 42% with little variation, if any within the species, the range for *E. coli* is about 4.0% (Table 1). Likewise, the intraspecies range of 16S-rDNA sequence is conservatively about 1% for *E. coli* whereas it is less than that between the primate *H. sapiens* compared with the house mouse, *Mus musculus*, a mammalian species in the rodent order. Furthermore, we now know that some strains of *E. coli* have a genome size of 4.6 Mb whereas others are greater than 5.5 Mb, a striking difference of more than 20%. It is highly unlikely that the human genome range is greater than 1%.

This comparison clearly shows that the bacterial species is much broader than that of its hosts. Indeed, one could argue on this basis that the bacterial species is comparable to that of an animal family or perhaps even an animal order.

Some microbiologists have pointed out that the bacterial species concept based on DNA/DNA reassociation is arbitrary (Staley, 1997) and artificial (Ward, 1998) in that a single strain is selected, i.e., one that happens to be available in culture, which is designated as the type strain of a species. The type strain is then deposited in a culture collection as a reference for DNA hybridization analyses and phenotypic comparisons with closely related organisms. The immobilized deep-frozen pure culture is no longer an entity that is evolving in its normal community.

This brief review of the current status of species and speciation raises two important issues:

- The only species concept that promises to be universally applicable to all organisms is the PSC.
- Speciation in *Bacteria* and *Archaea* is not occurring at the level of the currently defined species, it is occurring at the level of strains.

SPECIFIC RECOMMENDATION: THE PHYLOSPECIES

The two issues raised above need to be resolved before we further consider bacterial speciation. The PSC is especially appealing because it would enable biologists to have a universal concept for a species. However, because of practical reasons, I do not advocate that the current bacterial species concept or definition be eliminated at this time. First, an immediate change would have a confounding impact on the practice of microbiology in fields such as medicine, plant pathology, and industrial microbiology. In addition, it is not critical to change the species definition immediately even though it might be justified scientifically or because taxonomists who are interested in speciation and biodiversity would welcome this change. Nonetheless, a PSC can be considered to be a goal for microbial taxonomists in the not too distant future.

In the meantime, bacteriologists who are interested in speciation need to look at this process at the level of strains. In particular, I am proposing the following term: phylospecies. The term phylospecies is coined to describe those microorganisms, in particular *Bacteria* and *Archaea*, that, according to the phylogenetic species concept, form a monophyletic clade at a fundamental level and that are occupying, living, and evolving in a specific niche. Admittedly, this is a subjective definition because bacteriologists know so little about the niche of specific organisms where the evolutionary process is occurring. As researchers accumulate more data, better bounds on the phylospecies should become apparent.

Also, it is important to note that it would not be essential to name these new phylospecies even though they could be described, at least until there is a better understanding of how they are related to current taxa. New names, unless carefully considered, will cause a great deal of confusion until the broader scientific community of microbiologists understand and concur with the implications of a revised taxonomy. Perhaps specific numbers, such as *Pseudomonas putida,* ps (for phylospecies) A101 could designate a strain from a specific niche. Indeed, it may be preferable to have a dual system of classification of *Bacteria* and *Archaea* during this transitional period as discussed below.

Because the important events of evolution occur at the level of strains, in particular in the bacterial world, rather than at the level of the currently defined bacterial species, we use the term speciation to denote the process by which phylospecies originate.

I first discuss the environmental factors that may drive the speciation of an organism, and then I treat the intrinsic responses of the organism in its environment that affect a speciation event.

ENVIRONMENTAL (OR EXTRINSIC) FACTORS

Geography

Although allopatric speciation is the major factor thought to account for plant and animal speciation, until recently there has been no clear evidence that geographic separation is responsible for speciation in bacteria (see chapter 22). Certain eukaryotic microbiologists have vehemently argued that all free-living eukaryotic microorganisms are cosmopolitan (Finlay, 2002; Finlay and Esteban, chapter 21). The dismissal of the possibility of endemism of eukaryotic microorganisms likely lies in the common use of the morphospecies concept to distinguish species from one another. The rather simple morphological features of these microorganisms may also belie the true nature of endemism, which may occur at the level of strains or phylospecies. Clearly much more research is needed to address this important issue as its solution is necessary to help microbiologists better resolve the total number of microbial species and hence their biodiversity. Furthermore, knowledge of endemism is important in determining whether any microbial species are threatened with extinction because it is the endemic species whose habitats are threatened that are most likely to become extinct (Staley, 1997).

Why is it thought that geography may not play a major role in the speciation of bacteria? In addition to the lack of evidence in support of its occurrence, three factors seem to be important and are discussed below.

Broad bacterial species definition

As discussed above, the breadth of the bacterial species means that it is more likely to have cosmopolitan bacterial species than endemic species in comparison with plants and animals. Consider the following absurd example. If the bacterial species

definition were applied to primates, then primates would be considered to comprise a single species. Furthermore, gorillas and orangutans would not be considered threatened because they would be the same species as humans, which have a cosmopolitan distribution.

Dispersal

Bacteria are hardy and readily dispersed by winds and animals from one location on Earth to another. If a novel strain is selected for in one geographic area, it can be transferred to another area by various means of dispersal such as wind, animals, water currents, etc. Whether the strain can survive the dispersal process will depend on its hardiness under the conditions of transit. Certainly, endospores have a much higher likelihood of survival in comparison with a more fragile species that has no dormant stage. This may explain why some bacterial species, e.g., *Bacillus,* are regarded as being cosmopolitan.

When the novel strain arrives at a new site, it cannot survive unless it can colonize the habitat. However, if it is well suited to the habitat it may multiply and displace other strains of the same species that are less well adapted. Cohan (2002) discusses this possibility in some detail. Although one might expect that the existing strains should be better suited in the environment, the new strain may carry genes, such as those that confer resistance to inhibitors in the environment or that allow it to survive and successfully compete with the indigenous microbiota.

In addition, if the new strain can transfer genes to the existing population, those that are beneficial may be selected for in the native population as discussed next. Ready genetic exchange and rapid growth rates mean that favorable traits of transported bacteria can be incorporated into recipient organisms in the new environment making them homogeneous with populations thousands of kilometers away.

Chemical and Physical Factors

So, what factors, other than geography, can play a major role in bacterial speciation? Consider the following two points. First, although we do not know the nature of early life forms, we believe that they originated from a gene pool with a common evolutionary ancestry (Woese, 2002). Second, enormous physiological and nutritional differences exist among extant species of *Bacteria* and *Archaea.* From this, it can be inferred that the original organisms that have evolved and diversified to become the numerous species that exist today each has its own resources (energy sources, required nutrients, and lives within its specific temperature and pH range, etc.) and microhabitat. In short, the physical and chemical factors of environments have established conditions that have resulted in the diversification of bacterial life. Genetic variation has, over hundreds of millions of years, resulted in the speciation of all the different types of microbial life that rely on the multitude of various energy sources and physical conditions of the many available environments.

The upshot of this is that the physical and chemical features of the environment have clearly been a major driving force for speciation of bacteria. This particular type of speciation is closest to parapatric speciation of plants and animals where localized differences in the chemical and physical milieu have resulted in the selection of novel species through genetic variation and natural selection.

Biological Factors

Because chemical factors can play a major role in speciation, then biochemical factors that are due to the activities of other organisms should do so as well. Indeed, many examples of this are known. For example, chitin-degrading bacteria have been selected in environments such as the marine and soil environments when chitin-producing animals and fungi abound, respectively.

Cospeciation is another clear example of a speciation process occurring in bacteria that is controlled by external biological factors. A number of examples of cospeciation exist. One of the most notable and interesting examples is that of *Buchnera*, a bacterial genus that is an endosymbiont of aphids. *Buchnera* spp. live in a special organ inside the insect called the bacteriosome. The 16S and 18S rDNA sequences of *Buchnera* and its aphid host, respectively, provide evidence of cospeciation (Moran and Baumann, 2000). That is, as the aphid lineages have diverged to form new species, *Buchnera* has speciated at the same time thereby forming a phylogenetic tree, which is the mirror image of the host tree (Moran et al., 1993).

Buchnera are members of the gamma-Proteobacteria, and their ancestors had genome sizes of about 4.0 Mb. In contrast, *Buchnera* genomes are only 650 kb in size. The small size of their genome is accounted for by the process of reductive evolution. *Buchnera* species have lost genes that they no longer need for an independent life. They have become slave species for the aphid. They provide the aphid with amino acids by the special metabolic pathways they have retained and, in turn, have a habitat where they live and are provided with nutrients from the host.

Cospeciation of *Buchnera*, in this case, is assured by reproductive isolation. The entrapped *Buchnera* are not able to exchange genes with other bacteria, and therefore their phylogeny can be inferred directly from their gene sequences.

Do similar patterns occur in other bacterial symbionts? What about commensal bacteria such as those that inhabit the intestinal tract of warm-blooded animals? No evidence exists that cospeciation has occurred in any of these associations, however, not much work has been dedicated to following this possibility. Nonetheless, an interesting example of a possible cospeciation process might be occurring in the oral commensal bacterium *Simonsiella*. Recent evidence from our lab indicates that different species of *Simonsiella* inhabit humans, dogs, and sheep in a phylogenetic tree based on 16S rDNA sequences (Hedlund and Staley, 2002). The fourth group of strains from cats is also monophyletic and genetically sufficiently different to comprise a new species, but is phenotypically very similar to the dog species and has therefore not yet been named. In order to determine whether speciation has occurred at finer taxonomic levels, that is, within animal genera and species rather than orders and families, strains from additional more closely related animal species will need to be studied phylogenetically.

From a biodiversity standpoint, as well as from the viewpoint of speciation, it would be very helpful to know more about this speciation process. For example, if it could be shown that each mammalian species has its own *Simonsiella* phylospecies, then there would be more than 5,000 species of *Simonsiella*, as there are about 5,000 species of mammals. Importantly, these 5,000 phylospecies of the genus *Simonsiella* would then rival the total number of all bacterial species that have been described. Again, this would support the contention made above that the current bacterial species is far more broadly defined than that of plants and animals.

INTRINSIC FACTORS

What are the intrinsic biological factors that affect bacterial speciation? Certainly the ability of bacteria to acquire and express genes from different species is important. The remarkable genetic fluidity of the prokaryotic organism is due to several factors including their haploid nature, their ready ease of transfer of genetic material between organisms (termed horizontal gene transfer), and their huge population sizes and rapid growth rates. Each of these is considered individually below as well as constraints on genetic fluidity, followed by a section on genome size.

Haploidy and Mutation

One of the distinctive features of prokaryotic organisms is their haploid genome, which most commonly comprises a single chromosome. In most bacteria the genome is circular; however, in some, such as certain members of the spirochetes, it is linear. Prokaryotes may also contain plasmids, which have additional genetic information that is potentially available to the host organism. In contrast to prokaryotes, eukaryotic genomes contain multiple chromosomes that are commonly in the paired, or diploid, configuration.

In contrast to diploid organisms, haploid organisms are ideally suited for the ease of expression of genes. There is no recessive or dominant gene involved, because only a single gene exists. Thus, if a novel gene can be introduced into a recipient organism and the gene can be transcribed and translated, it will be expressed.

Because of their haploid nature, bacteria are particularly responsive to the evolutionary processes of mutation and selection. If a mutant strain is better suited to survive than its parental genotype, it will rapidly displace the parental type in a habitat (Palys et al., 1997). Therefore, the haploid nature of bacteria likely plays a major role that results in the rapid evolution of prokaryotes.

Horizontal Gene Transfer

Horizontal (or lateral) gene transfer is the process whereby genes from one species are transferred and expressed in another species. This process is entirely different from vertical gene transfer that occurs in the establishment of hereditary lines of animals.

Genetic transfer in bacteria is commonplace. Genes can be transferred between bacteria by several mechanisms including conjugation, transduction, and transformation. Each of these processes is well understood and is not discussed further here.

Gene transfer occurs most commonly between organisms that are closely related to one another. However, genetic material can be transferred between bacteria that are not very closely related to one another at all. For example, evidence indicates that gene transfer has occurred between the *Archaea* and the *Bacteria*, e.g., the transfer of genes involved in one-carbon metabolism and between methanogens and methanotrophs (Chistoserdova et al., 1998). Although these types of event occur much less frequently than transfer within phylogenetic groups,

there is good evidence from genomic studies that this must have happened in many different organisms (Ochman et al., 2000). Recently the selfish operon theory has been developed that discusses how genes are transferred in operons, clusters of genes that comprise a functional unit such as a biochemical pathway (Lawrence and Roth, 1996).

Thus, bacteria are quite different from eukaryotic organisms that have introduced barriers to genetic transfer. Even organisms that are closely related to one another such as different species of primates are not genetically compatible. Indeed, this difference, which is now being overcome by genetic engineering, comprises an important feature that distinguishes prokaryotic from eukaryotic species and their speciation processes.

Population Sizes and Rapid Growth Rates

Another distinctive feature of bacteria is that their species have gigantic populations. It is not unusual for a particular species to have 10,000 or more cells in a gram of soil or milliliter of water. Thus, a kilogram of soil or a liter of water contains 10 million members, and a garden or a lake, can contain several billion members, more than the population of humans on the planet.

Furthermore, cells in natural environments multiply rapidly. For example, doubling times of 6 to 12 hours are commonplace for aquatic bacteria (e.g., Poindexter et al., 2000). The huge population sizes combined with their haploidy, rapid growth rates, and mutation can result in rapid changes in the natural selection of a strain of a species over a short period of time. Population purges can occur rapidly as described by Palys et al. (1997).

Constraints on Genetic Fluidity

If factors causing genetic fluidity were the only ones responsible for bacterial evolution, it is doubtful that we would see any branches in the tree of life at all. Instead, it would appear as a tangled web of genetic and phenotypic incoherence. In actuality, it is striking that most branches in the tree of life contain phenotypes with a common phylogenetic ancestry. The *Cyanobacteria*, the only phylogenetic group in which oxygenic photosynthesis has evolved, is one example of this. Likewise the *Spirochetes*, with their unique axial filament and helical structure, are another. Indeed, many major phylogenetic groups (divisions, kingdoms) show similar patterns. Furthermore, there is evidence from other conserved genes such as ATPase and elongation factors that support, in general, the robust nature of the tree of life.

Therefore, other factors must be acting to constrain the genetic fluidity of bacteria. In particular, the niche must play a central role in bacterial speciation. Indeed, the best possibility for relating the bacterial species concept to that of plants and animals is through ecological features, i.e., each species' distinctive niche, and in particular its habitat. However, even here confusion exists. For example, some claim that two or more species are responsible for a single function in an ecosystem, a phenomenon referred to as functional redundancy. It is true that in broader activities such as sulfate reduction and photosynthesis, one would find a guild representing several species or even phyla to be responsible for a single activity. However, I would argue that each of these species has unique abilities and functions and therefore they are not redundant. Indeed, the redundancy concept may be merely an indication of our ignorance of the true interactions. After all, how can it be argued that a species or ecospecies is unnecessary when it actually exists?

Genome Size

Bacteria are noted for their small genomes. Indeed, some are less than one megabase in size. At the other extreme, some are as large as 10 Mb. Thus, the genome size among prokaryotes ranges over about 20-fold. It is not clear why there is a limit on the size of prokaryotic genomes.

Why is genome size important in evolution and hence speciation? First, it poses a constraint on the number of genes an organism can have. The maximum number of genes for a bacterium with a large genome is about 10,000. With a maximum limit on the size of a genome, a physical constraint is placed on its genetic capability and the repertoire of the organism's enzymes. If there were no constraint, or even if the maximum size were 20,000 genes, then the pattern of bacterial evolution may have been much different. For example, if an organism had 30,000 genes, then they would, in theory, be capable of carrying out many additional activities compared with one with only 2,000 or 5,000 genes. This would argue for fewer species carrying out many more activities, rather than the apparent existing situation in which there are more species, each of which carries out fewer activities. The limit on genome size may also serve as a constraint to the ready transferability of genetic material between organisms. For example, how can an organism accept additional genes if it has already saturated its capacity for genetic material and accepting additional genes would jeopardize its ability to maintain its niche?

What determines the size of a bacterial genome? Several major factors seem to be at play:

- free-living versus symbiotic life style,
- simplicity of metabolism,
- energy availability,
- complexity of life cycle, and
- the DNA processing ability constrains the upper limit to about 10 Mb

Small genomes are encountered in bacterial symbionts and pathogens. These organisms rely on a host organism for much of their sustenance. Thus, many have undergone evolutionary reduction. One example of this is the endosymbiont of aphids, the bacterial genus *Buchnera*. The ancestors of this genus had a genome of about 4 Mb, but now after 150 to 250 million years of evolution, it is only 650 kb. Another example is the genus *Chlamydia* that are obligate intracellular pathogens. Their genome size is also small, 1.22 Mb. The smallest known genome, 480 kb, is that of *Nanoarchaeum equitans*, a newly described archaeon that parasitizes other *Archaea*.

For free-living bacteria, other factors are important, such as the energetics of their metabolism (Table 2). Anaerobic organisms such as methanogens have small genomes ranging in size from 1 to 3 Mb. In contrast, aerobic organisms such as *Pseudomonas* spp. have larger genomes, about 6 Mb. Enteric bacteria, which are facultative aerobes, have an intermediate genome size of about 4 to 6 Mb. The largest genomes

Table 2. Bacterial genome size related to symbiosis and energy yield (approximated by type and simplicity[a] of metabolism) and complexity of life cycle

Bacterial group	Genome size
Obligate symbionts and pathogens	
Buchnera spp.	650 kb
Chlamydia trachomatis	1.22 Mb
Free-living or commensal with simple life cycles	
Obligate anaerobes	
Methanococcus jannaschii	1.66 Mb
Methanobacterium thermoautotrophicum	1.75 Mb
Chlorobium tepidum	2.1 Mb
Clostridium acetobutylicum	4.1 Mb
Facultative aerobes	
E. coli	4.6–5.5 Mb
Bacillus subtilis	4.2 Mb
Rhodobacter sphaeroides	4.4 Mb
Rhodopseudomonas palustris	5.46 Mb
Obligate aerobes	
Pseudomonas aeruginosa	6.30 Mb
Synechocystis sp.	3.57 Mb
Free-living aerobes with complex life cycles	
Myxococcus sp.	ca. 8 Mb
Streptomyces coelicolor	8.7 Mb

[a] Organisms such as hydrogen autotrophs have a simple metabolism and therefore require somewhat fewer genes than organisms such as photoautotrophs.

are found in aerobic organisms that undergo complex life cycles such as the myxobacteria and the actinobacteria. These bacteria have genome sizes of about 8 to 9 Mb. Their metabolism is much more efficient as they are able to derive energy from the complete oxidation of organic materials.

The rough correlation between energy yield and genome size indicates that there is a cost to having genetic material for an organism. A bacterial species cannot afford to carry an unlimited number of genes. Furthermore, it suggests that many anaerobic species may be on the borderline in terms of producing enough ATP to sustain a large genome. If so, then there are consequences to this limit. First, they may be producing only enough genes to sustain their frugal existence. Second, they cannot be as flexible as aerobic species in incorporating new genetic features. Third, these species would not be expected to evolve as quickly as aerobic organisms because they have so little room for additional genes. Furthermore, any genetic change that does occur may result in displacing them from their niche.

The importance of an upper limit to the genome size is paramount, especially when it relates to the evolution of anaerobic prokaryotes. This implies that external genetic material cannot be incorporated into the genome of an organism that is at its upper limit unless some of its existing chromosome is lost. Each gene in the genome is important. If genetic material from another organism is incorporated and displaces any critical gene, then the recipient organism will be selected against because it will lose its niche.

FINAL COMMENTS AND RECOMMENDATIONS

The broad definition of the bacterial species based on DNA/DNA hybridization has impeded the microbial ecologist and taxonomist in understanding speciation, the microbial niche, and biodiversity (Staley, 1997, 1999). I therefore propose the establishment a dual system of taxonomy for the foreseeable future. In addition to the currently defined species based on DNA/DNA hybridization, I propose the introduction of the phylospecies concept for *Bacteria* and *Archaea* that is defined in this chapter. In the phylospecies concept, the organism that is undergoing speciation in the environment is currently considered as an ecovar or ecotype. Newly described phylospecies do not have to have a name, at least at this time; however, a phylospecies description must provide information about the niche of the organism and the gene and protein sequences that provide the evidence that the phylospecies exists.

There are two primary consequences of establishing the phylospecies concept for *Bacteria* and *Archaea*. First, it is consistent with the PSC that applies to all organisms and therefore places the taxonomy of microorganisms, particularly *Bacteria* and *Archaea*, on a similar intellectual footing as other organisms. Second, it will enable microbiologists interested in microbial ecology and taxonomy to address the important questions relating to niche, speciation, and biodiversity.

From the standpoint of those interested in biodiversity, a PSC for microorganisms that would result in the naming of many additional phylospecies or species would help correct the longstanding, misleading belief by some that microorganisms are not diverse because there are so few species (Mayr, 1998; Staley, 2002). Moreover, it would stimulate more microbiologists, both eukaryotic and prokaryotic, who are interested in evolution, microbial ecology, taxonomy, and biodiversity to experimentally investigate the process of speciation.

Acknowledgments. I am grateful for the helpful comments from Cheryl Jenkins, Jonathan Miller, and Sujatha Srinivasan. I especially appreciate the intensive reviews and suggestions of Brian Hedlund and Brian Oakley. This work was supported in part by our Sea Grant research in which we are investigating the biodiversity of marine polyaromatic hydrocarbon-degrading bacteria.

REFERENCES

Brenner, D. J., J. T. Staley, and N. R. Krieg. 2000. Classification of prokaryotic organisms and the concept of speciation. In D. Boone, R. Castenholz, and G. Garrity (ed.), *Bergey's Manual of Systematic Bacteriology*, 2nd ed. The Williams & Wilkins Co., Baltimore, Md.

Chistoserdova, L., J. A. Vorholt, R. K. Thauer, and M. E. Lidstrom. 1998. C1 transfer enzymes and coenzymes linking methylotrophic bacteria and methanogenic Archaea. *Science* 281:99–102.

Cohan, F. M. 2002. What are bacterial species? *Annu. Rev. Microbiol.* 56:457–487.

Craycraft, J. 1989. Speciation and ontology: the empirical consequences of alternative species concepts for understanding patterns and processes of differentiation, p. 28–59. *In* D. Otte and J. A. Endler (ed.), *Speciation and Its Consequences*. Sinauer Associates, Inc., Sunderland, Mass.

Eldredge, N., and S. J. Gould. 1972. Punctuated equilibria: an alternative to phyletic gradualism, p. 82–115. *In* T. J. M. Schapf (ed.), *Models in Paeleobiology*. Freeman, Cooper, and Co., San Francisco, Calif.

Finlay, B. J. 2002. Global dispersal of free-living microbial eukaryotic species. *Science* 296:1061–1063.

Freeman, S., and J. C. Herron. 1998. *Evolutionary Analysis*. Prentice-Hall, Upper Saddle River, N. J.

Hedlund, B. P., and J. T. Staley. 2002. Phylogeney of the genus *Simonsiella* and other members of the Neisseriaceae. *Int. J. Syst. Evol. Microbiol.* 52:1377–1382.

Lawrence, J. G., and J. R. Roth. 1996. The selfish operon theory. *Genetics* 143:1843–1860.

Mayr, E. 1942. *Systematics and the Origin of Species*. Columbia University Press, New York, N.Y.

Mayr, E. 1998. Two empires or three? *Proc. Natl. Acad. Sci. USA* 95:9720–9723.

Moran, N. A., and P. Baumann. 2000. Bacterial endosymbionts in animals. *Curr. Opin. Microbiol.* 31:270–275.

Moran, N. A., M. A. Munson, P. Baumann, and A. Ishikawa. 1993. A molecular clock in endosymbiotic bacteria is calibrated using the insect hosts. *Proc. R. Soc. London Ser. B* 253:167–171.

Nelson, G. J., and N. I. Platnick. 1981. *Systematics and Biogeography: Cladistics and Vicariance.* Columbia University Press, New York, N. Y.

Ochman, H., J. G. Lawrence, and E. A. Groisman. 2000. Lateral gene transfer and the nature of bacterial innovation. *Nature* 405:299–304.

Palys, T., L. K. Nakamura, and F. M. Cohan. 1997. Discovery and classification of ecological diversity in the bacterial world: the role of DNA sequence data. *Int. J. Syst. Bacteriol.* 47:1145–1156.

Poindexter, J. S., K. P. Pujara, and J. T. Staley. 2000. *In situ* reproductive rate of freshwater *Caulobacter* spp. *Appl. Environ. Microbiol.* 66:4105–4111.

Rosen, D. E. 1978. Vicariant patterns and historical explanation in biogeography. *Syst. Zool.* 27:159–188.

Rosselló-Mora, R., and R. Amann. 2001. The species concept for prokaryotes. *FEMS Microbiol. Rev.* 25:39–67.

Sibley, C. G., and J. E. Ahlquist. 1987. DNA hybridization evidence of hominid phylogeny: results from an expanded data set. *J. Mol. Evol.* 26:99–121.

Staley, J. T. 1997. Biodiversity: are microbial species threatened? *Curr. Opin. Biotechnol.* 8:340–345.

Staley, J. T. 1999. Bacterial biodiversity: a time for place. *ASM News* 65:681–687.

Staley, J. T. 2002. A microbiological perspective of biodiversity. *In Biodiversity of Microbial Life: Foundation of Earth's Biosphere.* John Wiley & Sons. New York, N.Y.

Ward, D. M. 1998. A natural species concept for prokaryotes. *Curr. Opin. Microbiol.* 1:271–277.

Wayne L. G., D. J. Brenner, R. R. Colwell, P. A. D. Grimont, O. Kandler, M. I. Krischevsky, L. H. Moore, W. E. C. Moore, R. G. E. Murray, and E. Stackebrandt, M. P. Starr and H. G. Trüper. 1987. Report of the ad hoc committee on reconciliation of approaches to bacterial systematics. *Int. J. Syst. Bacteriol.* 37:463–464.

Woese, C. R. 2002. On the evolution of cells. *Proc. Natl. Acad. Sci. USA* 99:8742–8747.

Microbial Diversity and Bioprospecting
Edited by Alan T. Bull
© 2004 ASM Press, Washington, D.C.

Chapter 5

Approaches to Identification

FERGUS G. PRIEST

The practice of identification was clarified by the late Samuel T. Cowan when he deconstructed taxonomy into his trinity of classification, nomenclature, and identification (Cowan, 1965). In so doing he emphasized that identification could not take place without a prior classification because it involves the matching of an unknown organism with one that has been previously characterized and placed in a classification scheme. He consequently began to focus the minds of taxonomists on the construction of classifications that would enable effective identification. The situation has not changed today; we still depend on the foundations of robust and stable classifications for accurate identification.

The classification can be a general-purpose classification designed to provide a comprehensive catalog of all bacteria, or it can be a special-purpose classification covering a restricted range of organisms that are found in a particular environment or possess particular properties. Organisms encountered only in dairies, clinical situations, or water would represent such special-purpose classifications. Ideally the special-purpose classifications will conform to the larger general classification and simply represent a subset of that classification.

Currently, classifications are generally held as computer databases to enable their rapid and effective interrogation for identification purposes. Such databases may not be structured as traditional hierarchical classifications, although they will probably be founded on such. In this chapter I summarize various approaches to identification of microorganisms with emphasis on prokaryotes. However, virtually all the methods can be adapted to microbial eukaryotes.

APPROACHES TO IDENTIFICATION

Microorganisms can be analyzed at various levels to gain information suitable for constructing databases and effecting identification. The highest level is the genome and its direct expression as RNA. Sequence analysis of various genes provides for stable classifications and accurate identification, which have become the cornerstone of modern phylogenetic taxonomy. Nucleic acid hybridization also offers identification possibilities; indeed, chromosomal DNA hybridization forms the basis of the generally accepted species definition in bacterial systematics (see Rosselló-Mora and Kämpfer, Chapter 3), that members of the same species should hybridize more than 70% with minimal mismatch as displayed by reduction in melting temperature ($\Delta T_m < 5$°C [Wayne et al., 1987]).

The genetic information is expressed as proteins. Electrophoretic anaysis of whole-cell proteins has been an effective approach to classification and identification. Other cell components can be analyzed by a range of techniques applied to either whole cells or particular cell extracts. For example, cellular and membrane lipids can be profiled using gas chromatography (GC), or whole cells can be volatilized and the products detected by mass spectrometry (MS).

Morphology and physiology are the classical levels at which most conventional identification is done. The presence or absence of particular enzymes or metabolic pathways are typically characterized using commercial kits, and, again, computerized databases can be developed to enable rapid identification. In the remainder of this chapter I review the application of these various methods and the practicalities of using them in the context of environmental microbiology.

NUCLEIC-ACID-BASED IDENTIFICATION

Ribosequencing (Phylotyping)

Bacterial phylogenetic classification is based on sequence analysis of the small-subunit (ssu) 16S rRNA molecule or its genes. Over 20,000 ssu RNA gene se-

Fergus G. Priest • School of Life Sciences, John Muir Building, Heriot-Watt University, Riccarton, Edinburgh EH14 4AS, Scotland.

quences have now been deposited, either in general nucleic acid sequence databases or in specialist ribosomal RNA databases such as the rRNA Database Project (RDP [Maidak et al., 2001]). These databases provide the most comprehensive opportunities currently available for identification, and their exploitation for identification has been termed ribosequencing or phylotyping. Universal primers for amplification of the complete ssu rRNA gene or sections of it are well established, and generally a partial gene sequence of 300 to 500 bp is sufficient for identification purposes. Once the sequence has been obtained, it is submitted to a BLAST search to one of the publicly available websites such as the site for the National Center for Biotechnology Information (http://www.ncbi.nlm.nib.gov). A 99 to 100% similarity with an entry in the database can be accepted as a strong indication of identification. This process has been commercialized as the MicroSeq 500 Bacterial Identification System (Perkin-Elmer Biosystems, Foster City, Calif.) with dedicated software, database, and kits to perform molecular biology. Typical time to obtain an identification is 15 to 18 hours. rDNA sequencing is not restricted to bacteria but has also been used for identification of fungi and yeasts (Valente et al., 1999; Cappa and Cocconcelli, 2001).

One of the key decisions for ribosequencing is the choice of the most suitable region of the ssu RNA gene to be sequenced. It is best to target 500 bases of the most variable region for maximum discrimination. This can be located differently in different microorganisms. We examined the range of lactobacilli in Scotch whisky fermentations using the first 500 bases of the gene, but in some cases almost 900 bases were needed to obtain an accurate distinction of closely related bacteria (Simpson et al., 2001). Similarly, the 5-end region of the gene is useful for discrimination of *Bacillus* species (Goto et al., 2000), corynebacteria (Tang et al., 2000), and mycobacteria (Baldus Patel et al., 2000), whereas the distal region was used for identification of various clinical microorganisms (Trotha et al., 2001). Scrutiny of an alignment of the complete ssu rRNA genes for a given group of organisms would enable the most discriminatory region to be chosen. A particularly exciting development is the use of real-time DNA sequencing based on pyrosequencing to provide the DNA sequence for database searching (Unnerstad et al., 2001).

There are two problems with this otherwise straightforward procedure. First, even a perfect match over the complete 16S sequence may not always provide an unequivocal answer, for example, *Lactobacillus casei* and *L. paracasei* are very difficult to distinguish by 16S rRNA sequence alone (Simpson et al., 2001), and some species such as *Bacillus sub-tilis* and *B. mojavensis* have diverged so recently that their ssu RNA genes are identical even though DNA hybridization and fatty acid composition place them in separate species (Roberts et al., 1994). In such a situation the BLAST search will not provide a single hit, and more detailed characterization will be necessary for specific identification. The second problem is determining a level of similarity at which to accept identification. Given the conservation of the 16S rRNA gene, at least 99% similarity seems to be a commonly accepted score for identification (Drancourt et al., 2000; Simpson et al., 2001). In a recent study of 177 environmental and clinical isolates, 139 (78.5%) could be identified at 99 to 100% and 159 (90%) at 97% or above (Drancourt et al., 2000). Such high levels of similarity leave no room for inaccuracies in sequence determination. In summary, this is a rapid, powerful, and effective method for identification of microorganisms in general. Running the MicroSeq system has been estimated at $84 per test including kits, reagents, and labor (Tang et al. 2000), but this can be dramatically reduced by avoiding kits and using publicly accessible databases.

Hybridization

DNA hybridization has played a pivotal role in bacterial systematics and provided one of the first molecular methods for identification. A key decision is the target for a specific hybridization reaction.

If the organism has a particular and specific trait, hybridization reactions are valuable for identifying members of the taxon from the environment, foods, or clinical samples. For example, we used a DNA probe aimed at the insect toxic crystal protein genes of *B. sphaericus* to identify novel strains on isolation plates in a colony blot format (Aquino de Muro and Priest, 1994), and the toxin genes of *B. anthracis* are a popular target for identification of this bacterium (e.g., Ellerbrook et al., 2002). Such a procedure could be developed for screening for particular genes from a variety of microorganisms.

The most popular probe target by far is the ssu rRNA molecule or its larger counterpart. These are present in high copy number (10^4 to 10^5 molecules per bacterial cell), thus greatly improving the sensitivity of the hybridization. Within these genes are regions that are highly variable and differ significantly between species, whereas other areas are more conserved and suitable for identification at the generic level (reviewed by Amann and Ludwig, 2000). This has resulted in the development of sets of rRNA gene probes discriminatory at various levels from domain to species via kingdom, order, and genus (Amman et al., 1995; Stahl, 1995).

Large-format hybridization procedures are particularly attractive for identification applications. The original reverse dot blot hybridization is when the total DNA isolated from a particular environment (soil, water, milk, etc.) is labeled and hybridized with dots of DNA probes from known organisms bound to a membrane. Any reassociated DNA is subsequently detected and indicates the presence of the bacterium in the original sample. This approach was originally developed for identification of sulfate-reducing bacteria in oil field samples (Voordouw et al., 1991) and has been adopted for various applications such as identification of enterococci in drinking water (Behr et al., 2000). This technique is the forerunner of the hybridization micro- and macroarrays that have now been adopted for bacterial identification.

A high-density microarray of ssu rDNA probes comprising over 30,000 20-mer oligonucleotides complementary to a subalignment of sequences in the RDP has been prepared by Affymetrix (Santa Clara, Calif.). The oligonucleotides are arranged in hierarchical order so that a given organism should hybridize at the appropriate taxonomic levels such as kingdom, genus, species, etc. The chip and standard software were able to correctly match ssu rDNA amplicons with corresponding sequences in the RDP database for 15 of 17 strains of gram-positive and gram-negative bacteria and were able to cope with organisms isolated from air as test samples (Wilson et al., 2002). More restricted oligonucleotide microarrays have been developed for identification of sulfate-reducing bacteria (the SRP-PhyloChip) comprising 132 16S rRNA targeted oligonucleotide (18-mer) probes (Loy et al., 2002) and for identification of human intestinal bacteria (Wang et al., 2002). The ability to distinguish closely related bacteria, such as *B. anthracis*, *B. cereus*, and *B. thuringiensis*, using an array is shown in Color Plate 1.

An alternative approach, referred to as oligonucleotide fingerprinting, is to immobilize cloned rDNA on the chip or membrane and hybridize with a range of oligonucleotide probes designed to be specific for various taxa. The pattern of hybridization of the clones with the 27 probes is a fingerprint that enables assignment of the clones (or bacteria) to a species. In their development of this procedure, Valinsky et al. (2002) fingerprinted 1,536 clones derived from soil. The fingerprints were used to prepare an unweighted pair group method with arithmetic averages (UPGMA) dendrogram from which they recovered 766 clusters of approximate species rank. These represented five major taxa of gram-positive and gram-negative bacteria. An example of the reactions of the 27 probes with a set of clones is shown in Fig. 1A

with hybridization of a single probe with all 1,536 clones in Fig. 1B. The further refinement of approaches such as these may allow for cost-effective identification of all organisms in various ecosystems.

Ribotyping and Related Approaches

Ribotyping (also referred to as riboprinting) in its original concept involved restriction fragment length polymorphism (RFLP) patterns of ssu and/or large subunit rRNA genes and their operons. The attraction of the method lies in the relatively simple fingerprints and the use of a universal probe, the ssu rRNA gene. Bacterial chromosomes contain variable numbers of rRNA operons (generally between 2 and 10). Restriction enzyme digestion of total chromosomal DNA, agarose gel electrophoresis, Southern blotting, and hybridization to a ssu rDNA probe can produce 10 or more bands depending on the specificity of the enzymes used (i.e., 4-bp cutting enzymes would reveal more complex patterns than 6-bp enzymes). These RFLPs are sometimes species specific but usually strain specific. The technical complexity of producing reproducible RFLPs has been alleviated with the introduction of the RiboPrinter (Qualicon, Wilmington, Del.) in which automated hardware and software provide reproducible fingerprints for comparison with databases and accurate identification. Ribotyping has been adopted in several food and beverage applications, for example, for tracing *Pediococcus* strains in breweries (Barney et al., 2001) and food poisoning bacteria in foods (De Cesare and Manfreda, 2002). Indeed the manufacturer has demonstrated that the RiboPrinter is close to 100% accurate at identifying genus and species.

Some PCR-Based Procedures

The exquisite specificity of nucleic acid hybridization is exploited in the PCR as a popular approach to microbial identification. The oligonucleotide probe aimed at the ssu rRNA in hybridization format is obviously easily adapted as a PCR primer to effect identification. In general, a PCR is as specific as a labeled probe reaction, and the execution of the hybridization is simpler because no labeling and detection are involved. PCR-based identifications of bacteria based on specific oligonucleotides targeted to the 16 S rRNA genes are far too numerous to catalog here. Probe design has been covered in recent reviews (Amman et al., 1995; Stahl, 1995; Amman and Ludwig, 2000), and specialist websites such as ARB are invaluable tools to assist with the process (for ARB, one of the most comprehensive tools for rRNA tar-

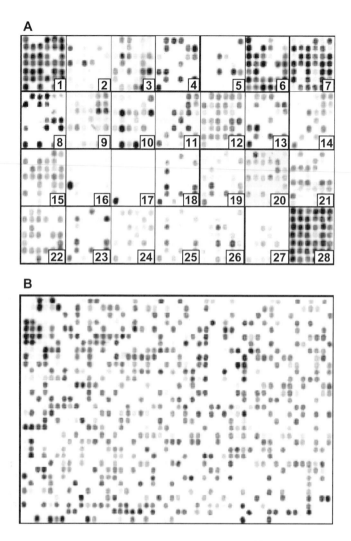

Figure 1. Arrayed bacterial rDNA clones on a nylon membrane hybridized with ^{32}P-labeled DNA oligonucleotide probes. (A) The full set of 27 discriminating probes and reference probe 28, each hybridized to a common set of clones; (B) a single probe (discriminating probe 4) hybridized to all 1,536 soil rDNA clones used in the study. Analysis with all 27 probes produced a hybridization fingerprint for every clone (from Valinsky et al., 2002 with permission).

geted probe design, see http://www.mikro.biologie. tu-meunchen.de).

Whereas regular PCR requires agarose gel electrophoresis for visualization of the products, real-time PCR enables the reaction to be followed as it proceeds. Thus, product formation can be determined by using fluorescent primers and monitoring the increase in fluorescence that is due to the inclusion of primers into amplicons. A LightCycler that contains an internal fluorimeter to measure the fluorescent products is required for this procedure. Real-time PCR assays targeting rRNA genes provide simultaneous detection, identification, and quantification of organisms in complex mixtures. The LightCycler has been used in numerous situations, for example, in the clinical lab for detection of *Campylobacter* (Logan et al., 2001) and *Chlamydia* (Mygind et al., 2001), and in

an environmental context for detection and identification of thermoanaerobes of the genera *Fervidobacterium* and *Caloramator* (Connolly and Patel, 2002) and *B. anthracis* (Ellerbrock et al., 2002).

PCR provides numerous typing procedures that can be adopted for bacterial identification with varying levels of sophistication, reproducibility, and specificity. A comprehensive review of the various nucleic-acid-based fingerprinting techniques used in bacterial systematics has been provided by Vaneechoutte (1996).

TECHNIQUES FOR EXAMINING PROTEINS

Sodium dodecyl sulfate-polyacrylamide gel electrophoresis of whole-cell protein extracts has found major application in bacterial classification and

Table 1. Features of some whole-cell fingerprinting methods[a]

Method	Destructive?	Sample size (μg)	Typical no. of cells	Reproducibility	Typical dimensionality
PyMS	Yes	~50	10^6–10^7	Poor	150
MALDI-TOF-MS	Yes	~1	10^5–10^6	Mediocre	~100 protein peaks
ESI-MS	Yes	~5	10^5–10^6	Good	100–3,000
FTIR	No (although sample is dried)	~50	10^6–10^7	Excellent	~1,000

[a] Adapted from Goodacre et al. (1998).

identification. In this approach, the complex patterns produced from one-dimensional electrophoresis of total cell extracts are scanned with a densitometer. The quantitative data are amenable to analysis by computer to provide clusters or species fingerprints that can be used as a database for identification purposes (Pot et al., 1994). Whole-cell protein electrophoretic patterns are most discriminatory at the species, and sometimes the subspecies, level. Although the method has been largely superseded by nucleic-acid-based techniques, sodium dodecyl sulfate-polyacrylamide gel electrophoresis is efficient for analysis of large numbers of organisms and is applicable to all organisms provided that an efficient cell disruption method is available. Its main application is for preliminary clustering of large numbers of isolates from a particular environment or as part of a polyphasic taxonomic study (Vandamme et al., 1996).

METHODS BASED ON CELL COMPOSITION

Numerous analytical techniques can be applied to the study of the cellular composition of microorganisms, but few have emerged as viable methods for analyzing cell constituents or whole organisms for purposes of classification and identification. These include fatty acid analysis by GC and whole-cell analyses by MS and Fourier-transform infrared (FTIR) spectroscopy.

Fatty Acid Analysis

GC analysis of fatty acid methyl esters (FAMEs) is a general-purpose technique for chemotaxonomic analysis of bacteria that forms the basis of the commercial Sherlock Microbial Identification System (MIDI Inc., Newark, Del.). Standardized procedures for cultivation of the microorganisms and preparation of the FAME samples are important. Bacteria are grown for 24 h as spread plates, and the cells are saponified in a strong base. The solution is then acidified, and the liberated fatty acids are methylated to increase their volatility for GC. The methyl esters are extracted and separated on a capillary GC column. The system is completed by a computer for recording data and software to match the FAME profiles against a reference library. The commercially available libraries cover yeasts as well as gram-negative and gram-positive genera, but the technique could be applied to any microorganism with individual reference libraries constructed according to interests. The production of FAME profiles takes approximately 60 to 90 min (60-min sample preparation time).

The technique of FAME profiling has wide-ranging potential as an identification tool and has been used for identification of various organisms such as nitrogen-fixing bacteria (Jarvis et al., 1996), soil communities (Carpenter Boggs et al., 1998; Ibekwe and Kennedy, 1998), and organisms with specific traits such as hydrocarbon utilization (Daane et al., 2001). One disadvantage of this system is the lack of resolution between some closely related species.

Fourier-Transform Infrared Spectroscopy

FTIR spectroscopy is one of a group of methods that can be applied directly to whole cells (Table 1). The infrared spectrum of any compound expresses a unique fingerprint, and it is this characteristic that allows infrared spectroscopy to be used for identification by comparison with spectral data libraries. The recent microbiological applications rely on FTIR spectra of whole cells that measure the vibrations of bonds within functional groups. FTIR spectroscopy probes the total composition of a given organism in a single experiment providing selectivity to a very high level, for example, at the strain level (Timmins et al., 1998a). Moreover, because whole cells are being tested, complicated and time-consuming preparation of cell constituents is avoided. Once a database of reference spectra has been established, reliable identification of most microorganisms including yeasts (Timmins et al., 1998b) and bacteria (Goodacre et al., 1998; Obereuter et al., 2002) can be readily achieved from a single colony. In a recent development of this technique, microcolonies (70 to 250 μm in diameter) grown on solid medium are transferred to a light microscope and spectra are recorded by a spectrometer coupled to the microscope. This FTIR microspectroscopy enables identification from environmental samples within a day, so long as the original microcolonies are well separated on the plate (Wenning et al., 2002).

Mass Spectrometry

Various forms of MS have been used for microbial identification. Initially, pyrolysis, which is the controlled thermal degradation of a compound in an inert atmosphere, was coupled to MS (PyMS). Pyrolysis of microbial biomass produces a complex mixture of low-molecular-weight volatile compounds that can be analyzed by MS to provide a fingerprint of the organism. Attractions of PyMS as a characterization technique are the fast analysis times and the ability to analyze a single microbial colony taken directly from the surface of a culture plate. The method is capable of discriminating organisms at taxonomic levels below species and even of providing fingerprints of individual strains (Timmins et al., 1998a). A major concern about the use of PyMS for identification of microorganisms relates to the long-term stability of the mass spectrometers used and the problem this may pose for the construction of reference libraries of key spectra. These problems can be largely circumvented by always including suitable reference strains with the organisms to be identified.

Alternatively, a matrix-assisted laser desorption ionization–time-of-flight (MALDI-TOF) mass spectrometer can be used. In this technique, a sample of cells is placed in a MALDI-TOF mass spectrometer and pulsed with a laser light. Macromolecules from the surface of the organism are desorbed, mass analyzed, and the results reported as a mass spectrum (Van Baar, 2000; Lay, 2001). This mass fingerprint is reproducible and specific, allowing new strains to be matched against libraries of known organisms. This procedure has been commercialized as the MicrobeLynx system (www.micromass.co.uk). A refinement of this approach enabling characterization of noncultured organisms involves PCR amplification of ssu rDNA and subsequent analysis by MALDI-TOF MS (von Wintzingerode et al., 2002).

Electrospray ionization (ESI) MS has generally been used for analysis of cell components, and its application to microbial identification based on whole-cell fingerprints has been limited. Nevertheless, coupled with a time-of-flight analyzer, ESI MS provides reproducible spectral information that can be used to discriminate at the strain level (Vaidyanathan et al., 2001).

METHODS BASED ON PHYSIOLOGY

The miniaturization of traditional biochemical and physiological tests and their availability as commercial kits have eased and improved microbial identification over the past 30 years. A diversity of kits is now available aimed principally, but not exclusively, at organisms of clinical importance. For the identification of members of the *Enterobacteriaceae* and other gram-negative bacteria, there are several systems available of which the API20E (www.biomerieux-USA.com) is the most well known. Results are generally available in 24 h or less depending on the bacterium. The main advantages for such kits come from ease of use and standardization of reagent manufacture, which result in reproducible identifications. bioMérieux also markets general-purpose kits such as the API50CH that enable the preparation of carbohydrate assimilation patterns for any organism for which a suitable inoculation and incubation protocol can be established.

Biolog (www.biolog.com) has introduced a methodology for characterization and identification based on carbon substrate utilization revealed by a tetrazolium-based indicator of respiration. Carbon sources specific for gram-negative bacteria, gram-positive bacteria, or filamentous fungi are provided in standard 96-well microtiter dishes. Inoculation of the 96 wells with a microbial suspension (by hand or with a dedicated autoinoculator) and incubation for 4 to 16 h provide a biochemical fingerprint of the microorganism for the 95 substrates (1 control well). These patterns can be read by eye or with an automated reader and compared with a database using standard computer algorithms. Biolog has one of the most extensive databases for microbial identification, making this system attractive to the environmental microbiologist.

CONCLUDING REMARKS

Given the bewildering array of approaches to microbial identification available to the 21st century microbiologist, how to choose? Much depends on the numbers of organisms to be identified and the resources available for purchase of dedicated equipment.

rDNA sequence determination must be a strong contender for any laboratory with DNA sequencing facilities available but with no resources for dedicated technology. PCRs are generally routine and, with modern large-capacity sequencers, 96 strains (from a microtiter plate) can be sequenced readily. Indeed the technology is amenable to automation if very large numbers of strains are to be processed. Databases for identification are free to use, and phylogenetic information is accumulated that is often predictive of primary or secondary product formation. A drawback can be the need to isolate DNA from strains for use as a PCR template because whole-cell extracts often amplify poorly. Moreover, data analysis and checking sequence traces can be very time-consuming. It is essential that only high-quality sequence information be used because a handful of errors in a partial sequence of 500 bases will result in strains that cannot be identified with confidence; there is nothing more frustrating than large numbers of strains that identify at 98 to 99% with the database. I prefer sequence-based molecular techniques for identification as they have the accuracy of a digital system (individual bases) rather than an analogue system (bands in a gel or on a filter).

Large-format hybridization techniques in the form of macro- and microarrays may hold possibilities for the future, particularly when large numbers of culturable and nonculturable organisms from particular environmental niches are to be characterized. However, these methods are currently in development and remain the preserve of specialist laboratories.

A chemotaxomic approach can be attractive to the laboratory needing to identify colonies from isolation plates rapidly and with minimum complication. FAME profiles with a dedicated machine such as the Sherlock Microbial Identification System (www. midi-inc.com) can be processed at a rate of 45 samples per day (90 with a dual tower) and require additional sample preparation (generally a technician can process about 75 samples per day). Data analysis is straightforward leading to this being a popular choice. MALDI-TOF MS has been developed for microbial identification by Micromass as the MicrobeLynx system and has the advantages of rapid sample preparation (a few minutes starting with a colony on a plate) and extensive databases for identification. Financial investment in a MALDI-TOF MS, however, is considerable. Finally, FTIR spectroscopy is one of the few chemotaxonomic methods that provides nondestructive identification in quick and convenient format (up to 400 samples per hour [Goodacre et al., 1998]).

Phenotypic methods must not be forgotten. They still have their place, especially for the nonspecialist laboratory having infrequent identification requirements. It seems likely that, for all the achievements of modern molecular biology and analytical chemistry, traditional phenotypic tests, probably in the form of API trays or Biolog plates will be with us for awhile.

Acknowledgments. I am grateful to Royston Goodacre and David Stahl for material for inclusion in Table 1 and Color Plate 1, respectively.

REFERENCES

Amann, R., and W. Ludwig. 2000. Ribosomal RNA-targeted nucleic acid probes for studies of microbial ecology. *FEMS Microbiol. Rev.* 24:555–565.

Amann, R. I., W. Ludwig, and K.-H. Schleifer. 1995. Phylogenetic identification and in situ detection of individual microbial cells without cultivation. *Microbiol. Rev.* 59:143–169.

Aquino de Muro, M., and F. G. Priest. 1994. A colony hybridization procedure for the identification of mosquitocidal strains of *Bacillus sphaericus* on isolation plates. *J. Invertebr. Pathol.* 63:310–333.

Baldus Patel, J., D. G. B. Leonard, X. Pan, J. M. Musser, R. E. Berman, and I. Nachamkin. 2000. Sequence-based identification of *Mycobacterium* species using the Microseq 500 16S rDNA bacterial identification system. *J. Clin. Microbiol.* 38: 246–251.

Barney, M., A. Volgyi, A. Navarro, and D. Ryder. 2001. Riboprinting and 16S rRNA gene sequencing for identification of brewery *Pediococcus* isolates. *Appl. Environ. Microbiol.* 67: 553–560.

Behr, T., C. Koob, M. Schedl, A. Mehlen, H. Meier, D. Knopp, E. Frahm, U. Obst, K. H. Schleifer, R. Niessner, and W. Ludwig. 2000. A nested array of rRNA targeted probes for the detection and identification of enterococci by reverse hybridization. *Syst. App. Microbiol.* 23:563–572.

Cappa, F., and P. S. Cocconcelli. 2001. Identification of fungi from dairy products by means of 18S rRNA analysis. *Int. J. Food Microbiol.* 69:157–160.

Carpenter Boggs, L., A. C. Kennedy, and J. P. Reganold. 1998. Use of phospholipid fatty acids and carbon source utilization to track microbial community succession in developing compost. *Appl. Environ. Microbiol.* 64:4062–4064.

Connolly, G. R., and B. K. C. Patel. 2002. Development of fluorescent adjacent hybridization probes and their application in real-time PCR for the simultaneous detection and identification of *Fervidobacterium* and *Caloramator. Int. J. Syst. Evol. Microbiol.* 52:1837–1843.

Cowan, S. T. 1965. Principles and practice of bacterial taxonomy. *J. Gen. Microbiol.* 39:143–155.

Daane, L. L., I. Harjono, G. J. Zylstra, and M. M. Haggblom. 2001. Isolation and characterization of polycyclic aromatic hydrocarbon-degrading bacteria associated with the rhizosphere of salt marsh plants. *Appl. Environ. Microbiol.* 67:2683–2691.

De Cesare, A., and G. Manfreda. 2002. Use of the automated ribotyping for epidemiological investigations. *Ann. Microbiol.* 52:181–190.

Drancourt, M., C. Bollet, A. Carlioz, R. Martelin, J.-P. Gayral, and D. Raoult. 2000. 16S ribosmal sequence analysis of a large collection of environmental and clinical unidentifiable bacterial isolates. *J. Clin. Microbiol.* 38:3623–3630.

Ellerbrook, H., H. Nattermann, M. Özel, I. Beutin, B. Appel, and G. Pauli. 2002. Rapid and sensitive identification of pathogenic and apathogenic *Bacillus anthracis* by real-time PCR. *FEMS Microbiol. Lett.* 214:51–59.

Goodacre, R., E. M. Timmins, R. Burton, N. Kaderbhai, A. M. Woodward, D. B. Kell, and P. J. Rooney. 1998. Rapid identification of urinary tract infection bacteria using hyperspectral whole-organism fingerprinting and artificial neural networks. *Microbiology* 144:1157–1170.

Goto, K., T. Omura, Y. Hara, and Y. Sadaie. 2000. Application of the partial 16S rDNA sequence as an index for rapid identification of species in the genus *Bacillus. J. Gen. Appl. Microbiol.* 46:1–8.

Ibekwe, A. M., and A. C. Kennedy. 1998. Phospholipid fatty acid profiles and carbon utilization patterns for analysis of microbial community structure under field and greenhouse conditions. *FEMS Microbiol. Ecol.* **26**:151–163.

Jarvis, B. D. W., S. Sivakumaran, S. W. Tighe, and M. Gillis. 1996. Identification of *Agrobacterium* and *Rhizobium* species based on cellular fatty acid composition. *Plant Soil* **184**:143–158.

Lay, J. O. 2001. MALDI-TOF-MS of bacteria. *Mass Spectrom. Rev.* **20**:172–194.

Liu, W. T., A. D. Mirzabekov, and D. A. Stahl. 2001. Optimization of an oligonucleotide microchip for microbial identification studies: a non-equilibrium dissociation approach. *Environ. Microbiol.* **3**:619–629.

Logan, J. M. J., K. J. Edwards, N. A. Saunders, and J. Stanley. 2001. Rapid identification of *Campylobacter* spp. by melting peak analysis of biprobes in real-time PCR. *J. Clin. Microbiol.* **39**:2227–2232.

Loy, A., A. Lehner, N. Lee, J. Adamczyk, H. Meier, J. Ernst, K.-H. Schleifer, and M. Wagner. 2002. Oligonucleotide microarray for 16S rRNA gene-based detection of all recognized lineages of sulfate-reducing prokaryotes on the environment. *Appl. Environ. Microbiol.* **68**:5064–5081.

Maidak, B. L., J. R. Cole, T. G. Lilburn, C. T. J. Parker, P. R. Saxman, P. J. Farris, G. M. Garrity, G. J. Olsen, T. M. Schmidt, and J. M. Tiedje. 2001. The RDP-II (Ribosomal Database Project). *Nucleic Acids Res.* **29**:82–85.

Mygind, T., S. Birklund, E. Falk, and G. Christiansen. 2001. Evaluation of real-time quantitative PCR for identification and quantification of *Chlamydia pneumoniae* by comparison with immunohistochemistry. J. Microbiol. Methods. **46**:241–251.

Obereuter, H., H. J. Charzinski, and S. Scherer. 2002. Identification of coryneform bacteria and related taxa by Fourier transform infrared spectroscopy (FT-IR). *Int. J. Syst. Evol. Microbiol.* **52**:91–100.

Pot, B., P. Vandamme, and K. Kersters. 1994. Analysis of electrophorestic whole organisms fingerprints, p. 493–521. *In* M. Goodfellow and A. G. O'Donnell (ed.), *Chemical Methods in Prokaryotic Systematics.* Wiley, Chichester, United Kingdom.

Roberts, M. S., L. K. Nakamura, and F. M. Cohan. 1994. *Bacillus mojavensis* sp. nov., distinguishable from *Bacillus subtilis* by sexual isolation, divergence in DNA sequence and differences in fatty acid composition. *Int. J. Syst. Bacteriol.* **44**:256–264.

Simpson, K. L., B. Pettersson, and F. G. Priest. 2001. Characterization of lactobacilli from Scotch malt whisky distilleries and description of *Lactobacillus ferintoshensis* sp. nov., a new species isolated from malt whisky fermentations. *Microbiology* **147**:1007–1016.

Stahl, D. A. 1995. Application of phylogenetically based hybridization probes to microbial ecology. *Mol. Ecol.* **4**:535–542.

Tang, Y.-W., A. Von Graevenitz, M. G. Waddington, M. K. Hopkins, D. H. Smith, H. Li, C. P. Kolbert, S. O. Montbomery, and D. H. Persing. 2000. Identification of coryneform bacterial isolates by ribosomal DNA sequence analysis. *J. Clin. Microbiol.* **38**:1676–1678.

Timmins, E. M., D. E. Quain, and R. Goodacre. 1998a. Differentiation of brewing yeast strains by pyrolysis mass spectrometry and Fourier transform infrared spectroscopy. *Yeast* **14**:885–893.

Timmins, E. M., S. A. Howell, B. K. Alsberg, W. C. Noble, and R. Goodacre. 1998b. Rapid differentiation of closely related *Candida* species and strains using pyrolysis-mass spectrometry and

Fourier transform-infrared spectroscopy. *J. Clin. Microbiol.* **36**:367–374.

Trotha, R., T. Hanck, W. König, and B. König. 2001. Rapid ribosequencing—an effective diagnostic tool for detecting microbial infection. *Infection* **29**:12–16.

Unnerstad, H., H. Ericsson, A. Alderborn, W. Tham, M.-L. Danielsson-Tham, and J. G. Mattsson. 2001. Pyrosequencing as a method for grouping of *Listeria monocytogenes* strains on the basis of single-nucleotide polymorphisms. *Appl. Environ. Microbiol.* **67**:5339–5342.

Vaidyanathan, S., J. J. Rowland, D. B. Kell, and R. Goodacre. 2001. Discrimination of aerobic endospore-forming bacteria via electrospray-ionization mass spectrometry of whole cell suspensions. *Anal. Chem.* **73**:4134–4144.

Valente, P., J. Ramos, and O. Leoncini. 1999. Sequencing as a tool in yeast molecular taxonomy. *Can. J. Microbiol.* **45**:949–958.

Valinsky, L., G. Della Vedova, A. J. Scupham, S. Alvey, A. Figueroa, B. Yin, J. Hartin, M. Chroback, D. E. Crowley, T. Jiang, and J. Borneman. 2002. Analysis of bacterial community composition by oligonucleotide fingerprinting of rRNA genes. *Appl. Environ. Microbiol.* **68**:3243–3250.

Van Baar, B. L. 2000. Characterization of bacteria by matrix assisted laser desorption/ionisationand electrospray mass spectrometry. *FEMS Microbiol. Rev.* **24**:193–219.

Vandamme, P., B. Pot, M. Gillis, P. De Vos, K. Kersters, and J. Swings. 1996. Polyphasic taxonomy, a consensus approach to bacterial systematics. *Microbiol. Rev.* **60**:407–438.

Vaneechoutte, M. 1996. DNA fingerprinting techniques for microorganisms; a proposal for classification and nomenclature. *Mol. Biotechnol.* **6**:115–142.

von Wintzingerode, F., S. Bocker, C. Schlotelburg, N. H. L. Chiu, N. Storm, C. Jurinke, C. R. Cantor, U. B. Gobel, and D. van den Boom. 2002. Base-specific fragmentation of amplified 16S rRNA genes analyzed by mass spectrometry: a tool for rapid bacterial identification. *Proc. Natl. Acad. Sci. USA* **99**:7039–7044.

Voordouw, G., J. K. Voordouw, R. R. Karkhoff-Schweizer, P. M. Fedorak, and D. W. S. Westlake. 1991. Reverse sample genome probing, a new technique for identification of bacteria in environmental samples by DNA hybridization, and its application to the identification of sulfate-reducing bacteria in oil field samples. *Appl. Environ. Microbiol.* **57**:3070–3078.

Wang, R.-F., M. L. Beggs, L. H. Robertson, and C. E. Cerniglia. 2002. Design and evaluation of oligonucleotide-microarray method for the detection of human intestinal bacteria in fecal samples. *FEMS Microbiol. Lett.* **213**:175–182.

Wayne, L. G., D. J. Brenner, R. R. Colwell, P. A. D. Grimont, O. Kandler, M. I. Krichevsky, W. F. C. Moore, R. G. E. Murray, E. Stackebrandt, M. P. Starr, and H. G. Trüper. 1987. Report of the ad hoc committee on reconciliation of approaches to bacterial systematics. *Int. J. Syst. Bacteriol.* **37**:463–464.

Wenning, M., H. Seiler, and S. Scherer. 2002. Fourier-transform infrared microspectroscopy, a novel and rapid tool for identification of yeasts. *Appl. Environ. Microbiol.* **68**:4717–4721.

Wilson, K. H., W. J. Wilson, J. L. Radosevich, T. Z. DeSantis, V. S. Viswanathan, T. A. Kuczmarski, and G. L. Andersen. 2002. High-density microarray of small-subunit ribosomal DNA probes. *Appl. Environ. Microbiol.* **68**:2535–2541.

Microbial Diversity and Bioprospecting
Edited by Alan T. Bull
© 2004 ASM Press, Washington, D.C.

Chapter 6

Eukaryotic Diversity—a Synoptic View

LAURA A. KATZ

OVERVIEW

The diversity of eukaryotic microbes is still relatively unexplored, particularly in extreme environments and for the smallest eukaryotes. Estimates of the numbers of clades of eukaryotes vary dramatically and range up to as many as 200 lineages, of which plants, animals, and fungi represent just three clades (Corliss, 2002; Patterson, 1999). This review presents an overview on eukaryotic relationships, describes major innovations within eukaryotes, and illustrates these innovations through examples from major clades.

DEFINING EUKARYOTES

All eukaryotes contain both a nucleus and a cytoskeleton. Although named for the presence of the nucleus (eu = true; karyon = kernel or seed), it is the presence of the cytoskeleton and its related proteins that has allowed the dramatic morphological variation found among eukaryotic cells. While many eukaryotes also contain additional internalized organelles such as mitochondria, dictyosomes, and chloroplasts, these organelles are not found in all eukaryotes. For example, there are numerous lineages of eukaryotes including chytrids, *Entamoeba,* and some ciliates that have lost their mitochondria (Embley, 2002; Roger, 1999; Roger and Silberman, 2002).

Evolutionary Relationships Among Eukaryotes Remain Unclear

Despite growing morphological and molecular data sets from some microbial eukaryotes, the scaffolding of the eukaryotic portion of the tree of life remains obscure. Problems in determining homology and modeling evolutionary transitions have hampered attempts to reconstruct eukaryotic relationships by morphology alone. Attempts to reconstruct eukaryotic relationships using molecular data have also proven controversial as analyses of different gene genealogies that include broadly sampled eukaryotes often result in conflicting topologies (Katz, 1999; Philippe and Adoutte, 1998). Inadequate taxonomic sampling surely contributes to the instability of gene genealogies as many known eukaryotic groups remain understudied and additional clades of eukaryotes are still being discovered.

Reconstructing eukaryotic relationships is further complicated by the fact that eukaryotic genomes are chimeric, containing genes acquired through lateral transmission as well as vertical descent (Katz, 1999, 2002). Recent models of the origin of eukaryotes frequently invoke fusion events between various prokaryotes, both archaea and bacteria, to explain the chimeric nature of eukaryotic genomes (reviewed in Katz, 2002). For example, the most recent version of the serial endosymbiosis theory presented by Margulis et al. (2000) argues that eukaryotes evolved through symbioses involving an archaeon, a spirochaete, and an alphaproteobacterium. Such models predict that eukaryotic genes should trace back to particular prokaryotic lineages. However, analyses of multiple gene genealogies indicate that eukaryotes have acquired genes through lateral transfers from a broad diversity of donor lineages (Katz, 2002). These data are inconsistent with models that invoke a small number of fusion or symbiotic events to explain the origin of eukaryotic genomes. Instead, these data imply that continuous lateral gene transfers into eukaryotic genomes have occurred.

In the future, we must elucidate the extent of lateral gene transfers in the genomes of microbial eukaryotes, improve our sampling of both taxa and genes, and apply more sophisticated analytical techniques if we are to discover the evolutionary relationships among all eukaryotes. A hypothesis for eukaryotic relationships is shown in Fig. 1 but should be

Laura A. Katz • Department of Biological Sciences, Smith College, Northampton, MA 01063 and Program in Organismic and Evolutionary Biology, University of Massachusetts-Amherst, Amherst, MA 01003.

interpreted with caution as many nodes are likely to change in the future. This figure is loosely based on a multigene analysis of protein coding genes (Baldauf et al., 2000).

Search for Early Eukaryotes

Not surprisingly, given their complex evolutionary history, the quest to uncover the earliest extant lineage of eukaryotes has proven very difficult. The limited fossil record of early eukaryotes provides little guidance. Precambrian fossils, including the earliest fossil eukaryotes from ~1.7 to 2.1 billion years ago, are of unknown taxonomic affinity, as they are distinct from extant eukaryotes (Knoll, 1992). These "acritarchs" may represent early experiments in eukaryotic life. Although there are reports of Precambrian fossilized organisms resembling red algae, brown algae, green algae, chytrids, ciliates, and testate amoebae, many of these fossils are controversial (e.g., Butterfield, 2000). The oldest convincing fossils for most eukaryotic lineages that have a fossil record start in the Cambrian period.

Molecular genealogies often place the parasitic lineages *Giardia* and *Trichomonas* at the base of the eukaryotic tree (Baldauf et al., 2000; Keeling and Palmer, 2000; Roger, 1999). However, rates of evolution are elevated in these lineages making their taxonomic position suspect (e.g., Dacks et al., 2002). Moreover, these potentially early diverging eukaryotes have virtually no fossil record. Although recently the eukaryotic tree has been rooted using a single-gene fusion (Stechmann and Cavalier-Smith, 2002), such single-gene approaches have been misleading in the past. Additional data are required to determine the extant lineages that represent the earliest-diverging eukaryotes.

Eukaryotic Metabolism

In contrast to the dramatic metabolic diversity found among prokaryotes, eukaryotes contain relatively few metabolic strategies, and the limited metabolic diversity that exists has often been acquired through lateral gene transfers or endosymbiosis. Although the nature of the host cell is unknown, the morphology and genomes of mitochondria indicate that this organelle is descended from a proteobacterial endosymbiont (Gray, 1999; Lang et al., 1999). Similarly, photoautotrophic eukaryotes evolved through the endosymbiotic acquisition of plastids (see below). Furthermore, increasing data on the nature of metabolic genes in some anaerobic eukaryotes suggest that the transition from aerobic to anaerobic lifestyles may have occurred alongside the acquisition of prokaryotic metabolic genes. For example, the iron hydrogenase gene in the anaerobic eukaryotes *Entamoeba histolytica* (the causative agent of amoebic dysentery) and *Nyctotherus ovalis* (a ciliate) may have been acquired by independent lateral gene transfers (Horner et al., 2000, 2002).

Eukaryotes Have Become Photosynthetic Numerous Times

Eukaryotes have become photosynthetic multiple times through the acquisition of primary, secondary, and perhaps even tertiary or quaternary endosymbionts. Although the details of these symbiotic events remain unclear, current data are consistent with a single primary symbiotic event between a heterotrophic eukaryote and a cyanobacterium (Archibald and Keeling, 2002; Delwiche, 1999; McFadden, 2001). The photosynthetic descendants of the primary symbiosis event are a clade containing glaucocystophytes, red algae, green algae, and plants. The glaucocystophytes are the only eukaryotes in which the wall of the chloroplast still contains peptidoglycan, a cross-linked sugar and amino acid structure found in the cell walls of many bacteria. The predominantly marine red algae are marked by diverse life cycles that include calcifying stages in some species and complex alternation of generations in others. Similarly, the ~20,000 species of nonplant green algae are marked by diverse forms of mitosis, multiple origins of multicellularity, and complex morphological development in large, multinucleated cells. Evidence to support the claim that these lineages resulted from primary symbiosis includes the structure of plastids in these lineages, which tend to be contained in double membranes (as opposed to three, four, or even more membranes found in other photosynthetic eukaryotes) as well as genealogical analyses of chloroplast genes that place these lineages basal to other photosynthetic eukaryotes (Delwiche, 1999; McFadden, 2001).

The remaining lineages of eukaryotes, including brown algae, diatoms, dinoflagellates, cryptomonads, euglenids, and chlorarachniophytes, evolved through secondary (or greater) symbioses in which heterotrophic eukaryotes became photosynthetic by engulfing either a red or green alga (Archibald and Keeling, 2002; Delwiche, 1999; McFadden, 2001). Such secondary endosymbioses are best documented in two nonsister lineages of eukaryotes, the chlorarachniophytes and cryptomonads (McFadden, 2001). In addition to chloroplasts surrounded by multiple membranes, both of these photosynthetic lineages contain nucleomorphs, remnant nuclei from the eukaryotic symbiont. For example, in addition to the host nuclear and mitochondrial genomes, cryptomonads contain a plastid genome and a remnant nuclear genome from the symbiont. Analyses of gene sequences from the nucleus and nucleomorph of cryptomonads indicate that this photosynthetic lineage evolved when a heterotrophic flagellate engulfed a red alga (Maier et al., 2000; McFadden, 2001).

REPRESENTATIVE LINEAGES

Although much of the structure of the eukaryotic portion of the tree remains to be elucidated, several major clades have emerged. Support for these lineages varies as some clades have been discovered by gene sequencing and are supported by only a few morphological characters whereas other clades are supported by both molecular and morphological analyses. Rather than listing all known eukaryotic lineages, I focus on representatives of five major clades—alveolates, heterokonts, euglenozoa, opisthokonts, and mycetozoans—as well as a few groups of uncertain taxonomic position—foraminifera, diplomonads, parabasalids (Fig. 1). Further description of these

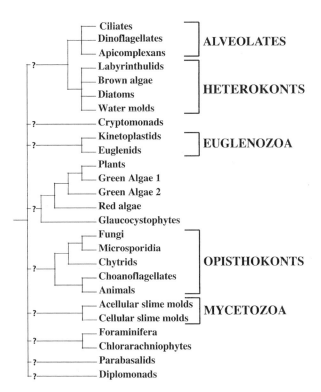

Figure 1. Hypothesis for eukaryotic relationships derived from multigene analyses of Baldauf and colleagues (2000). Question marks represent unresolved nodes. Green algae are paraphyletic. Many branches are likely to change as additional data are analyzed.

major groups, plus a list of taxa not discussed, can be found in Corliss (2002) and Patterson (1999).

Alveolates

The alveolates are a well-defined clade that emerges from many gene genealogies and includes three major lineages: the ciliates, apicomplexans, and dinoflagellates. All alveolates contain alveolar sacs that are located beneath the cell membrane. However, the function of these sacs is not known for most lineages.

Ciliates

Ciliates (Fig. 2a and b) are defined by two features: (1) the presence of cilia, rows of eukaryotic flagella, during at least some portion of their life cycle, and (2) nuclear dualism, the presence of two distinct genomes within the same cell. Nuclear dualism appears to have evolved at least twice in microbial eukaryotes, once at the origin of ciliates and once within the foraminifera. Within ciliates, where nuclear dualism has been best studied, the two distinct genomes develop from a single zygotic nucleus. The germ line micronucleus is transcriptionally inactive and divides by mitosis. In contrast, chromosomes in the somatic macronucleus, the site of virtually all transcription, are processed through fragmentation, elimination of internal sequences, and amplification of remaining chromosomes (reviewed in Katz, 2001). The extent of processing varies among lineages, but in at least two classes, the spirotrichs and phyllopharyngeans, the macronuclei end up with gene-sized chromosomes. In the most extreme cases, found in some spirotrich ciliates, the zygotic chromosomes are processed to generate approximately 25,000 to 30,000 unique chromosomes in the macronucleus, each of which is then amplified roughly 1,000 times. Not surprisingly, ciliate macronuclei divide by a poorly understood form of amitosis.

Apicomplexans

The ~4,000 species of apicomplexans are defined by the presence of the apical complex, a set of structures at the anterior end of the cells that enable the organisms to penetrate host cells (Levine, 1988). All three major groups of apicomplexans are parasites and have a spore stage (hence their former inclusion within the sporozoa); the Gregarinidea are both intracellular and extracellular parasites, whereas the Coccidea and Haematozoea are predominantly intracellular parasites. The life cycles of apicomplexans include multiple morphological forms and sometimes

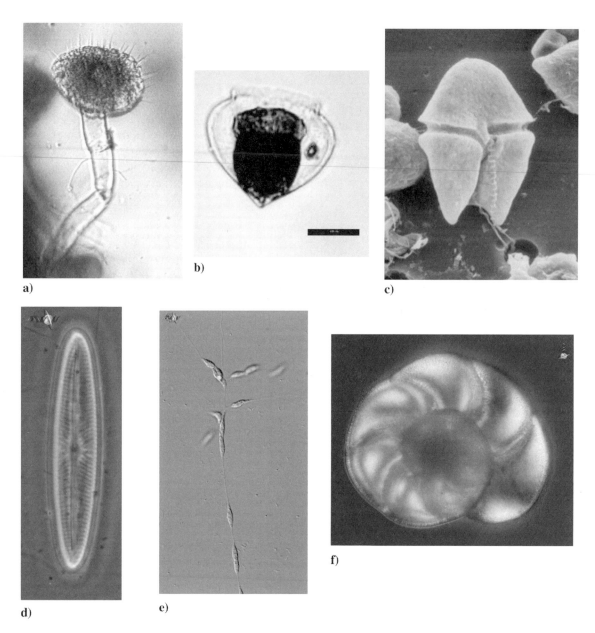

Figure 2. Microbial eukaryotes: (a) the ciliate *Ephelota* sp. (L. A. Katz); (b) the ciliate *Metacylis angulata*, scale bar at 30 μm (G. McManus, University of Connecticut); (c) the dinoflagellate *Gymnodinium sanguineum* (M. Farmer, University of Georgia); (d) a pennate diatom (www.mbl.edu/microscope); (e) *Labyrinthula* sp. with ectoplasmic net (D. J. Patterson, L. A. Zettler, and V. Edgcomb, www.mbl.edu/microscope); (f) the foraminiferan *Ammonia* sp. (M. Farmer and D. J. Patterson, www.mbl.edu/microscope).

multiple hosts. Some apicomplexans can divide by schizogony, a process in which a single multinucleated cell rapidly divides into many single-nucleated cells.

Perhaps the best-known apicomplexans are in the genus *Plasmodium*, including *P. falciparum*, a causative agent of malaria. Malaria remains one of the most devastating diseases throughout much of the developing world, despite decades of dedicated effort aimed at eradication. The reasons for the difficulty in preventing and curing malaria are multifold and include the complex life cycle of *P. falciparum*, in which at least 10 morphological stages pass through two hosts, anopheline mosquitoes and some mammals. The recent completion of the genome sequence of *P. falciparum* (Gardner et al., 2002; Nelson, chapter 25) may speed up the development of a vaccine against malaria.

Although not predicted by the parasitic life histories of apicomplexans, members of this group contain a remnant plastid now termed the apicoplast (Howe, 1992; Marechal and Cesbron-Delauw, 2001). This remnant plastid encodes genes necessary for fatty acid synthesis (Howe, 1992; Marechal and Cesbron-Delauw, 2001) and may serve as a target for drugs developed against malaria (Ralph et al., 2001).

Dinoflagellates

The third group of alveolates, the dinoflagellates (*deino* = whirling), is marked by numerous unique features (Fig. 2c) (Taylor, 1987). For example, the evolution of photosynthesis within dinoflagellates involved multiple losses of plastids as well as acquisitions of secondary, tertiary, and perhaps even quaternary endosymbionts (Saldarriaga et al., 2001; Yoon et al., 2002). Mitosis is unique in dinoflagellates as some lineages retain their nuclear envelope during karyokinesis (a trait also present in parabasalids) and attach chromosomes to microtubules through tunnel-like structures that form in the dividing nucleus. Furthermore, some dinoflagellates have lost their histone proteins and instead have continuously condensed chromosomes associated with novel basic proteins (Taylor, 1987). Dinoflagellates are important members of marine and freshwater food webs, the causative agents of many red tides, and the photosynthetic symbionts of cnidarians in coral reefs.

Heterokonts

The heterokonts, also called stramenopiles, are a diverse group of eukaryotes that include brown algae, diatoms, labyrinthulids, and water molds (Baldauf et al., 2000; Cavalier-Smith, 2002; Patterson, 1999). The monophyly of this group is supported by many gene genealogies as well as by the presence of hairlike projections, mastigonemes, on flagella when they occur. Furthermore, the plastids of the photosynthetic heterokonts have extra membranes associated with the endoplasmic reticulum, termed the chloroplast endoplasmic reticulum (McFadden, 2001).

Brown algae

The predominantly marine brown algae, or phaeophytes, are major components of seaweed communities and include single-celled organisms as well as the giant kelp *Macrocystis* and the tropical *Sargassum*. Brown algae show an alternation of generation with multicellular diploid and haploid phases and only reproductive cells are flagellated.

Diatoms

The silica-containing diatoms are divided into two major groups: the bilaterally symmetric pennate diatoms (Fig. 2d) and the radially symmetric centric diatoms. Because of their hard parts, diatoms have an extensive fossil record dating back to the Cretaceous period, about 100 million years ago. There are an estimated 10,000 species of diatoms, which is about twice the number of mammal species. The only flagellated stages of diatoms are the male gametes of centric diatoms; the remaining diatoms are either free floating or glide slowly through their substrate. The molecular properties of diatoms are only beginning to be explored.

Labyrinthulids

The roughly 50 species of labyrinthulids, also known as slime nets, create a complex network of excreted material, an ectoplasmic net, through which amoeboid cells travel (Fig. 2e). Once considered part of the lower fungi, this clade is united with the heterokonts based on a combination of ultrastructural and molecular data. One species, *Labyrinthula macrocystis*, is believed to be responsible for extensive periods of death of the marine plant eelgrass (*Zostera marina*) such as occurred on the eastern coast of the United States in the 1930s (Muehlstein et al., 1991).

Water molds

The 700 species of oomycetes, or water molds, are another group of heterotrophic microbial eukaryotes that formerly were considered to be lower fungi. The diplontic water molds include flagellated stages with mastigonemes, a characteristic of heterokonts. Like the apicomplexans, water molds have the ability to generate large, multinucleated cells that can rapidly go through a series of divisions to generate multiple, single-nucleated zoospores. Some oomycetes parasitize plants and animals: contrary to popular belief, the Great Famine of 1846 that resulted from the rapid destruction of the potato crop in Ireland was caused by the oomycete *Phytophthora infestans*, not by a fungus.

Euglenozoa

The euglenozoa include two major lineages, the euglenids and the kinetoplastids, whose sister status is supported by both ultrastructural and molecular analyses (Taylor, 1999). Members of the euglenozoa have mitochondria with discoid cristae and many contain two flagella, one leading and one trailing, which emerge from a pocket at the anterior end of the cell (Patterson, 1999; Taylor, 1999).

Euglenids

The euglenids include both heterotrophic and secondarily photosynthetic flagellates that are common in many freshwater environments. The striped pellicle that characterizes euglenids results from complex proteinaceous stripes underlying the cell membrane (Leander et al., 2001). Some euglenids can contract and expand their cells dramatically through a process known as metaboly. The plastids of photosynthetic euglenids are surrounded by at least three membranes, with the third membrane possibly resulting from a food vacuole that engulfed a photosynthetic eukaryote during secondary endosymbiosis. The green color and presence of chlorophyll *b* suggest that euglenid plastids are derived from a green algal symbiont.

Kinetoplastids

The second clade within the euglenozoa, the kinetoplastids, contains both parasitic and free-living flagellates (Maslov et al., 2001). This group is named because of the unusual structure of mitochondria found in its members. Instead of the simple circular DNA present in the mitochondria of many eukaryotes, the mitochondrial DNA of kinetoplastids is arranged in complex networks of maxicircles and minicircles. Some maxicircles contain incomplete copies of mitochondrial proteins that require correcting through RNA editing (Simpson et al., 2000). In its most extreme form, guide sequences in minicircles template the alteration of more than 80% of the nucleotides required to make a full-length mitochondrial mRNA.

The genome sequence of the kinetoplastid parasite *Leishmania major*, the causative agent of the "flesh-eating" disease leishmaniasis, is under way. Analysis of sequences on the first chromosome fully sequenced reveals a surprising pattern of absolute strand polarity, in which stretches of 30 to 50 genes are found on one strand with no intervening genes present on the other strand (McDonagh et al., 2000).

Opisthokonts

The opisthokonts include several microbial lineages (e.g., choanoflagellates and microsporidians) as well as two predominantly macroscopic clades (animals and fungi [Baldauf, 1999; Cavalier-Smith, 2002]). Opisthokonts, a clade supported by many gene genealogies, derive their name from the placement of flagella at the posterior (opistho) end of the

cells (Cavalier-Smith, 2002). Further evidence for the monopoly of this group comes from an 11-amino-acid insertion found in the elongation factor 1-alpha gene of all opisthokonts (Baldauf, 1999).

Choanoflagellates

The choanoflagellates (or collared flagellates) are morphologically similar to choanocyte cells of sponges and have been demonstrated by morphological and molecular data to be the sister lineage to animals (Leadbeater and Kelly, 2001; Snell et al., 2001). Choanoflagellates can be stalked or stalkless and can live either as single cells or in colonies. The life cycles of most choanoflagellates remain unknown.

Microsporidians

Perhaps one of the most surprising groups within the opisthokonts is the microsporidia. Microsporidians, single-celled eukaryotes whose life cycle includes a spore stage and amoeboid stages, are causative agents of some diarrheas in mammals (Mathis, 2000). Although the microsporidia were originally thought to be amitochondriate, a remnant mitochondrion has recently been found in one genus of microsporidian (Williams et al., 2002). Analyses of the small-subunit rDNA gene from numerous microsporidians have placed these lineages toward the base of the eukaryotic tree as one of the earliest lineages of eukaryotes. However, analyses of numerous protein-coding genes coupled with reanalysis of small-subunit rDNA sequences have moved this lineage to either within or basal to the fungi (e.g., Hirt et al., 1999). The sequence of the genome of one microsporidian, *Encephalitozoon cuniculi*, has been completed and reveals a dramatically compact genome (roughly 2,000 genes) that represents the smallest known nuclear genome (Katinka et al., 2001).

Fungi

Chytrids, a potential sister lineage to "true" fungi, are predominantly aquatic and include flagellated stages in their life cycles (Taylor et al., in press). In contrast, true fungi, such as mushrooms and yeasts, have lost their flagellated stages. Chytrids cause diseases in numerous animal lineages and may be responsible for amphibian declines in Central America, including the now-extinct golden toad. Fossil chytrids may date back to the Precambrian period.

The true fungi have classically been divided based on their reproductive stages into the ascomycete (sac fungi), basidiomycetes (club fungi), and zygomycetes (bread molds), with the deuteromycetes representing asexual lineages. Although the monophyly of the zygomycetes is not supported by molecular analyses, the ascomycetes and basidiomycetes appear to be monophyletic as well as sister taxa (Bruns et al., 1993; Taylor et al., in press). Parasitism has evolved multiple times in fungi, including the 7,000 species of rusts (basidomycetes) that attack numerous plants, the numerous parasitic zygomycetes that affect invertebrates, and the ascomycetes *Pneumocystis carinii* and *Candida albicans*, causative agents of pneumonia and vaginitis, respectively (Taylor et al., in press). Although increasing molecular data may resolve relationships within and between major fungal groups, it is possible that frequent lateral gene transfers may ultimately make it difficult to reconstruct fungal phylogeny (Rosewich and Kistler, 2000).

Mycetozoans

The mycetozoans, or slime molds, are characterized by complex life cycles that include multicellular fruiting bodies. There are two major types of slime molds: cellular and acellular. The monophyly of these slime molds is supported by many, but not all, gene genealogies (Baldauf and Doolittle, 1997; Baldauf et al., 2000; Sogin and Silberman, 1998).

Cellular slime molds

The cellular slime molds are predominantly haplont and exist in several distinct life stages: single-celled amoebae, slugs, fruiting bodies, and cysts (Bonner 1998, 2000). Single-celled amoebae feeding on bacteria in the soil will aggregate to form multicellular "slugs" upon starvation. These slugs then travel through the soil until eventually they differentiate into a multicelled fruiting body. Differentiation of cells to form a fruiting body has been best studied in members of the genus *Dictyostelium*, where some cells transform into stalk cells whereas others transform into spores from which amoebae emerge under the proper environmental signals. Studying development in *Dictyostelium* sheds light on the evolution of multicellularity in the mycetozoans (Bonner 1998, 2000).

Acellular slime molds

In contrast, the life cycle of the acellular slime molds, including the well-studied genus *Physarum*, is marked by the growth of large cells in which karyokinesis occurs in the absence of cytokinesis. These predominantly diplont organisms form plasmodia, large multinucleated cells, from which multicelled fruiting bodies develop. The plasmodium of an acellular slime mold can grow to be greater than 20 m in diameter.

Foraminifera

The chambered shells constructed by foraminifera have left an impressive fossil record dating back 550 million years ago (Lee and Anderson, 1991). Foraminifera (Fig. 2f) can be multinucleated and large, with fossilized forams reaching sizes greater than 10 cm in diameter. Foraminifera have bidirectional cytoplasmic movement within their pseudopodia. Metabolism varies among foraminifera with some species feeding on bacteria, microbial eukaryotes, and even small animals. Other foraminifera harbor photosynthetic dinoflagellates or diatoms that they farm as a way of acquiring complex carbons. Life cycles in foraminifera are complex with alternating haploid and diploid stages.

Parabasalids

The parabasalids represent a group of potentially early-diverging eukaryotes. Parabasalids are named for the association of their unusual Golgi apparatus with kinetosomes (basal bodies of eukaryotic flagella [Keeling and Palmer, 2000; Patterson, 1999]). Parabasalids lack mitochondria and instead contain hydrogenosomes, hydrogen-producing organelles that are also found in chytrids and some ciliates (Embley et al., 1997; Hackstein et al., 1999; Müller, 1993). Most well known among the parabasalids is *Trichomonas vaginalis*, the causative agent of the sexually transmitted disease trichomoniasis. Other parabasalids live as symbionts in the guts of termites and wood-eating cockroaches where they, along with their symbiotic bacteria, digest the wood that the insect eats. The taxonomic placement of the parabasalids is controversial, as many genes in this lineage appear to be evolving rapidly, generating long branches in genealogies and weak support for the placement of this group.

Diplomonads

The diplomonads are named for their characteristic symmetrical body shapes that look like two flagellates attached to one another. Hence, diplomonads contain two nuclei and two sets of flagella. Most extant diplomonads are parasites, including the well-studied *Giardia lamblia*, and the few free-living diplomonads appear to be derived from parasitic ancestors. Numerous gene genealogies place *Giardia* and its relatives at the base of the eukaryotic tree (e.g., Baldauf et al., 2000), creating a conundrum as to how a predominantly parasitic group could be early-diverging eukaryotes. There are at least two explanations for this apparent conundrum: (i) the parasitic nature of extant diplomonads could be derived or (ii) the relatively fast rates of evolution in many diplomonad genes could create long branch attraction that draws the diplomonads to the base of the tree.

There are numerous other unusual features of diplomonads including hypervariable karyotypes. Genome evolution in *Giardia* is remarkable in that significant segments (10–100 kb) of chromosomes are rearranged during cell division (Upcroft and Upcroft, 1999). This high rate of rearrangements generates variable karyotypes among strains of *Giardia* species.

LOOKING INTO THE FUTURE

Collecting data on both the genomes and the cell biology of diverse lineages is necessary to accurately describe the origin and diversification of eukaryotes. In the near future, comparisons of multiple gene genealogies are required to outline relationships among eukaryotes and to provide a framework for placing newly discovered lineages. However, to fully characterize the diversity of eukaryotes will require concerted efforts in constructing gene genealogies, analyzing cell structures, and searching for new lineages of eukaryotes.

Acknowledgment. This work is supported by a National Science Foundation CAREER grant to L.A.K. (NSF 0079325).

REFERENCES

Archibald, J. M., and P. J. Keeling. 2002. Recycled plastids: a "green movement" in eukaryotic evolution. *Trends Genet.* **18:**577–584.

Baldauf, S. L. 1999. A search for the origins of animals and fungi: comparing and combining molecular data. *Am. Nat.* **154:**S178–S188.

Baldauf, S. L., and W. F. Doolittle. 1997. Origin and evolution of the slime molds (Mycetozoa). *Proc. Natl. Acad. Sci. USA* **94:**12007–12012.

Baldauf, S. L., A. J. Roger, I. Wenk-Siefert, and W. F. Doolittle. 2000. A kingdom-level phylogeny of eukaryotes based on combined protein data. *Science* **290:**972–977.

Bonner, J. T. 1998. The origins of multicellularity. *Integ. Biol.* **1:**27–36.

Bonner, J. T. 2000 *First Signals: The Evolution of Multicellular Development.* Princeton University Press, Princeton, N.J.

Bruns, T. D., R. Vilgalys, S. M. Barns, D. Gonzalez, D. S. Hibbett, D. J. Lane, L. Simon, S. Stickel, T. M. Szaro, W. G. Weisburg, and M. L. Sogin. 1993. Evolutionary relationships within the fungi: analysis of nuclear small subunit rRNA sequences. *Mol. Phylogenet. Evol.* **1:**231–241.

Butterfield, N. J. 2000. Bangiomorpha pubescens n. gen., n. sp.: implications for the evolution of sex, multicellularity, and the Mesoproterozoic/Neoproterozoic radiation of eukaryotes. *Paleobiology* **26:**386–404.

Cavalier-Smith, T. 2002. The phagotrophic origin of eukaryotes and phylogenetic classification of protozoa. *Int. J. Syst. Evol. Microbiol.* **52:**297–354.

Corliss, J. O. 2002. Biodiversity and biocomplexity of the protists and an overview of their significant roles in maintenance of our biosphere. *Acta Protozool.* **41:**199–219.

Dacks, J. B., A. Marinets, W. F. Doolittle, T. Cavalier-Smith, and J. M. Logsdon. 2002. Analyses of RNA polymerase II genes from free-living protists: phylogeny, long branch attraction, and the eukaryotic big bang. *Mol. Biol. Evol.* **19:**830–840.

Delwiche, C. F. 1999. Tracing the tread of plastid diversity through the tapestry of life. *Am. Nat.* **154:**S164–S177.

Embley, T. M. 2002. Anaerobic eukaryotes and their archaebacterial endosymbionts. *Environ. Microbiol.* **4:**15–16.

Embley, T. M., D. A. Horner, and R. P. Hirt. 1997. Anaerobic eukaryote evolution: hydrogenosomes as biochemically modified mitochondria? *Trends Ecol. Evol.* **12:**437–441.

Gardner, M. J., N. Hall, E. Fung, O. White, M. Berriman, R. W. Hyman, J. M. Carlton, A. Pain, K. E. Nelson, S. Bowman, I. T. Paulsen, K. James, J. A. Eisen, K. Rutherford, S. L. Salzberg, A. Craig, S. Kyes, M. S. Chan, V. Nene, S. J. Shallom, B. Suh, J. Peterson, S. Angiuoli, M. Pertea, J. Allen, J. Selengut, D. Haft, M. W. Mather, A. B. Vaidya, D. M. A. Martin, A. H. Fairlamb, M. J. Fraunholz, D. S. Roos, S. A. Ralph, G. I. McFadden, L. M. Cummings, G. M. Subramanian, C. Mungall, J. C. Venter, D. J. Carucci, S. L. Hoffman, C. Newbold, R. W. Davis, C. M. Fraser, and B. Barrell. 2002. Genome sequence of the human malaria parasite *Plasmodium falciparum*. *Nature* **419:**498–511.

Gray, M. W. 1999. Evolution of organellar genomes. *Curr. Opin. Genet. Dev.* **9:**678–687.

Hackstein, J. H., A. Akhmanova, B. Boxma, H. R. Harhangi, and F. G. Voncken. 1999. Hydrogenosomes: eukaryotic adaptations to anaerobic environments. *Trends Microbiol.* **7:**441–447.

Hirt, R. P., J. M. Logsdon, B. Healy, M. W. Dorey, W. F. Doolittle, and T. M. Embley. 1999. Microsporidia are related to fungi: evidence from the largest subunit of RNA polymerase II and other proteins. *Proc. Natl. Acad. Sci. USA* **96:**580–585.

Horner, D. S., P. G. Foster, and T. M. Embley. 2000. Iron hydrogenases and the evolution of anaerobic eukaryotes. *Mol. Biol. Evol.* **17:**1695–1709.

Horner, D. S., B. Heil, T. Happe, and T. M. Embley. 2002. Iron hydrogenases—ancient enzymes in modern eukaryotes. *Trends Biochem. Sci.* **27:**148–153.

Howe, C. J. 1992. Plastid origin of an extrachromosomal DNA molecular from *Plasmodium*, the causative agent of malaria. *J. Theor. Biol.* **158:**199–205.

Katinka, M. D., S. Duprat, E. Cornillot, G. Metenier, F. Thomarat, G. Prensier, V. Barbe, E. Peyretaillade, P. Brottier, P. Wincker, F. Delbac, H. El Alaoui, P. Peyret, W. Saurin, M. Gouy, J. Weissenbach, and C. P. Vivares. 2001. Genome sequence and gene compaction of the eukaryote parasite *Encephalitozoon cuniculi*. *Nature* **414:**450–453.

Katz, L. A. 1999. The tangled web: gene genealogies and the origin of eukaryotes. *Am. Nat.* **154:**S137–S145.

Katz, L. A. 2001. Evolution of nuclear dualism in ciliates: a reanalysis in light of recent molecular data. *Int. J. Syst. Evol. Microbiol.* **51:**1587–1592.

Katz, L. A. 2002. Lateral gene transfers and the evolution of eukaryotes: theories and data. *Int. J. Syst. Evol. Microbiol.* **52:**1893–1900.

Keeling, P. J., and J. D. Palmer. 2000. Phylogeny—parabasalian flagellates are ancient eukaryotes. *Nature* **405:**635–637.

Knoll, A. H. 1992. The early evolution of eukaryotes: a geological perspective. *Science* **256:**622–627.

Lang, B. F., M. W. Gray, and G. Burger. 1999. Mitochondrial genome evolution and the origin of eukaryotes. *Annu. Rev. Genet.* **33:**351–397.

Leadbeater, B., and M. Kelly. 2001. Evolution of animals—choanoflagellates and sponges. *Water Atmos.* **9:**9–11.

Leander, B. S., R. P. Witek, and M. A. Farmer. 2001. Trends in the evolution of the euglenid pellicle. *Evolution* 55:2215–2235.

Lee, J. J., and R. A. Anderson. 1991. *Biology of Foraminifera.* Academic Press, San Diego, Calif.

Levine, N. D. 1988. *The Protozoan Phylum Apicomplexa.* CRC Press, Inc., Boca Raton, Fla.

Maier, U. G., S. E. Douglas, and T. Cavalier-Smith. 2000. The nucleomorph genomes of cryptophytes and chlorarachniophytes. *Protist* 151:103–109.

Marechal, E., and M. F. Cesbron-Delauw. 2001. The apicoplast: a new member of the plastid family. *Trends Plant Sci.* 6: 200–205.

Margulis, L., M. F. Dolan, and R. Guerrero. 2000. The chimeric eukaryote: origin of the nucleus from the karyomastigont in amitochondriate protists. *Proc. Natl. Acad. Sci. USA* 97:6954–6959.

Maslov, D. A., S. A. Podtipaev, and J. Lukes. 2001. Phylogeny of the kinetoplastida: taxonomic problems and insights into the evolution of parasitism. *Mem. Inst. Oswaldo Cruz* 96:397–402.

Mathis, A. 2000. Microsporidia: emerging advances in understanding the basic biology of these unique organisms. *Int. J. Parasitol.* 30:795–804.

McDonagh, P. D., P. J. Myler, and K. Stuart. 2000. The unusual gene organization of Leishmania major chromosome 1 may reflect novel transcription processes. *Nucleic Acids Res.* 28:2800–2803.

McFadden, G. I. 2001. Primary and secondary endosymbiosis and the origin of plastids. *J. Phycol.* 37:951–959.

Muehlstein, L. K., D. Porter, and F. T. Short. 1991. *Labyrinthula zosterae* sp. nov., the causative agent of wasting disease of Eelgrass, *Zostera marina. Mycologia* 83:180–191.

Müller, M. 1993. The hydrogenosome. *J. Gen. Microbiol.* 139: 2879–2889.

Patterson, D. J. 1999. The diversity of eukaryotes. *Am. Nat.* 154:S96–S124.

Philippe, H., and A. Adoutte. 1998. The molecular phylogeny of Eukaryota: solid facts and uncertainties, p. 25–56. *In* G. H. Coombs, K. Vickerman, M. A. Sleigh, and A. Warren (ed.), *Evolutionary Relationships Among Protozoa.* Kluwer Academic Publishers, Dordrecht, The Netherlands.

Ralph, S. A., M. C. D'Ombrain, and G. I. McFadden. 2001. The apicoplast as an antimalarial drug target. *Drug Resist. Updates* 4:145–151.

Roger, A. J. 1999. Reconstructing early events in eukaryotic evolution. *Am. Nat.* 154:S146–S163.

Roger, A. J., and J. D. Silberman. 2002. Cell evolution: mitochondria in hiding. *Nature* 418:827–829.

Rosewich, U. L., and H. C. Kistler. 2000. Role of horizontal gene transfer in the evolution of fungi. *Annu. Rev. Phytopathol.* 38:325–363.

Saldarriaga, J. F., F. J. R. Taylor, P. J. Keeling, and T. Cavalier-Smith. 2001. Dinoflagellate nuclear SSU rRNA phylogeny suggests multiple plastid losses and replacements. *J. Mol. Evol.* 53:204–213.

Simpson, L., O. H. Thiemann, N. J. Savill, J. D. Alfonzo, and D. A. Maslov. 2000. Evolution of RNA editing in trypanosome mitochondria. *Proc. Natl. Acad. Sci. USA* 97:6986–6993.

Snell, E. A., R. F. Furlong, and P. W. H. Holland. 2001. Hsp70 sequences indicate that choanoflagellates are closely related to animals. *Curr. Biol.* 11:967–970.

Sogin, M., and J. D. Silberman. 1998. Evolution of the protists and protistan parasites from the perspective of molecular systematics. *Int. J. Parasitol.* 28:11–20.

Stechmann, A., and T. Cavalier-Smith. 2002. Rooting the eukaryote tree by using a derived gene fusion. *Science* 297:89–91.

Taylor, F. J. R. 1987. *The Biology of Dinoflagellates.* Blackwell Scientific, Boston, Mass.

Taylor, F. J. R. 1999. Ultrastructure as a control for protistan molecular phylogeny. *Am. Nat.* 154:S125–S136.

Taylor, J. W., J. Spatafora, K. O'Donnell, F. Lutzoni, T. James, D. S. Hibbett, D. Geisser, T. D. Bruns, and M. Blackwell. 2002. The fungi. *In* J. Cracraft and M. Donoghue (ed.), *Assembling the Tree of Life.* Oxford University Press, New York, N.Y.

Upcroft, P., and J. A. Upcroft. 1999. Organization and structure of the Giardia genome. *Protist* 150:17–23.

Williams, B. A. P., R. P. Hirt, J. M. Lucocq, and T. M. Embley. 2002. A mitochondrial remnant in the microsporidian *Trachipleistophora hominis. Nature* 418:865–869.

Yoon, H. S., J. D. Hackett, and D. Bhattacharya. 2002. A single origin of the peridinin- and fucoxanthin-containing plastids in dinoflagellates through tertiary endosymbiosis. *Proc. Natl. Acad. Sci. USA* 99:11724–1729.

III. MICROBIAL ECOLOGY: THE KEY TO DISCOVERY

PREAMBLE

In the past decade microbiology, to a considerable extent, has become polarized into traditional culture-dependent and modern culture-independent studies. In chapter 9, Kornelia Smalla deliberates on the need for culture-independent approaches to the exploration of microbial diversity and discusses the emerging techniques that are enabling functions to be ascribed to particular ribotypes. Increasingly, investigations are embracing both culture-dependent and culture-independent approaches with a view to establishing what factors affect culturability (or dormancy) and how to maximize the recovery of exploitable biology—organisms and/or genes. But, of course, the dichotomy is apparent rather than real, as the impact of molecular biology on resuscitation and culturability is beginning to reveal. Thus, Douglas Kell and his colleagues, the discoverers of the resuscitation-promoting factor, have shown that it is widely distributed in the division *Actinobacteria* (chapter 10), while the growing realization of genome downsizing may be providing clues for the development of new cultivation strategies and media.

Culture-dependent microbiology probably suffers from being time and effort intensive and empirical, yet its importance to basic science and to biotechnology should not be underestimated. This is why developments directed to the recovery of organisms are and will continue to be crucially important. And here, overelaboration in terms of techniques development may not always be the most successful approach to adopt; rather, we might revisit some of the simple but intelligent methodologies pioneered by 19th century bacteriologists that were characterized by diligence and patience. Impressive endorsement of such thinking has recently been demonstrated by Sait et al. (2002), who have brought into culture significant numbers of globally distributed but previously uncultured phylogenetically novel bacteria, including representatives of *Acidobacteria* and *Actinobacteria*. This success was achieved by the simple expedient of using minimal agar supplemented with xylan and long incubation times to cultivate soil bacteria. Extinction culture (see Fry, chapter 8, and Button, chapter 16) and ecosystem mimicry also illustrate the innovative thinking that has been brought to bear on the problem of culturing difficult-to-recover but often dominant members of microbial communities. Extinction culture also exposes a certain reluctance to adopt proven but demanding techniques. Similarly, a greater focus on the analysis of microbial communities and their interactions undoubtedly would yield potent clues to

novel cultivation strategies. Syntrophy in particular is likely to be a very common phenomenon within microbial communities (Bull and Slater, 1982), implying that certain members will not be brought into axenic culture because they rely on the metabolism of others (Overmann and van Gemerden, 2000) and until the basis of syntrophic associations are defined. The complexity of microbial interactions is further increased by cell-cell communications that occur between the same and different species (Molin et al., 2000). An understanding of such signaling systems has clear implications for a wide range of biotechnology search and discovery activities (see chapter 36).

In this section of the book two critical questions are posed, the first of which is *how* do we look for novel microbial diversity? The culture-dependent versus culture-independent approach has already been introduced, but superimposed upon it is the often neglected matter of scale and, arising from scale, the problem of effective sampling of the environment. Hence the second question: *where* is novel diversity to be found? And secondarily, is it worthwhile to bioprospect in so-called exotic biomes and habitats? I proffer a few arguments in support of this strategy (chapter 7). An exhaustive examination of biomes and habitats in this context is neither possible nor necessary; instead, four have been selected to illustrate current trends and discoveries. Soil, arguably the most intensively studied habitat regarding microbial diversity, is included here to demonstrate how the intervention of new technology (metagenomics) is providing novel prospecting opportunities. Similarly, the application of molecular (and to a lesser extent culture) methods has led to a major reappraisal of the diversity of marine bacterioplankton, while the development of ocean and terrestrial drilling projects has enabled microbiologists to probe hitherto totally inaccessible territory and to open up the exploration of deep biospheres (see Parkes and Wellsbury, chapter 12, and websites devoted to subsurface microbiology in Wackett, 2002). John Priscu and Brent Christner (chapter 13) remind us that, overall, Earth's biosphere is cold and that therefore it is appropriate to focus on microbial diversity that has evolved under this dominant environmental condition.

The term "extremophile" is used to describe life under extreme conditions of temperature, pressure, pH, salinity, oxygen availability, and in a variety of toxic or stress environments such as those exposed to radiation or heavy metals. Again, microbial diversity and activity in a selection of environments are used to portray extremophile research and some options for biotechnological exploitation. Semantic arguments can readily surface about what constitutes an extremophile, and some readers may take issue with my decision to include oligotrophs. I acknowledge the possible incongruity of this position (but include oligotrophs here largely as a matter of convenience), as well as the somewhat arbitrary definition of extremophiles by reference to just a single environmental parameter. Given the impressive range of anaerobic microorganisms, Guy Fauque and Bernard Ollivier (chapter 17) decided to focus on sulfur-reducing bacteria in order to emphasize the enormous metabolic diversity that can reside in a single category of anaerobes. Although not considered in this book, it is important to recognize that anaerobic eukaryotes also comprise a large and diverse collection of parasitic and free-living taxa, some of which contain hydrogenosomes and support methanogenic archaea and/or bacterial endosymbionts (Embley 2002; Embley and Hirt, 1998).

This section concludes with a return to a particularly intimate and globally significant expression of organism interaction defined as symbiosis. Here the selection of case studies demonstrates that symbioses are universal with respect to marine and terrestrial ecosystems; they are a common feature in the largest single group of organisms (arthropods), they are established in animals and in plants; and as Peter Jeffries (chapter 20) comments, they probably have been a key driver of evolutionary processes.

Throughout much of the discussion in this section of the book the question of the distribution—both local and global—of microorganisms has been raised. This is a question that has broad scientific interest and also major significance for microbial prospecting: it is our subject of enquiry in section IV.

REFERENCES

Bull, A. T., and J. H. Slater. 1982. *Microbial Interactions and Communities.* Academic Press, London, United Kingdom.

Embley, T. M. 2002. Anaerobic eukaryotes and their archaebacterial endosymbionts. *Environ. Microbiol.* 4:15–16.

Embley, T. M., and R. P. Hirt. 1998. Early branching eukaryotes? *Curr. Opin. Genet. Dev.* 8:624–629.

Molin, S., A. T. Nielsen, A. Heydorn, T. Tolker-Nielsen, and C. Sternberg. 2000. Environmental microbiology at the end of the second millenium. *Environ. Microbiol.* 2:6–7.

Overmann, J., and H. van Gemerden. 2000. Microbial interactions involving sulfur bacteria: implications for the ecology and evolution of bacterial communities. *FEMS Microbiol. Rev.* 24:591–599.

Sait, M., P. Hugenholtz, and P. H. Janssen. 2002. Cultivation of globally distributed soil bacteria from phylogenetic lines previously only detected in cultivation-independent surveys. *Environ. Microbiol.* 4:654–666.

Wackett, L. P. 2002. Subsurface microbiology. An annotated selection of World Wide Web sites relevant to the topics in *Environmental Microbiology. Environ. Microbiol.* 4:430–431.

Microbial Diversity and Bioprospecting
Edited by Alan T. Bull
© 2004 ASM Press, Washington, D.C.

Chapter 7

How To Look, Where To Look

ALAN T. BULL

SO MANY ORGANISMS

Supersaturated Coexistence

A distinctive feature of microbial communities is that their diversity can be extraordinarily high, not infrequently orders of magnitude greater that the diversity of macrobial ecosystems (see chapter 2). The "paradox of the plankton" advanced by Hutchinson (1961) challenges microbiologists to account for the situation wherein the number of coexisting species exceeds the number of limiting resources and, with particular reference to plankton, in a relatively isotrophic, unstructured habitat. The ambience of soils and sediments is different from that of aquatic habitats, and the high diversity is sustained at small scales in a complex and dynamic chemical environment located within a heterogeneous physical milieu. On the basis of microbial community diversity profiles derived for surface, vadose, and saturated zone soils, Zhou et al. (1997) identified noncompetitive, competitive, and intermediate diversity patterns, respectively, in the three soil types. Various hypotheses have been erected in order to explain such noncompetitive and competitive patterns and the overall megadiversity of microbial communities. Tiedje et al. (1999) argued that spatial isolation and/or resource heterogeneity could be critical determinants of diversity patterns, and subsequently Zhou et al. (2002) have reported that uniform diversity profiles characterized the surface soil communities from which they concluded that competition did not determine community structure.

The original paradox of the plankton arose from an assumption that multiple plankters occupied a common ecological niche. Subsequently it was shown that even in such an apparently homogeneous medium a variety of ecological niches occur at the micro and pico scales as a consequence of resource competition, plasticity, cell shape, and physical constraints (Steinberg and Geller, 1993). The state of "supersaturated coexistence" that defines the coexisting species-limiting resource paradox may be maintained by competition for such resources such that nonequilibrium conditions are generated (Huisman and Weissing, 1999). Although this model has been criticized (Schippers et al., 2001), more recent simulations by Huisman and his colleagues add credence to the hypothesis that competitive chaos promotes biodiversity. Rather than using random species parameters (e.g., half-saturation constant, specific growth rate, resource content), Huisman et al. (2001) modeled on the basis of "plausible physiological trade-offs" and observed oscillations between equilibrium and nonequilibrium dynamics—and hence normal and supersaturated coexistence—in 10 million-day simulations. Supersaturation values of 40 species on three resources occurred frequently and peaked at 100 species. A somewhat similar consumer and resource model developed by Vandermeer et al. (2002) with which to investigate the consequences of increasing competition pressure showed that more species could coexist when competition intensity was higher. Neither of these groups has yet defined the ecological forces that are responsible for such supersaturated coexistence.

Antibiosis is a widespread function of microorganisms, and in a recent thought-provoking paper, Czaran et al. (2002) developed a game theory model suggesting that antibiotic interactions within microbial communities also may be an effective means of maintaining supersaturated existence. The model incorporated killer, sensitive, and resistant strains and an ability for the system to evolve by allowing several distinct systems of toxin production, sensitivity and resistance, and mutation and recombination. The simulation revealed that local interference of resource competition by antibiosis produced a quasistable equilibrium in which the indefinite coexistence of large numbers of different strains or species could be maintained.

Alan T. Bull • Research School of Biosciences, University of Kent, Canterbury, Kent CT2 7NJ, United Kingdom.

Models of the type discussed above provide the theoretical bases of testable hypotheses. For example, a number of groups have started to use simple microcosms to explore the importance of spatial isolation on two-species competition (Xia et al., 2001), whereas Huisman et al. (2001) point to experimental support for some of their multispecies competition simulations. The challenge now is to design experiments that can discriminate between these models for supersaturated coexistence, or enable more realistic ones to be formulated.

Scale Matters

It is increasingly clear that scale effects, both spatial and temporal, are principal determinants of species richness and need to be recognized when attempting to quantify microbial diversity, when addressing basic problems in microbial ecology and microbiogeography (see section IV), and when maximizing returns from biotechnology search and discovery programs. Issues relating to scale are much more familiar to macrobiologists than to microbiologists where a hierarchy of factors extending over local, landscape, regional, continental, and global scales have been identified that facilitate understanding of biodiversity distribution (Willis and Whittaker, 2002). Such a hierarchical framework implies that biodiversity processes are nested both spatially and temporally but, as Willis and Whittaker emphasize, variables that best account for species richness at a local spatial scale or recent time scale may not be the same as those operating at a regional spatial scale or a long time scale. These are challenging problems for microbiologists for whom scale matters also at the submicro level: thus the local spatial scale with which microbiologists must contend is 4 to 5 orders of magnitude less than that concerning macrobiologists (1 to 10 m^2), and it has been argued that the prokaryotic diversity in 100-cm^3 soil compares with the regional or γ-diversity of macroorganisms (Torsvik et al., 2002). A recent elegant example of microbial community structure at the microenvironment scale was reported by Tankéré et al. (2002) who combined high-resolution chemical measurements with molecular profiling to reveal fine-scale distributions of the microbiota at water-sediment interfaces. In the wider context, the use of species accumulation curves is one way forward and offers a practical means of assessing microbial species richness over the whole range of spatial scales.

Temporal changes in microbial population composition also can occur, often very rapidly, and may be induced by the biota and/or driven by external forces. Such temporal variability is dramatically illustrated by the diversity of cultivated bacterioplankton isolated from an eutrophic lake (Jaspers et al., 2001). Community fingerprints (enterobacterial repetitive intergenic consensus-PCR) of October, April, and May samplings were temporally unique, and, although *Cytophaga-Flavobacterium* dominated April communities, not a single strain of this group was isolated one month later. Moreover, even isolates obtained at one sampling time and having identical enterobacterial repetitive intergenic consensus-PCR fingerprints were shown to differ in their physiology. Other recent studies of this type have focused on bacterial communities in cultivated soil and plant crop rhizospheres (Smit et al., 2001; Smalla et al., 2001).

Complex relationships between scale and species richness pertain both to terrestrial and to marine biomes, and patterns of diversity in the deep sea particularly are much more complicated than previously thought (Levin et al., 2001). Consider, for example, regional scale (100 to 1,000 km^2) in the deep sea: most studies of faunal species richness have been made at bathyal (to 2,000-m) depths, whereas knowledge of spatial variation in the abyssal (>4,000-m) depths is much less secure. Nevertheless, the remarkable diversity of deep-sea communities raises once again the paradox of coexisting species and the exploitation of a common limited resource, namely, organic detritus. Although the composition of soft-sediment communities can be related to sediment characteristics, explanations for such relationships remain to be established. Moreover, as Levin et al. have pointed out, correlation between species diversity and sediment heterogeneity does not imply causality: such relationships "may be spurious or reflect more important proximal factors." Thus, multiple forces shape patterns of diversity that are associated with large-scale variations in sediment, nutrient input, productivity, oxygen concentration, deep-sea hydrology, catastrophic events such as debris flows, and volcanic eruptions. Accordingly, Levin et al. (2001) advocate species accumulation curves with randomly pooled samples for comparing regional-scale species richness; and, because typically they will be lower than rarefaction curves, the differences between these curves will be a measure of the heterogeneity at the scale of the sampler. All of which sharply focuses attention on sampling methodology, a problem that is especially acute for establishing patterns of microbial species richness.

HOW TO LOOK

Sampling Strategies and Effort

All too frequently insufficient attention is paid to the design of environment sampling strategies and to evaluating the sampling effort. The sampling strategy clearly should be appropriate to the question being asked or the hypothesis being tested. In the context of biotechnology prospecting the principal objective will be to maximize the recovery of all, or of targeted, organisms or genes, and the strategies can be formulated as empirical inventories or correlations research (Totsche, 1995). Inventory projects are to provide information on a system and its components, namely, biodiversity richness, whereas correlations research aims to provide information on the relations between a set of variables, namely, the patterns of microbial communities related to their environment. For example, it may be possible to correlate microbial patterns with edaphic conditions, vegetation cover, or anthropogenic disturbance so that sampling might become more predictable.

The sampling strategy should be devised to yield the most representative, reliable, and indicative information, and, because it is unlikely that an entire site or habitat can be analyzed, sampling plans are usually based on a limited number of specimens (in some cases logistical constraints may limit the number to just one or a very few specimens from a given habitat, e.g., abyssal sediments). In general there is no decision recipe to guide the microbial prospector, although ecological understanding and instinct are valuable attributes. Totsche's review (1995) of spatial sampling and analysis is a helpful starting point for microbiologists, and two strategies in particular commend themselves: (i) rigid or fixed sampling plans and (ii) algorithm sampling plans. Rigid sampling plans can be used for detection purposes and include proportionate sampling (each equalsized environmental parcel is sampled the same number of times as every other parcel of a screen, e.g., orthogonal and nonorthogonal screens) and nonproportionate sampling (randomized sampling such as Latin square design and polar sampling plans). Algorithm sampling plans are recommended for establishing the origins and distributions of organisms or perturbations in the environment. It may be the rule to take as many samples as possible depending on time and other logistical constraints, and this may be mandatory when heterogeneous habitats are being sampled. Alternatively it is expedient to analyze composite samples, and ideally these should consist of equal quantities of separate samples that are mixed thoroughly. When sampling for microbial symbionts, it may be necessary to collect large environmental samples; for example, the deep-sea fauna is sparsely distributed (Levin et al., 2001) and large samples are essential for assessing symbiont diversity. We have seen above that temporal variations exert major influences on microbial community composition, and the only way to obtain complete inventories on species richness is to take repeated samples throughout the year.

In order to estimate the effectiveness of the sampling effort we return again to the species (operational taxonomic unit) accumulation procedure. Specifically the microbial area-species curve provides a scientific basis to guide the discovery of microbial diversity and hence novel exploitable biology. An instructive illustration of the effort required for determining species richness is the recent study of African soil microarthropods (Noti et al., 2003). These authors analyzed two sampling sites in three habitats (barren soil, termitaria, plant patches), each containing three ecosystems (forest, woodland, savanna) on four occasions over 2 years. From each of these 72 locations, 10 soil cores were extracted. The specimen data set exceeded 22,000 individuals, and various nonparametric estimators (see chapter 2) were applied to assess species richness. Even this high sampling effort was insufficient to reach the asymptote of the total species accumulation curve, which is indicative of complete species inventory. Nevertheless, the analysis clearly revealed which ecosystem had the greatest α-diversity and which contained the highest numbers of exclusive species, information that is invaluable in the context of developing conservation measures. In this context, Tiedje (1995) has posed a number of important questions that impact significantly on bioprospecting programs: (i) What are the dominant but as yet uncultured microorganisms in nature? (ii) Which are the rare microorganisms and how can they be accessed? (iii) Is it possible to construct a comprehensive understanding of microbial distribution, for example, can an *Arthrobacter* landscape be predicted? Tiedje also draws attention to the "degree of difference among environments and the degree of difference among the microbes occupying these different environments" and the level of taxonomic resolution that is meaningful and manageable for the bioprospector.

As we have emphasized previously, many of the economically exploitable properties of microorganisms are expressed at the intraspecific level: discrimination at such a taxonomic level no longer presents problems because high-throughput dereplication can readily be achieved using chemometric and molecular fingerprinting techniques (Colquhoun et al., 2000; Seguritan and Rohwer, 2001; Brandao et al., 2002).

Recovery of Organisms

One, if not the most, depressing statement with which to confront a microbiologist runs along the lines that only 0.1 to 5% of microorganisms detected in the environment can be cultured. Resort to phrases such as "unculturable microorganisms," "viable but nonculturable organisms," and "as yet uncultured microorganisms" only serve to obfuscate or excuse the generally felt frustration. Yet most microbiologists will empathize with Steve Giovannoni's response that "Nothing beats actually having the organism in culture" or concur with Zinder (2002) that "the frequent use of the term 'uncultivable' [has] a chilling effect on considerable efforts needed to culture organisms from natural habitats." Despite the spectacular advances in the molecular detection and circumscription of microorganisms and functional genomics, organisms in culture are essential for providing an understanding of microbial interactions, pathogenesis, phenotypic variability, and, in the present context, for delivering biotechnological innovation. As Tiedje (1995) has observed, interesting organisms found through cultivation will also be targets for analyzing the information contained in their DNA, whereas ribosomal sequence evaluation on uncultured organisms may prove useful in improving isolation and cultivation methods for such organisms. The problems inherent in isolation and characterizing environmentally significant microorganisms notwithstanding, steady progress has been made as exemplified by the wide range of extremophiles, entomopathogenic microbes, nitrogen-fixing symbionts, and heterotrophs (Bull et al., 2000). As Zinder (2002) opines, the crisis carries the seeds of a revival for cultural microbiology.

This discussion is not intended as a vade mecum for microbial isolation and cultivation; various manuals and texts, such as Alef and Nannipieri (1995) on methods for soils and sediments, are available for specialist reference. Rather, the object of this brief statement is to set up some signposts that may aid more effective cultural microbiology and to prelude chapter 8. In parenthesis, it must be noted that, as more microorganisms are brought into culture, the infrastructure and funding of culture collections will require urgent strengthening (see section VII). Section III of this book considers microbial ecology as the key to discovery, a position that is especially consonant with culturability. Three approaches are presented here: habitat simulation, technological innovation, and taxonomic databases.

Habitat simulation

Microbiologists have long made crude concoctions and mimicked the ambient conditions of given environments in efforts to culture organisms. In turn such empiricism has led to more sophisticated attempts to simulate habitats. The success of Kaeberlein et al. (2002) in isolating previously uncultivated bacteria from intertidal sediments well illustrates this approach. Organisms were separated from sediment particles and suspended in agar made with seawater and housed in sealed diffusion chambers; the chambers were laid on top of blocks of sediment and incubated in seawater aquaria. Up to 40% of the inoculated organisms formed microcolonies, but the majority of these did not continue to grow after passage to petri dishes. The ones that did continue to grow on passage appeared to be mixed cultures, and the authors invoked specific signaling mechanisms to explain this behavior. A potential breakthrough in attempts to grow arbuscular mycorrhizal fungi (AMF) independent of their plant hosts may be indicated by the recent findings of Hildebrandt et al. (2002). Plant growth-promoting rhizosphere bacteria are known to enhance the colonization of plants by AMF, but these authors have isolated a strain of *Paenibacillus validus* that specifically facilitates the growth in vitro of *Glomus intraradicis* up to the stage of spore formation in the absence of plant cells. To date, neither the conditions under which such daughter spores can regerminate and produce competent infections nor the mechanism by which *Paenibacillus* supports growth of the AMF have been elucidated.

Paradoxically, disruption of the ecological integrity of niches also may be necessary for the recovery of microorganisms from the environment. Thus the need to overcome persistent associations of microbes with soil and sediment particles is a serious limitation to their isolation. Of the numerous methods used to disrupt these associations (Bull et al., 2000), a dispersion-differential centrifugation procedure (Hopkins et al., 1991) significantly improved recoveries of actinomycetes from soils (MacNaughton and O'Donnell, 1994; Atalan et al., 2000) and marine sediments (Mexson, 2001).

Technological innovation

Technological advances that can or promise to promote more efficient recovery of microorganisms include manipulative procedures such as optical tweezers (Morishima et al., 1998), robotics, atomic force microscopy (Boyd et al., 2002), the use of density gradient centrifugation, and cultivation methods such as continuous and dilution-to-extinction cultures. Density gradient centrifugation has been applied to great effect for isolating an anaerobic ammonia oxidizing plectomycete, the so-called "missing lithotroph," following careful electron microscope scrutiny of multispecies biofilms (Strous et al., 1999). Various continuous culture systems have been used successfully for isolating environmental organisms and particularly for recovering diverse interacting communities (see Veldkamp, 1977; Bull, 1983). Regretably, the demise of microbial physiology in general has been accompanied by a decline in the deployment of continuous culture methods.

In a similar way, the adoption of extinction dilution culture developed by Button et al. (1993) has been slow despite its proven effectiveness for isolating novel bacteria. This latter technique has an important advantage in providing a means for isolating organisms that are abundant in the environment yet are outcompeted in conventional enrichment culture by kinetically more-versatile organisms. For isolating oligotrophic marine bacteria, Button recommended the use of unamended sterilized seawater, monitoring the developing populations over at least a 9-week period, and assessing growth with sensitive techniques such as epifluorescence microscopy and flow cytometry. Examples of the successful use of dilution culture are few, but the work of Schut et al. (1997) on the marine ultramicrobacterium *Sphingomonas alaskensis* RB2256, Button et al. (1998) on *Cycloclasticus oligotrophus*, and Eguchi et al (2001) on *S. alaskensis* strain AFO1 are model investigations of this type. The ability of strain AFO1 to grow on solid media developed only after serial passage in liquid medium for 12 months. Recently the use of dilution culture has produced significant increases in the culturability of planktonic lake bacteria (Bussmann et al., 2001) and of soil bacteria (Janssen et al., 2002). The results of the latter investigation carry some important messages: (i) the use of solidified rather than liquid media may yield better recoveries of organisms, (ii) long incubation times are likely to be required to allow maximum recovery (see also the Button references above), and (iii) a large proportion of the isolates were the first known cultured representatives of novel lineages in the divisions *Acidobacteria, Actinobacteria, Proteobacteria,* and *Verrucomicrobia.*

Taxonomic databases

Innumerable medium formulations and sample pretreatments have been advocated for the selective isolation of microorganisms, but the ingredients and conditions most usually have been chosen empirically, hence the basis of such selectivity is obscure. However, the development of computer-assisted procedures now enables selective isolation media to be designed and evaluated in a rational way. Numerical taxonomic databases containing extensive information on the nutritional, physiological, and inhibitory sensitivity profiles of the microbial taxa are ideal resources for the formulation of new selective media designed to isolate rare and novel organisms of biotechnological importance (Bull et al., 1992, 2000).

Molecular Detection

The application of molecular phylogenetic methods in the 1980s (Stahl et al., 1984; Pace et al., 1986), based on rRNA gene sequences, paved the way for culture-independent assessments of microbial diversity, while also confirming microscopic observations that only a small proportion of total populations were being cultivated. It is now commonplace to analyze microbial communities via PCR amplification of small-subunit rRNA gene primers that can be designed for various taxonomic levels. Alternatives to the cloning approach to microbial community profiling include PCR-denaturing gradient gel electrophoresis (Muyzer et al., 1993) and single-strand conformation polymorphism (Lee et al., 1996; Stach et al., 2002). The use of such techniques has revealed spectacular patterns of diversity even in previously well-studied habitats such as the human gut (Suau et al., 1999) and the phyllospheres of crop plants (Yang et al., 2001). The success of the molecular detection strategy depends on the quality of the DNA extracted from the environment (see Akkermans et al., 1995; Courtois et al., 2001, and references therein). Moreover, DNA-based analyses of microbial communities may be subject to a number of pitfalls (see Bull et al., 2000, for references).

In recent years molecular detection methods have evolved for more sophisticated characterization to be made of uncultured natural diversity. Thus, fluorescent in situ hybridization, bacterial artificial chromosomes and metagenomics, and oligonucleotide microarray technology now enable simultaneous detection of different organisms, activity in addition to the presence of organisms in a habitat to be determined, and biochemical pathways of uncultured organisms to be reconstructed. However, in a number of instances these methods have yet to be rigorously tested and validated with environmental samples, for example, with regard to sensitivity, selectivity, and quantitation in the case of microarrays (Zhou and Thompson, 2002).

Judging Success

Increasingly, culture-independent methods are being used as the first or only recourse for evaluating the composition of microbial communities. So, have cultural approaches become irrelevant in this context and for biotechnology prospecting? First, it is important to point out that congruence between molecular and cultivation detection methods can be poor. Thus our work on deep-sea actinomycetes illustrates this inconsistency: in one series of experiments, numbers of culturable actinomycetes from bathyal, abyssal, and hadal sites in the northwest Pacific Ocean ranged from 1.6×10^4 to 3.4×10^2 CFU g^{-1} wet sediment (Colquhoun et al., 1998), yet 16S rDNA clones libraries obtained from the same sites, and in some cases identical sediment samples (Li et al., 1999; Urakawa et al., 1999), failed to reveal the presence of cultured actinomycete taxa, or, indeed of any actinomycete signatures. Second, relatively few twin-track studies have been made in which cultivation and direct recovery of 16S rRNA gene sequences have been used to assess the microbial diversity of natural bacterial communities. Such comparative studies are essential because both culture and 16S rDNA cloning are subject to biases that can distort community composition. Although a somewhat mixed picture emerges from the available comparative studies of natural microbial ecosystems, carefully planned and performed investigations have produced close matching between phylogenetic and culture data sets. Good examples of the twin-track approach are the studies of rice paddy soil bacteria (Hengstmann et al., 1999) and bacterial communities at hypersaline sites in the Red Sea (Eder et al., 2001). Dilution culture in a medium that simulated resources and conditions of the paddy soil was used to isolate numerically abundant bacteria. The good congruence between the cultural and 16S rDNA data sets indicated that the former populations represented the dominant and functionally relevant bacteria in the paddy habitat. The study of Eder et al. (2001) is notable for the careful seawater sampling system that was deployed and the use of optical tweezers to separate individual bacteria from the enriched communities. One obvious conclusion arising from a review of these studies is that species (or operational taxonomic unit) accumulation curve analyses provide an effective means of judging the efficacy of culture-dependent and culture-independent approaches to inventorying natural microbial communities.

The question remains: are there microorganisms that cannot be cultured, or, more specifically, cannot be brought into pure culture. Pace (2000) makes the important point that many organisms are unlikely to be brought into pure culture "because they are syntrophic, relying absolutely on the activities of other organisms." In some cases an understanding of genome downsizing may provide vital clues to why certain organisms prove to be persistently "unculturable." For example, comparison of the *Mycobacterium leprae* and *M. tuberculosis* genomes reveals a 26% downsizing of the unculturable leprosy bacillus (Cole et al., 2001); notably many key metabolic functions including siderophore synthesis; parts of the oxidative, microaerophilic, and anaerobic respiratory chains; and numerous catabolic sequences and their regulation have been lost.

We have concluded previously (Bull et al., 2000) "that both innovative cultural procedures and culture-independent methods have a role to play in unravelling the full extent of (microbial) diversity in natural habitats. Although the two approaches sometimes provide different assessments of relative community diversity, the discrepancies may be attributed to sampling different subsets of the microbial community and to limitations inherent in each of the two approaches. In addition, highlighting consistent relationships between environments based on the dual approach may be highly habitat-dependent due to the limited ability of a single cultural method to survey the full extent of the bacterial communities and the influence of bacterial physiology *in situ* on the success of cultivation in the laboratory." Ironically, the "holistic view of the phenotype owes much to advances in molecular biology, and yet molecular biology is reductionism in the extreme . . . . Precisely how to avoid the pitfalls of extreme reductionism while at the same time benefiting from the tremendous insights made possible by molecular biology is not clear" (Rainey, 2000). Thus, it is essential to reiterate that physiological and metabolic properties of microorganisms cannot be extrapolated from rDNA (or other gene) data alone as the compeling example of perchlorate-reducing bacteria makes clear (Achenbach and Coates, 2000).

WHERE TO LOOK

In order to find novel microorganisms, all one has to do is look in one's own backyard. At first sight such an admonishment is hard to refute given the fact that intensive investigation, using new molecular methods, is revealing bountiful diversity in previously intensively studied ecosystems. The case may appear well proven with reference to soils, wastewater treatment systems, and animal guts: for example, the discovery of anaerobic ammonia oxidizers (see chapter 9), novel *Betaproteobacteria* related to *Rhodocyclus*, and glycogen- and phosphorus-accumulating *Gammaproteobacteria* (Wagner and Loy, 2002) in sewage works and the very high proportion of phylotypes identified in pig gastrointestinal tracts having a less than 97% 16S rDNA sequence similarity to any sequences in the databases (Leser et al., 2002). Consequently it is necessary to raise the question of why look in more exotic locations for diverse microorganisms? There are a number of rejoinders to this question, among which are the following:

1. The unique metabolic profiles of related organisms obtained from different environments. For example, the fungal genus *Phoma* contains many terrestrial as well as obligate marine species. Using diode array and thermospray mass spectrometry, Osterhage et al. (2000) showed that many metabolites synthesized by marine *Phoma* species were similar to those synthesized by terrestrial relatives; however, both groups were distinguished by their unique metabolic products, indicating that these marine isolates can be regarded as "valuable sources of new natural products."

2. The well-founded principle of biodiversity "hot spots" may be applicable to microorganisms. Such hot-spot localities were defined by Norman Myers as areas of great species diversity and/or endemism that are under particular threat of destruction (Myers, 1988; Myers et al., 2000). The original 18 hot spots defined on the basis of vegetation characteristics contained about 50,000 endemic plant species and encompassed only 0.5% of the total land area. Currently we have little or no data by which to recognize microbiolgical hot spots, but the reality of such a concept may be approached by way of symbiotic associations; coral reefs provide an excellent test bed for this hypothesis. Here restricted-range faunal species are clustered into distinct centers of endemism, the 10 richest of which cover only 0.012% of the world's oceans (Roberts et al., 2002). The excellent recent survey of tropical Australian sponges (Hooper et al., 2002) offers a unique opportunity for exploring microbial hot-spot diversity.

3. Many ecosystems and biomes remain poorly investigated in terms of their microbiology. These systems vary from the local (e.g., endophytes and other symbionts) to the global (e.g., ocean and deep biosphere) scales. The case of symbionts is beautifully illustrated by research on insects in Breznak's laboratory (Lilburn et al., 1999) (see chapter 19) and by the recent explorations of endophytic fungi (Kursar et al., 1999; Brady et al., 2000). The exploration of the deep biospheres has started quite recently (see Parkes and Wellsbury, chapter 12) and opens up a remarkable window not only onto the archaeology of microorganisms (Inagaki et al., 2001, 2002) but to accessing archaeic DNA for possible biotechnological exploitation.

4. The comparative phylogeographic approach to biodiversity. Taberlet (1998) analyzed the available information on phylogeographic studies (admittedly on macroorganisms) in order to deduce some general rules by which intraspecific biodiversity can be governed. Taberlet opined that when designing a biotechnology sampling strategy, two guidelines could be recommended: the more widespread the sites sampled, the greater the intraspecific variation can be expected; and special attention should be paid to sampling in refugia from geologically cold periods because of their intrinsically high intraspecific varia-

tion. This proposition warrants investigation in terms of microbial genetic diversity.

5. Curiosity-driven research—arguably the most potent catalyst for discovering novelty. In this context, Tiedje's (1995) espousal of tackling an All Taxa Biodiversity Inventory that would include microorganisms deserves attention, although it would present the most taxing and novel challenge to microbiologists.

The individual chapters that follow in this section provide the detailed evidence for the assertions described above and also lead logically to the wider discussion of biogeography and endemism that is taken up in section IV.

REFERENCES

Achenbach, L. A., and J. D. Coates. 2000. Disparity between bacterial phylogeny and physiology. *ASM News* **66:**714–715.

Akkermans, A. D. L., J. D. van Elsas, and F. J. de Bruijn (ed.). 1995. *Molecular Microbial Ecology Manual.* Kluwer Academic Publishers, Dordrecht, The Netherlands.

Alef, K., and P. Nannipieri (ed.). 1995. *Quality Control and Quality Assurance in Applied Soil Microbiology and Biochemistry.* Academic Press Ltd., London, United Kingdom.

Atalan, E., G. P. Manfio, A. C. Ward, R. M. Kroppenstedt, and M. Goodfellow. 2000. Biosystematic studies on novel streptomycetes from soil. *Antonie Leeuwenhoek* **77:**337–353.

Boyd, R. D., J. Verran, M. V. Jones, and M. Bhakoo. 2002. Use of the atomic force microscope to determine the effect of substratum surface topography on bacterial adhesion. *Langmuir* **18:**2342–2346.

Brady, S. F., M. M. Wagenaar, M. P. Singh, J. E. Janso, and J. Clardy. 2000. The cytosporones, new octaketide antibiotics isolated from an endophytic fungus. *Org. Lett.* **2:**4043–4046.

Brandao, P. F. B., M. Torimura, R. Kurane, and A. T. Bull. 2002. Dereplication for biotechnology screening: PyMS analysis and PCR-RFLP-SSCP (PRS) profiling of 16S rRNA genes of marine and terrestrial actinomycetes. *Appl. Microbiol. Biotechnol.* **58:**77–83.

Bull, A. T. 1983. Continuous culture for production, p. 405–437. *In* A. Hollaender, A. I. Laskin, and P. Rogers (ed.), *Basic Biology of New Developments in Biotechnology.* Plenum Press, New York, N.Y.

Bull, A. T., M. Goodfellow, and J. H. Slater. 1992. Biodiversity as a source of innovation in biotechnology. *Annu. Rev. Microbiol.* **42:**219–257.

Bull, A. T., A. C. Ward, and M. Goodfellow. 2000. Search and discovery strategies for biotechnology: the paradigm shift. *Microbiol. Mol. Biol. Rev.* **64:**573–606.

Bussmann, I., B. Philipp, and B. Schink. 2001. Factors influencing the cultivability of lake water bacteria. *J. Microbiol. Methods* **47:**41–50.

Button, D. K., F. Schut, P. Quang, R. Martin, and B. R. Robinson. 1993. Viability and isolation of marine bacteria by dilution culture: theory, procedures, and initial results. *Appl. Environ. Microbiol.* **59:**881–891.

Button, D. K., B. R. Robertson, P. W. Lepp, and T. M. Schmidt. 1998. A small, dilute-cytoplasm, high-affinity, novel bacterium isolated by extinction culture and having kinetic constants compatible with growth at ambient concentrations of dissolved nutrients in seawater. *Appl. Environ. Microbiol.* **64:**4467–4476.

Cole, S. T., K. Eiglmeier, J. Parkhill, K. D. James, N. R. Thomson, P. R. Wheeler, N. Honoré, T. Garnier, C. Churcha, D. Harris et al. 2001. Massive gene decay in the leprosy bacillus. *Nature* 409:1007–1011.

Colquhoun, J. A., J. Mexson, M. Goodfellow, A. C. Ward, K. Horikoshi, and A. T. Bull. 1998. Novel rhodococci and other mycolate actinomycetes from the deep sea. *Antonie Leeuwenhoek* 74:27–40.

Colquhoun, J. A., J. Zulu, M. Goodfellow, K. Horikoshi, A. C. Ward, and A. T. Bull. 2000. Rapid characterisation of deep-sea actinomycetes for biotechnological screening programmes. *Antonie Leeuwenhoek* 77:359–367.

Courtois, S., A. Frostegård, P. Göransson, G. Depret, P. Jeannin, and P. Simonet. 2001. Quantification of bacterial subgroups in soil: comparison of DNA extracted directly from soil or from cells previously released by density gradient centrifugation. *Environ. Microbiol.* 3:341–439.

Czaran, T., R. F. Hoekstra, and L. Pagie. 2002. Chemical warfare between microbes promotes biodiversity. *Proc. Natl. Acad. Sci. USA* 99:786–790.

Eder, W., L. L. Jahnke, M. Schmidt, and R. Huber. 2001. Microbial diversity of the brine-seawater interface of the Kebrit Deep, Red Sea, studied via 16S rRNA gene sequences and cultivation methods. *Appl. Environ. Microbiol.* 67:3077–3085.

Eguchi, M., M. Ostrowski, F. Fegatella, J. Bowman, D. Nichols, T. Nishino, and R. Cavicchioli. 2001. *Sphingomonas alaskensis* strain AF01, an abundant oligotrophic ultramicrobacterium from the North Pacific. *Appl. Environ. Microbiol.* 67:4945–4954.

Hengstmann, U., K.-J. Chin, P. H. Janssen, and W. Liesack. 1999. Comparative phylogenetic assignment of environmental sequences of genes encoding 16S rRNA and numerically abundant culturable bacteria from an anoxic rice paddy soil. *Appl. Environ. Microbiol.* 65:5050–5058.

Hildebrandt, U., K. Janetta, and H. Bothe. 2002. Towards growth of arbuscular mycorrhizal fungi independent of a plant host. *Appl. Environ. Microbiol.* 68:1919–1924.

Hooper, J. N. A., J. A. Kennedy, and R. J. Quinn. 2002. Biodiversity "hotspots," patterns of richness and endemism, and taxonomic affinities of tropical Australian sponges (Porifera). *Biodivers. Conserv.* 11:851–885.

Hopkins, D. W., S. J. MacNaughton, and A. G. O'Donnell. 1991. A dispersion and differential centrifugation technique for representatively sampling microorganisms from soil. *Soil Biol. Biochem.* 23:217–225.

Huisman, J., and F. J. Weissing. 1999. Biodiversity of plankton by species oscillations and chaos. *Nature* 402:407–410

Huisman, J., A. M. Johansson, E. O. Folmer, and F. J. Weissing. 2001. Towards a solution of the plankton paradox: the importance of physiology and life history. *Ecol. Lett.* 4:408–411.

Hutchinson, G. E. 1961. The paradox of the plankton. *Am. Nat.* 95:137–145.

Inagaki, F., K. Takai, T. Komatsu, T. Kanamatsu, K. Fujioka, and K. Horikoshi. 2001. Archaeology of Archaea: geomicrobiological record of Pleistocene events in a deep-sea subseafloor environment. *Extremophiles* 5:385–392.

Inagaki, F., Y. Sakihama, A. Inoue, C. Kato, and K. Horikoshi. 2002. Molecular phylogenetic analyses of reverse-transcribed bacterial rRNA obtained from deep-sea cold seep sediments. *Environ. Microbiol.* 4:277–286.

Janssen, P. H., P. S. Yates, B. E. Grinton, P. M. Taylor, and M. Sait. 2002. Improved culturability of soil bacteria and isolation in pure culture of novel members of the Divisions *Acidobacteria, Actinobacteria, Proteobacteria,* and *Verrucomicrobia. Appl. Environ. Microbiol.* 68:2391–2396.

Jaspers, E., K. Nauhaus, H. Cypionka, and J. Overmann. 2001. Multitude and temporal variability of ecological niches as indicated by the diversity of cultivated bacterioplankton. *FEMS Microbiol. Ecol.* 36:153–164.

Kaeberlein, T., K. Lewis, and S. S. Epstein. 2002. Isolating "uncultivable" microorganisms in pure culture in a simulated natural environment. *Science* 296:1127–1129.

Kursar, T. A., T. L. Capson, P. D. Coley, D. G. Corley, M. B. Gupta, L. A. Harrison, E. Ortega-Barria, and D. M. Windsor. 1999. Ecologically guided bioprospecting in Panama. *Pharmaceut. Biol.* 37:114–126.

Lee, D. H., Y. G. Zo, and S. J. Kim. 1996. Nonradioactive method to study genetic profiles of natural bacterial communities by PCR-single-strand-conformation polymorphism. *Appl. Environ. Microbiol.* 62:3112–3120.

Leser, T. D., J. Z. Amenuvor, T. K. Jensen, R. H. Lindecrona, M. Boye, and K. Møller. 2002. Culture-independent analysis of gut bacteria: the pig gastrointestinal tract microbiota revisted. *Appl. Environ. Microbiol.* 68:673–690.

Levin, L. A., R. J. Etter, M. A. Rex, A. J. Gooday, C. R. Smith, J. Pineda, C. T. Stuart, R. R. Hessler, and D. Pawson. 2001. Environmental influences on regional deep-sea species diversity. *Annu. Rev. Ecol. Syst.* 32:51–93.

Li, L., C. Kato, and K. Horikoshi. 1999. Bacterial diversity in deep-sea sediments from different depths. *Biodiver. Conserv.* 8:659–677.

Lilburn, T. G., T. M. Schmidt, and J. A. Breznak. 1999. Phylogenetic diversity of termite gut spirochaetes. *Environ. Microbiol.* 1:331–345.

MacNaughton, S. J., and A. G. O'Donnell. 1994. Tuberculostearic acid as a means of estimating the recovery (using dispersion and differential centrifugation) of actinomycetes from soil. *J. Microbiol. Methods* 20:69–77.

Mexson, J. 2001. Selective isolation and taxonomic analysis of the genus *Micromonospora.* Ph.D. thesis. University of Kent at Canterbury, United Kingdom.

Morishima, K., F. Arai, T. Fukuda, and H. Matsuura. 1998. Screening of single *Escherichia coli* in a microchannel system by electric field and laser tweezers. *Anal. Chim. Acta* 365:273–278.

Muyzer, G., E. C. de Waal, and A. G. Uitterlinden. 1993. Profile of complex microbial populations by denaturing gradient gel electrophoresis analysis of polymerase chain reaction-amplified genes encoding for 16S rRNA. *Appl. Environ. Microbiol.* 59:695–700.

Myers, N. 1988. Threatened biotas: "hotspots" in tropical forests. *Environmentalist* 8:1–20.

Myers, N., R. A. Mittermeier, C. G. Mittermeier, G. A. B. da Fonseca, and J. Kent. 2000. Biodiversity hotspots for conservation priorities. *Nature* 403:853–858.

Noti, M.-I., H. M. André, X. Ducarme, and P. Lebrun. 2003. Diversity of soil oribatid mites (Acari: Oribatida) from high Katanga (Dem. Rep. Congo): a multiscale and multifactor approach. *Biodivers. Conserv.* 12:767–785.

Osterhage, C., M. Schwibbe, G. M. König, and A. D. Wright. 2000. Differences between marine and terrestrial *Phoma* species as determined by HPLC-DAD and HPLC-MS. *Phytochem. Anal.* 11:288–294.

Pace, N. R. 2000. Community interactions: towards a natural history of the microbial world. *Environ. Microbiol.* 2:7–8.

Pace, N. R., D. A. Stahl, D. J. Lane, and G. J. Olsen. 1986. The analysis of natural microbial populations by ribosomal RNA sequences. *Microb. Ecol.* 9:1–56.

Rainey, P. 2000. An organism is more than its genotype. *Environ. Microbiol.* 2:8–9.

Roberts, C. M., C. J. McClean, J. E. N. Veron, J. P. Hawkins, G. R. Allen, D. E. McAllister, C. G. Mittermeir, F. W. Schueler, M. Spalding, F. Wells, C. Vynne, and T. B. Werner. 2002. Marine biodiversity hotspots and conservation priorities for tropical reefs. *Science* 295:1280–1284.

Schippers, P., A. M. Verschoor, M. Vos, and W. M. Mooij. 2001. Does "supersaturated coexistence" resolve the "paradox of the plankton"? *Ecol. Lett.* **4:**404–407.

Schut, F., J. C. Gottschal, and R. A. Prins. 1997. Isolation and characterisation of the marine ultramicrobacterium *Sphingomonas* sp. strain RB2256. *FEMS Microbiol. Rev.* **20:**363–369.

Seguritan, V., and F. Rohwer. 2001. FastGroup: a program to dereplicate libraries of 16S rDNA sequences. *BMC Bioinformatics* **2:**9.

Smalla, K., G. Wieland, A. Buchner, A. Zock, J. Parzy, S. Kaiser, N. Roskot, H. Heuer, and G. Berg. 2001. Bulk and rhizosphere soil bacterial communities studied by denaturing gradient gel electrophoresis: plant-dependent enrichment and seasonal shifts revealed. *Appl. Environ. Microbiol.* **67:**4742–4751.

Smit, E., P. Leeflang, S. Gommans, J. van den Broek, S. van Mil, and K. Wernars. 2001. Diversity and seasonal fluctuations of the dominant members of the bacterial soil community in a wheat field as determined by cultivation and molecular methods. *Appl. Environ. Microbiol.* **67:**2284–2291.

Stach, J. E. M., and R. G. Burns. 2002. Enrichment versus biofilm culture: a functional and phylogenetic comparison of polycyclic aromatic hydrocarbon-degrading microbial communities. *Environ. Microbiol.* **4:**169–182.

Stahl, D. A., D. J. Lane, G. J. Olsen, and N. R. Pace. 1984. Analysis of hydrothermal vent-associated symbionts by ribosomal-RNA sequences. *Science* **224:**409–411.

Steinberg, C. E. W., and W. Geller. 1993. Biodiversity and interactions within pelagic nutrient cycling and productivity, p. 43–64. *In* E.-D. Schulze and H. A. Mooney (ed.), *Biodiversity and Ecosystem Function.* Springer-Verlag, Berlin, Germany.

Strous, M., J. A. Fuerst, E. H. M. Kramer, S. Logemann, G. Muyzer, K. T. Van de Pas-Schooner, R. Webb, J. Gijs Kuenen, and M. S. M. Jeiten. 1999. Missing lithotroph identified as a new planctomycete. *Nature* **400:**446–449.

Suau, A., R. Bonnet, M. Sutren, J. J. Godon, G. R. Gibson, M. D. Collins, and J. Dore. 1999. Direct analysis of genes encoding 16S rRNA from complex communities reveals many novel molecular species within the human gut. *Appl. Environ. Microbiol.* **65:**4799–4807.

Taberlet, P. 1998. Biodiversity at the intraspecific level: the comparative phylogeographic approach. *J. Biotechnol.* **64:**91–100.

Tankéré, S. P. C., D. G. Bourne, F. L. L. Muller, and V. Torsvik. 2002. Microenvironments and microbial community structure in sediments. *Environ. Microbiol.* **4:**97–105.

Tiedje, J. M. 1995. Approaches to the comprehensive evaluation of prokaryotic diversity of a habitat, p. 73–87. *In* D. Allsopp, R. R. Colwell, and D. L. Hawksworth. (ed.), *Microbial Diversity and Ecosystem Function.* CABI Publishing, Wallingford, United Kingdom.

Tiedje, J. M., J. C. Cho, A. Murray, D. Treves, B. Xia, and J. Zhou. 1999. Soil teeming with life: new frontiers for soil science, p. 393–412. *In* R. M. Rees, B. C. Ball, C. D. Campbell, and C. A. Watson (ed.), *Sustainable Management of Soil Organic Matter.* CABI Publishing, Wallingford, United Kingdom.

Torsvik, V., L. Øvreås, and T. F. Thingstad. 2002. Prokaryotic diversity—magnitude, dynamics, and controlling factors. *Science* **296:**1064–1066.

Totsche, K. 1995. Quality–project design–spatial sampling, p. 5–51. *In* K. Alef. and P. Nannnipieri. (ed.), *Quality Control and Quality Assurance in Applied Soil Microbiology and Biochemistry.* Academic Press Ltd., London, United Kingdom.

Urakawa, H., K. Kita-Tsukamoto, and K. Ohwada. 1999. Microbial diversity in marine sediments from Sagami Bay and Tokyo Bay, Japan, as determined by 16S rRNA gene analysis. *Microbiology* **145:**3305–3315.

Vandermeer, J., M. A. Evans, P. Foster, T. Höök, M. Reiskind, and M. Wund. 2002. Increased competition may promote species coexistence. *Proc. Natl. Acad. Sci. USA* **99:**8731–8736.

Veldkamp, H. 1977. Ecological studies with the chemostat. *Adv. Microbial Ecol.* **1:**59–94.

Wagner, M., and A. Loy. 2002. Bacterial community composition and function in sewage treatment systems. *Curr. Opin. Biotechnol.* **13:**218–227.

Willis, K. J., and R. J. Whittaker. 2002. Species diversity—scale matters. *Science* **295:**1245–1248.

Xia, B. C., D. S. Treves, J. Z. Zhou, and J. M. Tiedje. 2001. Soil microbial community diversity and driving mechanisms. *Prog. Nat. Sci.* **11:**818–824.

Yang, C.-H., D. E. Crowley, J. Borneman, and N. T. Keen. 2001. Microbial phyllosphere populations are more complex than previously realized. *Proc. Natl. Acad. Sci. USA* **98:**3889–3894.

Zhou, J., and D. K. Thompson. 2002. Challenges in applying microarrays to environmental studies. *Curr. Opin. Biotechnol.* **13:**204–207.

Zhou, J. Z., M. E. Davey, J. B. Figueras, E. Rivkina, D. Gilichinsky, and J. M. Tiedje. 1997. Phylogenetic diversity of a bacterial community determined from Siberian tundra soil DNA. *Microbiology* **143:**3913–3919.

Zhou, J. Z., B. C. Xia, D. S. Treves, L.-Y. Wu, T. L. Marsh, R. V. O'Neill, A. V. Palumbo, and J. M. Tiedje. 2002. Spatial and resource factors influencing high microbial diversity in soil. *Appl. Environ. Microbiol.* **68:**326–334.

Zinder, S. H. 2002. The future for culturing environmental organisms: a golden era ahead? *Environ. Microbiol.* **4:**14–15.

Microbial Diversity and Bioprospecting
Edited by Alan T. Bull
© 2004 ASM Press, Washington, D.C.

Chapter 8

Culture-Dependent Microbiology

JOHN C. FRY

TRADITIONAL APPROACHES TO STUDYING MICROBIAL DIVERSITY

Microbiologists have been using culture techniques to study the diversity of bacteria in natural habitats for well over 150 years. They have been doing this mainly by the use of selective enrichment and plating on agar with a wide range of general and selective media. This has led to the wide range of pure cultures stored in culture collections from natural habitats such as soil, seawater, lakes and rivers, and man-made environments such as waste treatment works and polluted soils. Once isolated, the physiology of these bacteria was investigated and they were described and given names and attempts were made to put them within a taxonomic framework to aid in the identification of further isolates.

Many microbiologists have focused on investigating bacteria that contribute to biogeochemical pathways in their searches to illuminate bacterial biodiversity. This has led to extensive study of particular groups of bacteria with specific physiology (Fry, 1987). Therefore, organisms such as nitrifiers, sulfate-reducing bacteria, methanogens, methane-oxidizing bacteria are well studied and explored, but general heterotrophs have received less attention. Similarly, habitats with strong gradients and extreme environmental conditions have been investigated to elucidate both the diversity present and the physiological interactions that occur. Therefore, we understand reasonably well the populations of bacteria in habitats such as marine and freshwater surface sediments, the interfaces in eutrophic and meromictic lakes, alkaline lakes, hot springs, and microbial mats. Furthermore, we are beginning to understand diversity in deep-sea hydrothermal vents, artificial high-salt ponds used for salt extraction (salterns), and extremely cold terrestrial habitats such as arctic tundra.

Pollutants have also been a source of biotechnological interest for microbiologists. Many pollutant-degrading bacteria have been isolated and studied. These include bacteria that degrade aromatic and aliphatic compounds, especially pesticides, herbicides, and other compounds on governmental priority lists for removal from the environment.

From about 1950 to 1985 many microbiologists isolated collections of bacteria from different environments, usually on agar plates (see Color Plate 2a to c). They then used numerical taxonomy to compare the isolates, identify them, and improve the taxonomy of the major groups. With this approach, extensive information was collected about the diversity of bacteria in nature. For example, a comparison of nine studies of this type from aquatic habitats (Fry, 1987) found that the most common genera were *Flavobacterium, Pseudomonas, Vibrio* (all found in seven studies out of nine; 78%), *Aeromonas* (67%), and *Alcaligenes* (56%).

THE NEED FOR MORE PURE CULTURES TO STUDY DIVERSITY

These traditional approaches to studying bacterial diversity have led to much confusion and uncertainty, mainly because of the inherent problems of classifying bacteria using physiology and morphology. The development of molecular approaches to taxonomy over the past 2 decades is helping to resolve these difficulties. In particular, analysis of rRNA gene sequences is rapidly providing a strong phylogenetic framework. This taxonomic structure seems secure as it is based on strong evolutionary principles. Furthermore, it allows new bacteria to be placed within this structure, and putatively identified, by simply sequencing most of the 16S rRNA gene (rDNA) (at least 1,300 bp). Since 1990, microbial ecologists have been studying the diversity of bacteria from many environments, not by culturing them, but by isolating community DNA, amplifying the 16S

John C. Fry • Cardiff School of Biosciences, Cardiff University, Main Building, Park Place, Cardiff CF10 3TL, United Kingdom.

rDNA, cloning the fragments, and sequencing the clones. This culture-independent approach has shown that most of the bacteria in culture collections, and that grow on conventional media, are not the most abundant in natural habitats. Studying collections of cultures without reference to the equivalent 16S rDNA clone libraries will tell us little about the diversity of the most abundant bacteria. Accordingly, many scientists are currently arguing the necessity for isolating the "as-yet-uncultured" bacteria that dominate habitats (e.g., Hugenholtz et al., 1998; Giovannoni and Rappé, 2000; Stackebrandt and Tindall, 2000; Breznak, 2002). Therefore, it is imperative to obtain more pure cultures of these dominant bacteria so that we can study the physiology of ecologically relevant bacteria and understand the forces driving natural populations.

The basic problem seems to be that many numerically abundant bacteria grow more slowly than their less-dominant colleagues on most laboratory media. This idea is supported by the greater success of microcolony techniques for estimating viability (Fry and Zia, 1982; Bartscht et al., 1999) than traditional plating or most-probable-number (MPN) methods. Indeed some abundant bacteria may not grow on conventional media at all. This leads to isolation of only the faster-growing, less-numerous populations with conventional approaches. For this reason, in this chapter I concentrate on some examples of the latest techniques being successfully exploited for isolating dominant but uncultured bacteria from a variety of habitats (see Table 1 for a summary). For those wishing to use the traditional methods outlined above, there are numerous reviews (e.g., Bull and Slater, 1982; Austin, 1988; Grigorova and Norris, 1990) and original papers to consult.

MEDIA AND DETECTING THE TARGET BACTERIA

Regardless of what isolation method is used, the choice of an appropriate medium is crucial when culturing bacteria in nature. There are many books that provide lists of media for growing environmental bacteria (e.g., Austin, 1988; Holt, 1994, and other books in this series; Atlas, 1995). The most comprehensive publication is *The Prokaryotes* (Balows et al., 1992), which describes groups of bacteria and their ecology and media for isolating and growing them. Rich media such as nutrient agar and Tryptone soya broth agar (with >8 g of organic compounds per liter) give lower counts than media with less nutrients, such as casein peptone starch (CPS) agar and R2A (<4 g of organic compounds per liter). These lower-nutrient media have been designed specifically for counting environmental bacteria and give a wider range of colonial morphologies than high-nutrient media. Many bacteria from natural habitats are thought to be oligotrophs, and so media designed to isolate these bacteria (Fry, 1990) have even lower organic nutrient content ($\approx$10 mg of organic compounds per liter).

The target bacteria being isolated have to be identified in some way. Although features such as colony and cell morphology can help, modern molecular methods such as PCR and hybridization based on rRNA and rDNA sequences (e.g., see some chapters in Paul, 2001; also see Smalla, Chapter 9) are more effective for tracking target bacteria during isolation.

PLATING METHODS

Suzuki et al. (1997) compared collections of isolates from R2A plates with 16S rDNA sequences from a clone library from the same seawater sample. They found that most of the cultured isolates were novel bacteria, but few were closely related to the most abundant bacteria from the clones. Similar studies have confirmed these results for other habitats, and so other methods have been used to isolate the most abundant bacteria. However, this finding, although common, is not universal. Pinhassi et al. (1997) made a collection of 48 isolates obtained over a year by plating Baltic Sea water on Zobell agar at 15°C for 10 to 15 days. They then used this collection to screen community DNA from the same seawater by whole-genome DNA hybridization and found that the isolates represented 7 to 69% of the total bacterial count. Although their isolates where widely spread phylogenetically, they were not from the predominant groups normally isolated from seawater (Giovannoni and Rappe, 2000).

Direct plating on low-nutrient plates has been successful in isolating the most abundant seawater bacteria in some cases. For example, Gonzalez and Moran (1997) found that up to 40% of the colonies isolated from coastal water in the United States were members of the abundant *Roseobacter* clade from the *Alphaproteobacteria* (16% of all marine rDNA clones [Giovannoni and Rappe, 2000]). They used plates containing 10 mg of peptone per liter and 5 mg of yeast extract per liter in membrane-filtered Sargasso Sea water incubated for 35 days at 15°C. Hybridization with a specific 16S probe showed that this group made up 10 to 25% of the bacteria from coastal seawater and 54 to 86% of a clone library from the same sites. Similar results have been found in the North Sea (e.g., Eilers et al., 2000), although members of other abundant groups were rarely cul-

Table 1. Some examples of methods used for the isolation of abundant *Bacteria* from natural habitats[a]

Method	Habitat	Summary of techniques used[a]	Isolate(s)	Phylogeny	Abundance	Reference
Plating	North Sea	Synthetic seawater + organics, $\approx$10 mg liter^{-1}; 37 days at 15°C + colony excision	Strain KT71	*Gammaproteobacteria*	8% of total count in North Sea	Eilers et al. (2001)
Plating	Georgia coast	Low-nutrient agar + 15 mg of organics + aged seawater; 35 days at 15°C	*Roseobacter* spp.	*Roseobacter* clade, *Alphaproteobacteria*	16% of marine 16S rDNA clones	Gonzalez and Moran (1997)
Plating	Australian soil	NB diluted 100 times + gellan gum, 84 days at 25°C	Various	*Acidobacteria, Verrucomicrobia*	Common in soil rDNA libraries	Janssen et al. (2002)
Plating	River Elbe aggregates	18 media, very low nutrient best + 100 mg of organics per liter; 28 days at 26°C	*Matsuebacter chitosanotabidus*	*Betaproteobacteria*	>1% of total count	Bockelmann et al. (2000)
Enrichment	Rice paddy soil	Anoxic MPN with rich organic media (>1 g liter^{-1}); 90 days at 25°C	$\approx$9 abundant strains	*Bacteriodetes, Verrucomicrobia,* and clostridia	>5% of total count	Chin et al. (1999)
Enrichment	Gold mine aquifer	900 MPN trials, inorganic medium + vitamins, 5% O$_2$ at 50–70°C	Strains HGM-K1 and pHAuB-D	*Aquificales*	$\approx$30% of total count	Takai et al. (2002)
Extinction culture	Alaska fjord water	Sterile seawater + <1 mgC of organics per liter; 21–56 days at 10°C	*Cycloclasticus oligotrophus*	*Gammaproteobacteria*	Not tested	Button et al. (1993)
Extinction culture	Alaska fjord water	Sterile seawater + <1 mgC of organics per liter; 21–56 days at 10°C	*Sphingomonas alaskensis*	*Sphingomonas* clade, *Alphaproteobacteria*	Uncertain	Schut et al. (1993)
Extinction culture	Oregon coast	Inorganic + 70 mg of organics + vitamins per liter; 23 days at 15°C, FISH screening	*Pelagibacter ubique* (proposed)	SAR11 clade, *Alphaproteobacteria*	26% of marine 16S rDNA clones	Rappé et al. (2002)

[a] "Organics" is used throughout this table to mean a mixture of organic carbon compounds; "<1 mgC of organics per liter" means <1 mg of organic carbon per liter.

tured. In another study (Eilers et al., 2001), slowly growing bacteria (see Color Plate 2d for examples of small, slowly growing colonies) on very-low-nutrient agar plates incubated for 36 days were studied by completely removing rapidly growing colonies by excision. They found 10-fold more colonies of *Alphaproteobacteria* developing later during incubation, but these were not the most abundant phylotypes found. In particular the *Roseobacter* isolates were not those in their clone libraries. However, one isolate (KT71) that developed after 15 days of incubation was a strain of *Gammaproteobacteria* belonging to the abundant NOR5 group and made up 8% of the total microscopic count.

The success of very low-nutrient media at isolating some abundant bacteria has been confirmed in other habitats. In bulk soil from an Australian pasture, nutrient broth diluted 100 times and solidified with gellan gum incubated at 25°C for 84 days resulted in viable counts that were 14% of the total microscopic count (Janssen et al., 2002). Furthermore, several of the isolates identified by partial 16S rDNA sequencing belonged to hitherto uncultured clusters of abundant soil bacteria in the *Verrucomicrobia* and *Acidobacteria*. Similarly, Bockelmann et al. (2000) compared 18 media for growing bacteria in aggregates from the River Elbe. They found that many more of the *Betaproteobacteria* that were most abundant (54% of the total count) grew on the very-low-nutrient media used than on richer media such as R2A. In addition, their *Betaproteobacteria* isolate 21 (*Matsuebacter chitosanotabidus*) made up at least 1% of the total count.

Therefore, the most successful isolation approaches for numerically abundant types of bacteria by plating on solid media use very-low-nutrient media (5 to 80 mg of organic components per liter) incubated for up to 84 days. However, although some numerically dominant bacteria can be isolated in this way, many are not.

Plating has also been used to isolate new, abundant bacteria from some extreme environments. For example, close relatives of the recently discovered, extremely halophilic bacterium, *Salinibacter ruber*, belonging to a novel deeply branching group of the *Bacteroidetes* phylum, make up 5 to 25% of the total prokaryotic community in Spanish saltern ponds. This organism was isolated as red colonies that grew on R2A medium with 20 to 25% salt at 37°C (Anton et al., 2002).

ENRICHMENT AND MICROMANIPULATION

One way of isolating abundant natural bacteria by enrichment is to examine the bacteria growing in the highest-dilution tubes during MPN enumeration experiments. This has been a successful technique for investigating the diversity of physiologically specialist bacteria, such as nitrifying bacteria, for many years and has recently been applied to grow bacteria revealed as numerically abundant by 16S rDNA cloning methods. For example, Chin et al. (1999) used this approach to investigate anoxic rice paddy field soil. Several isolates constituted more than 5% of the total direct count using about 1 g of xylan, pectin, or a mixture of seven mono- and disaccharides per liter as substrates. These included members of the *Verrucomicrobia*, *Bacteroidetes*, and gram-positive bacteria that belonged to groups closely related to abundant 16S rDNA clones from the same rice paddy soil. The most likely reason for the success of this approach was that rapidly growing but less-abundant isolates were diluted out in the terminal MPN tubes.

A similar approach has been used to culture lake water bacteria (Bartscht et al., 1999). A new synthetic freshwater medium was used to mimic natural lake water with a salts mixture in HEPES buffer containing trace vitamins and about 40 mg of organic substrates per liter. The MPN cultures were grown in microtiter plates using seven 10-fold dilutions and six replicates. When compared with direct microscopy, MPN counts gave up to 7% culturability (range of 0.2 to 25%). The R2A medium gave much lower MPN counts, perhaps due to the inhibitory effects of the phosphate buffer. Glucose, mixtures of four fatty acids (140 mg liter^{-1}), and 28-fold-diluted YPG medium all gave similar MPN counts, as did YPG alone, which contained yeast extract (0.2%), peptone (0.1%), and glucose (0.2%). This approach was extended by the addition of 10 μM cAMP to the MPN tubes using artificial brackish seawater with about 50 mg of a glucose and amino acid mixture per liter (Bruns et al., 2002). This gave culturability averaging about 15% (range of 2 to 100%) from the Baltic Sea. However, organisms cultured from two of these tubes were not the most abundant bacteria in the seawater, although they were *Alphaproteobacteria* from the *Rhodobacteriaceae*.

Therefore, the MPN approach needs further refinement before it will work in all habitats. In fact, research following the growth of *Thermus* isolates during enrichment from hot pools in New Zealand has shown that the abundant phylotypes in the original habitat can be overgrown by less-abundant 16S rDNA variants (Saul et al., 1999). Similarly, work isolating thermophilic *Archaea* from hydrothermal vents has shown that small variations in growth temperature and carbon source generated great taxonomic diversity between the enrichments (Wery et al., 2002); these authors further concluded that high-nutrient media were unlikely to isolate novel deep-sea thermophiles.

Enrichment has been especially successful at isolating naturally abundant thermophiles in the *Aquificales* from a hot-subsurface aquifer from a gold mine. Takai et al. (2002) tested 900 enrichment culture conditions to show that three media were successful. They isolated pure cultures of the dominant 16S rDNA phylotype of target bacteria by diluting the successful enrichments until no more grew; the pure cultures grew in the terminal enrichments. The successful media were based on $NaHCO_3$, Na_2SiO_3, and NH_4Cl with a vitamin mixture incubated at 65°C supplied with a gas mixture of N_2-CO_2 with 5% O_2 with some other nonessential supplements. This study emphasizes the large range of enrichment conditions that sometimes have to be tested to culture naturally abundant bacteria.

Micromanipulation can also be a valuable aid for isolating target bacteria, especially when fluorescent in situ hybridization (FISH) with phylogenetic probes is used to visualize the target bacteria. In this way Huber et al. (1995) used a strongly focused infrared laser ("optical tweezers") to separate a new hyperthermophilic archaeum from a hot pool in Yellowstone National Park. This organism grew in distinctive aggregates in enrichments dominated by filamentous bacteria and had been previously identified from in situ phylogenetic analyses of 16S rDNA. After separation by micromanipulation, the aggregates were successfully grown in pure culture anaerobically when incubated at 83°C for 1 week. Morphology without FISH probes is sometimes enough to target isolation. For example, several morphologically distinct filamentous bacteria, common in activated sludge wastewater treatment systems, have been isolated by micromanipulation (Kämpfer, 1997). Similarly, abundant filamentous *Bacteria* belonging to the green non-sulfur division have been isolated from thermophilic methanogenic sludge granules (Sekiguchi et al., 2001). This was achieved by removing filamentous outgrowths from the granules identified as the target green non-sulfur cells by FISH with a phylogenetic probe. The slow-growing target cells were grown in pure culture after several attempts from further high-dilution enrichments on media containing sucrose or glucose plus yeast extract incubated anaerobically for 14 days at 55°C.

EXTINCTION CULTURE

This method is also called dilution culture and dilution to extinction. It involves diluting water samples with filter-sterilized water until only a few bacteria remain and then growing the cells in either the unamended water or by adding small amounts of organic substrates to culture them. The method was first described by Button et al. (1993), and has recently been described elsewhere (Button et al., 2001; see also Button, chapter 16). The approach must not be confused with the MPN methods described above, where the term dilution to extinction is also used to refer to bacteria growing in the terminal tubes.

Button's method was carried out as follows. First, the total bacteria in the sample was counted microscopically. Then the sample was diluted with the filter-sterilized sample water so that very small numbers of bacteria remained. The diluted samples were then incubated with or without very small amounts of carbon source in 60-ml vials at 10°C, which was very close to the in situ temperature. The target inoculum in each vial was between 0.1 and 10^3 bacteria. The developing populations were then scored by microscopic examination over 3 to 8 weeks; positive vials were those in which at least 100 bacteria could be counted, equivalent to $>10^4$ bacteria per ml. In the original trials of the method (Button et al., 1993) with seawater from Resurrection Bay, Gulf of Alaska, all 11 experiments gave cultures.

Bacteria grew with extinction culture when no nutrients were added to the seawater, but populations increased up to 0.4×10^6 to 4.1×10^6 cells ml^{-1} when 0.01 to 1.0 mg of C per liter was added, although growth was inhibited at 5 mg liter^{-1} or more. There was no sign of wall growth in negative tubes, indicating that false negatives did not occur. When assessed by flow cytometry, the bacteria that developed were small (0.002 to 0.1 μm^3) and contained small amounts of DNA (1 to 8 fg per cell); these values were fairly typical of the marine bacteria observed before incubation. Bacteria in the vials with the highest inocula grew faster than those in vials with the lowest inocula, indicating that only the lower inocula cultured the most abundant bacteria. Mathematical analysis showed that almost all the original marine bacteria were viable (about 60% of direct counts). Viability decreased with increasing inoculum size, which supports the hypothesis that rare bacteria overgrow the most abundant small cells in tubes with high inocula. In this method dilution is not exact, so numbers of bacteria in the tubes will vary stochastically, and this will influence the number of pure cultures that result. Analysis showed that the number of pure cultures could be expected to be between one and six from 20 tubes but could never exceed 36%.

Using the methodology described above, Button et al. (1993) obtained three cultures, two of which were cultured for 3 years. One of these cultures, strain RB1, has been extensively studied physiologically and identified as *Cycloclasticus oligotrophus* (Button et al., 1998). This is a small rod-shaped bacterium (ca. 0.65 by 0.35 μm; volume, 0.08 μm^3) that reaches populations of 10^5 bacteria per ml in unamended seawater. However, with 100 mg of acetate incorporated per liter, it can form turbid cultures in liquid and small, smooth, translucent colonies on agar after 10 days of incubation at 20°C. It is closely related to *C. pugetii,* which is in the *Gammaproteobacteria* (Garrity et al., 2002).

Schut et al. (1993) extended the extinction dilution approach to isolate 37 strains from Resurrection Bay and the North Sea. Cultures from the highest dilutions generated single-cell types that were mostly rod shaped, but the lower dilutions grew mixed cultures of larger cells. The cultures of the smallest rod-shaped cells could not be grown on high-nutrient media (8 g of complex organic substrates per liter), but all could grow on synthetic seawater with 2 mg of Casamino Acids per liter. Plating the cultures onto agar after storage for 6 to 12 months at 5 to 8°C gave very small colonies after 32 to 42 days of incubation at 20°C; these were only just visible to the naked eye and were best observed at ×25 magnification. Both seawater salts agar with 10 mg of mixed carbon sources per liter and Zobell 2214E agar (seawater salts with 5 g of peptone per liter and 1 g of yeast extract per liter) gave the highest colony counts (1 × 10^5 to 10 × 10^5 CFU ml^{-1}). These counts were 80% of the direct counts, showing that the cultures were facultatively oligotrophic. Once again, these bacteria were extremely small (volume, 0.05 to 0.09 μm^3; diameter, 0.2 to 0.5 μm; length, 0.5 to 3.0 μm), equivalent in size to most natural marine bacteria and starved cells of other marine isolates showing the reduction-division starvation response. This small size is maintained even on most rich laboratory media, so they are true "ultramicrobacteria." They also grow on Trypticase soy agar with 0.5% salt aerobically at 28°C as convex, yellowish colonies when the cells are larger (0.8 × 2 to 3 μm) than on other media (Vancanneyt et al., 2001). Furthermore, they are stress resistant and do not show the typical reduction-division as a starvation-survival response and so are almost certainly well adapted to growth in the sea. Seven of these isolates have now been assigned to the new species *Sphingomonas alaskensis,* with isolate RB2256 as the type strain (Vancanneyt et al., 2001). They are phylogenetically related but not physiologically similar to other *Sphingomonas* spp. and are *Alphaproteobacteria* (Garrity et al., 2002). Recently another strain of *S. alaskensis* AFO1 has been isolated by extinction culture from the North Pacific (Eguchi et al., 2001). This strain is very similar in physiology to RB2256, which supports both the widespread occurrence of this species in the oceans and its high degree of adaptation to the marine environment.

The SAR11 clade of the *Alphaproteobacteria* is one of the most abundant bacteria groups in the oceans, accounting for 26% of all rDNA sequences isolated from seawater (Giovannoni and Rappé, 2000). It also occurs worldwide and features in almost all clone libraries from coastal and oceanic waters. In August 2002, Rappé et al. (2002) reported the isolation of 11 cultures of this bacterium from Oregon coast seawater by extinction culture. They diluted and grew the samples in two media: first, a sterile Pacific Ocean water supplemented with NH_4Cl and KH_2PO_4 (both at 0.1 μM), and second, a similar medium supplemented with a mixture of carbon sources (10 mg each of glucose, ribose, succinic acid, pyruvic acid, glycerol, and *N*-acetyl glucosamine per liter): ethanol (20 mg liter^{-1}), and a vitamin mixture. The water samples were diluted so that about 22 cells were in each of 288 1-ml cultures in microtiter plates, which were incubated for 23 days at 15°C. Culture wells were tested for replicating bacteria and SAR11 cells by arraying small volumes on polycarbonate membranes. The arrays were stained with a DNA stain and four FISH probes for the SAR11 group, before screening with epifluorescence microscopy and image analysis, to achieve a limit of detection of about 200 cells in 100 μl of culture. Ten pure cultures of SAR11 bacteria and eight mixed cultures were obtained. These cultures belonged to two different phylogenetic groups of the SAR11 clade and were 99% similar to over 30 16S rDNA clones isolated from diverse locations. Both media grew and isolated the SAR11 strains equally well, but growth was inhibited by 10 mg of proteose peptone per liter. The cultured cells are very small, curved rods (diameter, 0.1 to 0.2 μm; length, 0.4 to 0.9 μm; volume, 0.01 μm^3) and are among the smallest bacteria in culture. The organism has been named "*Candidatus* Pelagibacter ubique."

PROSPECTS FOR THE FUTURE

The recent isolation of bacteria from the abundant, widespread, and elusive marine SAR11 clade is clearly the greatest success so far in isolating dominant bacteria from nature. This confirms the earlier studies that strongly suggested that extinction culture could be used to isolate some of the most important bacteria from oligotrophic waters. It seems highly likely that more attempts will be made to isolate other not-yet-cultured groups from marine and freshwater habitats. However, the research summarized in this chapter demonstrates that other methods such as enrichment, micromanipulation, and plating can be successful in other habitats. It is likely that a mixture of high-throughput methods to screen large numbers of cultures using many different tactics, coupled with effective olionucleotide probes to target specific groups of bacteria, will be the general approach of choice in the future. The development of methods such as arraying, robotics, and automated microscopic screening needs to be applied to bacterial isolation. Medical microbiology shows us that almost any bacterium can be grown if enough time and effort are applied. There are still whole divisions of *Bacteria* (Hugenholtz et al., 1998) and *Archaea* (DeLong and Pace, 2001) for which there are no cultivated representatives. It is only by growing these organisms and studying their physiology that we will fully understand nature's biogeochemical cycles. As Zinder (2002) has recently said, the "golden era of culturable microbiology" lies ahead.

REFERENCES

Anton, J., A. Oren, S. Beniloch, F. Rodriguez-Valera, R. Amann, and R. Rossello-Mora. 2002. *Salinibacter ruber* gen. nov., sp nov., a novel, extremely halophilic member of the Bacteria from saltern crystallizer ponds. *Int. J. Syst. Evol. Microbiol.* 52:485–491.

Atlas, R. M. 1995. *Handbook of Media for Environmental Microbiology.* CRC Press, Boca Raton, Fla.

Austin, B. (ed.). 1988. *Methods in Aquatic Bacteriology.* John Wiley, Chichester, United Kingdom.

Balows, A., H. G. Truper, W. H. Dworkin, and K.-H. Schleifer (ed.). 1992. *The Prokaryotes: A Handbook on the Biology of Bacteria: Ecophysiology, Isolation, Identification, Applications,* 2nd ed, vol. 1–4. Springer-Verlag, Berlin, Germany.

Bartscht, K., H. Cypionka, and J. Overmann. 1999. Evaluation of cell activity and of methods for the cultivation of bacteria from a natural lake community. *FEMS Microbiol. Ecol.* 28:249–259.

Bockelmann, U., W. Manz, T. R. Neu, and U. Szewzyk. 2000. Characterization of the microbial community of lotic organic aggregates ("river snow") in the Elbe River of Germany by cultivation and molecular methods. *FEMS Microbiol. Ecol.* 33:157–170.

Breznak, J. A. 2002. A need to retrieve the not-yet-cultured majority. *Environ. Microbiol.* 4:4–5.

Bruns, A., H. Cypionka, and J. Overmann. 2002. Cyclic AMP and acyl homoserine lactones increase the cultivation efficiency of heterotrophic bacteria from the central Baltic Sea. *Appl. Environ. Microbiol.* 68:3978–3987.

Bull, A. T., and J. H. Slater (ed.). 1982. *Microbial Interactions and Communities,* vol. 1. Academic Press, London, United Kingdom.

Button, D., F. Schut, P. Quang, R. Martin, and B. Robertson. 1993. Viability and isolation of marine bacteria by dilution culture: theory, procedures, and initial results. *Appl. Environ. Microbiol.* 59:881–891.

Button, D. K., B. R. Robertson, P. W. Lepp, and T. M. Schmidt. 1998. A small, dilute-cytoplasm, high-affinity, novel bacterium isolated by extinction culture and having kinetic constants compatible with growth at ambient concentrations of dissolved nutrients in seawater. *Appl. Environ. Microbiol.* 64:4467–4476.

Button, D. K., B. K. Robertson, and P. Quang. 2001. Isolation of oligobacteria, p. 161–173. *In* J. H. Paul (ed.), *Marine Microbiology, Methods in Microbiology,* vol. 30. Academic Press, San Diego, Calif.

Chin, K. J., D. Hahn, U. Hengstmann, W. Liesack, and P. H. Janssen. 1999. Characterization and identification of numerically abundant culturable bacteria from the anoxic bulk soil of rice paddy microcosms. *Appl. Environ. Microbiol.* 65:5042–5049.

DeLong, E. E., and N. R. Pace. 2001. Environmental diversity of *Bacteria* and *Archaea. Syst. Biol.* 50:470–478.

Eguchi, M., M. Ostrowski, F. Fegatella, J. Bowman, D. Nichols, T. Nishino, and R. Cavicchioli. 2001. *Sphingomonas alaskensis* strain AFO1, an abundant oligotrophic ultramicrobacterium from the North Pacific. *Appl. Environ. Microbiol.* 67:4945–4954.

Eilers, H., J. Pernthaler, F. O. Glockner, and R. Amann. 2000. Culturability and *in situ* abundance of pelagic bacteria from the North Sea. *Appl. Environ. Microbiol.* 66:3044–3051.

Eilers, H., J. Pernthaler, J. Peplies, F. O. Glockner, G. Gerdts, and R. Amann. 2001. Isolation of novel pelagic bacteria from the German Bight and their seasonal contributions to surface picoplankton. *Appl. Environ. Microbiol.* 67:5134–5142.

Fry, J. C. 1987. Functional roles of major groups of bacteria associated with detritus, p. 83–122. *In* D. J. W. Moriarty and R. S. V. Pullin (ed.), *Detritus and Microbial Ecology in Aquaculture,* vol. 14, ICLARM Conference Proceedings. International Centre for Living Aquatic Resources Management, Manila, Philippines.

Fry, J. C. 1990. Oligotrophs, p. 93–116. *In* C. Edwards (ed.), *Microbiology of Extreme Environments.* Open University Press, Milton Keynes, United Kingdom.

Fry, J. C., and T. Zia. 1982. Viability of heterotrophic bacteria in fresh-water. *J. Gen. Microbiol.* 128:2841–2850.

Garrity, G. M., M. Winters, A. W. Kuo, and D. B. Searles. 2002. *Taxonomic Outline of the Procaryotes. Bergey's Manual of Systematic Bacteriology.* Release 2, 2nd ed. Springer-Verlag, New York, N.Y.

Giovannoni, S. J., and M. S. Rappé. 2000. Evolution, diversity, and molecular ecology of marine prokaryotes, p. 47–84. *In* D. L. Kirchmann (ed.), *Microbial Ecology of the Oceans.* Wiley, New York, N.Y.

Gonzalez, J. M., and M. A. Moran. 1997. Numerical dominance of a group of marine bacteria in the alpha-subclass of the class *Proteobacteria* in coastal seawater. *Appl. Environ. Microbiol.* 63:4237–4242.

Grigorova, R., and J. R. Norris (ed.). 1990. Techniques in microbial ecology p. 627. *In Methods in Microbiology,* vol. 22. Academic Press, London, United Kingdom.

Holt, J. G. (ed.). 1994. *Bergey's Manual of Determinative Bacteriology,* 9th ed., p. 787. Williams & Wilkins, Baltimore, Md.

Huber, R., S. Burggraf, T. Mayer, S. M. Barns, P. Rossnagel, and K. O. Stetter. 1995. Isolation of a hyperthermophilic archaeum predicted by *in-situ* RNA analysis. *Nature* 376:57–58.

Hugenholtz, P., B. M. Goebel, and N. R. Pace. 1998. Impact of culture-independent studies on the emerging phylogenetic view of bacterial diversity. *J. Bacteriol.* 180:4765–4774.

Janssen, P. H., P. S. Yates, B. E. Grinton, P. M. Taylor, and M. Sait. 2002. Improved culturability of soil bacteria and isolation in pure culture of novel members of the divisions. *Acidobacteria, Actinobacteria, Proteobacteria,* and *Verrucomicrobia. Appl. Environ. Microbiol.* 68:2391–2396.

Kämpfer, P. 1997. Detection and cultivation of filamentous bacteria from activated sludge. *FEMS Microbiol. Ecol.* 23:169–181.

Paul, J. H. (ed.). 2001. Marine microbiology, p. 666. *In Methods in Microbiology,* vol. 30. Academic Press, San Diego, Calif.

Pinhassi, J., U. L. Zweifel, and A. Hagstrom. 1997. Dominant marine bacterioplankton species found among colony-forming bacteria. *Appl. Environ. Microbiol.* 63:3359–3366.

Rappé, M. S., S. A. Connon, K. L. Vergin, and S. J. Giovannoni. 2002. Cultivation of the ubiquitous SAR11 marine bacterioplankton clade. *Nature* **418:**630–633.

Saul, D. J., R. A. Reeves, H. W. Morgan, and P. L. Bergquist. 1999. *Thermus* diversity and strain loss during enrichment. *FEMS Microbiol. Ecol.* **30:**157–162.

Schut, F., E. de Vries, J. Gottschal, B. Robertson, W. Harder, R. Prins, and D. Button. 1993. Isolation of typical marine bacteria by dilution culture: growth, maintenance, and characteristics of isolates under laboratory conditions. *Appl. Environ. Microbiol.* **59:**2150–2160.

Sekiguchi, Y., H. Takahashi, Y. Kamagata, A. Ohashi, and H. Harada. 2001. *In situ* detection, isolation, and physiological properties of a thin filamentous microorganism abundant in methanogenic granular sludges: a novel isolate affiliated with a clone cluster, the Green Non-Sulfur Bacteria, subdivision I. *Appl. Environ. Microbiol.* **67:**5740–5749.

Stackebrandt, E., and B. J. Tindall. 2000. Appreciating microbial diversity: rediscovering the importance of isolation and characterization of microorganisms. *Environ. Microbiol.* **2:**9–10.

Suzuki, M. T., M. S. Rappé, Z. W. Haimberger, H. Winfield, N. Adair, J. Strobel, and S. J. Giovannoni. 1997. Bacterial diversity among small-subunit rRNA gene clones and cellular isolates from the same seawater sample. *Appl. Environ. Microbiol.* **63:**983–989.

Takai, K., H. Hirayama, Y. Sakihama, F. Inagaki, Y. Yamato, and K. Horikoshi. 2002. Isolation and metabolic characteristics of previously uncultured members of the order *Aquificales* in a subsurface gold mine. *Appl. Environ. Microbiol.* **68:**3046–3054.

Vancanneyt, M., F. Schut, C. Snauwaert, J. Goris, J. Swings, and J. Gottschal. 2001. *Sphingomonas alaskensis* sp. nov., a dominant bacterium from a marine oligotrophic environment. *Int. J. Syst. Evol. Microbiol.* **51:**73–79.

Wery, N., M.-A. Cambon-Bonavita, F. Lesongeur, and G. Barbier. 2002. Diversity of anaerobic heterotrophic thermophiles isolated from deep-sea hydrothermal vents of the Mid-Atlantic Ridge. *FEMS Microbiol. Ecol.* **41:**105–114.

Zinder, S. H. 2002. The future for culturing environmental organisms: a golden era ahead? *Environ. Microbiol.* **4:**14–15.

Microbial Diversity and Bioprospecting
Edited by Alan T. Bull
© 2004 ASM Press, Washington, D.C.

Chapter 9

Culture-Independent Microbiology

KORNELIA SMALLA

WHY ARE CULTIVATION-INDEPENDENT TECHNIQUES NEEDED?

The phenomenon that only a small proportion of bacteria can form colonies when traditional plating techniques are used (Amann et al., 1995) was first described by Staley and Konopka (1985) as the "great plate anomaly." The proportion of the hidden bacterial diversity largely depends on the kind of environmental sample. Based on microscopic studies it is estimated that in seawater, freshwater, mesotrophic lakes, estuarine waters, activated sludge, sediments, and soil, 0.0001 to 0.1%, 0.25%, 0.1 to 1%, 0.1 to 3%, 1 to 15%, 0.25%, and 0.3%, respectively, of the total bacterial cells are accessible to traditional cultivation (Amann et al., 1995). It seems that the more oligotrophic an environmental sample, the higher the proportion of bacteria that do not grow under standard cultivation conditions. Different approaches to isolate bacteria that do not easily form colonies under standard cultivation conditions are described in chapter 8.

A further limitation of the cultivation-based studies of microbial communities is that under environmental stress bacteria can enter a state called "viable but nonculturable," and again these bacteria would not be accessible to traditional cultivation techniques (Roszak and Colwell, 1987; Oliver, 2000). Thus, monitoring the presence of pathogens or genetically modified inoculants, for example, exclusively by cultivation-based methods can be misleading and might have potential health implications.

The development of methods to extract nucleic acids directly from environmental samples that should be representative for the microbial genomes present in such samples opened a new dimension for the study of microbial communities. The analysis of DNA can provide information on the structural diversity of environmental samples or on the presence or absence of certain genes (e.g., antibiotic resistance genes or plasmid replicon-related sequences) or of introduced bacteria. However, in general, the analysis of DNA does not allow conclusions on the metabolic activity of members of the bacterial or fungal community or on gene expression. This information might be obtained from RNA analysis. Other approaches for the study of metabolically active bacteria are based on the incorporation of bromadeoxyuridine (BrdU) or ^{13}C into the nucleic acids of growing cells. Because the basis of most cultivation-independent approaches in microbiology (except whole-cell in situ hybridization or exogenous isolation of mobile genetic elements) is DNA or RNA that is extracted from environmental matrices, the following section is devoted to nucleic extraction.

NUCLEIC ACID EXTRACTION FROM ENVIRONMENTAL SAMPLES

More than 20 years after the first paper on the extraction of DNA from soil had been published by Torsvik (1980), obtaining nucleic acids from environmental matrices that are suitable for molecular analysis remains a challenge. This is mirrored by the large number of publications on this subject (for a review, see Van Elsas et al., 2000). None of the protocols seem suitable for all kinds of environmental matrices, in particular for soils and sediments originating from contaminated sites. Only recently have commercial kits, e.g., for DNA extraction from soil, become available that are indeed a major breakthrough in view of a simplification and miniaturization of this crucial method for many cultivation-independent analysis methods.

Two principal approaches to recover nucleic acids from environmental matrices exist. The first approach pioneered by Ogram et al. (1987) is based on

Kornelia Smalla • Federal Biological Research Centre for Agriculture and Forestry, Institute for Plant Virology, Microbiology and Biosafety, Messeweg 11–12, 38104 Braunschweig, Germany.

direct or in situ lysis of microbial cells in the presence of the environmental matrix (e.g., soil, sediments, or plant material), followed by separation of the nucleic acids from matrix components and cell debris, which by far is the most frequently used method. The advantage of the direct nucleic acid extraction approach is that it is less time-consuming and that a much higher DNA yield is achieved (Van Elsas et al., 2000). However, directly extracted DNA often contains considerable amounts of coextracted substances such as humic acids that do interfere with the subsequent molecular analysis. Furthermore, a considerable proportion of directly extracted DNA might originate from nonbacterial sources or free DNA.

In the second approach, pioneered by Torsvik (1980) and Holben et al. (1988), the microbial fraction is recovered before the cells are lysed and the DNA is extracted and purified. The indirect DNA extraction approach might preferably be used when problematic environmental matrices are to be analyzed, or when cloning large DNA fragments (e.g., to generate bacterial artificial chromosome [BAC] libraries) from soil or sediment DNA where a high proportion of DNA of bacterial origin is crucial. To concentrate the microbial fraction from aquatic samples, centrifugation or filtration through membranes is used. The recovery of the bacterial fraction from soils or sediments commonly involves repeated homogenization and differential centrifugation steps (Hopkins et al., 1991), as originally suggested by Faegri et al. (1977). However, protocols differ considerably with respect to the solutions used to break up soil colloids and dislodge surface-attached cells that adhere to surfaces by various bonding mechanisms such as polymers, electrostatic forces, and water bridging and with different strengths (Bakken and Lindahl, 1995; Van Elsas et al., 2000). Homogenization is usually achieved by shaking suspensions with gravel or blending steps. Although a complete dislodgment of cells seems to be impossible, it is important that cells that are bound to the surface with different degrees of strength are released with similar efficiency. A clear advantage of the indirect approach is that the nucleic acids recovered are less contaminated with coextracted humic acids and DNA of nonbacterial origin.

The efficient disruption of the bacterial and fungal cell walls is crucial for the recovery of representative DNA that reflects the genomes of microbes present in an environmental sample and their relative abundance (Moré et al., 1994; Miller et al., 1999). Cell lysis can be achieved by mechanical cell disruption and by enzymatic or chemical disintegration of cell walls, and a combination of these methods is often used. High-molecular-weight DNA is an important criterion when evaluating and comparing

different protocols because sheared DNA can cause PCR artifacts and is not suitable for direct cloning of large DNA fragments. To obtain large DNA fragments is particularly important for the generation of soil DNA BAC libraries. The complete removal of coextracted humic acids is critical for the subsequent molecular analysis because humic acids were shown to interfere with DNA hybridization, restriction enzyme digestions, and PCR amplification (Tebbe and Vahjen, 1993). The degree of coextracted humic substances strongly depends on the environmental matrix, e.g., the soil type. To reduce the complexity of bacterial communities, a protocol to fractionate the DNA on the basis of guanine-plus-cytosine (G+C) content by using bisbenzimidazole and equilibrium density gradient ultracentrifugation was developed by Holben and Harris (1995). Bisbenzimidazole binds to adenine and thymidine and thus changes the buoyant density of DNA corresponding to its G+C content.

RNA EXTRACTION FROM ENVIRONMENTAL MATRICES

In contrast to direct and indirect DNA extraction protocols that are widely used in cultivation-independent microbiology, methods for RNA extraction are less frequently used, and considerable efforts are needed to ensure the absence of RNases. Because of the short half-life of bacterial mRNA, an unbiased recovery of total RNA is still a major methodological challenge. Different protocols aimed at the simultaneous extraction of RNA and DNA have recently been published (Duarte et al., 1998; Griffiths et al., 2000; Hurt et al., 2001; Weinbauer et al., 2002). Important criteria for the quality and suitability of RNA extraction protocols are the yield, the integrity, and the purity of the RNA. Thus, methods for efficiently removing humic substances and residual DNA without partial degradation of the RNA are required for the reliable use of RNA for microbial community analysis.

MOLECULAR ANALYSIS OF NUCLEIC ACIDS OBTAINED FROM ENVIRONMENTAL SAMPLES

The analysis of nucleic acids extracted directly from environmental samples provides information on the structural composition of microbial communities as well as the presence or absence of functional genes independent from traditional cultivation techniques. If RNA is used, the expression of certain genes in response to certain environmental stimuli can be studied. The growing database of partial or complete microbial sequences is essential for the development of tools such as primers or probes. Many of the common methods in cultivation-independent microbiology are based on PCR.

Phylogenetic Markers Used in Microbiology

The most frequently analyzed phylogenetic markers are the 16S rRNA genes. By comparing ribosomal RNA (rRNA) sequences, Woese (1987) developed a sequence-based phylogenetic tree. The rRNA molecule offers great potential as a phylogenetic marker because it is universal to all forms of life, it is structurally and functionally conserved, and it contains regions with different degrees of relatedness. Another considerable advantage is the rapidly growing database of rRNA gene sequences. A disadvantage for its use in cultivation-independent microbiology is that bacteria possess different numbers of rRNA operons which might reflect different ecological strategies of bacteria (Klappenbach et al., 2000) and that sequence heterogeneity of the different operons might occur (Nübel et al., 1996; von Wintzingerode et al., 1997). Recently, alternative markers such as the gene coding for the σ factor *rpoB* have been suggested because this gene is present only as a single copy and was shown to allow a higher resolution between species (Dahllöf et al., 2000). Furthermore, a number of functional genes have been employed successfully as phylogenetic markers. Examples of functional genes that are good phylogenetic markers because they reflect the 16S rDNA-based phylogeny are those encoding ammonia monooxygenase (*amoA*) and particulate methane monooxygenase (*pmoA*) (for a review see Liesack and Dunfield, 2002).

The first attempts to characterize the structural composition of microbial communities independent from cultivation targeted the 5S rRNA that was directly extracted from environmental samples. Because of its size (about 120 nucleotides), the information content was relatively small, and thus this approach could be used only for less complex environments. The use of the 16S rRNA gene in microbial ecology was suggested by Pace and coworkers (Pace et al., 1986). One approach uses the direct cloning of community DNA, identification of an rRNA gene containing clones by probing, and subsequent sequencing of the 16S rRNA gene (Ward et al., 1990). Although this omits PCR biases, screening of shotgun libraries of random DNA fragments for the presence of RNA genes by hybridization is rather laborious.

PCR, Cloning, and Sequencing Approach

The highly conserved nature of the rRNA can be used to design primers that are universal and thus anneal to sequences from all kinds of organisms in an environmental sample (Pace, 1997). Cloning of 16S rDNA fragments amplified from community DNA or RNA by PCR and sequencing opened a new era of cultivation-independent microbiology. The PCR-based amplification of 16S rDNA sequences from community DNA was first applied by Giovannoni et al. (1990) to analyze the picoplankton of the Sargasso Sea. The cloning and sequencing of PCR amplicons from community DNA or RNA became one of the most important tools in cultivation-independent microbiology. This approach has been applied to a wide variety of environments and has revealed substantial bacterial diversity with minor or no representation among cultured bacteria (Hugenholtz et al., 1998). The wide application of the cloning and sequencing approach of PCR-amplified 16S rRNA genes or fragments for more than one decade has led to an increase in the number of identified bacterial divisions to 40 or more. The majority of divisions has been retrieved from different environments. However, only 23 of the presently recognized 40 bacterial divisions are represented by isolates. In contrast to the four divisions (the *Proteobacteria*, the *Actinobacteria*, the low G+C gram-positives, and the *Cytophagales*, which are well represented by cultured bacteria) others such as the *Acidobacteria,* the *Verrucomicrobia*, or the green non-sulfur bacteria, which are frequently retrieved from 16S rDNA clone libraries, are represented only by very few isolates. Clones of 16S rDNA amplified from DNA or reverse-transcribed RNA from geographically distant sites often showed highly similar or identical sequences. The analysis of 16S rDNA clone libraries from different soils revealed an enormous diversity that confirmed predictions based on DNA reassociation studies (Torsvik et al., 1990). Although a comparison of the results from different studies is often problematic because of variations in the sampling strategy, the sample type (soil type, land use, climate), and the molecular methods used (nucleic acid extraction protocol, primers targeting different conserved regions of the 16S rDNA), some general observations can be made. Usually, little redundancy among sequenced 16S rDNA from soil is found (McCaig et al., 1999). The presence of duplicate sequences in soil libraries might be an indication of a reduced diversity, as proposed by Felske et al. (1998). In several soil studies *Alphaproteobacteria* belonged to the most abundant ribotypes whereas in other soils *Actinomycetes* were the most abundant (McCaig et al., 1999). Nogales et al.

(2001) assessed the bacterial diversity of polychlorinated biphenyl-polluted soils from clone libraries generated from rRNA or rDNA. A good correspondence of the community composition in both types of libraries and a high bacterial diversity was found even in polychlorinated biphenyl-contaminated soil.

Several studies have utilized reverse transcription-PCR (RT-PCR) to investigate the expression and diversity of different functional genes (e.g., *merA*, *nahA*, *amoA*, *pmoA*, *nirS*, or *nirK*) in various environments (for a review, see Liesack and Dunfield, 2002). Nogales et al. (2002) successfully used the RT-PCR-based approach to study the diversity and expression of five key enzymes involved in bacterial denitrification in river sediments. However, although detection of mRNA in environmental RNA by RT-PCR is a potentially powerful tool, several shortcomings might affect the results. In addition to the well-known biases of PCR from environmental nucleic acids (von Wintzingerode et al., 1997), the most critical factor is the stability of the transcripts. It seems that this approach can currently be successfully applied only for transcripts expressed at a high level (Nogales et al., 2002).

Recently, with cloning large fragments of environmental DNA and subsequent screening of the metagenomic libraries, it has become feasible to explore the functional potential of microbial communities independently of cultivation. This approach has been successfully applied to recover DNA coding for novel enzymes or antibiotics directly from soil DNA (Rondon et al., 2000; Gillespie et al., 2002). However, despite the vast potential of this approach, several methodological challenges remain to be solved (for more information, see chapter 14).

Molecular Fingerprinting Techniques

To study spatial and temporal variation of microbial communities in relation to environmental factors, which are due to perturbation or experimental treatment, multiple sample analysis is essential. For this purpose, approaches based on cloning and sequencing of 16S rDNA fragments PCR amplified from community DNA or on characterization of bacterial isolates are too labor-intensive and time-consuming, especially for terrestrial habitats of a large microbial diversity. More appropriate are fingerprinting techniques such as denaturing or temperature gradient gel electrophoresis (DGGE; TGGE) (Muyzer et al., 1993; Muyzer and Smalla, 1998), single-strand conformation polymorphism (SSCP) (Schwieger and Tebbe, 1998), or terminal restriction fragment analysis of 16S ribosomal genes or fragments amplified from total community DNA (Liu

et al., 1997; Osborn et al., 2000). The advantage of a community analysis based on the rRNA genes is that not only fingerprints are generated but also information on the phylogenetic affiliation of dominant community members is possible. The DGGE or SSCP techniques that were originally developed to detect point mutations in medical research had to be adopted to use in microbial ecology (Heuer et al., 2001; Tebbe et al., 2001). DGGE, SSCP, and terminal restriction-fragment-length polymorphism (T-RFLP) are ideal for the analysis of PCR products amplified from genes such as the 16S rRNA genes composed of conserved and variable regions. PCR products of the same length but of different sequences can be separated by DGGE according to the melting behavior of the DNA or by means of SSCP because of their conformation polymorphism. For T-RFLP the 16S rDNA fragments are amplified with primers labeled with fluorochromes from total community DNA. The PCR products are subsequently digested with restriction enzymes. Because of the sequence differences in the variable regions, terminal restriction fragments of different lengths are obtained. The fingerprints of 16S rDNA fragments amplified from community DNA or RNA by PCR or RT-PCR with primers annealing to regions conserved within the bacteria are an example of the most predominant ribotypes. It is estimated that a population must represent about 1% of the total community to be detectable in a fingerprint. Nested or seminested PCR approaches with taxa-specific primers were successfully used to analyze less-abundant ribotypes or to reduce the complexity of the pattern (Fig. 1) (Heuer et al., 1997; Gomes et al., 2001; Boon et al., 2002). Prominent bands can be excised and used for sequence determination in order to obtain further information about the phylogeny of the dominating ribotypes. The detection level of PCR-amplified ribosomal fragments might vary for several reasons, such as different numbers of rRNA operons or mismatches to the primer used for amplification. More than one species population might be hidden behind one band, thus leading to an underestimation of the diversity. Thus, the number and intensity of bands (DGGE, SSCP) or peaks (T-RFLP) does not necessarily correspond to the relative abundance of a species. Despite several pitfalls of PCR-based rRNA analysis, profiling of microbial communities by DGGE, SCCP, or T-RFLP proved to be a powerful method, allowing a cultivation-independent analysis of large numbers of samples. Molecular fingerprints have been applied in many studies to follow shifts of bacterial communities, e.g., during plant growth development (see Fig. 2), or to show plant-dependent diversity in the rhizosphere of different crop plants (Schwieger and Tebbe,

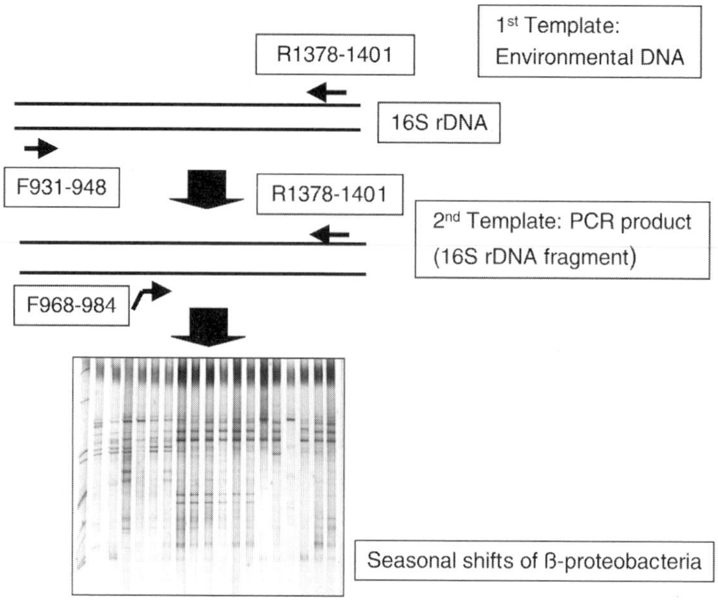

1st Template: Environmental DNA

R1378-1401

16S rDNA

F931-948

R1378-1401

2nd Template: PCR product (16S rDNA fragment)

F968-984

Seasonal shifts of ß-proteobacteria

Figure 1. DGGE and TGGE analysis of specific taxa to dissect complex communities and to detect less-abundant ribotypes. Experimental approach for the analysis of patterns of *Betaproteobacteria* (ß-proteobacteria in the figure): in the first PCR a forward primer specific for *Betaproteobacteria* is used in combination with a universal primer to amplify *Betaproteobacteria* 16S rDNA from community DNA. The amplicons are used in a second PCR with a G+C-clamped bacterial primer in combination with a universal primer.

2000; Yang and Crowley, 2000; Duineveld et al., 2001; Smalla et al., 2001). Molecular fingerprinting techniques have also been used to analyze the diversity of functional genes (Liesack and Dunfield, 2002). More information becomes available when linking 16S rRNA gene or functional gene-based molecular fingerprints with the analysis of clones and isolates obtained from the same samples.

Quantitative Dot Blot Hybridization and Gene Arrays

DNA or RNA extracted from different environmental samples can be spotted on membranes and hybridized with phylogenetic probes. A large set of phylogenetic probes with different degrees of specificity has been described. Quantitative dot or slot blot

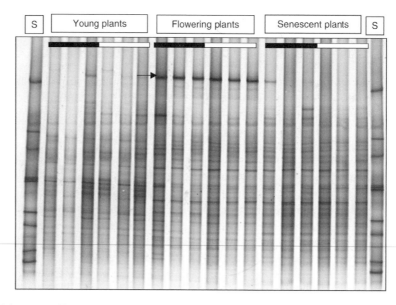

S | Young plants | Flowering plants | Senescent plants | S

Figure 2. Seasonal dynamics of bacterial communities in the potato rhizosphere as revealed by DGGE analysis of 16S rDNA fragments amplified from community DNA. Nontransgenic (light gray) and transgenic (black) T4-lysozyme expressing Désirée. The arrow indicates *Serratia ficaria*. (Reprinted from Heuer et al. 2002.)

hybridization with phylogenetic probes has been used in several studies to analyze the structural composition of various different environments. The relative abundance of different taxa can be determined by hybridization of dot-blotted RNA with a specific probe and a universal probe and is calculated as the ratio of their respective hybridization signals. Coextracted humic substances and DNA have recently been shown to affect RNA hybridization results with oligonucleotide probes. Different regions of the rRNA are differently susceptible to the attack of RNases, which might result in a partial loss of probe or primer target sites. Such a partial RNA degradation is particularly critical when the RNA is used for quantitative hybridization (Alm and Stahl, 2000; Alm et al., 2000).

Nucleic acid microarrays that are a further development of conventional membrane-based technology represent the latest advance in molecular technology that offers enormous advantages of multiplex detection. Thousands of genes can be simultaneously assessed by using large sets of probes fixed on glass slides. Thus, microarrays are ideal to study the sequence diversity of 16S rDNA genes and of functional genes in environmental samples. However, the performance of microarray hybridization in environmental microbiology studies has to be carefully evaluated, and obviously a number of technological challenges need to be solved before this technique can be explored in hybridization with nucleic acids from complex environmental samples. A microarray with 100 functional genes was used by Wu et al. (2001) to systematically study the specificity, sensitivity, and quantification of microarray hybridization with DNA from complex environmental samples. When environmental DNA is used without prior PCR amplification, this seems to be the most difficult challenge. The level of detection is 1,000- to 10,000-fold lower than with PCR amplification (Wu et al., 2001). Another critical parameter is specificity, which is largely determined by the nucleotide composition of mismatches and the mismatch composition. The DNA microarray technology was used by Koizumi et al. (2002) to characterize anaerobic toluene and ethylbenzene-degrading microbial consortia. In this study, optimal washing conditions were established by determining the melting profiles of all probe target duplexes. RNA transcribed in vitro was used to increase the sensitivity. A so-called PhyloChip with 132 16S rRNA gene probes specific for all known lineage of sulfate-reducing prokaryotes was tested (Loy et al., 2002). In this study, 16S rDNA and dissimilatory bisulfite reductase PCR amplificates were fluorescently labeled and used for hybridization. The sulfate-reducing prokaryote PhyloChip results were confirmed by several cultivation-independent methods.

ATTEMPTS TO LINK MICROBIAL COMMUNITY STRUCTURE WITH FUNCTION

A major criticism of most cultivation-independent studies of microbial communities is that, despite the fact that exciting new insights into the structural diversity of microbial communities are provided, actually little is known of the function of these populations. Recently, different methodological approaches have been developed that seem to have a considerable potential to link information on metabolic activity or the involvement in biogeochemical transformation processes to certain ribotypes. Two of these approaches are making use of the incorporation of stable isotopes or BrdU, which allows a subsequent separation from the bacterial fraction that is not labeled. Furthermore, the fluorescence in situ hybridization (FISH) technique in combination with microautoradiography (MAR) has been used to demonstrate the involvement of uncultured organisms in transformation processes.

The BrdU Method

BrdU is a structural analogue of thymidine. The uptake of [^{3}H]thymidine has routinely been used for measuring the in situ growth of bacteria in different environments. Recently, BrdU incubation was used to detect metabolically active bacteria in microbial communities from lake water (Urbach et al., 1999) and in soil (Borneman, 1999; Yin et al., 2000). The procedure consists of four steps: incubation of environmental bacteria (soil or microbial fraction) followed by the extraction of DNA directly from the environmental sample or the microbial fraction and immunocapture of DNA that incorporates BrdU by using magnetic beads covered with anti-BrdU antibody. The BrdU approach was used by Borneman (1999) to study soil bacterial communities responding to certain environmental stimuli such as the addition of glucose. Limitations of the BrdU method might be nonspecific binding of DNA to the magnetic beads that did not incorporate BrdU, or the possibility that not all bacteria are capable of taking up BrdU. Although the majority of bacteria are supposed to take up and incorporate [^{3}H]thymidine, this is not fully studied for its analogue BrdU.

The Stable Isotope Technique

The stable isotope technique is a promising strategy in particular for cultivation-independent identification of organisms responsible for certain in situ biogeochemical transformation processes (Boschker and Middelburg, 2002). This technique was first used to identify ^{13}C-enriched phospholipid fatty acid signature profiles, e.g., in experiments aiming to identify microbial populations responsible for acetate oxidation in sediments (Boschker et al., 1998). Recently, Radajewski et al. (2000) showed that ^{13}C-DNA produced during growth of microbes on a ^{13}C-enriched carbon source can be separated from ^{12}C by density gradient centrifugation. When this approach was applied to study methanol-utilizing microorganisms in soil, the involvement of *Alphaproteobacteria* and *Acidobacterium* lineages could be demonstrated. A major limitation of the method might be the dilution of the labeled substrate before its incorporation (Radajewski et al., 2000) because this will reduce the proportion of ^{13}C-labeled DNA. A further limitation of this approach might be the DNA synthesis rate in situ, which reflects the replication of bacterial cells. Because RNA synthesis rates are higher than those of DNA, Manefield et al. (2002) proposed that RNA could serve as a more responsive biomarker for the stable isotope labeling. In the study by Manefield et al. (2002), ^{13}C-labeled phenol was added to a phenol-degrading microbial community in an aerobic bioreactor. Stable isotope-labeled RNA was obtained after density gradient centrifugation and analysis of reverse-transcribed RNA by DGGE. Sequencing of a dominant band revealed that a dominant member of the microbial community of the reactor belonged to the genus *Thauera*.

In Situ Hybridization of Whole Cells

FISH allows the study of complex microbial communities at the level of individual cells. The strategy is to fix cells in their environmental setting and to hybridize with specific fluorescently labeled oligonucleotide probes targeting the 16S or 23S rRNA gene and to quantify the fluorescence signal by image processing. This technique offers the unique chance to study the composition and spatial organization of bacterial communities in situ. When used in combination with flow cytometry, quantitative information on the relative abundance of certain bacteria can be obtained. FISH has become an important tool to monitor bacteria, e.g., in biofilms, in sewage flocs, or plant-associated bacteria. FISH studies have revealed that surface-associated bacterial communities are highly structured and that, within microcolonies, cells with various different metabolic states might coexist. The use of FISH in combination with microsensors can provide information on chemical gradients and the phylogeny of bacteria. Ramsing et al. (1993) were the first who used this combination to study the vertical distribution of sulfate-reducing bacteria in biofilms of

trickling filters of a sewage plant. They showed that hydrogen sulfide and sulfate-reducing bacteria were detectable in the anoxic zone of the biofilm. In another study of nitrifying communities in biofilms, the participation of three different ammonium-oxidizing populations with different spatial localization in the biofilm was revealed (Gieseke et al., 2001). Recently, a combination of FISH with rRNA-targeted oligonucleotide probes and MAR has been suggested as a new tool to link structure and function in microbial ecology studies (Lee et al., 1999). This method can be used to follow the in vivo uptake of radioactively labeled substrates in microbial communities characterized by rRNA-targeted probes. When MAR was applied to the study of the activity and phylogenetic affiliation of iron reducers in activated sludge, it was found that all cells able to consume [^{14}C]acetate under iron-reducing conditions were targeted by the eubacterial probe EUB338, but only 30% of these were hybridized with the probes tested, leaving the majority unidentified (Nielsen et al., 2002). A combination of FISH and MAR was also applied to the in situ characterization of *Nitrosospira*-like nitrite-oxidizing bacteria in wastewater (Daims et al., 2001). However, when complex bacterial communities from environmental samples are analyzed by FISH with rRNA-targeted probes, several technical problems and potential artifacts might be encountered. Bacteria in less-nutrient-rich environments have a low ribosome content, which affects the sensitivity of detection. A prerequisite for hybridization of labeled oligonucleotide with rRNA is that the different kinds of cells be sufficiently permeable to allow the probe to enter the cell. Thus, insufficient permeability of bacterial cells as well as in situ accessibility of the rRNA can be a limiting factor. The accessibility of probe target sites of the rRNA varies considerably because of the higher-order structure of the ribosome (Fuchs et al., 2001).

Marker and Reporter Genes

Marker gene technology has become an extremely valuable tool in microbial ecology to study the spatial distribution of marked bacteria and their metabolic activity and to follow gene transfer (Stoltzfus et al., 2000). The most commonly used marker or reporter genes are those that code for the green fluorescent protein (GFP), firefly luciferase, bacterial luciferase, β-galactosidase, β-glucuronidase, or catechol 2,3-dioxygenase (Leveau and Lindow, 2002). Marker genes expressed from a constitutive promoter can be used as a kind of tagging for bacteria that allows their fate and spatial distribution in complex microbial communities to be followed. The marker gene present in an environmental sample can be detected independently from cultivation by PCR from total community DNA or RNA. Furthermore, the proteins encoded by the different reporter genes allow easy and specific identification of tagged bacteria that are able to form colonies on solid nutrient media.

However, most exciting is the possibility of following tagged bacteria in situ. Very few reporter genes enable in situ observation. For in situ detection, GFP is most frequently used because of its easy detection in individual cells. The polypeptide chain contains a fluorochrome that can be excited with blue light without the requirement of externally added substrates. The in situ observation of GFP-tagged single cells by microscopy allows insights in the spatial context. A disadvantage of the wild-type GFP is its high stability in bacteria, which makes conclusions on the actual metabolic state difficult. Thus, the development of new variants of GFP with reduced stability was a prerequisite for using this reporter to monitor growth rates (Andersen et al., 1998). RNA synthesis rates could be used as an indicator of growth activity (Molin and Givskov, 1999). When the reporter gene is fused to an environmentally or metabolically responsive promoter in situ, studies on changes of the respective stimulus can be done. The application of unstable GFP variants has shown how heterogeneous the metabolic state of cells within microcolonies in biofilms or of plant-associated cells is and how relevant a micrometer distance is for perceiving certain environmental stimuli by the microbe. To monitor growth activities in biofilms, the reporter gene coding for instable GFP was used in fusion with an rRNA promoter. The use of GFP-tagged rhizosphere bacteria has uncovered not only the spatial localization of plant-associated bacteria but also their metabolic state. Studies with GFP-tagged *Pseudomonas fluorescens* in the rhizosphere of barley showed that only a minor part of the population is active while the majority of cells have properties typical for starved cells (Normander et al., 1999). For in situ monitoring during colonization of wheat, Unge and Jansson (2001) used the biocontrol strain *P. fluorescens* SBW25 tagged with GFP *luxAB*. In contrast to the findings of Normander et al. (1999), *P. fluorescens* SBW25 was found to be metabolically active on all parts of the plant and occurred primarily on the seeds.

CULTIVATION-INDEPENDENT STUDY OF THE HORIZONTAL GENE POOL

Only recently, the importance of horizontal gene exchange for bacterial adaptation and for suc-

cessful colonization of ecological niches has been fully appreciated. Currently, mobile genetic elements (MGEs) are recognized as an important and essential component that promotes bacterial diversity. The lack of information on the distribution of MGEs such as plasmids in environmental bacteria is partly due to the fact that only a minor proportion of bacteria are accessible to cultivation techniques. The development of tools for studying the prevalence and transfer of MGEs independently from the culturability of their hosts is an important prerequisite to gain a more complete picture of the horizontal gene pool. The PCR-based detection of MGE-specific sequences in community DNA was first used by Götz et al. (1996). Primers targeting replicon-specific sequences were designed on the basis of sequenced broad-host-range (BHR) plasmids. In combination with Southern blot hybridization, a specific and sensitive monitoring of large numbers of environmental samples became possible. Direct capturing of MGEs by means of so-called exogenous plasmid isolation techniques (Bale et al., 1988; Hill et al., 1992) and the PCR-based amplification of MGEs have allowed researchers to gain new insights into the spread of MGEs in bacteria originating from different environments. MGEs conferring selectable traits such as mercury or antibiotic resistance have been acquired from a wide range of environmental samples in gram-negative recipients functioning as a genetic sink. The capture of degradative genes resident on MGEs has been demonstrated as well (Top et al., 1995). In several of the studies, increased transfer frequencies were observed when the environmental sample was previously exposed to pollutants. Many of the plasmids captured in biparental matings displayed a broad host range (BHR plasmids) and the capacity to mobilize $tra^- mob^+$ MGEs. In addition to IncP-1 plasmids, which were frequently obtained with the exogenous isolation approach, new classes of BHR plasmids have been identified. The PCR-based detection of MGEs has been used for monitoring a wide range of environments for the incidence of MGEs, and hot spots with high abundance of MGEs could be identified.

The group of S. Molin (Denmark) developed zygotic induction systems that allow online tracking of plasmid transfer in mixed communities (Christensen et al., 1996, 1998). The principle is simple: the plasmid under investigation is marked with a GFP reporter gene that is repressed in the donor cell but will be expressed in transconjugant cells. With this approach it was shown that plasmid transfer in biofilms occurred only between actively growing cells. Nutrient diffusion gradients in biofilms seem to allow active cell growth and plasmid transfer only on the sur-

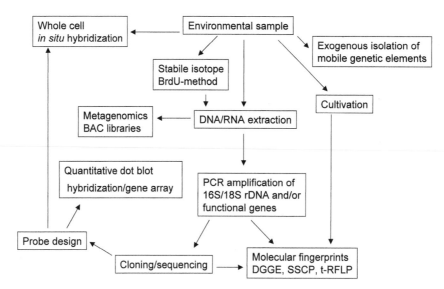

Figure 3. A polyphasic approach is required for the analysis of microbial communities.

face of microcolonies. This system allowed Dahlberg et al. (1998) to monitor transfer of the plasmid pBF1::*gfp* from *P. putida* to the total marine bacterial community directly in seawater. A variety of morphologically different indigenous cells exhibited green fluorescence, which indicated that these cells had received the plasmid, confirming the broad host range of pBF1. Considering the low cell densities in seawater, the extent of gene transfer observed was quite astonishing. Sequencing of newly isolated MGEs is an important basis for the development of probes and primers as well as for the use of gene arrays and proteomics. Recently, Smalla and Sobecky (2002) proposed that a polyphasic approach should be taken to assess the horizontal gene pool.

CONCLUSION AND OUTLOOK

The field of cultivation-independent microbiology has rapidly advanced over the past few years. The technological developments briefly described in this chapter have brought us the opportunity to study the enormous complexity of natural microbial communities in more comprehensive and complete terms. The use of molecular tools has dramatically changed our view of the microbial world. Microbial taxonomy has been revolutionized (Pace, 1997). However, most importantly, it has become possible to study microbes in their natural settings. Microbial community structures are studied using multiphasic approaches by combining various methods described here (see Fig. 3). Although the kinds of bacterial populations present in an environmental sample are still explored

best by the cloning of 16S rDNA genes or other phylogenetic markers, the temporal and spatial distribution of ribotypes can be followed best by molecular fingerprints. Information on the localization of respective ribotypes and their metabolic activities can be provided by whole-cell in situ hybridization. In addition, reporter genes are a powerful tool to study how microbes perceive their surroundings and how their metabolic activity relates to their spatial distribution. High-density DNA arrays that will allow monitoring gene content and expression—although still a methodological challenge—will provide new insights into complex microbial communities by linking information on structure and function. The advances of genomics strongly affect our understanding of microbes. In the future, advanced protein detection methods will become more important in addition to gene arrays in cultivation-independent microbiology.

REFERENCES

Alm, E. W., and D. A. Stahl. 2000. Critical factors influencing the recovery and integrity of rRNA extracted from environmental samples: use of an optimized protocol to measure depth-related biomass distribution in freshwater sediments. *J. Microbiol. Methods* **40:**153–162.

Alm, E. W., D. Zheng, and L. Raskin. 2000. The presence of humic substances and DNA in RNA extracts affects hybridization results. *Appl. Environ. Microbiol.* **66:**4547–4554.

Amann, R. I., W. Ludwig, and K.-H. Schleifer. 1995. Phylogenetic identification and in situ detection of individual microbial cells without cultivation. *Microbiol. Rev.* **59:**143–169.

Andersen, J. B., C. Sternberg, L. K. Poulsen, S. P. Bjorn, M. Givskov, and S. Molin. 1998. New unstable variants of green fluorescent protein for studies of transient gene expression in situ. *Appl. Environ. Microbiol.* **64:**2240–2246.

Bakken, L. R., and V. Lindahl. 1995. Recovery of bacterial cells from soil, p. 13–27. *In* J. T. Trevors and J. D. Elsas (ed.), *Nucleic Acids in the Environment*. Springer-Verlag, Berlin, Germany.

Bale, M. J., M. J. Day, and J. C. Fry. 1988. Novel method for studying plasmid transfer in undisturbed river epilithon. *Appl. Environ. Microbiol.* **54**:2756–2758.

Boon, N., W. de Windt, W. Verstraete, and E. M. Top. 2002. Evaluation of nested PCR-DGGE (denaturing gradient gel electrophoresis) with group-specific 16S rRNA primers for the analysis of bacterial communities from different wastewater treatment plants. *FEMS Microbiol. Ecol.* **39**:101–112.

Borneman, J. 1999. Culture-independent identification of microorganisms that respond to specified stimuli. *Appl. Environ. Microbiol.* **65**:3398–3400.

Boschker, H. T. S., and J. J. Middelburg. 2002. Stable isotopes and biomarker in microbial ecology. *FEMS Microbiol. Ecol.* **1334**: 1–12.

Boschker, H. T. S., S. C. Nold, P. Wellsbury, D. Bos, W. de Graaf, R. Pel, R. J. Parkes, and T. E. Cappenburg. 1998. Direct linking of microbial populations to specific biogeochemical processes by ^{13}C-labelling of biomarkers. *Nature* **392**:801–804.

Christensen, B. B., C. Sternberg, and S. Molin. 1996. Bacterial plasmid conjugation on semi-solid surfaces monitored with the green fluorescent protein (Gfp) from *Aequorea victoria* as a marker. *Gene* **173**:59–65.

Christensen, B. B., C. Sternberg, J. B. Andersen, L. Eberl, S. Moeller, M. Givskov, and S. Molin. 1998. Establishment of new traits in a microbial biofilm community. *Appl. Environ. Microbiol.* **64**:2247–2255.

Dahlberg, C., M. Bergström, and M. Hermansson. 1998. In situ detection of high levels of horizontal plasmid transfer in marine bacterial communities. *Appl. Environ. Microbiol.* **64**:2670–2675.

Dahllöf, I., H. Baillie, and S. Kjelleberg. 2000. rpoB-based microbial community analysis avoids limitations inherent in 16S rRNA gene intraspecies heterogeneity. *Appl. Environ. Microbiol.* **66**:3376–3380.

Daims, H., J. L. Nielsen, P. H. Nielsen, K.-H. Schleifer, and M. Wagner. 2001. In situ characterization of *Nitrosospira*-like nitrite-oxidizing bacteria active in wastewater treatment plants. *Appl. Environ. Microbiol.* **67**:5273–5284.

Duarte, G. F., A. S. Rosado, L. Seldin, A. C. Keijzer-Wolters, and J. D. van Elsas. 1998. Extraction of ribosomal RNA and genomic DNA from soil for studying the diversity of the indigenous microbial community. *J. Microbiol. Method.* **32**:21–29.

Duineveld, B. M., G. A. Kowalchuk, A. Keijzer, J. D. van Elsas, and J. A. van Veen. 2001. Analysis of bacterial communities in the rhizosphere of chrysanthemum via denaturing gradient gel electrophoresis of PCR-amplified 16S rRNA as well as DNA fragments coding for 16S rRNA. *Appl. Environ. Microbiol.* **67**:172–178.

Fægri, A., V. L. Torsvik, and J. Goksøyr. 1977. Bacterial and fungal activities in soil: separation of bacteria and fungi by a rapid fractionated centrifugation technique. *Soil Biol. Biochem.* **9**:105–112.

Felske, A., A. Wolterink, R. van Lis, and A. D. L. Akkermans. 1998. Phylogeny of the main bacterial 16S rRNA sequences in Drentse A grassland soils (The Netherlands). *Appl. Environ. Microbiol.* **64**:871–879.

Fuchs, B. M., K. Syutsubo, W. Ludwig, and R. Amann. 2001. *In situ* accessibility of *Escherichia coli* 23S rRNA to fluorescently labeled probes. *Appl. Environ. Microbiol.* **67**:961–968.

Gieseke, A., U. Purkhold, M. Wagner, R. Amann, and A. Schramm. 2001. Community structure and activity dynamics of nitrifying bacteria in a phosphate-removing biofilm. *Appl. Environ. Microbiol.* **67**:1351–1362.

Gillespie, D. E., S. F. Brady, A. D. Bettermann, N. P. Cianciotto, M. R. Liles, M. R. Rondon, J. Clardy, R. M. Goodman, and J. Handelsman. 2002. Isolation of antibiotics turbomycin A and B from a metagenomic library of soil microbial DNA. *Appl. Environ. Microbiol.* **68**:4301–4306.

Giovannoni, S. J., T. B. Britschgi, C. L. Moyer, and K. G. Field. 1990. Genetic diversity in Sargasso Sea bacterioplankton. *Nature* **345**:60–63.

Gomes, N. C. M., H. Heuer, J. Schönfeld, R. Costa, L. Hagler-Mendonca, and K. Smalla. 2001. Bacterial diversity of the rhizosphere of maize *(Zea mays)* grown in tropical soil studied by temperature gradient gel electrophoresis. *Plant Soil* **232**: 167–180.

Götz, A., R. Pukall, E. Smit, E. Tietze, R. Prager, H. Tschäpe, J. D. van Elsas, and K. Smalla. 1996. Detection and characterization of broad-host-range plasmids in environmental bacteria by PCR. *Appl. Environ. Microbiol.* **62**:2621–2628.

Griffiths, R. I., A. S. Whiteley, A. G. O'Donnell, and M. J. Bailey. 2000. Rapid method for coextraction of DNA and RNA from natural environments for analysis of ribosomal DNA- and rRNA-based microbial community composition. *Appl. Environ. Microbiol.* **66**:5488–5491.

Heuer, H., R. M. Kroppenstedt, J. Lottmann, G. Berg, and K. Smalla. 2002. Effects of T4 lysozyme release from transgenic potato roots on bacterial rhizosphere communities are negligible relative to natural factors. *Appl. Environ. Microbiol.* **68**: 1325–1335.

Heuer, H., M. Krsek, P. Baker, K. Smalla, and E. M. H. Wellington. 1997. Analysis of actinomycete communities by specific amplification of genes encoding 16S rRNA and gel-electrophoretic separation in denaturing gradients. *Appl. Environ. Microbiol.* **63**:3233–3241.

Heuer, H., G. Wieland, J. Schönfeld, A. Schönwälder, N. C. M. Gomes, and K. Smalla. 2001. Bacterial community profiling using DGGE or TGGE analysis, p. 177–190. *In* P. Rouchelle (ed.), *Environmental Molecular Microbiology: Protocols and Applications*. Horizon Scientific Press, Wymondham, United Kingdom.

Hill, K. E., A. J. Weightman, and J. C. Fry. 1992. Isolation and screening of plasmids from the epilithon which mobilize recombinant plasmid pD10. *Appl. Environ. Microbiol.* **58**:1292–1300.

Holben, W. E., and D. Harris. 1995. DNA-based monitoring of total bacterial community structure in environmental samples. *Mol. Ecol.* **4**:627–631.

Holben, W. E., J. K. Jansson, B. K. Chelm, and J. M. Tiedje. 1988. DNA probe method for the detection of specific microorganisms in the soil bacterial community. *Appl. Environ. Microbiol.* **54**:703–711.

Hopkins, D. W., S. J. Macnaughton, and A. G. O'Donnell. 1991. A dispersion and differential centrifugation technique for representatively sampling microorganisms from soil. *Soil Biol. Biochem.* **23**:217–225.

Hugenholtz, P., B. M. Goebel, and N. R. Pace. 1998. Impact of culture-independent studies on the emerging phylogenetic view of bacterial diversity. *J. Bacteriol.* **180**:4765–4774.

Hurt, R. A., X. Qiu, L. Wu, Y. Roh, A. V. Palumbo, J. M. Tiedje, and J. Zhou. 2001. Simultaneous recovery of RNA and DNA from soils and sediments. *Appl. Environ. Microbiol.* **67**:4495–4503.

Klappenbach, J. A., J. M. Dunbar, and T. M. Schmidt. 2000. rRNA operon copy number reflects ecological strategies of bacteria. *Appl. Environ. Microbiol.* **66**:1328–1333.

Koizumi, Y., J. J. Kelly, T. Nakagawa, H. Urakawa, S. El-Fantroussi, S. Al-muzaini, M. Fukui, Y. Urushigawa, and D. Stahl. 2002. Parallel characterization of anaerobic toluene- and ethylbenzene-degrading microbial consortia by PCR-denaturing gradient gel electrophoresis, RNA-DNA membrane hybridization, and DNA microarray technology. *Appl. Environ. Microbiol.* **68:**3215–3225.

Lee, N., P. H. Nielsen, K. H. Andreasen, S. Juretschko, J. L. Nielsen, K.-H. Schleifer, and M. Wagner. 1999. Combination of fluorescent in situ hybridization and microautoradiography—a new tool for structure-function analyses in microbial ecology. *Appl. Environ. Microbiol.* **65:**1289–1297.

Leveau, J. H. J., and S. E. Lindow. 2002. Bioreporters in microbial ecology. *Curr. Opin. Microbol.* **5:**259–265.

Liesack, W., and P. F. Dunfield. 2002. Biodiversity in soils: use of molecular methods for its characterization, p. 528–544. *In* G. Bitton (ed.), *Encyclopedia of Environmental Microbiology.* Wiley & Sons Inc., New York, N. Y.

Liu, W.-T., T. L. Marsh, H. Cheng, and L. J. Forney. 1997. Characterization of microbial diversity by determining terminal restriction fragment length polymorphisms of genes encoding 16S rRNA. *Appl. Environ. Microbiol.* **63:**4516–4522.

Loy, A., A. Lehner, N. Lee, J. Adamczyk, H. Meier, J. Ernst, K.-H. Schleifer, and M. Wagner. 2002. Oligonucleotide microarray for 16S rRNA gene-based detection of all recognized lineages of sulfate-reducing prokaryotes in the environment. *Appl. Environ. Microbiol.* **68:**5064–5081.

Manefield, M., A. S. Whiteley, R. I. Griffiths, and M. J. Bailey. 2002. RNA stable isotope probing, a novel means of linking microbial community function to phylogeny. *Appl. Environ. Microbiol.* **68:**5367–5373.

McCaig, A. E., L. A. Glover, and J. Prosser. 1999. Molecular analysis of bacterial community structure and diversity in unimproved and improved upland grass pastures. *Appl. Environ. Microbiol.* **65:**1721–1730.

Miller, D. N., J. E. Bryant, E. L. Madsen, and W. C. Ghiorse. 1999. Evaluation and optimization of DNA extraction and purification procedures for soil and sediment samples. *Appl. Environ. Microbiol.* **65:**4715–4724.

Molin, S., and M. Givskov. 1999. Application of molecular tools for in situ monitoring of bacterial growth activities. *Environ. Microbiol.* **1:**383–391.

Moré, M. I., J. B. Herrick, M. C. Silva, W. C. Ghiorse, and E. L. Madsen. 1994. Quantitative cell lysis of indigenous microorganisms and rapid extraction of microbial DNA from sediment. *Appl. Environ. Microbiol.* **60:**1572–1580.

Muyzer, G., and K. Smalla. 1998. Application of denaturing gradient gel electrophoresis (DGGE) and temperature gradient gel electrophoresis (TGGE) in microbial ecology. *Antonie Leeuwenhoek* **73:**127–141.

Muyzer, G., E. C. de Waal, and A. G. Uitterlinden. 1993. Profiling of complex microbial populations by denaturing gradient gel electrophoresis analysis of polymerase chain reaction-amplified genes encoding for 16S rRNA. *Appl. Environ. Microbiol.* **59:**695–700.

Nielsen, J. L., S. Juretschko, M. Wagner, and P. H. Nielsen. 2002. Abundance and phylogenetic affiliation of iron reducers in activated sludge as assessed by fluorescence in situ hybridization and microautoradiography. *Appl. Environ. Microbiol.* **68:**4629–4636.

Nogales, B., E. R. B. Moore, E. Llobet-Brossa, R. Rossello-Mora, R. Amann, and K. N. Timmis. 2001. Combined use of 16S ribosomal DNA and 16S rRNA to study the bacterial community of polychlorinated biphenyl-polluted soil. *Appl. Environ. Microbiol.* **67:**1874–1884.

Nogales, B., K. N. Timmis, D. B. Nedwell, and A. M. Osborn. 2002. Detection and diversity of expressed denitrification genes in estuarine sediments after reverse transcriptase-PCR amplification from mRNA. *Appl. Environ. Microbiol.* **68:**5017–5025.

Normander, B., N. B. Hendriksen, and O. Nybroe. 1999. Green fluorescent protein-marked *Pseudomonas fluorescens:* localiza-

tion, viability, and activity in the natural barley rhizosphere. *Appl. Environ. Microbiol.* **65:**4646–4651.

Nübel, U., B. Engelen, A. Felske, J. Snaidr, A. Wiesenhuber, R. I. Amann, W. Ludwig, and H. Backhaus. 1996. Sequence heterogeneities of genes encoding 16S rRNAs in *Paenibacillus polymyxa* detected by temperature gradient gel electrophoresis. *J. Bacteriol.* **178:**5636–5643.

Ogram, A., G. S. Sayler, and T. J. Barkay. 1987. DNA extraction and purification from sediments. *J. Microbiol. Methods* **7:**57–66.

Oliver, J. D. 2000. Problems in detecting dormant (VBNC) cells and the role of DNA elements in this response, p. 1–15. *In* J. K. Jansson, J. D. van Elsas, and M. J. Bailey (ed.), *Tracking Genetically-Engineered Microorganisms.* Eurekah, Austin, Tex.

Osborn, A. M., E. R. B. Moore, and K. N. Timmis. 2000. An evaluation of terminal-restriction fragment length polymorphism (T-RFLP) analysis for the study of microbial community structure and dynamics. *Environ. Microbiol.* **2:**39–50.

Pace, N. R. 1997. A molecular view of microbial diversity and the biosphere. *Science* **276:**734–740.

Pace, N. R., D. A. Stahl, D. L. Lane, and G. J. Olsen. 1986. The analysis of natural microbial populations by rRNA sequences. *Adv. Microb. Ecol.* **9:**1–55.

Radajewski, S., P. Ineson, N. R. Parekh, and J. C. Murrell. 2000. Stable-isotope probing as a tool in microbial ecology. *Nature* **403:**646–649.

Ramsing, N. B., M. Kühl, and B. B. Jørgensen. 1993. Distribution of sulfate-reducing bacteria, O_2, and H_2S in photosynthetic biofilms determined by oligonucleotide probes and microelectrodes. *Appl. Environ. Microbiol.* **59:**3840–3849.

Rondon, M. R., P. R. August, A. D. Bettermann, S. F. Brady, T. H. Grossman, M. R. Liles, K. A. Loiacono, B. A. Lynch, I. A. MacNeil, M. S. Osburne, J. Clardy, J. Handelsman, and R. M. Goodman. 2000. Cloning the soil metagenome: a strategy for accessing the genetic and functional diversity of uncultured microorganisms. *Appl. Environ. Microbiol.* **66:**2541–2547.

Roszak, D. B., and R. R. Colwell. 1987. Survival strategies of bacteria in the natural environment. *Microbiol. Rev.* **51:**365–379.

Schwieger, F., and C. C. Tebbe. 1998. A new approach to utilize PCR-single strand-conformation polymorphism for 16S rRNA gene-based microbial community analysis. *Appl. Environ. Microbiol.* **64:**4870–4876.

Schwieger, F., and C. C. Tebbe. 2000. Effect of field inoculation with *Sinorhizobium meliloti* L33 on the composition of bacterial communities in rhizospheres of a target plant (*Medicago saliva*) and a non-target plant (*Chenopodium album*)-linking of 16S rRNA gene-based single-strand conformation polymorphism community profiles to the diversity of cultivated bacteria. *Appl. Environ. Microbiol.* **66:**3556–3565.

Smalla, K., and P. Sobecky. 2002. The prevalence and diversity of mobile genetic elements in environmental bacteria assessed with new tools. *FEMS Microbiol. Ecol.* **42:**165–175.

Smalla, K., G. Wieland, A. Buchner, A. Zock, J. Parzy, S. Kaiser, N. Roskot, H. Heuer, and G. Berg. 2001. Bulk and rhizosphere soil bacterial communities studied by denaturing gradient gel electrophoresis: plant-dependent enrichment and seasonal shifts revealed. *Appl. Environ. Microbiol.* **67:**4742–4751.

Staley, J. T., and A. Konopka. 1985. Measurement of *in situ* activities of nonphotosynthetic microorganisms in aquatic and terrestrial habitats. *Annu. Rev. Microbiol.* **39:**321–346.

Stoltzfus, J. R., J. K. Jansson, and F. J. de Bruijn. 2000. Using green fluorescent protein (GFP) as biomarker or bioreporter for bacteria, p. 101–116. *In* J. K. Jansson, J. D. van Elsas, and M. J. Bailey (ed.), *Tracking Genetically-Engineered Microorganisms.* Eurekah, Austin, Tex.

Tebbe, C. C., and W. Vahjen. 1993. Interference of humic acids and DNA extracted directly from soil in detection and transformation of recombinant DNA from bacteria and a yeast. *Appl. Environ. Microbiol.* **59:**2657–2665.

Tebbe, C. C., A. Schmalenberger, S. Peters, and F. Schwieger. 2001. Single-strand conformation polymorphism (SSCP) for microbial community analysis, p. 161–175. *In* P. Rouchelle (ed.), *Environmental Molecular Microbiology: Protocols and Applications.* Horizon Scientific Press, Wymondham, United Kingdom.

Top, E. M., W. E. Holben, and L. J. Forney. 1995. Characterization of diverse 2,4-dichlorophenoxyacetic acid-degradative plasmids isolated from soil by complementation. *Appl. Environ. Microbiol.* **61:**1691–1698.

Torsvik, V. 1980. Isolation of bacterial DNA from soil. *Soil Biol. Biochem.* **12:**15–21.

Torsvik, V., J. Goksøyr, and F. L. Daae. 1990. High diversity in DNA of soil bacteria. *Appl. Environ. Microbiol.* **56:**782–787.

Unge, A., and J. Jansson. 2001. Monitoring population size, activity, and distribution of *gfp-luxAB*-tagged *Pseudomonas fluorescens* SBW25 during colonization of wheat. *Microbiol. Ecol.* **41:**290–300.

Urbach, E., K. L. Vergin, and S. J. Giovannoni. 1999. Immunochemical detection and isolation of DNA from metabolically active bacteria. *Appl. Environ. Microbiol.* **65:**1207–1213.

Van Elsas, J. D., K. Smalla, and C. C. Tebbe. 2000. Extraction and analysis of microbial community nucleic acids from environmental matrices, p. 29–51. *In* J. K. Jansson, J. D. van Elsas, and

M. J. Bailey (ed.), *Tracking Genetically-Engineered Microorganisms.* Eurekah, Austin, Tex.

von Wintzingerode, F., U. B. Göbel, and E. Stackebrandt. 1997. Determination of microbial diversity in environmental samples: pitfalls of PCR-based rRNA analysis. *FEMS Microbiol. Rev.* **21:**213–229.

Ward, D. M., M. M. Bateson, R. Weller, and A. L. Ruff-Roberts. 1990. 16S rRNA sequences reveal numerous microorganisms in a natural community. *Nature* **345:**63–65.

Weinbauer, M. G., I. Fritz, D. F. Wenderoth, and M. G. Höfle. 2002. Simultaneous extraction from bacterioplankton of total RNA and DNA suitable for quantitative structure and function analyses. *Appl. Environ. Microbiol.* **68:**1082–1087.

Woese, C. R. 1987. Bacterial evolution. *Microbiol. Rev.* **51:**221–271.

Wu, L., D. K. Thompson, G. Li, R. A. Hurt, J. M. Tiedje, and J. Zhou. 2001. Development and evaluation of functional gene arrays for detection of selected genes in the environment. *Appl. Environ. Microbiol.* **67:**5780–5790.

Yang, C.-H., and D. E. Crowley. 2000. Rhizosphere microbial community structure in relation to root location and plant iron nutritional status. *Appl. Environ. Microbiol.* **66:**345–351.

Yin, B., D. Crowley, G. Sparovek, W. J. de Melo, and J. Borneman. 2000. Bacterial functional redundancy along a soil reclamation gradient. *Appl. Environ. Microbiol.* **66:**4361–4365.

Microbial Diversity and Bioprospecting
Edited by Alan T. Bull
© 2004 ASM Press, Washington, D.C.

Chapter 10

Resuscitation of "Uncultured" Microorganisms

Douglas B. Kell, Galya V. Mukamolova, Christopher L. Finan, Hongjuan Zhao,
Royston Goodacre, Arseny S. Kaprelyants, and Michael Young

These germs—these bacilli—are transparent bodies. Like glass. Like water. To make them visible you must stain them. Well, my dear Paddy, do what you will, some of them won't stain; they won't take cochineal, they won't take any methylene blue, they won't take gentian violet, they won't take any colouring matter. Consequently, though we know as scientific men that they exist, we cannot see them.

Sir Ralph Bloomfield-Bonington.
The Doctor's Dilemma. George Bernard Shaw.

It is by now well known that the number, and probably the nature (Torsvik et al., 1990a, b, 1996), of microorganisms visible in—and whose activity may often be demonstrated within—a natural environmental or clinical sample is often orders of magnitude greater than the number of cells or propagules that can be isolated and brought into culture therefrom (e.g., Alvarez-Barrientos et al., 2000; Amann et al., 1995; Barer et al., 1993, 1998; Barer and Harwood, 1999; 1998; Biketov et al., 2000; Bogosian and Bourneuf, 2001; Bull et al., 2000; Dobrovol'skaya et al., 2001; Domingue and Woody, 1997; Fredricks and Relman, 1996; Gangadharam, 1995; Gao and Moore, 1996; Head et al., 1998; Kaprelyants et al., 1993, 1996, 1994, 1999; Kaprelyants and Kell, 1993; Kell et al., 1998; Kell and Young, 2000; McDougald et al., 1998; Mukamolova et al., 1995a and 1995b; Postgate, 1976, 1969; Relman, 1999; Rondon et al., 1999; Schut et al., 1997; Smith et al., 2002; Tiedje and Stein, 1999; Votyakova et al., 1994; Watts et al., 1999; Wayne, 1994). What is less than clear, however, is whether these ostensibly "unculturable" cells have permanently lost culturability (i.e., are effectively dead), are killed by (or simply unable to grow on) our standard isolation media, or are in a dormant state from which we might recover them if only we

knew how (Barer et al., 1998; Kaprelyants et al., 1993, 1999).

Two important definitions are immediately necessary. First, as noted by Postgate (1976, 1969), we always equate viability and culturability. The consequences of this are at least twofold: (i) phrases such as "viable but not culturable" are to be seen as an oxymoron (Barer et al., 1993, 1998; Barer and Harwood, 1999), and (ii) the definition means that a property such as viability or culturability is not an innate property of a microbe but an operational definition or property. In other words, the (apparent) property of a microbial cell of interest depends not only on the cell itself but on the experimental manipulations we perform to assess its state, as with the Schrödinger's cat paradox (Kell et al., 1998; Primas, 1981). Because we have reviewed this elsewhere at some length recently (Barer et al., 1998; Kell et al., 1998), and it is likely that most or all cells termed viable but not culturable are in fact simply dead or at least irreversibly nonculturable (Bogosian, et al., 2000, 1998; Bogosian and Bourneuf, 2001; Nystrom, 2001; Smith et al., 2002), we do not discuss this specific aspect further. The second definition is that of dormancy, which we define (Kaprelyants et al., 1993) as a reversible state of low metabolic activity, in which cells can persist for extended periods without division. This often corresponds to a state in which cells are not "alive" in the sense of being able to form a colony when plated on a suitable solid medium, but one in which they are not "dead" in that, when conditions are more favorable, they can revert to a state of "aliveness" as so defined. However, the mycobacterial literature refers to a state of latency in which cells also persist for extended periods without net multiplication and possibly without division (Chaisson, 2000; Domingue and Woody, 1997; Flynn and Chan, 2001; Parrish et al., 1998;

Douglas B. Kell • Department of Chemistry, Faraday Building, Sackville Street, UMIST, P.O. Box 88, Manchester M60 IQD, United Kingdom. **Galya V. Mukamolova, Christopher L. Finan, Hongjuan Zhao, Royston Goodacre, and Michael Young** • Institute of Biological Sciences, University of Wales, Aberystwyth, Aberystwyth SY23 3DD, United Kingdom. **Arseny S. Kaprelyants** • Bakh Institute of Biochemistry, Leninskii Prospekt 33, 117071 Moscow, Russia.

Phyu et al., 1998; van Pinxteren et al., 2000; Wayne, 1994). Such cells may or may not be dormant by the above definition, but there is evidence that they are at least metabolically active (Höner zu Bentrup and Russell, 2001; McKinney et al., 2000). In particular, we note here the important incorporation of the adjective "reversible" in the definition of dormant and dormancy. Finally, we distinguish our use of the phrase "uncultured" organism (when we refer to a microbial strain that can be detected in a natural environment, usually by molecular means, but that is not yet cultured) from "nonculturable" organism (which refers to an organism that has been cultured in a laboratory but that has entered a physiological state in which it is incapable of growth in conditions that normally support its growth).

Because this review is about the resuscitation of uncultured or nonculturable microbes, and such microbes that are successfully resuscitated must by definition have been dormant or latent, we concentrate on this issue, particularly with reference to the actinobacteria that are the source of most of the bioactive secondary metabolites of industrial or applied interest.

LOSS OF CULTURABILITY IN LABORATORY CULTURES—BASIC EXPERIMENTAL ISSUES

Consider the technically undemanding but intellectually rather interesting experiment (taken from Kaprelyants and Kell, 1993) shown in Fig. 1. Part A shows the loss of culturability of *Micrococcus luteus* as a decrease in the plate count at more or less constant total count. On the basis of this type of observation alone, we do not know whether the cells that have lost culturability are dormant or dead. Part B shows what at first sight appears to be the resuscitation of most of these previously unculturable cells (which could not form a colony in a plate count assay) as an increase in culturable (plate) count. However, that fact alone still does not in fact allow one to claim that those ostensibly resuscitating cells were dormant. This is because the noise on the total count (let us charitably say ±10%) is such that the increase in plate count between 32 and 55 h is entirely within the noise of the total (microscopic) count and thus could easily have been due to regrowth (multiplication) of the cells that were already culturable at 22 h; indeed, by the last data point at 58 h most or all of the cells in the culture appear to have initiated regrowth. The clear conclusion is that the presence initially of more than one physiological class of cell confounds the simple analysis based on culturable and total counts alone.

The additional evidence that suggests that the increase in culturable count during the period 32 to 55 h in Fig. 1B is due to resuscitation comes from several sources. First, the kinetics of the increase in culturable count are far more rapid than the known doubling time of *M. luteus* in this medium; the increase in culturable cell count could not be due to growth of the initially viable cells. Second, the fraction of cells that can resuscitate (as seen, for example, at 55 h) before a measurable increase in plate count is about 30% of the total, i.e., significantly greater than the noise in the total cell count. Third, the morphology of the different physiological types of cells allows them to be discriminated, most easily by size; the increase in large cells is exactly matched by the decrease in small

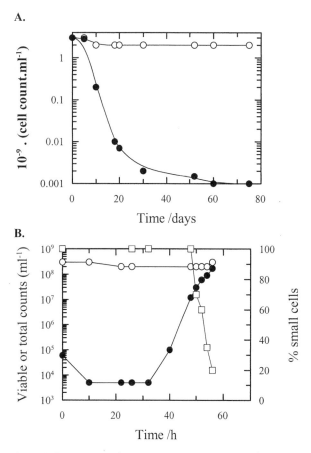

Figure 1. Dormancy and resuscitation in *Micrococcus luteus*. (A) Changes in the viable and total counts of *M. luteus* grown in batch culture and subjected to starvation. Cells were grown and starved, and the total (microscopic) and viable (plate) counts were measured as described by Kaprelyants and Kell (1993). Time zero corresponds to the onset of the stationary phase. (B) Changes in viable and total counts and the percentage of small cells during resuscitation of a starved culture of *M. luteus*. Cells were starved for 75 days, incubated with penicillin G for 10 hours, washed, and resuscitated by the addition of growth medium as described by Kaprelyants and Kell (1993). Total counts, open circles; viable counts, closed circles; percentage of small cells (<0.5 μm in diameter), squares.

cells as the small cells resuscitate (rather than there simply being an increase in the number of large cells).

However, the crucial piece of evidence comes from the use of the most-probable-number (MPN) technique. In the MPN technique, samples of cells are diluted seriatim into a broth that supports their multiplication and whether there is growth is subsequently scored via turbidity. The pattern observed can be compared with the (Poisson) distribution expected from standard tables, and the MPN of cells originally present is then calculated. The advantage of this approach is that the properties of the cells can be determined individually, without the influence of any other cells (or activators or inhibitors) that might be present in the initial medium (Smith et al., 2002). In the system shown in Fig. 1, it was expected that a dormant cell would score as culturable in the MPN assay, as the liquid medium into which the samples were diluted—which was the same as that employed in the previous experiment and identical in each case—should have permitted any dormant cells (ostensibly the vast majority) to resuscitate and then grow. The curious finding, however, was that it did not (Kaprelyants et al., 1994), although the above evidence had already shown that such cells were indeed dormant (i.e., metabolically inactive) (Kaprelyants and Kell, 1993). The big difference in the experiments, however, was that in the experiment of Fig. 1B culturable cells were present during the resuscitation, but in the MPN experiments they were not. This led us to opine that the presence of the culturable cells was necessary not for their own (re)growth but for the production of a substance necessary for the resuscitation of the dormant cells (Kaprelyants et al., 1994; Votyakova et al., 1994). This could be (and was) tested simply by adding sterile supernatant from a culture of viable cells in the MPN experiment. For different cultures, the presence of appropriate concentrations of supernatant increased the culturable count of starved cells in an MPN assay by 3 to 5 orders of magnitude: the supernatant contained a resuscitation-promoting factor (Rpf) produced by culturable cells.

PHEROMONES

Pheromones are substances produced by an organism that have specific effects on other organisms of the same species; although the presence of pheromones in prokaryotes was not widely recognized at the time when the existence of Rpf was proposed (notwithstanding a prescient review by Stephens [1986]), this was a clear example of pheromonal activity. Our own short survey in 1995 (Kell et al., 1995) noted that there was a tendency for

molecules such as lactones to be used in gram-negative bacteria whereas gram-positive bacteria often used oligopeptides (often produced from larger precursor proteins). Although there are now many more examples, the basic trend remains unchanged (Fuqua and Greenberg, 1998; Kleerebezem and Quadri, 2001; Kleerebezem et al., 1997; Lazazzera, 2001; Lazazzera and Grossman, 1998).

RPF: A BACTERIAL CYTOKINE FAMILY AND ITS BIOLOGY

Given the resuscitation assay for the M. luteus Rpf, we were able to purify it to homogeneity and thus to characterize it (Mukamolova et al., 1998a). The first surprising fact about it was first that it turned out to be a protein of 220 amino acids (Molecular weight of 19,148.5) and, so far as is still known, the protein is not cleaved (apart from the removal of its signal sequence during secretion) to produce activity. In addition, the protein served not only to resuscitate dormant cells but was required for the growth of normal (viable) cells from which it can be removed by washing (Mukamolova et al., 1998a, 1999). This is the activity of a cytokine (Callard and Gearing, 1994; Heath, 1993) or proteinaceous growth factor. The Rpf was extremely potent, this version being active at picomolar concentrations (Mukamolova et al., 1998a, 1999) (and any such estimates are underestimates as not all molecules added will be active; in addition there is a cell-wall binding motif similar to that in lysM [Bateman and Bycroft, 2000] that may sequester it unproductively). Finally, recombinant Rpf cloned in Escherichia coli (a host lacking any genes coding for any Rpf homologs and also devoid of any background activity) showed that Rpf alone was the active substance (Kaprelyants et al., 1999; Mukamolova et al., 1998a).

There is a highly conserved Rpf domain (motif) of some 70 amino acids near the N terminus (Kaprelyants et al., 1999; Kell and Young, 2000) of the M. luteus Rpf (Fig. 2), and there are by now some 144 known homologs containing this domain (Table 1). The main evidence comes from systematic genome sequencing programs. All examples of proteins containing this domain come from the actinomycetes or actinobacteria, i.e., the high-G+C clade of gram-positive bacteria; indeed we know of no actinomycete that has fewer than one. M. luteus seems to have one homolog only (Mukamolova et al., 2002a) (not two as originally suspected [Mukamolova et al., 1998a]), whereas Mycobacterium tuberculosis (Cole et al., 1998) and Streptomyces coelicolor (Bentley et al., 2002) each have five homologs, one (encoded by

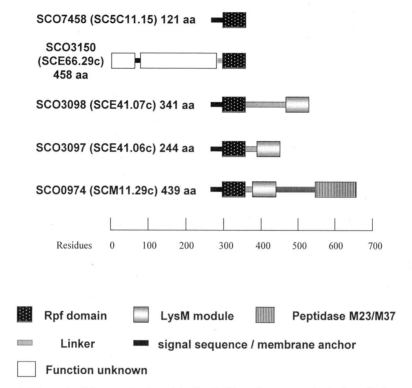

Figure 2. Schematic of the organization of the five Rpf homologs present in the *S. coelicolor* genome.

Rv1009 in *M. tuberculosis*) with a membrane anchor (Fig. 3).

Because the molecule is so potent, it follows (on thermodynamic grounds practically inevitably) (i) that it is the conserved Rpf domain that is responsible for activity and (ii) that there should be at least some level of interspecies cross reactivity. Both predictions are fulfilled (Kaprelyants et al., 1999; Mukamolova et al., 2002b; see also Freeman et al., 2002). In addition, Rpf can increase the culturable count of *M. tuberculosis* obtainable from macrophages ex vivo by several orders of magnitude (Biketov et al., 2000). Similar statements are true for the fast-growing *Rhodococcus rhodochrous* and the slow-growing *M. tuberculosis* in vitro, which adopted a small, coccoid

Table 1. Some organisms that have been shown to contain Rpf-like genes, with the number of homologs known to date

Organism	No. of genes
Micrococcus luteus	1
Corynebacterium glutamicum	2
C. diphtheriae	3
Mycobacterium tuberculosis	5
M. bovis	5
M. leprae	3
M. avium	4
Streptomyces coelicolor	5
Streptomyces avermitilis	6
Saccharopolyspora erythraea	4

morphology upon lengthy starvation and where cultures appeared to lose completely the ability to multiply on agar plates (Shleeva et al., 2002). All of the above facts make it clear that a basic tenet of microbiology—one cell, one culture—does not hold in its purest form because, in the absence of any cell-wall sequestration of Rpf (which is now obviously seen as potentially beneficial), a cell cannot propagate merely upon being placed in a medium that normally sustains its growth and reproduction (Kaprelyants and Kell, 1996).

The important biological questions to be asked of Rpf and its homologs include the following:

- Is it necessary for growth and/or resuscitation in vivo as well as in vitro? The answer to this appears to be in the affirmative, at least ex vivo for cells isolated from macrophages (Biketov et al., 2000), although the crucial and difficult truly in vivo experiments (e.g., in tissues obtained from infected hosts [see, e.g., Hernandez-Pando et al., 2000]) have yet to be performed.

- When is it expressed? Expression increases rapidly soon after inoculation of stationary phase cells into fresh growth medium, is maximal during lag and early exponential phase, decreases during late-exponential phase, and ceases in stationary phase (Mukamolova et al.,

```
SCO3150   VSRTQFEPAETYTAYTAYDEAYEAHEAYEAHEAYEAHETYERPAPYDLHSAPTLPYGGTY   60
SCO3150   PSAPPPVHEPTVPALPRQSDGRAERRRRARGAGRADASLRRLVPRALVVAFLAGGTTAFV   120
SCO3150   AKDKAVELTVDGSPRTLHTFADDVSELLAEEGVQVGAHDVIAPAPGTPLTSGEDVTVHHG   180
SCO3150   RPVLLTLDGHRRQVWTTAGTVAGALRQLGVRTQGAYLSTGPSRRIGREGLALVVRTERVV   240
SCO3150   TVMADGRTRTVRTNAATVGEVVEEAGITLRGEDTTSVPATGFPRDGQTVTVLRITGSQEV   300

MlutRpf   ------------------------------MDTMTLFTTSATRSRRATASIVA   23
SCO3150   REDPIPFDERRAEDASLFRGTEVVQEAGRPGLRRTTYALRTVNGVRQKPRRLRTEVVREP   360
SCO7458   ---------------------------MICRRNDRQSDVRGSRGRRIRTAAVTL   27
SCO3098   --------------------------MLSGNGRHRRPRQAPALVVAAG   22
SCO3097   --------------------------MLFSGKGKHRRPSKATRVIAVAG   23
SCO0974   --------------------------MAVRGRHRRYQPNRINRASLTVT   23

MlutRpf   GMTLAGAAAVGFSAPAQAATVDTWDRLAECESNGTWDINTGNGFYGGV-QFTLSSWQAVG   82
SCO3150   SPRIVRVGTRPRPASVHGADHLNWQGLAACESGGRADAVDPSGTYGGLYQFDSATWHGLG   420
SCO7458   VAATALGATGEAVAAPSAPLRTDWDAIAACESSGNWQANTGNGYYGGL-QFARSSWIAAG   86
SCO3098   VTGSAIAIPLLGATGAHAADSTNWDQVAECETGGAWSQNSGNGYYGGL-QLSQDAWEQYG   81
SCO3097   VTGAAVAAPLMAAGNASAATASEWDAVAQCESGGNWSINTGNGYYGGL-QFSASTWAAYG   82
SCO0974   AGGAGLALPLVGTGTAHAADAATWDKVAACESTDDWDINTGNGYYGGL-QFTQSTWEAFG   82
                             *  .* #*. * *  *.***.***. **   .*  *

MlutRpf   GEG---YPHQASKAEQIKRAEILQDLQGWGAWPLCSQKLGLTQADADAGDVDATEAAPVA   139
SCO3150   GEG---RPEDASAAEQTYRAQKLYVRSGADAWPHCGARLRE.................   458
SCO7458   GLKYAPRADLATRGEQIAVAERLARLQGMSAW-GCA.....................   121
SCO3098   GLDYAPSADQASRSQQIRIAEKIHASQGIAAWPTCGLLAGLGNGSGGTGDGSGAAGDGAS   141
SCO3097   GTQYASTADQASKSQQIQIAEKVLAGQGKGAWPVCGTGLSGAAYTGGGSEGSGSGSSEGS   142
SCO0974   GTRYAPRADLATREQQIAGAEKVLDTQGPGAWPVCSERAGLTRGGDPPDIRPAGSAAPAQ   142
          *            . *.. *. *... ** *** #

MlutRpf   VERTATVQRQSAADEAAAEQAAAAEQAVVAEAE-------------------------   172
SCO3098   EGSDASGEQDTTKSSESPATTETPESSQSSESSGSSETPESTSGASSSSPSPSSSPSSSD   201
SCO3097   QSQSSGSTAERSTEQKASRSAERPAAEPKAEKPAAKKKTVTTPTGKKV-----------   196
SCO0974   KTSDSVKDVQPQTTPQSRAGKAR----------------------------------   165

MlutRpf   -------------------------------------------------------   173
SCO3098   APSDGSSGASGDSSDGAGQSAKPDTSTESDPSGSAEPQGTEGSSGSGKHRGGSADEGATG   261
SCO3097   -------------------------------------------------------   196
SCO0974   -------------------------------------------------------   166

MlutRpf   ----------------------TIVVKSGDSLWTLANEYEVEGGWTALYEANKGAVS-   207
SCO3098   EGRTDPASGRHASRDGGEREAGDGRYVVRTGDSLWAIADSLDVDGGWHALYADNETVVGA   321
SCO3097   ------------------EKGDGEYKVVKGDTLSSIAEEHDVKGGWAKLFKLNDDIVD-   230
SCO0974   ----------------------MYTVVTGDTLSGIADTHEVRGGWQRLYEANRSAIGS   201

MlutRpf   DAAVIYVGQELVLPQA.............................   223
SCO3098   DPDHILPGQTLTVTGESGEK.........................   341
SCO3097   DADLIYPGQQLHLK...............................   244
SCO0974   DPDLILPGQRLSLRGQGTTRAPGAEAGRRQDEQQPQRDKQRQKQQDKQQKQHRKEQKQEQ   261

SCO0974   KQEPKQEPKEQKRQEQKQQEQRKAPKESSSDSGKAKAAGKATAHRAVVAPVDAATGTPYH   321
SCO0974   QAGSSWSKGYHTGVDFPVPTGTSVKSVADGRVVSAGWGGSYGYQVVVRHGDGRYSQYAHL   381
SCO0974   SAISVKSGQSVGVGQRLGRSGSTGNVTGPHLHFEVRTGPGFGSDVDPVAYLRAGGVRI..   439
```

Figure 3. Sequence and alignment of *M. luteus* Rpf with the five Rpf-like gene products found in *S. coelicolor*.

2002a). We do not yet know much about the relative controls on mRNA and protein stability and sequestration that will determine its persistence postproduction.

- Does inhibiting its activity inhibit growth of target actinobacteria? Yes, (polyclonal) antibodies raised against Rpf inhibit its activity, the inhibition being overcome by adding extra Rpf (Mukamolova et al., 2002a).

- What are the mechanisms by which actively growing cells (i) become dormant and (ii) resuscitate from the dormant state? The extent of our ignorance of these matters is well illustrated by an excellent recent review (Barer and Harwood, 1999) in which the authors suggest two possible mechanisms for the transition to dormancy (in *M. luteus*). On the one hand, it

might be an ordered developmental program (as is recognized for more obviously developmental processes such as sporulation [Errington, 1996; Losick and Dworkin, 1999] or the stationary phase in gram-negative organisms [Kjelleberg, 1993; Kolter et al., 1993]). On the other, it might simply be a gratuitous and graceful degradation from a state of normal activity and culturability, for which the loss of any number of different functions might be responsible (McDougald et al., 1998). Similarly, the return from dormancy to culturability (resuscitation) would involve either a reproducible and ordered program of gene expression or a more general and stochastic repair and recovery process. Thus far, we have little information about this. We strongly suspect that dormancy and resuscitation of *M. luteus* are indeed both active and programmed processes. This is suggested by the comparatively coherent timings of the loss and gain of metabolic and biochemical functions (as indicated, for example, by rhodamine 123 uptake) by individual cells in a bacterial population, as observed by flow cytometry (Kaprelyants and Kell, 1993; Mukamolova et al., 1995b).

- Is there a receptor or target for Rpf? Although this is an attractive and almost compelling assumption, nothing is yet known about this. The potency of the bioactivity observed means that possibly only a few molecules per cell are necessary for its activity, and novel methods of detection of binding (e.g., Haupts et al., 2000; Rudiger et al., 2001) may be required.

- What is the actual biochemical role of Rpf? This is as yet unknown. However, so far as the basic phenomenology is concerned, Rpf does have the ability to cause cells that are in what would normally be seen as a nutritionally adequate medium to multiply under conditions in which its absence does not (and which indeed may cause cells to die). In this sense it does indeed exhibit the behavior of a cytokine (Kaprelyants et al., 1999; Kell and Young, 2000; Mukamolova et al., 1998a and b) and as such may contribute to the process of bacterial cell-cycle progression (a topic about which we are also remarkably ignorant in gram-positive bacteria).

RESUSCITABILITY OF BACTERIA TAKEN FROM ENVIRONMENTAL SAMPLES

All of the foregoing leads to the view that it is at least reasonable that the uncultured status of actinobacteria from the environment may be due, at least in part, to the fact that they have indeed become dormant (i.e., only reversibly nonculturable) and thus might be resuscitable given the right nutritional conditions, including the presence of Rpf or an equivalent bioactivity. Certainly what evidence there is shows that most actinobacteria that are culturable will grow on the same general types of media as judged merely by nutritional composition, and in large measure the uncultured actinobacteria observable by molecular means are phylogenetically close enough to cultured clades (McVeigh et al., 1996). The problem, then, is not of culturability but of bringing these strains into culture, where, to paraphrase McLuhan (McLuhan and Fiore, 1971), the medium is

the message. Thus, there is evidence that the concentration of nutrients necessary to resuscitate is much lower than that needed for optimal growth (MacDonell and Hood, 1982; Mukamolova et al., 1998b; Shleeva et al., 2002), given that these organisms would normally be considered copiotrophs (Schut et al, 1997). Indeed, the shock of adding specific nutrients to a laboratory culture whose growth was previously limited by such nutrients can cause substrate-accelerated death (see Poindexter, 1987; Postgate, 1967).

AN EVOLUTIONARY CODA

If individual microbes make cell signals that can resuscitate other organisms, as we have shown, this raises a number of evolutionary issues. First, and most obvious, is what is in it for the producer organisms? Genetical kinship theory (Hamilton, 1963, 1964) shows that, provided the benefit of an altruistic action to the recipient times the degree of its genetic relatedness to the producer exceeds the cost to the producer, the action is then selected (Kell et al., 1995), and because the degree of relatedness in clonally propagating bacteria is 1, this selection is likely.

Second, albeit related, the question then arises as to how we might account (in terms of evolutionary selection) for the cross reactivity we have seen for Rpf between organisms from entirely different species or indeed genera, whose degree of kinship is thus negligible. Here the answer is that the competing organisms are not normally associated spatially with the producers, most obviously where we contrast, for example, pathogenic mycobacteria in the lungs of hosts with harmless soil-living streptomycetes, but even at the small-scale level where the existence of very localized microenvironments in the soil is recognized (Bakken and Olsen, 1987).

Another ostensible conundrum relates to how it could make sense to use peptides or proteins as signaling elements under conditions (in the soil or in the sea) in which one would suppose that a large amount of proteolytic activity was present. The probable answer here comes from the potency of the Rpf systems, where the effective concentrations are far below the K_m (Michaelis constant) of known proteases.

Next, although it has been rather commonplace to assume that the humble bacteria are rather homogeneous, especially in axenic cultures in which they are presumed to be genetically identical, phenotypic differentiation (e.g., to produce sentinels [Postgate, 1995] or an insurance policy [Koch, 1987]) is now recognized as widespread (Davey and Kell, 1996;

Sumner and Avery, 2002), and thus we see an important linkage between evolutionary selection and epigenetic phenomena. As to genetic change, perhaps the most interesting recent development in this field, following the original article by Cairns and colleagues (Cairns, et al., 1988), is the recognition that a small fraction of the cells in nongrowing cultures of *E. coli* enter a hypermutable state that can help them escape their condition (Bull et al., 2001; Foster, 1999; McKenzie et al., 2000; Rosenberg, 1997), a phenomenon (the benefits of hypermutation) also observed in chemostats (Riley et al., 2001; Sniegowski et al., 1997), in pathogens in vivo (Oliver et al., 2000), in experimental directed evolution (Zaccolo and Gherardi, 1999), and in evolutionary computation in silico (Corne et al., 2002; Oates et al., 2000).

CONCLUDING REMARKS

It is clear that the vast majority of microbes observable by molecular means have never been cultured. Indeed, a number of biotechnology companies have been started on the premise that, to exploit the biosynthetic potential of such microbes, it will be much easier to express their DNA in other hosts than to try and bring them into culture. Certainly the complexity of natural ecosystems makes their study difficult. However, two facts are pertinent: (i) all known actinobacteria possess genes encoding at least one member of the Rpf family of bacterial cytokines, and (ii) we have demonstrated in the laboratory that such organisms can enter a reversible state of dormancy when starved, and that Rpf can be used to resuscitate them. This makes one optimistic that Rpf—or small molecules mimicking its activity—might be of utility in increasing massively the number of microbes that have been cultured and, thus, the number of useful bioactive substances available to the natural products scientist.

Acknowledgments. We thank the Biotechnology and Biological Sciences Research Council and the National Environmental Research Council for financial support.

REFERENCES

Alvarez-Barrientos, A., J. Arroyo, R. Canton, C. Nombela, and M. Sanchez-Perez. 2000. Applications of flow cytometry to clinical microbiology. *Clin. Microbiol. Rev.* **13**:167–195.

Amann, R. I., W. Ludwig, and K. H. Schleifer. 1995. Phylogenetic identification and *in situ* detection of individual microbial cells without cultivation. *Microbiol. Rev.* **59**:143–169.

Bakken, L. R., and R. A. Olsen. 1987. The relationship between cell size and viability of soil bacteria. *Microb. Ecol.* **13**:103–114.

Barer, M. R., and C. R. Harwood. 1999. Bacterial viability and culturability. *Adv. Microb. Physiol.* **41**:93–137.

Barer, M. R., L. T. Gribbon, C. R. Harwood, and C. E. Nwoguh. 1993. The viable but non-culturable hypothesis and medical microbiology. *Rev. Med. Microbiol.* **4**:183–191.

Barer, M. R., A. S. Kaprelyants, D. H. Weichart, C. R. Harwood, and D. B. Kell. 1998. Microbial stress and culturability: conceptual and operational domains. *Microbiology* (UK) **144**:2009–2010.

Bateman, A., and M. Bycroft. 2000. The structure of a LysM domain from *E. coli* membrane-bound lytic murein transglycosylase D (MltD). *J. Mol. Biol.* **299**:1113–1119.

Bentley, S. D., K. F. Chater, A.-M. Cerdeno-Tarraga, G. L. Challis, N. R. Thomson, K. D. James, D. E. Harris, M. A. Quail, H. Kieser, D. Harper, A. Bateman, S. Brown, G. Chandra, C. W. Chen, M. Collins, A. Cronin, A. Fraser, A. Goble, J. Hidalgo, T. Hornsby, S. Howarth, C.-H. Huang, T. Kieser, L. Larke, L. Murphey, K. Oliver, S. O'Neil, E. Rabbinowitsch, M.-A. Rajandream, K. Rutherford, S. Rutter, K. Seeger, D. Saunders, S. Sharp, R. Squares, S. Squares, K. Taylor, T. Warren, A. Wietzorrek, J. Woodward, B. G. Barrell, J. Parkhill, and D. A. Hopwood. 2002. Complete genome sequence of the model actinomycete *Streptomyces coelicolor* A3(2). *Nature* **417**:141–147.

Biketov, S., G. V. Mukamolova, V. Potapov, E. Gilenkov, G. Vostroknutova, D. B. Kell, M. Young, and A. S. Kaprelyants. 2000. Culturability of *Mycobacterium tuberculosis* cells isolated from murine macrophages: a bacterial growth factor promotes recovery. *FEMS Immunol. Med. Microbiol.* **29**:233–240.

Bogosian, G., and E. V. Bourneuf. 2001. A matter of bacterial life and death. *EMBO Rep.* **2**:770–774.

Bogosian, G., P. J. L. Morris, and J. P. O'Neil. 1998. A mixed culture recovery method indicates that enteric bacteria do not enter the viable but nonculturable state. *Appl. Environ. Microbiol.* **64**:1736–1742.

Bogosian, G., N. D. Aardema, E. V. Bourneuf, P. J. L. Morris, and J. P. O'Neil. 2000. Recovery of hydrogen peroxide-sensitive culturable cells of Vibrio vulnificus gives the appearance of resuscitation from a viable but nonculturable state. *J. Bacteriol.* **182**:5070–5075.

Bull, A. T., A. C. Ward, and M. Goodfellow. 2000. Search and discovery strategies for biotechnology: the paradigm shift. *Microbiol. Mol. Biol. Rev.* **64**:573–606.

Bull, H. J., M. J. Lombardo, and S. M. Rosenberg. 2001. Stationary-phase mutation in the bacterial chromosome: recombination protein and DNA polymerase IV dependence. *Proc. Natl. Acad. Sci. USA* **98**:8334–8341.

Cairns, J., J. Overbaugh, and S. Miller. 1988. The origin of mutants. *Nature* **335**:142–145.

Callard, R., and A. Gearing. 1994. *The Cytokine Facts Book*. Academic Press, London, United Kingdom.

Chaisson, R. W. 2000. New developments in the treatment of latent tuberculosis. *Int. J. Tuber. Lung Dis.* **4**:S176–S181.

Cole, S. T., R. Brosch, J. Parkhill, T. Garnier, C. Churcher, D. Harris, S. B. Gordon, K. Eiglmeier, S. Gas, C. E. Barry, F. Tekaia, K. Badcock, D. Basham, D. Brown, T. Chillingworth, R. Connor, R. Davies, K. Devlin, T. Feltwell, S. Gentles, N. Hamlin, S. Holroyd, T. Hornby, K. Jagels, A. Krogh, J. McLean, S. Moule, L. Murphy, K. Oliver, J. Osborne, M. A. Quail, M. A. Rajandream, J. Rogers, S. Rutter, K. Seeger, J. Skelton, R. Squares, S. Squares, J. E. Sulston, K. Taylor, S. Whitehead, and B. G. Barrell. 1998. Deciphering the biology of *Mycobacterium tuberculosis* from the complete genome sequence. *Nature* **393**:537–544.

Corne, D. W., M. J. Oates, and D. B. Kell. 2002. On fitness distributions and expected fitness gains of parallelised mutation operators: implications for high mutation rates and rate adaptation in parallel evolutionary algorithms, p. 132–141. *In* J. J. Merelo Guervós, P. Adamidis, H.-G. Beyer, J.-L. Fernández-Villacañas, and H.-P. Schwefel (ed.), *Parallel Problem Solving from Nature—PPSN VII*. Springer, Berlin, Germany.

Davey, H. M., and D. B. Kell. 1996. Flow cytometry and cell sorting of heterogeneous microbial populations: the importance of single-cell analysis. *Microbiol. Rev.* **60:**641–696.

Dobrovol'skaya, T. G., L. V. Lysak, G. M. Zenova, and D. G. Zvyagintsev. 2001. Analysis of soil bacterial diversity: methods, potentiality, and prospects. *Microbiology* **70:**119–132.

Domingue, G. J., and H. B. Woody. 1997. Bacterial persistence and expression of disease. *Clin. Microbiol. Rev.* **10:**320–344.

Errington, J. 1996. Determination of cell fate in *Bacillus subtilis*. *Trends Genet.* **12:**31–34.

Flynn, J. L., and J. Chan. 2001. Tuberculosis: latency and reactivation. *Infect. Immun.* **69:**4195–4201.

Foster, P. L. 1999. Mechanisms of stationary phase mutation: a decade of adaptive mutation. *Annu. Rev. Genet.* **33:**57–88.

Fredricks, D. N., and D. A. Relman. 1996. Sequence-based identification of microbial pathogens—a reconsideration of Koch's postulates. *Clin. Microbiol. Rev.* **9:**18–33.

Freeman, R., J. Dunn, J. Magee, and A. Barrett. 2002. The enhancement of isolation of mycobacteria from a rapid liquid culture system by broth culture supernate of *Micrococcus luteus*. *J. Med. Microbiol.* **51:**92–93.

Fuqua, C., and E. P. Greenberg. 1998. Self perception in bacteria: quorum sensing with acylated homoserine lactones. *Curr. Opin. Microbiol.* **1:**183–189.

Gangadharam, P. R. J. 1995. Mycobacterial dormancy. *Tuber. Lung Dis.* **76:**477–479.

Gao, S. J., and P. S. Moore. 1996. Molecular approaches to the identification of unculturable infectious agents. *Emerg. Infect. Dis.* **2:**159–167.

Hamilton, W. D. 1963. The evolution of altruistic behaviour. *Am. Nat.* **97:**354–356.

Hamilton, W. D. 1964. The genetical evolution of social behaviour, I and II. *J. Theor. Biol.* **7:**1–52.

Haupts, U., M. Rüdiger, and A. J. Pope. 2000. Macroscopic versus microscopic fluorescence techniques in (ultra)-high-throughput screening. *Drug Discov. Today, HTS Suppl.* **1:**3–9.

Head, I. M., J. R. Saunders, and R. W. Pickup. 1998. Microbial evolution, diversity, and ecology: a decade of ribosomal RNA analysis of uncultivated microorganisms. *Microb. Ecol.* **35:**1–21.

Heath, J. K. 1993. *Growth Factors*. IRL Press, Oxford, United Kingdom.

Hernandez-Pando, R., M. Jeyanathan, G. Mengistu, D. Aguilar, H. Orozco, M. Harboe, G. A. W. Rook, and G. Bjune. 2000. Persistence of DNA from *Mycobacterium tuberculosis* in superficially normal lung tissue during latent infection. *Lancet* **356:**2133–2138.

Höner zu Bentrup, K., and D. G. Russell. 2001. Mycobacterial persistence: adaptation to a changing environment. *Trends Microbiol.* **9:**597–605.

Kaprelyants, A. S., and D. B. Kell. 1993. Dormancy in stationary-phase cultures of *Micrococcus luteus*: flow cytometric analysis of starvation and resuscitation. *Appl. Environ. Microbiol.* **59:**3187–3196.

Kaprelyants, A. S., and D. B. Kell. 1996. Do bacteria need to communicate with each other for growth? *Trends Microbiol.* **4:**237–242.

Kaprelyants, A. S., J. C. Gottschal, and D. B. Kell. 1993. Dormancy in nonsporulating bacteria. *FEMS Microbiol. Rev.* **104:**271–286.

Kaprelyants, A. S., G. V. Mukamolova, and D. B. Kell. 1994. Estimation of dormant *Micrococcus luteus* cells by penicillin lysis and by resuscitation in cell-free spent medium at high dilution. *FEMS Microbiol. Lett.* **115:**347–352.

Kaprelyants, A. S., G. V. Mukamolova, H. M. Davey, and D. B. Kell. 1996. Quantitative analysis of the physiological heterogeneity within starved cultures of *Micrococcus luteus* using flow cytometry and cell sorting. *Appl. Environ. Microbiol.* **62:**1311–1316.

Kaprelyants, A. S., G. V. Mukamolova, S. S. Kormer, D. H. Weichart, M. Young, and D. B. Kell. 1999. Intercellular signalling and the multiplication of prokaryotes: bacterial cytokines. *Symp. Soc. Gen. Microbiol.* **57:**33–69.

Kell, D. B., and M. Young. 2000. Bacterial dormancy and culturability: the role of autocrine growth factors. *Curr. Opin. Microbiol.* **3:**238–243.

Kell, D. B., A. S. Kaprelyants, and A. Grafen. 1995. On pheromones, social behaviour and the functions of secondary metabolism in bacteria. *Trends Ecol. Evol.* **10:**126–129.

Kell, D. B., A. S. Kaprelyants, D. H. Weichart, C. L. Harwood, and M. R. Barer. 1998. Viability and activity in readily culturable bacteria: a review and discussion of the practical issues. *Antonie Leeuwenhoek* **73:**169–187.

Kjelleberg, S. (ed.). 1993. *Starvation in Bacteria*. Plenum Press, New York, N.Y.

Kleerebezem, M., and L. E. Quadri. 2001. Peptide pheromone-dependent regulation of antimicrobial peptide production in Gram-positive bacteria: a case of multicellular behavior. *Peptides* **22:**1579–1596.

Kleerebezem, M., L. E. N. Quadri, O. P. Kuipers, and W. M. deVos. 1997. Quorum sensing by peptide pheromones and two-component signal-transduction systems in Gram-positive bacteria. *Mol. Microbiol.* **24:**895–904.

Koch, A. L. 1987. The variability and individuality of the bacterium, p. 1606–1614. *In* F. C. Neidhardt, K. B. Low, B. Magasanik, M. Schaechter, and H. E. Umbarger (ed.), *Escherichia coli and Salmonella typhimurium: Cellular and Molecular Biology*, vol. 2. American Society for Microbiology, Washington, D.C.

Kolter, R., D. A. Siegele, and A. Tormo. 1993. The stationary phase of the bacterial life cycle. *Annu. Rev. Microbiol.* **47:**855–874.

Lazazzera, B. A. 2001. The intracellular function of extracellular signaling peptides. *Peptides* **22:**1519–1527.

Lazazzera, B. A., and A. D. Grossman. 1998. The ins and outs of peptide signaling. *Trends Microbiol.* **6:**288–294.

Losick, R., and J. Dworkin. 1999. Linking asymmetric division to cell fate: teaching an old microbe new tricks. *Genes Devel.* **13:**377–381.

MacDonell, M. T., and M. A. Hood. 1982. Isolation and characterization of ultramicrobacteria from a gulf coast estuary. *Appl. Environ. Microbiol.* **43:**566–571.

McDougald, D., S. A. Rice, D. Weichart, and S. Kjelleberg. 1998. Nonculturability: adaptation or debilitation? *FEMS Microbiol. Ecol.* **25:**1–9.

McKenzie, G. J., R. S. Harris, P. L. Lee, and S. M. Rosenberg. 2000. The SOS response regulates adaptive mutation. *Proc. Natl. Acad. Sci. USA* **97:**6646–6651.

McKinney, J. D., K. H. zu Bentrup, E. J. Munoz-Elias, A. Miczak, B. Chen, W. T. Chan, D. Swenson, J. C. Sacchettini, W. R. Jacobs, and D. G. Russell. 2000. Persistence of *Mycobacterium tuberculosis* in macrophages and mice requires the glyoxylate shunt enzyme isocitrate lyase. *Nature* **406:**735–738.

McLuhan, M., and Q. Fiore. 1971. *The medium is the Massage*. Penguin Books, London, United Kingdom.

McVeigh, H. P., J. Munro, and T. M. Embley. 1996. Molecular evidence for the presence of novel actinomycete lineages in a temperate forest soil. *J. Ind. Microbiol.* **17:**197–204.

Mukamolova, G. V., A. S. Kaprelyants, and D. B. Kell. 1995a. Secretion of an antibacterial factor during resuscitation of dormant cells in *Micrococcus luteus* cultures held in an extended stationary phase. *Antonie Leeuwenhoek* 67:289–295.

Mukamolova, G. V., N. D. Yanopolskaya, T. V. Votyakova, V. I. Popov, A. S. Kaprelyants, and D. B. Kell. 1995b. Biochemical changes accompanying the long-term starvation of *Micrococcus luteus* cells in spent growth medium. *Arch. Microbiol.* 163:373–379.

Mukamolova, G. V., A. S. Kaprelyants, D. I. Young, M. Young, and D. B. Kell. 1998a. A bacterial cytokine. *Proc. Natl. Acad. Sci. USA* 95:8916–8921.

Mukamolova, G. V., N. D. Yanopolskaya, D. B. Kell, and A. S. Kaprelyants. 1998b. On resuscitation from the dormant state of *Micrococcus luteus*. *Antonie Leeuwenhoek* 73:237–243.

Mukamolova, G. V., S. S. Kormer, D. B. Kell, and A. S. Kaprelyants. 1999. Stimulation of the multiplication of *Micrococcus luteus* by an autocrine growth factor. *Arch. Microbiol.* 172:9–14.

Mukamolova, G. V., O. A. Turapov, K. Kazaryan, M. Telkov, A. S. Kaprelyants, D. B. Kell, and M. Young. 2002a. The *rpf* gene of *Micrococcus luteus* encodes an essential secreted growth factor. *Mol. Microbiol.* 46:611–621.

Mukamolova, G. V., O. A. Turapov, D. I. Young, A. S. Kaprelyants, D. B. Kell, and M. Young. 2002b. A family of autocrine growth factors in *Mycobacterium tuberculosis*. *Mol. Microbiol.* 46:623–635.

Nystrom, T. 2001. Not quite dead enough: on bacterial life, culturability, senescence, and death. *Arch. Microbiol.* 176:159–164.

Oates, M., D. Corne, and R. Loader. 2000. A tri-phase multimodal evolutionary search performance profile on the "hierarchical if and only if" problem, p. 339–346. *In* D. Whitley, D. Goldberg, E. Cantú-Paz, L. Spector, I. Parmee, and H.-G. Beyer (ed.), *Proceedings of GECCO-2000*. Morgan Kaufmann, San Francisco, Calif.

Oliver, A., R. Canton, P. Campo, F. Baquero, and J. Blazquez. 2000. High frequency of hypermutable *Pseudomonas aeruginosa* in cystic fibrosis lung infection. *Science* 288:1251–1253.

Parrish, N. M., J. D. Dick, and W. R. Bishai. 1998. Mechanisms of latency in *Mycobacterium tuberculosis*. *Trends Microbiol.* 6:107–112.

Phyu, S., T. Mustafa, T. Hofstad, R. Nilsen, R. Fosse, and G. Bjune. 1998. A mouse model for latent tuberculosis. *Scand. J. Infect. Dis.* 30:59–68.

Poindexter, J. S. 1987. Bacterial responses to nutrient limitation. *Symp. Soc. Gen. Microbiol.* 41:283–317.

Postgate, J. R. 1967. Viability measurements and the survival of microbes under minimum stress. *Adv. Microbiol. Physiol.* 1:1–23.

Postgate, J. R. 1969. Viable counts and viability. *Methods Microbiol.* 1:611–628.

Postgate, J. R. 1976. Death in microbes and macrobes, p. 1–19. *In* T. R. G. Gray and J. R. Postgate (ed.), *The Survival of Vegetative Microbes*. Cambridge University Press, Cambridge, United Kingdom.

Postgate, J. R. 1995. Danger of sleeping bacteria. *The (London) Times*, Nov. 13, p. 19.

Primas, H. 1981. *Chemistry, Quantum Mechanics and Reductionism*. Springer, Berlin, Germany.

Relman, D. A. 1999. The search for unrecognized pathogens. *Science* 284:1308–1310.

Riley, M. S., V. S. Cooper, R. E. Lenski, L. J. Forney, and T. L. Marsh. 2001. Rapid phenotypic change and diversification of a soil bacterium during 1000 generations of experimental evolution. *Microbiology* 147:995–1006.

Rondon, M. R., R. M. Goodman, and J. Handelsman. 1999. The Earth's bounty: assessing and accessing soil microbial diversity. *Trends Biotechnol.* 17:403–409.

Rosenberg, S. M. 1997. Mutation for survival. *Curr. Opin. Genet. Dev.* 7:829–834.

Rudiger, M., U. Haupts, K. J. Moore, and A. J. Pope. 2001. Single-molecule detection technologies in miniaturized high throughput screening: binding assays for G protein-coupled receptors using fluorescence intensity distribution analysis and fluorescence anisotropy. *J. Biomol. Screen.* 6:29–37.

Schut, F., R. A. Prins, and J. C. Gottschal. 1997. Oligotrophy and pelagic marine bacteria: facts and fiction. *Aquat. Microbial. Ecol.* 12:177–202.

Shleeva, M. O., K. Bagramyan, M. V. Telkov, G. V. Mukamolova, M. Young, D. B. Kell, and A. S. Kaprelyants. 2002. Formation and resuscitation of "non-culturable" cells of *Rhodococcus rhodochrous* and *Mycobacterium tuberculosis* in prolonged stationary phase. *Microbiology* 148:1581–1591.

Smith, R. J., A. T. Newton, C. R. Harwood, and M. R. Barer. 2002. Active but nonculturable cells of *Salmonella enterica* serovar Typhimurium do not infect or colonize mice. *Microbiology* 148:2717–2728.

Sniegowski, P. D., P. J. Gerrish, and R. E. Lenski. 1997. Evolution of high mutation rates in experimental populations of *E. coli*. *Nature* 387:703–705.

Stephens, K. 1986. Pheromones among the prokaryotes. *CRC Crit. Rev. Microbiol.* 13:309–334.

Sumner, E. R., and S. V. Avery. 2002. Phenotypic heterogeneity: differential stress resistance among individual cells of the yeast *Saccharomyces cerevisiae*. *Microbiology* 148:345–351.

Tiedje, J. M., and J. L. Stein. 1999. Microbial biodiversity: strategies for its recovery, p. 682–692. *In* R. M. Atlas, G. Cohen, C. L. Hershberger, W.-S. Hu, D. H. Sherman, R. C. Willson, and J. H. D. Wu (ed.), *Manual of Industrial Microbiology and Biotechnology*, 2nd ed. American Society for Microbiology, Washington, D.C.

Torsvik, V., J. Goksøyr, and F. L. Daae. 1990a. High diversity in DNA of soil bacteria. *Appl. Environ. Microbiol.* 56:782–787.

Torsvik, V., K. Salte, R. Sorheim, and J. Goksøyr. 1990b. Comparison of phenotypic diversity and DNA heterogeneity in a population of soil bacteria. *Appl. Environ. Microbiol.* 56:776–781.

Torsvik, V., R. Sorheim, and J. Goksøyr. 1996. Total bacterial diversity in soil and sediment communities—a review. *J. Ind. Microbiol.* 17:170–178.

van Piuxteren, L. A. H., J. P. Cassidy, B. H. C. Smedegaard, E. M. Agger, and P. Andersen. 2000. Control of latent *Mycobacterium tuberculosis* infection is dependent on CD8 T cells. *Eur. J. Immunol.* 30:3689–3698.

Votyakova, T. V., A. S. Kaprelyants, and D. B. Kell. 1994. Influence of viable cells on the resuscitation of dormant cells in *Micrococcus luteus* cultures held in extended stationary phase. The population effect. *Appl. Environ. Microbiol.* 60:3284–3291.

Watts, J. E. M., A. S. Huddleston-Anderson, and E. M. H. Wellington. 1999. Bioprospecting, p. 631–641. *In* R. M. Atlas, G. Cohen, C. L. Hershberger, W.-S. Hu, D. H. Sherman, R. C. Willson, and J. H. D. Wu (ed.), *Manual of Industrial Microbiology and Biotechnology*, 2nd ed. American Society for Microbiology, Washington, D.C.

Wayne, L. G. 1994. Dormancy of *Mycobacterium tuberculosis* and latency of disease. *Eur. J. Clin. Microbiol. Infect. Dis.* 13:908–914.

Zaccolo, M., and E. Gherardi. 1999. The effect of high-frequency random mutagenesis on *in vitro* protein evolution: a study on TEM-1 β-lactamase. *J. Mol. Biol.* 285:775–783.

Microbial Diversity and Bioprospecting
Edited by Alan T. Bull
© 2004 ASM Press, Washington, D.C.

Chapter 11

Soils—the Metagenomics Approach

JO HANDELSMAN

THE HISTORY OF SOIL BIOLOGY

Soil is a living organ of the Earth. Until recent times, large and small cultures throughout the world recognized this truth, expressing it through a deep spiritual relationship with the soil. The ancient Greeks worshipped soil through the goddesses Gaia and Demeter, the Germans through the goddess Ertha, and the Native Americans through Mother Earth, and the Old Testament describes the soil as a source of healing: "The Lord hath created medicines out of the earth; and he that is wise will not abhor them." (*Ecclesiasticus,* XXXVIII, 4.)

Advances in science, medicine, and agriculture during the 20th century provided numerous reasons to continue to venerate the soil, well beyond the basic necessity of food production. Discoveries during this century about soil bacteria led to development of life-saving drugs, including antibiotics, antitumor agents, and immunosuppressants. And yet, just as these advances grew more impressive, urbanization and other global economic influences tended to disassociate many people from an intimate association with the environment. Gradually, we lost appreciation of our reliance on the soil, an ignoble decline typified by common use of the term "dirt." But that trend may be about to be altered by the renaissance in soil biology that presents the soil as one of the last great frontiers available for discovery and bioprospecting. What we are poised to discover, in the form of knowledge and medicinal and agricultural chemistry, may well restore the soil to a revered position in the now-global society. This renaissance is driven by the tools of molecular biology, which enable exploration of the life and chemistry of the soil in new ways. One emerging tool that draws together traditional soil biology and modern molecular biology is metagenomics, which entails analysis of the collective genomes of an assemblage of microorganisms.

The past 100 years of microbiological research has yielded experimental evidence verifying the historical belief that soil is the site of many life-sustaining processes. Microorganisms active in the soil are largely responsible for the biogeochemical cycles that support life on Earth (Paul and Clark, 1996). Microorganisms mediate the nitrogen cycle, providing the only biological route for atmospheric nitrogen to enter living systems through nitrogen fixation, oxidizing it further through nitrification, and recycling it into the atmosphere by denitrification. Similarly, microorganisms play a central role in the carbon and sulfur cycles and in the oxidative and reductive transformation of metals such as iron and mercury. Soil microorganisms also influence the health of plants and animals. Although they tend to receive more attention for the diseases they incite, microorganisms are likely more significant for their role in maintaining the health of plants and animals by protecting them from attack by other microorganisms, providing vitamins and other nutrients, and influencing developmental processes.

Although it is easy to celebrate these positive and essential activities, microorganisms exist within a carefully balanced natural system of checks and balances, and when disrupted through human agency can produce harmful environmental effects. One classic example is acid mine drainage, a microbially mediated process that generates a toxic waste typified by acid production, generating pH's below 1 or 0, and high soluble metal concentrations that can pollute waterways (Edwards et al., 1999; Schleper et al., 1995). The disruption of the Earth and exposure of coal or copper mines to oxygen sets the stage for the microbial activity that produces this highly damaging waste.

The biological processes in soil are executed in a complex physical and chemical environment (Fig. 1). The physical structure of soil is varied and dynamic. Soil particles vary in size from 1 to 1,000 μm, and the

Jo Handelsman • Department of Plant Pathology, University of Wisconsin, Madison, WI 53706.

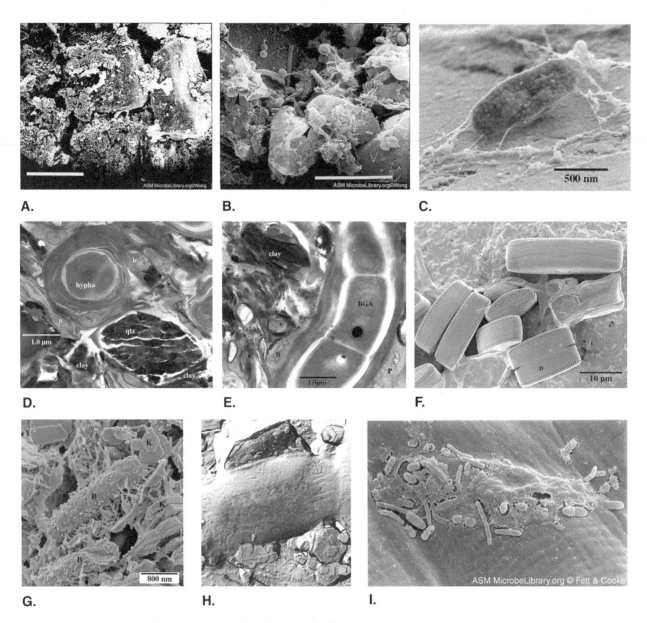

Figure 1. Microorganisms in soil, on minerals, and on plant surfaces.
(A) and (B) Scanning electron micrographs of the naturally occurring biofilm on sand grains in the clog mat of a septic system infiltration mound. The biofilm is composed of mineral particles, a variety of microorganisms, and a network of slime, or glycocalyx, that binds the microorganisms and particles together. Image (A) scale bar is 150 μm. Image (B) scale bar is 4.3 μm. Copyright Amy C. Lee Wong. Licensed for use, ASM MicrobeLibrary (linked to http://www.microbelibrary.org). (C) Unidentified bacterium attached to a feldspar surface by extracellular polymers. High-resolution, low-voltage cryoscanning electron micrograph of high-pressure frozen, freeze-fracture/sublimed culture sample. (D) Microbial soil assemblage consisting of extensive extracellular polymer networks (p), anhedral clay minerals (clay), quartz (qtz), bacteria (b), and fungi. Energy filtered transmission electron micrograph of ultrathin section of high-pressure frozen, freeze-substituted, undisturbed soil sample. (E) Microbial soil assemblage consisting of extensive extracellular polymer networks (P), anhedral clay minerals (clay), bacteria (B), and filamentous cyanobacteria (BGA). Energy-filtered transmission electron micrograph of ultrathin section of high-pressure frozen, freeze-substituted, undisturbed soil sample. (F) Diatoms (D), cyanobacteria (C), and fungal hyphae (F) inhabit the surface of a quartz grain from a spring seep sandstone outcrop near Mount Horeb, Wisconsin. High-resolution, low-voltage scanning electron micrograph. (G) *Pseudomonas fluorescens* (B) bound to euhedral kaolinite (K) by extracellular polysaccharides (arrows). High-resolution, low-voltage cryoscanning electron micrograph of high-pressure frozen, freeze-fracture/sublimed culture sample. (H) *Escherichia coli* (B) attached to euhedral kaolinite (k) by f-type sex pili (P). Transmission electron micrograph of a propane cryojet-frozen, freeze-etch Pt replica. (Micrographs in panels C to H courtesy of William W. Barker, The College of Letters and Science, University of Wisconsin-Madison.) (I) Micrograph of naturally occurring biofilm on a plant surface (alfalfa sprout hypocotyl). The size of the bacteria ranged from 0.4 μm in diameter (cocci) to 0.2 × 1.7 μm (rods). Copyright Peter Cooke and William Fett. Licensed for use, ASM Microbe Library (linked to http://www.microbelibrary.org).

moisture content and chemistry of the soil affect particle size. Particle size, in turn, affects water flow, gas exchange, and temperature gradients in soil. The chemical composition of soil is derived from a combination of its geologic and biologic origins. The bedrock contributes to the inorganic mineral fraction, and geologic events influence the ion exchange capacity, weathering rates, and particle size. The plants that grow in a soil and the microorganisms that live on them and decompose them create the organic fraction of soil, which is highly complex and changes rapidly. The soil, then, is a dynamic interplay of physical forces with chemical substrate and the life that grows in it. It changes over short and long time scales. The moisture content, structure, and nutrient solubility of a soil can change dramatically in the course of a day. Over years and centuries, the organic and mineral fractions can change substantially. Many of these changes are driven by microorganisms; all of them affect the structure and function of the microbial communities in soil (Brady and Weil, 2002).

It is this dynamic complexity of soil that has challenged and excited soil biologists for more than a century. The questions about soil are complex, the opportunities are profound, and the influence of new knowledge is expansive, constituting a broad frontier that intersects with many areas of science. Although we know vastly more about soil now than in the 19th century, the unknown is far greater than the known. The interplay between the known and the unknown and the integration of diverse sciences to dissect soil biology is reflected in the careers, discoveries, and beliefs of some of the most prominent microbiologists of the modern era. The impact of soil biology on human welfare has been molded by the integration of sciences by scientists of diverse origins and beliefs. Selman Waksman is remembered, for instance, for his discovery of streptomycin in the 1940s, hailed at the time as the ultimate weapon against the "White Plague," as tuberculosis was then known. Waksman was a biologist interested in the ecology of soil microorganisms, and as early as 1916 began publishing both a taxonomy of soil fungi and his findings regarding the effects of protozoa on populations of soil bacteria. He did not set out to find a medical miracle, but his explorations in basic soil biology led directly to the discovery of the actinomycete antibiotics that became that miracle (Waksman and Curtis, 1916; Waksman and Foster, 1937).

Waksman was a scientist who valued knowledge and appreciated how much was known about soil biology. His optimistic celebration of "the known" is sharply juxtaposed against the awe with which some of his contemporaries approached "the unknown." Waksman believed that the microbiology of the soil could be described in its entirety relatively quickly: "A large body of information has accumulated that enables us to construct a clear picture . . . of . . . the microscopic population of the soil . . ." (Waksman and Starkey, 1931).

Waksman's contemporary, Francis Clark, by contrast, considered the true nature of soil to be unknowable because of the undiscovered extent and myriad functions of microorganisms. Clark, a medical microbiologist, reminded his readers of the salient observations of the 18th-century botanist Linnaeus, who suggested that the overwhelming complexity of the microbial world might be beyond the reach of tidy classification: "Linnaeus . . . recognized the existence of microscopic forms of life but skirted a taxonomic quagmire by simply placing all microbes in a group designated 'Chaos'" (Paul and Clark, 1989).

This contrast, between what is known or unknown, knowable or unknowable, introduces one of the most dramatic recent advances in soil science. For decades, soil microbiology has concentrated on cultured organisms. These studies, conducted through the essential but limiting screen of the petri dish, have produced significant advances. The advent of new molecular and microscopic methods has fostered a growing realization, however, that there is much to be known about soil microorganisms that cannot be elucidated through traditional approaches because the vast majority of soil microorganisms cannot be cultured. Thus, separating that unknown universe from Linnaeus' centuries-old description of "chaos" requires new methods.

SOIL BIOLOGY AND THE ORIGINS OF METAGENOMICS

Only a small minority of the microorganisms living in soil are readily culturable on standard media. Microscopic studies, analysis of DNA complexity, and studies of species richness with culture-independent methods indicate that between 0.1 and 1% of the viable prokaryotes in soil grow in a standard culturing experiment (Torsvik et al., 1990, 1996; Whitman et al., 1998). The most remarkable aspect of these studies is that the as-yet-uncultured bacteria are highly diverse. Reassociation of DNA from the bacterial component in soil indicates that the complexity of the bacterial community is comprised of at least 4,000 distinct genomes, which exceeds the complexity estimated by culturing by 200-fold (Torsvik et al., 1990). PCR amplification, cloning, and sequencing of rRNA genes from soil have led to a partial understanding of the species diversity (Pace, 1996; Hugenholtz et al., 1998) (Fig. 2). Phylogenetic analysis indicates that the

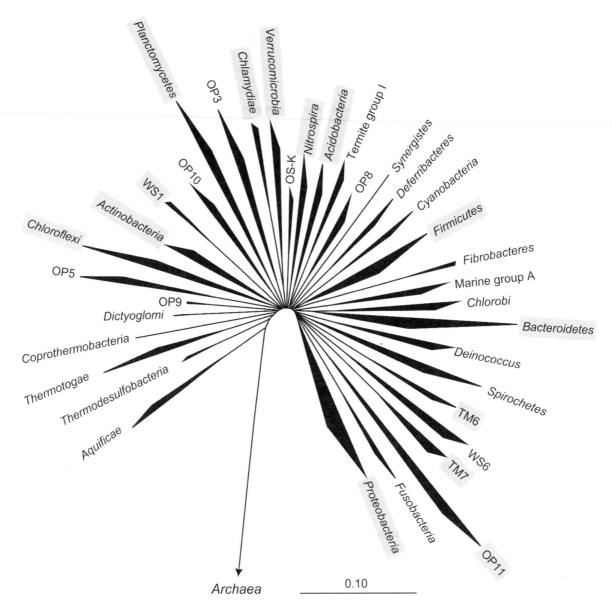

Figure 2. Phylogenetic diversity of microbial life. Phyla that have been detected in soil are indicated by shading.

approximately 2,400 sequences from soil contain few sequences that are identical to each other, span 13 phyla of *Bacteria*, contain clades of *Archaea* that, until recently, were unknown in terrestrial environments (Bintrim et al., 1997), and contain groups that diverge deeply from any cultured isolates (Borneman et al., 1996; Liles et al., 2003) (Fig. 2). The experimental evidence makes it abundantly clear that there are many thousands of species of prokaryotes yet to be discovered in soil. It is likely that these microorganisms mediate soil processes that have not been described, form novel associations with plant roots and soil animals, and produce chemicals that could be of benefit in medicine and agriculture.

Although molecular methods of analysis have improved the quality of our image of the soil micro-

bial community by unmasking the as-yet-uncultured members, the inability to decipher further the secrets of that community has both tantalized and frustrated soil microbiologists. The 16S rRNA gene analysis provides little functional information about the membership of uncultured communities, as phylogeny rarely predicts function in microbial taxa. Therefore, the contemporary quest for a fuller understanding of the entire biology of the soil hinges on approaches that can link phylogeny with function by integrating information about both cultured and uncultured microorganisms, as well as species abundance, nutrient cycling, signal molecules, antibiotics, and other small diffusible compounds.

Excitement about the potential for understanding the undiscovered forms of life indicated by 16S rRNA gene sequence analysis, coupled with vexation at the lack of available technology to delve into the biology of the newly described taxa, propelled the development of the approach of metagenomics (also known as environmental or community genomics), which enables genomic analysis of uncultured organisms. The idea of directly cloning DNA from the environment was suggested by Pace et al. (1985) and Pace (1986) and was developed independently by microbiologists studying soil and seawater. It was first applied successfully to analysis of oceanic microbial communities (Schmidt et al., 1991; Stein et al., 1996) and subsequently to soil communities (Handelsman et al., 1998; Rondon et al., 2000).

METAGENOMICS AS AN EXPERIMENTAL STRATEGY

Application of genomics to uncultured communities in soil involves extracting DNA directly from the soil, cloning it into a suitable vector, and transforming it into a culturable host cell (Fig. 3). Some libraries of metagenomic DNA have been constructed in fosmids and bacterial artificial chromosomes to maintain large fragments of DNA, which are necessary to capture complete biosynthetic pathways or to study genome organization. Alternatively, some libraries have been constructed in expression vectors that maintain small fragments linked to a promoter known to express genes in the host organism, facilitating functional analysis by increasing the chance of obtaining expression of foreign genes. It is the functional expression studies that facilitate the link between phylogeny and function, lending unique power to metagenomics analysis.

FUNCTIONAL DIVERSITY IN SOIL

Little is known about the relationship between phylogenetic diversity and functional diversity. A premise of much of the metagenomic work conducted on soil microbial communities is that the diversity of species will be reflected in the diversity of physiology and biochemistry. The estimates of species diversity suggest that analysis of soil microbial communities will require libraries containing large amounts of DNA to access the diverse genomes. If the estimates of soil microbial diversity are correct, and if the genomes of the unknown species are similar to the sequenced genomes, an average of 5 Mb, then 1 million clones, each containing 50 kb of DNA would be required to provide single-fold coverage of the genomes of the estimated 10,000 different species in a gram of

soil. To date, none of the libraries constructed approach this size, so substantial work remains to be done.

Access to minor species in the community will provide a challenge in the advanced stages of metagenomic analysis. If complete coverage of the soil "metagenome" is to be approached, then methods will need to be developed both to avoid repeatedly cloning DNA from the abundant organisms and to gain access to DNA from rare components of the community. There has been little redundancy detected in the libraries constructed thus far, suggesting that a complete census of the soil community is a distant prospect with current methods.

The relative abundance of culturable and as-yet-uncultured organisms has not been well described. Therefore, a future challenge is to determine the contribution of DNA from culturable and uncultured organisms to metagenomic libraries and elucidate the genetic and functional relationships among the organisms that can be dissected in the laboratory and those that cannot.

LINKING PHYLOGENY AND FUNCTION

Extracting the full value of metagenomic libraries requires systematically applying functional, sequence, and phylogenetic analyses. By identifying a phylogenetic marker on a clone that expresses a functional gene product or contains sequences with significant similarity to genes of known function, the linkage of phylogeny and the function of uncultured microorganisms can be accomplished (Stein et al., 1996). A striking example of this type of discovery is the identification of a clone derived from seawater that contained a 16S rRNA gene whose sequence indicated that it was of bacterial origin (Beja et al., 2000a, 2000b). Sequence analysis of the regions flanking the 16S rRNA gene revealed a new photorhodopsin gene, which was then expressed in *Escherichia coli* and shown to have light-harvesting capability. All microbial photorhodopsins discovered prior to this study were found in *Archaea*, and *Bacteria* were assumed not to contain this light-harvesting strategy. The bacteriorhodopsin discovery in a bacterial genome provides a new paradigm for linking phylogeny and function in uncultured microorganisms.

Although much can be inferred from sequence analysis of the genomes of uncultured microorganisms, sequence-driven study is limited to the functional knowledge about the sequences currently in the genetic databases. Functional analysis has the potential to expand the range of known functions and elucidate functions of genes with no known homologs.

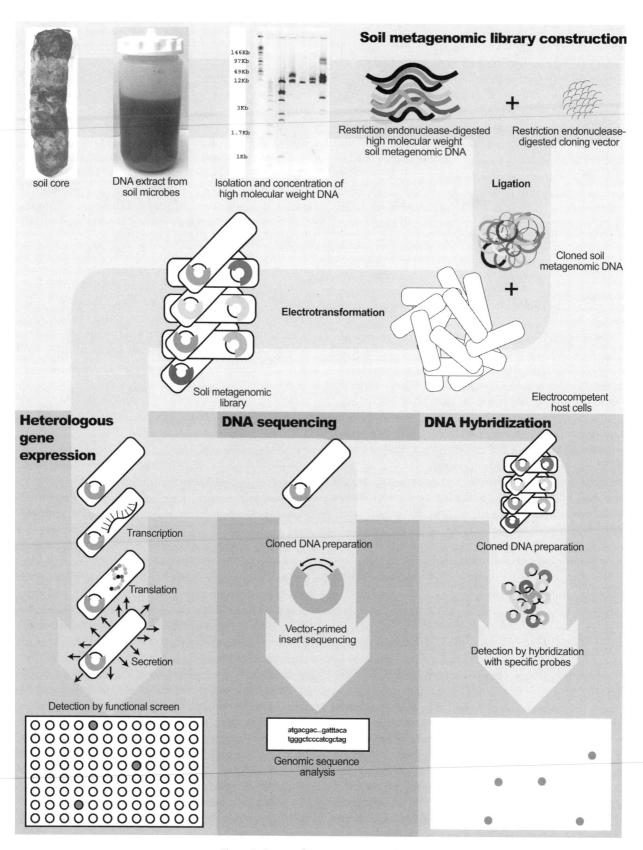

Figure 3. Strategy for metagenomic analysis.

Metagenomic libraries from soil have provided a rich source of new functions, which have enriched our knowledge of microbial biochemistry and contributed significantly to the genetic databases.

BIOLOGICAL INSIGHTS INTO THE SOIL FROM METAGENOMICS

Metagenomic analysis has added to our knowledge of proteins and small molecules encoded in the genomes of soil microorganisms. Soil metagenomic libraries produce diverse antibiotics. Biosynthetic pathways encoded by metagenomic DNA and expressed in *E. coli* direct the production of violacein, terragine, turbomycin, indirubin, fatty dienic alcohols, and the acyl tyrosines (Fig. 4). The synthesis of some of these antibiotics is encoded by a single gene whereas some require a complex operon for expression. Violacein, for example, requires coordinated expression of seven genes for its synthesis. The cloning of this pathway demonstrates the feasibility of heterologous expression of pathways from anonymous microorganisms from soil. Some of the antibiotics discovered are novel, such as the acyl tyrosines, terragine, and turbomycin, further illustrating the utility of metagenomics for bioprospecting. New insights into the evolution of antibiotic biosynthetic pathways may emerge as even the genes for synthesis of known antibiotics cloned in the metagenomic libraries diverged deeply from related genes from cultured organisms, suggesting that there is substantial genetic diversity to be discovered in known genes in uncultured organisms. In addition to new antibiotics, new enzymes, an antiporter, and antibiotic resistance determinants have been discovered (Table 1). A striking feature of the studies leading to these discoveries is the low frequency of clones expressing any given phenotype in metagenomic libraries. For example, one library consisting of 25,000 clones representing 1 Gb of DNA contained 3 that expressed detectable antimicrobial activity (Gillespie et al., 2002). Two screens for lipolytic activity detected 1 positive out of 730,000 clones and 3 positives out of 286,000 clones, respectively (Henne et al., 2000). The most feasible metagenomic studies thus involve selections or enrichments, rather than screens, for the phenotype of interest, negating the need to test thousands of clones individually. Robotics and high-throughput screening will accelerate the pace of discovery of genes for which there is no selection.

The emerging paradigm of metagenomic analysis is to screen libraries for 16S rRNA genes or other phylogenetic markers and sequence the flanking DNA to identify functions that can be recognized by sequence similarity, or to identify functionally active clones by selection or screen and then sequence the DNA flanking the function genes to identify phylogenetic markers (Sandler et al., 1999; Schleper et al., 1997). This approach will yield data linking phylogeny and function that will be used to design culturing strategies that are driven by the physiology of the organisms to be cultured.

CHALLENGES AND LIMITATIONS IN METAGENOMIC ANALYSIS

Although metagenomic libraries have yielded interesting information about biological systems, there are many barriers to extracting the full range of information that is locked in the genomes of the uncultured microbial world. First, some *Bacteria* and *Archaea* are recalcitrant to lysis and their DNA is not represented in the libraries. Some genes or their products will be unstable or toxic to the host cell and therefore will not be maintained in the libraries. Many genes will be derived from organisms that are phylogenetically distant from *E. coli* or another culturable species that is used as host to the library, and thus functional analysis will not detect these genes. The most interesting new genes are those that do not have high sequence similarity to those in the databases, and therefore expression and functional analysis are essential for discerning the nature of the gene product.

One goal of metagenomic analysis is to reconstruct genomes of uncultured organisms. This will be particularly challenging in soil communities because they are so complex. In addition to the large number of species, currently estimated at 10^4 or more per g of soil, there is likely microheterogeneity among genomes within a species. Recent work on *Cenarchaeum symbiosum*, a species that forms an intimate associate with a sponge, showed substantial variation among genomes within a local population of the bacterium (Schleper et al., 1998). Therefore, layers of complexity may make it impossible to reconstruct a genome, as is possible in simpler communities, but association of genes, their functions, and genomic organization with particular taxa will be a fruitful area of endeavor in soil metagenomics. Once the task of reassembling 5 to 10 genomes of uncultured organisms simultaneously has been accomplished in a simpler habitat, it will be easier to predict whether overcoming the technical and computational challenges of reconstructing genomes from soil is possible with today's expertise.

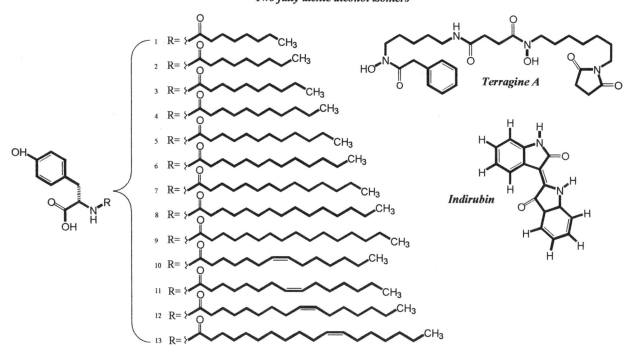

Turbomycin A

Turbomycin B

Violacein

Deoxyviolacein

Two fatty dienic alcohol isomers

Terragine A

Indirubin

Long-chain N-Acyl Amino Acid Antibiotics

Figure 4. Small molecules discovered in metagenomic libraries.

Table 1. Genes and gene products discovered in metagenomic libraries constructed with DNA extracted directly from soil

Gene and/or function	Relevant characteristic	Reference(s)
N-Acyl amino acid compounds	Novel antibiotics	Brady and Clardy (2000)
Violacein	Antibiotic known in cultured organisms	Brady et al. (2001)
Turbomycin A and B	Tri-aryl cation antibiotics	Gillespie et al. (2002)
Indirubin and related compounds	Known compound; some compounds have antimicrobial activity	MacNeil et al. (2001)
Terragine A and related small molecules	Novel antibiotics	Wang et al. (2000)
Kanamycin resistance gene	New acetyltransferase homolog	Courtois et al. (2003)
Fatty dienic alcohol	Novel compounds	Courtois et al. (2003)
Na$^+$/H$^+$ antiporter	New homolog	Majernik et al. (2001)
4-Hydroxybutyrate utilization enzymes	New homologs	Henne et al. (1999)
Amylases	Diverse homologs	Rondon et al. (2000)
Lipolytic activity	New homologs	Henne et al. (2000), Rondon et al. (2000)
Hemolytic activity	Diverse homologs	Rondon et al. (2000)
DNase	New homolog	Rondon et al. (2000)
16S rRNA	Diverse sequences	Rondon et al. (2000)

FUTURE DIRECTIONS IN METAGENOMICS

Metagenomic analysis depends on technical innovations that will provide improved access to DNA, heterologous gene expression, and DNA sequence analysis. To access DNA from a wider range of organisms, we need methods that facilitate aggressive bacterial cell lysis while protecting the integrity of the DNA. We need strategies to increase representation of rare members of the community by removing the DNA from the abundant organisms. This might involve antibodies that remove certain bacterial cell types prior to cell lysis or methods for removing the genomes after lysis, although the latter will be challenging because of the conservation among all bacterial genomes.

Heterologous gene expression will likely be one of the areas of greatest improvement in metagenomic analysis, largely because of the sophisticated understanding of the elements in gene regulation that has emerged from the past few decades of study. Gene expression can be enhanced by transferring clones to a variety of organisms that represent a range of G+C content, preferred codon usage, and protein secretion strategies. This requires shuttle vectors that replicate in *E. coli* and a host cell with gene expression machinery that differs substantially from that of *E. coli*, such as *Bacillus* (a low G+C gram positive) or *Streptomyces* (a high G+C gram positive).

Alternatively, heterologous gene expression could be enhanced by using a host cell modified to express a broader range of genes. To enhance transcription, genes encoding sigma factors from organisms closely related to the uncultured clades of interest can be cloned into the host cell. To enhance translation, a broader range of tRNAs can be supplied by introducing a range of genes encoding tRNAs that may be rare in the host cell but are typical of other cells. *E coli* carrying genes for tRNAs typically required by organisms with G+C- or A+T-rich genomes are currently available and ready to be tested for their efficacy in enhancing functional expression in metagenomic libraries (Stratagene, BL21-Condon Plus Competent Cells, 2002, http://www.stratagene.com/).

Heterologous expression of many of the genes cloned in metagenomic libraries is likely to be poor even if measures are taken to enhance gene expression. Therefore, in order to capture most of the expressed genes in functional screens, highly sensitive screens are essential. Detection of gene products in assays that amplify an initial signal will increase the rate of discovery. Assays in which the detection system is inside the cell carrying the cloned DNA will enhance detection as local concentrations of gene products may be higher inside the cell than following secretion into the surrounding medium. Highly sensitive detectors and small-volume assays that maintain high local concentration of products from a clone will likely increase the rate of finding active clones.

CONCLUSIONS

Metagenomics offers a new look by accessing the as-yet-unculturable microorganisms that represent the majority of life in soil. Metagenomic analysis of soil is challenging because soil is such a complex environment, containing diverse organisms in a dynamic matrix. The biological, chemical, and physical properties of soil all contribute to the technical difficulties of the analysis, and thus numerous obstacles to cloning and analyzing the metagenome of soil remain to be overcome. The potential is vast. Discovery of metabolic networks and small molecules from the cultured microorganisms from soil has been unparalleled in any other environment, and therefore it is likely that the potential for discovery from the uncultured community is similarly enormous. Because most of the uncultured life forms in soil appear to be new species and many represent new genera, there is much to learn about the fundamental functioning of soil microbial communities, and these communities have already yielded new enzymes and antibiotics. Technical advances in DNA recovery, gene expression, and functional analysis will enhance the rate of discovery and make possible productive prospecting of soil for the medicinal, agricultural, and industrial chemicals.

Acknowledgments. I thank Zakee Sabree and Tina Matta for preparing the figures and Ed DeLong for helpful discussions. The preparation of the manuscript was supported by the Howard Hughes Medical Institute, National Science Foundation grant MCB-0132085, and the University of Wisconsin-Madison.

REFERENCES

Beja, O., L. Aravind, E. V. Koonin, M. Suzuki, A. Hadd, L. P. Nguyen, S. B. Jovanovich, C. M. Gates, R. A. Feldman, J. L. Spudich, E. N. Spudich, and E. F. DeLong. 2000a. Bacterial rhodopsin: evidence for a new type of phototrophy in the sea. *Science* **289:**1902–1906.

Beja, O., M. T. Suzuki, E. V. Koonin, L. Aravind, A. Hadd, L. P. Nguyen, R. Villacorta, M. Amjadi, C. Garrigues, S. B. Jovanovich, R. A. Feldman, and E. F. DeLong. 2000b. Construction and analysis of bacterial artificial chromosome libraries from a marine microbial assemblage. *Environ. Microbiol.* **2:**516–529.

Bintrim, S. B., T. J. Donohue, J. Handelsman, G. P. Roberts, and R. M. Goodman. 1997. Molecular phylogeny of Archaea from soil. *Proc. Natl. Acad. Sci. USA* **94:**277–282.

Borneman, J., P. W. Skroch, K. M. O'Sullivan, J. A. Palus, N. G. Rumjanek, J. L. Jansen, J. Nienhuis, and E. W. Triplett. 1996. Molecular microbial diversity of an agricultural soil in Wisconsin. *Appl. Environ. Microbiol.* **62:**1935–1943.

Brady, N. C., and R. R. Weil. 2002. *The Nature and Properties of Soils,* 13th ed. Prentice-Hall, Upper Saddle River, N.J.

Brady, S. F., and J. Clardy. 2000. Long-chain N-acyl amino acid antibiotics isolated from heterologously expressed environmental DNA. *J. Am. Chem. Soc.* **122:**12903–12904.

Brady, S. F., C. J. Chao, J. Handelsman, and J. Clardy. 2001. Cloning and heterologous expression of a natural product biosynthetic gene cluster from eDNA. *Org. Lett.* **3:**1981–1984.

Courtois, S., C. M. Cappellano, M. Ball, E.-X. Francou, P. Normand, G. Helynck, A. Martinez, S. J. Kolvek, J. Hopke, M. S. Osburne, P. R. August, R. Nalin, M. Guérineau, P. Jeannin, P. Simonet, and J. L. Pornodet. 2003 Recombinant environmental libraries provide access to microbial diversity for drug discovery from natural products. *Appl. Environ. Microbiol.* **69:**49–55.

Edwards, K. J., T. M. Gihring, and J. F. Banfield. 1999 Interdependence of microbial populations and environmental conditions at an extreme acid mine drainage environment. *Appl. Environ. Microbiol.* **65:**3627–3632.

Gillespie, D. E., S. F. Brady, A. D. Bettermann, N. P. Cianciotto, M. R. Liles, M. R. Rondon, J. Clardy, R. M. Goodman, and J. Handelsman. 2002. Isolation of antibiotics turbomycin A and B from a metagenomic library of soil microbial DNA. *Appl. Environ. Microbiol.* **68:**4301–4306.

Handelsman, J., M. R. Rondon, S. Brady, J. Clardy, and R. M. Goodman. 1998. Molecular biology provides access to the chemistry of unknown soil microbes: a new frontier for natural products. *Chem. Biol.* **5:**R245–R249.

Henne, A., R. Daniel, R. A. Schmitz, and G. Gottschalk. 1999. Construction of environmental DNA libraries in *Escherichia coli* and screening for the presence of genes conferring utilization of 4-hydroxybutyrate. *Appl. Environ. Microbiol.* **65:**3901–3907.

Henne, A., R. A. Schmitz, M. Bomeke, G. Gottschalk, and R. Daniel. 2000. Screening of environmental DNA libraries for the presence of genes conferring lipolytic activity on *Escherichia coli. Appl. Environ. Microbiol.* **66:**3113–3116.

Hugenholtz, P., B. M. Goebel, and N. R. Pace. 1998 Impact of culture-independent studies on the emerging phylogenetic view of bacterial diversity. *J. Bacteriol.* **180:**4765–4774.

Liles, M. R., B. F. Manske, S. B. Bintrim, J. Handelsman, and R. M. Goodman. 2003. A census of rRNA genes and linked genomic sequences within a soil metagenomic library. *Appl. Environ. Microbiol.* **69:**2684–2691.

MacNeil, I. A., T. C. Minor, P. R. August, T. H. Grossman, K. A. Loiacono, et al. B. A. Lynch, T. Philips, S. Narula, R. Sundaramoorthi, A. Tyler, T. Aldredge, H. Long, M. Gilman, and M. S. Osburne. 2001. Expression and isolation of antimicrobial small molecules from soil DNA libraries. *J. Mol. Microbiol. Biotechnol.* **3:**301–308.

Majernik, A., G. Gottschalk, and R. Daniel. 2001. Screening of environmental DNA libraries for the presence of genes conferring Na$^+$(Li$^+$)/H$^+$ antiporter activity on *Escherichia coli:* characterization of the recovered genes and the corresponding gene products. *J. Bacteriol.* **183:**6645–6653.

Pace, N. R., D. A. Stahl, D. J. Lane, and G. J. Olsen. 1985. Analyzing natural microbial populations by rRNA sequences. *ASM News* **51:**4–12.

Pace, N. R. 1996. New perspective on the natural microbial world: molecular microbial ecology. *ASM News* **62:**463–470.

Paul, E. A., and F. E. Clark. 1989. *Soil Microbiology and Biochemistry.* Academic Press, New York, N.Y.

Paul, E. A., and F. E. Clark. 1996. *Soil Microbiology and Biochemistry,* 2nd ed. Academic Press, New York, N.Y.

Rondon, M. R., P. R. August, A. D. Bettermann, S. F. Brady, T. H. Grossman, M. R. Liles, K. A. Loiacono, B. A. Lynch, I. A. MacNeil, M. S. Osburne, J. Clardy, J. Handelsman, and R. M. Goodman. 2000. Cloning the soil metagenome: a strategy for accessing the genetic and functional diversity of uncultured microorganisms. *Appl. Environ. Microbiol.* **66:**2541–2547.

Sandler, S. J., P. Hugenholtz, C. Schleper, E. F. DeLong, N. R. Pace, and A. J. Clark. 1999. Diversity of *radA* genes from cultured and uncultured *Archaea:* comparative analysis of putative RadA proteins and their use as a phylogenetic marker. *J. Bacteriol.* **181:**907–915.

Schleper, C., G. Pühler, B. Kühlmorgen, and W. Zillig. 1995 Life at extremely low pH. *Nature* **375:**741–742.

Schleper, C., R. V. Swanson, E. J. Mathur, and E. F. DeLong. 1997. Characterization of a DNA polymerase from the uncultivated psychrophilic archaeon *Cenarchaeum symbiosum. J. Bacteriol.* **179:**7803–7811.

Schleper, C., E. F. DeLong, C. M. Preston, R. A. Feldman, K. Y. Wu, and R. V. Swanson. 1998. Genomic analysis reveals chromosomal variation in natural populations of the uncultured psychrophilic archaeon *Cenarchaeum symbiosum. J. Bacteriol.* **180:**5003–5009.

Schmidt, T. M., E. F. DeLong, and N. R. Pace. 1991. Analysis of a marine picoplankton community by 16S rRNA gene cloning and sequencing. *J. Bacteriol.* **173:**4371–4378.

Stein, J. L., T. L. Marsh, K. Y. Wu, H. Shizuya, and E. F. DeLong. 1996. Characterization of uncultivated prokaryotes: isolation and analysis of a 40-kilobase-pair genome fragment from a planktonic marine archaeon. *J. Bacteriol.* **178:**591–599.

Torsvik, V., J. Goksøyr, and F. L. Daae. 1990. High diversity in DNA of soil bacteria. *Appl. Environ. Microbiol.* **56:**782–787.

Torsvik, V., R. Sorheim, and J. Goksøyr. 1996. Total bacterial diversity in soil and sediment communities—a review. *J. Ind. Microbiol.* **17:**170–178.

Whitman, W. B., D. C. Coleman, and W. J. Wiebe. 1998. Prokaryotes: the unseen majority. *Proc. Natl. Acad. Sci. USA* **95:**6578–6583.

Waksman, S. A., and R. E. Curtis. 1916. The actinomyces of the soil. *Soil Sci.* **1:**99–134.

Waksman, S. A., and J. W. Foster. 1937. Associative and antagonistic effects of microorganisms: II. Antagonistic effects of microorganisms grown on artificial substrates. *Soil Sci.* **43:** 69–76.

Waksman, S. A., and R. L. Starkey. 1931. *The Soil and the Microbe.* John Wiley, New York, N.Y., p. viii.

Wang, G.-Y.-S., E. Graziani, B. Waters, W. Pan, X. Li, J. McDermott, G. Meurer, G. Saxona, J. Andersen, and J. Davies. 2000. Novel natural products from soil DNA libraries in a Streptomycete host. *Org. Lett.* **2:**2401–2404.

Microbial Diversity and Bioprospecting
Edited by Alan T. Bull
© 2004 ASM Press, Washington, D.C.

Chapter 12

Deep Biospheres

R. JOHN PARKES AND PETE WELLSBURY

Large bacterial populations are present in surface soils and sediments where they efficiently degrade deposited organic matter and thus recycle nutrients that are essential for continued photosynthetic primary production. This microbial filter is so efficient that in marine sediments <0.5% of water column primary production is preserved (Hedges and Keil, 1995). The remaining organic material presumably survives because of its recalcitrance. Hence, despite organic matter accumulating over geological time scales in sedimentary deposits and rocks creating the largest global reservoir of organic matter (Berner and Lasaga, 1989), these geosphere environments have been considered to be devoid of life. In addition, conditions become more extreme with increasing depth. Pressure and temperature (average 30°C/km) increase, porosity and permeability decrease, the more energy efficient electron acceptors are removed (Froelich et al., 1979), and certain nutrients become limited. There was little surprise, therefore, when early research into bacteria in subsurface environments found that viable bacteria were restricted to the top few meters of sediment (e.g., Morita and Zobell [1955] found no bacteria below ~8 m). Although bacteria were found in deeper layers at some locations, these were generally dismissed as contamination during sampling or as reactivated dormant organisms (Zobell, 1938).

Since that time, researchers have demonstrated that only a small proportion of all bacteria in the environment can be cultured (e.g., 0.0001 to 10% of the microscopically detectable cells [Jannasch and Jones, 1959]). Hence, the absence of viable bacteria does not mean the absence of all bacteria. This together with the isolation of bacteria that are able to grow under an amazing range of conditions (e.g., temperature −10 to 113°C, pressure >1,000 atm, pH 0 to 10.5, salt 0 to 5 mM [Rothschild and Mancinelli 2001; see also chapters 14 to 16]) caused researchers to reconsider the possibility of subsurface microbial life. The application of modern microbial ecological techniques to investigate subsurface environments from the mid 1980s has slowly provided evidence for the presence of deep biosphere microorganisms. Positive results from a range of different approaches used on the same sample has produced convincing evidence for a deep biosphere in a range of different environments: terrestrial subsurface (Ghiorse and Wobber, 1989), deep aquifers (McMahon and Chapelle, 1991), deep marine sediments (Parkes et al., 1994), oil reservoirs (Stetter et al., 1993; Haridon et al., 1995), terrestrial and marine basalts (Stevens and McKinley, 1995; Fisk et al., 1998), confined cretaceous rocks (Krumholz et al., 1997) and granites (Pedersen, 1997), ancient halite (Vreeland et al., 2000), sandstone (Colwell et al., 1997) and carbonate deposits (Martens et al., 1991), caves (Smith et al., 2002), and gold mines (Takai et al., 2001). Some of these environments are now reviewed in detail.

MARINE SEDIMENTS AND ROCKS

The presence of active bacteria in deep marine sediments has been investigated extensively as part of the International Ocean Drilling Program. Some 16 different sites have been analyzed using a range of complementary approaches, including direct microscopic and viable counts, radiotracer activity measurements, bacterial molecular genetic and biomarker analysis, and pure culture isolation (Parkes et al., 1994, 1995, 2000; Barnes et al., 1998; Marchesi et al., 2001). Consistent demonstration of the presence of bacteria by this range of approaches and their correspondence with geochemical and stable isotopic data has provided unequivocal evidence for a deep bacterial biosphere in marine sediments. Although bacterial populations decrease exponentially with depth (Fig. 1), bacterial populations are still considerable ($2.76 \times 10^6/cm^3$) at the mean ocean sediment

R. John Parkes • Department of Earth Sciences, University of Cardiff, Main Building, P.O. Box 914, Cardiff CF10 3YE, United Kingdom.   **Pete Wellsbury** • Department of Earth Sciences, University of Bristol, Bristol BS8 1RJ, United Kingdom.

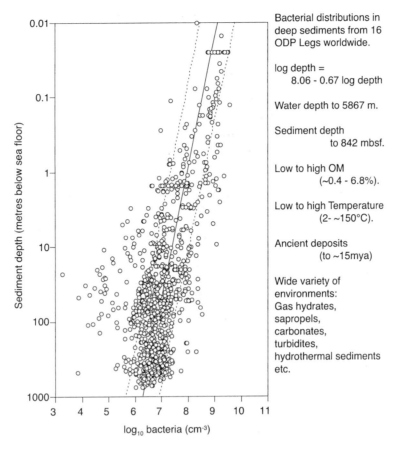

Figure 1. Bacterial distributions in deep marine sediments. The solid line is the regression line and the dotted lines on either side are the 95% prediction limits. ODP, Ocean Drilling Program; OM, organic matter.

depth of 500 m below the seafloor (mbsf) (Chester 1990). This decrease in bacteria (97%) presumably reflects the preferential removal of the most degradable components of organic matter during burial, and thus the remaining organic matter becomes increasingly recalcitrant. Therefore, it is surprising that any organic matter is still degradable after millions of years of burial. It has been estimated that the global amount of bacterial organic carbon in marine sediments represents only a tiny proportion of the total amount of organic carbon present (0.004%); remarkably, however, it is equivalent to 10% of all living carbon in the surface biosphere (Parkes et al., 1994). Hence, the subsurface marine biosphere is a quantitatively important ecosystem on Earth, and its discovery has significantly increased the total known biosphere on Earth.

Although bacterial populations decrease with depth at most locations, there are several sites where there are subsurface peaks in bacterial populations and activity. These occur where geochemical, depositional, or other changes result in increases in subsurface energy sources and include geothermal methane (Cragg et al., 1992), brine or sulphate incursions

(Parkes et al., 1990; Cragg et al., 1999; Mather and Parkes, 2000), high organic carbon (Cragg et al., 1998), gas hydrate sediments (Cragg et al., 1996; Wellsbury et al., 2000), and subsurface fluid flow in hydrothermal systems (Parkes et al., 2000). These situations provide clear evidence for the in situ viability of subsurface bacteria and are deep biosphere hot spots. An extreme example of this are gas hydrate sediments (Kvenvolden, 1995), especially toward and below the base of the hydrate stability zone (bottom simulating seismic reflector, where free methane gas and fluid flow occur), where bacterial processes are greatly stimulated (Fig. 2) (Wellsbury et al., 2000). Stimulation is so great that, for some bacterial activities, rates are greater at depth than at the sediment surface. These include methanogenesis, and hence active bacterial methane formation may play an important role in gas hydrate formation. Interestingly, the organic acid acetate, which is an important bacterial substrate, increases considerably beneath the hydrate zone. As gas hydrates contain twice as much carbon as other fossil fuels (Kvenvolden, 1988), elevated bacterial activity in association with these deposits may have global significance.

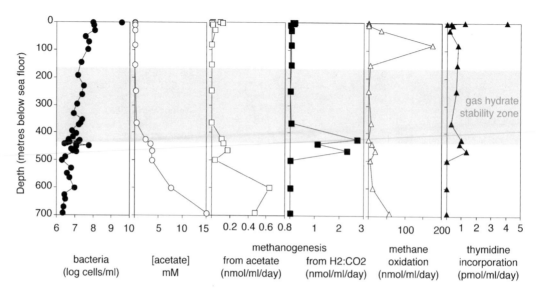

Figure 2. Bacterial populations and activity in gas hydrate sediments from Blake Ridge, Atlantic Ocean (modified from Wellsbury et al., 2000). Thymidine incorporation into bacterial DNA is a measure of growth rates.

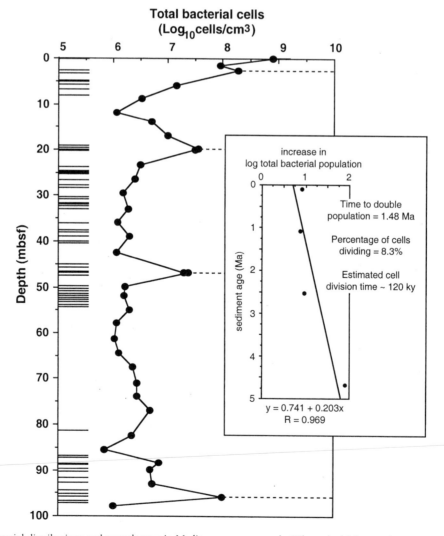

Figure 3. Bacterial distributions and growth rates in Mediterranean sapropels. Where the high organic matter sapropel layers (horizontal bars on the depth axis) were sampled (dashed horizontal lines), bacterial populations increased significantly (from Parkes et al., 2000). ky, thousands of years.

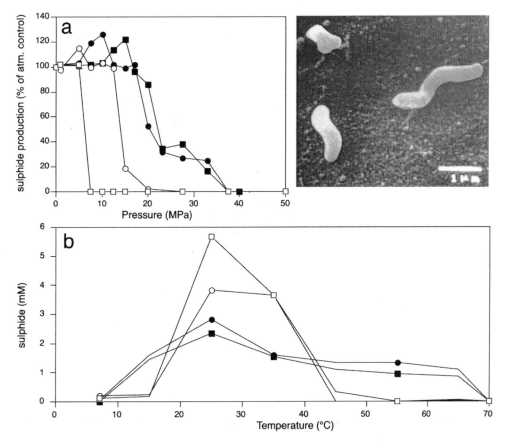

Figure 4. (a) Pressure and (b) temperature characteristics of the deep marine sediment bacterium *D. profundus* (photomicrograph) strain 80-55 [●] and strain 500-1 [■], compared with other *Desulfovibrios: D. desulfuricans* (○) and *D. salexigens* (□) (Bale, et al., 1997).

Bacterial populations also increase in zones of high organic matter. One example of this is in sapropel layers that are discrete zones of elevated organic matter (up to 30%) where bacterial populations are significantly elevated (Fig. 3) (Cragg et al., 1998) compared with the lower organic matter surrounding sediments (0 to 3%). This elevation in bacterial populations increases with the depth and age of the sapropel, suggesting that bacteria are continuously growing in the sapropel during burial. As the sapropels are accurately dated, the elevation in bacterial populations can be used to calculate growth rates. Average doubling times for dividing cells are a remarkable ~120,000 years. Although there are some uncertainties concerning the accuracy of this estimate—for example, the death rate is unknown, and a much smaller proportion of the microscopically determined dividing cell count may be actively dividing—the order of magnitude of the estimate demonstrates that bacteria in deep sediments are growing very slowly, which is consistent with their

utilization of highly recalcitrant and ancient organic matter. Under these conditions, any energy can be channeled to maintenance rather than growth, which would help to explain the surprisingly high biomasses in deep sediments (Fig. 1).

Bacterial populations also occur in the basement rock beneath sediments, where they are associated with basalt weathering (Fisk et al., 1998). Although it is not clear what energy sources they are utilizing, weathering continues over millions of years and, hence again, bacterial activity must be remarkably slow in this "geosphere" habitat.

Little is known about the physiology of bacteria in deep marine sediments as they have proved difficult to isolate (Barnes et al., 1998). However, one bacterium that has been isolated, *Desulfovibrio profundus* (Bale et al., 1997), indicates that deep biosphere bacteria are probably well adapted to their subsurface habitat, as its pressure optimum for activity corresponded with its in situ pressure (Fig. 4). In addition, *D. profundus* has an unusually wide temperature optimum (15 to 65°C), although the significance of this remains unclear. A number of possibly unique deep bacteria have also been detected by molecular genetic approaches (Rochelle et al., 1994) as well as gene sequences closely related to cultured organisms (Marchesi et al., 2001).

PETROLEUM AND COAL RESERVOIRS

Although the presence of bacteria in deep oil reservoirs had been suspected for some time (Bastin, 1926), the consensus view was that the deep subsurface was sterile. Thus, any reservoir bacteria must have been recently introduced from the surface environment, such as by drilling or other oil production operations. However, this still left the puzzle of how surface bacteria could effectively adapt so quickly to extreme subsurface conditions. In the past 10 years research has demonstrated that bacteria can be widespread in oil reservoirs and that some may be indigenous to the formations (Magot et al., 2000; Haridon et al., 1995). For example, high concentrations of hyperthermophiles have been found in oil reservoirs 3,000 m below the seabed in the North Sea and below the permafrost surface of the North Slope of Alaska (Stetter et al., 1993). Enrichments of sulfide-producing bacteria grew at 85 and 103°C, temperatures close to those in situ. Several enrichments were also able to grow anaerobically on crude oil as the sole carbon and energy source. These results suggest that bacteria may be responsible for in-reservoir souring (sulfide formation) and oil biodegradation even at these high temperatures. The role of deep bacteria in oil biodegradation is underlined by the observation that burial of the reservoir to high temperatures (80 to 90°C) may sterilize some reservoirs, or remove important oil-degrading organisms, thus preventing biodegradation, even if the reservoir is subsequently uplifted to cooler temperatures (Wilhelms et al., 2001). Petroleum reservoirs may be another example of a deep biosphere hot spot, especially as it has been estimated that these reservoirs can discharge between 3 and 16 kg of bacteria per reservoir per day. Hence, ironically, the deep oil reservoir biosphere may be responsible for inoculating the surface biosphere, rather than the other way around.

Microbial utilization of hydrocarbons probably involves interacting groups of syntrophic, anaerobic organisms, and hence populations in oil reservoirs may be complex. For example, a seven-membered consortium has been described that metabolizes saturated hydrocarbons (e.g., hexadecane) to methane (Zengler et al., 1999). This involves acetogenic bacteria that metabolize hexadecane to acetate and hydro-

gen, which are then utilized by methanogens. Bacterial methane formation from hydrocarbons may be a significant source of natural gas in oil reservoirs. In some coal fields there is also considerable bacterial methane formation. Most notable is the San Juan Coalfield, the most prolific coal bed gas basin in the world (Scott et al., 1994), where subsurface groundwater flow transports bacteria and probably nutrients through permeable rock strata to the coal beds. The bacteria metabolize hydrocarbons and other organic compounds in the coal to produce secondary biogenic gases, predominantly methane and carbon dioxide. Continuous biogenic gas production is essential to maintain the exceptional gas production of this field. Hence, the methane in nonconventional gas resources, such as coal beds and organic-rich shales, which has previously been attributed to largely thermogenic processes, may contain substantial quantities of biogenic gas.

TERRESTRIAL

In addition to the above, environments where deep bacteria have been found include Coastal Plain deposits (Ghiorse and Wobber 1989), deep aquifers (McMahon and Chapelle 1991), continental oil reservoirs (Haridon et al. 1995), Columbia River basalts (Stevens and McKinley 1995; Fisk et al. 1998), confined Cretaceous rocks (Krumholz et al. 1997) and granites (Pedersen 1997), ancient halite (Vreeland et al. 2000), sandstone and carbonate deposits (Colwell et al. 1997), caves (Smith et al. 2002), and gold mines (Takai et al. 2001). Depths exceed 3,000 m and a wide range of different bacterial groups have been detected, either by enrichment and isolation or by molecular genetic approaches (Chandler et al., 1997). Rates of bacterial activity have also been measured, these are among the lowest rates ever recorded (Chapelle and Lovely, 1990) and are 10^3 to 10^5 times lower than in modern surface sedimentary environments (Lovley and Chapelle, 1995). Bacterial distributions and activity differ between water-saturated and unsaturated formations and between different sediment types. Lower bacterial activity, for example, has been found in clays compared with sand layers (McMahon and Chapelle, 1991). This is surprising, as sands generally have lower organic matter content than clays. A probable explanation is that lower permeability restricts diffusion of electron acceptors into the clay layers, which restricts organic matter metabolism to inefficient fermentation (Fig. 5). This restricted diffusion, combined with the recalcitrance of the buried organic matter, results in low bacterial activity and populations. However, organic acids from this fermentation slowly diffuse out into the more permeable sandy lay-

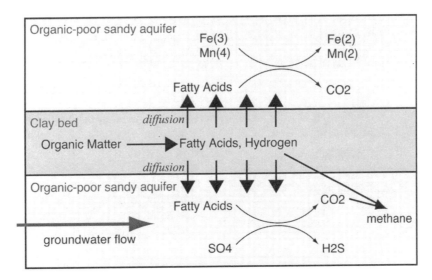

Figure 5. Model of bacterial metabolic interactions between clay and sand layers in deep terrestrial formations.

ers, where they can be respired efficiently because of the presence of electron acceptors supplied in this more open system. Again, the surprising aspect is that these reactions are apparently still occurring in Cretaceous age formations, both in permeable formations and consolidated rocks (Krumholz et al., 1997). Bacteria have been detected in fluid inclusions in Permian halite crystals from 569 m in an evaporite deposit in New Mexico (Vreeland et al., 2000). The age of this formation was 250 million years, and it is suggested that the bacteria had been entombed in the inclusion since its formation. These results have profound implications, such as the potential immortality of bacterial spores, the biochemistry that enables this, and the implications of long-term survival for transport of life between planets (Parkes, 2000), all of which have elicited considerable debate (Nickle et al., 2002). Subsequently, bacteria have been found in even older formations (425 million years ago [mya]) (Fish et al., 2002). The exact age of these bacteria may be in some doubt, but they add to the growing evidence for the presence of bacteria in geological formations.

Bacteria are also present in igneous rocks such as granites (Pedersen, 1997) and basalts (Stevens and McKinley, 1995), but due to the absence of organic matter it is more difficult to determine their energy sources. If these rocks are fractured, and there is groundwater flow, this could provide some energy sources. However, these substrates are likely to be removed in shallower zones, leaving little for deeper formations. However, H_2 generation from waterrock weathering and other reactions has been suggested as a possible geosphere energy source in such environments (Pedersen, 2000). For example, in the deep (3- to 5-km) crystalline rock aquifers of the Columbia River Basalt Group there are high concentrations of

CH_4 (up to 160 mM) and H_2 (up to 60 μM). Furthermore, the $\delta^{13}C$ of methane suggests that it is a product of bacterial methanogenesis from $H_2:CO_2$ and that this forms the base of a unique deep, autotrophically driven bacterial ecosystem (Stevens and McKinley, 1995). This anaerobic community is entirely independent of photosynthesis and hence the surface biosphere. Experiments have shown that weathering of these basalt rocks produces H_2 for the methanogens, although recently questions have been raised about the environmental applicability of these experiments (Anderson et al., 1998, 2001; Stevens and McKinley, 2000). However, despite the criticism, this aquifer environment demonstrates the potential for a bacterial ecosystem supported by abiotic processes, i.e., H_2 generation. Similarly, autotrophic methanogens and homoacetogens, which utilize subterranean hydrogen and bicarbonate, have also been detected at a 446-m depth in granitic aquifers in Sweden (Kotelnikova and Pedersen, 1997).

COLD DEEP BIOSPHERE

Glaciers cover 11% of the modern-day land surface; however, they present physiological challenges to organisms in terms of extremes of temperature, pressure, pH, desiccation, salinity, and the availability of nutrients and energy sources, and hence were considered devoid of life. A number of factors indicate that subglacial environments may be suitable microbial habitats (Sharp et al., 1999) (Fig. 6). First, the glacier bed is insulated from temperature fluctuations across the freezing point, and water is present in basal ice, at ice grain boundaries, in subglacial sediments, and subglacial cavities and channels. Second, the

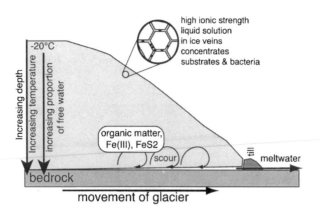

Figure 6. Glacial biosphere habitat.

production of fine-grained flour by glacier abrasion produces geochemically reactive sites for bacterial metabolism and suitable energy sources. Subglacial sediments frequently contain trace quantities of organic carbon, either from bedrock and/or soils overridden during glacial advance, and metal sulphides, both of which are bacterial energy sources. Increases in microbial populations with increased sediment concentration in subglacial environments are consistent with this energy supply (Sharp et al., 1999). NO_3^-, SO_4^{2-}, and Fe(III) introduced to glacial water by dissolution processes may also be utilized by microbes as electron acceptors. Third, basal melting and continued flow of meltwater over the bed serve to supply nutrients (e.g., N, P, S, Fe), dissolved gases, and particulate material within the microbial habitat. In addition, surface-derived meltwater may supplement the subglacial nutrient supply with snowmelt-derived NO_3^- and NH_4, particularly during the early melt season, if appropriate hydrological connections exist. Fourth, soluble and particulate impurities are routinely excluded from the ice lattice as it forms, resulting in a matrix of discreet ice crystals permeated by a network of connecting liquid-filled veins (Price, 2000). These veins contain elevated levels of common atmospherically and terrestrially derived impurities (H_2SO_4, HNO_3, HCl, NaCl, NH_4, and dissolved organic carbon), potentially providing nutrients and energy sources for bacteria. The resultant depression of the freezing point by these high ion concentrations and the ordering of the water molecules ensures that liquid water, considered a prerequisite for life, can be present in the vein network down to temperatures of $-35°C$.

The presence and activity of microbes in subglacial environments have now been demonstrated directly at several glaciers. Large bacterial populations were found in meltwater at two alpine valley glaciers, Tsanfleuron and the Haut Glacier d'Arolla,

with the highest counts displayed by subglacial borehole waters (10^6/ml) and debris-rich basal ice and till (10^7/ml) (Sharp et al., 1999). Analysis of active populations of psychrophilic bacteria from subglacial meltwater and basal ice samples from John Evans Glacier also showed that these organisms could be metabolically and morphologically diverse (Skidmore et al., 2000). Enriched bacteria included aerobic heterotrophs, anaerobic nitrate, and sulfate reducers and methanogens.

Analysis of a portion of the Vostok ice core, which is thought to contain frozen water derived from Lake Vostok, Antarctica (a body of liquid water located beneath about 4 km of glacial ice and isolated for ~1 million years), has also been shown to contain bacterial populations (Karl et al., 1999; Priscu et al., 1999). Between 10^2 and 10^4 bacterial cells per milliliter and low concentrations of potential growth nutrients were present. The gram-negative bacterial cell biomarker lipopolysaccharide was also detected at concentrations consistent with the cell enumeration data, which suggested a predominance of gram-negative bacteria in the ice. At least a portion of the microbial assemblage was active, as determined by the respiration of [^{14}C]acetate and glucose substrates during incubation at 3°C and 1 atm (see also Priscu and Christner, chapter 13).

SIGNIFICANCE OF THE DEEP BIOSPHERE

An initial estimate of the magnitude of the deep biosphere surprisingly indicates that the vast majority of bacteria on Earth reside in the subsurface (~97%), with ocean sediments being the largest intraterrestrial habitat (Whitman et al., 1998). These bacteria, however, grow extremely slowly with average division times between 1,000 and 2,000 years, which is consistent with their low-energy, deep habitat. It may be that only a small portion of the population is active at any one moment in time, thus allowing slightly increased growth for the active component. Extremely slow growth rates may be possible in the subsurface as low pore size (down to <0.2 μm) limits eukaryotic and other grazers. However, it is still difficult to understand how energy sources are available over millions of years and kilometer depths. A possible mechanism of activating organic matter during burial is by temperature increases, which are a characteristic of all deep environments due to the Earth's thermal gradient (Wellsbury et al., 1997). This has been shown to stimulate the formation of acetate from sedimentary organic matter during laboratory heating experiments, and it is proposed that other substrates associated with thermal alteration of organic matter could

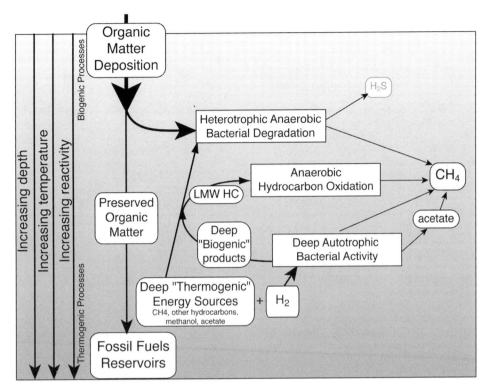

Figure 7. Schematic diagram of possible deep-biosphere energy sources and bacterial: thermogenic interactions in deep marine sediments, including increasing reactivity of buried components as temperature increases with depth (Wellsbury et al., 1997). LMW HC, low-molecular-weight hydrocarbons.

also be produced at lower temperatures and be utilized by bacteria (Fig. 7). When organic matter concentrations and bioavailability are high, as at Blake Ridge, deep acetate (Fig. 2) and dissolved organic carbon formation (Egeberg and Barth, 1998) can be greatly stimulated in situ. This temperature activation may also occur for iron minerals (Stevens and McKinley, 2000) and stimulate deep H_2 formation and hence an autotrophically driven deep biosphere (Chapelle et al., 2002).

The presence of a deep biosphere on Earth that can obtain energy from subsurface geochemical processes has clear implications for the possibility of life on other planets, whether these are hot or cold. However, we still need to know much more about our own deep biosphere as there are implications for the origin of life, biodiversity, and biosphere:geosphere interactions (e.g., climate feedbacks, gas hydrates) in addition to potential biotechnological applications such as microbial-enhanced oil recovery, petroleum and gas formation (Gold, 1999), subsurface bioremediation, and as a source of novel enzymes and compounds.

Acknowledgments. We acknowledge the Natural Environment Research Council and the European Union for funding aspects of this research. We also thank Michala Pettitt for contributions to the "Cold deep biosphere" section.

REFERENCES

Anderson, R. T., F. H. Chapelle, and D. R. Lovley. 1998. Evidence against hydrogen-based microbial ecosystems in basalt aquifers. *Science* **281**:976–977.

Anderson, R. T., F. H. Chapelle, and D. R. Lovley. 2001. Comment on "Abiotic controls on H-2 production from basalt-water reactions and implications for aquifer biogeochemistry." *Environ. Sci. Technol.* **35**:1556–1557.

Bale, S. J., K. Goodman, P. A. Rochelle, J. R. Marchesi, J. C. Fry, A. J. Weightman, and R. J. Parkes. 1997. *Desulfovibrio profundus* sp. nov., a novel barophilic sulfate-reducing bacterium from deep sediment layers in the Japan Sea. *Int. J. Sys. Bacteriol.* **47:**515–521.

Barnes, S. P., S. D. Bradbrook, B. A. Cragg, J. R. Marchesi, A. J. Weightman, J. C. Fry, and R. J. Parkes. 1998. Isolation of sulfate-reducing bacteria from deep sediment layers of the Pacific Ocean. *Geomicrobiol. J.* **15:**67–83.

Bastin, E. S. 1926. The problem of the natural reduction of sulphates. *Bull. Am. Assoc. Petrol. Geol.* **10:**1270–1299.

Berner, R. A., and A. C. Lasaga. 1989. Modeling the geochemical carbon cycle. *Sci. Am.* **260:**74–81.

Chandler, D. P., S. M. Li, C. M. Spadoni, G. R. Drake, D. L. Balkwill, J. K. Fredrickson, and F. J. Brockman. 1997. A molecular comparison of culturable aerobic heterotrophic bacteria and 16S rDNA clones derived from a deep subsurface sediment. *FEMS Microbiol. Ecol.* **23:**131–144.

Chapelle, F. H., and D. R. Lovley. 1990. Rates of microbial-metabolism in deep coastal-plain aquifers. *Appl. Environ. Microbiol.* **56:**1865–1874.

Chapelle, F. H., K. O'Neil, P. M. Bradley, B. A. Methe, S. A. Ciufo, L. L. Knobel, and D. R. Lovley. 2002. A hydrogen-based subsurface microbial community dominated by methanogens. *Nature* **415**: 312–315.

Chester, R. 1990. *Marine Geochemistry.* Unwin Hyman Ltd, London, United Kingdom.

Colwell, F. S., et al. 1997. Microorganisms from deep, high temperature sandstones: constraints on microbial colonization. *FEMS Microbiol. Rev.* **20**:425–435.

Cragg, B. A., S. M. Harvey, J. C. Fry, R. A. Herbert, and R. J. Parkes. 1992. Bacterial biomass and activity in the deep sediment layers of the Japan Sea, Hole 798B, p. 761–776. *In* K. A. Pisciotto, J. C. Ingle, Jr., M. T. Bryemann, J. Barron, et al. (ed.), *Proceeding of the Ocean Drilling Program Scientific Research,* vol. 127/128. Ocean Drilling Program, College Station, Tex.

Cragg, B. A., R. J. Parkes, J. C. Fry, A. J. Weightman, P. A. Rochelle, and J. R. Maxwell. 1996. Bacterial populations and processes in sediments containing gas hydrates (ODP Leg 146: Cascadia Margin). *Earth Planet. Sci. Lett.* **139**:497–507.

Cragg, B. A., K. M. Law, A. Cramp, and R. J. Parkes. 1998. The response of bacterial populations to sapropels in deep sediments of the Eastern Medierranean (Site 969), p. 303–307. *In* A. H. F. Robertson, K.-C. Emeis, C. Richter, and A. Camerlenghi (ed.), *Proceedings of the Ocean Drilling Program Scientific Research,* Vol. 160. Ocean Drilling Program, College Station, Tex.

Cragg, B. A., K. M. Law, G. M. O'Sullivan, and R. J. Parkes. 1999. Bacterial profiles in deep sediments of the Alboran Sea, Western Mediterranean (Sites 976–978), p. 433–438. *In* R. Zahn, M. C. Comas, and A. Klaus (ed.), *Proceedings of the Ocean Drilling Program Scientific Results,* vol. 161. Ocean Drilling Program, College Station, Tex.

Egeberg, P. K., and T. Barth. 1998. Contribution of dissolved organic species to the carbon and energy budgets of hydrate bearing deep sea sediments (Ocean Drilling Program Site 997 Blake Ridge). *Chem. Geol.* **149**:25–35.

Fish, S. A., T. J. Shepherd, T. J. McGenity, and W. D. Grant. 2002. Recovery of 16S ribosomal RNA gene fragments from ancient halite. *Nature* **417**:432–436.

Fisk, M. R., S. J. Giovannoni, and H. Furnes. 1998. Alteration of oceanic volcanic glass: textural evidence of microbial activity. *Science* **281**:978–980.

Froelich, P. N., G. P. Klinkhammer, M. L. Bender, N. A. Luedtke, G. R. Heath, D. Cullen, P. Dauphin, D. Hammond, B. Hartmen, and V. Maynard. 1979. Early oxidation of organic matter in pelagic sediments of the eastern equatorial Atlantic: suboxic diagenesis. *Geochim. Cosmochim. Acta* **43**:1075–1090.

Ghiorse, W. C., and F. J. Wobber. 1989. Special issue on deep subsurface microbiology. *Geomicrobiol. J.* **7**:1–135.

Gold, T. 1999. *The Deep Hot Biosphere.* Springer-Verlag New York, N.Y.

Haridon, S. L., A. Reysenbach, P. Glenat, D. Prieur, and C. Jeanthon. 1995. Hot subterranean biosphere in continental oil reservoir. *Nature* **377**:223–224.

Hedges, J. I., and R. G. Keil. 1995. Marine chemistry discussion paper. Sedimentary organic matter preservation: an assessment and speculative synthesis. *Mar. Chem.* **4**:81–115.

Jannasch, H. W., and G. E. Jones. 1959. Bacterial populations in seawater as determined by different methods of enumeration. *Limnol. Oceanogr.* **4**:128–139.

Karl, D. M., D. F. Bird, K. Bjorkman, T. Houlihan, R. Shackelford, and L. Tupas. 1999. Microorganisms in the accreted ice of Lake Vostock, Antarctica. *Science* **286**:2144–2147.

Kotelnikova, S. and K. Pedersen. 1997. Evidence for methanogenic Archaea and homoacetogenic Bacteria in deep granitic rock aquifers. *FEMS Microbiol. Rev.* **20**:339–349.

Krumholz, L. R., J. P. Mckinley, G. A. Ulrich, and J. M. Suflita. 1997. Confined subsurface microbial communities in Cretaceous rock. *Nature* **386**:64–66.

Kvenvolden, K. A. 1988. Methane hydrate—a major reservoir of carbon in the shallow geosphere. *Chem. Geol.* **71**:41–51.

Kvenvolden, K. A. 1995. A review of the geochemistry of methane in natural gas hydrate. *Org. Geochem.* **23**:997–1008.

Lovley, D. R., and F. H. Chapelle. 1995. Deep subsurface microbial processes. *Rev. Geophys.* **33**:365–381.

Magot, M., B. Ollivier, and B. K. C. Patel. 2000. Microbiology of petroleum reservoirs. *Antonie Leeuwenhoek Int. J. Gen. Mol. Microbiol.* **77**:103–116.

Marchesi, J. R., A. J. Weightman, B. A. Cragg, R. J. Parkes, and J. C. Fry. 2001. Methanogen and bacterial diversity and distribution in deep gas hydrate sediments from the Cascadia Margin as revealed by 16SrRNA molecular analysis. *FEMS Microbiol. Ecol.* **34**:221–228.

Martens, C. S., J. P. Chanton, and C. K. Paull. 1991. Biogenic methane from abyssal brine seeps at the base of the Florida escarpment. *Geology* **19**:851–854.

Mather, I. D., and R. J. Parkes. 2000. Bacterial populations in sediments of the Eastern Flank of the Juan de Fuca Ridge, Sites 1026 and 1027, p. 161–165. *In* A. Fisler, E. E. Davis, and C. Escutia (ed.), *Proceedings of the Ocean Drilling Program Scientific Results,* vol. 168. Ocean Drilling Program, College Station, Tex.

McMahon, P. B., and F. H. Chapelle. 1991. Microbial production of organic acids in aquitard sediments and its role in aquifer geochemistry. *Nature* **349**:233–235.

Morita, R. Y., and C. E. Zobell. 1955. Occurrence of bacteria in pelagic sediments collected during the Mid-Pacific Expedition. *Deep-Sea Res.* **3**:66–73.

Nickle, D. C., G. H. Learn, M. W. Rain, J. I. Mullins, and J. E. Mittler. 2002. Curiously modern DNA for a "250 million-year-old" bacterium. *J. Mol. Evol.* **54**:134–137.

Parkes, R. J. 2000. A case of bacterial immortality? *Nature* **407**:844–845.

Parkes, R. J., B. A. Cragg, J. C. Fry, R. A. Herbert, and J. W. T. Wimpenny. 1990. Bacterial biomass and activity in deep sediment layers from the Peru Margin. *Phil. Trans. R. Soc.* London Ser. A **331**:139–153.

Parkes, R. J., et al. 1994. Deep bacterial biosphere in Pacific-Ocean sediments. *Nature* **371**:410–413.

Parkes, R. J., B. A. Cragg, S. J. Bale, K. Goodman, and J. C. Fry. 1995. A combined ecological and physiological approach to studying sulfate reduction within deep marine sediment layers. *J. Microbiol. Methods* **23**:235–249.

Parkes, R. J., B. A. Cragg, and P. Wellsbury. 2000. Recent studies on bacterial populations and processes in subseafloor sediments: a review. *Hydrogeol. J.* **8**:11–28.

Pedersen, K. 1997. Microbial life in deep granitic rock. *FEMS Microbiol. Rev.* **20**:399–414.

Pedersen, K. 2000. Exploration of deep intraterrestrial microbial life: current perspectives. *FEMS Microbiol. Lett.* **185**:9–16.

Price, P. B. 2000. A habitat for psychrophiles in deep Antarctic ice. *Proc. Nat. Acad. Sci. USA* **97**:1247–1251.

Priscu, J. C., et al. 1999. Geomicrobiology of subglacial ice above Lake Vostok, Antarctica. *Science* **286**:2141–2144.

Rochelle, P. A., B. A. Cragg, J. C. Fry, R. J. Parkes, and A. J. Weightman. 1994. Effect of sample handling on estimation of bacterial diversity in marine-sediments by 16s ribosomal-RNA gene sequence-analysis. *FEMS Microbiol. Ecol.* **15**:215–225.

Rothschild, L. J., and R. L. Mancinelli. 2001. Life in extreme environments. *Nature* **409**:1092–1101.

Scott, A. R., W. R. Kaiser, and J. W. B. Ayers. 1994. Thermogenic and secondary biogenic gases, San Juan Basin, Colorado and New Mexico—implications for coalbed gas productivity. *Am. Assoc. Pet. Geol. Bull.* **78**:1186–1209.

Sharp, M., J. Parkes, B. Cragg, I. J. Fairchild, H. Lamb, and M. Tranter. 1999. Widespread bacterial populations at glacier beds and their relationship to rock weathering and carbon cycling. *Geology* **27**:107–110.

Skidmore, M. L., J. M. Foght, and M. J. Sharp. 2000. Microbial life beneath a high Arctic glacier. *Appl. Environ. Microbiol.* **66**:3214–3220.

Smith, S., F. Whitaker, R. Parkes, P. Smart, P. Beddows, and S. Bottrell. 2002. Active dolomitisation by saline groundwaters in the Yucatan Peninsula, Mexico. *In* J. B. Marten, C. M. Wicks, and I. D. Sasaowsky (ed.), *Hydrogeology and Biology of Post-Paleozoic Carbonate Aquifers*, Karst Waters Institute Special Publication 7. Karst Waters Institute, Charles Town, W.V.

Stetter, K. O., et al. 1993. Hyperthermophilic Archaea are thriving in deep North-Sea and Alaskan oil-reservoirs. *Nature* **365**:743–745.

Stevens, T. O., and J. P. McKinley. 1995. Lithoautotrophic microbial ecosystems in deep basalt aquifers. *Science* **270**:450–454.

Stevens, T. O., and J. P. McKinley. 2000. Abiotic controls on H-2 production from basalt-water reactions and implications for aquifer biogeochemistry. *Environ. Sci. Technol.* **34**:826–831.

Takai, K., D. P. Moser, M. DeFlaun, T. C. Onstott, and J. K. Fredrickson. 2001. Archaeal diversity in waters from deep South African gold mines. *Appl. Environ. Microbiol.* **67**:5750–5760.

Vreeland, R. H., W. D. Rosenzweig, and D. W. Powers. 2000. Isolation of a 250 million year old halotolerant bacterium from a primary salt crystal. *Nature* **407**:897–900.

Wellsbury, P., K. Goodman, T. Barth, B. A. Cragg, S. P. Barnes, and R. J. Parkes. 1997. Deep marine biosphere fuelled by increasing organic matter availability during burial and heating. *Nature* **388**:573–576.

Wellsbury, P., K. Goodman, B. A. Cragg, and R. J. Parkes. 2000. The geomicrobiology of deep marine sediments from Blake Ridge containing methane hydrate (Sites 994, 995 and 997), p. 379–391. *In* C. K. Paull, R. Matsumoto, P. J. Wallace, and W. P. Dillon (ed.), *Proceedings of the Ocean Drilling Program Scientific Results*, vol. 164. Ocean Drilling Program, College Park, Tex.

Whitman, W. B., D. C. Coleman, and W. J. Wiebe. 1998. Prokaryotes: the unseen majority. *Proc. Nat. Acad. Sci. USA* **95**:6578–6583.

Wilhelms, A., S. R. Larter, I. Head, P. Farrimond, R. di-Primio, and C. Zwach. 2001. Biodegradation of oil in uplifted basins prevented by deep-burial sterilization. *Nature* **411**:1034–1037.

Zengler, K., H. H. Richnow, R. Rossello-Mora, W. Michaelis, and F. Widdel. 1999. Methane formation from long chain alkanes by anaerobic microorganisms. *Nature* **401**:266–269.

Zobell, C. E. 1938. Studies on the bacterial flora of marine bottom sediments. *J. Sed. Petrol.* **8**:10–18.

Microbial Diversity and Bioprospecting
Edited by Alan T. Bull
© 2004 ASM Press, Washington, D.C.

Chapter 13

Earth's Icy Biosphere

John C. Priscu and Brent C. Christner

EARTH'S COLD BIOSPHERE

Earth's biosphere is cold, with 14% being polar and 90% (by volume) cold ocean at <5°C. More than 70% of Earth's freshwater occurs as ice, and a large portion of the soil ecosystem (~20%) exists as permafrost. There is even evidence that bacteria proliferate in high-altitude supercooled cloud droplets (Sattler et al. 2001), using organic acids and alcohols for growth. Microorganisms in sea ice were noted by early sailors and first studied as a scientific curiosity by Bunt (1964). At maximum extent, sea ice covers some 5% of the Northern Hemisphere and 8% of the Southern Hemisphere and accounts for about 67% of the Earth's ice cover. Despite the global significance of sea ice, only recently has it been explored for novel microorganisms (Bowman et al., 1997; Junge et al., 2002; Thomas and Dieckmann, 2002). Recent studies of microbial diversity in polar oceans have shown a predominance of *Archaea* in the subgroup *Chrenarchaeota* (DeLong et al., 1994). Previous to this discovery, *Archaea* were thought to be confined to thermal systems. Paleoclimate records for the past 500,000 years have shown that the surface temperature on Earth has fluctuated drastically, with four major glaciations occurring during this period (Petit et al., 1999). Strong evidence also exists showing that the Earth was completely ice covered during the Paleoproterozoic and Neoproterozoic periods (Kirschvink, 1992; Hoffman et al., 1998). New discoveries of microbial life in cold (−5°C) and saline lakes (Franzman et al., 1997; Priscu et al., 1999a; Takacs et al. 2001), permanent lake ice (Priscu et al., 1998; Fritsen and Priscu, 1998; Psenner et al., 1999), glacial ice (Christner et al., 2000, 2001; Skidmore et al., 2000), and polar snow (Carpenter et al., 2000) are extending the bounds of our biosphere. The recent description of potential bacterial life in Lake Vostok (Priscu et al., 1999b; Karl et al., 1999; Christner et al., 2001) and the discovery of at least 100 other Antarctic subglacial lakes extend the known boundaries for life on Earth even further. Even with the spatial and temporal records for icy systems on Earth, little is known about the psychrophilic or psychrotolerant microorganisms that inhabit them. Despite the mounting evidence for microbial life in frozen ecosystems, many textbooks limit their definitions of the biosphere to the region between the outer portion of the geosphere and the inner portion of the atmosphere, neglecting icy habitats. Clearly, we must extend the bounds of what is currently considered the Earth's biosphere to include icy systems.

COLD EXTRATERRESTRIAL LIFE?

Studies of Earthly ice-bound microbes are also relevant to the evolution and persistence of life on extraterrestrial bodies. This is particularly evident because the average temperature of the Universe is just a few degrees above absolute zero. During the transition from a clement environment to an inhospitable environment on Mars, liquid water may have progressed from a primarily liquid phase to a solid phase, and the Martian surface would have eventually become ice covered (Wharton et al., 1995). Evidence from Martian orbiter laser altimeter images has revealed that water ice exists at the poles and below the surface of Mars (Boynton et al., 2002; Malin and Carr, 1999), and studies of Martian meteorites have inferred that prokaryotes were once present (Thomas-Keprta et al., 2002). Habitats in polar ice may serve as a model for life on Mars (Priscu et al. 1998, 1999a, 1999b; Paerl and Priscu, 1998; Thomas and Dieckmann, 2002) as it cooled and may assist us in our search for extinct or extant life on Mars today. Biochemical traces of life or even viable microorganisms may well be protected from destruction if deposited within polar perennial

---

John C. Priscu and **Brent C. Christner** • Department of Land Resources and Environmental Sciences, 304 Leon Johnson Hall, Montana State University, Bozeman, MT 59717.

ice or frozen below the planet's surface. During high obliquity, increases in the temperature and atmospheric pressure at the northern pole of Mars (McKay and Stoker, 1989; Malin and Carr, 1999) could result in the discharge of liquid water that might create environments with ecological niches similar to those inhabited by microorganisms in terrestrial polar and glacial regions. Periodic effluxes of hydrothermal heat to the surface could move microorganisms from the Martian subterranean, where conditions may be more favorable for extant life (McKay, 2001). The annual partial melting of the ice caps might then provide conditions compatible with active life or at least provide water in which these microorganisms may be preserved by subsequent freezing (McKay and Stoker, 1989; Clifford et al., 2000). The microfossils and chemical signatures of potential biological origin that were recently discovered in Alan Hills meteorite ALH84001 reinvigorated the debate over the possibility of life on Mars (McKay et al., 1996; Thomas-Keprta et al., 2002). However, such circumstantial evidence will require confirmation by scientific missions to explore and study the frozen surface of Mars.

Surface ice on Europa, one of the moons of Jupiter, appears to exist in contact with subsurface liquid water (Greenburg et al., 2000; Kivelson et al., 2000). Geothermal heating and the tidal forces generated by orbiting Jupiter are thought to maintain a 50- to 100-km-deep liquid ocean on Europa with perhaps twice the volume of the Earth's ocean (Chyba and Phillips, 2001), but beneath an ice shell at least 3 to 4 km thick (Turtle and Pierazzo, 2001). Cold temperatures (<128 K [Orton et al., 1996]) combined with intense levels of radiation would appear to preclude the existence of life on the surface, and the zone of habitability (i.e., where liquid water is stable) may be present only kilometers below the surface where sunlight is unable to penetrate (Chyba and Hand, 2001). Europa's surface appears strikingly similar to terrestrial polar ice floes, suggesting that the outer shell of ice is periodically exchanged with the underlying ocean. The ridges in the crust and the apparent rafting of dislocated pieces imply that subterranean liquid water flows up through stress-induced tidal cracks, which may then offer provisional habitats at shallow depth for photosynthesis or other forms of metabolism (Gaidos and Nimmo, 2000; Greenberg et al., 2000). Gaidos et al. (1999) argue that without a source of oxidants, Europa's subsurface ocean would be destined to reach chemical equilibrium, making biologically dependent redox reactions thermodynamically impossible. However, the surface is continually bombarded with high-energy particles, producing molecular oxygen and peroxides, as well as formaldehyde and other organic carbon sources (Chyba, 2000; Chyba and

Hand, 2001); and it is conceivable that Europan microbial life might subsist without employing photosynthetic or chemoautotrophic lifestyles. In this scenario, mixing between the crust and the subsurface need be the only mechanism required to provide organics and oxygen at levels sufficient to support life (Chyba, 2000). Tidal heat generation and electrolysis could also provide sources of energy that could be coupled to bioenergetic redox reactions (Greenberg et al., 2000). The vast network of Antarctic subglacial lakes that lie ~4 km beneath the permanent ice sheet provides an earthly analog in the search for life on Europa and a model system to develop the noncontaminating technologies that will be required to sample Europa.

EVOLUTION OF LIFE ON A FROZEN EARTH

The biology of permanently cold environments on our own planet has received relatively little investigation. Similar to their high-temperature counterparts, frozen ecosystems are dominated by microorganisms. Expectations of commercial applications and interest in the early evolution of life have led many researchers to examine microorganisms, including cyanobacteria, in thermal systems. Based on the deep-rooted phylogeny of thermophilic bacteria and archaea in the tree of life (i.e., small-subunit rRNA phylogeny), in concert with extensive geothermal activity during the early evolution of our planet, it is generally thought that life on Earth evolved in a hot environment (Huber et al., 2000; Pederson, 1997). Recent considerations about the evolution of life, however, have suggested that a "hot start" was probably not the only alternative for the origin of life. Although there are strong arguments for a thermal origin of life based on small-subunit ribosomal RNA (16S/18S rRNA) phylogenetic relationships, the validity of this relationship is questioned by researchers who believe that phylogenies are strongly biased by the use of just a single gene for the construction of the tree of life. If lateral gene transfer is common among all prokaryotic organisms (Nelson et al., 1999), then a hierarchical universal classification is difficult or impossible (Pennisi, 1998, 1999), and evolutionary patterns must be reassessed (Doolittle, 1999). A hot origin of life is also not supported by new results from phylogenetic trees based on genes that do not code for rRNA, chemical experiments with alternative structure for the nucleic acid backbone (Eschenmoser, 1999), considerations about the thermal stability of basic molecules found in all organisms, and statistical analysis of the G+C content of DNA (Galtier et al., 1999). Adaptation to life in hot environments may even have occurred late in evolutionary time (Balter, 1999).

Although much more research is required to determine whether life originated in hot or cold environments, it is highly probable that cold environments have acted as a refuge for life during major glaciations. Recent evidence has indicated that, around 600 million years ago during the Neoproterozoic, early microbes endured an ice age with such intensity that even the tropics froze over (Hoffman et al., 1998; Hoffman and Schrag, 2000; Schrag and Hoffman, 2001). According to this hypothesis, known as the Snowball Earth Hypothesis, the Earth would have been completely ice covered for 10 million years or more, with ice thickness exceeding 1 km. Only the deepest oceans would have contained liquid water. One of the primary criticisms of the Snowball Earth Hypothesis is that the thick ice cover over the world ocean would cut off the supply of sunlight to organisms in the seawater below and thereby eliminate photosynthesis and all life associated with photosynthetic carbon production. Others have concluded that global-scale freezing would extinguish all surface life (Williams et al., 1998). Only the hardiest of microbes would have survived this extreme environmental circumstance, and perhaps icy refuges or hot springs on the seafloor and terrestrial surface may have served as oases for life during these lengthy crises. Hoffman and Schrag (2000), Vincent and Howard-Williams (2001), and Vincent et al. (2002) suggest that photosynthetic cyanobacteria and bacteria, similar to those found in the permanent ice covers of contemporary polar systems, may have acted as an icy refuge during this period, until postsnowball melting introduced conditions suitable for activity in terrestrial and marine habitats. The resultant high concentration of microbes in these icy environments would favor intense chemical and biological interactions between species, which could entice the development of symbiotic associations, and perhaps influence eukaryotic development through evolutionary time (Vincent et al., 2002). Although this "density speeds evolution" theory has been considered primarily in the context of thermal microbial mat communities (Margulis and Sagan, 1997), it is possible that ice-bound habitats also provided opportunities for microbial evolution and the acquired biological innovations may have triggered the Cambrian explosion, which occurred immediately after the last snowball Earth event (Knoll, 1994; Hoffman and Schrag, 2000; Kirschvink et al., 2000). Ironically, the deeprooted phylogeny of thermophilic species, generally interpreted as evidence for the origin of life under hot circumstances, may instead be the consequence of an evolutionary "bottleneck" imposed by the extremes of multiple snowball Earth events (Kirschvink et al., 2000).

It is clear that a great diversity of icy environments make up Earth's cold biosphere. Given the space constraints of this book, we dedicate the remainder of this chapter to describing research conducted in our own laboratories on the newly discovered life associated with permanent Antarctic lake ice, glaciers and ice sheets (polar and temperate), and subglacial Antarctic lakes.

PERMANENT ANTARCTIC LAKE ICE

The McMurdo Dry Valleys form the largest ($\sim$4,000-km^2) ice-free expanse of land on the Antarctic continent. Meteorological conditions in the dry valleys reveal the extreme conditions that organisms must overcome to survive (Priscu et al., in press; Doran et al., 2002a, 2002b, in press). Surface air temperatures average $-27.6°C$, and there are on average only about 6.2 degree-days per year above freezing (temperature above freezing times the number of days above freezing each year [Doran et al., 2002]). These conditions produce the only permanently ice-covered lakes on Earth. During studies on the biogeochemistry of nitrous oxide in McMurdo Dry Valleys lakes, Priscu and coworkers (Priscu et al., 1996; Priscu, 1997) observed a peak in nitrous oxide associated with a sediment layer 2 m beneath the surface of the 4-m-thick ice cover of Lake Bonney (Priscu et al., 1998). This observation together with elevated levels of chlorophyll *a*, particulate organic carbon, particulate organic nitrogen, ammonium, and dissolved organic carbon (DOC) (Wing and Priscu, 1993) led to the hypothesis that phototrophic and heterotrophic microorganisms were present within the lake ice and were metabolically active. Subsequent research showed that adequate liquid water was produced during summer (Fritsen et al., 1998; Adams et al., 1998) to support an active prokaryotic ecosystem within the ice consisting of cyanobacteria and a diversity of bacterial species (e.g., Priscu et al., 1998; Paerl and Priscu, 1998; Pinckney and Paerl, 1996; Gordon et al., 2000).

The phylogenetic diversity of bacteria and cyanobacteria colonizing sediment particles at a depth of 2.5 m in the permanent ice cover of Lake Bonney was characterized by analyses of 16S rRNA genes amplified from environmental DNA (Gordon et al. 1996, 2000). An rRNA gene clone library of 198 clones was made and characterized by sequencing and oligonucleotide probe hybridization. The library was dominated by representatives of the cyanobacteria, proteobacteria, and Planctomycetales, but also contained diverse clones representing the *Acidobacterium* and *Holophaga* division, the Green Non-Sulfur division, and the *Actinobacteria* (Fig. 1). Of the cyanobacterial

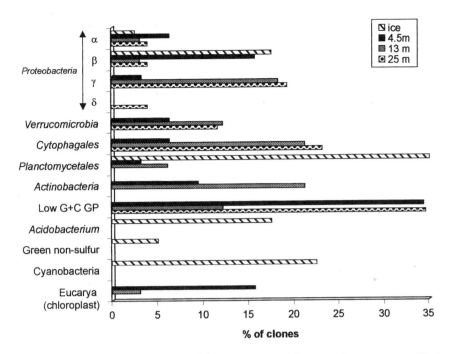

Figure 1. Lake Bonney 16S rDNA summary comparing lake ice sequences with water column sequences. The ice sample was collected about 2 m beneath the surface of the 4-m-thick permanent ice cover; the 4.5- and 13-m samples were from the east lobe, and the 25-m sample was from the west lobe of Lake Bonney. See Priscu *et al.* (1997) for hydrographic characteristics of the water column of these lake basins and Priscu et al. (1998) for details of the ice column. GP, gram positive.

gene clusters characterized, only one was closely (>97% similarity) affiliated with a well-characterized cyanobacterial species, *Chamaesiphon subglobosus*. The remaining cyanobacterial gene clusters were less than 93% similar to any characterized sequences in public databases although they resembled the *Leptolyngbya* sp. and *Phormidium* sp. Oligonucleotide probes made from three lake ice cyanobacterial clusters were used to screen environmental 16S rDNA samples obtained from the terrestrial (soil and stream) environment in the vicinity of Lake Bonney and Lake Fryxell. The probes designed to hybridize to cyanobacterial 16S rRNA genes effectively hybridized to each sample, indicating that the cyanobacterial sequences present in the lake ice of Lake Bonney are also found in terrestrial cyanobacterial mat samples. Molecular characterization (PCR amplification of the *nifH* fragments) of the *nifH* gene (encoding for the highly conserved Fe-protein subunit) of nitrogenase in lake ice sediments from Lake Bonney also demonstrated the presence of a diverse diazotrophic assemblage (Olson et al., 1998). The *nifH* analysis suggested that phototrophic cyanobacteria and heterotrophic microorganisms have the potential to fix atmospheric nitrogen when liquid water is present in the ice cover. The expression of nitrogenase was confirmed by the acetylene reduction assay for nitrogenase activity (Grue et al., 1996; Olson et al., 1998; Paerl and Priscu,

1998). Sequence analysis in concert with physiological data indicates that the cyanobacterial (and bacterial) community within the lake ice is dominated by organisms that did not evolve in the lake ice ecosystem. Instead, the strong katabatic winds common to the region act to disperse microorganisms in the desert environment and provide the biological seed for the lake ice microbial assemblage (Gordon et al., 2000; Priscu et al., 1998). Different prokaryotic sequence data from selected depths within the water column of Lake Bonney further corroborate this contention (Fig. 1), confirming that distinctly different microbial communities exist within the ice and water column. Although cyanobacteria, *Planctomyces*, and *Acidobacteria* dominate the assemblages within lake ice, the most frequently identified water column species are members of the *Verrucomicrobia*, *Cytophagales*, and low-G+C gram-positive bacteria; none of the water column groups were detected in the ice cover. Preliminary analysis of the 16S rDNA sequences obtained at depth in Lake Bonney revealed that ~70% of the clones obtained had high identity with species from marine and lake habitats, with more than half of these sequences being most similar to isolates and clones from polar marine or lake environments (J. P. Zehr and J. C. Priscu, unpublished). The occurrence of related phylotypes from geographically diverse but predominantly cold aquatic environments argues that these species probably have physiologies adapted for survival, persistence, and activity at low temperature.

A majority of the cyanobacterial and bacterial activity within the dry valley lake ice was associated with sediment aggregates, as opposed to individual microorganisms embedded in the ice matrix. A core of Lake Bonney ice collected from 0.2 to 0.3 m showed considerable differences between clean and sediment-laden layers based on epifluorescence counts of material stained with SYBR gold, a probe specific for DNA (Chen et al., 2001). Based on the SYBR gold-staining study, the average bacterial and viruslike particle (VLP) densities in clear ice were 2.29×10^3 cells ml^{-1} and 1.23×10^4 VLPs ml^{-1}, respectively (Lisle and Priscu, in press). The sediment-laden portion of this ice core had a bacterial density of 1.15×10^4 cell ml^{-1} and a VLP count of 2.77×10^4 ml^{-1}. The virus:bacteria ratio was 5.37 for the clear ice and 2.41 for the sediment sections of the core. These ratios are within the range

of those seen in more temperate climates that include eutrophic and oligotrophic waters (Wommack and Colwell, 2000) but are lower than those in the water columns of freshwater lakes of Signey Island, Antarctica (Wilson et al., 2000), and Lake Hoare, Antarctica (Kepner et al., 1998). There were on average 5.02-fold more bacteria and 2.25-fold more VLPs in the Lake Bonney ice core sample that included the sediment than in the clear ice section of the core. The virus:bacteria ratio for sediment-laden ice was less than half of that observed in the clear ice, implying that there are fewer viruses per bacterium in the sediment-containing section of the core. It remains unclear if the VLPs were bacteriophage or cyanobacteriophage. However, the presence of viral particles in the ice indicates that phage may play a major role in genetic transfer and overall survival of prokaryotes in the ice.

Figure 2 shows the lake ice assemblage on several scales. Microautoradiographic studies reveal that

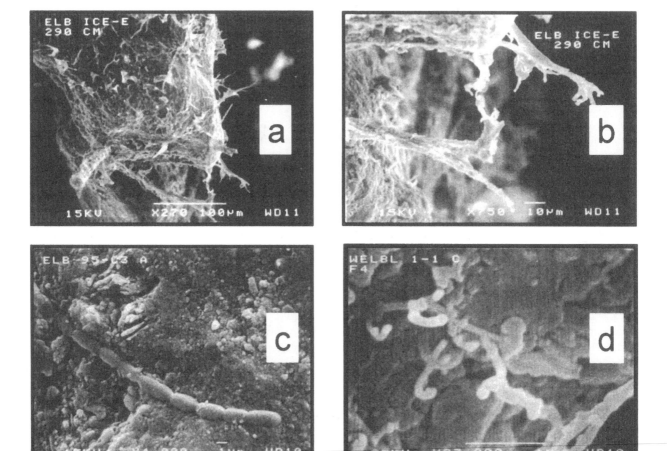

Figure 2. Scanning electron microscope (SEM) images of microbial assemblages collected 2 m beneath the surface of the east lobe Lake Bonney ice cover. (a) and (b) represent low- and high-magnification images of cyanobacterial filaments attached to lithogenic material; (c) a single cyanobacterial filament attached to a surface, (d) small unknown organic filaments attached to a surface. Images were obtained by cryogenic SEM (JEOL-6100 SEM with an Oxford Instruments cryogenic preparation stage) on particles captured by 0.2-μm filtration of melted ice.

both bacterial and cyanobacterial activities were closely associated with sediment particles, corroborating experimental results (Priscu et al., 1998). Microautoradiographs also indicated that virtually all of the incorporation of radiolabeled organic substrates was mediated by nonautofluorescent (nonchlorophyll-containing) bacterial-size rods (0.5 to 1 μm in length) and filaments (0.5 μm wide) closely associated with aggregates, whereas $^{14}CO_2$ incorporation was limited to filamentous cyanobacteria (Paerl and Priscu, 1998). Heterotrophic bacteria were attached to soil particles and associated with cyanobacterial colonies and aggregates. These observations are similar to those reported for temperate and tropical cyanobacteria-dominated systems (Paerl and Pinckney, 1996). Tetrazolium salt reduction assays further revealed that, when melting occurs, localized O_2 consumption associated with aggregates is sufficient to create reduced microzones. These microzones are associated with regions colonized by bacteria and cyanobacteria, suggesting that they may be potential sites for O_2-sensitive processes such as atmospheric nitrogen fixation (Olson et al., 1998; Paerl and Priscu, 1998). Pinkney and Paerl (1996) showed that cyanobacterial and bacterial biomass and activities were heterogeneously distributed among aggregates, promoting the development of O_2 and, possibly, other biogeochemical gradients. Biogeochemical zonation and diffusional O_2 and nutrient concentration gradients likely result from microscale patchiness in microbial metabolic activities (i.e., photosynthesis, respiration). These gradients, in turn, promote metabolic diversity and differential photosynthetic and heterotrophic growth rates.

Phototrophy, heterotrophy, and diazotrophy (N_2 fixation) can occur simultaneously in ice-aggregate microbial communities. Mineralization of particulate organic carbon and particulate organic nitrogen is highly dependent on organic matter availability, the main source being cyanobacterial photosynthesis. Therefore, close spatial proximity of heterotrophs to phototrophs is essential for completion of carbon, nitrogen, and phosphorus cycling. The paucity of higher trophic levels in the ice (e.g., protozoans) magnifies the importance of microbial interactions within the ice assemblage and amplifies the role played by viruses in terms of microbial survival and possibly diversity. Clearly, the spatial and temporal relationships within the ice produce a microbial consortium that is of fundamental importance for initiating, maintaining, and optimizing essential life-sustaining production and nutrient transformation processes (Priscu et al., in press). The close spatial and temporal coupling of metabolites within the microbial consortium appears to be essential for the microbes to survive and replicate

in what has been characterized as "the edge of life" (Paerl and Priscu, 1998). Data on microbial activity for the ice assemblage indicate that metabolic complementation among functionally diverse, but structurally simple, prokaryotic consortia along microscale biogeochemical gradients is a unique and effective strategy for meeting the requirements of life in what appears to be an otherwise inhospitable environment.

CRYOCONITE HOLES

Cryoconite holes form as windblown particulates accumulate on the surface of a glacier, are warmed by the Sun, and melt into the ice, producing a cylindrical basin of liquid water (Fig. 3). Cells released from the melted glacial ice and deposited attached to airborne particulates inoculate these environments with viable organisms. Primary production by algae and cyanobacteria supply sufficient reduced carbon and nutrients to support complex microbial and invertebrate communities, and cryoconite hole ecosystems occur globally in Arctic (Gerdel and Drouet, 1960; De Smet and Van Rompu, 1994; Grøngaard et al., 1999; Mueller et al., 2001), Antarctic (Wharton et al., 1981; Christner et al., 2003b), and alpine glaciers (Kohshima, 1989; Takeuchi et al., 2000). During the austral summer in the McMurdo Dry Valleys, the 24 daylight hours and increased temperature enable liquid water to exist on the glacial surface, and the melting process is greatly accelerated in cracks and depressions within the ice that collect heat-absorbing sediments (Wharton et al., 1985). Under these circumstances, aquatic communities based on algal and cyanobacterial photosynthetic primary production develop, but are destined to refreeze, and presumably become inactive, through the cold, dark winter months.

Although dominated by microorganisms, cryoconite holes are one of the few environments in the dry valleys inhabitable by metazoan life, and the resident rotifer, tardigrade, and nematode species have the ability to differentiate under adverse conditions into metabolically dormant forms and, as such, could possibly also survive within a cryoconite sediment that is completely frozen (Spmme, 1996). Every cryoconite hole formed is unique and therefore may support a novel and discrete ecosystem. However, based on results from a phylogenetic survey of a cryoconite hole on the Canada glacier (Christner et al., 2003b), these ecosystems are inhabited by species very similar to those in adjacent microbial mat and lake ice communities in this polar desert environment (Priscu et al., 1998; Gordon et al., 2000). Thus, particulates blown onto the glacier from adja-

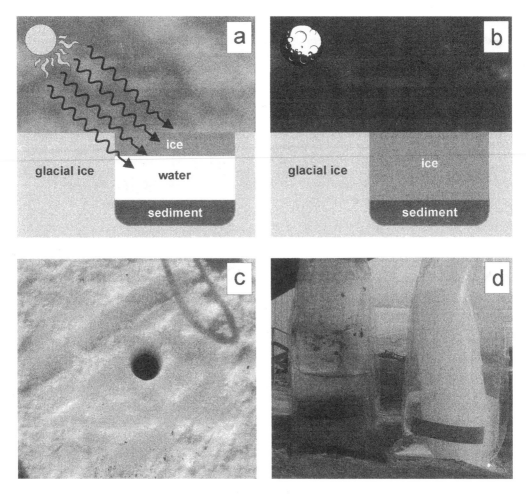

Figure 3. The cryoconite hole environment in the McMurdo Dry Valleys. In summer, sediment collects on glacial surfaces, and exposure to solar irradiation produces (a) melt pools within the ice, which may subsequently freeze on the surface (b) and completely freeze during the winter. The cryoconite hole illustrated in (c) was located on the Canada glacier and was completely frozen when sampled in January 2001. (d) A comparison of cores retrieved from the cryoconite hole (left) with a core from the adjacent glacial ice. Note the dense layer of sediment and organic material present within the bottom 5 cm of the cryoconite hole core.

cent locations are responsible for seeding cryoconite holes with biological material. Although these environments become completely frozen during the Antarctic winter, upon summer warming and glacial melting, the surviving members of these communities might serve in reverse to ensure the reseeding of surrounding environments. The notion that cryoconite holes serve as biological refuges in this very cold, essentially desert environment warrants more detailed investigation.

LIFE IN GLACIAL ICE

Snowfall accumulates into continental ice sheets in the polar regions and globally at high altitudes (Fig. 4). Depending on the topological nature of the accumulation environment, high-elevation ice fields are termed valley or alpine glaciers and termed icecaps when a flat bedrock surface or volcanic crater is completely covered in ice. The expansive icecaps of Greenland and Antarctica cover ~10% of Earth's terrestrial surface with ice and contain ~70% of the fresh water on the planet (Patterson, 1994). Earth's climate is currently in an interglacial stage of a 100,000-year cycle, caused largely by episodic changes in the planet's axial tilt and ellipticity of its orbit around the Sun. During the last glacial maximum 18,000 years ago, sea levels were $\approx$120 m lower than today and the north polar icecap advanced to cover 5 million square kilometers, blanketing what is now Canada and half of the United States (Hughes, 1998).

Archived chronologically within glacial ice are samples of the atmospheric constituents from different times in the past, including biological material such as insects, plant fragments, seeds, pollen grains, fungal spores, viruses, and bacteria (Abyzov, 1993; Abyzov et al., 1998; Dancer et al., 1997; Castello et al., 1999; Willerslev et al., 1999; Christner et al., 2000, 2003; Zhang et al., 2001). Ice cores extending thousands of meters below the glacial surface can represent hundreds of thousands of years of snowfall accumulation, and the assemblages of microorganisms immured chronologically within a core are species that were distributed in the atmosphere at different times in history. Studies indicate that the topography, local and global environmental conditions, and proximity of ecosystems contributing biological particles to a particular air mass influence the concentration and diversity of airborne microorganisms (Lighthart and Shaffer, 1995; Giorgio et al., 1996; Fuzzi et al., 1997; Marshall and Chalmers, 1997). Consistent with this, ice core samples from nonpolar, high-altitude glaciers contain a greater number and variety of culturable bacterial species than polar ices, and, similarly, the highest recoveries from polar ice cores were obtained from Antarctic regions adjacent to exposed soils and rock surfaces of the McMurdo Dry Valleys complex (Christner et al., 2000). Hence, increased microbial deposition occurs in glaciers contiguous to

environments that supply airborne rock grains and soils, which presumably serve to transport and protect attached microorganisms.

Aerosolized microorganisms can travel large distances on atmospheric currents, often in a viable but dormant state. Remarkably, some air conditions actually provide a medium for growth, and microbial metabolism has been detected in fog particles (Fuzzi et al., 1997) and supercooled clouds (Sattler et al., 2001). For an airborne microorganism deposited in glacial ice to retain viability, the stress associated with desiccation, solar irradiation, freezing, an extended period of no growth, and subsequent thawing must not result in a lethal level of unrepairable cellular damage. It is therefore not surprising that many of the species that are isolated form spores, structures known to confer resistance to environmental abuses. Many also have thick cell walls or polysaccharide capsules and resist repeated cycles of freezing and thawing. Regardless of the ice cores' geographical source, related but not identical species are frequently recovered. Interestingly, members of the bacterial genera *Sphingomonas*, *Acinetobacter*, and *Arthrobacter* are commonly isolated from glacial samples (Christner

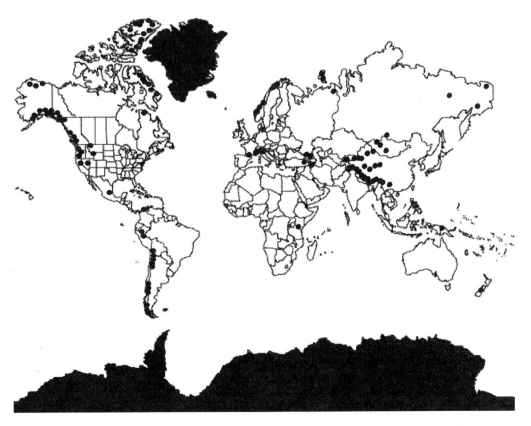

Figure 4. Global locations of existing glacial ice sheets and caps (denoted by shading). At each geographical location, the nearest terrestrial or marine ecosystem that would most likely contribute the majority of airborne particles are very different.

et al., 2000, 2001, 2003a), and these are also the most frequently isolated genera in enrichment surveys of terrestrial subsurface environments (Balkwill et al., 1997). As such, these genera would appear to contain species that can survive for extended times under low-nutrient, nongrowth conditions, and similar survival strategies may be in effect in deep ice and subsurface situations.

Viable cells and nucleic acids remain preserved for hundreds of thousands of years in glacial ice (Abyzov, 1993; Christner et al., 2003a). Although many glacial isolates appear to possess features that might enhance their survival while dormant (i.e., ability to form spores), the thermodynamic reality is that, in the absence of metabolic activity, cells must incur a significant amount of macromolecular damage over such long periods of time. Temperature and the hydration of nucleic acids and protein strongly influence the rate of depurination and L-amino acid racemization, respectively (Lindahl, 1993; Bada et al., 1994). Amino acids in amber have retarded racemization rates, with the observed stereochemical preservation attributed to the anhydrous nature of amber (Poinar et al., 1996), and this could also pertain to ice. It is also possible that entrapped microbes carry out a slow rate of metabolism which allows repair of macromolecular damage, but not growth. Thin veins of liquid water between ice crystals could potentially provide a microbial habitat within apparently solid ice (Mader et al., 1992a,b; Price, 2000). Studies of permafrost (Rivkina et al., 2000), surface snow (Carpenter et al., 2000), and frozen bacterial suspensions (Fig. 5) have demonstrated low levels of metabolic activity at subzero temperatures. Based on the minimum required input to avoid microbial carbon loss, Price (2000) calculated that the vein environment contains sufficient carbon and nutrients to support a small population of cells (10 to 10^2 cells ml^{-1}) for hundreds of thousands of years. Indirect evidence for microbial activity in glacial ice was obtained when analysis of the air bubbles in cores from Vostok Station, Antarctica, and Sajama, Bolivia, revealed isotopic fractionation profiles consistent with in situ microbiological production of nitrous oxide and methane, respectively (Sowers, 2001; T. Sowers, personal communication). Geochemical anomalies attributed to microbial activity in Greenland ice have also been reported (Souchez et al., 1995, 1998), and this issue must now be experimentally addressed, particularly with respect to the interpretation of paleoclimate records obtained from ice cores.

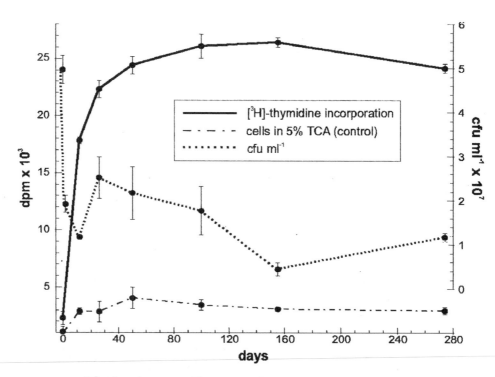

Figure 5. Incorporation of [^{3}H]thymidine into trichloroacetic acid (TCA)-precipitable material and the number of CFU mL^{-1} for the glacial isolate *Psychrobacter* sp. Trans 1 after 9 months at −15°C. Cells in logarithmic growth were suspended in distilled water with 1 μCi of [^{3}H]thymidine, frozen rapidly at −70°C, and incubated at −15°C for an extended period. Under these circumstances, cells were able to conduct a low level of macromolecular synthesis, but this activity was not sufficient for reproductive growth. For more details, see Christner 2002.

SUBGLACIAL LAKES

Much attention is currently focused on the exciting possibility that the subglacial environments of Antarctica may harbor microbial ecosystems under thousands of kilometers of ice, isolated from the atmosphere for as long as the continent has been glaciated (20 to 25 million years, [Naish et al., 2001]). The discovery of more than 70 subglacial lakes in central Antarctica during the early 1970s (Siegert et al. 1996; Siegert, 2000) went relatively unnoticed by the biological scientific community; however, curiosity about the nature of these environments has recently intensified as a result of the discovery of other large subglacial lakes (McKay et al., in press; Tobacco et al., 1998; Tikku et al., 2002) and increasing international interest in the largest subglacial lake, Lake Vostok. The freshwater in Lake Vostok originates from the overlying ice sheet, which melts near the shoreline of the lake and at the ice-water interface in the north and is thought to have a relatively high dissolved oxygen content supplied from air bubbles released from overlying glacial ice (Lipenkov and Barkov, 1998; Lipenkov and Istomin, 2001; Lipenkov et al. 2000). Lake water refreezes (accretes) at the base of the ice sheet in the central and southern regions, removing water from the lake (Kapitsa et al., 1996; Jouzel et al., 1999; Siegert et al., 2001; Bell et al., 2002). Hence, constituents in the accretion ice should reflect those in the actual lake water in a proportion equal to the partitioning that occurs when the water freezes (Priscu et al., 1999; Karl et al., 1999, Siegert et al., 2001). Vostok ice coring reached a record depth of 3,623 m in 1998, but due to concerns regarding contamination, drilling stopped at ~120 m above the lake-ice interface; the deepest part of the core recovered 150 m of accretion ice (Petit et al., 1999). The presence of microorganisms within Lake Vostok accretion ice has now been confirmed independently by at least three laboratories (Priscu et al., 1999b; Karl et al., 1999; Christner et al., 2001). Molecular profiling of accreted ice microbes using 16S rDNA techniques (Priscu et al., 1999b; Christner et al., 2001) showed close agreement with present-day surface microbiota. Phylotypes have mapped closely to extant members of the *Alpha*- and *Betaproteobacteria* and to *Actinomycetes* (the latter also isolated in Vostok glacial ice [Abyzov et al., 1998]). If the accreted ice microbes are representative of the lake microbiota, this would imply that microbes within Lake Vostok do not represent an evolutionarily distinct subglacial biota (Siegert et al., 2003). The time scale of isolation within Lake Vostok (~20 to 25 million years) is not long in terms of prokaryotic evolution compared with its 3.7×10^9-year history on Earth, and studies of species divergence of other

prokaryotes have shown that species-level divergence may take ~100 million years (Lawrence and Ochman, 1998). However, other mechanisms of genetic change (such as recombination and mutator genes) could allow more rapid alteration of organism phenotype allowing for adaptation to conditions within Lake Vostok (Page and Holmes, 1998). An alternative scenario is that glacial meltwater entering the lake forms a lens overlying the Vostok water column. If so, the microbes discovered within accretion ice would likely have spent little time within the actual lake water itself (few, if any, cell divisions occurring) before being frozen within the accretion ice. The microbes within the main body of the lake below such a freshwater lens may have originated primarily from basal sediments and rocks, and if so, their period of isolation may be adequate for significant evolutionary divergence, particularly given the potential selection pressures that exist within the subglacial environment.

A recent report on microbial diversity in Lake Vostok accretion ice, using PCR-based analyses of 16S rDNA in accretion ice (Bulat et al., 2002), has revealed three phylotypes closely related to DNA signatures representative of thermophiles. One of them is a known extant chemolithoautotroph identified previously in hot springs and capable of obtaining energy by oxidizing H_2S at reduced O_2 tension. Two other taxa are not identified in the current databases, but showed relatedness to bacteria associated with hydrothermal vents and associated surface sediments. Evidence for the presence of thermophiles is supported by the recent interpretation by French scientists of He^3/He^4 data from accretion ice (Petit et al., 2002). This interpretation now implies that there may be extensive faulting beneath Lake Vostok resulting in hydrothermal activity in the southern part of the lake. If this emerging picture is correct, Lake Vostok could harbor a unique assemblage of organisms fueled by chemical energy. Although it seems inevitable that viable microorganisms from the overlying glacial ice and in sediment scoured from bedrock adjacent to the lake are regularly seeded into the lake, the question remains whether these or preexisting microorganisms have established a flourishing community within Lake Vostok. If a microbial ecosystem were found to exist within the water or sediment of this subsurface environment, it would represent one of the most extreme and unusual environments on Earth.

Abyzov et al. (1998) measured bacterial cell densities ranging from 1.2×10^3 to 8.3×10^3 in Vostok glacial ice between 1,665 and 2,750 m and showed that density was positively correlated with atmospheric microparticles within the ice. The highest cell and atmospheric particle densities occurred during a glacial period indicating that paleoclimate

Table 1. Summary of the bacterial cell number and organic carbon contribution from Antarctic subglacial lakes and the Antarctic and Greenland ice sheets[a]

Parameter	Antarctica			Greenland ice sheet	Both poles, lakes + ice sheet	Global		
	Lakes	Ice sheet	Total			Fresh waters	Open ocean	Soils
Cell number ($\times 10^{25}$)	1.20	8.84	10.04	0.77	10.81	13.1	10,100	26,000
Cell-C (Pg) ($\times 10^{-3}$)	0.33	2.44	2.77	0.21	2.99	3.63	2,790	26,000
DOC (Pg)	0.01	3.31	3.32	0.29	3.61	NA	NA	NA
Cell-C+DOC (Pg)	0.02	3.32	3.34	0.29	3.62	NA	NA	NA

[a] Carbon concentrations are in petagrams (Pg = 10^{15} g). Ice sheet DOC was estimated assuming an ice concentration of 0.11 mg liter^{-1}; cell number and cell-C for the Greenland ice sheet assume bacterial densities similar to those measured in the Antarctic ice sheet. Global estimates for cell number and cell carbon (cell-C) are from Whitman et al. (1998). Freshwaters represent all surface rivers and lakes, excluding subglacial lakes. NA, not computed.

has a major role in the distribution and perhaps type of microorganisms found in the ice. Priscu et al. (1999b) used epifluorescence microscopy of DNA-stained cells and scanning electron microscopy to measure bacterial cell densities of 2.8×10^3 and 3.6×10^4, respectively, in Vostok accretion ice (3590 m). Priscu et al. further reported DOC concentrations in core 3590 of 0.51 mg l^{-1}. Based on these values and estimates of partitioning coefficients for the water to ice phase change, they estimated that the water in Lake Vostok had bacterial cell concentrations of 10^5 to 10^6 ml^{-1} and a DOC concentration of 1.2 mg l^{-1}. Using data on the volume of the Antarctic ice sheet and subglacial lakes (Siegert, 2000) in concert with published bacterial volume-to-carbon conversion factors (Riemann and Spndergaard, 1986), we estimated the cell number and carbon content within the Antarctic ice sheet and Antarctic subglacial lakes (Table 1). Our estimates assume that concentrations of bacterial cells and DOC in all subglacial lakes are similar to those estimated for Lake Vostok. Based on these calculations, subglacial lakes contain about 12% of the total cell number and cell carbon with respect to the pools associated with the Antarctic continent (e.g., subglacial lakes plus ice sheet). The number of prokaryotic cells we estimate for subglacial lakes plus the ice sheet (10.04×10^{25} cells) is about 50-fold higher than that estimated for prokaryotic cells in Antarctic sea ice (2.2×10^{24} cells) and about 25-fold higher than sea ice in both polar regions combined (4×10^{24} cells) (Whitman et al., 1998). The prokaryotic carbon content of subglacial lakes plus the ice sheet is about 76% of that estimated for all of Earth's liquid surface fresh waters (rivers plus lakes, excluding Antarctic subglacial lakes) but more than 3 orders of magnitude less than that in the open ocean and soils. The number of cells we estimate for Antarctic subglacial lakes alone (1.20×10^{25}) approaches that reported by Whitman et al. (1998) for the Earth's surface freshwater lakes and rivers combined (13.1×10^{25}). To our knowledge, no data on cell density have been published for the Greenland ice sheet. Assuming that prokaryotic cell densities in the Greenland ice sheet are the same as those measured in the Vostok ice core, we estimate that Greenland adds about 8% to the carbon pools we estimated for the Antarctic ice sheet. Our seminal estimates of the number of prokaryotes and organic carbon associated with polar ice and Antarctic subglacial lakes are clearly tentative and should be refined once additional data become available. These calculations, however, do imply that polar ice, particularly Antarctic ice, contains an organic carbon reservoir that should be considered when addressing issues concerning global carbon dynamics.

EVIDENCE FOR COLD-ADAPTED MICROBIAL SPECIES

Molecular-based approaches to microbial ecology yield data that measure the natural evolutionary relationships between microorganisms. As such, a phylogenetic comparison of the species inhabiting similar environments provides a way to examine biogeographical relationships—an essential prerequisite to determine global biodiversity and resolve the ecological role of species distributed throughout the biosphere. A number of studies have examined bacterial and archaeal biogeography in soil, microbial mat, hot spring, hydrothermal vent, oil reservoir, and polar seawater and ice environments (Stetter et al., 1993; Garcia-Pichel et al., 1996; Fulthorpe et al., 1998; Staley and Gosink, 1999; Reysenbach and Shock, 2002; Van Dover et al., 2001; Hollibaugh et al., 2002). Cosmopolitan microorganisms are found globally

(Stetter et al., 1993; Garcia-Pichel et al., 1996; Fulthorpe et al., 1998; Hollibaugh et al., 2002), and Staley and Gosink (1999) hypothesized that endemism results from the inability of certain species to survive dissemination through air or water to other locations suitable for colonization.

Figure 6 illustrates the phylogenetic relatedness, based on 16S rDNA identity, between bacteria recovered in our laboratories and by others (Benson et al., 2000) from Antarctica and permanently cold nonpolar locales. As indicated, these psychrophilic and psychrotrophic isolates originate from locations ranging from aquatic and marine ecosystems to terrestrial soils and glacial ice, with little in common between these environments except that all are permanently cold or frozen. The isolation of related species from such diverse frozen environments argues that clades in these bacterial genera evolved under cold circumstances and likely possess similar strategies to survive freezing and remain active at low temperature. Although not possible through analysis of a single gene, a polyphasic approach could reveal patterns of conserved inheritance and divergence from a common ancestor or identify parallel evolutionary pathways. Such information, coupled with a dedicated effort to further investigate microbial diversity within the planet's frozen realms, will provide the perspective necessary to understand the evolution and ecological impacts of microbial ecosystems residing within Earth's icy biosphere.

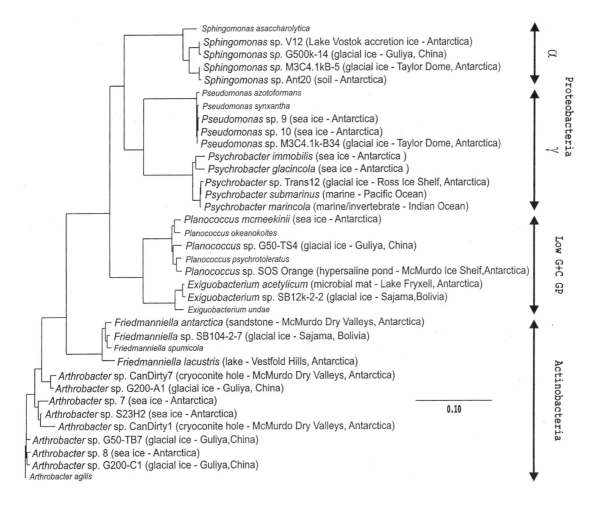

Figure 6. Phylogenetic analysis of bacteria obtained in microbiological surveys of permanently cold and frozen environments. Isolates from cold habitats are shown in bold, followed by the source environment and geographical location. The 16S rDNA sequences corresponding to nucleotides 27-1492 of the *Escherichia coli* 16S rDNA were aligned based on secondary structure and used to construct this neighbor-joining tree. The scale bar represents 0.1 fixed substitutions per nucleotide position. GP, gram positive.

CONCLUDING REMARKS

From a microbial perspective, our planet's zone of habitation (the biosphere) extends from high into the atmosphere to the inner depths of the Earth, where temperatures rise with depth to exceed those assumed possible for known carbon-based life, to the bottom of ice sheets where temperatures rarely exceed 0°C. Based on information gathered over the past 5 years, it is clear that the role of permanently cold ecosystems in global ecology must be reassessed and included in formal definitions of Earth's biosphere. As such we propose that the Earth's cryosphere and associated sub-ice lakes should be included as biospheric components of our planet. Cryosphere is defined here as that portion of the Earth's surface where water is in a solid form as snow or ice. Water in its solid form includes sea ice, freshwater ice, snow, glaciers, ice sheets, and frozen ground.

Examining permanently ice-covered habitats and microorganisms preserved for extended periods within ice is also relevant to astrobiological discussions of past or present life on Mars or in the subsurface ocean of Europa and the concept that planetary bodies may not be biologically isolated. We are rapidly reaching a point in our search for the origins of life on Earth that studies must be extended beyond our own planet. Clearly our efforts will be enhanced if we increase our sample size beyond one. Such remote and seemingly inconsequential frozen environments may harbor as yet undiscovered microbial ecosystems that could shed light on the natural history and evolution of life on a frozen Earth, as well as other icy planets and moons in the solar system.

REFERENCES

Abyzov, S. S. 1993. Microorganisms in the Antarctic ice, p. 265–295. *In* E. I. Friedmann (ed.), *Antarctic Microbiology.* Wiley-Liss, Inc., New York, N.Y.

Abyzov, S. S., I. N. Mitskevich, and M. N. Poglazova. 1998. Microflora of the deep glacier horizons of central Antarctica. *Microbiology* (Moscow) 67:66–73.

Adams, E. E., J. C. Priscu, C. H. Fritsen, S. R. Smith, and S. L. Brackman. 1998. Permanent ice covers of the McMurdo Dry Valley Lakes, Antarctica: bubble formation and metamorphism. *In* J. C. Priscu (ed.), *Ecosystem Dynamics in a Polar Desert: The McMurdo Dry Valleys, Antarctica. Ant. Res. Ser.* 72:281–296.

Bada, J. L., X. S. Wang, H. N. Poinar, S. Paabo, and G. O. Poinar. 1994. Amino acid racemization in amber-entombed insects: implications for DNA preservation. *Geochim. Cosmochim. Acta* 58:3131–3135.

Balkwill, D. L., R. H. Reeves, G. R. Drake, J. Y. Reeves, F. H. Crocker, M. B. King, and D. R. Boone. 1997. Phylogentic characterization of bacteria in the subsurface microbial culture collection. *FEMS Microbiol. Rev.* 20:201–216.

Balter, M. 1999. Did life begin in hot water? *Science* 280:31.

Bell, R. E., M. Studinger, A. A. Tikku, G. K. C. Clarke, M. M. Gutner, and C. Meertens. 2002. Origin and fate of Lake Vostok water refrozen to the base of the East Antarctic ice sheet. *Nature* 416:307–310.

Benson, D. A., I. Karsch-Mizrachi, D. J. Lipman, J. Ostell, B. A. Rapp, and D. L. Wheeler. 2000. GenBank. *Nucl. Acids Res.* 28:15–18.

Bowman, J. P., S. A. McCammon, M. V. Brown, D. S. Nichols, and T. A. McMeekin. 1997. Diversity and association of psychrophilic bacteria in Antarctic sea ice. *Appl. Environ. Microbiol.* 63:3068–3078.

Boynton, W. V., W. C. Feldman, S. W. Squyres, T. H. Prettyman, J. Bruckner, L. G. Evans, R. C. Reedy, R. Starr, J. R. Arnold, D. M. Drake, P. A. J. Englert, A. E. Metzger, I. Mitrofanov, J. I. Trombka, C. d'Uston, H. Wanke, O. Gasnault, D. K. Hamara, D. M. Janes, R. L. Marcialis, S. Maurico, I. Mikheeva, G. J. Taylor, R. Tokar, and C. Shinohara. 2002. Distribution of hydrogen in the near surface of Mars: evidence for subsurface ice deposits. *Science* 297:81–85.

Bulat, S. A., I. A. Alekhina, M. Blot, J.-R. Petit, D. Waggenbach, V. Y. Lipenkov, D. Raynaud, and V. V. Lukin. 2002. Thermophiles microbe signature in Lake Vostok, Antarctica. American Geophysical Union Spring 2002 Meeting. Washington, D.C.

Bunt, J. S. 1964. Primary productivity under sea ice in Antarctic waters. 2. Influence of light and other factors on photosynthetic activities of Antarctic marine microalgae. *Antarct. Res.* 1:27–31.

Carpenter, E. J., S. Lin, and D. G. Capone. 2000. Bacterial activity in South Pole snow. *Appl. Environ. Microbiol.* 66:4514–4517.

Castello, J. D., S. O. Rogers, W. T. Starmer, C. M. Catranis, L. Ma, G. D. Bachand, Y. Zhao, and J. E. Smith. 1999. Detection of tomato mosaic tobamovirus RNA in ancient glacial ice. *Polar Biol.* 22:207–212.

Chen, F., J. Lu, B. Binder, Y. Liu, and R. Hodson. 2001. Application of digital image analysis and flow cytometry to enumerate marine viruses stained with SYBR Gold. *Appl. Environ. Microbiol.* 67:539–545.

Christner, B. C. 2002. Incorporation of DNA and protein precursors into macromolecules by bacteria at −15°C. *Appl. Environ. Microbiol.* 68:6435–6438.

Christner, B. C., E. Mosley-Thompson, L. G. Thompson, V. Zagorodnov, K. Sandman, and J. N. Reeve. 2000. Recovery and identification of viable bacteria immured in glacial ice. *Icarus* 144:479–485.

Christner, B. C., E. Mosley-Thompson, L. G. Thompson, and J. N. Reeve. 2001. Isolation of bacteria and 16S rDNAs from Lake Vostok accretion ice. *Environ. Microbiol.* 3:570–577.

Christner, B. C., E. Mosley-Thompson, L. G. Thompson, and J. N. Reeve. 2003a. Recovery of bacteria from ancient ice. *Environ. Microbiol.* 5:433–436.

Christner, B. C., B. H., Kvitko, and J. N. Reeve. 2003b. Molecular identification of bacteria and eukarya inhabiting an Antarctic cryoconite hole. *Extremophiles* 7:177–183.

Chyba, C. F. 2000. Energy for microbial life on Europa. *Nature* 403:381–382.

Chyba, C. F., and K. P. Hand. 2001. Life without photosynthesis. *Science* 292:2026–2027.

Chyba, C. F., and C. B. Phillips. 2001. Possible ecosystems and the search for life on Europa. *Proc. Natl. Acad. Sci. USA* 98:801–804.

Clifford, S. M., D. Crisp, D. A. Fisher, K. E. Herkenhoff, S. E. Smrekar, P. C. Thomas, D. D. Wynn Williams, R. W. Zurek, J. R. Barnes, B. G. Bills, E. W. Blake, et al. 2000. The state and future of Mars polar science and exploration. *Icarus* 144:210–242.

Dancer, S. J., P. Shears, and D. J. Platt. 1997. Isolation and characterization of coliforms from glacial ice and water in Canada's high Arctic. *J. Appl. Microbiol.* 82:597–609.

DeLong, E. F., K. Y. Wu, B. B. Prezelin, and R. V. M. Jovine. 1994. High abundance of Archaea in Antarctic marine picoplankton. *Nature* 371:695–697.

De Smet, W. H., and E. A. Van Rompu. 1994. Rotifera and tardigrada from some cryoconite holes on a Spitsbergen (Svalbard) glacier. *Belg. J. Zool.* 124:27–37.

Doolittle, W. F. 1999. Phylogenetic classification and the universal tree. *Science* 284:2124–2128.

Doran, P. T., J. C. Priscu, W. B. Lyons, J. E. Walsh, A. G. Fountain, D. M. McKnight, D. L. Moorhead, R. A. Virginia, D. H. Wall, G. D. Clow, C. H. Fritsen, C. P. McKay, and A. N. Parsons. 2002a. Antarctic climate cooling and terrestrial ecosystem response. *Nature* 415:517–520.

Doran, P. T., C. P. McKay, G. D. Clow, G. L. Dana, A. G. Fountain, T. Nylen, and W. B. Lyons. 2002b. Valley floor climate observations from the McMurdo Dry Valleys, Antarctica, 1986–2000. *J. Geophys. Res.* 107(D24, 4772):1–12.

Eschenmoser, A. 1999. Chemical etiology of nucleic acid structure. *Science* 284:2118–2124.

Franzmann, P. D., Y. Liu, D. L. Balkwill, H. C. Aldrich, E. Conway de Marcario, and D. R. Boone. 1997. *Methanogenium frigidum* sp. nov., a psychrophilic, H_2-using methanogen from Ace Lake, Antarctica. *Int. J. Syst. Bacteriol.* 47:1068–1072.

Fritsen, C. H., and J. C. Priscu. 1998. Cyanobacterial assemblages in permanently ice covers on Antarctic lakes: distribution, growth rate, and temperature response of photosynthesis. *J. Phycol.* 34:587–597.

Fritsen, C. H., E. E. Adams, C. M. McKay, and J. C. Priscu. 1998. Permanent ice covers of the McMurdo Dry Valley Lakes, Antarctica: liquid water content. *In* J. C. Priscu (ed.), *Ecosystem Dynamics in a Polar Desert: The McMurdo Dry Valleys, Antarctica. Antarct. Res. Ser.* 72:269–280.

Fulthorpe, R. R., A. N. Rhodes, and J. M. Tiedje. 1998. High levels of endemicity of 3-chlorobenzoate-degrading soil bacteria. *Appl. Environ. Microbiol.* 64:1620–1627.

Fuzzi, G., P. Mandrioli, and A. Perfetto. 1997. Fog droplets—an atmospheric source of secondary biological aerosol particles. *Atmos. Environ.* 31:287–290.

Gaidos, E. J., and F. Nimmo. 2000. Tectonics and water on Europa. *Nature* 405:637.

Gaidos, E. J., K. H. Nealson, and J. L. Kirschvink. 1999. Life in ice-covered oceans. *Science* 284:1631–1633.

Galtier, N., N. Tourasse, and M. Gouy. 1999. A non-hyperthermophilic common ancestor to extant life forms. *Science* 283:220–222.

Garcia-Pichel, F., L. Prufert-Bebout, and G. Muyzer. 1996. Phenotypic and phylogenetic analyses show *Microcoleus chthonoplastes* to be a cosmopolitan cyanobacterium. *Appl. Environ. Microbiol.* 62:3284–3291.

Gerdel, R. W., and F. Drouet. 1960. The cryoconite of the Thule Area, Greenland. *Trans. Am. Microsc. Soc.* 79:256–272.

Giorgio, C. D., A. Krempff, H. Guiraud, P. Binder, C. Tiret, and G. Dumenil. 1996. Atmospheric pollution by airborne microorganisms in the city of Marseilles. *Atmos. Environ.* 30:155–160.

Gordon, D. A., B. Lanoil, S. Giovannoni, and J. C. Priscu. 1996. Cyanobacterial communities associated with mineral particles in Antarctic lake ice. *Antarct. J. US* 31:224–225.

Gordon, D. A., J. C. Priscu, and S. Giovannoni. 2000. Distribution and phylogeny of bacterial communities associated with mineral particles in Antarctic lake ice. *Microb. Ecol.* 39:197–202.

Greenberg, R., P. Geissler, B. R. Tufts, and G. V. Hoppa. 2000. Habitability of Europa's crust: the role of tidal-tectonic processes. *J. Geophys. Res.* 105:17551–17562.

Grøngaard, A., P. J. A. Pugh, and S. J. McInnes. 1999. Tardigrades, and other cryoconite biota, on the Greenland ice sheet. *Zool. Anz.* (Germany) 238:211–214.

Grue, A. M., C. H. Fritsen, and J. C. Priscu. 1996. Nitrogen fixation within permanent ice covers on lakes in the McMurdo Dry Valleys, Antarctica. *Antarct. J. US* 2:218–220.

Hoffman, P. F., and D. P. Schrag. 2000. Snowball Earth. *Sci. Am.* 282:68–75.

Hoffman, P. F., A. J. Kaufman, G. P. Halverson, and D. P. Schrag. 1998. A neoproterozoic snowball Earth. *Science* 281:1342–1346.

Hollibaugh, J. T., N. Bano, and H. W. Ducklow. 2002. Widespread distribution in polar oceans of a 16S rRNA gene sequence with affinity to *Nitrosospira*-like ammonia-oxidizing bacteria. *Appl. Environ. Microbiol.* 68:1478–1484.

Huber, R., H. Huber, and K. O. Stetter. 2000. Towards the ecology of hyperthermophiles: biotopes, new isolation strategies and novel metabolic properties. *FEMS Microbiol. Rev.* 24:615–623.

Hughes, T. J. 1998. *Ice Sheets.* Oxford University Press Inc., New York, N.Y.

Jouzel, J., J. R. Petit, R. Souchez, N. I. Barkov, V. Y. Lifenkov, D. Raymond, M. Stievenard, N. I. Vassiliev, V. Verbeke, and F. Vimeux. 1999. More than 200 meters of lake ice above subglacial Lake Vostok, Antarctica. *Science* 286:2138–2141.

Junge, K., F. Imhoff, T. Staley, and J. W. Deming. 2002. Phylogenetic diversity of numerically important Arctic sea-ice bacteria cultured at subzero temperatures. *Microb. Ecol.* 43:315–328.

Kapitsa, A. P., J. K. Ridley, G. deQ. Robin, M. J. Siegert, and I. A. Zotikov. 1996. A large deep freshwater lake beneath the ice of central East Antarctica. *Nature* 381:684–686.

Karl, D. M., D. F. Bird, K. Björkman, T. Houlihan, R. Shackelford, and L. Tupas. 1999. Microorganisms in the accreted ice of Lake Vostok, Antarctica. *Science* 286:2144–2147.

Kepner, R. L., R. A. Wharton, Jr., and C. A. Suttle. 1998. Viruses in Antarctic lakes. *Limnol. Oceanogr.* 43:1754–1761.

Kirschvink, J. L. 1992. Late Proterozoic low-latitude global glaciation: the Snowball Earth, p. 51–52. *In* J. W. Schopt, C. Klein, and D. Des Maris (ed.), *The Proterozoic Biosphere: A Multidisciplinary Study.* Cambridge University Press, Cambridge, United Kingdom.

Kirschvink, J. L., E. J. Gaidos, L. E. Bertani, N. J. Beukes, J. Gutzmer, L. N. Maepa, and R. E. Steinberger. 2000. Paleoproterozoic snowball Earth: extreme climatic and geochemical global change and its biological consequences. *Proc. Natl. Acad. Sci. USA* 97:1400–1405.

Kivelson, M. G., K. K. Khurana, C. T. Russell, M. Volwerk, R. J. Walker, and C. Zimmer. 2000. Galileo magnetometer measurements: a stronger case for a subsurface ocean at Europa. *Science* 289:1340–1343.

Knoll, A. H. 1994. Proterozoic and early Cambrian protists: evidence for accelerating evolutionary tempo. *Proc. Natl. Acad. Sci. USA* 91:6743–6750.

Kohshima, S. 1989. Glaciological importance of microorganisms in the surface mud-like material and dirt layer particles of the Chongce Ice Cap and Gozha Glacier, West Kunlun Mountains, China. *Bull. Glacier Res.* (Japan) 7:59–65.

Lawrence, J. G., and H. Ochman. 1998. Molecular archaeology of the *Escherichia coli* genome. *Proc. Natl. Acad. Sci. USA* 95:9413–9417.

Lighthart, B., and B. T. Shaffer. 1995. Airborne bacteria in the atmospheric surface layer: temporal distribution above a grass seed field. *Appl. Environ. Microbiol.* 61:1492–1496.

Lindahl, T. 1993. Instability and decay of the primary structure of DNA. *Nature* 362:709–715.

Lipenkov, V. Y., and N. I. Barkov. 1998. Internal structure of the Antarctic ice sheet as revealed by deep core drilling at Vostok station, p. 31–35. *In Lake Vostok Study: Scientific Objectives and Technological Requirements. Abstracts of an International*

Workshop (24 to 26 March 1998). Arctic and Antarctic Research Institute, St. Petersburg, Russia.

Lipenkov, V. Y., and V. A. Istomin. 2001. On the stability of air clathrate-hydrate crystals in subglacial Lake Vostok. *Mater. Glyatsiol. Issled.* [Data Glaciol. Stud.] **91**:138–149.

Lipenkov, V. Y., N. I. Barkov, and A. N. Salamatin. 2000. Istoriya klimata i oledeneniya Antarktidy po rezul'tatam izucheniya ledanogo kerna so stantsii Vostok [The history of climate and glaciation of Antarctica from results of the ice core study at Vostok Station]. *Probl. Arktiki Antarkt.* [Probl. Arctic Antarct.] **72**:197–236.

Lisle, J. T., and J. C. Priscu. The occurrence of lysogenic bacteria and microbial aggregates in the lakes of the McMurdo Dry Valleys, Antarctica. *Microb. Ecol.*, in press.

Mader, H. 1992a. Observations of the water-vein system in polycrystalline ice. *J. Glaciol.* **38**:333–347.

Mader, H. 1992b. The thermal behaviour of the water-vein system in polycrystalline ice. *J. Glaciol.* **38**:359–374.

Malin, M. C., and M. H. Carr. 1999. Groundwater formation of Martian valleys. *Nature* **397**:589–591.

Margulis, L., and D. Sagan. 1997. *Micro-Cosmos: Four Billion Years of Microbial Evolution*, p. 304. University of California Press, Berkeley, Calif.

Marshall, W. A., and M. O. Chalmers. 1997. Airborne dispersal of Antarctic algae and cyanobacteria. *Ecography* **20**:585–594.

McKay, C. P. 2001. The deep biosphere: lessons for planetary exploration, p. 315–327. *In* J. K. Fredrickson and M. Fletcher (ed.), *Subsurface Microbiology and Biogeochemistry*, Wiley-Liss Inc., New York, N.Y.

McKay, C. P., K. P. Hand, P. T. Dolan, D. T. Anderson, and J. C. Priscu. Clathrate formation and the fate of noble and biologically useful gases in Lake Vostok, Antarctica. *Geophys. Res. Lett.*, in press.

McKay, C. P., and C. R. Stoker. 1989. The early environment and its evolution on Mars: implications for life. *Rev. Geophys.* **27**:189–214.

McKay, D. S., E. K. Gibson, K. L. Thomas-Keptra, H. Vali, S. Romanek, S. J. Clemett, X. D. F. Chillier, C. R. Maechling, and N. Zare. 1996. Search for past life on Mars: possible relic biogenic activity in martian meteorite ALH84001. *Science* **273**:924–930.

Mueller, D. R., W. F. Vincent, W. H. Pollard, and C. H. Fritsen. 2001. Glacial cryoconite ecosystems: a bipolar comparison of algal communities and habitats, p. 173–197. *In* J. Elster, J. Seckbach, W. F. Vincent, and O. Lhotsky (ed.), *Algae and Extreme Environments; Ecology and Physiology. Proceedings of the International Conference, 11 to 16 September 2000, Trebon, Czech Republic.* J. Cramer, Berlin, Germany.

Naish, T. R., K. J. Woolfe, P. J. Barrett, G. S. Wilson, C. Atkins, S. M. Bohaty, C. J. Bücker, M. Claps, F. J. Davey, G. B. Dunbar, A. G. Dunn, C. R. Fielding, F. Florindo, M. J. Hannah, D. M. Harwood, S. A. Henrys, L. A. Krissek, M. Lavelle, J. van der Meer, W. C. McIntosh, F. Niessen, S. Passchier, R. D. Powell, A. P. Roberts, L. Sagnotti, R. P. Scherer, C. P. Strong, F. Talarico, K. L. Verosub, G. Villa, D. K. Watkins, P. N. Webb, and T. Wonik 2001. Orbitally induced oscillations in the East Antarctic ice sheet at the Oligocene/Miocene boundary. *Nature* **413**:719–723.

Nelson, K. E., R. A. Clayton, S. R. Gill, M. L. Gwinn, R. J. Dodson, D. H. Haft, E. K. Hickey, J. D. Peterson, W. C. Nelson, K. A. Ketchum, L. McDonald, T. R. Utterback, J. A. Malek, K. D. Linher, M. M. Garrett, A. M. Stewart, M. D. Cotton, M. S. Pratt, C. A. Phillips, D. Richardson, J. Heidelberg, G. G. Sutton, R. D. Fleischmann, J. A. Eisen, O. White, S. L. Salzberg, H. O. Smith, J. C. Venter, and C. M. Fraser. 1999. Evidence for lateral gene transfer between Archaea and Bacteria from genome sequence of *Thermotoga maritime*. *Nature* **399**:323–328.

Olson, J. B., T. F. Steppe, R. W. Litaker, and H. W. Paerl. 1998. N_2-fixing microbial consortia associated with the ice cover of Lake Bonney, Antarctica. *Microb. Ecol.* **36**:231–238.

Orton, G. S., J. R. Spencer, L. D. Travis, T. Z. Martin, and L. K. Tamppari. 1996. Galileo photopolarimeter-radiometer observations of Jupiter and the Galilean satellites. *Science* **274**:389–391.

Paerl, H. W., and J. L. Pinckney. 1996. Ice aggregates as a microbial habitat in Lake Bonney, dry valley lakes, Antarctica: nutrient-rich micro-ozones in an oligotrophic ecosystem. *Antarct. J. US* **31**:220–222.

Paerl, H. W., and J. C. Priscu. 1998. Microbial phototrophic, heterotrophic, and diazotrophic activities associated with aggregates in the permanent ice cover of Lake Bonney, Antarctica. *Microb. Ecol.* **36**:221–230.

Page, R. R. M., and E. C. Holmes. 1998. *Molecular Evolution: A Phylogenetic Approach*, p. 352. Blackwell Science, Oxford, United Kingdom.

Patterson, W. S. B. 1994. *The Physics of Glaciers*, 3rd ed. Elsevier Science Inc., Tarrytown, N.Y.

Pederson, K. 1997. Microbial life in deep granitic rock. *FEMS Microbiol. Rev.* **20**:399–414.

Pennisi, E. 1998. Genome data shake tree of life. *Science* **280**:672–674.

Pennisi, E. 1999. Is it time to uproot the tree of life? *Science* **284**:1305–1307.

Petit, J.-R., J. Jouzel, D. Raynaud, N. I. Barkov, J. M. Barnola, I. Basile, M. Benders, J. Chappellaz, M. Davis, G. Delaygue, M. Dolmotte, V. M. Dotlyakov, M. Legrand, V. Y. Lipendoc, C. Lorius, L. Pepin, C. Ritz, F. Saltzman, and M. Stievenard. 1999. Climate and atmospheric history of the past 420,000 years from the Vostok ice core, Antarctica. *Nature* **399**: 429–436.

Petit, J.-R., C. Ritz, P. Jean Baptiste, R. Souchez, V. Y. Lipenkov, and A. Salamatin. 2002. Hot spots in Lake Vostok? American Geophysical Union Spring 2002 Meeting. Washington, D.C.

Pinckney, J. L., and H. W. Paerl. 1996. Lake ice algal phototroph community composition and growth rates, Lake Bonney, Dry Valley Lakes, Antarctica. *Antarct. J. US* **31**:215–216.

Poinar, H. N., M. Hoss, J. L. Bada, and S. Paabo. 1996. Amino acid racemization and the preservation of ancient DNA. *Science* **272**:864–866.

Price, B. P. 2000. A habitat for psychrophiles in deep Antarctic ice. *Proc. Natl. Acad. Sci. USA* **97**:1247–1251.

Priscu, J. C. 1997. The biogeochemistry of nitrous oxide in permanently ice-covered lakes of the McMurdo Dry Valleys, Antarctica. *Glob. Change Biol.* **3**:301–305.

Priscu, J. C., M. T. Downes, and C. P. McKay. 1996. Extreme super-saturation of nitrous oxide in a permanently ice-covered Antarctic Lake. *Limnol. Oceanogr.* **41**:1544–1551.

Priscu, J. C., C. H. Fritsen, E. E. Adams, S. J. Giovannoni, H. W. Paerl, C. P. McKay, P. T. Doran, D. A. Gordon, B. D. Lanoil, and J. L. Pinckney. 1998. Perennial Antarctic lake ice: an oasis for life in a polar desert. *Science* **280**:2095–2098.

Priscu, J. C., C. F. Wolf, C. D. Takacs, C. H. Fritsen, J. Laybourn-Parry, E. C. Roberts, and W. Berry Lyons. 1999a. Carbon transformations in the water column of a perennially ice-covered Antarctic Lake. *Bioscience* **49**:997–1008.

Priscu, J. C., E. E. Adams, W. B. Lyons, M. A. Voytek, D. W. Mogk, R. L. Brown, C. P. McKay, C. D. Takacs, K. A. Welch, C. F. Wolf, J. D. Kirschtein, and R. Avci. 1999b. Geomicrobiology of subglacial ice above Lake Vostok, Antarctica. *Science* **286**:2141–2144.

Priscu, J. C., C. H. Fritsen, E. E. Adams, H. W. Paerl, J. T. Lisle, J. E. Dore, C. F. Wolf, and J. Milucki. Perennial Antarctic lake ice: a refuge for cyanobacteria in an extreme environment. *In* S. O. Rogers and J. Castello (ed.), *Life in Ancient Ice*. Princeton University Press, Princeton, N. J., in press.

Psenner, R., B. Sattler, A. Willie, C. H. Fritsen, J. C. Priscu, M. Felip, and J. Catalan. 1999. Lake ice microbial communities in alpine and Antarctic lakes. p. 17–31. *In* P. Schinner and R. Margesin (ed.), *Adaptations of Organisms to Cold Environments.* Springer-Verlag, New York.

Reysenbach, A. L., and E. Shock. 2002. Merging genomes with geochemistry in hydrothermal ecosystems. *Science* **296:**1077–1082.

Riemann, B., and M. Søndergaard. 1986. *Carbon Dynamics in Eutrophic, Temperate Lakes.* Elsevier, Amsterdam, The Netherlands.

Rivkina, E. M., E. I. Friedmann, C. P. McKay, and D. A. Gilichinsky. 2000. Metabolic activity of permafrost bacteria below the freezing point. *Appl. Environ. Microbiol.* **66:**3230–3233.

Sattler, B., H. Puxbaum, and R. Psenner. 2001. Bacterial growth in supercooled cloud droplets. *Geophys. Res. Lett.* **28:**239–242.

Schrag, D. P., and P. F. Hoffman. 2001. Life, geology and snowball Earth. *Nature* **409:**306.

Siegert, M. J. 2000. Antarctic subglacial lakes. *Earth-Sci. Rev.* **50:**29–50.

Siegert, M. J., J. A. Dowdeswell, M. R. Gorman, and N. F. McIntyre. 1996. An inventory of Antarctic subglacial lakes. *Antarct. Sci.* **8:**281–286.

Siegert, M. J., R. Kwok, C. Mayer, and B. Hubbard. 2000. Water exchange between subglacial Lake Vostok and the overlying ice sheet. *Nature* **403:**643–646.

Siegert, M. J., J. C. Ellis-Evans, M. Tranter, C. Mayer, J.-R. Petit, A. Salamatin, and J. C. Priscu. 2001. Physical, chemical and biological processes in Lake Vostok and other Antarctic subglacial lakes. *Nature* **414:**603–609.

Siegert, M. J., M. Tranter, J. C. Ellis-Evans, J. C. Priscu, and W. B. Lyons. 2003. The hydrochemistry of Lake Vostok and the potential for life in Antarctic subglacial lakes. *Hydro. Process.* **17:**795–814.

Skidmore, M. L., J. M. Foght, and M. J. Sharp. 2000. Microbial life beneath a high Arctic glacier. *Appl. Environ. Microbiol.* **66:**3214–3220.

Sömme, L. 1996. Anhydrobiosis and cold tolerance in tardigrades. *Eur. J. Entomol.* **93:**349–357.

Souchez, R., M. Janssens, M. Lemmens, and B. Stauffer. 1995. Very low oxygen concentration in basal ice from Summit, Central Greenland. *Geophys. Res. Lett.* **22:**2001–2004.

Souchez, R., A. Bouzette, H. B. Clausen, S. J. Johnsen, and J. Jouzel. 1998. A stacked mixing sequence at the base of the Dye 3 core. *Geophys. Res. Lett.* **25:**1943–1946.

Sowers, T. 2001. The N_2O record spanning the penultimate deglaciation from the Vostok ice core. *J. Geograph. Res.* **106:**31903–31914.

Staley, J. T., and J. J. Gosink. 1999. Poles apart: biodiversity and biogeography of sea ice bacteria. *Annu. Rev. Microbiol.* **53:**189–215.

Stetter, K. O., R. Huber, E. Blochl, M. Kurr, R. D. Eden, M. Fielder, H. Cash, and I. Vance. 1993. Hyperthermophilic archaea are thriving in deep North Sea and Alaskan oil reservoirs. *Nature* **365:**743–745.

Takacs, C. D., J. C. Priscu, and D. McKnight. 2001. Bacterial dissolved organic carbon demand in McMurdo Dry Valley lakes, Antarctica. *Limnol. Oceanogr.* **46:**1189–1194.

Takeuchi, N., S. Kohshima, Y. Yoshimura, K. Seko, and K. Fujita. 2000. Characteristics of cryoconite holes on a Himalayan glacier, Yala Glacier central Nepal. *Bull. Glaciol. Res.* (Japan) **17:**51–59.

Thomas, D. N., and G. S. Dieckmann. 2002. Antarctic sea ice—a habitat for extremophiles. *Science* **295:**641–644.

Thomas-Keprta, K. L., S. J. Clemett, D. A. Bazylinski, J. L. Kirschvink, D. S. McKay, S. J. Wentworth, H. Valli, E. K. Gibson, Jr., and C. S. Romanek. 2002. Magnetofossils from ancient Mars: a robust biosignature in the martian meteorite ALH84001. *Appl. Environ. Microbiol.* **68:**3663–3672.

Tikku, A. A., R. E. Bell, and M. Studinger. 2002. Lake Concordia: a second Significant Lake Beneath the East Antarctic Ice Sheet. American Geophysical Union 2002 Spring Meeting, Washington, D.C.

Tobacco, I. E., A. Passerini, F. Corbelli, and M. Gorman. 1998. Determination of the surface and bed topography at Dome C, East Antarctica. *J. Glaciol.* **44:**185–190.

Turtle, E. P., and E. Pierazzo. 2001. Thickness of a Europan ice shell from impact crater simulations. *Science* **294:**1326–1328.

Van Dover, C. L., S. E. Humphris, D. Fornari, C. M. Cavanaugh, R. Collier, S. K. Goffredi, J. Hashimoto, M. D. Lilley, A. L. Reysenbach, T. M. Shank, K. L. Von Damm, A. Banta, R. M. Gallant, D. Götz, D. Green, J. Hall, T. L. Harmer, L. A. Hurtado, P. Johnson, Z. P. McKiness, C. Meredith, E. Olson, I. L. Pan, M. Turnipseed, Y. Won, C. R. Young III, and R. C. Vrijenhoek. 2001. Biogeography and ecological setting of Indian Ocean hydrothermal vents. *Science* **294:**818–823.

Vincent, W. F., J. A. E. Gibson, R. Pienitz, and V. Villenueve. 2000. Ice shelf microbial ecosystems in the high Arctic and implications for life on snowball Earth. *Naturwissenshaften* **87:**137–141.

Vincent, W. F., and C. Howard-Williams. 2001. Life on snowball Earth. *Science* **287:**2421.

Vincent, W. F., J. A. E. Gibson, R. Pienitz, V. Villeneuve, P. A. Broady, P. B. Hamilton, and C. Howard-Williams. 2002. Ice shelf microbial ecosystems in the High Arctic and implications for life on Snowball Earth. *Naturwissenschaften* **87:**137–141.

Wharton, R. A., Jr., W. C. Vinyard, B. C. Parker, G. M. Simmons, Jr., and K. G. Seaburg. 1981. Algae in cryoconite holes on Canada Glacier in southern Victoria Land, Antarctica. *Phycologia* **20:**208–211.

Wharton, R. A., Jr., C. P. McKay, G. M. Simmons, Jr., and B. C. Parker. 1985. Cryoconite holes on glaciers. *Bioscience* **35:**499–503.

Wharton, R. A., Jr., R. A. Jamison, M. Crosby, C. P. McKay, and J. W. Rice, Jr. 1995. Paleolakes on Mars. *J. Paleolimn.* **13:**267–283.

Whitman, W. B., D. C. Coleman, and W. J. Wiebe. 1998. Prokaryotes: the unseen majority. *Proc. Natl. Acad. Sci. USA* **95:**6578–6583.

Willerslev, E., A. J. Hansen, B. Christensen, J. P. Steffensen, and P. Arctander. 1999. Diversity of Holocene life forms in fossil glacier ice. *Proc. Natl. Acad. Sci. USA* **96:**8017–8021.

Williams, D. M., J. F. Kasting, and L. A. Frakes. 1998. Low-latitude glaciation and rapid changes in the earth's obliquity explained by obliquity-oblateness feedback. *Nature* **396:**453–455.

Wilson, W. H., D. Lane, D. A. Pearce, and J. S. Ellis-Evans. 2000. Transmission electron microscope analysis of virus-like particles in freshwater lakes of Signy Island, Antarctica. *Polar Biol.* **23:**657–660.

Wing, K. T., and J. C. Priscu. 1993. Microbial communities in the permanent ice cap of Lake Bonney, Antarctica: relationships among chlorophyll *a*, gravel and nutrients. *Antarct. J. US* **28:**246–249.

Wommack, E., and R. Colwell. 2000. Virioplankton: viruses in aquatic ecosystems. *Microbiol. Mol. Biol. Rev.* **64:**69–114.

Zhang, X., T. Yao, X. Ma, and N. Wang. 2001. Analysis of the characteristics of microorganisms packed in the ice core of Malan Glacier, Tibet. *Sci. China* (Series D) **44:**165–170.

Microbial Diversity and Bioprospecting
Edited by Alan T. Bull
© 2004 ASM Press, Washington, D.C.

Chapter 14

Extremophiles: pH, Temperature, and Salinity

CONSTANTINOS E. VORGIAS AND GARABED ANTRANIKIAN

From the anthropocentric point of view, microorganisms that are able to survive and grow optimally at temperatures below 10°C and above 40°C, at a pH below 5.0 or above 8.0, at a pressure above 1 atm, and at a salt concentration of more than 30 g/liter are defined as extremophiles. There are many environments around the world that are extreme. These are geothermal areas with high temperature, polar regions with temperatures around the freezing point of water, deep oceans with very high pressure, and acid or alkaline springs with low or high pH, respectively. As conditions become increasingly demanding, extreme environments become exclusively populated by microorganisms belonging to prokaryotes. It is very likely that eukaryotic organisms were partially unable to adapt and survive under extreme conditions because of their cellular complexity. The realization that extreme environments harbor different kinds of prokaryote lineages has resulted in a complete reassessment of our concept of microbial evolution and has given considerable impetus to extremophile research (Leuschner and Antranikian, 1995; Ladenstein and Antranikian, 1998; Niehaus et al., 1999; Horikoshi and Grant, 1998, Rohschild and Mancmelli, 2001).

During the past 20 years, rapidly growing research activities focused on the elucidation of the basic rules that govern these extreme microorganisms have been conducted all over the world. This new field was strongly supported by industry and academia because it became obvious that the extremophilic organisms provide a unique resource for a variety of biomolecules such as enzymes and compounds with high potential for applications in the biotechnological industry. Furthermore, the finding of novel biocatalysts will allow the development of more efficient and environmentally friendly industrial processes. In this chapter we focus on a general description of the ecology, the general properties, and some examples of the biotechnological application of microorganisms that are able to grow optimally under very low or very high temperature, extreme pH, and high salinity. Such extremophiles can be found in terrestrial and marine environments all over the world and particularly in exotic ecological niches such as polar regions, solfataric fields, soda lakes, and abyssal hypothermal vents.

MICROORGANISMS LIVING AT EXTREME PH

Acidophiles

Solfataric fields are the most common biotopes for microorganisms that prefer to live at high temperature and acidic conditions. Solfataric soils consist of two layers with distinguished features. The upper layer is aerobic and has an ochre color that is due to the presence of ferric iron, and the lower layer is anaerobic and appears rather blackish-blue due to the presence of ferrous iron. The chemical composition of the two layers determines the variety of microorganisms that dominate in this environment.

Thermophilic acidophiles belonging to the genera *Sulfolobus*, *Acidianus*, *Thermoplasma*, and *Picrophilus*, with growth optima between 60 and 90°C and a pH of 0.7 to 5.0, are usually found in the aerobic upper layer of solfataric fields. These microorganisms maintain their intracellular pH between 4 and 6.5. On the other hand, slightly acidophilic or neutrophilic anaerobes such as *Thermoproteus tenax* or *Methanothermus fervidus* have been isolated from the lower layer of solfataric fields. *Thermoplasma* spp. (growth optima, pH 2 and 60°C) have been found in hot springs, solfataras, and coal refuse piles (Horikoshi and Grant, 1998). While their phylogenetic relatives *Picrophilus* spp. have been found in

Constantinos E. Vorgias • National and Kapodistrian University of Athens, Faculty of Biology, Department of Biochemistry-Molecular Biology, Panepistimiopolis-Zographou, 15701 Athens, Greece. **Garabed Antranikian** • Technical University Hamburg-Harburg, Technical Microbiology, Kasernenstrasse 12, 21073 Hamburg, Germany.

solfataras and are so far the most extreme acidophiles because they grow at a pH close to 0, *P. oshimae* and *P. torridus* are both aerobic, heterotrophic *Archaea* that grow optimally at 60°C and pH 0.7 and utilize various biopolymers such as starch and proteins as carbon sources.

Members of the genus *Sulfolobus* are strict aerobes growing either autotrophically, heterotrophically, or facultative heterotrophically. During autotrophic growth, S^0, S^{2-}, and H_2 are oxidized to sulfuric acid or water as end products. *Sulfolobus metallicus* and *S. brierley* are able to grow by oxidation of sulfidic ores. A dense biofilm of these microorganisms is responsible for the microbial ore leaching process, in which heavy metal ions such as Fe^{2+}, Zn^{2+}, and Cu^{2+} are solubilized. Several hyperthermophilic acidophiles have been assigned to the genera *Metallosphaera* (growth range, 50 to 80°C, pH 1 to 4.5), *Acidianus* (growth range, 60 to 95°C, pH 1.5 to 5), and *Stygioglobus* (growth range, 57 to 89°C, pH 1 to 5.5) (Stetter, 1996) and to *Thiobacillus caldus* (T_{max}, 55°C), isolated by Hallberg and Lindström (1994).

Alkaliphiles

Alkaliphiles, which grow at high pH values, are widely distributed throughout the world and require alkaline environments and sodium ions for growth. They have been found in carbonate-rich springs and alkaline soils, where the pH is usually 10.0 or higher, whereas their intracellular pH is maintained around 8.0. In these places, several species of cyanobacteria and *Bacillus* are normally abundant and provide organic matter for diverse groups of heterotrophs. Sodium ion-dependent uptake of nutrients has been reported in alkaliphiles. Many alkaliphiles require various nutrients for growth. A few alkaliphilic *Bacillus* strains can grow in simple minimal media containing glycerol, glutamic acid, and citric acid (Horikoshi and Grant, 1998). In general, the cultivation temperature is in the range of 20 to 55°C.

Haloalkaliphiles are microorganisms that have been isolated from alkaline hypersaline lakes and can grow in alkaline media containing 20% NaCl. A typical habitat where alkaliphilic microorganisms can be isolated is the soda lakes in the Rift Valley of Kenya. Similar lakes have been found in a few other places on Earth that are highly alkaline with pH values between 11.0 and 12.0.

Two thermoalkaliphilic bacteria, *Anaerobranca gottschalkii* and *Anaerobranca horikoshii*, have been isolated from Lake Bogoria in Kenya and from Yellowstone National Park, respectively. The new isolates represent a new line within the *Clostridium/*

Table 1. Microorganisms living at extreme pHs

Microorganisms	Optimal growth	
	°C	pH
Acidophilic microorganisms		
Sarcina ventriculi	37	4.0
Thiobacillus ferrooxidans	37	2.5
Alicyclobacillus acidocaldarius	55	2.0–6.0
Picrophilus oshimae	60	0.7
Picrophilus torridus	60	0.7
Thermoplasma acidophilum	60	2.0
Sulfolobus acidocaldarius	75	2.5
Acidianus infernus	75	2.0
Alkaliphilic microorganisms		
Many cyanobacteria	(25–37)	6.0–8.0
Spirulina spp.		8.0–10.0
Chromatium sp.		8.5
Bacillus spp.		11.5
Anaerobranca gottschalkii	55	9.5
Thermococcus alcaliphilus	85	9.0
Thermococcus acidoaminivorans	85	9.0

Bacillus subphylum (Prowe and Antranikian, 2001; Wiegel, 1998). The two archaeal thermoalkaliphiles identified to date are *Thermococcus alcaliphilus* and *Thermococcus acidoaminivorans*, both growing at 85°C and with a pH of 9.0. Table 1 summarizes the most studied acidophilic and alkaliphilic microorganisms.

MICROORGANISMS GROWING AROUND THE FREEZING POINT OF WATER

Many parts of the world rarely reach temperatures above 5°C, and numerous bacteria, yeast, unicellular algae, and fungi have successfully colonized cold environments (see chapter 13). These psychrophilic microorganisms are able to grow at temperatures around 0°C and have developed adaptation mechanisms at various levels (enzymes, membranes, etc.). Most of the cold-adapted microorganisms have been isolated and characterized from the Arctic and Antarctic seawaters. It is noteworthy that, despite the harsh conditions, the density of bacterial cells in the Antarctic oceans is as high as the regular density in temperate waters (Deming, 2002; Russel and Hamamoto, 1998).

The microorganisms that are able to grow at or close to the freezing point of water can be divided into two main groups: psychrophiles and psychrotolerants.

Psychrophilic microorganisms are defined by an optimum temperature for growth at about 15°C, a maximum growth temperature at about 20°C, and a minimum temperature for growth at 0°C or lower. Psychrophiles can be found in permanently cold environments such as the deep sea, glaciers, mountain regions, or soils, and in fresh or saline waters associated with cold-blooded animals such as fish or crustaceans. Psychrotolerant microorganisms generally do not grow at 0°C and have optimum and maximum growth temperatures above 20°C (Morita, 1975). In general, psychrophiles have significantly narrower growth temperature ranges and lower optimum or maximum growth temperatures compared with psychrotolerant microorganisms.

Because psychrotolerant microorganisms are able to grow above 20°C, they could be pathogenic for humans and animals. A notable example is *Listeria monocytogenes*, which has a broad growth temperature between 0 and 40°C. Several other psychrophilic microorganisms are pathogenic, such as *Yersinia enterocolitica* and *Erwinia* spp. Several food-spoilage psychrotolerant bacteria have been identified, including *Brochothrix thermosphacta*, *Pseudomonas fragi*, *Bacillus cereus*, and *Clostridium botulinum*. Table 2 lists a few selected psychrophilic microorganisms.

The current studies on psychrophilic microorganisms are not so extensive compared with the other extremophiles. The biotechnological potential of psychrophiles has not fully emerged, and it is necessary to stress efforts to this area in order to fill this gap.

A systematic investigation has been undertaken to describe the microbial diversity and to quantify the microorganism in cold environments. Most of the studies, based on molecular biology techniques, have been concentrated on benthic communities while the sea-ice ecosystems have been relatively neglected (but see chapter 13). Bowman and Bowman (2001) have investigated bacterial diversity in sea ice using direct cloning of 16S rDNA sequences amplified from sea-ice DNA. Most of the bacteria identified were classified in four phylogenetic groups: *Alphaproteobacteria*, *Betaproteobacteria*, *Gammaproteobacteria*, and the Cytophaga-Flavobacterium-Bacteroides division, as were supported by cultivation data. Many clones detected in these groups demonstrated high similarity to cultured strains isolated from Arctic or Antarctic sea ice or from other polar habitats. Surprisingly, no *Archaea* were detected using the universal archaeal-specific primers in any of the investigated sea-ice samples.

At the biochemical level, a systematic investigation has been carried out in order to understand the rules governing the molecular mechanisms for adaptation to low temperatures. These fundamental aspects are closely associated with a strong biotechnological interest aimed at the exploitation of these microorganisms and their cell components such as membranes, polysaccharides, and enzymes (Russell, 2000).

MICROORGANISMS GROWING AROUND THE BOILING POINT OF WATER

Hyperthermophilic microorganisms are adapted to grow optimally at high temperatures (60 to 108°C) and have been isolated from their most common biotopes, i.e., volcanic and hydrothermal vent systems, solfataric fields, neutral hot springs, and submarine hot vents. The most interesting are submarine hydrothermal systems that are situated in shallow and abyssal depths. They consist of hot fumaroles, springs, sediments, and deep-sea vents with temperatures up to 400°C ("black smokers") (Stetter, 1996). Shallow marine hydrothermal systems have been detected at the beaches of Vulcano, Naples, and Ischia (Italy); Sao Miguel (Azores); and Djibouti (Africa). The best-studied deep-sea hydrothermal systems are (i) the Guaymas Basin (depth, 1,500 m) and the East Pacific Rise (depth, 2,500 m), both off the coast of Mexico; (ii) the Mid-Atlantic Ridge (depth, 3,700 m); and (c) the Okinawa Trough (depth, 1,400 m) (Grote et al., 1999; Jeanthon et al., 1999, 1998; Gonzáles et al., 1998; Canganella et al., 1998).

The genera *Pyrococcus*, *Pyrodictium*, *Igneococcus*, *Thermococcus*, *Methanococcus*, *Archaeoglobus*, and *Thermotoga* are the major members of the shallow as well as deep-sea hydrothermal systems (Jeanthon et al., 1998). So far, members of the genus *Methanopyrus* have been found only at greater depths, whereas *Aquifex* was isolated exclusively from shallow hydrothermal vents (Stetter, 1998). Recently, interesting biotopes of extreme and hyperthermophiles were discovered in deep, geothermally heated oil reservoirs around 3,500 m below the bed of the North Sea and the permafrost soil of north Alaska (Lien et al., 1998; Stetter et al., 1983).

Moderate thermophiles are microorganisms that are capable of growing optimally at temperatures

Table 2. Microorganisms growing at low temperature

Microorganisms	Optimal growth(°C)
Vibrio sp.	<20
Micrococcus criophilus	<20
Arthrobacter glacialis	<20
Vibrio psychroerythreus	<20
Carnobacterium sp.	<20
Antarctic strain TABS	<20
Alteromonas haloplanctis A23	<20
Moritella marina	<20
Psychrobacter sp. TA137	<20
Bacillus psychrosaccharolyticus	<20
Moraxella sp. TA144	<20
Psychrobacter immobilis B10	<20
Aquaspirillum arcticum	<20

between 50 and 60°C. Most of these microorganisms belong to many different taxonomic groups of eukaryotic and prokaryotic microorganisms such as protozoa, fungi, algae, streptomycetes, and cyanobacteria, which comprise mainly mesophilic species. It can be assumed that moderate thermophiles, which are closely related phylogenetically to mesophilic organisms, may be secondarily adapted to life in hot environments. Extreme thermophiles, which grow optimally between 60 and 80°C, are widely distributed among the genera *Bacillus*, *Clostridium*, *Thermoanaerobacter*, *Thermus*, *Fervidobacterium*, and *Thermotoga*.

Hyperthermophiles are represented among all the deepest and shortest lineages, including the genera *Aquifex* and *Thermotoga* within the *Bacteria* and *Pyrodictium*, *Pyrobaculum*, *Thermoproteus*, *Desulfurococcus*, *Sulfolobus*, *Methanopyrus*, *Pyrococcus*, *Thermococcus*, *Methanococcus*, and *Archaeoglobus* within the *Archaea*. Interestingly, the majority of the hyperthermophiles isolated to date belong to the archaeal domain of life, and no eukaryotic organism has been found that can grow at the boiling point of water.

The relative abundance of *Archaea* and *Bacteria* in high-temperature environments was, until recently, mainly studied by cultivation-based techniques. Because of the frequent isolation of *Archaea* from these habitats, it was assumed that Archaea dominate the high-temperature biotopes (Baross and Deming, 1995; Stetter et al., 1990). Recently, the application of molecular biological methods revealed a quite different picture. Slot-blot hybridizations of rRNA utilizing oligonucleotide probes targeting the 16S rRNA of Archaea and Bacteria revealed that bacteria seem to be the major population of the microbial community along a thermal gradient at a shallow submarine hydrothermal vent near Milos Island, Greece (Sievert et al., 2000). Bacteria made up at least 78% (mean, 95%) of the prokaryotic rRNA. Along the steepest temperature gradient, the proportion of archaeal rRNA increased. Nevertheless, even in the hottest sediment layer, archaeal rRNA made up only around 12% of the prokaryotic rRNA. These results suggest that archaea may generally be of lower abundance in hot environments than could be assumed from cultivation-based experiments. However, the factors that allow bacteria to dominate in high-temperature habitats, that were once believed to be the territory of archaea, remain unknown. Most of these microorganisms that can be found in low-salinity and submarine environments are strict anaerobes. Terrestrial solfataric fields found in Italy or Iceland harbor members of the genera *Pyrobaculum*, *Thermoproteus*, *Thermofilum*, *Desulfurococcus*, and *Methanothermus*. *Pyrobaculum islandicum* and *Thermoproteus tenax*

are able to grow chemolithoautotrophically, gaining energy by anaerobic reduction of S^0 by H_2. In contrast to these strictly anaerobic microorganisms, *Pyrobaculum aerophilum* and *Aeropyrum pernix* are able to use oxygen as a final electron acceptor. *Methanothermus fervidus*, on the other hand, is highly sensitive toward oxygen and can survive only in low-redox environments at temperatures between 65 and 97°C. Some microorganisms from marine environments such as members of the genera *Archaeoglobus*, *Methanococcus*, and *Methanopyrus* are able to grow chemolithoautotrophically, gaining energy by the reduction of SO_4^{2-} by H_2 (*Archaeoglobus lithothrophicus* and *A. fulgidus*) or by the reduction of CO_2 by H_2 (*Methanococcus jannaschii*, *Methanopyrus kandleri*). Other members of the hyperthermophilic genera *Staphylothermus*, *Pyrococcus*, *Thermococcus*, and *Pyrodictium* are adapted to marine environments (NaCl concentration of about 30 g/liter). Most of them gain energy by fermentation of polysaccharides, peptides, amino acids, and sugars (Stetter, 1996). Consequently, such thermophilic microorganisms have been found to be producers of interesting polymer-degrading enzymes of industrial relevance (Niehaus et al., 1999; Vorgias et al., 2000).

Among the bacterial domain of life, members of the genera *Aquifex* and *Thermotoga* represent the deepest phylogenetic branches. Within the latter genus, *Thermotoga maritima* and *T. neapolitana* are the most thermophilic species with a maximal growth temperature of about 90°C. The representatives of the order of *Thermotogales* also provide a resource of unique thermoactive enzymes. Some representative microorganisms living at high temperature are presented in Table 3.

Table 3. Microorganisms growing at elevated temperatures

Microorganisms	Optimal growth (°C)
Moderate thermophiles (50–60°C)	
Bacillus acidocaldarius	50
Bacillus stearothermophilus	55
Extreme thermophiles (60–80°C)	
Thermus aquaticus	70
Thermoanaerobacter ethanolicus	65
Clostridium thermusulfurogenes	60
Fervidobacterium pennivorans	75
Hyperthermophiles (80–110°C)	
Thermotoga maritima	90
Aquifex pyrophilus	85
Archeoglobus fulgidus	83
Methanopyrus kandleri	88
Sulfolobus sulfataricus	88
Thermococcus aggregans	88
Pyrobaculum islandicum	100
Pyrococcus furiosus	100
Pyrodictium occultum	105
Pyrolobus fumarii	106

MICROORGANISMS GROWING AT EXTREME SALINITY

Halophiles are *Bacteria* and *Archaea* that grow optimally at NaCl concentrations above the average concentration of seawater (>0.6 M NaCl). Generally, halophilic microorganisms are classified as moderate halophiles for those that can grow at salt concentrations between 0.4 and 3.5 M NaCl and as extreme halophiles for those that require NaCl concentrations above 2 M for growth (Grant et al., 1998). Halophiles have been mainly isolated from saline lakes, such as the Great Salt Lake in Utah (salinity, >2.6 M) and from evaporitic lagoons and coastal salterns with NaCl concentrations between 1 and 2.6 M (Grant et al., 1998). Saline soils are less well explored. Bulk salinity measurements of 1.7 to 3.4 M NaCl have been reported for saltern soils (Ventosa and Bieto, 1995). Saline soils constitute less-stable biotopes than hypersaline waters because they are subjected to periodic significant dilution during rainy periods. There is no doubt that almost all hypersaline habitats harbor significant populations of specifically adapted microorganisms. However, it remains unclear what substrates might be available for growth in these biotopes. Hypersaline lakes often contain up to 1 g of dissolved organic carbon per liter. In many of these lakes, primary producers such as cyanobacteria, anoxygenic phototrophic bacteria, and algae may be the main sources of organic compounds (Grant et al., 1998; Jones et al., 1998; Kamekura, 1998).

In a study of aerobic heterotrophs in a marine saltern it has been shown that bacterial halophiles were predominant up to 2 M NaCl. Above this concentration, archaeal halophiles become predominant, almost to the exclusion of *Bacteria*. Halophilic primary producers mainly belong to the cyanobacteria and anoxygenic phototrophic sulfurbacteria. The former often thrive in eutrophic salterns, forming large floating mats. The latter group, on the other hand, grows either in anaerobic sediments or in the water column where they are responsible for the characteristic red color of high-salinity habitats. The range of heterotrophic *Bacteria* comprises proteobacteria, actinomycetes, and gram-positive rods and cocci. Fermentative anaerobes as well as sulfur oxidizers, sulfate reducers, and nitrate reducers are also present and give rise to the assumption that all kind of metabolic features can be found in high-salinity environments. Halophilic bacteria do not belong to one homogeneous group but rather fall into many bacterial taxa in which the capability to grow at high salt concentrations is a secondary adaptation.

The term "halobacteria" refers to the red-pigmented, extremely halophilic *Archaea*, members of the family *Halobacteriaceae* and the only family in the order *Halobacteriales* (Ventosa and Bieto, 1995). Most halobacteria require 1.5 M NaCl in order to grow and to retain the structural integrity of the cell. Halobacteria can be distinguished from halophilic bacteria by their archaeal characteristics, in particular the presence of ether-linked lipids (Ross et al., 1981). Most halobacteria are colored red or orange due to the presence of carotenoids, but some species are colorless, and those with gas vesicles form opaque, white, or pink colonies. A purple hue can be seen in halobacteria that form the bacteriorhodopsin-containing purple membrane (Grant et al., 1998). Halobacteria are the most halophilic organisms known so far and form the dominant microbial population when hypersaline waters approach saturation (Ventosa et al., 1998). Interestingly, the reddening caused by halobacterial blooms has an impact on the evaporation rates in salterns. It is known that the carotenoid pigments of halobacteria trap solar radiation, thus increasing the ambient temperature and evaporation rates (Madigan and Oren, 1999). Some representative microorganisms living at high salt concentration are listed in Table 4.

The singular physiology of halophilic microorganisms that have to cope with 4 M ion concentration inside and outside the cell has evolved potentially interesting enzymes that can function under conditions of low-water activity that could be imposed by compounds other than salts, for example, solvents. Interestingly, halophilic and marine halo-

Table 4. Microorganisms living at high salt concentrations

Halophilic microorganisms	NaCl concentration (M) required for growth		
	Maximum	Minimum	Optimum
Dunaliella spp.	0.3		5.0
Clostridium halophilum	0.15	0.6	6.0
Haloanaerobium praevalens	0.8	2.2	4.3
Halobacterium sp.		2.0	5.5
Halobacterium denitrificans	1.5	2.5	4.5
Haloferax vulcanii	1.0	1.5	3.0
Methanohalobium evestigatum			4.3

tolerant bacteria produce and/or accumulate organic osmolytes (compatible solutes, e.g., ectoin, hydroxyectoin, trehalose) for osmotic equilibrium. These metabolically compatible hygroscopic compounds not only protect living cells in a low-water environment but also exhibit an enzyme-stabilizing effect in vitro against a variety of stress factors such as heating, freezing, urea, and other denaturants (Margesin and Schinner, 2001).

BIOCATALYSIS UNDER EXTREME CONDITIONS

Extremophiles in general have been considered as a group of microorganisms with biotechnological potential. It is clear that hyperthermophiles represent a huge, almost unexplored potential for novel applications in modern biotechnology. Applications of hyperthermophiles in biotechnology are in an early phase of growth. At present there are several known examples but these represent only a small portion of the real potential of extremophilic biomolecules (Herbert, 1992). A breakthrough in applications of hyperthermophilic enzymes occurred in the late 1980s with the application of heat-stable DNA polymerase for DNA amplification by PCR. This was a milestone in the field of basic and applied biological research (Frey and Suppmann, 1995). Several thermophilic hydrolases, such as proteases, lipases, amylases, and xylanases, have the potential to be incorporated in existing industrial processes toward a more economical and environmentally friendly procedure (Sunna et al., 1997).

Several enzymes from hyperthermophiles have been purified and characterized (Niehaus et al., 1999; Vorgias et al., 2000). As a general rule, they show an extraordinary heat stability even in vitro. *Pyrococcus woesei* harbors an amylase that is active even at 130°C for more than 30 min (Koch et al., 1991). Within the bacterial domain, *Thermotoga maritima* MSB8 possesses an extremely heat-resistant, membrane-associated xylanase that is optimally active at 105°C (Winterhalter and Liebl, 1995). Even complex enzymes such as DNA-dependent RNA polymerases or glutamate dehydrogenases show a remarkable heat stability.

Because extremophilic microorganisms are able to survive under unusual conditions, it is clear that they have to develop strategies to withstand these conditions. For the case of the extreme thermophilic organisms, the cell components have to be either intrinsically heat resistant or stabilized by other components within the cells. The molecular basis of thermostability is a very exciting and difficult issue and is still under intensive investigation (Ladenstein and Antranikian, 1998; Stetter, 1998; Vetriani et al., 1998). Up to now a solid mechanistic basis for thermostabilization has been achieved, and one is able to understand but not yet predict thermostabilization mechanisms. The basic principles of heat stabilization of several thermostable enzymes have been established, and it has become clear that each protein has found its own strategy for stabilization through the evolution (Jaenicke and Bohm, 1998).

The best known example of a biotechnological product derived from the studies on halophiles is the small molecules known as compatible solutes. Halophilic and/or halotolerant bacteria produce, among others, the ectoine-type osmolytes (2-methyl-1,4,5,6-tetrahydropyrimidine derivatives) that represent the most abundant class of stabilizing solutes (De Costa et al., 1998). The extrinsic stabilization effect of ectoines and other compatible solutes is most likely based on solvent-modulating properties of these compounds. Osmolytes have considerable potential as effective stabilizers for the hydration shell of proteins and hence could be highly efficient against stress conditions for biomolecules and therefore stabilizers that might be suitable for sensitive vaccines or industrial enzymes functioning under extreme conditions (Galinski, 1993). Recently, it has been shown that several hyperthermophilic microorganisms are also able to produce a variety of compatible solutes that were found to be effective in enzyme stabilization (De Costa et al., 1998). Because halophiles can work at high salt concentrations (up to 4 M), their enzymes might be capable of working under low-water conditions and open the possibility to function in nonaquatic solvents. Although there is considerable interest in halophiles, to date they have little impact in the commercial field.

The main industrial application of alkaliphilic enzymes is in the detergent industry, which accounts for nearly 30% of the total worldwide enzyme production. Alkaline enzymes have been used in the hide-dehairing process that is usually carried out at a pH between 8 and 10. Several other alkaliphilic hydrolases such as amylases, xylanases, pullulanases, and cyclomaltodextrin glucanotransferases (CGTases) have been isolated and are under intensive investigation for their potential industrial application (Horikoshi and Grant, 1998).

The possible applications of cold-active enzymes are as detergents for cold washing (proteases, lipases, cellulases) and food additives such as polyunsaturated flavor-modifying agents. In the dairy industry, β-galactosidase is applied to reduce the amount of lactose in milk, which is responsible for lactose intolerance in approximately two-thirds of the world's population. The clarification of fruit juices is achieved by the addition of cold-active pectinases. Further applications of cold-active enzymes are found in food processing such as cheese manufacturing; tenderizing meat by proteases; and improving the baking process by the addition of amylases, proteases, and xylanases. Cold-active enzymes are also used in biosensors for environmental applications and for cleaning contact lenses (Russell and Hamamoto, 1998).

Some of the advantages of using psychrophiles and their enzymes in biotechnological applications are the rapid termination of the process by moderate heat treatment, higher yields of thermosensitive components, modulation of the (stereo-) specificity of enzyme-catalyzed reactions, cost saving by elimination of expensive heating and cooling process steps, and finally the capacity for online monitoring under environmental conditions (Gerday et al., 2000; Feller et al., 1996).

REFERENCES

Baross, J. A., and J. W. Deming. 1995. Growth at high temperatures: isolation and taxonomy, physiology, and ecology, p. 169–217. In D. M. Karl (ed.), The Microbiology of Deep-Sea Hydrothermal Vents. CRC Press, Boca Raton, Fla.

Bowman, M. V., and J. P. Bowman. 2001. A molecular phylogenetic survey of sea-ice microbial communities. FEMS Microbiol. Ecol. 35:267–275.

Canganella, F., W. J. Jones, A. Gambacorta, and G. Antranikian. 1998. Thermococcus guaymasensis sp. nov. and Thermococcus aggregans sp. nov., two novel thermophilic archaea isolated from the Guaymas Basin hydrothermal vent site. Int. J. Syst. Bacteriol. 48:1181–1185.

De Costa, M. S., H. Santos, and E. A. Galinski. 1998. An overview of the role and diversity of compatible solutes in Bacteria and Archaea. Adv. Biochem. Eng. Biotechnol. 61:117–153.

Deming, J. W. 2002. Psychrophiles and polar regions. Curr. Opin. Microbiol. 5:301–309.

Demirjian, D. C., F. Moris-Varas, and C. S. Cassidy. 2001. Enzymes from extremophiles. Curr. Opin. Chem. Biol. 5:144–151.

Feller, G., E. Narinx, J. L. Arpigny, M. Aittaleb, E. Baise, S. Genicot, and C. Gerday. 1996. Enzyme from extremophilic mocroorganisms. FEMS Microbiol. Rev. 18:189–202.

Frey, B., and B. Suppmann. 1995. Demonstration of the Expand PCR System's greater fidelity and higher yields with a lacI-based fidelity assay. Biochemica 2:34–35.

Galinski, E. A. 1993. Compatible solutes of halophilic eubacteria: molecular principles, water-soluble interactions, stress protection. Experientia 49:487–496.

Gerday, C., M. Aittaleb, M. Bentahir, J. P. Chessa, P. Claverie, T. Collins, S. D'Amico, J. Dumont, G. Garsoux, D. Georlette, A. Hoyoux, T. Lonhienne, M. A. Meuwis, and G. Feller. 2000. Cold-adapted enzymes: from fundamentals to biotechnology. Trends Biotechnol. 18:103–107.

Gonzáles, J. M., Y. Masuchi, F. T. Robb, J. W. Ammerman, D. L. Maeder, M. Yanagibayashi, J. Tamaoka, and C. Kato. 1998. Pyrococcus horikoshii sp. nov., a hyperthermophilic archaeon isolated from a hydrothermal vent at the Okinawa Trough. Extremophiles 2:123–130.

Grant, W. D., R. T. Gemmell, and T. J. McGenity. 1998. Halophiles, p. 93–133. In K. Horikoshi and W. D. Grant (ed.), Halophiles. Wiley-Liss, New York, N.Y.

Grote, R., L. Li, J. Tamoaoka, C. Kato, K. Horikoshi, and G. Antranikian. 1999. Thermococcus siculi sp. nov., a novel hyperthermophilic archaeon isolated from a deep-sea hydrothermal vent at the Mid-Okinawa Trough. Extremophiles 3:55–62.

Hallberg, K. B., and E. B. Lindström. 1994. Characterization of Thiobacillus caldus sp. nov., a moderately thermophilic acidophile. Microbiology 140:3451–3456.

Herbert, R. A. 1992. A perspective on the biotechnological potential of extremophiles. Trends Biotechnol. 110:395–402.

Horikoshi, K., and W. D. Grant (ed.). 1998. Extremophiles—Microbial Life in Extreme Environments. Wiley-Liss, New York, N.Y.

Jaenicke R, and G. Bohm. 1998. The stability of proteins in extreme environments. Curr. Opin. Struct. Biol. 8:738–748.

Jeanthon, C., S. L'Haridon, A. L. Reysenbach, M. Vernet, P. Messner, U. B. Sleytr, and D. Prieur. 1998. Methanococcus infernus sp. nov., a novel hyperthermophilic lithotrophic methanogen isolated from a deep-sea hydrothermal vent. Int. J. Syst. Bacteriol. 48:913–919.

Jeanthon, C., S. L.'Haridon, A. L. Reysenbach, E. Corre, M. Vernet, P. Messner, U. B. Sleytr, and D. Prieur. 1999. Methanococcus vulcanius sp. nov., a novel hyperthermophilic methanogen isolated from East Pacific Rise, and identification of Methanococcus sp. DSM 4213T as Methanococcus fervens sp. nov. Int. J. Syst. Bacteriol. 49:583–589.

Jones, B. E., W. D., Grant, A. W. Duckworth, and G. G. Owenson. 1998. Microbial diversity of soda lakes. Extremophiles 2:191–200.

Kamekura, M. 1998. Diversity of extremely halophilic bacteria. Extremophiles 2:289–295.

Koch, R., K. Spreinat, K. Lemke, and G. Antranikian. 1991. Purification and properties of a hyperthermoactive alpha-amylase from the archaeobaterium Pyrococcus woesei. Arch. Microbiol. 155:572–578.

Ladenstein, R., and G. Antranikian. 1998. Proteins from hyperthermophiles: stability and enzymatic catalysis close to the boiling point of water. Adv. Biochem. Eng. Biotechnol. 61:37–85.

Leuschner, C., and G. Antranikian. 1995. Heat-stable enzymes from extremely thermophilic and hyperthermophilic microorganisms. World Microbiol. Biotechnol. 11:95–114.

Lien, T., M. Madsen, F. A. Rainey, and N. K. Birkeland. 1998. Petrotoga mobilis sp. nov., from a North Sea oil-production well. Int. J. Syst. Bacteriol. 48:1007–1013.

Madigan, M. T., and A. Oren. 1999. Thermophilic and halophilic extremophiles. Curr. Opin. Microbiol. 2:265–269.

Margesin, R., and F. Schinner. 2001. Potential of halotolerant and halophilic microorganisms for biotechnology. Extremophiles 5:73–83.

Morita, R. Y. 1975. Psychrophilic bacteria. Bacteriol. Rev. 39:144–167.

Niehaus, F., C. Bertoldo, M. Kahler, and G. Antranikian. 1999. Extremophiles as a source of novel enzymes for industrial application. Appl. Microbiol. Biotechnol. 51:711–729.

Prowe, S. G., and G. Antranikian. 2001. Anaerobrances gottschalkii sp. nov., a new hermoalkaliphilic bacterium that grows anaerobically at high pH and temperature. Int. J. Syst. Evol. Microbiol. 51:457–465.

Ross, H. N. M., M. D. Collins, B. J. Tindall, and W. D. Grant. 1981. A rapid method for detection of archaeabacterial lipids in halophilic bacteria. J. Gen. Microbiol. 123:75–80.

Rothschild, L. J., and R. L. Mancinelli. 2001. Life in extreme environments. Nature 22:1092–1101.

Russell, N. J. 2000. Toward a molecular understanding of cold activity of enzymes from psychrophiles. *Extremophiles* 4:83–90.

Russell, N. J. and T. Hamamoto. 1998. Psychrophiles, p. 25–47. *In* K. Horikoshi and W. D. Grant (ed.), *Extremophiles—Microbial Life in Extreme Environments,* Wiley-Liss, New York, N.Y.

Sievert, S. M., W. Ziebis, J. Kuever, and K. Sahm. 2000. Relative abundance of Archaea and Bacteria along a thermal gradient of a shallow-water hydrothermal vent quantified by rRNA slot-blot hybridization. *Microbiol* 146:1287–1293.

Stetter, K. O. 1996. Hyperthermophilic procaryotes. *FEMS Microbiol. Rev.* 18:149–158.

Stetter, K. O. 1998. Hyperthermophiles: isolation, classification and properties, p. 1–24. *In* K. Horikoshi and W. D. Grant (ed.), *Extremophiles—Microbial Life in Extreme Environments.* Wiley-Liss, New York, N.Y.

Stetter, K. O., H. König, and E. Stackebrandt. 1983. *Pyrodictium* gen. nov., a new genus of submarine disc-shaped sulfur reducing archaebacteria growing optimally at 105°C. *Syst. Appl. Microbiol.* 4:535–551.

Stetter, K. O., G. Fiala, G. Huber, R. Huber, and A. Segerer. 1990. Hyperthermophilic microorganisms. *FEMS Microbiol. Rev.* 75:117–124.

Sunna, A., M. Moracci, M. Rossi, and G. Antranikian. 1997. Glycosyl hydrolases from hyperthermophiles. *Extremophiles* 1:2–13.

Ventosa, A., and J. J. Bieto. 1995. Biotechnological applications and potentialities of halophilic microorganisms. *World J. Microbiol. Technol.* 11:85–94.

Ventosa, A., M. C. Marquez, M. J. Garabito, and D. R. Arahal. 1998. Moderately halophilic gram-positive bacterial diversity in hypersaline environments. *Extremophiles* 2:297–304.

Vetriani, C., D. L. Maeder, N. Tolliday, K. S. P. Yip, T. J. Stillman, K. L. Britton, D. W. Rice, H. H. Klump, and F. T. Robb. 1998. Protein thermostability above 100°C: a key role for ionic interactions. *Proc. Nat. Acad. Sci. USA* 95:12300–12305.

Vorgias, C. E., and G. Antranikian. 2000. Glycosyl hydrolases from extremophiles, p. 313–339. *In* Doyle (ed.), *Glycomicrobiology.* Kluwer Academic, New York, N.Y.

Wiegel, J. 1998. Anaerobic alkalithermophiles, a novel group of extremophiles. *Extremophiles* 2:257–267.

Winterhalter, C., and W. Liebl. 1995. 2 Extremely thermostable xylamases of the hyperthermophilic bacterium *Thermotoga maritima* MSB8. *Appl. Environ. Microbiol.* 61:1810–1815.

Microbial Diversity and Bioprospecting
Edited by Alan T. Bull
© 2004 ASM Press, Washington, D.C.

Chapter 15

Extremophiles: Pressure

Fumiyoshi Abe, Chiaki Kato, and Koki Horikoshi

The ocean, with an average depth of 3,800 m and therefore a pressure of 38 MPa (0.1 megapascals [MPa] = 1 bar = 0.9869 atmosphere [atm]) comprises approximately 70% of the biosphere. It has been suggested that life originated in the deep sea some 3.5 to 4 billion years ago. Therefore, hydrostatic pressure would have been a very important stimulus for the early stages of life. Recently, scientists have proposed that life might have originated in deep-sea hydrothermal vents, and thus it seems possible that high-pressure-adapted mechanisms of gene expression, protein synthesis, or metabolism could represent features present in early forms of life. Thus, the study of deep-sea microorganisms may not only enhance our understanding of specific adaptations to abyssal and hadal ocean realms, but also may provide valuable insights into the origin and evolution of all life.

ZoBell and Johnson (1949) first coined the term "barophile" (more recently termed "piezophile"), and ZoBell and Morita (1957) obtained the first evidence of piezophilic growth in mixed microbial cultures recovered from the deep sea. The first isolate of pressure-adapted bacteria was reported by Yayanos et al. in 1979. Subsequently, many psychrophilic piezophiles with various optimal growth pressures have been isolated and characterized physiologically and genetically. Deep-sea hydrothermal vents are interesting sources of novel isolates, many of which were discovered in the course of investigations into the origin of life. Thermophilic microorganisms have also been examined physiologically under high-pressure conditions. Thus, studies on the effects of pressure on microorganisms have mainly been performed using two types of microorganism, psychrophilic piezophiles and thermophilic piezophiles. Several species of these microorganisms are listed in Table 1.

PHYSIOLOGICAL EFFECTS OF HIGH PRESSURE

The basis of all pressure effects arises from a single influence, that is, the change in system volume that accompanies a biological reaction. When a reaction is accompanied by a volume increase, it is inhibited by increasing pressure. When a reaction is accompanied by a volume decrease, it is enhanced by increasing pressure. For example, dissociation of ribosome subunits is significantly enhanced by increasing pressure because solvation of charged groups of protein molecules is accompanied by a volume decrease (Pande and Wishnia, 1986). The equilibria $H_2PO_4 \leftrightarrow HPO_4^- + H^+$ and $H_2CO_3 \leftrightarrow HCO_3^- + H^+$ shift to the generation of protons with increasing pressure. The intracellular pH in the yeast *Saccharomyces cerevisiae* was measured at high pressure using pH-sensitive fluorescent probes. Both the cytoplasm and the vacuole are acidified upon application of high pressure in a manner dependent on the production of carbon dioxide through glycolysis (Abe and Horikoshi, 1998). Thus, maintenance of the cytoplasm at neutral pH may be one mechanism required for piezoadaptation.

The structure of lipid bilayers is particularly sensitive to increasing pressure. Increasing pressure enhances the order of hydrocarbon chains and raises the transition temperature of membranes, thereby leading to a decrease in membrane fluidity (Macdonald, 1987). There is considerable evidence that an increased proportion of unsaturated fatty acids in membrane lipids is strongly correlated with bacterial growth profiles under high pressure as well as at low temperature (DeLong and Yayanos, 1985). A higher proportion of unsaturated fatty acids would help maintain favorable fluidity and viscosity of biological membranes under high pressure.

Fumiyoshi Abe and Koki Horikoshi • The DEEPSTAR Group, Japan Marine Science and Technology Center (JAMSTEC), Yokosuka 237-0061, Japan. **Chiaki Kato** • Department of Marine Ecosystems Research, Japan Marine Science and Technology Center (JAMSTEC), Yokosuka 237-0061, Japan.

Table 1. Investigations of the effects of high pressure on piezophiles[a]

Organism	Experimental conditions	Investigation	Reference
Psychrophilic piezophiles			
Colwellia hadaliensis BNL-1	75–94 MPa at 2°C	Physiology	Deming et al. (1988)
Moritella japonica DSK1	50 MPa at 15°C	Physiology	Nogi et al. (1998b)
Moritella yayanosii DB21MT-5	80 MPa at 10°C	Membrane lipids	Kato et al. (1998)
Photobacterium profundum SS9	28 MPa at 9°C	Gene expression, membrane protein, membrane fatty acid	Bartlett et al. (1989) Welch and Bartlett (1996) Welch and Bartlett (1998) Allen et al. (1999)
Shewanella benthica strains	50–70 MPa at 10–15°C	Physiology, gene expression, membrane fatty acid	Kato et al. (1998) Qureshi et al. (1998a) Qureshi et al. (1998b) Kato and Nogi (2001)
Shewanella violacia DSS12	30 MPa at 8°C	Gene expression, respiratory system	Kato et al. (1996) Kato et al. (1997) Tamegai et al. (1998) Nakasone et al. (1998) Ikegami et al. (2000) Yamada et al. (2000)
Other unidentified strains	41–62 MPa at 3°C	Physiology	DeLong et al. (1997)
Thermophilic piezophiles			
Methanococcus jannaschii	75 MPa at 86 and 90°C, 51 MPa at 90°C[b]	Pressure stabilization of hydrogenase Methanogenesis	Miller et al. (1988) Hei and Clark (1994)
Palaeococcus ferrophilus DMJ	30 MPa at 84°C	Physiology	Takai et al. (2000)
Pyrococcus abyssi GE5	20–40 MPa at 73–112°C[b]	Protein synthesis	Marteinsson et al. (1997)
Pyrococcus sp. GB-D	20 MPa at 103°C[b]	Physiology	Jannasch et al. (1992)
Thermococcus peptonophilus OG1	45 MPa at 90 and 95°C	Physiology	Canganella et al. (1997)
Thermococcus sp. ES1	22 MPa at 91°C[c], 22 MPa at 103°C[c]	Physiology	Pledger et al. (1994)

[a] Organisms whose optimal growth pressures have not been determined are also included.
[b] Examinations were performed under the indicated conditions.
[c] The condition is not optimal for growth, but piezophily was exhibited.

Recently, two extremely piezophilic bacteria, *Shewanella* sp. strain DB21MT-2 and *Moritella* sp. strain DB21MT-5, were isolated from the Challenger Deep in the Mariana Trench (Kato et al., 1998). The optimal pressure for growth of DB21MT-2 and DB21MT-5 is 70 and 80 MPa, respectively. Strain DB21MT-2 contains a significantly higher proportion of the monounsaturated fatty acid octadecenoic acid (18:1) compared with the type strain of *Shewanella benthica*. The other strain, DB21MT-5, contains a high proportion of tetradecenoic acid (14:1) compared with the type strain of *Moritella marinus*. Although there was no significant difference in the proportion of the polyunsaturated fatty acid eicosapentaenoic acid (20:5) between the two *Shewanella* strains or that of docosahexaenoic acid (22:6) between the two *Moritella* strains, the proportions in each instance were high (8 to 18% of total fatty acids) (Kato et al., 1998). Genetic analysis has shown that monounsaturated, but not polyunsaturated, fatty acids in membrane lipids are required for growth at high pressure and low temperature in the moderate piezophile *Photobacterium profundum* SS9 (Allen et al., 1999). As polyunsaturated fatty acids potentially increase the fluidity of cell membranes, it is worthwhile to analyze the significance of these fatty acids in growth or survival under more severe temperature and pressure conditions.

THE ROLE OF MEMBRANE PROTEINS UNDER PRESSURE CONDITIONS

Because the effect of pressure on membrane structures is considerable, the function of membrane proteins must also be affected by increasing pressure. Focusing on the respiratory system, biochemical analyses of respiratory-chain components in the piezophile *S. benthica* DB172F have resulted in the isolation of two pressure-regulated *c*-type cytochromes, *c*-551 (membrane bound) and *c*-552 (cytoplasmic), and a *ccb*-type quinol oxidase (membrane bound) (Qureshi et al., 1998a,b). Cytochrome *c*-551 is constitutively synthesized at a pressure from 0.1 to 60 MPa, whereas cytochrome *c*-552 is synthesized only at 0.1 MPa (Qureshi et al., 1998a). However, the amount of the *ccb*-type quinol oxidase appears to increase depending on the pressure applied during growth and reaches a peak at 60 MPa (Qureshi et al., 1998b). A low level of N,N,N',N'-tetramethyl-p-phenylenediamine (TMPDH2)-oxidase activity suggests that cells grown at 60 MPa lack the bc_1 complex and cytochrome c oxidase. A possible model for the pressure-regulated respiratory system in *S. benthica* DB172F is shown in Fig. 1. It is likely that a cytochrome c-dependent oxidase functions to reduce oxygen at 0.1 MPa, whereas at high pressure, a quinol oxidase functions to reduce oxygen using quinol as a substrate. It should be noted that the respiratory chain in strain DB172F appears to be compacted or abbreviated under high pressure compared with that at atmospheric pressure.

Photobacterium fundum SS9 has been investigated genetically in great detail, particularly the syn-

thesis of the outer-membrane proteins OmpH and OmpL. OmpH is most abundant when SS9 is grown at its pressure optimum of 28 MPa, whereas OmpL is produced in the greatest quantity at 0.1 MPa (Bartlett et al., 1989; Welch and Bartlett, 1996). OmpH is thought to function as a nutrient transporter in nutrient-limited environments such as the deep sea. As suggested by the results of a study of mutants defective in OmpH/OmpL regulation (Welch and Bartlett, 1998), the transmembrane proteins ToxR and ToxS appear to be involved in pressure sensing in strain SS9. ToxR is a dimeric transmembrane protein, the cytoplasmic face of which binds those genes under its direct control. The activity of ToxR is dependent on dimerization and is modulated by ToxS. Elevated pressure usually promotes the dissociation of multimeric proteins. It has been postulated that pressure may affect the

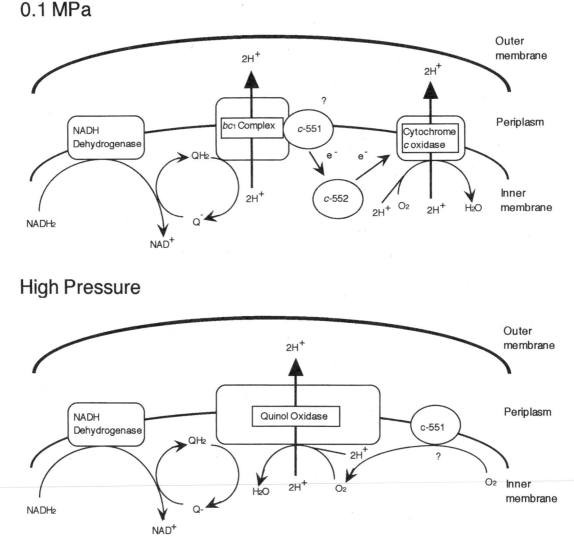

Figure 1. A model for the electron transport system in *S. benthica* DB172F. The respiratory system seems to be abbreviated under high pressure compared with 0.1 MPa. Q, quinone; QH2, quinol.

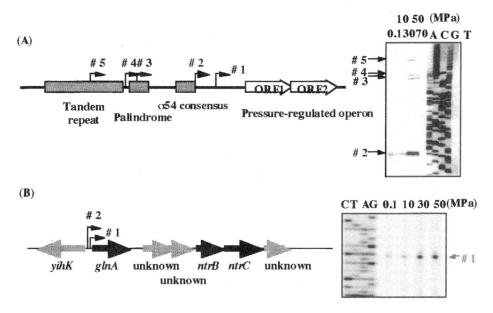

Figure 2. Diagrammatic representation of the pressure-regulated genes in *S. violacea* DSS12. (A) Pressure-regulated operon. Arrow #2 shows the transcription controlled by the sigma 54 factor. (B) Glutamine synthetase operon. Arrow #1 shows the transcription controlled by the sigma 54 factor.

multimeric structure of the ToxR-ToxS proteins in the cell membrane, either by a direct effect on its conformation or by a structural alteration of the lipid bilayer. It would be interesting to measure ToxR-ToxS association or dissociation directly under high pressure, in combination with analysis of the expression of such nutrient transporters as OmpH and OmpL.

The significance of nutrient availability was genetically documented in *S. cerevisiae*, the growth of which is inhibited by high pressure. Under high-pressure conditions, the uptake of tryptophan via the high-affinity tryptophan permease Tat2 is impaired and the expression of Tat2 is downregulated, leading to growth arrest (Abe and Horikoshi, 2000). These results are consistent with those of transport studies in marine bacteria, which indicate that the rate of uptake of many substrates, including glutamate and acetate, is low at a pressure up to 18 MPa (Jannasch and Taylor, 1984). Interestingly, the addition of excess tryptophan or overexpression of Tat2 protein enables *S. cerevisiae* cells to grow at 25 MPa (Abe and Horikoshi, 2000). These results suggest that the uptake of tryptophan is one of the most pressure-sensitive processes in living yeast cells.

TRANSCRIPTIONAL REGULATION UNDER PRESSURE CONDITIONS

The moderately piezophilic *S. violacea* DSS12 grows optimally at 30 MPa and 8°C, but also at 0.1 MPa and 8°C (Nogi et al., 1998a). Recently, a pressure-regulated operon containing two unknown open reading frames was cloned from this strain. The promoter is activated by growth under high pressure. This operon has five transcription initiation sites and is controlled at the transcriptional level by elevated pressure (Fig. 2) (Nakasone et al., 1998). At 0.1 MPa, most transcripts from the operon coincide with those from initiation site 2 (Fig. 2A). The consensus sequence for the RNA polymerase sigma factor, sigma 54, was found upstream from this operon, and the *S. violacea* sigma 54 binds to this region. The *glnA* operon is one of the operons under the control of sigma 54. This operon contains a gene *(glnA)* encoding glutamine synthetase, which is involved in nitrogen metabolism. Transcription of the *glnA* operon is also controlled by elevated pressure conditions in *S. violacea*. These results suggest that sigma 54 may play an important role in pressure-regulated transcription in piezophilic bacteria, particularly in nitrogen metabolism.

Sigma 54 and a two-component regulatory system composed of the bacterial signal-transducing protein NtrB and the bacterial enhancer-binding protein NtrC have been investigated in *S. violacea* DSS12. The *S. violacea* NtrC protein specifically recognizes two NtrC-binding consensus sites found in the *S. violacea glnA* operon. Autophosphorylation of NtrB, followed by transphosphorylation of the phosphorylated NtrB to NtrC, is characteristic of the two-component regulatory system. When the autophosphorylation activity was examined under low-temperature conditions at 0 to 10°C, in vitro autophosphorylation occurred with NtrB from *S. violacea*, but not with its

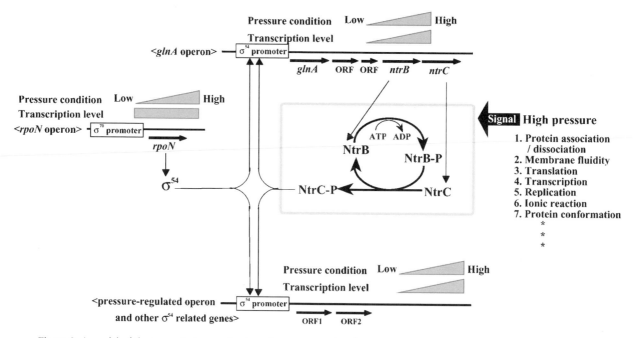

Figure 3. A model of the transcription mechanism of pressure-regulated gene expression in piezophilic bacterium, *S. violacea* DSS12.

homologue from *Escherichia coli*, suggesting that this piezophilic bacterium adapts to the psychrosphere (low-temperature environment) and may have evolved a low-temperature-adapted system in the deep-sea environment (Nakasone et al., 2002). The sigma 54 in *S. violacea* is expressed at a constant level under both atmospheric and high-pressure conditions, and the level of NtrC at high pressure is greater than that at 0.1 MPa. A possible model depicting the pressure-regulated *glnA* operon in *S. violacea* under the control of the NtrB-NtrC system is shown in Fig. 3. In this model, NtrB functions as a pressure sensor via its autophosphorylation in response to increasing pressure. Then the phosphate is transferred to NtrC followed by the activation of the sigma 54-dependent promoter. Detailed studies on the structure and the function of NtrB will be necessary to validate this model.

CONCLUSIONS

Considering the large amount of background information available on the effects of high pressure on physicochemical equilibria, biochemical and biophysical reactions, and protein structure, research on piezophiles is expected to progress more mechanistically. Most of the research on piezophiles to date has focused on psychrophilic piezophiles. This is reasonable because most deep oceans are low-temperature environments. However, mesophilic and thermophilic

piezophiles are also available for biotechnological investigation and use, as in the case of other extremophiles (Abe and Horikoshi, 2001).

The integration of basic knowledge of the effects of pressure on biochemical or biophysical reactions and knowledge acquired through recent advances in molecular biology and entire genome sequencing information will enable us to understand the mechanisms of piezoadaptation as well as the microbial diversity in the deep sea. Some recent reviews have been published (Kato and Bartlett, 1997; Bartlett, 1999, 2002; Abe et al., 1999; Abe and Horikoshi, 2001. We sincerely hope that more researchers will become interested in investigating the effects of high pressure in piezophiles.

REFERENCES

Abe, F., and K. Horikoshi. 1998. Analysis of intracellular pH in the yeast *Saccharomyces cerevisiae* under elevated hydrostatic pressure: a study in baro-(piezo-)physiology. *Extremophiles* 2:223–228.

Abe, F., and K. Horikoshi. 2000. Tryptophan permease gene *TAT2* confers high-pressure growth in *Saccharomyces cerevisiae*. *Mol. Cell. Biol.* 20:8093–8102.

Abe, F., and K. Horikoshi. 2001. The biotechnological potential of piezophiles. *Trends Biotechnol.* 19:102–108.

Abe, F., C. Kato, and K. Horikoshi. 1999. Pressure-regulated metabolism in microorganisms. *Trends Microbiol.* 7:447–452.

Allen, E. E., D. H. Facciotti, and D. H. Bartlett. 1999. Monounsaturated but not polyunsaturated fatty acids are required for growth of the deep-sea bacterium *Photobacterium profundum* SS9 at high pressure and low temperature. *Appl. Environ. Microbiol.* 65:1710–1720.

Bartlett, D. H. 1999. Microbial adaptations to the psychrosphere/peizosphere. *J. Mol. Microbiol. Biotechnol.* **1**:93–100.

Bartlett, D. H. 2002. Pressure effects on in vivo microbial processes. *Biochim. Biophys. Acta* **1595**:367–381.

Bartlett, D. H., M. Wright, A. A. Yayanos, and M. Silverman. 1989. Isolation of a gene regulated by hydrostatic pressure. *Nature* **342**:572–574.

Canganella, F., J. M. Gonzalez, M. Yanagibayashi, C. Kato, and K. Horikoshi. 1997. Pressure and temperature effects on growth and viability of the hyperthermophilic archaeon *Thermococcus peptonophilus*. *Arch. Microbiol.* **168**:1–7.

DeLong, E. F., and A. A. Yayanos. 1985. Adaptation of the membrane lipids of a deep-sea bacterium to changes in hydrostatic pressure. *Science* **228**:1101–1103.

DeLong, E. F., D. G. Franks, and A. A. Yayanos. 1997. Evolutionary relationship of cultivated psychrophilic and barophilic deep-sea bacteria. *Appl. Environ. Microbiol.* **63**:2105–2108.

Deming, J. W., L. K. Somers, W. L. Straube, D. G. Swartz, and M. T. Macdonnel. 1988. Isolation of an obligatory barophilic bacterium and description of a new genus, *Colwellia* gen. nov. *Syst. Appl. Microbiol.* **10**:152–160.

Hei, D. J., and D. S. Clark. 1994. Pressure stabilization of proteins from extreme thermophiles. *Appl. Environ. Microbiol.* **60**: 932–939.

Ikegami, A., K. Nakasone, S. Fujii, C. Kato, R. Usami, and K. Horikoshi. 2000. Cloning and characterization of the gene encoding RNA polymerase sigma factor σ^{54} of deep-sea piezophilic *Shewanella violacea*. *Biochim. Biophys. Acta* **1491**: 315–320.

Jannasch, H. W., and C. D. Taylor. 1984. Deep-sea microbiology. *Annu. Rev. Microbiol.* **38**:487–514.

Jannasch, H. W., C. O. Wirsen, and S. J. Molyneaux. 1992. Comparative physiological studies on hyperthermophilic archaea isolated from deep-sea hot vent with emphasis on *Pyrococcus* strain GB-D. *Appl. Environ. Microbiol.* **58**:3472–3481.

Kato, C., and D. H. Bartlett. 1997. The molecular biology of barophilic bacteria. *Extremophiles* **1**:111–116.

Kato, C., and Y. Nogi. 2001. Correlation between phylogenetic structure and function: examples from deep-sea *Shewanella*. *FEMS Microbiol. Ecol.* **35**:223–230.

Kato, C., H. Tamegai, A. Ikegami, R. Usami, and K. Horikoshi. 1996. Open reading frame 3 of the barotolerant bacterium strain DSS12 is complementary with *cyd*D in *Escherichia coli*: *cyd*D functions are required for cell stability at high pressure. *J. Biochem.* **120**:301–305.

Kato, C., A. Ikegami, M. Smorawinska, R. Usami, and K. Horikoshi. 1997. Structure of genes in a pressure-regulated operon and adjacent regions from a barotolerant bacterium strain DSS12. *J. Mar. Biotechnol.* **5**:210–218.

Kato, C., L. Li, Y. Nogi, Y. Nakamura, J. Tamaoka, and K. Horikoshi. 1998. Extremely barophilic bacteria isolated from the Mariana Trench, Challenger Deep, at a depth of 11000 meters. *Appl. Environ. Microbiol.* **64**:1510–1513.

Macdonald, A. G. 1987. The role of membrane fluidity in complex processes under high pressure, p. 207–223. *In* R. E. Marquis, A. M. Zimmerman, and H. W. Jannasch (ed.), *Current Perspectives in High Pressure Biology.* Academic Press, London, United Kingdom.

Marteinsson, V. T., P. Moulin, J.-L. Birrien, A. Gambacorta, M. Vernet, and D. Prieur. 1997. Physiological responses to stress conditions and barophilic behavior of the hyperthermophilic vent archaeon *Pyrococcus abyssi*. *Appl. Environ. Microbiol.* **63**:1230–1236.

Miller, J. F., N. N. Shah, C. M. Nelson, J. M. Ludlow, and D. S. Clark. 1988. Pressure and temperature effects on growth and methane production of the extreme thermophile *Methanococcus jannaschii*. *Appl. Environ. Microbiol.* **54**: 3039–3042.

Nakasone, K., A. Ikegami, C. Kato, R. Usami, and K. Horikoshi. 1998. Mechanisms of gene expression controlled by pressure in deep-sea microorganisms *Extremophiles* **2**:149–154.

Nakasone, K., A. Ikegami, H. Kawano, C. Kato, R. Usami, and K. Horikoshi. 2002. Transcriptional regulation under pressure conditions by RNA polymerase sigma54 factor with a two-component regulatory system in *Shewanella violacea*. *Extremophiles* **6**:89–95.

Nogi, Y., C. Kato, and K. Horikoshi. 1998a. Taxonomic studies of deep-sea barophilic *Shewanella* strains and description of *Shewanella violacea* sp. nov. *Arch. Microbiol.* **170**:331–338.

Nogi, Y., C. Kato, and K. Horikoshi. 1998b. *Moritella japonica* sp. nov., a novel barophilic bacterium isolated from a Japan Trench sediment. *J. Gen. Appl. Microbiol.* **44**:289–295.

Pande, C., and A. Wishnia. 1986. Pressure dependence of equilibria and kinetics of *Escherichia coli* ribosomal subunit association. *J. Biol. Chem.* **261**:6272–6278.

Pledger, R. J., B. C. Crump, and J. A. Baross. 1994. A barophilic response by two hyperthermophilic, hydrothermal vent *Archaea*: an upward shift in the optimal temperature and acceleration of growth rate at supra-optimal temperatures by elevated pressure. *FEMS Microbiol. Ecol.* **14**:233–242.

Qureshi, M. H., C. Kato, and K. Horikoshi. 1998a. Purification of two pressure-regulated c-type cytochromes from a deep-sea barophilic bacterium. *Shewanella* sp. strain DB-172F. *FEMS Microbiol. Lett.* **161**:301–309.

Qureshi, M. H., C. Kato, and K. Horikoshi. 1998b. Purification of a ccb-type quinol oxidase specifically induced in a deep-sea barophilic bacterium, *Shewanella* sp. strain DB-172F. *Extremophiles* **2**:93–99.

Takai, K., A. Sugai, T. Itoh, and K. Horikoshi. 2000. *Palaeococcus ferrophilus* gen. nov., sp. nov., a barophilic, hyperthermophlic archaeon from a deep-sea hydrothermal vent chimney. *Int. J. Syst. Evol. Microbiol.* **50**:489–500.

Tamegai, H., C. Kato, and K. Horikoshi. 1998. Pressure-regulated respiratory system in barotolerant bacterium, *Shewanella* sp. strain DSS12. *J. Biochem. Mol. Biol. Biophys.* **1**:213–220.

Welch, T. J., and D. H. Bartlett. 1996. Isolation and characterization of the structural gene for OmpL, a pressure-regulated porin-like protein from the deep-sea bacterium *Photobacterium* species strain SS9. *J. Bacteriol.* **178**:5027–5031.

Welch, T. J., and D. H. Bartlett. 1998. Identification of a regulatory protein required for pressure-responsive gene expression in the deep-sea bacterium *Photobacterium* species strain SS9. *Mol. Microbiol.* **27**:977–985.

Yamada, M., K. Nakasone, H. Tamegai, C. Kato, R. Usami, and K. Horikoshi. 2000. Pressure regulation of soluble cytochromes *c* in a deep-sea piezophilic bacterium, *Shewanella violacea*. *J. Bacteriol.* **182**:2945–2952.

Yayanos, A. A., S. Dietz, and R. Van Boxtel. 1979. Isolation of a deep-sea barophilic bacterium and some of its growth characteristics. *Science* **205**:808–810.

ZoBell, C. E., and F. H. Johnson. 1949. The influence of hydrostatic pressure on the growth and viability of terrestrial and marine bacteria. *J. Bacteriol.* **57**:179–189.

ZoBell, C. E., and R. Y. Morita. 1957. Barophilic bacteria in some deep-sea sediments. *J. Bacteriol.* **73**:563–568.

Microbial Diversity and Bioprospecting
Edited by Alan T. Bull
© 2004 ASM Press, Washington, D.C.

Chapter 16

Life in Extremely Dilute Environments: the Major Role of Oligobacteria

D. K. BUTTON

Oligobacteria are bacteria adapted to growth in dilute conditions. Their main food source is dissolved photosynthetically produced organic material. The major products are carbon dioxide, minerals, and cellular biochemicals. Most nutrition is heterotrophic. Oligobacteria evolved to sustain at nutrient concentrations that are small. These small nutrient concentrations are of their own doing in that they are responsible for the depletion. Small cell size provides much surface area but limited space for DNA so that even though usually only one double-stranded molecule of DNA is contained, it must be very small as well. This limits information content. Copeotrophs, on the other hand, have larger cell size, larger genomes, and more cytoplasmic constituents. Most can survive periods of starvation to populate rich food sources and also defend against hostile influences. The properties of oligobacteria are based largely on copeotrophs such as *Escherichia coli* because oligobacteria have been unavailable in sufficient mass for investigation.

Phytoplankton collect nutrients and reduce them to concentrations that are small as well. They are obliged to concentrate both light energy and mineral nutrients. Some, such as *Synechococcus* and *Prochlorococcus*, are very small giving a large surface-to-volume ratio for mineral collection. This ability is smaller than for oligobacteria because phytoplankton must synthesize their own organics as well. Competition strategies appear different as phytoplankton maintain a "bloom and bust" distribution with occasional fast growth in mineral-rich conditions and limited predation, followed by periods of mineral replenishment by oligobacteria. Our focus here is on the oligobacteria.

PROPERTIES

The problem of oligobacterial growth in natural water systems is one of nutrient acquisition. This requirement dominates oligobacterial properties as summarized in Table 1. The characteristically small size and large DNA content of marine bacteria that results from eons of competition in dilute environments is shown by the DNA/particle mass cytogram of surface water organisms (Fig. 1). Deep-water organisms have almost indistinguishable properties according to this technology, in contrast to some observations by microscopy.

Growth rates are inherently slower for oligobacteria than for commonly cultured bacteria because of limited numbers of enzymes in their dilute cytoplasm. Rates measured are generally near upper limits for oligobacteria because the focus is often on photic zone processes in active regions. Most exist in cold, deep, and extremely oligotrophic conditions. Rates are unknown, but estimates from oxygen consumption and from nutrient concentrations and kinetics below the rates are very small. The calculation of rates in the predominant less-active waters are further compromised by uncertainty about the distribution of metabolic activity among the population, whether uniform or strongly graded. Among the best evidence for continued activity throughout the extremely oligotrophic deep ocean is uniform populations of about 10^4/ml with properties similar to those above and a constant virus/bacteria ratio throughout (Fuhrman, 2000). Phage life times are only hours, and dead vegetative bacteria probably would autolyze as well, so the bacteria, consistent with surface water autoradiography data (Karner and Fuhrman 1997), are probably alive.

D. K. Button • Institute of Marine Science and Department of Chemistry and Biochemistry, University of Alaska, Fairbanks, AK 99775.

Table 1. Oligobacterial composition

Parameter	*Escherichia coli*	Bacterioplankton	Units	Reference
Growth rate	1	0.1	g produced per g present per h	Indirect calculations
Diameter	1.5	0.5	µm	Calculated
Cell mass, dry	255	18.5	fg/cell	Button and Robertson (2000)
Cell mass	1,740	102	fg/cell	Velimirov and Walenta-Simon (1992)
Volume	1.71	0.09	µm	Velimirov and Walenta-Simon (1992)
Area and dry weight	15	55	µm^2/µg	Robertson et al. (1998)
Cell density	1.1	1.07	g/cm^2	Robertson et al. (1998)
Dry weight	28.6	14.18	%	Robertson et al. (1998)
Transporters	304	30–100	Types per cell	Meidanis et al. (2002)
Protein content	55	53	% solids	Ingraham et al. (1983)
RNA content	20	?	% solids	Kirchman (2000)
DNA content	11	2	fg/cell	Button and Robertson (2000)
DNA content	1	12	% dry weight	Button and Robertson (2000)
Genome size	4.7	1.9	Mb	Rudd et al. (1990)
Lipid content	2–9	?	% dry weight	Kirchman (2000)

Cell volumes and solids content for *E. coli* are reasonably certain because of the ease in growing large amounts of biomass and analytical methods such as Coulter counting that have the necessary sensitivity. However, values assigned to oligobacteria have been quite variable. Values here are by flow cytometry and agree with refractive indices and with carbon content by CHN analysis and by equilibrium sedimentation. Volumes taken from extinction culture *Cycloclasticus oligotrophus* were just large enough to determine by

Coulter counter impedance. This organism grows well on hydrocarbons but uses acetate slowly, perhaps by diffusion. Thus, large quantities could be grown without substrate-accelerated death. No other marine bacteria have yet been examined in this way. But the small solids content and dilute cytoplasm are consistent with the advantages of a large surface-to-dry-mass ratio and room for DNA replication.

DNA content and dry weight can be measured directly on in situ populations. Chromosome size

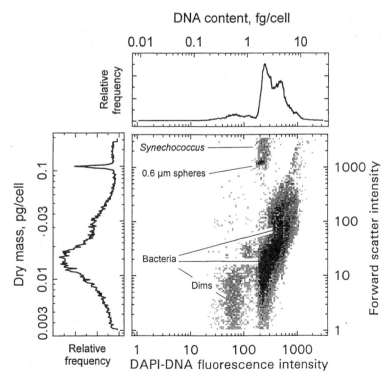

Figure 1. Distribution of the DNA content and mass per cell in a surface water sample analyzed by flow cytometry. Redrawn from Button and Robertson (2000). DAPI, 4′,6′-diamidino-2-phenylindole.

varies only by a factor of about 3, from a little less than 2 fg/cell to a little more than 4. Many organisms in the environment have only a single pair of DNA molecules and are therefore 1n, whereas actively growing organisms often have two and a few have more. The smaller cells are more than 10% DNA dry weight, a large enough value to question whether a smaller size limit has been reached for these already small-genome organisms. This small genome size is of particular interest. It appears to correlate with cultur-ability, and free-living representatives are not com-monly cultivated. Clearly some normal information content is absent because these organisms have the added problem of transporting and metabolizing nu-merous substrates along with other normal functions. Loss of metabolic control is one possibility because the chemical composition of the aquatic environment is somewhat constant, and bacterioplankton could rely on kinetic control instead.

Other aspects of cell composition are less certain. The RNA content of bacterioplankton is unknown but values may be very small because few ribosomes, the main RNA-containing constituent, are required for very slow-growing organisms. *E. coli* can have 1,000 ribosomes, but environmental organisms usu-ally produce their ribosomes with only a single ribo-somal operon. One ribosome contains about 3 ag or about 0.01% of an oligobacterium's dry weight.

The number of transporter types is probably large. *E. coli* has the largest number of major trans-porters for a sequenced gram-negative bacterium (Meidanis et al., 2002). Pathogens generally have fewer; for example, the plant-infecting *Xyllella fas-tidiosa* has 75. The amount of major transporter DNA per megabase for sequenced microorganisms is somewhat constant at 2.4 to 7.1%, so that the total number increases with genome size, and bacterio-plankton have small genomes. Because the simultane-ous transport of numerous substrates is necessary for growth at the observed rates, in situ nutrient concen-trations, and specific affinities, it is necessary for the organisms to either have a large number of permease types or ones that are fairly nonspecific. An estimate of the number of transporter copies per transporter type is of the order of 100 from the dioxygenase con-tent of *C. oligotrophus* (Button et al., 1998); how-ever, the number of transporter copies per cell has yet to be measured for any organism.

Substrates

All enzymatically formed organics are necessar-ily used because none accumulate. Major compounds include the hundred or so macromolecular monomers and metabolic intermediates of organisms. Steady-state concentrations of monomers such as common amino acids and sugars are near 1 μg/liter in active waters. Some substrates are used slowly, such as ex-oskeleton and cell-wall components constructed with unusual stereochemistry to discourage enzymatic at-tack, so their concentrations increase relative to their rate of supply. These combine to form part of the dis-solved organic material. This mixture is the remains of structural material after having been worked over by the microflora and totals about 1 mg/liter. It is quite susceptible to photolysis by UV light, and pas-sage through the photic zone renders portions avail-able to the microflora. Nonpolar substrates such as hydrocarbons and terpenes are reasonably abundant because many coastal organisms can produce them, and they are analytically detectable.

Products

Under the best conditions in aerobic systems, metabolized substrates are split about equally be-tween cell material and carbon dioxide. As growth slows, yields are uncertain as endogenous metabolism takes its toll. Substrate quality affects yield as well. Those rich in hydrogen such as hydrocarbons pro-duce larger yields because the hydrogen can be oxi-dized to produce energy. Substrate molecular com-plexity reduces yield. Often oxidation is incomplete, and hydrocarbons are particularly prone to the liber-ation of partly oxidized products such as 2-hydroxy muconic semialdehyde. As oxygen is depleted, organ-ics can serve as electron acceptors, and these reduced products are liberated as well. Some products such as dimethyl sulfide and methane along with carbon dioxide are greenhouse gases, compounds that affect global climate. Quorum sensors are also liberated as regulators of population. Particularly interesting are various lactones liberated by bacteria, and some dra-matically increase the culturability of marine bacteria (Bruns et al., 2002).

Speciation

Although speciation is thoroughly addressed elsewhere in this volume (chapter 4), the following is intended as a background to oligotrophy. Only about half the prokaryotes isolated by current techniques belong to known phyla; most belong to two main groups, the *Bacteria* and the *Archaea*. Many of the isolated organisms belong to the gamma subclass of the *Proteobacteria;* α, β, δ, and ε complete the lin-eages in this class. Examples of the 17 other major lineages are *Cytophagales*, green sulfur, cyanobacte-ria, *Flexistipes*, and *Aquificales* (Giovannoni and Rappé, 2000). The other major group of microorganisms is the *Archaea*. Sometimes called the third kingdom, this ancient lineage is biochemically unique due to an absence of thymine in tRNA, ether-linked and branched fatty acids in their cell membranes, and occasionally methane production. Many archaeons are thermophiles; hot, cold, deep, and very salty waters are rich in this group. These adaptations result in part from the ability to produce a diverse variety of lipids that facilitate the function of membrane proteins.

Phylogenetics is based on differences in DNA sequences that code for ribosomal DNA. This DNA is relatively ancient and stable so that lineages can be traced. However, the cosmopolitan distribution of oligobacteria among the various phyla suggest that metabolic patterns could have since evolved and that metabolic strategies may have converged.

Most aquatic forms prefer their native environment to that of the laboratory, and only 0.01 to 0.1% will grow by conventional methods such as agar spread plate. New cultivation technology such as extinction culturing (Button et al., 1993) has fostered the isolation of new species (Fig. 2). Extinction culturing involves determining the total population to calculate the required dilution. This is done by counting fluorescently stained organisms on membrane filters or by flow cytometry. The latter is particularly accurate because both organism size, from light scatter intensity and associated Mie theory, and DNA content by fluorescence labeling of their DNA, are evaluated for each organism. This avoids interference from viruses that have sufficient DNA for visible fluorescence and from cytoplasm-free ghosts or other small particles free of DNA. Use of 30 substrate species simultaneously by the average organism gives the potential for 10^{10} species if each prefers a different combination. Yet DNA fragment analysis suggests that far fewer normally coexist in detectible

numbers; only 5 or 10 are sometimes detected by fragment analysis.

Viability

True viabilities are arguably near 100% depending on phage infectivity. Autoradiography shows that nearly all are metabolically active, improved isolation techniques generally yield larger viabilities, and the very small fraction is as active as the large (Button and Robertson, 2000). Cultivation in liquid media such as filtered and sterilized seawater, perhaps in the presence of allelochemicals such as lactones, in communication with natural seawater through diffusion membranes, and in the absence of large concentrations of substrate, is helping to increase apparent values. Noteworthy is the fact that the method favors large values in that an occasional false positive can greatly inflate results.

ACTIVITY CONTROL BY SUBSTRATE CONCENTRATION

The major direct control of oligobacterial activity is the concentration of utilizable nutrients at the surface of the cell. These nutrients are mostly small biochemically common molecules. The organic monomers are either directly liberated from phytoplankton or enzymatically hydrolyzed by extracellular enzymes in the guts of grazers, in fecal particles, and on solid surfaces. Photolysis is also a significant mechanism of generating metabolically labile particles in the euphotic zone.

Microbial kinetics is a discipline that includes treatment of the rate of nutrient acquisition by transthreptic organisms. Transthreptic organisms are across-feeding or surface feeders that live by concentrating organic nutrients across their cell surfaces and include yeasts, molds, and bacteria, including some mixotrophs that use sunlight as well. The paradigm for nutrient transport is Michaelis-Menten. However, deriving from surface catalysis, it has limited ability to describe microbial processes. Changes in metabolic capacity of microorganisms that stem from changes in enzyme content generate ambiguity in concepts of organism affinity for substrates. Inductive and bioenergetic changes in organism activity that change with the concentration of substrate are difficult to incorporate into the paradigm. Neither does it provide a good way to compare the activity of organisms across species or community lines. However, the process of nutrient acquisition can be described from first principles in a step-by-step manner that has come to be known as specific affinity theory (Button, 1998).

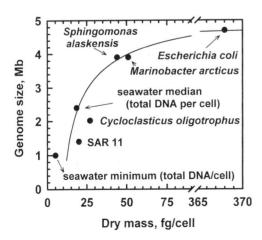

Figure 2. Dry mass, genome size profiles of bacterial isolates along with total DNA mass in surface seawater.

Kinetics

To help understand oligobacterial strategies, we can formulate an understanding of component processes. Main components (Fig. 3) are the concentration of substrate S that is accumulated by transporters of type T to concentration S_c in the internal pools. E_A and E_B are enzyme concentrations for members of the pathways able to transform the substrates, and X is the biomass that is formed. The initial rate-affecting process for transport is the frequency of effective collisions between external substrate S and energized or active permeases T that are N_a in number. The transporters are taken as limited in type and general in specificity as found in *Caulobacter crescentus* and to avoid uncertainties in the frequency of recollisions due to cage (Berg and Purcell, 1977) and surface effects (Astumian and Chock, 1985). Some mechanisms differ in these details. For example, paraplasmic binding proteins can first collect the substrate for efficient transfer across the inner-cell membrane by transporters. For hydrocarbons, the substrate first reacts with a cytoplasmic oxygenase where it is trapped. But the basic kinetics and energetics remain constant. With all permeases active and in the absence of saturation, the rate of transport per unit biomass v is

$$v = a_S^\circ S \tag{1}$$

where a_S° is the base-specific affinity for substrates. Specific affinity is a second-order rate constant in that it specifies rate as the product of two species, substrate molecules, and cells. Because rate is specified as per unit biomass in this case, it can be called a pseudo second-order constant. It takes on different units if rate is specified differently. If specified in terms of cell number, the units would be liters per cell per unit time. Conversion to the previous units thus involves multiplying by the average mass per cell. Thus, specific affinity can be reported in any convenient conserved property of the microbiota. But some property, for example, bacterial carbon or nitrogen, must be specified.

For oligotrophic systems, the portion of permeases that are active becomes significant. Common

transport mechanisms such as symport transport depend on the availability of protons or other ions such as sodium that can be exchanged for protons. Then the number of active permeases, N_a

$$N_a = \frac{N \Delta p^n}{k^n + \Delta p^n} \tag{2}$$

is taken as dependent on the transmembrane potential, often the proton motive force Δp, as it activates the total number of transporters T according to a constant k and a positive cooperativity. The factor n gives a sigmoidal relationship with a foot or threshold and truncation of the number of active permeases when all N transporters become active. The constants of this curve remain speculative due to difficulties in dealing with the small substrate concentrations involved and interference by cosubstrates, but the presence of a requirement for a membrane potential during rapid metabolism is certain. The thermodynamics of generating the concentration gradient observed in metabolic pools requires energy input. Also, starved bacteria effect normal rates of transport only in the presence of a usable source of energy.

With ample substrate and permeases, the proton motive force maximizes at a commonly measured value Δp_{max}. But if S is scarce it depends on the concentration of the substrate as well. This relationship is also taken as sigmoidal. Then the proton motive force from the actively transported substrate Δp_T is

$$\Delta p_T = \frac{\Delta p_{max} S^n}{K_{\Delta p} S^n}, \tag{3}$$

where $K_{\Delta p}$ and n are constants. S and Δp_T are interdependent, exponentially slowing growth in low-energy environments such as the deep sea and the emergence of growth following spring blooms. Both small and nonpolar molecules can enter organisms by diffusion alone, in which case equation 3 is that portion of the proton potential obtained from active transport. That obtained from diffusion Δp_D is $\Delta p_D = a_{SD} S_D$, where S_D is a diffusion substrate. Although a_{SD} is unknown, it may be significant in extremely oligotrophic systems. Uptake is given by the number of active permeases and the residence time τ of the substrate in them during transport. When S becomes large enough to reduce the frequency of success of collisions between substrate and permeases (Button et al., 1998), the specific affinity is a reduced a_S to the parenthetical term in equation 4 and

$$v = \left(\frac{N_a c}{1 + S c \tau} \right) S, \tag{4}$$

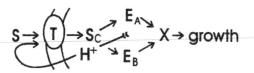

Figure 3. Model for the nutrient-limited rate of transport and growth by an oligobacterium.

where c is the collection constant,

$$c = \frac{5 \, D \, M \, r_S^2}{2 \, r_X^4} \tag{5}$$

Here M is the molecular weight of S mobilized according to the diffusion constant D, and r_S and r_X are the cell and transporter site radii. Cytoplasmic substrate concentration S_c is obtained from the uptake rate according to a linear increase in S_c with the uptake rate v in the same way that gas pressure in a leaky chamber responds to the input rate of a gas:

$$S_C = v \, L \tag{6}$$

so that S_c reflects v as a net transport rate. S_c is transformed to cell material by enzyme sequences E_A, E_B.... The mean amount of enzyme in each pathway step can be calculated from temperature-corrected maximal velocities V_m for cytoplasmic enzymes. The steady-state V_m generally increases with S and

$$V_i = V_m(1 + K_i S), \tag{7}$$

where V_i is the induced value of $V_m \cdot K_i$ can be obtained from observed increases in maximal velocities with induction for the system in question to give the rate of transport

$$v = \frac{V_j S_C}{K_C + S_C}. \tag{8}$$

Thus, organism activity is specified in terms of its cytoarchitecture, a property dependent on ambient concentrations of nutrients.

Endogenous Effects

In addition, growth with respect to uptake rate v varies because of the effect of endogenous losses on cell yield and increases the threshold in S for growth as calculated below.

Conclusions from the Calculations

The effect of substrate concentration on specific affinity is shown in Fig. 4. The effect of energy costs serves to minimize the specific affinity at concentrations of substrate that are small as shown in Fig. 4A. When obtained from the initial slope of the Michaelis-Menten equation for transport, the specific affinity is maximal at $S = 0$ to give a base value $a°_S = a_{Smax}$. However, the combination of equations

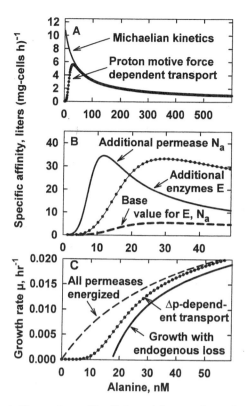

Figure 4. Changes in specific affinity and the rate of growth calculated for *Marinobacter arcticus* according to specific affinity theory.

2 through 8 show minima at both zero and large concentrations of substrate, with the maxima a_{Smax} at $S > 0$. Then active transport can occur only in the presence of sufficient substrate for the number of active permeases N_a to be above zero, and a threshold for active transport is specified. Panel B of Fig. 4 shows that the specific affinity can be increased by increasing either the spatial density of transporters near the surface of the cell or by increasing the concentration of cytoplasmic enzymes involved in the transport of the substrate. The specific affinity is also increased with larger concentrations of transported substrate in the cytoplasmic pools due to effects on the cytoplasmic enzymes. Endogenous requirements are unknown for oligobacteria, and a rate constant of 2.7×10^{-3}/h was used in calculations of growth rate compared with 0.07 to 0.3/h for *E. coli* (Tempest and Neijssel, 1987). Values subtract from the amount of material available for growth. With this small value, more than a year is required to consume all cell components, yet an obvious threshold in S for growth remains. Figure 4C shows that the rate of growth is further reduced at small concentrations according to the membrane potential available for transport and the amount of energy loss from endogenous processes.

Variations in a_s

Growth limitation by sub-microgram per liter concentrations of substrate is not easily evaluated for in situ processes. However, credible values for specific affinity can be obtained from uptake rates of added radiolabeled substrate, values that reflect real affinities at ambient substrate levels irrespective of variances and values for saturation curves. Background substrate is accounted for because it dilutes the ambient and accumulated substrate by equal amounts. Thus, specific affinity, v/S from equation 1, gives a numerical index of microbial activity and one that decreases from maximal levels only when saturation becomes significant. Typical values for oligotrophic isolates and copeotrophs and in situ values are shown in Table 2.

The Multisubstrate Requirement

Stimulation of growth is commonly observed by the addition of cosubstrates (Law and Button, 1986), and affinities are increased in starved cells as well (Schmidt and Alexander, 1985). However, in typical aquatic environments specific affinities for substrates added in combination are not different from those added separately if the total concentration added remains constant. Equation 9 expresses the total rate of transported nutrient from n substrates according to their associated specific affinities.

$$v = \sum_{i=1}^{n} a_S S_i. \tag{9}$$

Then total flux is a function only of the number, concentration, and associated specific affinity for each substrate. With S at 0.5 µg/liter for $n = 20$ substrates with a maximal (July) specific affinity of 1.5 liters/mg cells per h in Harding Lake, the flux is $(0.5 \times 10^{-6}$ g/liter) (20) $(1.5 \times 10^3$ liters/g of cells per h) = 0.15 g of substrate/g of cells per h. The growth rate is $\mu = vY$, where Y is the yield. Taking the yield as 2.5 g of cells wet/g of substrate used, the growth rate is $\mu = 0.015 \times 2.5 = 0.0375$/h for a doubling time of ln 2/0.0375 = 18 h. Then, use of 20 amino acids plus 20 sugars or other substrates would be needed for a reasonably rapid but a likely growth rate of 2 per day. Winter growth rates in the Alaskan environment would be far slower, with specific affinities that are lower by a factor of 70 and a substrate concentration that is lower by a factor of 10 according to amino acids data.

Oligotrophic Capacity

Affinities generally increase with an inverse Michaelis constant so that log-specific affinity versus $-\log K_m$ is a straight line of positive slope (Button, 1991), with copeotrophs at the bottom and oligotrophs at the top. Large Michaelis constants reflect concentrations at a half-maximal uptake rate that are large because the organisms require the ability to process large quantities of substrate quickly. This can be done by using large quantities of transporters and enzymes in tandem. The burden of nutrient processing is then spread among many proteins so that increased external concentrations of substrate are needed to keep them all occupied. Most oligotrophs, on the other hand, must build many permeases for substrate collection. Because large nutrient concentrations are rarely faced, only small numbers of each sequence are

Table 2. Typical values for specific affinity

Organism	Substrate	Specific affinity (liters [mg cells h]$^{-1}$)	Reference
Cycloclasticus oligotrophus	Toluene	47.4	Button et al. (1998)
Escherichia coli	Glucose	7.3	Senn et al. (1994)
Flavobacterium sp. S12	Starch	1.6	van der Kooij (1990)
Summer polynya water	Mixed amino acids	1.1	Yager and Deming (1999)
Sphingomonas alaskensis	Alanine	0.4	Schut et al. (1993)
Sphingomonas alaskensis	Glucose	0.2	Schut et al. (1993)
Saccharomyces cerevisiae	Gucose	6×10^{-4}	Coons et al. (1995)
Vibrio parahaemolyticus	Glucose	3.0×10^{-4}	Sarker et al. (1994)
Harding Lake surface water from near Fairbanks, Alaska			
January	Mixed amino acids	0.12	Unpublished, this laboratory
April	Mixed amino acids	0.07	Unpublished, this laboratory
July	Mixed amino acids	0.5	Unpublished, this laboratory
Gulf of Alaska surface seawater			
May	Leucine	6×10^{-3}	Unpublished, this laboratory
May with warming	Leucine	100×10^{-3}	Unpublished, this laboratory
December	Glutamate	4.4×10^{-3}	Unpublished, this laboratory
December	Mixed amino acids	4.3×10^{-3}	Unpublished, this laboratory

required. This reduces capacity, which decreases the size of the Michaelis constant because saturation can be achieved with a smaller substrate flux. Oligobacteria that are best able to grow on single substrates, such as *C. oligotrophus*, have specific affinities of near 47 liters/g of cells per h and K_ω values of about 1 μg/liter (10^{-9} g/liter). The product of log a_S and $-\log K_m$ is 9, which reflects a large oligotrophic capacity as compared with copeotrophs, where the value approaches unity.

In natural environments, Michaelis constants are small due to limited enzyme content. Measured as the uptake of a single substrate, affinities are also small because flux depends on many substrates collected in concert. If a group of substrates is added to measure the affinity and the total concentration remains constant, the affinity will remain constant, but the nutrient flux will have increased in proportion to the number of substrates added.

Functionality

The carbon cycle is shown as a flywheel in Fig. 5. The arrow width approximates the relative flux as taken from Stomm (2000). Bacterioplankton are responsible for a major part of the remineralization, providing N, P, C, and other minerals for phytoplankton growth. Flywheel mass, the amount of material in the various components, is set by the amount of these minerals in the system. The cycling rate depends on the kinetics of their capture by phytoplankton. The rate of formation of organic material is particularly patchy in natural water systems with regional

and seasonal blooms of phytoplankton. Bacterial populations are quite constant, in contrast, usually in the range of 10^5 to 10^6 organisms per ml. They provide capacitance in the sense of a maintained reserve for the consumption of organic carbon by responding to surges in carbon fixation with increased activity. Then blooms initiated can be extended in time. The result is a large increase in biomass over the cycles because the faster-growing phytoplankton can increase before the smaller herbivores have time to meet the new resource with new demand. The result is biomass maximization from available resources, a combined principle of thermodynamics and biological kinetics, that stems from niche filling, because the more efficient consumers produce the most potential for reproduction.

In active systems such as surface water, the bacterial population is controlled by bacterivore feeding ability at the point where clearance efficiency is significantly limited by their collection rate. Following the blooms and associated input of organics, a small decrease in bacterial population probably causes the bacterivore population to subside. But the bacteria are able to alter their consumption rate of organics by perhaps 2 orders of magnitude and yet sustain a presence. These organisms can then provide a rapid response to the organics from a phytoplankton bloom.

Details of this turn-on remain unclear. Specific affinities in springtime water samples are greatly increased by warming, so ready reserve capacity is certain and can proceed in the absence of nutrient amendments. These low-activity-inducible populations can be found in seasonal blooms at high latitude, so there can be a lag between the bloom in phytoplankton and the bloom in bacterial activity in cold water for reasons unknown. DNA restriction analysis suggests some population shifts over the seasons, and different populations have different metabolic capacities at different temperatures due to changes in tertiary structure. Much of metabolic control is usually maintained with multiple effectors, and although the kinetic and thermodynamic control described here may be incomplete, it appears reasonable as a primary influence. Low non-bloom activity extends to the bottom on the oceans, and this turn-on may also accompany upwelling.

CONCLUSIONS

Oligobacteria are uniquely adapted to the extremely dilute organic nutrient conditions of the marine and other aquatic environments. They are small in size to minimize surface-to-volume ratio and to minimize nutrient depletion in their immediate surroundings. Cytoarchitecture is further diluted, minimizing nutritional costs and is dependent on the col-

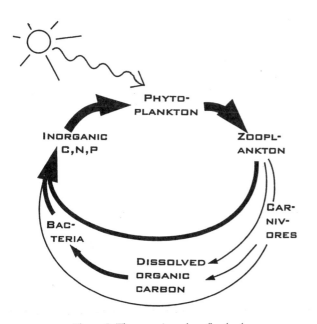

Figure 5. The aquatic carbon flywheel.

lection rate of nutrients. Small size limits chromosome size as a matter of space for DNA replication. Metabolic flexibility is limited, and growth of all but the larger representatives in the laboratory is traditionally problematic, perhaps due to an absence of metabolic control. Most cultivated oligotrophs resist growth above 10^6 organisms per ml (Rappé et al., 2002). Their properties are sufficiently unique that, despite their dominant abundance, physiological, molecular biological, and cytochemical properties remain uninvestigated. Some specialists that focus on a single class of nutrients such as hydrocarbons have a very high affinity for them. Most, however, do not; they rely instead on having only moderate affinity for specific organics, but are able to use many simultaneously, thereby increasing nutrient flux. One result is competition for a rather common nutrient pool and the coexistence of fewer species than otherwise might exist. The ability to concentrate many nutrients in this way appears to be cosmopolitan so that successful members of the oligobacteria have coevolved across many genetic lineages.

REFERENCES

Astumian, R. D., and P. B. Chock. 1985. Interfacial reaction dynamics. *J. Phys. Chem.* **89**:3477–3482.

Berg, H. C., and E. M. Purcell. 1977. Physics of chemoreception. *Biophys. J.* **20**:193–219.

Bruns, A., H. Cypionka, and R. Overmann. 2002. Cyclic amp and acyl homoserine lactones increase the cultivation efficiency of heterotrophic bacteria from the central Baltic Sea. *Appl. Environ. Microbiol.* **68**:3978–3987.

Button, D. K. 1991. Biochemical basis for whole-cell uptake kinetics: specific affinity, oligotrophic capacity, and the meaning of the Michaelis constant. *Appl. Environ. Microbiol.* **57**:2033–2038.

Button, D. K. 1998. Nutrient uptake by microorganisms according to kinetic parameters from theory as related to cytoarchitecture. *Microbiol. Mol. Biol. Rev.* **62**:636–645.

Button, D. K., and B. R. Robertson. 2000. Effect of nutrient kinetics and cytoarchitecture on bacterioplankter size. *Limnol. Oceanogr.* **45**:499–505.

Button, D. K., F. Schut, P. Quang, R. M. Martin, and B. Robertson. 1993. Viability and isolation of typical marine oligobacteria by dilution culture: theory, procedures and initial results. *Appl. Environ. Microbiol.* **59**:881–891.

Button, D. K., B. R. Robertson, T. Schmidt, and P. Lepp. 1998. A small, dilute-cytoplasm, high-affinity, novel bacterium isolated by extinction culture that has kinetic constants compatible with growth at measured concentrations of dissolved nutrients in seawater. *Appl. Environ. Microbiol.* **64**:4467–4476.

Coons, D. M., R. B. Boulton, and L. F. Bisson. 1995. Computer-assisted nonlinear regression analysis of the multicomponent glucose uptake kinetics of *Saccharomyces cerivisiae*. *J. Bacteriol.* **177**:3251–3258.

Fuhrman, J. 2000. Impact of viruses on bacterial processes, p. 327–350. *In* D. L. Kirchman (ed.), *Microbial Ecology of the Oceans*. Wiley-Liss, New York, N.Y.

Giovannoni, S., and M. Rappé. 2000. Evolution, diversity, and molecular ecology of marine prokaryotes, p. 47–84. *In* D. Kirchman (ed.), *Microbial Ecology of the Oceans*. Wiley-Liss, New York, N.Y.

Ingraham, J., O. Maaloe, and F. C. Neidhardt. 1983. *Growth of the Bacterial Cell*. Sinauer Assoc., Inc., Sunderland, Mass.

Karner, M., and J. A. Fuhrman. 1997. Determination of active marine bacterioplankton: a comparison of universal 16S rRNA probes, autoradiography and nucleoid staining. *Appl. Environ. Microbiol.* **63**:1208–1213.

Kirchman, D. L. 2000. Uptake and regeneration of inorganic nutrients by marine heterotrophic bacteria, p. 261–288. *In* D. Kirchman (ed.), *Microbial Ecology of the Oceans*. Wiley-Liss, New York, N.Y.

Law, A. T., and D. K. Button. 1986. Modulation of the affinity of a marine pseudomonad for toluene and benzene by hydrocarbon exposure. *Appl. Environ. Microbiol.* **51**:469–476.

Meidanis, J., D. V. Braga, and S. Verjovski-Almeida. 2002. Whole-genome analysis of transorters in the plant pathogen *Xylella fastiodiosa*. *Microbiol. Mol. Biol. Rev.* **66**:272–299.

Rappé, M. S., S. A. Connon, K. L. Vergin, and S. J. Giovannoni. 2002. Cultivation of the ubiquitous SAR11 marine bacterium clade. *Nature* **418**:630–632.

Robertson, B. R., D. K. Button, and A. L. Koch. 1998. Determination of the biomasses of small bacteria at low concentration in a mixture of species with forward light scatter measurements by flow cytometry. *Appl. Environ. Microbiol.* **64**:3900–3909.

Rudd, K. E., W. Miller, J. Ostell, and D. A. Benson. 1990. Alignment of *Escherichia coli* K12 DNA sequences to a genomic restriction map. *Nucleic Acids Res.* **18**:313–321.

Sarker, R. I., W. Ogawa, M. Tsuda, S. Tanaka, and T. Tsuchiya. 1994. Characterization of a glucose transport system in *Vibrio parahaemolyticus*. *J. Bacteriol.* **176**:7378–7382.

Schmidt, S. K., and M. Alexander. 1985. Effects of dissolved organic carbon and second substrates on the biodegradation of organic compounds at low concentrations. *Appl. Environ. Microbiol.* **49**:822–827.

Schut, F., E. DeVries, J. C. Gottschal, B. R. Robertson, W. Harder, R. A. Prins, and D. K. Button. 1993. Isolation of typical marine bacteria by dilution culture: growth, maintenance, and characteristics of isolates under laboratory conditions. *Appl. Environ. Microbiol.* **59**:2150–2160.

Senn, H., U. K. Lendenmann, M. Snozzi, G. Hamer, and T. Egli. 1994. The growth of *Escherichia coli* in glucose-limited chemostat cultures: a re-examination of the kinetics. *Biochim. Biophys. Acta* **1201**:424–436.

Strom, S. L. 2000. Bacterivory: interactions between bacteria and their grazers, p. 351–386. *In* D. Kirchman (ed.), *Microbial Ecology of the Oceans*. Wiley-Liss, New York, N.Y.

Tempest, D. W., and O. M. Neijssel. 1987. Growth yield and energy distribution, p. 797–806. *In* F. C. Neidhardt (ed.), *Escherichia Coli and Salmonella Typhimurium*. American Society for Microbiology, Washington, D.C.

van der Kooij, D. 1990. Assimilable organic carbon (AOC) in drinking water, p. 57–84. *In* G. A. McFeters (ed.), *Drinking Water Microbiology*. Springer-Verlag, New York, N.Y.

Velimirov, B., and M. Walenta-Simon. 1992. Seasonal changes in specific growth rates, production and biomass of a bacterial community in the water column above a Mediterranean seagrass system. *Mar. Ecol. Prog. Ser.* **80**:237–248.

Yager, P. L., and J. W. Deming. 1999. Pelagic microbial activity in an arctic polynya: testing for temperature and substrate interactions using a kinetic approach. *Limnol. Oceanogr.* **44**:1882–1893.

Chapter 17

Anaerobes: the Sulfate-Reducing Bacteria as an Example of Metabolic Diversity†

GUY FAUQUE AND BERNARD OLLIVIER

Life in the absence of oxygen has become possible because anaerobes can gain energy by using various metabolical strategies in anaerobiosis, especially fermentative pathways. In addition to anaerobes ability to perform fermentation, their use of inorganic (e.g., heavy metals, sulfur or nitrogen compounds, carbon dioxide, etc.) and organic electron acceptors (e.g. fumarate) is also widespread. During the past two decades, it has also been shown that microorganisms formerly known as strict anaerobes were capable of using oxygen under microaerobic conditions. In this case, oxygen serves as an electron acceptor or can be used to detoxify the immediate environment of anaerobes. Therefore, the discovery of such relationships between anaerobes and oxygen has fundamentally modified the concept of anaerobic microorganisms that will probably disclose in the near future other metabolical properties involving oxygen. Metabolical activities performed by anaerobic sulfate-reducing bacteria (SRB) may exemplify what are the metabolical options for anaerobes to live in the absence and in the presence of oxygen. In an attempt to briefly illustrate what is anaerobic life, we focus on the metabolism of SRB, which may have an autotrophic, lithoautotrophic, heterotrophic, or respiration type of life under anaerobiosis.

Many groups of microorganisms utilize a wide variety of inorganic and organic compounds as terminal electron acceptors for anaerobic respiration. Among these final electron acceptors, inorganic sulfur compounds such as elemental sulfur and polysulfide (Fauque et al., 1980, 1994; Hedderich et al., 1999; Widdel and Pfennig, 1992), sulfate and sulfite (Akagi, 1995; Fauque, 1995; Widdel and Hansen, 1992), and thiosulfate and polythionates (Le Faou et al., 1990) play important roles. SRB are a unique physiological group of microorganisms sharing the ability of utiliz-

ing sulfate for their growth, with the production of large amounts of hydrogen sulfide (Fauque, 1995; Smith, 1993; Widdel, 1988). The physiology and the taxonomy of SRB have undergone enormous changes in the past 20 years, and almost 40 genera of sulfate reducers have been characterized. SRB include a broad range of prokaryotes currently divided into four phylogenetic groups based on 16S rRNA sequence analysis: gram-positive spore-forming sulfate reducers, gram-negative mesophilic and thermophilic sulfate reducers belonging to the domain *Bacteria,* and thermophilic archaeal sulfate reducers (Castro et al., 2000; Stackebrandt et al., 1995).

BIOENERGETIC MECHANISMS OF DISSIMILATORY SULFATE REDUCTION

Dissimilatory sulfate reduction in SRB is enzymologically distinct from assimilatory sulfate reduction, which occurs in green plants, algae, and other groups of prokaryotes (Peck, 1993; Peck and Lissolo, 1988). Dissimilatory sulfate reduction in *Desulfovibrio* spp. is linked to electron transport-coupled phosphorylation because substrate-level phophorylation is inadequate to support their growth (Peck, 1960). The SRB belonging to the genus *Desulfovibrio* possess a number of unique physiological and biochemical characteristics such as the requirement for ATP to reduce inorganic sulfate (Peck, 1959), the cytoplasmic localization of enzymes (APS reductases and bisulfite reductases) involved in the pathway of respiratory sulfate reduction (Kremer et al., 1988), the periplasmic localization of some hydrogenases (Fauque et al., 1988), and the abundance of multihemic *c*-type cytochromes (LeGall and Fauque, 1988; Fauque et al., 1991).

Guy Fauque and Bernard Ollivier • UR 101 Extrêmophiles, Institut de Recherche pour le Développement, IFR-BAIM, Universités de Provence et de la Méditerranée, ESIL, Case 925, 163 avenue de Luminy, F-13288 Marseille Cedex 09, France.
†We dedicate this chapter to the memory of Dr. Jean LeGall (1932–2003) for his great contribution to our knowledge of the biochemistry and microbiology of dissimilatory sulfate reduction.

Sulfate Activation and Its Reduction to Bisulfite

Owing to its chemical inertia, sulfate first needs to be activated by consumption of ATP (Peck, 1959). The ATP sulfurylase (EC 2.7.7.4; ATP sulfate adenylyltransferase) forms APS (adenylyl sulfate) and PP$_i$ (pyrophosphate) from sulfate and ATP:

$$SO_4^{2-} + ATP + 2H^+ \rightarrow APS + PP_i$$
$$\Delta G^{\circ\prime} = +46 \text{ kJ/mol}$$

ATP sulfurylase has been purified from crude extracts of two species of the genus *Desulfovibrio*: *D. gigas* and *D. desulfuricans* ATCC 27774; it is a novel metalloprotein containing zinc and cobalt (Gavel et al., 1998). Because the formation of PP$_i$ is thermodynamically unfavorable, the reaction needs to be pulled to completion by a second enzyme, an inorganic pyrophosphatase (EC 3.6.1.1; pyrophosphate phosphohydrolase), which hydrolyzes PP$_i$ according to this reaction:

$$PP_i + H_2O \rightarrow 2P_i$$
$$\Delta G^{\circ\prime} = -22 \text{ kJ/mol}$$

The reduction of APS to AMP and bisulfite, catalyzed by APS reductase, is the first redox reaction and is more exergonic than the pyrophosphate cleavage:

$$APS + H_2 \rightarrow HSO_3^- + AMP + H^+$$
$$\Delta G^{\circ\prime} = -69 \text{ kJ/mol}$$

APS reductase (EC 1.8.99.2) has been purified and characterized from several species of *Desulfovibrio* (Lampreia et al., 1994) and from the hyperthermophilic archaebacterial sulfate reducer *Archaeoglobus fulgidus* (Lampreia et al., 1991). APS reductase is a cytoplasmic iron-sulfur flavoprotein constituting 2 to 3% of soluble proteins and containing one flavin adenine dinucleotide per molecule and eight iron atoms arranged as two different [4Fe-4S] clusters (Lampreia et al., 1994). APS reductase and dissimilatory bisulfite reductase are present in a ratio close to 1. The specific electron donor required for the reduction of APS to bisulfite is yet unknown.

Bisulfite Reduction to Sulfide

The reduction of bisulfite to sulfide must compensate for the energy investment of sulfate activation and yield additional ATP for growth. The standard free-energy change of sulfite reduction to sulfide with molecular hydrogen as electron donor is -174 kJ/mol. Thermodynamically, this could allow the regeneration of at least two ATP. The pathway of bisulfite reduction to sulfide is somewhat controversial,

and two mechanisms have been proposed. The first is the six-electron reduction of bisulfite to H$_2$S in one step, catalyzed by bisulfite reductase, without the formation of free intermediates. In the second, also called the trithionate pathway, bisulfite is reduced to sulfide in three steps via the free intermediates trithionate and thiosulfate (Akagi, 1995; Cypionka, 1995). Arguments for and against a trithionate pathway have been discussed (LeGall and Fauque, 1988), but only the isolation of mutants that will be altered with respect to one or both reductase activities would provide definitive information on the bisulfite reduction mechanism.

Two types of bisulfite reductases exist in SRB. Their substrate is actually the bisulfite ion as deduced from the acidic optimum pH of these two enzymes. The first type comprises the low-spin bisulfite reductases, also called assimilatory-type sulfite reductases, which have a low molecular mass (27 kDa) and only one polypeptide chain and contain a single [4Fe-4S] center coupled to a siroheme (Moura and Lino, 1994). The physiological significance of the low-spin sulfite reductases is still not understood. The second type is constituted by the high-spin bisulfite reductases (EC 1.8.99.1), which have a large molecular mass (around 200 kDa) and a complex structure with at least two different polypeptides in an $\alpha_2\beta_2$ tetramer containing siroheme and [4Fe-4S] clusters. So far, four different enzymes belonging to this class have been purified and characterized from different genera of sulfate reducers (LeGall and Fauque, 1988). Desulfoviridin and desulforubidin belong mainly to the genera *Desulfovibrio* and *Desulfomicrobium*, desulfofuscidin to the thermophilic genus *Thermodesulfobacterium*, and P-582 to the spore-forming genus *Desulfotomaculum* (Fauque et al., 1990, 1991; Hatchikian, 1994; Moura et al., 1988a; Widdel, 1988). These four enzymes differ mainly by the behavior of their siroheme moieties, their major optical absorption spectra, and electron paramagnetic resonance (EPR) spectra.

Hydrogen Metabolism

Molecular hydrogen is a key intermediate in the metabolic interactions of a wide variety of microorganisms. Hydrogen plays a central role in the energy metabolism of sulfate reducers of the genus *Desulfovibrio*, which can either produce or utilize hydrogen depending on the growth conditions (Widdel, 1988). Hydrogenases (hydrogen: oxidoreductase EC.1.12) constitute a class of redox proteins that are very diversified in their structure and active center composition; they catalyze the reversible oxidation of the dihydrogen molecule to protons and electrons (Fauque, 1989; Fauque et al., 1988; Moura et al.,

1988b). Sulfate reducers of the genus *Desulfovibrio* contain three classes of hydrogenases ([Fe], [NiFe], [NiFeSe]), which differ in their metal structure and subunit composition, localization, amino acid sequence, gene structure, immunological reactivities, and catalytic properties (Fauque et al., 1988, 1991). Two mechanisms have been proposed for the formation of a proton gradient in *Desulfovibrio* species: a vectorial electron transport linked to the oxidation of molecular hydrogen by hydrogenases and a proton translocation coupled to the reduction of specific substrates (LeGall and Fauque, 1988). In the first mechanism, also called the obligate hydrogen cycling, H_2 is formed inside the cell by a cytoplasmic hydrogenase; then it diffuses into the periplasmic space where it is oxidized by a periplasmic hydrogenase (Peck and Odom, 1984; Peck et al., 1987). This process generates a membrane potential and a transmembrane pH gradient without pumping protons across the cell membrane. This mechanism is somewhat controversial even if recent observations are in favor of this model (Noguera et al., 1998; Steger et al., 2002). In the second mechanism, sulfate reducers pump protons across the cell membrane, performing the classical type of vectorial proton translocation. *D. desulfuricans* Essex was shown to generate a proton motive force by classical proton translocation using the reductant pulse method (Cypionka, 1995).

OTHER PROCESSES OF ENERGY CONSERVATION BY SULFATE REDUCERS

SRB can utilize processes of energy conservation other than dissimilatory sulfate reduction; they can use electrons acceptors other than sulfate, sulfite, or thiosulfate and are also capable of fermentative growth by dismutation or disproportionation of organic and inorganic compounds.

Fermentation of Inorganic Sulfur Compounds

A novel type of energy metabolism involving fermentation of inorganic sulfur compounds has been described in *Desulfovibrio sulfodismutans* (Bak and Pfennig, 1987). This species is able to conserve energy for growth under strictly anaerobic conditions by disproportionation (or dismutation) of sulfite or thiosulfate to sulfate and sulfide according to the following reactions:

$$4\,SO_3^{2-} + H^+ \rightarrow 3\,SO_4^{2-} + HS^-$$
$$\Delta G^{\circ\prime} = -235 \text{ kJ/mol}$$
$$S_2O_3^{2-} + H_2O \rightarrow SO_4^{2-} + HS^- + H^+$$
$$\Delta G^{\circ\prime} = -21.9 \text{ kJ/mol}$$

Eight of 19 sulfate-reducing strains tested were able to perform at least one of these two inorganic fermentations (Kramer and Cypionka, 1989). The free-energy change of thiosulfate dismutation is low (-21.9 kJ/mol) and does not always permit growth. The capacity of sulfite and thiosulfate disproportionation is constitutively expressed. The enzymes required for the dismutations appear to be the same as for sulfate reduction (Kramer and Cypionka, 1989). Evidence has been shown that during inorganic sulfur compounds fermentations, sulfate is formed via APS reductase and ATP sulfurylase, but not by sulfite oxidoreductase. Reversed electron transport is necessary to enable the reduction of sulfite or thiosulfate with the electrons derived from APS reductase (Kramer and Cypionka, 1989)

Fermentation of Organic Substrates

Some *Desulfovibrio* species ferment malate and fumarate with the formation of succinate, acetate, and carbon dioxide (Widdel, 1988). In the absence of sulfate, many species belonging to the genera *Desulfovibrio*, *Desulfobacterium*, *Desulfococcus*, *Desulfotomaculum*, and *Desulfobulbus* ferment pyruvate with acetate, carbon dioxide, and hydrogen appearing as major end products of metabolism (Widdel, 1988). *Desulfovibrio aminophilus* ferments pyruvate, peptone, Casamino Acids, glycine, serine, threonine, and cysteine (Baena et al., 1998). Some genera of sulfate reducers can also carry out a propionic fermentation that has been studied in detail in *Desulfobulbus propionicus* (Widdel, 1988). Several strains of *D. desulfuricans* were reported to ferment choline with the formation of acetate, ethanol, and trimethylamine. Four unidentified saccharolytic gram-negative nonsporulating mesophilic sulfate-reducing strains fermented fructose, sucrose or glucose to acetate, carbon dioxide, and hydrogen (Joubert and Britz, 1987). *Desulforibio fructosovorans* can ferment fructose to succinate and acetate with the production of small amounts of ethanol (Ollivier et al., 1988). Fermentation of lactate to acetate, carbon dioxide, and hydrogen has been reported with *Desulfovibrio vulgaris* Marburg (Pankhania et al., 1988), even if this reaction does not normally allow growth as lactate oxidation to pyruvate ($E^{\prime\circ} = -190$ mV) requires energy-dependent reverse electron transport, probably catalyzed by a membrane-bound enzyme. Sulfate reducers of the genus *Desulfovibrio* that cannot grow by fermentation of lactate, ethanol, or choline may grow with these substrates in the absence of sulfate in syntrophic cocultures with hydrogen-scavenging methanogenic bacteria (Widdel and Hansen, 1992).

Reduction of Elemental Sulfur, Nitrate, and Oxygen

Reduction of elemental sulfur

An important contribution to the knowledge of the biological sulfur cycle has been the discovery that some microorganisms utilize colloidal or elemental sulfur as a terminal electron acceptor. Several genera of the domain *Archaea* and *Bacteria* may gain energy for growth by a dissimilatory reduction of elemental sulfur to hydrogen sulfide in a respiratory-type metabolism (Fauque et al., 1991; Le Faou et al., 1990; Widdel, 1988; Widdel and Pfennig, 1992). The facultative sulfur-reducing bacteria, such as the SRB, utilize elemental sulfur as a respiratory substrate in the absence of other possible electron acceptors such as nitrate, nitrite, sulfate, sulfite, or thiosulfate. Even if most of sulfate reducers cannot grow by sulfur reduction, some thiophilic SRB, belonging to the genera *Desulfovibrio* and *Desulfomicrobium,* use elemental sulfur as an alternative electron acceptor (Biebl and Pfennig, 1977). Tetraheme cytochrome c_3 is the constitutive elemental sulfur reductase in some strains of *Desulfomicrobium* and *Desulfovibrio* from which the sulfur reductase activity can be purified with the tetrahemoprotein (Fauque et al., 1979; Fauque, 1994). An EPR spectroscopic study of tetraheme cytochrome c_3 from *Desulfomicrobium baculatum* Norway 4, in the presence of colloidal sulfur, suggests the need of a heme with low redox potential to reduce elemental sulfur (Cammack et al., 1984). A mechanism of attack of colloidal sulfur by the *Desulfomicrobium baculatum* Norway 4 tetraheme cytochrome c_3 has been proposed, and the reaction might involve insoluble S_8 molecules as intermediates (Cammack et al., 1984). Membranes isolated from *D. gigas* and *Desulfomicrobium baculatum* Norway 4 contained hydrogenase and *c*-type cytochromes and catalyzed the dissimilatory sulfur reduction. Membranes of *D. gigas* were able to couple esterification of orthophosphate to electron flow from hydrogen to elemental sulfur. A P/2e ratio of 0.1 was measured, and methyl viologen functioned as an uncoupling agent of phosphorylation (Fauque et al., 1980).

Dissimilatory reduction of nitrate

The dissimilatory reduction of nitrate or nitrite to ammonia (also called ammonification) can function as sole energy-conserving processes in some sulfate reducers. Nitrate is reduced to ammonia (with nitrite as intermediate) by a few strains belonging mainly to *D. desulfuricans, Desulfobacterium catecholicum,* and *Desulfobulbus propionicus* (Seitz and Cypionka, 1986; Widdel and Hansen, 1992). Depending on the organism, sulfate or nitrate may be the preferred electron acceptor. Nitrate reductase is inducible by nitrite or nitrate, whereas nitrite reductase is synthesized constitutively in *D. desulfuricans* Essex 6 (Seitz and Cypionka, 1986). Vectorial proton translocation during nitrate or nitrite reduction has been demonstrated with whole cells of *Desulfovibrio* species (Cypionka, 1995), and ATP synthesis coupled to the reduction of nitrite to ammonia was also obtained with membranes of *D. gigas* (Barton et al., 1983). The purification and the characterization of the respiratory nitrate and nitrite reductases has been reported only in *D. desulfuricans* ATCC 27774 grown on a lactate nitrate medium (Moura et al., 1997). This nitrate reductase is a periplasmic monomeric enzyme with a molecular mass of 74 kDa, containing one molybdenum atom and one [4Fe 4S] center by molecule, which exhibits EPR signals assigned to Mo(V) (Bursakov et al., 1995); it is the first dissimilatory nitrate reductase to have its crystal structure determined (Dias et al., 1999). The respiratory nitrite reductase (ammonia forming) is a membrane-bound heterooligomer cytochrome *c* containing two types of subunits of 62 and 19 kDa and forming a complex with a molecular mass of 750 kDa (Pereira et al., 1996; Liu et al., 1994). It was recently reported that this cytochrome *c* nitrite reductase also catalyzes the reduction of sulfite to sulfide (Pereira et al., 1996).

Fixation of molecular nitrogen has been demonstrated in species of the genera *Desulfobulbus, Desulfotomaculum, Desulfobacter,* and *Desulfovibrio* (Lespinat et al., 1987; Postgate et al., 1988; Widdel and Hansen, 1992).

Reduction of oxygen

Until 1980, sulfate reducers were thought to be strict anaerobes; then it was shown that they tolerate the transient presence of molecular oxygen (LeGall and Xavier, 1996; Cypionka, 2000). The capability of true aerobic respiration coupled to energy conservation was detected in some strains of *Desulfovibrio* spp., *Desulfococcus multivorans, Desulfobulbus propionicus,* and *Desulfobacterium autotrophicum* (Dilling and Cypionka, 1990). However, sulfate reducers usually do not grow with dioxygen as an electron acceptor. Aerobic respiration by SRB is microaerophilic and not sensitive to azide and cyanide (Dilling and Cypionka, 1990). SRB obviously contain terminal oxidases different from those of aerobic microorganisms. Different oxygen-reducing systems are present in *Desulfovibrio* species. NADH oxidase activity entirely responsible for the oxygen reduction to water was found in several *Desulfovibrio* species (Cypionka, 2000). In *D. desulfuricans, Desulfovibrio ter-*

mitidis, and *D. vulgaris,* oxygen reduction was coupled to proton translocation and ATP conservation. In these three species, periplasmic hydrogenase and tetraheme cytochrome c_3 play a major role in oxygen reduction (Cypionka, 2000). Sulfate reducers use superoxide reductases as one component of an alternative oxidative stress protection system that catalyzes reduction rather than disproportionation of superoxide to hydrogen peroxide. Two classes of superoxide reductases exist in sulfate reducers containing one (neelaredoxin) or two (rubredoxin oxidoreductase or desulfoferrodoxin) iron centers (Kurtz and Coulter, 2002). Recently, a membrane-bound terminal oxygen reductase of the cytochrome *bd* family was isolated and characterized from *D. gigas* and shown to completely reduce oxygen to water (Lemos et al., 2001).

METAL REDUCTION BY DISSIMILATORY SULFATE REDUCERS

In addition to their ability to use sulfur compounds as terminal electron acceptors, SRB have been demonstrated to reduce a wide range of heavy metals and radionuclides including Cr(VI), Mo(VI), Se(VI), Tc(VII), Fe(III), Mn(IV), As(V), Pd(II), and U(VI) (Lovley, 1993). Such abilities shared by several SRB make them useful candidates for bioremediation of environments contaminated by heavy metals. Metal reduction can be obtained not only by direct reduction but also by indirect chemical reduction when sulfate is supplied in the culture medium, thus favoring concomitant sulfide production and metal precipitation. *D. vulgaris* Hildenborough was the first SRB found to reduce U(VI) and Cr(VI) (Lovley et al., 1993; Lovley and Phillips, 1994). More recently, *D. desulfuricans* was also shown to reduce both these metals but also Mo(VI), Se(VI), and Tc(VII) (Tucker et al., 1998; Lloyd et al., 1999b). Technetium (VII) was also reduced by *D. fructosovorans* (De Luca et al., 2001). The use of Tc(VII) as a terminal electron acceptor with hydrogen as the preferred electron donor looks quite widespread within the SRB (Lloyd et al., 2001). In contrast, the same authors demonstrated that Cr(VI) reduction was restricted among the SRB tested and was not a hydrogen-dependent reaction, with lactate (but not hydrogen) used during this reductive process.

Members of the genus *Desulfotomaculum* were also found to be metal reducers. *Desulfotomaculum reducens* grows in the presence of Cr(VI), U(VI), Mn(IV), and Fe(III) as terminal electron acceptors and lactate as the electron donor (Tebo and Obraztsova, 1998), whereas *Desulfotomaculum auripigmentum* was shown to reduce arsenate [As(V)]

to arsenite [As(III)] (Newman et al., 1997). Metal-reducing capabilities were also demonstrated among members of the genus *Desulfomicrobium,* where the strain Ben-RB reduces As(V) (Macy et al., 2000). *Desulfomicrobium norvegicum* reduces Cr(VI) with a very high efficiency (Michel et al., 2001) and was proposed to be seeded in bioreactors for Cr(VI) industrial removal.

The use of heavy metals in lieu of sulfate as an electron acceptor by SRB is generally not coupled to growth. Energy conservation through metal reduction [Cr(VI)] has been demonstrated only for *Desulfotomaculum reducens* isolated from a heavy-metal-contaminated sediment (Tebo and Obraztsova, 1998). The direct reduction of heavy metals by SRB results from an enzymatic process that is still poorly understood. The tetraheme cytochrome c_3 of *D. vulgaris* Hildenborough has been demonstrated to catalyze the reductive precipitation of U(VI) and Cr(VI) (Lovley et al., 1993; Lovley and Phillips, 1994). Results with tetraheme cytochrome c_3 altered by site-directed mutagenesis indicate that negative redox potential hemes are crucial for metal reductase activity (Michel et al., 2001). Similar to the tetraheme cytochrome c_3, the [Fe] hydrogenase from *D. vulgaris* Hildenborough was shown to reduce Cr(VI) (Michel et al., 2001). The periplasmic hydrogenases of *D. desulfuricans* and *D. fructosovorans* were also shown to be associated with the Tc(VII) reductase activity of these two microorganisms (Lloyd et al., 1999a), thus suggesting that cytochromes and hydrogenases from SRB play an important role in metal reduction.

To date, few SRB have been tested for their ability to reduce heavy metals. It is therefore a good challenge (i) for microbiologists to isolate suitable SRB for bioremediation of soils and waters contaminated by heavy metals and (ii) for biochemists to further understand the enzymology of metal reduction by SRB in particular.

CONCLUDING REMARKS

SRB are a group of anaerobic prokaryotes which are unified by sharing the capacity to carry out dissimilatory sulfate reduction to sulfide as a major component of their bioenergetic processes. Despite this seeming physiological unity, sulfate reducers present a tremendous ecological, metabolic, morphological, and nutritional diversity. SRB contain a complex and diversified electron carrier system that obviously complicates bioenergetic studies of the process of dissimilatory sulfate reduction. During the past 20 years, progress has been made (i) in the purification and the biochemical characterization of novel types of enzymes and redox carriers involved in the dissimilatory sulfate metabolism of SRB and (ii) in the elucidation of the mechanisms for sulfate transport and degradation pathways. The use of a combined genetic and biochemical approach should enable answers to two remaining questions that concern the nature of the in vivo electron donors required for the reduction of APS and bisulfite and the importance of the trithionate pathway in the dissimilatory bisulfite reduction. As the enzymatic machinery adequate for aerobic respiration is present in bacteria classified as strict anaerobes, a revision of the classification of microorganisms in terms of their response to oxygen has to be done.

Acknowledgments. We gratefully thank P. Roger and J. L. Garcia for revising the manuscript.

REFERENCES

Akagi, J. M. 1995. Respiratory sulfate reduction, p. 89–111. *In* L. L. Barton (ed.), *Biotechnology Handbooks*, vol. 8, *Sulfate-Reducing Bacteria*. Plenum Press, New York, N.Y.

Baena, S., M.-L. Fardeau, M. Labat, B. Ollivier, J.-I. Garcia, and B. K. C. Patel. 1998. *Desulfovibrio aminophilus* sp. nov., a novel amino acid degrading and sulfate reducing bacterium from an anaerobic wastewater lagoon. *Syst. Appl. Microbiol.* **21:** 498–504.

Bak, F., and N. Pfennig. 1987. Chemolithotrophic growth of *Desulfovibrio sulfodismutans* sp. nov. by disproportionation of inorganic sulfur compounds. *Arch. Microbiol.* **147:**184–189.

Barton, L. L., J. LeGall, J. M. Odom, and H. D. Peck, Jr. 1983. Energy coupling to nitrite respiration in the sulfate-reducing bacterium *Desulfovibrio gigas. J. Bacteriol.* **153:**867–871.

Biebl, H., and N. Pfennig. 1977. Growth of sulfate-reducing bacteria with sulfur as electron acceptor. *Arch. Microbiol.* **112:** 115–117.

Bursakov, S., M.-Y. Liu, W. J. Payne, J. LeGall, I. Moura, and J. J. G. Moura. 1995. Isolation and preliminary characterization of a soluble nitrate reductase from the sulfate reducing organism *Desulfovibrio desulfuricans* ATCC 27774. *Anaerobe* **1:**55–60.

Cammack, R., G. Fauque, J. J. G. Moura, and J. LeGall. 1984. ESR studies of cytochrome c₃ from *Desulfovibrio desulfuricans* Norway 4: midpoint potentials of the four haems, and interactions with ferredoxin and colloidal sulphur. *Biochim. Biophys. Acta* **784:**68–74.

Castro, H. F., N. H. Williams, and A. Ogram. 2000. Phylogeny of sulfate-reducing bacteria. *FEMS Microbiol. Ecol.* **31:**1–9.

Cypionka, H. 1995. Solute transport and cell energetics, p. 151–184. *In* L. L. Barton (ed.), *Biotechnology Handbooks*, vol. 8, *Sulfate-Reducing Bacteria*. Plenum Press, New York, N.Y.

Cypionka, H. 2000. Oxygen respiration by *Desulfovibrio* species. *Annu. Rev. Microbiol.* **54:**827–848.

De Luca, G., P. De Philip, Z. Dermoun, M. Rousset, and A. Verméglio. 2001. Reduction of technetium (VII) by *Desulfovibrio fructosovorans* is mediated by the nickel-iron hydrogenase. *Appl. Environ. Microbiol.* **67:**4583–4587.

Dias, J. M., M. E. Than, A. Humm, R. Huber, G. P. Bourenkov, H. D. Bartunik, S. Bursakov, J. Calvete, J. Caldeira, C. Carneiro, J. J. G. Moura, I. Moura, and M. J. Romac. 1999. Crystal structure of the first dissimilatory nitrate reductase at 1.9 Å solved by MAD methods. *Structure* **7:**65–79.

Dilling, W., and H. Cypionka. 1990. Aerobic respiration in sulfate-reducing bacteria. *FEMS Microbiol. Lett.* **71:**123–128.

Fauque, G. 1989. Properties of [NiFe] and [NiFeSe] hydrogenases from methanogenic bacteria, p. 216–236. *In* M. S. Da Costa, J. C. Duarte, and R. A. D. Williams (ed.), *Microbiology of Extreme Environments and its Potential for Biotechnology*. Elsevier Applied Science, London, United Kingdom.

Fauque, G. D. 1994. Sulfur reductase from thiophilic sulfate-reducing bacteria. *Methods Enzymol.* **243:**353–367.

Fauque, G. D. 1995. Ecology of sulfate-reducing bacteria, p. 217–241. *In* L. L. Barton (ed.), *Biotechnology Handbooks*, vol. 8, *Sulfate-Reducing Bacteria*. Plenum Press, New York, N.Y.

Fauque, G., D. Herve, and J. LeGall. 1979. Structure-function relationship in hemoproteins: the role of cytochrome c₃ in the reduction of colloidal sulfur by sulfate-reducing bacteria. *Arch. Microbiol.* **121:**261–264.

Fauque, G. D., L. L. Barton, and J. LeGall. 1980. Oxidative phosphorylation linked to the dissimilatory reduction of elemental sulphur by *Desulfovibrio*, p. 71–86. *In Sulphur in Biology, Ciba Foundation Symposium 72*. Excerpta Medica, Amsterdam, The Netherlands.

Fauque, G., H. D. Peck, Jr., J. J. G. Moura, B. H. Huynh, Y. Berlier, D. V. DerVartanian, M. Teixeira, A. E. Przybyla, P. A. Lespinat, I. Moura, and J. LeGall. 1988. The three classes of hydrogenases from sulfate-reducing bacteria of the genus *Desulfovibrio. FEMS Microbiol. Rev.* **54:**299–344.

Fauque, G., A. R. Lino, M. Czechowski, L. Kang, D. V. DerVartanian, J. J. G. Moura, J. LeGall, and I. Moura. 1990. Purification and characterization of bisulfite reductase (desulfofuscidin) from *Desulfovibrio thermophilus* and its complexes with exogenous ligands. *Biochim. Biophys. Acta* **1040:**112–118.

Fauque, G., J. LeGall, and L. L. Barton. 1991. Sulfate-reducing and sulfur-reducing bacteria, p. 271–337. *In* J. M. Shively and L. L. Barton (ed.), *Variations in Autotrophic Life*. Academic Press Ltd., London, United Kingdom.

Fauque, G., O. Klimmek, and A. Kröger. 1994. Sulfur reductases from spirilloid mesophilic sulfur-reducing eubacteria. *Methods Enzymol.* **243:**367–383.

Gavel, O. Y., S. A. Bursakov, J. J. Calvete, G. N. George, J. J. G. Moura, and I. Moura. 1998. ATP sulfurylases from sulfate-reducing bacteria of the genus *Desulfovibrio*. A novel metalloprotein containing cobalt and zinc. *Biochemistry* **37:**16225–16232.

Hatchikian, E. C. 1994. Desulfofuscidin: dissimilatory, high-spin sulfite reductase of thermophilic, sulfate-reducing bacteria. *Methods Enzymol.* **243:**276–295.

Hedderich, R., O. Klimmek, A. Kroger, R. Dirmeier, M. Keller, and K. O. Stetter. 1999. Anaerobic respiration with elemental sulfur and with disulfides. *FEMS Microbiol. Rev.* **22:**353–381.

Joubert, W. A., and T. J. Britz. 1987. Isolation of saccharolytic dissimilatory sulfate-reducing bacteria. *FEMS Microbiol. Lett.* **48:**35–40.

Kramer, M., and H. Cypionka. 1989. Sulfate formation via ATP sulfurylase in thiosulfate- and sulfite-disproportionating bacteria. *Arch. Microbiol.* **151:**232–237.

Kremer, D. R., M. Veenhuis, G. Fauque, H. D. Peck, Jr., J. LeGall, J. Lampreia, J. J. G. Moura, and T. A. Hansen. 1988. Immunocytochemical localization of APS reductase and bisulfite reductase in three *Desulfovibrio* species. *Arch. Microbiol.* **150:**296–301.

Kurtz, D. M., and E. D. Coulter. 2002. The mechanism(s) of superoxide reduction by superoxide reductases in vitro and in vivo. *J. Biol. Inorg. Chem.* **7:**653–658.

Lampreia, J., G. Fauque, N. Speich, C. Dahl, I. Moura, H. G. Trüper, and J. J. G. Moura. 1991. Spectroscopic studies on APS reductase isolated from the hyperthermophilic sulfate-reducing archaebacterium. *Archaeoglobus fulgidus. Biochem. Biophys. Res. Commun.* **181:**342–347.

Lampreia, J., A. S. Pereira, and J. J. G. Moura. 1994. Adenylylsulfate reductases from sulfate-reducing bacteria. *Methods Enzymol.* **243:**241–260.

Le Faou, A., B. S. Rajagopal, L. Daniels, and G. Fauque. 1990. Thiosulfate, polythionates and elemental sulfur assimilation and reduction in the bacterial world. *FEMS Microbiol. Rev.* **75:**351–382.

LeGall, J., and G. Fauque. 1988. Dissimilatory reduction of sulfur compounds, p. 587–639. *In* A. J. B. Zehnder (ed.), *Biology of Anaerobic Microorganisms.* John Wiley & Sons, Inc., New York, N.Y.

LeGall, J., and A. V. Xavier. 1996. Anaerobes response to oxygen: the sulfate-reducing bacteria. *Anaerobe* **2:**1–9.

Lemos, R. S., C. M. Gomes, M. Santana, J. LeGall, A. V. Xavier, and M. Teixeira. 2001. The "strict" anaerobe *Desulfovibrio gigas* contains a membrane-bound oxygen-reducing respiratory chain. *FEBS Lett.* **496:**40–43.

Lespinat, P. A., Y. M. Berlier, G. D. Fauque, R. Toci, G. Denariaz, and J. LeGall. 1987. The relationship between hydrogen metabolism, sulfate reduction and nitrogen fixation in sulfate reducers. *J. Ind. Microbiol.* **1:**383–388.

Liu, M.-C., C. Costa, and I. Moura. 1994. Hexaheme nitrite reductase from *Desulfovibrio desulfuricans* ATCC 27774. *Methods Enzymol.* **243:**303–319.

Lloyd, J. R., J. Ridley, T. Khizniak, N. N. Lyalikova, and L. E. Macaskie. 1999a. Reduction of technetium by *Desulfovibrio desulfuricans:* biocatalyst characterization and use in a flowthrough bioreactor. *Appl. Environ. Microbiol.* **65:**2691–2696.

Lloyd, J. R., G. H. Thomas, J. A. Finley, J. A. Cole, and L. E. Macaskie. 1999b. Microbial reduction of technetium by *Escherichia coli* and *Desulfovibrio desulfuricans:* enhancement via the use of high-activity strains and effect of process parameters. *Biotechnol. Bioeng.* **66:**122–130.

Lloyd, J. R., A. N. Mabbett, D. R. Williams, and L. E. Macaskie. 2001. Metal reduction by sulphate-reducing bacteria: physiological diversity and metal specificity. *Hydrometallurgy* **59:**327–337.

Lovley, D. R. 1993. Dissimilatory metal reduction. *Annu. Rev. Microbiol.* **47:**263–290.

Lovley, D. R., and E. J. P. Phillips. 1994. Reduction of chromate by *Desulfovibrio vulgaris* and its c_3 cytochrome. *Appl. Environ. Microbiol.* **60:**726–728.

Lovley, D. R., P. K. Widman, J. C. Woodward, and E. J. P. Phillips. 1993. Reduction of uranium by cytochrome c_3 of *Desulfovibrio vulgaris. Appl. Environ. Microbiol.* **59:**3572–3576.

Macy, J. M., J. M. Santini, B. V. Pauling, A. H. O'Neill, and L. I. Sly. 2000. Two new arsenate/sulfate-reducing bacteria: mechanism of arsenate reduction. *Arch. Microbiol.* **173:**49–57.

Michel, C., M. Brugna, C. Aubert, A. Bernadac, and M. Bruschi. 2001. Enzymatic reduction of chromate: comparative studies using sulfate-reducing bacteria. Key role of polyheme cytochromes c and hydrogenases. *Appl. Microbiol. Biotechnol.* **55:**95–100.

Moura, I., and A. R. Lino. 1994. Low-spin sulfite reductases. *Methods Enzymol.* **243:**296–303.

Moura, I., S. Bursakov, C. Costa, and J. J. G. Moura. 1997. Nitrate and nitrite utilization in sulfate-reducing bacteria. *Anaerobe* **3:**279–290.

Moura, I., J. LeGall, A. R. Lino, H. D. Peck, Jr., G. Fauque, A. V. Xavier, D. V. DerVartanian, J. J. G. Moura, and B. H. Huynh. 1988a. Characterization of two dissimilatory sulfite reductases (desulforubidin and desulfoviridin) from the sulfate-reducing

bacteria. Mossbauer and EPR studies. *J. Am. Chem. Soc.* **110:**1075–1082.

Moura, J. J. G., I. Moura, M. Teixeira, A. V. Xavier, G. D. Fauque, and J. LeGall. 1988b. Nickel-containing hydrogenases, p. 285–314. *In* H. Sigel (ed.), *Metal Ions in Biological Systems,* vol. 23, *Nickel and Its Role in Biology.* Marcel Dekker, Inc., New York, N.Y.

Newman, D. K., E. K. Kennedy, J. D. Coates, D. Ahmann, D. J. Ellis, D. R. Lovley, and F. M. M. Morel. 1997. Dissimilatory arsenate and sulfate reduction in *Desulfotomaculum auripigmentum* sp. nov. *Arch. Microbiol.* **168:**380–388.

Noguera, D. R., G. A. Brusseau, B. E. Rittmann, and D. A. Stahl. 1998. A unified model describing the role of hydrogen in the growth of *Desulfovibrio vulgaris* under different environmental conditions. *Biotechnol. Bioeng.* **59:**732–746.

Ollivier, B., R. Cord-Ruwisch, E. C. Hatchikian, and J. L. Garcia. 1988. Characterization of *Desulfovibrio fructosovorans* sp. nov. *Arch. Microbiol.* **149:**447–450.

Pankhania, I. P., A. M. Spormann, W. A. Hamilton, and R. K. Thauer. 1988. Lactate conversion to acetate, CO_2 and H_2 in cells suspensions of *Desulfovibrio vulgaris* (Marburg): indications for the involvement of an energy driven reaction. *Arch. Microbiol.* **150:**26–31.

Peck, H. D., Jr. 1959. The ATP-dependent reduction of sulfate with hydrogen in extracts of *Desulfovibrio desulfuricans. Proc. Natl. Acad. Sci. USA* **45:**701–708.

Peck, H. D., Jr. 1960. Evidence for oxidative phosphorylation during the reduction of sulfate with hydrogen by *Desulfovibrio desulfuricans. J. Biol. Chem.* **235:**2734–2738.

Peck, H. D., Jr. 1993. Bioenergetics strategies of the sulfate-reducing bacteria, p. 41–76. *In* J. M. Odom and R. Singleton Jr. (ed.), *The Sulfate-Reducing Bacteria: Contemporary Perspectives.* Brock/Springer Series in Contemporary Bioscience. Springer-Verlag, New York, N.Y.

Peck, H. D., Jr., and T. Lissolo. 1988. Assimilatory and dissimilatory sulfate reduction: enzymology and bioenergetics, p. 99–132. *In* J. A. Cole and S. J. Ferguson (ed.), *The Nitrogen and Sulphur Cycles. Forty-Second Symposium of the Society for General Microbiology.* Cambridge University Press, Cambridge, United Kingdom.

Peck, H. D., Jr., and J. M. Odom. 1984. Hydrogen cycling in *Desulfovibrio:* a new mechanism for energy coupling in anaerobic microorganisms, p. 215–243. *In* Y. Cohen, R. W. Castenholz, and H. O. Halverson (ed.), *Microbial Mats Stromatolites.* Alan R. Liss, Inc., New York, N.Y.

Peck, H. D., Jr., J. LeGall, P. A. Lespinat, Y. Berlier, and G. Fauque. 1987. A direct demonstration of hydrogen cycling by *Desulfovibrio vulgaris* employing membrane-inlet mass spectrometry. *FEMS Microbiol. Lett.* **40:**295–299.

Pereira, I. C., I. A. Abreu, A. V. Xavier, J. LeGall, and M. Teixeira. 1996. Nitrite reductase from *Desulfovibrio desulfuricans* (ATCC 27774)—a heterooligomer heme protein with sulfite reductase activity. *Biochem. Biophys. Res. Commun.* **224:**611–618.

Postgate, J. R., H. M. Kent, and R. L. Robson. 1988. Nitrogen fixation by *Desulfovibrio,* p. 457–471. *In* J. A. Cole and S. J. Ferguson (ed.), *The Nitrogen and Sulphur Cycles. Forty-Second Symposium of the Society for General Microbiology.* Cambridge University Press, Cambridge, United Kingdom.

Seitz, H.-J., and H. Cypionka. 1986. Chemolithotrophic growth of *Desulfovibrio desulfuricans* with hydrogen coupled to ammonification of nitrate or nitrite. *Arch. Microbiol.* **146:**63–67.

Smith, D. W. 1993. Ecological actions of sulfate-reducing bacteria, p. 161–188. *In* J. M. Odom and R. Singleton, Jr. (ed.), *The Sulfate-Reducing Bacteria: Contemporary Perspectives.* Brock/

Springer Series in Contemporary Bioscience. Springer-Verlag, New York, N.Y.

Stackebrandt, E., D. A. Stahl, and R. Devereux. 1995. Taxonomic relationships, p. 49–87. *In* L. L. Barton (ed.), *Biotechnology Handbooks*, vol. 8, *Sulfate-Reducing Bacteria*. Plenum Press, New York, N.Y.

Steger, J. L., C. Vincent, J. D. Ballard, and L. R. Krumholz. 2002. *Desulfovibrio* sp. genes involved in the respiration of sulfate during metabolism of hydrogen and lactate. *Appl. Environ. Microbiol.* **68:**1932–1937.

Tebo, B. M., and A. Y. Obraztsova. 1998. Sulfate-reducing bacterium grows with Cr(VI),U(VI), Mn(IV), and Fe(III) as electron acceptors. *FEMS Microbiol. Lett.* **162:**193–198.

Tucker, M. D., L. L. Barton, and B. M. Thomson. 1998. Reduction of Cr, Mo, Se and U by *Desulfovibrio desulfuricans* immobilized in polyacrylamide gels. *J. Ind. Microbiol. Biotechnol.* **20:**13–19.

Widdel, F. 1988. Microbiology and ecology of sulfate- and sulfur-reducing bacteria, p. 469–585. *In* A. J. B. Zehnder (ed.), *Biology of Anaerobic Microorganisms*. John Wiley & Sons, Inc., New York, N.Y.

Widdel, F., and T. A. Hansen. 1992. The dissimilatory sulfate- and sulfur-reducing bacteria, p. 583–624. *In* A. Balows, H. G. Trüper, M. Dworkin, W. Harder, and K.-H. Schleifer (ed.), *The Prokaryotes: A Handbook on the Biology of Bacteria: Ecophysiology, Isolation, Identification, Applications*, 2nd ed. vol. I. Springer-Verlag, New York, N.Y.

Widdel, F, and N. Pfennig. 1992. The genus *Desulfuromonas* and other Gram-negative sulfur-reducing eubacteria, p. 3379–3389. *In* A. Balows, H. G. Trüper, M. Dworkin, W. Harder, and K.-H. Schleifer (ed.), *The Prokaryotes: A Handbook on the Biology of Bacteria: Ecophysiology, Isolation, Identification, Applications*, 2nd ed. vol. IV. Springer-Verlag, New York, N.Y.

Microbial Diversity and Bioprospecting
Edited by Alan T. Bull
© 2004 ASM Press, Washington, D.C.

Chapter 18

Microbes from Marine Sponges: A Treasure Trove of Biodiversity for Natural Products Discovery

RUSSELL T. HILL

Microbes associated with marine sponges are of interest in marine biotechnology for several reasons. Our knowledge of sponge microbiology has increased dramatically during the past 5 years as molecular techniques have been applied to the study of these microbial communities, and it is increasingly clear that these communities are highly diverse and include genera of particular interest for natural products screening. Sponge-associated microbes are therefore a resource for drug discovery.

For several decades, it has been known that sponges contain many bioactive compounds, including several compounds of great biomedical interest. Recent work has shown that in some cases these bioactive compounds may be produced by the sponge-associated microbes rather than by the sponges. As more compounds from marine sponges progress through clinical trials, a pressing need for large-scale pharmaceutical production exists. Harvesting of sponges from the natural environment as the sole source of these compounds is impractical. Culture of the producing bacterium in fermentation systems may provide an efficient method for production of the compound. Studies on the diversity of microbes associated with sponges and development of methods to culture additional sponge symbionts are therefore important in order to contribute to the future production of new pharmaceuticals.

In cases where pharmaceutically important compounds are produced by sponges rather than by symbiotic bacteria, it may be necessary to grow sponges by aquaculture in order to obtain an adequate supply of the compounds. The microbial communities associated with sponges are likely to be important for the health of the sponges. For this reason, it may be necessary to monitor changes in sponge-associated microbial communities when the sponges are brought into aquaculture systems.

SPONGES

Sponges belong to the phylum *Porifera* that has been estimated to comprise at least 15,000 extant species. Approximately 85% of these sponges belong to the class *Demospongiae* and the remainder belong to the *Hexactinellida* and the *Calcarea* (Hooper and Van Soest, 2002). Sponges are sessile metazoans found mainly in the marine environment and are filter feeders that use bacteria in the water column as a food source. Sponges generally have three cell layers. The outer layer is the pinacoderm that lines all external surfaces. An internal aquiferous canal system is lined by a layer of choanocyte cells that are flagellated and pump water through the sponge. The main part of the sponge body is contained between the pinacoderm and the choanoderm and is a gelatinous matrix termed the mesohyl. The mesohyl contains skeletal material composed of collagenous spongin fibers and siliceous or calcium carbonate spicules.

SPONGE MICROBIOLOGY

Sponge-microbe associations involve a diverse range of heterotrophic bacteria, cyanobacteria, facultative anaerobes, dinoflagellates, diatoms and, archaea. Bacteria within the intercellular sponge matrix and intracellular within sponge cells can occupy up to 60% of the sponge volume (Wilkinson, 1978b), although the proportion of symbiotic bacteria is highly variable between different sponge species. The cyanobacterial symbiont *Oscillatoria spongelliae* comprises approximately 50% of the cellular volume of the sponge *Dysidea herbacea* (Berthold et al., 1982). Densely packed morphologically diverse bacteria within sponge mesohyl are shown in Fig. 1. The total bacterial counts in sponge tissue samples can ex-

Russell T. Hill • Center of Marine Biotechnology, University of Maryland Biotechnology Institute, Columbus Center Suite 236, 701 East Pratt St., Baltimore, MD 21202-4031.

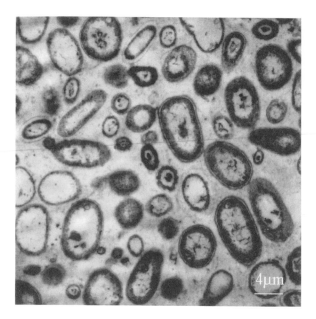

Figure 1. Densely packed bacteria within the mesohyl of *R. odorabile* (Webster and Hill, 2001).

ceed counts in the ambient seawater by between 1 and 3 orders of magnitude (Friedrich et al., 2001; Webster and Hill, 2001) and in one study ranged up to 8×10^9 cells/ml (Webster and Hill, 2001). On the other hand, some sponges, in particular hexactinellids, contain few bacteria within their tissue (Vacelet and Donadey, 1977). The sponge *Aplysina fistularis* was observed to contain remarkably few bacterial symbionts (Faulkner et al., 2000).

The microbes associated with sponges can be present as extracellular symbionts, either on the outer layers of sponges as exosymbionts or within the mesohyl as endosymbionts. Intracellular symbionts can be present within the sponge cells or even within the nuclei of sponge cells as intranuclear symbionts (Fuerst et al., 1999; Lee et al., 2001; Vacelet, 1970). Some extracellular bacteria and other microbes are likely to be cosmopolitan inhabitants of the seawater surrounding the sponge because sponges are filter feeders and take up microbes from the water column by phagocytosis as a food source (Vacelet, 1970). These extracellular microbes will be present within the choanocyte chambers and also transiently within the mesohyl prior to their digestion.

The mode whereby sponges acquire their symbionts is unclear, but there are two obvious possibilities, namely, acquisition by filtration from the ambient water and vertical transmission of symbionts from parent sponges through larvae. Sponges filter very large volumes of seawater. A 1-kg sponge can filter 24,000 liters of seawater per day (Vogel, 1977) and, assuming a typical bacterial cell density for the tropical coral reef environment of about 10^6 cells/ml, is ca-

pable of ingesting 2.4×10^{13} bacterial cells/day. This filtration therefore provides a huge stock of bacteria that can potentially be sequestered through the aquiferous system of the sponge. Any microbes that can resist digestion by the sponge as well as the sponge's immune response and are capable of growth in the microenvironment present in the sponge mesohyl have the potential to become established as symbionts (Wilkinson, 1987). In the case of the *Alphaproteobacteria* symbiont of the Great Barrier Reef sponge *Rhopaloeides odorabile*, densities of the symbiont within the sponge were about 10^4 to 10^6 cells/ml, yet this bacterium was never detected in the water surrounding the sponge. However, detection limits of the method employed to detect this bacterium in seawater were about 10^1 to 10^2 cells/ml. Even if strain NW001 were present at 1 cell/ml in the ambient water, well below the detection limit of the method employed in this study, *R. odorabile* would filter 2.4×10^7 cells of strain NW001 from the water column in a single day if the filtration rate of this sponge is similar to that established by Vogel (1977). Conversely, if strain NW001 were present at an average density of only 2.4 cells per 10,000 liters, one cell of this potential symbiont would still be acquired by the sponge each day. This illustrates that uptake by filtration can result in acquisition by a sponge of bacteria that are present at extremely low densities in the ambient water. Once taken up by the sponge, some bacteria destined to become symbionts may be adapted to prevent phagocytosis. Abundant bacterial morphotypes within sponges have been shown by electron microscopy to have thickened cell walls or slime capsules that may serve as barriers to digestion by sponge archaeocytes (Friedrich et al., 1999; Wilkinson et al., 1984). Bioactive compounds may also play a role in the selective retention of specific bacteria. The Red Sea sponge *Amphimedon viridis* contains bioactive pyridinium alkaloids, halitoxin and amphitoxin, that show selective activity against specific bacteria. These compounds were highly active against eight strains of bacteria isolated from the water column, whereas six bacterial strains isolated from the sponge were resistant to the compounds (Kelman et al., 2001)

The second possibility is that microbial symbionts are transmitted vertically through sponge larvae or during the process of asexual reproduction, in which gemmules, buds, or branches give rise to new adult sponges. Vertical transmission via asexual reproductive stages may allow for transmission of multiple bacteria and permit transmission of a greater diversity than the presumably more stringent transmission through the germ line (Hentschel et al., 2002).

Bacterial symbionts are thought to benefit sponges by supplying nutrients through direct incorporation of dissolved organic matter from seawater (Wilkinson and Garrone, 1980), translocating photosynthate from symbiotic cyanobacteria (Wilkinson, 1983), and assisting in chemical defense (Unson et al., 1994). Other possible roles of sponge symbionts include transportation of metabolites throughout the sponge mesohyl (Borowitzka et al., 1988) and the removal of waste products during periods when the sponges are not circulating water (Wilkinson, 1978a). It has even been speculated that polysaccharide-producing bacterial symbionts may contribute to the structural rigidity of sponges (Wilkinson et al., 1981). In only a few cases has the relationship between sponges and bacterial symbionts been experimentally demonstrated. More often, the nature of this relationship is suggested by circumstantial evidence. For example, Webster and Hill (2001) suggested that the role of the dominant culturable symbiont in the Great Barrier Reef sponge R. odorabile may be in nutrient uptake by the sponge, based on the location of this alphaproteobacterial symbiont around the choanocyte chambers and the ability of this strain to grow on medium in which amino acids and small polypeptides are the primary carbon sources. It would certainly be of great interest to determine unequivocally the role of additional sponge symbionts. Efforts in this direction would be aided by an ability to maintain sponges in aquaria or other in vitro culture systems so that controlled, manipulative, long-term experiments could be conducted. The nutritional requirements and environmental parameters for successful maintenance of sponges in captivity in closed systems are poorly understood (Osinga et al., 1999), although some progress in this regard has recently been made (Nickel et al., 2001; Osinga et al., 2001). The application of molecular approaches should assist greatly in determining the role of sponge symbionts. Fluorescent in situ hybridization (FISH) with specific (usually 16S rRNA) gene probes can be used to localize and identify individual symbiont cells within the sponge matrix (see Color Plate 3) (Friedrich et al., 1999; Webster and Hill, 2001; Webster et al., 2001b). This could be coupled with in situ PCR (Hodson et al., 1995) and in situ reverse transcription-PCR (Chen and Hodson, 2001) to detect presence and expression, respectively, of individual genes within bacterial symbionts, although these approaches would be technically very challenging within the complex sponge matrix due to the presence of PCR-inhibitory compounds and autofluorescence. Perhaps the most rapid progress could be made by taking a genomic approach. Genome sequencing of a known sponge symbiont (such as the Alphaproteobacteria found in R. odorabile [Webster and Hill, 2001]) could, by comparison with the genome of free-living marine Alphaproteobacteria, allow rapid identification of many genes involved in this symbiosis.

DIVERSITY OF MICROBES ASSOCIATED WITH SPONGES

Early work on the diversity of bacteria in sponges relied on culture-based studies. Sponge-associated microbes were shown to be markedly different from those in the water column (Santavy et al., 1990; Wilkinson et al., 1981). Studies of this type are limited by the small proportion of sponge-associated bacteria that can be cultured, generally <1% (Webster and Hill, 2001), although in the case of Ceratoporella nicholsoni, 3 to 11% of the total bacterial population in this sponge was culturable (Santavy et al., 1990). Molecular approaches are needed to answer these most basic questions about sponge-associated bacteria: (i) How diverse are the microbial communities associated with sponges? (ii) Are these communities stable in space and time? (iii) Do sponge-associated microbial communities include previously unknown microbes, including some unique to particular species of sponges? (iv) Are there ubiquitous microbes found in all sponges? This is an exciting time in sponge microbiology as the application of molecular techniques is now producing answers to some of these questions, at least for a handful of sponge species. As the microbial communities associated with larger numbers of sponges are studied, it will become possible during the next decade to draw some general conclusions about these microbial communities.

The first application of 16S rRNA community analysis to investigate the total microbial community associated with a marine sponge revealed previously undescribed Gammaproteobacteria and novel uncultivated strains in the lithistid sponge Discodermia (Lopez et al., 1999). A detailed study of the total microbial community associated with the Great Barrier Reef sponge R. odorabile first gave an indication of the remarkable bacterial diversity present in sponges. The community included representatives of the Betaproteobacteria, Gammaproteobacteria, Cytophaga/Flavobacterium, green sulfur bacteria, green non-sulfur bacteria, planctomycetes, low-G+C gram-positive bacteria, and Actinobacteria (Webster et al., 2001b). Aplysina aerophoba and Theonella swinhoei were chosen for bacterial community analysis because these sponges are distantly related and occur in different geographic locations (Hentschel et al., 2002). Comparative analysis of these communities and that of R. odorabile indicated that there is a uniform microbial community in sponges from different oceans. The majority of sponge-derived sequences were related to the *Acidobacteria* and the *Chloroflexi,* with abundant representatives also from the actinobacteria, *Alpha-, Beta-,* and *Gammaproteobacteria,* cyanobacteria, and *Nitrospira.* This important study revealed that a uniform microbial community, distinctly different from those in marine water or sediment, was present in marine sponges (Hentschel et al., 2002).

Molecular analysis of the total microbial community of an unidentified Indonesian sponge designated sponge 01IND 35 was recently undertaken (Hill et al., 2002). This sponge is of interest because it contains several promising bioactive alkaloid compounds known as manzamines (Yousaf et al., 2002). This analysis revealed the presence of members of the *Cytophaga,* acidobacteria, *Alphaproteobacteria, Gammaproteobacteria,* and *Nitrospira* (Fig. 2).

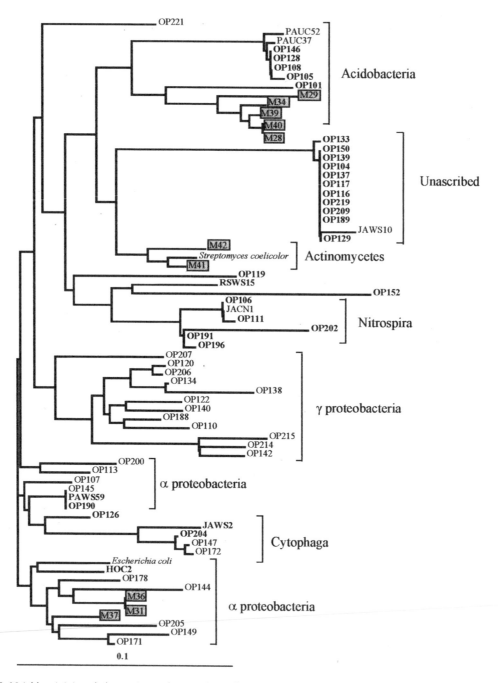

Figure 2. Neighbor-joining phylogenetic tree from analysis of about 500 bp of 16S rRNA gene sequence from clones obtained from the unidentified Indonesian sponge 01IND 35. The scale bar represents 0.1 substitutions per nucleotide position. Culturable isolates from sponge 35 are boxed. Sequences shown in bold are those whose nearest relatives, based on BLAST searches, are also from sponges.

The microbial community in sponge 01IND 35 includes a large assemblage of currently unascribed bacteria, most closely related to the uncultured strain JAWS10 (GenBank No. AF434968) from the sponge *T. swinhoei*. Interestingly, the closest culturable relatives to this assemblage were culturable actinomycete strains M41 and M42, indicating that it is possible that this unascribed group could be novel actinomycetes. In approximately 50% of cases, the closest relatives of the microbial community clone library from sponge 01IND 35 were sequences derived from other marine sponges, consistent with the finding of Hentschel et al. (2002) that there are specific sponge-associated groups of bacteria.

Archaea are also known to be present in marine sponges. The crenarchacote *Cenarchaeum symbiosum* is the single phylotype inhabiting the cold-water sponge *Axinella mexicana*. This symbiotic archaeon grows well within the sponge tissue at temperatures of 10°C, over 60°C below the growth temperature optimum of any cultivated species of *Crenarchaeota* (Preston et al., 1996), all of which are hyperthermophiles. No low-temperature crenarchaeote has yet been cultured, although it is now known that these microbes are ubiquitous in temperate marine waters. Genomic analyses of *C. symbiosum* revealed the presence of two closely related variants that were consistently found in the majority of *A. mexicana* individuals analyzed (Schleper et al., 1998). A crenarchaote related to *C. symbiosum* as well as a euryarchaeote were present in the tropical sponge *R. odorabile,* located primarily within the pinacoderm (Webster et al., 2001a). The role of these sponge-associated archaea within the sponge as well as their metabolic capabilities are completely unknown. Few archaea have been screened for production of bioactive compounds. It would be extremely interesting to screen these sponge-associated archaea but no progress has been made on growing these organisms in culture.

MICROBIAL DIVERSITY IN SPONGES AS A RESOURCE FOR NATURAL PRODUCTS DISCOVERY

The microbial communities associated with marine sponges are remarkably diverse and contain many novel bacteria (Hentschel et al., 2002; Lopez et al., 1999; Webster et al., 2001b) as discussed in detail above. Of particular importance for natural products discovery is the presence in marine sponges of groups of bacteria such as cyanobacteria and actinomycetes, with a good track record for production of bioactive compounds. Novel actinomycetes are an important common feature of the sponge communities of *R.*

odorabile (Webster et al., 2001b) and the unidentified Indonesian sponge 01IND 35 (Hill et al., 2002). Once the presence of actinomycetes was revealed in *R. odorabile* by 16S rRNA community analysis, considerable efforts were made to culture actinomycetes from this sponge, resulting in the successful isolation of three new actinomycete strains (Webster et al., 2001b). Novel actinomycetes have also been cultured from another Great Barrier Reef sponge, *Candidaspongia flabellata* (Burja and Hill, 2001). Because actinomycetes are of considerable interest for screening programs, efforts to culture additional sponge-associated actinomycetes are warranted. More generally, the tremendous diversity of bacteria in marine sponges will remain largely untapped as long as >99% of these bacteria remain uncultured and advances are needed in culturing methods for sponge-derived microbes. Promising results have been obtained by the addition of catalase and sodium pyruvate to isolation media (Olson et al., 2000), and the inclusion of sponge extracts in media may be beneficial (Webster et al., 2001b). Despite over 100 attempts to culture the filamentous theopalauamide-containing symbiont from *T. swinhoei* in liquid media, no growth was obtained. Finally, some colonies containing filamentous cells together with other bacteria were observed on agar media supplemented with aqueous sponge extract and sodium silicate (Schmidt et al., 2000). In some cases, it may prove to be extremely challenging to culture symbionts that have ancient, highly developed relationships with their sponge hosts.

PRODUCTION OF BIOACTIVE COMPOUNDS BY SPONGE-ASSOCIATED MICROBES

Many bioactive compounds of potential pharmaceutical importance have already been obtained from microbes isolated from marine sponges. The reviews by Faulkner (Faulkner [2002] and earlier reports in this series) give many additional examples of compounds from microbes isolated from sponges, and the review by Lee et al. (2001) tabulates 36 examples. In most of these cases, the relationship between the microbe and the sponge has not been studied in any detail and may not be a symbiotic relationship. Rather, sponges have been used simply as source material for isolation of microbes that have then been screened in drug discovery programs. A range of media can be used to attempt to grow as wide a range of sponge-associated microbes as possible. Alternatively, specific groups of microbes can be targeted as being of particular interest because they have a good track record in producing bioactive compounds (e.g., actinomycetes, fungi). Cells are then grown for extraction in organic solvents to obtain "extracts" for screening. Growth under different culture conditions frequently results in the production of a different suite of bioactive compounds. It can therefore be effective to grow each microbe under several different conditions, for example, in different liquid or solid media, in order to maximize the "chemical diversity" from that strain. The screening process is typically based on molecular targets in which the activity of each extract is evaluated against single targets, e.g., an enzyme-receptor binding assay. Detection of the interaction of interest may be by colorimetric, radiologic, or immunologic methodologies. Positive extracts are generally retested to confirm the initial results and are then subjected to chemical characterization. In the examples given here, sponges are used solely as sources of microbes for screening rather than attempts being made to isolate specific microbes that are producers of bioactive compounds previously characterized from the sponges themselves (this is discussed in the next section).

Fungi from sponges have been particularly productive. Jensen and Fenical (2000) give 10 examples of sponge-derived fungi that produce bioactive compounds. Recently, a *Microsphaeropsis* sp. from the Mediterranean sponge *A. aerophoba* was found to contain several tyrosine kinase inhibitors (Brauers et al., 2000).

Actinomycetes isolated from sponges have been sources of several promising compounds. The urauchimycins, novel antimycin antibiotics, were obtained from a *Streptomyces* sp. isolated from an unidentified sponge (Imamura et al., 1993). A *Microbacterium* sp. isolated from the sponge *Halichondria panicea* was the source of four glycoglyceropeptides, one of which showed antitumor activities (Wicke et al., 2000).

MICROBIAL SYMBIONTS OF SPONGES AND NATURAL PRODUCTS DISCOVERY

All natural products obtained from marine sources since 1977 have been reviewed by Faulkner (Faulkner [2002] and earlier reports in this series). During the year 2000, sponges were again the most studied of marine organisms, and a vast array of novel bioactive compounds have been isolated from these organisms. In almost all cases, it is not known whether these compounds are actually produced by the sponges or by sponge-associated microbes. There are very few cases where bioactive compounds found in sponges have been unequivocally shown to be of microbial origin. An often-cited example is the isolation of a *Micrococcus* sp. from the sponge *Tedania ignis* that produced three diketopiperazines (Stierle et al., 1988) previously isolated from extracts of this sponge (Schmitz et al., 1983). However, Faulkner et al. (2000) point out that the same diketopiperazines have been found in many other bacterial and fungal cultures and in other sponges and may simply be common metabolites that arise during the catabolism of proteins.

A group of Russian workers isolated a *Vibrio* sp. from the sponge *Dysidea* sp. that produced brominated diphenyl ethers previously isolated from this sponge (Elyakov et al., 1991). The peptide andrimid with some antibiotic activity against *Bacillus* sp., isolated from the sponge *Hyatella* sp., was found to be produced by a *Vibrio* sp. associated with this sponge (Oclarit et al., 1994).

In cases where a producing microbe can be isolated and production of the compound is maintained under culture conditions, there can be little argument that the symbiotic microbe is the source of the compound found in the sponge. However, it may be very difficult or impossible to culture the symbiont. The relationship between the sponge and the microbe may be sufficiently evolved that the microbe is an obligate symbiont. It would not be surprising if this occurred in at least some cases, considering that these relationships appear to have persisted for an extremely long time. Some sponge symbioses may have evolved approximately 500 million years ago (Wilkinson, 1987). If the producing microbe cannot be readily cultured, a range of approaches such as cell separations, electron microscopy, chemical analysis, and molecular biology techniques may be necessary in order to build a circumstantial case for the microbial production of the compound of interest. This approach can be particularly valuable if it facilitates the molecular characterization of the producing microbe by, for example, 16S rRNA gene sequence analysis. Additional culturing attempts can then be made on a more rational basis because the growth requirements can be deduced from those of closely related bacteria that may be more fully characterized.

This approach is exemplified by studies on the marine sponge *T. swinhoei* from Palau. This sponge contains a cytotoxic polyketide, swinholide A, and the bicyclic glycopeptide antifungal compound theopalauamide, among other compounds (Bewley and Faulkner, 1998). Bacteria associated with this sponge include unicellular cyanobacteria, filamentous bacteria, and unicellular bacteria. Cell separations were useful in the study of the symbionts of *T. swinhoei*. Swinholide A is likely to be a bacterial metabolite because this compound is associated with fractions of the unicellular bacteria from *T. swinhoei* and not with cyanobacterial, filamentous bacterial, or sponge cell fractions (Bewley et al., 1996). A single morphotype of a filamentous bacterium was present in a separate fraction that also contained the compound theopalauamide (Schmidt et al., 1998). Subsequent application of molecular approaches identified this filamentous bacterium as a novel deltaproteobacterium, related to myxobacteria (Schmidt et al., 2000). Extensive efforts to grow this filamentous bacterium in pure culture were unsuccessful although the bacterium was propagated in a mixed culture on agar plates (Schmidt et al., 2000). Although it cannot be totally proven that the filamentous bacteria are the source of theopalauamide unless production of the compound can be shown in a pure culture of the bacterial strain, a very strong circumstantial case has been made for production of theopalauamide by this symbiont.

Although there are few cases where there is strong evidence for bioactive compounds found in sponges of microbial origin, the converse is also true. In only a few cases have sponge-derived bioactive compounds been proven to be produced by sponge cells rather than by microbial symbionts. An example of strong circumstantial evidence for sponge cell production is the localization of the compound latrunculin B within the Red Sea sponge *Negombata magnifica*. The latrunculins are ichthyotoxic compounds found in several sponges, and their ecological role is presumably to deter feeding on the sponges by fish (Groweiss et al., 1983; Gulavita et al., 1992; Kakou et al., 1987). The location of latrunculin B in *N. magnifica* was investigated using specific antibodies and was shown to be mostly beneath the sponge cortex at the ectosome-endosome border, primarily in vacuoles within archeocytes and choanocytes (Gillor et al., 2000). It is not possible to completely eliminate the possibility that latrunculin B is produced by symbiotic bacteria and later localized within the sponge cells, but this appears unlikely because the bacterial symbionts were poorly labeled by the latrunculin B-specific antibody (Gillor et al., 2000).

In several studies using cell dissociation followed by separation on density gradients, compounds have been shown to be localized in sponge cell fractions. The antimitotic compound avarol from the sponge *Dysidea avara* was shown to be located in spherulous cells (Muller et al., 1986), specifically within choanocytes (Uriz et al., 1996). The tropical marine sponge *Amphimedon* sp. has the compound diisocyanoadociane localized in archeocytes and spherulous cells (Garson et al., 1992).

In some cases, unique characteristics of the metabolite of interest may aid in localization. The bromine in aerothionin and homoaerothionin assisted in localization of these compounds to spherulous cells

lining the exhalent channels of *A. fistularis* (Thompson et al., 1983), and in this case sponge cell production of the compounds is highly likely given the localization and the fact that *A. fistularis* contains very few bacterial cells and aerothionin and homoaerothionin are major metabolites of the sponge (Faulkner et al., 2000). In the case of the localization of the pyridoacridine alkaloid dercitamide in the sponge *Oceanapia sagittaria*, the natural fluorescence of dercitamide was used to detect this compound within cells. Confocal microscopy allowed precise localization of dercitamide in sponge cells containing spherical inclusions (Faulkner et al., 2000). Because these sponge cells are bacteria free, dercitamide is probably not produced by intracellular bacteria. It was hypothesized that dercitamide was biosynthesized and stored by the inclusional sponge cells rather than being synthesized by extracellular bacteria and transported to these storage cells (Salomon et al., 2001).

Another possibility is that bioactive metabolites found in sponges are the result of dietary uptake. Okadaic acid and derivatives are potent inhibitors of protein phosphatases 1 and 2A and are used as research tools (Kobayashi, 2000). Okadaic acid was first isolated from the sponge *Halichondria okadai* from Japan and is also present in the Caribbean sponge *Halichondria melanodocia* (Tachibana et al., 1981). Okadaic acid and derivatives are produced by the dinoflagellates *Prorocentrum lima* (Murakami et al., 1982) and *Prorocentrum concavum* (Dickey et al., 1990) and by *Dinophysis* spp. (Draisci et al., 1998). Since there does not appear to be a symbiotic relationship between these sponges and dinoflagellates, it is likely that the sponges feed on the dinoflagellates with which they coexist in shallow waters and acquire the okadaic acid by this dietary uptake (Faulkner et al., 2000). Dietary uptake is another method of acquisition of bioactive compounds that needs to be considered in marine invertebrates, particularly in filter feeders such as sponges.

The presence of related compounds in a variety of marine organisms of different phyla can be regarded as suggestive of production by symbiotic microbes. There are several examples where related macrolide compounds have been found in several sponges from different environments. Cinachyrolide A from the sponge *Cinachyra* sp. (Fusetani et al., 1993) and the spongiostatins found in *Spongia* sp. and *Spirastrella spinispirulifera* (Pettit et al., 1993) are closely related structurally to the altohyrtins found in *Hyrtios altum* from near Okinawa and *Haliclona* sp. collected off Amami Island, southwestern Japan (Kobayashi, 2000). This suggests that these compounds may be synthesized by symbiotic microorganisms (Kobayashi, 2000).

The example of dercitamide suggests that this interpretation is not always the correct one. More than 40 pyridoacridine alkaloid compounds related to dercitamide have been found in unrelated marine invertebrates, including ascidians, a mollusc, and a coelenterate in addition to sponges (Molinski, 1993). Similar biosynthetic pathways may have evolved in these different phyla, and the common presence of related pyridoacridine alkaloid may be an example of convergent evolution (Salomon et al., 2001) rather than being the result of a common microbial symbiont that is the producing organism.

Circumstantial evidence for the microbial origin of a sponge-derived compound may also be supplied by the structure of the compound. In some cases, sponge-derived molecules bear a close resemblance to compounds isolated from bacteria. For example, the cytotoxic depsipeptide arenastatin A isolated from the sponge *Dysidea arenaria* resembles the cryptophycins isolated from the cyanobacterium *Nostoc* sp. (Kobayashi, 2000). Synthetic cryptophycin 1 is in clinical trials.

In the case of callystatin A, circumstantial evidence for a microbial origin comes from both the structure of the compound and its occurrence in different marine invertebrates. Callystatin A, a potent cytotoxic polyketide isolated from the sponge *Callyspongia truncata*, has structural similarities to several antibiotics isolated from *Streptomyces* spp. Callystatin A has also been isolated from the sponge *Stelletta* sp. and from a tunicate that co-occurs with *C. truncata* at a site off the Goto Islands, southwestern Japan. This is suggestive of a possible microbial production of callystatin A by a microbe that is associated with both sponges and the tunicate (Kobayashi, 2000).

Faulkner and coauthors speculated that it is probably the exception rather than the rule that bioactive sponge metabolites are produced by symbiotic microbes, based on the increasing number of marine natural products that have been shown to be localized in sponge cells (Faulkner et al., 2000), some of which are discussed above. However, the enormous diversity of bacteria associated with sponges suggests a large metabolic capability of these symbionts. Furthermore, it is difficult to eliminate the possibility that a bioactive compound is being produced by symbiotic bacteria even if the compound is shown to be localized within sponge cells. So it may still prove to be the case that the majority of bioactive compounds found in sponges are produced by bacterial symbionts. Certainly, it is important to always consider this possibility when studying sponge-derived bioactive compounds of pharmaceutical importance. In some cases, careful microbiological investigation will be rewarded by the isolation of a microbe that produces the compound of interest.

THE SUPPLY PROBLEM

As with many compounds from marine sources, it can be problematic to obtain sufficient amounts of some of the compounds with pharmaceutical potential from marine sponges. Supply of the promising sponge-derived bioactive compounds halichondrin B and discodermolide is discussed to illustrate some of the issues involved in obtaining an abundant supply of compounds of interest. Halichondrin B (Fig. 3) is a bioactive compound found in several marine sponges. The anticancer properties of halichondrin B result from prevention of cell division by interaction with tubulin. Testing and development of halichondrin B has been constrained because of difficulties in obtaining a sufficient supply of the compound. The deepwater New Zealand sponge *Lissodendoryx* sp. is a good source of halichondrin B. In 2000, the world availability of halichondrin B was 300 mg (from a collec-

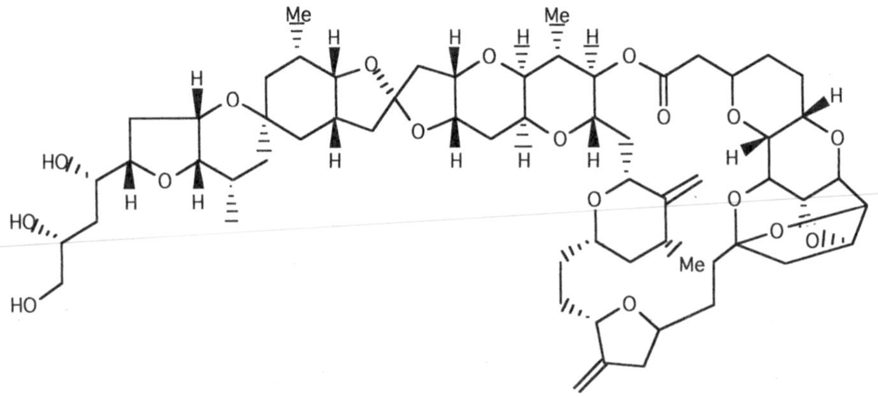

Figure 3. Structure of halichondrin B.

tion of 1 metric ton of *Lissodendoryx*), and the likely requirements for clinical trials alone would be about 10 g (Hart et al., 2000). Collection of the sponge from the wild is not a practical solution because the total biomass of the sponge has been estimated at only about 300 metric tons (Hart et al., 2000). One approach that has been taken to overcome this problem has been to grow the sponge *Lissodendoryx* sp. by aquaculture in New Zealand. Commercial production conditions have been simulated and confirmed that aquaculture of sponges would be a viable option for halichondrin B production (Hart et al., 2000). Another possible solution is to use the structure of halichondrin B as the basis for synthesis of smaller molecules that can be used as pharmaceuticals. Total synthesis of the halichondrin B molecule has been reported (Aicher et al., 1992), but the complexity of halichondrin B makes bulk synthesis impractical without a great deal of synthetic refinement (Hart et al., 2000). However, structurally simplified analogues of halichondrin B were recently synthesized that retained the ability of the "parent" compound to inhibit cell growth. The route to pharmaceutical exploitation of the halichondrin B structure may therefore lie in synthesis of biologically active analogues or subunits of this promising molecule (Hart et al., 2000). Preliminary cell separation experiments indicate that production of halichondrin B is probably by *Lissodendoryx* sponge cells rather than symbionts (Hart et al., 2000), so microbial production by isolation of symbionts may not be an option in this case.

Discodermolide is an anticancer compound extracted from the deepwater Caribbean lithistid sponge *Discodermia dissoluta*. Discodermolide has a mechanism of action similar to that of paclitaxel (Taxol); these compounds inhibit cell division by binding to microtubules and disrupting cell division. The structure of discodermolide appears to be related to that of the antibiotic ikarugamycin from a terrestrial *Streptomyces* sp. and the compound alteramide produced by an *Alteromonas* sp. isolated from the sponge *Halichondria okadai*, leading Kobayashi and Ishibashi (1993) to suggest that a microbial origin for discodermolide is very likely. However, there are as yet no reports of the successful isolation of a discodermolide-producing symbiont from *Discodermia dissoluta*. The solution to an abundant supply of discodermolide appears to be in the synthesis of this compound. Discodermolide has recently been synthesized on a sufficiently large scale to make future development of this drug possible (Smith et al., 1999). Isolation of a discodermolide-synthesizing symbiont remains of considerable scientific interest and may yet provide a more economic method for large-scale pharmaceutical production of discodermolide.

MICROBIOLOGICAL STUDIES OF MANZAMINE-CONTAINING SPONGES

Manzamines are alkaloid bioactive compounds first reported from the Okinawan sponge *Haliclona* sp. (Sakai et al., 1986). Manzamines have since been found in many additional sponge genera including *Prianos*, *Xestospongia*, *Pachypellina*, *Petrosia*, *Cribrochalina*, *Ircinia*, *Amphimedon*, and *Pellina*. Manzamines have great pharmaceutical potential, particularly as antimalarial drugs, with manzamine A exhibiting improved in vivo activity against malaria parasites over the currently used antimalarial drugs artemisin and chloroquine (Ang et al., 2000; El Sayed et al., 2001). Several manzamines, including manadomanzamines A and B, recently isolated from an Indonesian sponge, *Petrosiidae* sp., also show promising antimycobacterial activity (Hu et al., in press).

The isolation of manzamine alkaloids from many unrelated and geographically diverse sponge species strongly suggests a microbial origin for this important group of compounds. Isolation of symbiotic microbes that produce manzamines from manzamine-containing sponges may allow for the large-scale sustainable production of manzamines without the need for sponge harvest. In addition, it is possible that sponge-associated microbes may play a significant role in the bioconversion of manzamines to produce the large number of manzamine alkaloids found in sponges. For these reasons, an intensive microbiological investigation is under way on the microbiology of two manzamine-containing sponges from Indonesia. The total microbial communities associated with these sponges have been investigated by using 16S rRNA community analysis, and results for sponge 01IND 35 are discussed above and shown in Fig. 2. In addition, culturable isolates have been obtained from sponges 01IND 35 and 01IND 52 and include *Alphaproteobacteria*, actinomycetes, *Bacillus* sp., and *Staphylococcus* sp. (Peng et al., in press; Yousaf et al., 2002). All these culturable isolates are being screened for the ability to produce the manzamines found in these two sponges. Several groups, including acidobacteria, *Nitrospira*, and *Cytophaga*, shown to be present by the molecular analysis are not yet represented in the culturable assemblage obtained from these sponges. Cell enumeration by fluorescence microscopy indicated that only 0.2 to 0.7% of the microbes associated with the sponges were cultured in initial attempts to grow sponge-associated bacteria. Future culturing attempts will specifically target members of unrepresented groups now that they are known to be present in the total assemblage.

Variations in the microbial communities between different individuals of the same species have seldom been considered in sponge microbiology studies. It is possible that these variations could be a factor in explaining the differences in chemotype profiles that are sometimes observed between individuals of the same species. Denaturing gradient gel electrophoresis analysis (Muyzer et al., 1993) is being used to determine differences between individuals of sponges 01IND 35 and 01IND 52 and to compare the communities associated with these two species. Initial indications are that different individuals of sponge 01IND 35 contain many similar bacteria but there were also several bands that were present in one individual and not another. More important, there are common bacteria present in sponges 01IND 35 and 01IND 52, indicating that the same symbionts may be present in different sponge species. Because both species of sponge contain manzamines, these "shared" bacteria are candidates for bacterial production of manzamines. 16S rRNA gene sequencing of specific bands allows identification of the bacteria that are in common between the two sponges. These bacteria will be targeted in future culturing attempts from sponges 01IND 35 and 01IND 52. This finding of common bacteria in different sponge species is consistent with that of Hentschel et al. (2002) that components of the microbial community are uniform between sponges of different species.

SYMBIONTS FROM OTHER MARINE INVERTEBRATES

The importance of symbionts in marine invertebrates is certainly not limited to sponges. Many marine invertebrates harbor symbiotic microbes and may rely on these symbionts to produce chemicals that are used in defense. Symbiotic bacteria are implicated in the production of the promising marine bioactive compound bryostatin 1. Bryostatins are a family of cytotoxic macrolides derived from the Californian bryozoan *Bugula neritina*. This marine invertebrate attaches to underwater surfaces as a fouling organism. Bryostatin 1 is in clinical trials for treatment of various types of cancer, including non-Hodgkin's lymphoma, chronic lymphocytic leukemia, and metastatic melanoma. The mode of action of bryostatin 1 is modulation of the enzyme protein kinase C. A limiting factor in the development of this compound as a drug has been the difficulty in obtaining sufficient bryostatin 1. Harvesting of 13,000 kg of biomass of the bryozoan *B. neritina* was required to obtain 18 g of bryostatin for clinical trials, and this harvesting took almost 2 years (Mendola, 2000). Rod-shaped gram-negative bacteria are present in the pallial sinus (a sealed circular fold) in the larvae of *B. neritina* and in the funicular cords that connect zooids in the adults. The bacte-

rial symbiont of *B. neritina* has been intensively studied and is a candidate for the microbial production of bryostatins (Haygood et al., 1999). 16S rRNA gene sequence analysis indicated that there is a novel gammaproteobacterium present in the larvae, and the bacterium was named "*Candidatus Endobugula sertula*." A specific probe designed from this sequence bound to the bacteria in the larval pallial sinus in hybridization experiments (Davidson et al., 2001). Bryostatins are complex polyketides resembling bacterial secondary metabolites that are synthesized by enzymes with multiple active sites on a single polypeptide (type I polyketide synthases). Type 1 polyketide synthase gene fragments were cloned from DNA extracted from *B. neritina* containing "*Endobugula sertula*." Reduction in cell numbers of *E. sertula* resulted in the host *B. neritina* producing reduced levels of bryostatins. It was concluded that the symbiont *E. sertula* has the genetic potential to make bryostatins indicating that it may be possible to clone bryostatin genes from *B. neritina* directly and to use these to produce bryostatins in heterologous bacterial systems (Davidson et al., 2001). This approach is indicated since attempts to culture "*E. sertula*" have so far been unsuccessful. The fact that no metazoan enzymes responsible for synthesis of complex polyketides such as bryostatins were known initially suggested the possibility that these compounds are produced by bacteria and fungi (Haygood et al., 1999). Some marine dinoflagellates contain polyketides, but the enzymes responsible for their production are uncharacterized (Rein and Borrone, 1999). Interestingly, there is now a single report of a protist that encodes a putative polyketide synthase (Zhu et al., 2002).

Another extremely promising compound from the marine environment that is currently in clinical trials is ecteinascidin-743, one of several ecteinascidins from the Caribbean mangrove tunicate *Ecteinascidia turbinata*. This compound is a promising antitumor agent that acts by binding to the minor groove of DNA with a mode of action unique among the DNA alkylating agents with pharmaceutical potential (Rinehart, 2000). Isolation of purified ecteinascidins from *E. turbinata* took over a decade because of the very small amounts of the compounds present in the tunicate. Supply of ecteinascidin-743 is certainly a major stumbling-block limiting development of this compound. Large-scale supplies will need to be assured by growing *E. turbinata* in aquaculture systems or by laboratory synthesis of the compound (Mendola, 2000). One other possibility is that ecteinascidin-743 may be produced by a bacterial symbiont of *E. turbinata*. Similar tetrahydroisoquinoline alkaloids such as mimosamysin and saframycin have been isolated from *Streptomyces lavendulae*, leading Kobayashi and

Ishibashi (1993) to suggest that ecteinascidins may be biosynthesized by symbiotic microorganisms.

Considerable progress has been made in the aquacultural production of both bryostatin and ecteinascidin-743, showing that it is technically possible to produce these compounds in aquaculture systems (Mendola, 2000). These efforts have been very important in supplying material for clinical trials and in ensuring that pharmaceutical companies maintain some interest in these compounds. However, there are important advantages in isolation of symbionts that produce these compounds. If symbionts can be readily cultured and maintain production of the compounds under culture conditions, costs of production are likely to be low. Furthermore, there is the possibility that closely related symbionts could be isolated that produce additional byrostatins and ecteinascidins that may have enhanced therapeutic properties. If symbionts cannot be cultured, knowledge that symbionts are implicated in production of these compounds is still very valuable because it may facilitate cloning of genes encoding the compounds. Identification of symbionts by 16S rRNA gene sequence analysis may be useful in selection of appropriate heterologous bacterial systems for expression of biosynthetic genes for compound production. Separation of symbiotic bacteria from invertebrate source material may be possible in order to obtain DNA samples enriched in the target genes.

SPONGE HEALTH AND AQUACULTURE

In cases where aquaculture is used to produce marine bioactive compounds, it may still be useful to carefully study the microbiology of the system. *B. neritina* and *E. turbinata* have been grown under intensive aquaculture conditions (Mendola, 2000), and if this were done on a large scale, a considerable investment would be required. The possibility of disease, including microbial infection, would need to be considered and may warrant microbiological monitoring. Sponges grown in aquaculture, in either closed or open systems, may be susceptible to disease. The commercial bath sponge industry in the Mediterranean was decimated during a disease outbreak in 1986 to 1990, attributed to bacterial attack (Vacelet et al., 1994). The first isolation of a bacterium that causes sponge disease was recently reported (Webster et al., 2002). The etiological agent in this case was an *Alphaproteobacterium* that bored through the spongin fibers of the sponge *R. odorabile*. Interestingly, the resultant burrows closely resembled those observed by microscopy in diseased sponges during the 1986 to 1990 sponge disease outbreak.

Sponge health can be correlated with the presence of specific symbionts. In *R. odorabile*, the presence of the *Alphaproteobacterium* strain NW001 was investigated in 52 individual sponges from different sites on the Great Barrier Reef. In 50 of these individuals, strain NW001 was the dominant culturable bacterium present and the sponges appeared healthy, whereas in two sick individuals, this strain was absent or greatly reduced (Webster and Hill, 2001).

Consistent production of the desired metabolite may be affected by symbionts in the invertebrate, even in cases where the metabolite is produced by the invertebrate. This could be a direct effect in cases where necessary metabolic precursors can be supplied by a symbiont. Indirect effects could result if symbionts play an important role in the health of the invertebrate.

In *B. neritina* there are two genetic types, one occurring in deep water and one in shallow water. The two types have different strains of the symbiont "*E. sertula*," and only the deep type contains bryostatin 1. It is not known whether the difference in chemotype is due to the *B. neritina* strain difference or the presence of different "*E. sertula*" strains (Davidson and Haygood, 1999).

CONCLUSION

The study of microbial communities associated with marine sponges is at an exciting stage. It is now clear that these communities are extremely diverse and include many novel bacteria. Application of molecular techniques may now start to reveal the nature of the relationship between some of these symbiotic bacteria and their hosts. Bacteria-sponge symbioses are far more complex than monospecific symbioses such as the well-characterized symbiosis between the luminescent bacterium *Vibrio fischeri* and its host the bob-tail squid *Euprymna scolopes* (Ruby and Lee, 1998) and the *B. neritina*-"*E. sertula*" symbiosis (Haygood and Davidson, 1998; Haygood et al., 1999). The complexity of bacteria-sponge symbioses as well as their potential biotechnological importance certainly make them an intriguing and worthwhile area for study. The culture of additional sponge-associated symbionts warrants increased effort as this is the most direct method for large-scale production of important bioactive compounds from sponges that are actually produced by microbial symbionts. In some cases, it will prove to be impossible to isolate and culture the symbiont of interest. A combination of cell separations, FISH, and 16S rRNA gene sequencing may enable identification of the important symbiont and enrichment of symbiont cells from which DNA can be extracted. Genes encoding the bioactive compound can then be cloned into an appropriate heterologous host. This approach may not succeed, especially in cases where the compound of interest is present at very low concentrations and is produced by a symbiont that is a very small component of the total microbial community in the sponge. It may be necessary to take a "metagenomic" approach as has been developed for gene cloning from the microbial community in soil (Gillespie et al., 2002; Rondon et al., 2000; Handelsman, chapter 11). This needs a powerful selection method for the desired genes or products. In some cases, most likely when the metabolite is produced by sponge cells rather than symbionts, growth of sponge cells in cell culture systems or of sponges in aquaculture will be necessary. Microbiological studies will be valuable when sponges are grown in open or closed aquaculture systems to monitor for disease and to obtain healthy sponges that give consistent yields of compound. In general, the more that is understood about the microbiology of a sponge-symbiont system, the greater the likelihood of success in obtaining the sought-after bioactive compound.

Acknowledgments. I thank past graduate student Nicole Webster, current graduate students Olivier Peraud and Jayne Lohr, and postdoctoral associate Julie Enticknap for their efforts. Eric Schmidt is thanked for helpful discussions. Research on sponge microbiology in the Hill Laboratory has been funded by the VIRTUE Program (Wallenberg Foundation) and by Schering Plough Research Institute and is currently funded by the Microbial Observatories Program, National Science Foundation (MCB-0238515). This support is gratefully acknowledged. This is contribution no. 04-598 from the Center of Marine Biotechnology.

REFERENCES

Aicher, T. D., K. R. Buszek, F. G. Fang, C. J. Forsyth, S. H. Jung, Y. Kishi, P. M. Scola, D. M. Spero, and S. K. Yoon. 1992. Total synthesis of halichondrin B and norhalichondrin B. *J. Am. Chem. Soc.* 114:3162–3264.

Ang, K. K., M. J. Holmes, T. Higa, M. T. Hamann, and U. A. Kara. 2000. In vivo antimalarial activity of the beta-carboline alkaloid manzamine. A. *Antimicrob. Agents Chemother.* 44:1645–1649.

Berthold, R. J., M. A. Borowitzka, and M. A. Mackay. 1982. The ultrastructure of *Oscillatoria spongeliae*, the blue-green algal endosymbiont of the sponge *Dysidea herbacea*. *Phycologia* 21:327–335.

Bewley, C. A., and D. J. Faulkner. 1998. Lithistid sponges: star performers or hosts to the stars. *Angew. Chem. Int. Ed.* 37:2162–2178.

Bewley, C. A., N. D. Holland, and D. J. Faulkner. 1996. Two classes of metabolites from *Theonella swinhoei* are localized in distinct populations of bacterial symbionts. *Experientia* 52:716–722.

Borowitzka, M. A., R. Hinde, and F. Pironet. 1988. Carbon fixation by the sponge *Dysidea herbacea* and its endosymbiont *Oscillatoria spongeliae*, p. 151–155. *In* J. H. Choat, D. J. Barnes, M. A. Borowitzka, J. C. Coll, P. J. Davies, P. Flood, B. G. Hatcher, D. Hopley, P. A. Hutchings, D. Kingsey, G. R. Orme, M. Pichon, P. F. Sale, P. W. Sammarco, C. C. Wallace, C. R. Wilkinson, E. Wolanski, and O. Bellwood (ed.), *Proceedings of the 6th International Coral Reef Symposium*. Symposium Executive Committee, Townsville, Australia.

Brauers, G., R. A. Edrada, R. Ebel, P. Proksch, V. Wray, A. Berg, U. Grafe, C. Schachtele, F. Totzke, G. Finkenzeller, D. Marme, J. Kraus, M. Munchbach, M. Michel, G. Bringmann, and K. Schaumann. 2000. Anthraquinones and betaenone derivatives from the sponge-associated fungus *Microsphaeropsis* species: novel inhibitors of protein kinases. *J. Nat. Prod.* 63:739–745.

Burja, A. M., and R. T. Hill. 2001. Microbial symbionts of the Australian Great Barrier Reef sponge *Candidaspongia flabellata*. *Hydrobiologia* 461:41–47.

Chen, F., and R. E. Hodson. 2001. *In situ* PCR/RT-PCR coupled with in situ hybridization for detection of functional gene and gene expression in prokaryotic cells, p. 409–424. *In* J. Paul (ed.), *Methods in Marine Microbiology*. Academic Press, San Diego, Calif.

Davidson, S. K., and M. G. Haygood. 1999. Identification of sibling species of the bryozoan *Bugula neritina* that produce different anticancer bryostatins and harbor distinct strains of the bacterial symbiont "*Candidatus* Endobugula sertula." *Biol. Bull.* 196:273–280.

Davidson, S. K., S. W. Allen, G. E. Lim, C. M. Anderson, and M. G. Haygood. 2001. Evidence for the biosynthesis of bryostatins by the bacterial symbiont "*Candidatus* Endobugula sertula" of the bryozoan *Bugula neritina*. *Appl. Environ. Microbiol.* 67:4531–4537.

Dickey, R. W., S. C. Bobzin, D. J. Faulkner, F. A. Bencsath, and D. Andrzejewski. 1990. The identification of okadaic acid from a Caribbean dinoflagellate *Prorocentrum concavum*. *Toxicon* 28:371–377.

Draisci, R., L. Giannetti, L. Lucentini, C. Marchiafava, K. J. James, A. G. Bishop, B. M. Healy, and S. S. Kelly. 1998. Isolation of a new okadaic acid analogue from phytoplankton implicated in diarrhetic shellfish poisoning. *J. Chromatogr. A* 798:137–145.

El Sayed, K. A., M. Kelly, U. A. Kara, K. K. Ang, I. Katsuyama, D. C. Dunbar, A. A. Khan, and M. T. Hamann. 2001. New manzamine alkaloids with potent activity against infectious diseases. *J. Am. Chem. Soc.* 123:1804–1808.

Elyakov, G. B., T. Kuznetsova, V. V. Mikhailov, I. I. Maltsev, V. G. Voinov, and S. A. Fedoreyev. 1991. Brominated diphenyl ethers from a marine bacterium associated with the sponge *Dysidea* sp. *Experientia* 47:632–633.

Faulkner, D. J. 2002. Marine natural products. *Nat. Prod. Rep.* 19:1–48.

Faulkner, D. J., M. K. Harper, M. G. Haygood, C. E. Salomon, and E. W. Schmidt. 2000. Symbiotic bacteria in sponges: sources of bioactive substances, p. 107–119. *In* N. Fusetani (ed.), *Drugs from the Sea*. Karger, Basel, Switzerland.

Friedrich, A. B., H. Merkert, T. Fendert, J. Hacker, P. Proksch, and U. Hentschel. 1999. Microbial diversity in the marine sponge *Aplysina cavernicola* (formerly *Verongia cavernicola*) analyzed by fluorescence in situ hybridization (FISH). *Mar. Biol.* 134:461–470.

Friedrich, A. B., I. Fischer, P. Proksch, J. Hacker, and U. Hentschel. 2001. Temporal variation of the microbial community associated with the Mediterranean sponge *Aplysina aerophoba*. *FEMS Microbiol. Ecol.* 38:105–113.

Fuerst, J. A., R. I. Webb, M. J. Garson, L. Hardy, and H. M. Reiswig. 1999. Membrane-bounded nuclear bodies in a diverse range of microbial symbionts of Great Barrier Reef sponges. *Mem. Qld. Mus.* 44:193–203.

Fusetani, N., K. Shinoda, and S. Matsunaga. 1993. Cinachyrolide A: a potent cytotoxic macrolide possessing two spiroketals from marine sponge *Cinachyra*. *J. Am. Chem. Soc.* 115:3977–3981.

Garson, M. J., J. E. Thompson, R. M. Larsen, C. N. Battershill, P. T. Murphy, and P. R. Bergquist. 1992. Terpenes in sponge cell membranes: cell separation and membrane fractionation studies with the tropical marine sponge *Amphimedon* sp. *Lipids* 27:378–388.

Gillespie, D. E., S. F. Brady, A. D. Bettermann, N. P. Cianciotto, M. R. Liles, M. R. Rondon, J. Clardy, R. M. Goodman, and J. Handelsman. 2002. Isolation of antibiotics turbomycin a and B from a metagenomic library of soil microbial DNA. *Appl. Environ. Microbiol.* 68:4301–4306.

Gillor, O., S. Carmeli, Y. Rahamin, Z. Fishelson, and M. Ilan. 2000. Immunolocalization of the toxin latrunculin B within the Red Sea sponge *Negombata magnifica* (Demospongiae, Latrunculiidae). *Mar. Biotechnol.* 2:213–223.

Groweiss, A., U. Shmueli, and Y. Kashman. 1983. Marine toxins of *Latrunculia magnifica*. *J. Org. Chem.* 48:3512–3516.

Gulavita, N. K., S. P. Gunasekera, and S. A. Pomponi. 1992. Isolation of latrunculin A, 6,7-epoxylatrunculin A, fijianolide A, and euryfuran from a new genus of the family Thorectidae. *J. Nat. Prod.* 55:506–508.

Hart, J. B., R. E. Lill, S. J. H. Hichford, J. W. Blunt, and M. H. G. Munro. 2000. The halichondrins: chemistry, biology, supply and delivery, p. 134–153. *In* N. Fusetani (ed.), *Drugs from the Sea*. Karger, Basel, Switzerland.

Haygood, M. G., and S. K. Davidson. 1998. Bacterial symbionts of the bryostatin-producing bryozoan *Bugula neritina*, p. 281–284. *In* Y. Le Gal and H. O. Halvorson (ed.), *New Developments in Marine Biotechnology*. Plenum, New York, N.Y.

Haygood, M. G., E. W. Schmidt, S. K. Davidson, and D. J. Faulkner. 1999. Microbial symbionts of marine invertebrates: opportunities for microbial biotechnology. *J. Mol. Microbiol. Biotechnol.* 1:33–43.

Hentschel, U., J. Hopke, M. Horn, A. B. Friedrich, M. Wagner, J. Hacker, and B. S. Moore. 2002. Molecular evidence for a uniform microbial community in sponges from different oceans. *Appl. Environ. Microbiol.* 68:4431–4440.

Hill, R. T., O. Peraud, J. J. Enticknap, and M. T. Hamann. 2002. Molecular analysis of the microbial communities associated with marine sponges: importance for natural products discovery, p. 67–75. *In* Proceedings of the 2002 International Meeting of the Federation of Korean Microbiological Societies, 22 to 23 October 2002, Millenium Town, Chungcheongbuk-do, Korea. Federation of Korean Microbiology Societies, Korea.

Hodson, R. E., W. A. Dustman, R. P. Garg, and M. A. Moran. 1995. *In situ* PCR for visualization of microscale distribution of specific genes and gene products in prokaryotic communities. *Appl. Environ. Microbiol.* 61:4074–4082.

Hooper, J. N. A., and R. W. M. Van Soest. 2002. Systema Porifera: A Guide to the Classification of Sponges. Kluwer Academic/Plenum Publishers, New York, N.Y.

Hu, J.-F., M. T. Hamann, R. Hill, and M. Kelly. 2003. The manzamine alkaloids. *In* G. A. Cordell (ed.), *The Alkaloids*. Academic Press, San Diego, Calif.

Imamura, N., M. Nishijima, K. Adachi, and H. Sano. 1993. Novel antimycin antibiotics, urauchimycins A and B, produced by marine actinomycete. *J. Antibiot.* 46:241–246.

Jensen, P. R., and W. Fenical. 2000. Marine microorganisms and drug discovery: current status and future potential, p. 6–19. *In* N. Fusetani (ed.), *Drugs from the Sea*. Karger, Basel, Switzerland.

Kakou, Y., P. Crews, and G. J. Bakus. 1987. Dendrolasin and latrunculin A from the Fijian sponge *Spongia mycofijiensis* and an associated nudibranch *Chromodoris lochi*. *J. Nat. Prod.* 50:482–484.

Kelman, D., Y. Kashman, E. Rosenberg, M. Ilan, I. Ifrach, and Y. Loya. 2001. Antimicrobial activity of the reef sponge *Amphimedon viridis* from the Red Sea: evidence for selective toxicity. *Aquat. Microb. Ecol.* 24:9–16.

Kobayashi, J., and M. Ishibashi. 1993. Bioactive metabolites of symbiotic marine microorganisms. *Chem. Rev.* 93:8305–8308.

Kobayashi, M. 2000. Search for biologically active substances from marine sponges, p. 46–58. *In* N. Fusetani (ed.), *Drugs from the Sea*. Karger, Basel, Switzerland.

Lee, Y. K., J.-H. Lee, and H. K. Lee. 2001. Microbial symbiosis in marine sponges. *J. Microbiol.* 39:254–264.

Lopez, J. V., P. J. McCarthy, K. E. Janda, R. Willoughby, and S. A. Pomponi. 1999. Molecular techniques reveal wide phyletic diversity of heterotrophic microbes associated with *Discodermia* spp. (Porifera: Demospongia). *Mem. Qld. Mus.* 44:329–341.

Mendola, D. 2000. Aquacultural production of bryostatin 1 and ecteinascidin 743, p. 120–133. *In* N. Fusetani (ed.), *Drugs from the Sea*. Karger, Basel, Switzerland.

Molinski, T. F. 1993. Marine pyridoacridine alkaloids: structure, synthesis, and biological chemistry. *Chem. Rev.* 93:1825–1838.

Muller, W. E., B. Diehl-Seifert, C. Sobel, A. Bechtold, Z. Kljajic, and A. Dorn. 1986. Sponge secondary metabolites: biochemical and ultrastructural localization of the antimitotic agent avarol in *Dysidea avara*. *J. Histochem. Cytochem.* 34:1687–1690.

Murakami, Y., Y. Oshima, and T. Yasumoto. 1982. Identification of okadaic acid as a toxic component of a marine dinoflagellate *Prorocentrum lima*. *Bull. Jpn. Soc. Sci. Fish.* 48:69–72.

Muyzer, G., E. C. de Waal, and A. G. Uitterlinden. 1993. Profiling of complex microbial populations by denaturing gradient gel electrophoresis analysis of polymerase chain reaction-amplified genes coding for 16S rRNA. *Appl. Environ. Microbiol.* 59:695–700.

Nickel, M., S. Leininger, G. Proll, and F. Brümmer. 2001. Comparative studies on two potential methods for the biotechnological production of sponge biomass. *J. Biotechnol.* 92:169–178.

Oclarit, J. M., H. Okada, S. Ohta, K. Kaminura, Y. Yamaoka, T. Iizuka, S. Miyashiro, and S. Ikegami. 1994. Anti-bacillus substance in the marine sponge, *Hyatella* species, produced by an associated *Vibrio* species bacterium. *Microbios* 78:7–16.

Olson, J. B., C. C. Lord, and P. J. McCarthy. 2000. Improved recoverability of microbial colonies from marine sponge samples. *Microb. Ecol.* 40:139–147.

Osinga, R., J. Tramper, and R. H. Wijffels. 1999. Cultivation of marine sponges. *Mar. Biotechnol.* 1:509–532.

Osinga, R., R. Kleijn, E. Groenendijk, P. Niesink, J. Tramper, and R. H. Wijffels. 2001. Development of in vivo sponge cultures: particle feeding by the tropical sponge *Pseudosuberites* aff. *andrewsi*. *Mar. Biotechnol.* 3:544–554.

Pettit, G. R., Z. A. Cichacz, F. Gao, C. L. Herald, M. R. Boyd, J. M. Schmidt, and J. N. A. Hooper. 1993. Isolation and structure of spongistatin 1. *J. Org. Chem.* 58:1302–1304.

Preston, C. M., K. Y. Wu, T. F. Molinski, and E. F. DeLong. 1996. A psychrophilic crenarchaeon inhabits a marine sponge: *Cenarchaeum symbiosum* gen. nov., sp. nov. *Proc. Natl. Acad. Sci. USA* 93:6241–6246.

Rein, K. S., and J. Borrone. 1999. Polyketides from dinoflagellates: origins, pharmacology and biosynthesis. *Comp. Biochem. Physiol. B Biochem. Mol. Biol.* 124:117–131.

Rinehart, K. L. 2000. Antitumor compounds from tunicates. *Med. Res. Rev.* 20:1–27.

Rondon, M. R., P. R. August, A. D. Bettermann, S. F. Brady, T. H. Grossman, M. R. Liles, K. A. Loiacono, B. A. Lynch, I. A. MacNeil, C. Minor, C. L. Tiong, M. Gilman, M. S. Osburne, J. Clardy, J. Handelsman, and R. M. Goodman. 2000. Cloning the soil metagenome: a strategy for accessing the genetic and functional diversity of uncultured microorganisms. *Appl. Environ. Microbiol.* 66:2541–2547.

Ruby, E. G., and K.-H. Lee. 1998. The *Vibrio fischeri-Euprymna scolopes* light organ association: current ecological paradigms. *Appl. Environ. Microbiol.* **64:**805–812.

Sakai, R., T. Higa, C. W. Jefford, and G. Bernardinelli. 1986. Manzamine A: an antitumor alkaloid from a sponge. *J. Am. Chem. Soc.* **108:**6404–6405.

Salomon, C. E., T. Deerinck, M. H. Ellisman, and D. J. Faulkner. 2001. The cellular localization of dercitamide in the Palauan sponge *Oceanapia sagittaria. Mar. Biol.* **139:**313–319.

Santavy, D. L., P. Willenz, and R. R. Colwell. 1990. Phenotypic study of bacteria associated with the Caribbean sclerosponge, *Ceratoporella nicholsoni. Appl. Environ. Microbiol.* **56:**1750–1762.

Schleper, C., E. F. DeLong, C. M. Preston, R. A. Feldman, K. Y. Wu, and R. V. Swanson. 1998. Genomic analysis reveals chromosomal variation in natural populations of the uncultured psychrophilic archaeon *Cenarchaeum symbiosum. J. Bacteriol.* **180:**5003–5009.

Schmidt, E. W., C. A. Bewley, and D. J. Faulkner. 1998. Theopalauamide, a bicyclic glycopeptide from filamentous bacterial symbionts of the lithistid sponge *Theonella swinhoei. J. Org. Chem.* **63:**1254–1258.

Schmidt, E. W., A. Y. Obraztsova, S. K. Davidson, D. J. Faulkner, and M. G. Haygood. 2000. Identification of the antifungal peptide-containing symbiont of the marine sponge *Theonella swinhoei* as a novel δ-proteobacterium, "*Candidatus* Entotheonella palauensis." *Mar. Biol.* **136:**969–977.

Schmitz, F. J., D. J. Vanderah, K. H. Hollenbeak, C. E. L. Enwall, Y. Gopichand, P. K. Sengupta, M. B. Hossain, and C. van der Helm. 1983. Metabolites from the marine sponge *Tedania ignis*—a new atisanedial and several known diketopiperazines. *J. Org. Chem.* **48:**3941–3945.

Smith, A. B., III, M. D. Kaufman, T. J. Beauchamp, M. J. LaMarche, and H. Arimoto. 1999. Gram-scale synthesis of (+)-discodermolide. *Org. Lett.* **1:**1823–1826.

Stierle, A. C., J. H. I. Cardellina, and F. L. Singleton. 1988. A marine *Micrococcus* produces metabolites ascribed to the sponge *Tedania ignis. Experientia* **44:**1021.

Tachibana, K., P. J. Scheuer, Y. Tsukitani, H. Kikuchi, D. van Engen, J. Clardy, Y. Gopichand, and F. J. Schmitz. 1981. Okadaic acid, a cytotoxic polyether from two marine sponges of the genus *Halichondria. J. Am. Chem. Soc.* **103:**2469–2471.

Thompson, J. E., K. D. Barrow, and D. J. Faulkner. 1983. Localization of two brominated metabolites, aerothionin and homoaerothionin, in the spherulous cells of the marine sponge *Aplysina fistularis* (=*Verongia thiona*). *Acta Zool.* **64:**199–210.

Unson, M. D., N. D. Holland, and D. J. Faulkner. 1994. A brominated secondary metabolite synthesized by the cyanobacterial symbiont of a marine sponge and accumulation of the crystalline metabolite in the sponge tissue. *Mar. Biol.* **119:**1–11.

Uriz, M. J., X. Turon, J. Galera, and J. M. Tur. 1996. New light on the cell location of avarol within the sponge *Dysidea avara* (Dendroceratida). *Cell Tissue Res.* **285:**519–527.

Vacelet, J. 1970. Description de cellules à bactéries intranucleaires chez des eponges *Verongia. J. Microsc.* (Paris) **9:**333–346.

Vacelet, J., and C. Donadey. 1977. Electron microscope study of the association between some sponges and bacteria. *J. Exp. Mar. Biol. Ecol.* **30:**301–314.

Vacelet, J., E. Vacelet, E. Gaino, and M.-F. Gallissian. 1994. Bacterial attack of spongin skeleton during the 1986–1990 Mediterranean sponge disease, p. 355–362. *In* R. W. M. van Soest, T. M. G. van Kempen, and J. C. Braekman (ed.), *Sponges in Time and Space.* Balkema, Rotterdam, The Netherlands.

Vogel, S. 1977. Current-induced flow through living sponges in nature. *Proc. Natl. Acad. Sci. USA* **74:**2069–2071.

Webster, N. S., and R. T. Hill. 2001. The culturable microbial community of the Great Barrier Reef sponge *Rhopaloeides odorabile* is dominated by an α-Proteobacterium. *Mar. Biol.* **138:** 843–851.

Webster, N. S., J. E. M. Watts, and R. T. Hill. 2001a. Detection and phylogenetic analysis of novel crenarchaeote and euryarchaeote 16S rRNA gene sequences from a Great Barrier Reef sponge. *Mar. Biotechnol.* **3:**600–608.

Webster, N. S., K. J. Wilson, L. L. Blackall, and R. T. Hill. 2001b. Phylogenetic diversity of bacteria associated with the marine sponge *Rhopaloeides odorabile. Appl. Environ. Microbiol.* **67:**434–444.

Webster, N. S., A. P. Negri, R. I. Webb, and R. T. Hill. 2002. A spongin-boring α-proteobacterium is the etiological agent of disease in the Great Barrier Reef sponge *Rhopaloeides odorabile. Mar. Ecol. Prog. Ser.* **232:**305–309.

Wicke, C., M. Huners, V. Wray, M. Nimtz, U. Bilitewski, and S. Lang. 2002. Production and structure elucidation of glycoglycerolipids from a marine sponge-associated *Microbacterium* species. *J. Nat. Prod.* **63:**621–626.

Wilkinson, C. R. 1978a. Microbial associations in sponges. I. Ecology, physiology and microbial populations of coral reef sponges. *Mar. Biol.* **49:**161–167.

Wilkinson, C. R. 1978b. Microbial associations in sponges. III. Ultrastructure of the in situ associations of coral reef sponges. *Mar. Biol.* **49:**177–185.

Wilkinson, C. R. 1983. Net primary productivity in coral reef sponges. *Science* **219:**410–412.

Wilkinson, C. R. 1987. Significance of microbial symbionts in sponge evolution and ecology. *Symbiosis* **4:**135–146.

Wilkinson, C. R., and R. Garrone. 1980. Nutrition of marine sponges. Involvement of symbiotic bacteria in the uptake of dissolved carbon, p. 157–161. *In* D. C. Smith and Y. Tiffon (ed.), *Nutrition in the Lower Metazoa.* Pergamon Press, Oxford, United Kingdom.

Wilkinson, C. R., M. Nowak, B. Austin, and R. R. Colwell. 1981. Specificity of bacterial symbionts in Mediterranean and Great Barrier Reef sponges. *Microb. Ecol.* **7:**13–21.

Wilkinson, C. R., R. Garrone, and J. Vacelet. 1984. Marine sponges discriminate between food bacteria and bacterial symbionts: electron microscope radioautography and *in situ* evidence. *Proc. R. Soc. Lond. Sect. B* **220:**519–528.

Yousaf, M., K. A. El Sayed, K. V. Rao, C. W. Lim, J.-F. Hu, M. Kelly, F. Franzblau, O. Peraud, R. T. Hill, and M. T. Hamann. 2002. 12.34-Oxamanzamines, novel biocatalytic and natural products from manzamine producing Indo-Pacific sponges. *Tetrahedron* **58:**7397–7402.

Zhu, G., M. J. LaGier, F. Stejskal, J. J. Millership, X. Cai, and J. S. Keithly. 2002. *Cryptosporidium parvum:* the first protist known to encode a putative polyketide synthase. *Gene* **298:**79–89.

Chapter 19

Invertebrates—Insects

JOHN A. BREZNAK

Insects (phylum Arthropoda; class Insecta) are one of the most diverse groups of living creatures on Earth. First appearing in the late Carboniferous period about 300 million years ago, they have subsequently radiated into a group that now includes about 750,000 to several million species (Wilson, 1992; Pimm et al., 1995; Novotny et al, 2002). Equally diverse are the feeding habits and behaviors exhibited by the group as a whole, and there is virtually no terrestrial food source that escapes exploitation by one or more species. However, one property shared by all insects is their common association with microorganisms. The associations range from loose and nonspecific ones, in which the insect merely serves as an inadvertent carrier and distributor of microbes, to much tighter, highly interdependent and remarkably regulated symbiotic interactions, and they include literally all gradations in between these extremes.

Work by Buchner and others in the early to mid-1900s documented just how common symbiotic interactions between insects and microbes actually were (Buchner 1965). Such studies revealed the wide assortment of microbes that could engage in such associations, including various prokaryotes (*Bacteria*, and in recent studies *Archaea* as well) and eukaryotes (yeasts, filamentous fungi, and protozoa). Early studies also revealed the often striking anatomical and behavioral adaptations of insects to harbor the microbial partners in, on, or around them and to ensure transmission of microbial symbionts to offspring. One salient theme to emerge from early investigations was that distinct microbial symbionts, or communities of microbial symbionts, were especially common in insects that feed on restricted and/or relatively refractory food resources. Such diets are often deficient in nutrients such as amino acids (e.g., plant sap) or vitamins (e.g., animal blood), and lignocellulosic plant material is not only poor in nitrogenous compounds, vitamins, and sterols, but it is also difficult to digest.

Consequently, research into the bases for such symbioses has frequently been driven by hypotheses surrounding compensatory provision of missing nutrients or digestive enzymes by the microbial partner; and in many cases such notions have been validated. However, microbial symbionts are also involved in numerous other aspects of insect biology, e.g., detoxification of plant defensive secretions; production of insect behavior-modifying compounds; protection against microbial pathogens and pests; and alteration of host reproductive patterns. Accordingly, a major challenge in preparing this chapter has been to distill the enormous body of literature on insect-microbe interactions into a coherent and meaningful overview. To do this, I have often cited other reviews as a gateway to background literature on a particular topic, opting to focus here on some exciting recent discoveries, especially those in which biochemical, genetic, and molecular biological approaches have greatly improved our understanding of the symbioses and their evolution. Nevertheless, readers interested in digging deeper into the historical literature can get there from here. The theme of this chapter is to emphasize the biodiversity of the interacting partners and to suggest systems in which bioprospecting for novel and useful natural products, genes, enzymes, or organisms might be fruitful.

TERMINOLOGY

The terms symbiosis and symbiont and derivations thereof are used here in the original sense of de-Bary, meaning the permanent or semipermanent association of two dissimilar organisms (de Bary, 1879). More specific terms denoting the nature of the association (e.g., mutualistic, parasitic, etc.) are used where appropriate and where possible. However, it is worth noting that the nature of a particular symbiotic interaction can shift with environmental conditions,

John A. Breznak • *Department of Microbiology and Molecular Genetics, 6190 Biomedical and Physical Sciences, Michigan State University, East Lansing, MI 48824-4320.*

with the developmental stage of the insect, and with time (Smith and Douglas, 1987). Symbiotic microbes living internally are referred to as endosymbionts. Among endosymbionts are those occurring extracellularly (e.g., in the gut or in evaginations of the alimentary canal) or intracellularly. Insect cells containing microbial symbionts have been traditionally referred to as mycetocytes, even though this term implies that the intracellular symbiont is a yeast or filamentous fungus. The term bacteriocyte is being used more frequently when the intracellular symbiont is known to be a bacterium. Localized, organized aggregates of symbiont-containing cells surrounded by a membrane or sheath of epithelial cells are referred to as a mycetome or bacteriome, respectively. Symbiotic microbes living upon or otherwise outside an insect's body (e.g., in tunnels, galleries, or gardens within the nest) are termed ectosymbionts. In symbiotic relationships between microbes and insects, the insect partner has traditionally been referred to as the host and the microbial associate(s) as the symbiont.

MICROBIAL ENDOSYMBIONTS OF INSECTS (EXTRACELLULAR)

At least a dozen insect orders contain species that harbor microbial endosymbionts as a resident gut microbiota. These are Coleoptera (e.g., scarab beetles), Dictyoptera (cockroaches and termites), Diptera (e.g., craneflies, fruit flies), Ephemeroptera (e.g., mayflies), Heteroptera (e.g., kissing bugs, stink bugs), Hymenoptera (e.g., ants, bees), Lepidoptera (e.g., moths, butterflies), Megaloptera (e.g., dobsonflies, fishflies), Orthoptera (e.g., grasshoppers, crickets), Plectoptera (e.g., stoneflies), Trichoptera (e.g., caddisflies), and Zoraptera (summarized by Dasch et al. [1984]; Douglas [1992]; Cazemier et al. [1997]; Kane [1997]; Kaufman et al. [2000]; Brune [2003]). However, the density and diversity of the endosymbionts vary greatly, as does the nature and magnitude of their impact on the host. The largest microbial populations (attaining densities up to 10^9 to 10^{11} prokaryotic cells ml^{-1}) are typically found in the hindgut and are observed in insects like termites, scarab beetle larvae and cranefly larvae that feed on xylophagous or fiber-rich plant material or on residues derived from it (e.g., humus or detritus), as well as in omnivores like cockroaches, whose natural diets consist of a mixture of such materials (Nalepa and Bandi, 2000). In such insects, readily utilizable nutrients in the food are released and absorbed during passage through the midgut. Digestion of the more refractory components (e.g., cellulose and other polysaccharides, lignin moieties) may be initiated in the fore- and midgut via salivary and midgut enzymes, but then continues in the hindgut with the aid of a symbiotic gut microbiota.

Hindgut microbes densely colonize the gut wall and structures (e.g., cuticular spines) emanating from it, as well as the lumen proper (Fig. 1A). As inwardly

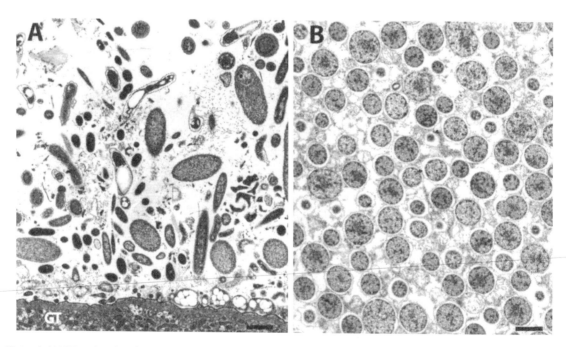

Figure 1. (A) Diversity of prokaryotic microbes located on and near the hindgut epithelium (GT) of the termite *Reticulitermes flavipes*. (B) Cells of *Buchnera aphidicola* within an aphid bacteriocyte (kindly supplied by P. Baumann). Bars: (A) 1 μm; (B) 5 μm.

diffusing oxygen is stripped out by the wall-associated microbiota (e.g., Brune et al. [1995]), a substantial fraction of dissimilatory activity may be effected fermentatively by the lumenal microbiota resulting in production of incompletely oxidized products such as organic acids (e.g., acetate, propionate, and butyrate), which are taken up and used by the insects as an energy source (Kane, 1997; Brune, 2003). Gaseous fermentation products such as H_2 and CO_2 are used by homoacetogenic bacteria resulting in production of additional acetate and/or by methanogenic archaea for the production of methane. The latter microbes are common in termites, cockroaches, and scarab beetles (Bayon, 1980; Hackstein and Stumm, 1994). The contribution of termites to global biogenic methane production has been estimated to be 4 to 10% (Sanderson, 1996; Sugimoto et al., 2000). The contribution from all insects is likely to be significantly higher.

In certain insects, protozoa are a significant component of the hindgut microbiota. A diversity of anaerobic, cellulolytic, and flagellate protozoa (members of the orders Oxymonadida, Trichomonadida, and Hypermastigida) is present in six (of seven) families of termites referred to as lower termites, which comprise about 25% of all known species of the suborder Isoptera, as well as in the hindgut of *Cryptocercus punctulatus*, a species of wood-eating cockroach closely related to termites phylogenetically (Yamin, 1979; Inoue et al., 2000). These protozoa endocytose wood particles into food vacuoles, in which cellulose hydrolysis occurs by protozoan cellulases, and then ferment the hydrolysis products mainly to acetate, CO_2, and H_2. Flagellate, ciliate, and amoeboid protozoa also occur in other species of cockroaches (Nalepa and Bandi, 2000). For example, significant populations of the ciliate *Nyctotherus ovalis* are present in hindguts of the American cockroach (*Periplaneta americana*), wherein they appear to contribute to cellulose digestion (Gijzen et al., 1994). When present, hindgut protozoa themselves can harbor bacteria and methanoarchaea either as ectosymbionts or endosymbionts (Gijzen et al., 1991; Breznak, 2000).

The impact of gut microbes on the vitality and behavior of their host varies among insects, as the following examples illustrate. In termites, the impact is profound, as the insects cannot survive without them. This dependency was once attributed mainly to provision, by microbes, of cellulases. However, it is now known that termites can synthesize and secrete their own cellulase components (Tokuda et al., 1999), and, apart from a significant contribution of cellulolytic activity from hindgut protozoa in so-called lower termites, the contribution of prokaryotic-derived enzymes to the cellulase repertoire is small, if any. The importance of prokaryotes may lie more in their roles in acetate production, nitrogen acquisition (e.g., N_2 fixation), nitrogen conservation (e.g., uric acid-N recycling), maintenance of a low redox potential, and prevention of colonization of the gut by pathogens (reviewed by Breznak [2000]). A more moderate and variable impact of gut microbes on insect nutrition is seen with crickets (*Acheta domesticus*), whereby the normally colonized insects digest relatively refractory diets (e.g., alfalfa) more efficiently than do those reared germ free (Kaufman et al., 1989; Kaufman and Klug, 1991), even though the latter grow well and reproduce. The gut microbiota of the desert locust *Schistocerca gregaria* is abundant, but simple, and consists mainly of *Enterobacteriaceae* derived from its food. Participation of gut microbes in fiber digestion is minimal, as the retention time of food is relatively short (Cazemier et al., 1997), and the insects can be reared germ free. Nevertheless, the gut microbiota of normally colonized locusts converts secondary plant chemicals to hydroxyquinone and dihydroxybenzoic acids, which provide colonization resistance against microbial pathogens, and to guaiacol and phenol, which are components of locust aggregation and cohesion pheromone (Dillon and Charnley, 2002). An uncertain and seemingly nonobligatory relationship exists between gut-associated *Enterobacteriaceae* and frugivorous dipterans, such as tephritid fruit flies (Drew and Lloyd, 1991). For example, despite the fact that that such bacteria (primarily *Pantoea* [*Enterobacter*] *agglomerans* and *Klebsiella oxytoca*) were always associated with the alimentary tract of *Rhagoletis pomonella*, the bacteria did not appear to be determinants of host plant specificity nor contribute directly to host nutrition in any discernible way (Howard et al., 1985). However, the bacteria in tephritids may themselves be a source of protein, and recent studies indicate that such bacteria are capable of degrading purines such as uric acid to volatile compounds (presumably ammonia) that are highly attractive to the flies (Lauzon et al, 2000). It may be that the bacteria have mainly exploited the insect for their own housing, nutrition, and dispersal.

The microbial diversity that exists among the gut microbiota of Earth's insects must be enormous, but it is not known with certainty. Up until the past decade or so, most of our information has come from cultivation-dependent approaches, which are widely acknowledged to severely underestimate microbial species diversity (Staley and Konopka, 1985). More recently, cultivation-independent, molecular biological approaches have begun to be used, and these are providing a much more illuminated view of the extant diversity (Hugenholtz et al., 1998; Amann, 2000). The most widely used method involves cloning and sequencing the small-subunit (SSU) rRNA obtained by reverse transcription of the rRNA itself, or inferring its sequence from the encoding gene (the SSU rDNA) cloned after PCR amplification. Most of such work done on insects has been on termites. Although a comprehensive analysis of the gut microbiota has not yet been made for a single termite species (a major undertaking in its own right), sufficient analyses of clone libraries have been done to suggest that microbial species diversity in individual termite species probably numbers in the hundreds, perhaps even more than a thousand. Given that there are about 2,000 known species of termites on Earth, whose biology, behavior, and nutritional ecology are highly diverse, the actual microbial diversity associated with this single suborder is likely to be quite large.

Most prevalent among termite gut bacterial SSU rDNA clones have been members of the *Betaproteobacteria*, *Gammaproteobacteria*, and *Deltaproteobacteria*, the low-G+C (*Clostridium*-related) and high-G+C gram-positive divisions, the *Bacteroides* subgroup of the *Cytophaga-Flavobacterium-Bacteroides* (CFB) division, and the *Spirochaetes* division (*Treponema* related) (Ohkuma and Kudo, 1996, 1998; Tokuda et al., 2000; Ohkuma et al., 2002). Although many of the *Proteobacteria* clones were closely related to cultivated members of the *Enterobacteriaceae* and sulfate-reducing bacteria, approximately two-third of all analyzed clones bear ≤90% similarity to SSU rDNA sequences of cultivated organisms. Considering that SSU rRNA sequence similarities ≤97% are taken to imply that two organisms are probably different species (Stackebrandt and Goebel, 1994), it seems clear that most of the termite gut bacterial community represents novel, but yet-uncultivated, species. Indeed, in a study with *Reticulitermes speratus*, 8 of 55 clones were so deeply divergent phylogenetically as to define a hitherto unknown bacterial division, referred to as Termite Group 1 (Ohkuma and Kudo, 1996). Phylogenetic analysis of *Bacteroides*-related SSU rDNA clones obtained from eight termite species revealed that they were distributed in five clades, two of which as yet contain no cultured representatives. One of the latter (clade V) comprised clones so far only found in termites (Ohkuma et al., 2002).

Spirochetes are a morphologically conspicuously and abundant component of the termite hindgut microbiota, accounting for as many as 50% of all prokaryotic cells present and residing primarily in the lumenal zone of the hindgut. A comprehensive analysis of over 300 SSU rDNA clones obtained from various termites revealed that all were related to the genus *Treponema*, but none were closely related (i.e., all bore ≤91% sequence similarity) to known treponemes (Lilburn et al., 1999). Conservative estimates suggested that at least 21 novel species of *Treponema* were present in hindguts of individual termite species. Some treponemal phylotypes from lower termites appear to form specific ectosymbiotic relationships with oxymonad protozoa (Iida et al., 2000). Recently, the first pure cultures of termite gut spirochetes were obtained and were shown to possess two properties hitherto unknown in the *Spirochaetes* division and that are important to termite nutrition, i.e., H_2/CO_2-acetogenesis (Leadbetter et al., 1999) and N_2 fixation (Lilburn et al., 2001). The phylogenetic diversity and physiology of termite gut spirochetes have been recently discussed (Breznak, 2002; Breznak and Leadbetter, 2002).

Most of the archaeal SSU rDNA clones obtained from termites have been phylogenetically related to known families of methanogens, although clones related to *Thermoplasmales* and *Crenarchaeota* have also been obtained (Friedrich et al. [2001] and references therein). Not surprisingly, most of the archaeal clones obtained so far form termite-specific phylogenetic clusters, of which some (as with spirochetes, above) are specifically associated with termite gut protozoa. To date, only three strains of methanoarchaea have been isolated from termites, but each represented a new species of *Methanobrevibacter* (Leadbetter and Breznak, 1996; Leadbetter et al., 1998). Methanogens representing a new genus, *Methanomicrococcus blatticola*, have been isolated from cockroaches (Sprenger et al., 2000).

MICROBIAL ENDOSYMBIONTS OF INSECTS (INTRACELLULAR)

A number of bacteria have evolved especially intimate relationships with insects, occurring as intracellular endosymbionts. Overall, such bacteria are not as phylogenetically diverse as the gut microbiota of insects, but they can exert profound effects on the survival, behavior, and reproductive patterns of their host (Dasch et al., 1984; Douglas, 1989; Kane and Pierce, 1994; O'Neill et al., 1997; Baumann et al., 2002). Most of the intracellular symbionts characterized to date belong to the *Alpha-*, *Beta-*, and *Gammaproteobacteria*, although a few are members of the CFB division. Only a few types have been cultured in vitro. The intracellular symbionts can be divided into two general groups based on their pattern of interaction with various insects and the effect they exert on their host: (i) the *Wolbachia* group, which induce reproductive anomalies in their hosts and may be regarded as parasites and (ii) the primary (bacteriocyte-associated) and secondary endosymbionts, which usually have beneficial or neutral effects on host fitness.

The *Wolbachia* Group

The *Wohlbachia* group is represented by *Wolbachia pipientis*, an alphaproteobacterium closely related to *Erlichia* and *Rickettsia*. The cells are dimorphic (irregular-shaped rods and cocci) and exist as small forms (0.25 to 0.5 µm diameter) and large forms (1 to 1.8 µm in diameter) within vacuoles surrounded by a host-derived membrane. Although primarily found in reproductive tissues, they have also been detected in other tissues. However, they do not occur in bacteriocytes or bacteriomes. Some infected insects (e.g., *Drosophila*) contain as many as 4×10^7 *Wolbachia* cells. The association of *Wolbachia* with insects is general in that the same or closely related strains can associate with a wide range of insect taxa. Hence, phylogenies of *Wolbachia* deduced from comparative sequences of SSU rDNA or *ftsZ* (a gene encoding a cell division-related protein) are not congruent with their hosts. This implies that a substantial amount of horizontal transfer has occurred, but the frequency and mode of such transfer are not entirely clear. Early estimates of the distribution of *Wolbachia* in neotropical arthropods, made by screening for the presence of *Wolbachia*-specific *ftsZ* genes in insect tissues, implied that about 17% of insect species were infected (Werren et al., 1995). More recent estimates, made by using a more sensitive screening technique (long PCR) targeting a *Wolbachia* surface protein-encoding gene (*wsp*), suggest that the frequency may be >70% of insect species (Jeyaprakash and Hoy, 2000).

Wolbachia are not required for host vitality, nor do they appear to make a contribution to the well-being of their hosts. Their effect on host fitness appears to be either neutral or negative. However, upon infection *Wolbachia* can induce a variety of reproductive alterations in their hosts, including cytoplasmic incompatibility, wherein a cross between an infected male and uninfected female, for example, results in embryo mortality; parthenogenesis, in which infected virgin females produce daughters; feminization, in which infected genetic males reproduce as females; and male killing, in which infected male embryos die while female embryos develop into infected females (Stouthamer et al., 1999). As *Wolbachia* is transferred to offspring only through infected eggs (not sperm), the net effect is to drive the infection through the population via the sex (females) mediating vertical transfer. None of the *Wolbachia* strains has been grown in pure culture.

Other endosymbionts capable of distorting the sex ratio of insect progeny include (i) *Arsenophonus nasoniae* (*Gammaproteobacteria*) in parasitic wasps (*Nasonia vitripennis*) (Gherna et al., 1991); (ii) the "sex ratio *Spiroplasma*" present in various species of *Drosophila* (Hackett et al., 1986); and (iii) a *Rickettsia*-like alphaproteobacterium and an unnamed member of the CFB division, both of which infect ladybird beetles (*Coleomegilla maculata* and *Adalia bipunctata*) (Werren et al., 1994; Hurst et al., 1997). Each of these induces the killing of male progeny and, like *Wolbachia*, can be considered to be parasitic in their hosts.

Primary (P) and Secondary (S) Endosymbionts

P-endosymbionts are mainly members of the *Gammaproteobacteria*; a few representatives are *Betaproteobacteria* or related to *Flavobacterium* within the CFB division. Most of them share several properties: (i) they occur in well-defined, relatively localized cells and tissues (bacteriocytes and bacteriomes); (ii) they occur in insects that feed on a relatively restricted diet deficient in one or more types of nutrient; (iii) they are essential to, or beneficial to, host survival; (iv) they are vertically transmitted to offspring via eggs and show a phylogeny that is congruent with their host, implying coevolution; and (v) some of them have genomes that exhibit a greatly reduced size, an AT-biased base composition, and accelerated sequence evolution (Moran and Baumann, 2000; Tamas et al., 2002). By contrast, S-endosymbionts do not occur by themselves in well-organized bacteriocytes within bacteriomes, although they can occur together with P-endosymbionts in bacteriocytes. They often occur in various cell and tissue types within the same insect or they may be entirely absent in some members of the species, implying that their existence in host lineages is more labile. The nature and extent of their impact on the host is, in most cases, uncertain. Like the P-endosymbionts, they are maternally inherited, but some appear to have been acquired horizontally. The nature and species distribution of some of the better characterized P- and S-endosymbionts are summarized in Table 1. Others can be found in the review by Dasch et al. (1984). This discussion will deal mainly with P-endosymbionts.

Among the plant phloem sap-sucking insects, the best-studied endosymbionts are the P-endosymbionts of aphids, which are currently grouped within the single species *Buchnera aphidicola*. They are spherical to oval cells 2 to 5 µm in diameter and possess a gram-negative type of cell wall architecture, which includes an inner (cytoplasmic) and outer membrane (Fig. 1B). The *Buchnera* genome (ca. 28 mol% G+C) is among

Table 1. Distribution of primary (P) and secondary (S) endosymbionts in insects[a]

Host category	Host food source	Symbiont (P or S)	Symbiont 16S rRNA group or other taxonomic designation
Order: Hemiptera, Superfamily: Aphidoidae (aphids)	Phloem sap	Buchnera aphidicola (P)	Gammaproteobacteria
		R, T, and U types (S)	Enterobacteriaceae
Superfamily: Psylloidae (psyllids)		Carsonella ruddii (P)	Gammaproteobacteria
		S-endosymbiont	Gammaproteobacteria
Family: Pseudococcidae (mealybugs)		Tremblaya princeps (P)	Betaproteobacteria
		S-endosymbiont	Gammaproteobacteria
Superfamily: Aleyrodoidae (whiteflies)		P-endosymbiont	Gammaproteobacteria
		S-endosymbiont	Enterobacteriaceae
Family: Cicadellidae (leafhoppers) Subfamily: Cicadellinae (sharpshooters)		Baumannia cicadellinicola (P)	Gammaproteobacteria
Order: Heteroptera Family: Reduviidae Subfamily: Triatominae (kissing bugs)	Vertebrate blood	Arsenophonus triatominarum (S)	Gammaproteobacteria
Order: Diptera Family: Muscidae Genus: Glossina (tsetse flies)		Wigglesworthia glossinidia (P) Sodalis glossinidius (S)	Gammaproteobacteria Enterobacteriaceae
Order: Coleoptera Family: Curculionidae Genus: Sitophilus (weevils)	Stored gain	P-endosymbiont	Gammaproteobacteria
Order: Hymenoptera Family: Formicidae Genus: Camponotus (carpenter ants)	Plant nectar, honeydew, detritus, and other sources	Blochmannia species (P)	Gammaproteobacteria
Order: Dictyoptera[b] Suborder: Blattaria (cockroaches)	Omnivorous	Blattabacterium cuenoti (P)	CFB[c] division
Cryptocercus punctulatus	Wood	Blattabacterium cuenoti (P)	CFB division
Suborder: Isoptera (termites) Mastotermes darwiniensis	Wood	Blattabacterium cuenoti (P)	CFB division

[a] Modified from Baumann et al. (2002).
[b] The phylogeny and systematics of Dictyoptera are still controversial, including the exact position of the wood-feeding cockroach, C. punctulatus, and the termite, M. darwiniensis, within the order (Nalepa and Bandi, 2000; Eggleton, 2001).
[c] Cytophaga-Flavobacterium-Bacteroides.

the smallest known for bacteria (0.64 Mb), presumably having this size through progressive gene losses from a larger ancestral genome, ultimately rendering cells obligate intracellular symbionts. Inasmuch as phloem sap is rich in sugars, but deficient in several essential amino acids, it was speculated that Buchnera supplemented the diet of aphids by providing essential amino acids to the host. This hypothesis has received experimental support, and the mechanisms that Buchnera have evolved to overproduce essential amino acids are fascinating. For example, Buchnera possess multiple copies of plasmids bearing tandem repeats of trpEG, the genes encoding anthranilate synthase, the first and rate-limiting enzyme specific to tryptophan biosynthesis. Likewise, Buchnera has multiple copies of plasmids containing genes (leuABCD) specific to leucine biosynthesis. Both tryptophan and leucine are essential for aphids.

Genes for biosynthesis of other amino acids required by the host are on the Buchnera chromosome, whose copy number can range from 50 to 200, varying with the life cycle of the host and the host's nutritional economy (Komaki and Ishikawa, 1999). By contrast, the Buchnera genome lacks genes for biosynthesis of those amino acids capable of being made by the host and supplied to Buchnera.

The endosymbiosis between Buchnera and aphids was apparently established at least 150 million years ago, but aside from genomic sequence evolution and inactivation and loss of individual genes, no chromosome rearrangements or gene acquisitions have occurred for the past 50 million years. This has apparently resulted from loss of phages, repeated sequences and recA, rendering Buchnera no longer a source of ecological innovation for its hosts (Tamas et al., 2002). In addition to

Buchnera, three different types of S-endosymbionts (so-called R, T, and U types) also occur in aphids: all fall within the *Enterobacteriaceae* (Sandstrom et al., 2001).

Endosymbionts of other plant sap-sucking insects have not been studied as extensively as *Buchnera,* but they presumably also function in host nutrition and are worth mention. The P-endosymbiont of psyllids (jumping plant lice) has been named *Carsonella ruddii* (Thao et al., 2000). Portions of its genome that have been sequenced have an exceptionally low G+C content (19.9 mol%), an almost complete absence of intergenic spaces, operon fusions, and several additional characteristics suggesting that the genome size is itself quite small (Clark et al., 2001b). Endosymbionts of mealybugs not only have a remarkable relationship with their host, but with each other as well. The P-endosymbiont is a betaproteobacterium named *Tremblaya princeps.* However, speroidal cells of *T. princeps* themselves harbor secondary S-endosymbionts that are members of the *Gammaproteobacteria* (von Dohlen et al., 2001)! This is the first known case of prokaryotic endosymbionts of other prokaryotes. Analyses of 4-kb genomic fragments containing 16S-23S rDNAs indicated that the P-endosymbionts were monophyletic and could be subdivided into five clusters, which were similar to the phylogenetic clusters inferred for the S-endosymbionts. This implies that the P-endosymbionts were infected multiple times with different ancestors of the S-endosymbionts, and once the association was established they were transmitted in mealybugs together (Thao et al., 2002). Whiteflies contain P- and S-endosymbionts, one or both of which are required for normal development and fecundity (Costa et al. 1993, 1997). Whiteflies are unusual in that they deliver an entire bacteriocyte containing endosymbionts to the egg for transmission (Costa et. al., 1996). Recently, endosymbionts of sharpshooters (a subgroup of leafhoppers) were examined by molecular methods and found to form a distinct clade within the gamma-3 subgroup of the *Proteobacteria.* Named "*Candidatus* Baumannia cicadellinicola," these bacteria also possess a relatively small genome (ca. 680 kb) and inhabit red-pigmented bacteriomes. Additional endosymbionts related to flavobacteria were found in yellow bacteriomes of one of the sharpshooter species examined (*Homalodisca coagulata*), as were bacteria related to *Wolbachia pipientis* in testicular tissues (Moran et al., 2003).

Blood-sucking insects contain both extracellular gut bacteria as well as intracellular endosymbionts, some of which are thought to supply B vitamins essential for host survival, either by secreting them to the host or, for gut bacteria, by being themselves digested as a source of these nutrients (Buchner, 1965; Dasch et al., 1984; Beard et al., 2002). Triatomine bugs (kissing bugs) include *Triatoma infestans,* a vector of *Trypanosoma cruzi,* which causes Chagas' disease in humans. *T. infestans* harbors a *Gammaproteobacteria* S-endosymbiont in various tissues and in its hemolymph. Cells of the S-endosymbiont are highly filamentous rods (1 to 1.5 μm in diameter × 15 μm or more in length) related to the sex ratio-altering parasites of wasps, i.e., *Arsenophonus nasoniae* (see above). They have been cultured *in vitro* in an *Aedes albopictus* (mosquito) cell line and named *Candidatus* Arsenophonus triatominarum (Hypsa and Dale, 1997). However, their role in the biology of *T. infestans* is unknown. By using PCR amplification as a detection technique, bacteria related to *Arsenophonus* have also been detected in pea aphids and citrus psyllids (Subandiyah et al., 2000; Tsuchida et al., 2002).

P-endosymbionts of tsetse flies (*Glossina* species), which transmit African trypanosomiases, are *Gammaproteobacteria* that have been named *Wigglesworthia glossinidia* (Aksoy, 1995). These obligate endosymbionts occupy a U-shaped bacteriome located in the anterior region of the gut, and they have a phylogeny that is congruent with their insect host, with which they have been in association for about 40 million years (Chen et al., 1999). The entire genome of *W. glossinidia* has been recently sequenced (Akman et al., 2002), and its analysis along with that of gene expression are consistent with the notion that *W. glossinidia* contributes to host nutrition and fecundity by vitamin and cofactor biosynthesis and possibly amino acid provision for larvae (Akman and Aksoy, 2001; Wernegreen, 2002). Tsetse flies also bear microaerophilic S-endosymbionts (*Sodalis glossinidius*), which are affiliated with the *Enterobacteriaceae* and which were the first endosymbionts to be isolated in pure culture from hemolymph (Dale and Maudlin, 1999). The genome size of *S. glossinidius* is about 2 Mb and has retained many biosynthetic genes, but many genes involved in energy metabolism and carbon compound assimilation are apparently missing, suggesting an adaptation to energy sources present only in blood (Akman et al., 2001). Interestingly, *S. glossinidius* appears to possess a type III secretion system, which is critical to their ability to establish an endosymbiosis in its host and to invade insect cells in vitro (Dale et al., 2001). As type III secretion systems are common in animal and plant pathogenic bacteria (Galan and Collmer, 1999), their discovery in *S. glossinidius* gives credence to the long-held (but until now, weakly supported) notion that mutualistic intracellular endosymbionts evolve from parasites, after establishment of vertical transmission and attenuation of virulence. More recently, homologues of type III secretion system genes have also been found in the (as yet unnamed *Gammaproteobacteria*) P-endosymbiont of grain weevils (*Sitophilus zeamais*), and the expression of these genes coincides with infection in developing weevils (Dale et al., 2002).

P-endosymbionts of *Camponotus* species (carpenter ants) are *Gammaproteobacteria* grouped in the candidate genus *Blochmannia* (*B. floridanus, B. herculeanus,* and *B. rufipes*) (Sauer et al., 2000). The endosymbionts are located primarily in midgut bacteriocytes and in oocytes and have cospeciated with their host (Sauer et al., 2002). Elimination of *Blochmannia* from adult insects had no apparent deleterious effects suggesting that if *Blochmannia* is important to the well-being of the ants, its impact may be greater during embryogenesis or larval development.

P-endosymbionts occur in all species of cockroaches examined and in one species of termite (*Mastotermes darwiniensis*). They consist of rods (1 μm × 1.6 to 9 μm) with a gram-negative type cell wall and are located in bacteriocytes within the fat body, a diffuse mass of abdominal tissue that is a site for intermediary metabolism and storage of glycogen and fats and which also contains cells (urocytes) for storage and excretion of uric acid (Dasch et al., 1984; Polver et al., 1986). The P-endosymbionts are currently assigned to a single species, *Blattabacterium cuenoti*, but they actually consist of a clade of *Flavobacterium*-related members within the CFB division (Bandi et al., 1994, 1995; Nalepa et al., 1997; Clark et al., 2001a). Within this clade, the phylogeny of individual strains of *B. cuenoti* is congruent with that of their host. It is interesting that *M. darwiniensis* is the only termite known to contain *B. cuenoti*. Owing to the close phylogenetic relationship between cockroaches and termites, with *Mastotermes* as the most basal termite group (reviewed by Eggleton [2001]), it seems likely that infection with an ancient *Blattabacterium* occurred 135 to 250 million years ago in an ancestor common to cockroaches and termites, but that subsequent loss of endosymbionts occurred in all termite lineages except that gaving rise to *Mastotermes* (Bandi et al., 1995).

The role of *Blattabacterium* in host vitality is still largely obscure. Radioactive tracer studies with cockroaches containing blattabacteria (but lacking gut bacteria) suggest that they play a role in essential amino acid synthesis (Dasch et al., 1984). However, the juxtaposition of bacteriocytes and urocytes in the fat body has long provoked the notion that blattabacteria are involved in uric acid degradation and mobilization, and the presence of xanthine dehydrogenase activity in crude extracts of blattabacteria harvested from oothecae support this idea (Wren and Cochran, 1987).

ECTOSYMBIONTS OF INSECTS

Ectosymbionts of insects include a wide array of bacteria, yeast, and filamentous fungi, many of which participate with the insects in varying degrees of intimacy in plant litter decomposition, or are consumed themselves as food or as a source of "acquired enzymes" for food digestion. Such interactions have been the topic of a number of earlier reviews (Anderson et al., 1984; Dasch et al., 1984; Martin, 1987; Dowd, 1992; Kaufman et al., 2000). This section will focus primarily on recent discoveries made in studies of fungus-cultivating ants, fungus-cultivating termites, and ambrosia and bark beetles.

Fungus-Cultivating (Attine) Ants

Ants of the tribe Attini contain over 200 species that inhabit tropical rain forests. They include the well known leaf-cutting ants that established an intimate relationship with *Basidiomycetes* fungi (mainly of the family *Lepiotaceae*) about 50 million years ago and which have coevolved with specific phylotypes of these fungi for about 23 million years (Chapela et al., 1994; Hinkle et al., 1994). The ants propagate their ectosymbionts in fungal gardens, on a substrate consisting of cut and well-masticated leaves of fresh vegetation from which the (antifungal) waxy coating has been scraped away by the ants. They then feed on the fungal mycelium that has developed in mature portions of the garden. Fungal enzymes acquired during such feeding are then distributed (via ant feces, after passage through the gut) into fresh portions of the substrate to enhance the initiation of growth of the fungus, which detoxifies plant anti-insect defense chemicals during growth (Martin, 1987; Hölldobler and Wilson, 1990). However, an important additional ectosymbiont has recently been discovered in this seemingly two-partner symbiosis. It is a *Streptomyces* sp. carried on genus-specific areas of the ant integument and transmitted vertically from parent to daughter nest. The streptomycete promotes the growth of the mutualistic fungus and at the same time selectively inhibits the growth of *Escovopsis* (anamorphic Hypocreales: Ascomycotina), a specialized virulent fungal parasite of attine fungus gardens (Currie et al., 1999). It does so by producing highly potent antibiotics that keep *Escovopsis* in check. Presumably, *Escovopsis* must have periodically developed resistance to the antibiotics, thereby putting selective pressure on the streptomycete to continually evolve new ones. However, little is yet known about the evolutionary history of these two organisms and how those two histories compare with that of the ants and their mutualistic fungus. Further aspects of the biology of this fascinating and complex interaction have been recently reviewed (Currie, 2001).

Fungus-Cultivating Termites (Subfamily *Macrotermitinae*)

Many species of termites are associated with fungi, and over 50 species of fungi have a beneficial influence on termite nutrition, survival, caste development, and nest construction (Rouland-Lèfevre, 2000). Most of the beneficial effects have been attributed to preliminary digestion of wood polymers (lignin and polysaccharides) by the fungus, an increase in nutritive value of the food resource imparted by the fungal mycelium, and detoxification by the fungus of plant chemicals. However, certain species of termites, like attine ants, have taken to cultivating certain fungi in well-tended, elaborate gardens, thereby evolving a whole suite of behavioral adaptations to ensure maintenance, growth, and transmission of their fungal symbionts. Fungus-cultivating termites belong to the subfamily *Macrotermitinae*, and their symbiotic fungus is represented by a single genus of basidiomycete fungi *Termitomyces*. The contribution of *Termitomyces* to termite nutrition appears to be multifold and includes an upgrade of nutritive value of plant litter harvested by the termites via fungal biomass, and it may also involve an enhancement of digestive capacity of the termites by enzymes acquired as a result of fungal browsing.

Although coevolution of *Termitomyces* and *Macrotermitinae* has been suggested, not until this past year has anything been known about the phylogeny of *Termitomyces* or their evolutionary relationship with *Macrotermitinae*. Three different laboratories approached this issue by examining (in aggregate) the sequences of nuclear-encoded SSU rDNA, internal transcribed spaces, and portions of the large subunit rDNA (Katoh et al., 2002; Rouland-Lefevre et al., 2002; Taprab et al., 2002). Initial results have led to several conclusions, which can be summarized as follows: (i) the fungal genus *Termitomyces* appears to be monophyletic, but contains at least eight species or subspecies; (ii) within a nest, the fungi comprising the mycelium and fruiting structures are usually identical and probably represent a monoculture; however, (iii) different phylogenetic types of *Termitomyces* can associate with a particular genus of termite, especially if the termites are collected from different geographical areas, implying a significant amount of horizontal transfer and/or the existence of free-living populations of cultivated fungi from which the strains

are acquired. Moreover, there was no apparent phylogenetic clustering of fungi within clades representing proposed functional contributions of the fungi to the symbiosis, i.e., noncontributors of digestive enzymes, contributors of oligo- and polysaccharidases, and contributors of polysaccharidases only (Rouland-Lefevre, et al. 2002).

Fungal Symbionts of Ambrosia Beetles and Bark Beetles

Ambrosia beetles and bark beetles comprise about 7,500 species distributed within the family *Curculionidae* (subfamilies *Scolytinae* and *Platypodinae*) that burrow into the phloem of trees for ovipositing and feeding. They are associated with ascomycete fungi of the order *Ophiostomatales* (an order of plant pathogenic fungi), which they carry into their host trees in glandular cuticular invaginations (mycangia) that maintain the fungal spores and mycelia in pure culture for inoculating into the tunnels and galleries. The ambrosia beetles exhibit obligate mutualisms with certain of the ophiostomatoid fungi (genera *Ambrosiella* and *Raffaelea*), which are vertically transmitted and which have become asexual polyphagous domesticates that serve as the primary food (ambrosia) of the beetles. As a consequence, individual species of ambrosia beetles can often use a wide variety of host taxa (Beaver, 1989). By contrast, bark beetles are host-specialized phloem feeders associated with *Ophiostoma* or other sexual free-living ambrosia fungi, whose primary role is to circumvent resinous defenses of trees by rapidly growing and blocking resin canals (Paine et al., 1997). Recent examination of the molecular systematics of the fungi (Cassar and Blackwell, 1996; Jones and Blackwell, 1998) and the beetles (Farrell et al., 2001) has revealed a number of interesting features of the life history of these associations: (i) the two genera *Ambrosiella* and *Raffaelea* are actually polyphyletic, having arisen at least five times; (ii) the ambrosia beetle habits have evolved repeatedly and are unreversed over a time period spanning 60 to 21 million years; (iii) bark beetles have shifted from ancestral association with conifers to angiosperms and back again several times, with each shift to angiosperms and conifers resulting in increased and decreased diversity, respectively. The adult habit of wood boring, and the ability of beetles to manage fungi as a source of food and protection of tree defenses, have undoubtedly played a major role in the ability of these insects to exploit one of Earth's most abundant sources of biomass.

BIOPROSPECTING WITHIN
MICROBIAL SYMBIONTS OF INSECTS

From this brief overview, it would appear that Earth's insects represent a fertile arena in which to bioprospect. Even if it is assumed that only 1% of all insect species have a unique microbial species associated with them, it is clear that insects constitute an enormous reservoir of novel and potentially exploitable microbes, microbial genes, and gene products. In some insect groups (e.g., termites, above), the microbial diversity in single species is quite large. Nevertheless, insect-associated microbes remain a resource that has been largely untapped (Gebhardt et al. [2002] and references therein). This is unfortunate, especially as the powerful tools of molecular biology can augment bioprospecting by rendering cultivation optional rather than an absolute necessity.

Among good starting points for exploration would be hot spots of microbial diversity, such as the gut microbiota of termites, cockroaches, and other insects. For example, Gebhardt and coworkers recently isolated an assortment of new natural products and antifungal and antibacterial antibiotics from bacilli isolated from guts of various insects (Gebhardt et al., 2002). Likewise, novel alkaliphilic bacteria capable of degrading biphenyl and polychlorinated biphenyls were isolated from termites. Using a cultivation-independent approach, Ohtoko and coworkers have recognized a rich diversity of novel cellulase enzymes in termite gut hypermastigote protozoa (Ohtoko et al., 2000). It is not unreasonable to think that large insert-containing BAC libraries prepared from complex gut microbiota DNA as source material could yield genes encoding a variety of useful natural products and enzymes, as has been initiated with the soil microbiota (see Handelsman, chapter 11). Another fertile area for exploration would seem to be insects associated with one or a few specific symbionts. Some factor(s) must exist to prevent intrusion of associations by undesired contaminants or pathogens, and production of antimicrobials by one or more of the partners would seem to be a selectable strategy, as seen with the streptomycete partners in the attine anti-fungus association. Perhaps analogous consorts exist elsewhere. Likewise, microbial symbionts of insects that feed on or live in food resources containing toxic substances may be a rich source of novel detoxification enzymes, as has been documented already for various insect-associated fungi (Dowd, 1992). Perhaps this review will inspire further explorations into some of these fascinating associations.

Acknowledgments. I thank Paul Baumann for providing the electron micrograph used for Fig. 1B. Figure 1A was prepared in collaboration with H. S. Pankratz, to whom I am also grateful.

REFERENCES

Akman, L., and S. Aksoy. 2001. A novel application of gene arrays: *Escherichia coli* array provides insight into the biology of the obligate endosymbiont of tsetse flies. *Proc. Nat. Acad. Sci. USA* 98:7546–7551.

Akman, L., R. V. M. Rio, C. B. Beard, and S. Aksoy. 2001. Genome size determination and coding capacity of *Sodalis glossinidius*, an enteric symbiont of tsetse flies, as revealed by hybridization to *Escehrichia coli* gene arrays. *J. Bacteriol.* 183:4517–4525.

Akman, L., A. Yamashita, H. Watanabe, K. Oshima, T. Shiba, M. Hattori, and S. Aksoy. 2002. Genome sequence of the endocellular obligate symbiont of tsetse flies, *Wigglesworthia glossinidia. Nat. Genet.* 32:402–407.

Aksoy, S. 1995. *Wigglesworthia* gen. nov. and *Wigglesworthia glossinidia* sp. nov., taxa consisting of the mycetocyte-associated, primary endosymbionts of tsetse flies. *Int. J. Syst. Bacteriol.* 45:848–851.

Amann, R. 2000. Who is out there? Microbial aspects of biodiversity. *Syst. Appl. Microbiol.* 23:1–8.

Anderson, J. M., A. D. M. Rayner, and D. W. H. Walton. 1984. *Invertebrate-Microbial Interactions.* Cambridge University Press, Cambridge, United Kingdom.

Bandi, C., G. Damiani, L. Magrassi, A. Grigolo, R. Fani, and L. Sacchi. 1994. Flavobacteria as intracellular symbionts in cockroaches. *Proc. R. Soc. Lond. Sci.* 257:43–48.

Bandi, C., M. Sironi, G. Damiani, L. Magrassi, C. A. Nalepa, U. Landani, and L. Sacchi. 1995. The establishment of intracellular symbiosis in an ancestor of cockroaches and termites. *Proc. R. Soc. Lond. Sci. B* 259:293–299.

Baumann, P., N. A. Moran, and L. Baumann. 2002. Bacteriocyte-associated endosymbionts of insects. *In* M. Dworkin (ed.), *The Prokaryotes* (on-line version). Springer-Verlag, New York, N.Y.

Bayon, C. 1980. Volatile fatty acids and methane production in relation to anaerobic carbohydrate fermentation in *Oryctes nasicornis* larvae (Coleoptera: Scarabaeidae). *J. Insect. Physiol.* 26:819–828.

Beard, C. B., C. Cordon-Rosales, and R. V. Durvasula. 2002. Bacterial symbionts of the *Triatominae* and their potential use in control of Chagas disease. *Annu. Rev. Entomol.* 47:123–141.

Beaver, R. A. 1989. Insect-fungus relationships in the bark and ambrosia beetles, p. 121–143. *In* N. Wilding, N. M. Collins, P. M. Hammond, and J. F. Webber (ed.), *Insect-Fungus Interactions.* Academic Press, London, United Kingdom.

Breznak, J. A. 2000. Ecology of prokaryotic microbes in the guts of wood-and litter-feeding termites, p. 209–231. *In* T. Abe, D. E. Bignell, and M. Higashi (ed.), *Termites: Evolution, Sociality, Symbiosis, Ecology.* Kluwer Academic, Dordrecht, The Netherlands.

Breznak, J. A. 2002. Phylogenetic diversity and physiology of termite hindgut spirochetes. *Integ. Comp. Biol.* 42:313–318.

Breznak, J. A., and J. R. Leadbetter. 2002. Termite gut spirochetes. *In* M. Dworkin (ed.), *The Prokaryotes.* [Online.] Springer-Verlag, New York, N.Y.

Brune, A. 2003. Symbionts aiding digestion, p. 1102–1107. *In* V. H. Resh and R. T. Cardé (ed.), *Encyclopedia of Insects.* Academic Press, Inc., New York, N.Y.

Brune, A., D. Emerson, and J. A. Breznak. 1995. The termite gut microflora as an oxygen sink: microelectrode determination of oxygen and pH gradients in guts of lower and higher termites. *Appl. Environ. Microbiol.* 61:2681–2687.

Buchner, P. 1965. *Endosymbiosis of Animals with Plant Microorganisms.* John Wiley & Sons, Inc., New York, N.Y.

Cassar, S., and M. Blackwell. 1996. Convergent origins of ambrosia fungi. *Mycologia* 88:596–601.

Cazemier, A. E., H. J. M. OpdenCamp, J. H. P. Hackstein, and G. D. Vogels. 1997. Fibre digestion in arthropods. *Comp. Biochem. Physiol. A* 118:101–109.

Chapela, I. H., S. A. Rehner, T. R. Schultz, and U. G. Mueller. 1994. Evolutionary history of the symbiosis between fungus-growing ants and their fungi. *Science* 266:1691–1694.

Chen, X. A., S. Li, and S. Aksoy. 1999. Concordant evolution of a symbiont with its host insect species: molecular phylogeny of genus *Glossina* and its bacteriome-associated endosymbiont, *Wigglesworthia glossinidia*. *J. Mol. Evol.* 48:49–58.

Clark, J. W., S. Hossain, C. R. Burnside, and S. Kambhampati. 2001a. Coevolution between a cockroach and its bacterial endosymbiont: a biogeographical perspective. *Proc. R. Soc. Lond. Ser. B* 268:393–398.

Clark, M. A., L. Baumann, M. L. L. Thao, N. A. Moran, and P. Baumann. 2001b. Degenerative minimalism in the genome of a psyllid endosymbiont. *J. Bacteriol.* 183:1853–1861.

Costa, H. S., D. E. Ulmann, M. W. Johnson, and B. E. Tabashnik. 1993. Antibiotic oxetetracycline interferes with Bemisia tabaci (Homoptera, Aleyrodidae) oviposition, development, and ability to induce squash silverleaf. *Ann. Entomol, Soc. Am.* 86:740–748.

Costa, H. S., N. C. Toscano, and T. J. Henneyberry. 1996. Mycetocyte inclusion in the oocytes of *Bemisia argentifolii* (Homoptera: Aleyrodidae). *Ann. Entomol. Soc. Am.* 89:694–699.

Costa, H. S., T. J. Henneberry, and N. C. Toscano. 1997. Effects of antibacterial materials on *Bemisia argentifolii* (Homoptera: Aleyrodidae) oviposition, growth, survival, and sex ratio. *J. Econ. Entomol.* 90:333–339.

Currie, C. R. 2001. A community of ants, fungi, and bacteria: a multilateral approach to studying symbiosis. *Annu. Rev. Microbiol.* 55:357–380.

Currie, C. R., J. A. Scott, R. C. Summerbell, and D. Malloch. 1999. Fungus-growing ants use antibiotic-producing bacteria to control garden parasites. *Nature* 398:701–704.

Dale C., and I. Maudlin. 1999. *Sodalis* gen. nov. and *Sodalis glossinidius* sp. nov., a microaerophilic secondary endosymbiont of the tsetse fly *Glossina morsitans morsitans*. *Int. J. Syst. Bacteriol.* 49:267–275.

Dale, C., S. A. Young, D. T. Haydon, and S. C. Welburn. 2001. The insect endosymbiont *Sodalis glossinidius* utilizes a type III secretion system for cell invasion. *Proc. Natl. Acad. Sci. USA* 98:1883–1888.

Dale, C., G. R. Plague, B. Wang, H. Ochman, and N. A. Moran. 2002. Type III secretion systems and the evolution of mutualistic endosymbiosis. *Proc. Natl. Acad. Sci. USA* 99:12397–12402.

Dasch, G. A., E. Weiss, and K. P. Chang. 1984. B. Endosymbionts of insects, p. 811–833. *In* N. R. Krieg, and J. G. Holt (ed.), *Bergey's Manual of Systematic Bacteriology*. Williams & Wilkins, Baltimore, Md.

de Bary, A. 1879 *Die Erscheinung der Symbiose*. Trubner, Strassburg, Austria.

Dillon, R., and K. Charnley. 2002. Mutualism between the desert locust *Schistocerca gregaria* and its gut microbiota. *Res. Microbiol.* 153:503–509.

Douglas, A. E. 1989. Mycetocyte symbiosis in insects. *Biol. Rev.* 64:409–434.

Douglas, A. E. 1992. Symbiotic microorganisms in insects, pp. 165–178. *In* J. Lederberg (ed.), *Encyclopedia of Microbiology*. Academic Press, Inc., San Diego, Calif.

Dowd, P. F. 1992. Insect fungal symbionts: a promising source of detoxifying enzymes. *J. Ind. Microbiol.* 9:149–161.

Drew, R. A. I., and A. C. Lloyd. 1991. Bacteria in the life cycle of tephritid fruit flies, p. 441–465. *In* P. Barbosa, V. A. Krischik, and C. G. Jones (ed.), *Microbial Mediation of Plant-Herbivore Interactions*. John Wiley & Sons, Inc., New York, N.Y.

Eggleton, P. 2001. Termites and trees: a review of recent advances in termite phylogenetics. *Insectes Sociaux* 48:187–193.

Farrell, B. D., A. S. Sequeira, B. C. O'Meara, B. B. Normark, J. H. Chung, and B. H. Jordal. 2001. The evolution of agriculture in beetles (Curculionidae: Scolytinae and Platypodinae). *Evolution* 55:2011–2027.

Friedrich, M. W., D. Schmitt-Wagner, T. Lueders, and A. Brune. 2001. Axial differences in community structure of *Crenarchaeota* and *Euryarchaeota* in the highly compartmentalized gut of the soil-feeding termite *Cubitermes orthognathus*. *Appl. Environ. Microbiol.* 67:4880–4890.

Galan, J. E., and A. Collmer. 1999. Type III secretion machines: bacterial devices for protein delivery into host cells. *Science* 284:1322–1328.

Gebhardt, K., et al. 2002. Screening for biologically active metabolites with endosymbiotic bacilli isolated from arthropods. *FEMS Lett.* 217:199–205.

Gherna, R. L., J. H. Werren, W. Weisburg, R. Cote, C. R. Woese, L. Mandeico, and D. J. Brenner. 1991. *Arsenophonus nasoniae* gen. nov., sp.nov., the causative agent of the son-killer trait in the parasitic wasp *Nasonia vitripennis*. *Int. J. Syst. Bacteriol.* 41:563–565.

Gijzen, H. J., C. A. M. Broers, M. Barughare, and C. K. Stumm. 1991. Methanogenic bacteria as endosymbionts of the ciliate *Nyctotherus ovalis* in the cockroach hindgut. *Appl. Environ. Microbiol.* 57:1630–1634.

Gijzen, H. J., C. van der Drift, M. Barugahare, and H. J. M. op den Camp. 1994. Effect of host diet and hindgut microbial composition on cellulolytic activity in the hindgut of the American cockroach, *Periplaneta americana*. *Appl. Environ. Microbiol.* 60:1822–1826.

Hackett, K. J., D. E. Lynn, D. L. Williamson, A. S. Ginsberg, and R. F. Whitcomb. 1986. Cultivation of the *Drosophila* sex-ratio *Spiroplasma*. *Science* 232:1253–1255.

Hackstein, J. H. P., and C. K. Stumm. 1994. Methane production in terrestrial arthropods. *Proc. Natl. Acad. Sci. USA* 91:5441–5445.

Hinkle, G., J. K. Wetterer, T. R. Schultz, and M. L. Sogin. 1994. Phylogeny of the attine ant fungi based on analysis of small subunit ribosomal RNA gene sequences. *Science* 266:1695–1697.

Hölldobler, B., and E. O. Wilson. 1990. *The Ants*. Belknap Press, Cambridge, Mass.

Howard, D. J., G. L. Bush, and J. A. Breznak. 1985. The evolutionary significance of bacteria associated with *Rhagoletis*. *Evolution* 39:405–417.

Hugenholtz, P., B. M. Goebel, and N. R. Pace. 1998. Impact of culture-independent studies on the emerging phylogenetic view of bacterial diversity. *J. Bacteriol.* 180:4765–4774.

Hurst, G. D. D., T. C. Hammarton, C. Bandi, T. M. O. Majerus, D. Bertrand, and M. E. N. Majerus. 1997. The diversity of inherited parasites of insects: the male-killing agent of the ladybird beetle *Coleomegilla maculata* is a member of the Flavobacteria. *Genet. Res.* 70:1–6.

Hypsa, V., and C. Dale. 1997. In vitro culture and phylogenetic analysis of "*Candidatus* Arsenophonus triatominarum," an intracellular bacterium from the triatomine bug, *Triatoma infestans*. *Int. J. Syst. Bacteriol.* 47:1140–1144.

Iida, T., M. Ohkuma, K. Ohtoko, and T. Kudo. 2000. Symbiotic spirochetes in the termite hindgut: phylogenetic identification of ectosymbiotic spirochetes of oxymonad protists. *FEMS Microbiol. Ecol.* 34:17–26.

Inoue, T., O. Kitade, T. Yoshimura, and I. Yamaoka. 2000. Symbiotic associations with protists, p. 275–288. *In* T. Abe, D. E. Bignell, and M. Higashi (ed.), *Termites: Evolution, Sociality, Symbioses, Ecology*. Kluwer Academic Publishers, Dordrecht, The Netherlands.

Jeyaprakash, A., and M. A. Hoy. 2000. Long PCR improves *Wolbachia* DNA amplification: *wsp* sequences found in 76% of sixty-three arthropod species. Insect *Mol. Biol.* 9:393–405.

Jones, K. G., and M. Blackwell. 1998. Phylogenetic analysis of ambrosia species in the genus *Raffaelea* based on 18S rDNA sequences. *Mycol. Res.* **102**:661–665.

Kane, M. D. 1997. Microbial fermentation in insect guts, pp. 231–265. *In* R. I. Mackie, and B. A. White (ed.), *Gastrointestinal Microbiology*. Chapman & Hall, New York, N.Y.

Kane, M. D., and N. E. Pierce. 1994. Diversity within diversity: molecular approaches to studying microbial interactions with insects, p. 509–524. *In* B. Schierwater, B. Streit, G. Wagner, and R. DeSalle (ed.), *Molecular Methods in Ecology and Evolution*. Birkhauser Verlag, Berlin, Germany.

Katoh, H., T. Miura, K. Maekawa, N. Shinzato, and T. Matsumoto. 2002. Genetic variation of symbiotic fungi cultivated by the macrotermitine termite *Odontotermes formosanus* (Isoptera: Termitidae) in the Ryukyu Archipelago. *Mol. Ecol.* **11**:1565–1572.

Kaufman, M. G., and M. J. Klug. 1991. The contribution of hindgut bacteria to dietary carbohydrate utilization by crickets (Orthoptera: Gryllidae). *Comp. Biochem. Physiol.* **98**:117–123.

Kaufman, M. G., M. J. Klug, and R. W. Merritt. 1989. Growth and food utilization parameters of germ-free house crickets, *Acheta domesticus*. *J. Insect Physiol.* **35**:957–967.

Kaufman, M. G., E. D. Walker, D. A. Odelson, and M. J. Klug. 2000. Microbial community ecology and insect nutrition. *Am. Entomol.* **46**:173–184.

Komaki, K., and H. Ishikawa. 1999. Intracellular bacterial symbionts of aphids posess many genomic copies per bacterium. *J. Mol. Evol.* **48**:717–722.

Lauzon, C. R., R. E. Sjogren, and R. J. Prokopy. 2000. Enzymatic capabilities of bacteria associated with apple maggot flies: a postulated role in attraction. *J. Chem. Ecol.* **26**:953–967.

Leadbetter, J. R., and J. A. Breznak. 1996. Physiological ecology of *Methanobrevibacter cuticularis* sp. nov. and *Methanobrevibacter curvatus* sp. nov., isolated from the hindgut of the termite *Reticulitermes flavipes*. *Appl. Environ. Microbiol.* **62**:3620–3631.

Leadbetter, J. R., L. D. Crosby, and J. A. Breznak. 1998. *Methanobrevibacter filiformis* sp. nov., a filamentous methanogen from termite hindguts. *Arch. Microbiol.* **169**:287–292.

Leadbetter, J. R., T. M. Schmidt, J. R. Graber, and J. A. Breznak. 1999. Acetogenesis from H_2 plus CO_2 by spirochetes from termite guts. *Science* **283**:686–689.

Lilburn, T. G., T. M. Schmidt, and J. A. Breznak. 1999. Phylogenetic diversity of termite gut spirochaetes. *Environ. Microbiol.* **1**:331–345.

Lilburn, T. G., K. S. Kim, N. E. Ostrom, K. R. Byzek, J. R. Leadbetter, and J. A. Breznak. 2001. Nitrogen fixation by symbiotic and free-living spirochetes. *Science* **292**:2495–2498.

Martin, M. M. 1987. *Invertebrate-Microbial Interactions: Ingested Fungal Enzymes in Arthropod Biology*. Comstock Publishing Associates, Ithaca, N.Y.

Moran, N. A., and P. Baumann. 2000. Bacterial endosymbionts in animals. *Curr. Opin. Microbiol.* **3**:270–275.

Moran, N. A., C. Dale, H. Dunbar, W.A. Smith, and H. Ochman. 2003. Intracellular symbionts of sharpshooters (Insecta: Hemiptera: Cicadellinae) form a distinct clade with a small genome. *Environ. Microbiol.* **5**:116–126.

Nalepa, C. A., and C. Bandi. 2000. Characterizing the ancestors: paedomorphosis and termite evolution, p. 53–73. *In* T. Abe, D. E. Bignell, and M. Higashi (ed.), *Termites: Evolution, Sociality, Symbioses, Ecology*. Kluwer Academic Publishers, Dordrecht, The Netherlands.

Nalepa, C., G. Byers, C. Bandi, and M. Sironi. 1997. Description of *Cryptocercus clevelandi* (Dictyoptera: Cryptocercidae) from the northwestern United States, molecular analysis of bacterial symbionts in its fat body, and notes on biology, distribution, and biogeography. *Ann. Entomol. Soc. Am.* **90**:416–424.

Novotny, V., Y. Basset, S. E. Miller, G. W. Weiblen, B. Bremer, L. Cizek, and P. Drozd. 2002. Low host specificity of herbivorous insects in a tropical forest. *Nature* **416**:841–844.

Ohkuma, M., and T. Kudo. 1996. Phylogenetic diversity of the intestinal bacterial community in the termite *Reticulitermes speratus*. *Appl. Environ. Microbiol.* **62**:461–468.

Ohkuma, M., and T. Kudo. 1998. Phylogenetic analysis of the symbiotic intestinal microflora of the termite *Cryptotermes domesticus*. *FEMS Microbiol. Lett.* **164**:389–395.

Ohkuma, M., S. Noda, Y. Hongoh, and T. Kudo. 2002. Diverse bacteria related to the *Bacteroides* subgroup of the CFB phylum within the gut symbiotic communities of various termites. *Biosci. Biotechnol. Biochem.* **66**:78–84.

Ohtoko, K., M. Ohkuma, S. Moriya, T. Inoue, R. Usami, and T. Kudo. 2000. Diverse genes of cellulase homologues of glycosyl hydrolase family 45 from the symbiotic protists in the hindgut of the termite *Reticulitermes speratus*. *Extremophiles* **4**:343–349.

O'Neill, S. L., A. A. Hoffmann, and J. H. Werren (ed). 1997. *Influential Passengers*. Oxford University Press, Oxford, United Kingdom.

Polver, P. P. D., L. Sacchi, L. Cima, A. Grigolo, and U. Laudani. 1986. Oxidoreductase distribution in the fat body and symbionts of the German cockroach *Blatella germanica*: a histochemical approach. *Cell. Mol. Biol.* **32**:701–708.

Paine, T. D., K. F. Raffa, and T. C. Harrington. 1997. Interactions among scolytid bark beetles, their associated fungi, and live host conifers. *Annu. Rev. Entomol.* **42**:179–206.

Pimm, S. L., G. J. Russell, J. L. Gittleman, and T. M. Brooks. 1995. Future of biodiversity. *Science* **269**:347–350.

Rouland-Lèfevre, C. 2000. Symbiosis with fungi, p. 289–306. *In* T. Abe, D. E. Bignell, and M. Higashi (ed.), *Termites: Evolution, Sociality, Symbioses, Ecology*. Kluwer Academic Publishers, Dordrecht, The Netherlands.

Rouland-Lefevre, C., N. M. Diouf, A. Brauman, and M. Neyra. 2002. Phylogenetic relationships in I (family Agaricaceae) based on the nucleotide sequence of ITS: a first approach to elucidate the evolutionary history of the symbiosis between fungus-growing termites and their fungi. *Mol. Phylogenet. Evol.* **22**:423–429.

Sanderson, M. G. 1996. Biomass of termites and their emissions of methane and carbon dioxide: a global database. *Global Biogeochem. Cycles* **10**:543–557.

Sandstrom, J. P., J. A. Russell, J. P. White, and N. A. Moran. 2001. Independent origins and horizontal transfer of bacterial symbionts of aphids. *Mol. Ecol.* **10**:217–228.

Sauer, C., E. Stackebrandt, J. Gadau, B. Holldobler, and R. Gross. 2000. Systematic relationships and cospeciation of bacterial endosymbionts and their carpenter ant host species: proposal of the new taxon *Candidatus* Blochmannia gen. nov. *Int. J. Syst. Evol. Microbiol.* **50**:1877–1886.

Sauer, C., D. Dudaczek, B. Holldobler, and R. Gross. 2002. Tissue localization of the endosymbiotic bacterium "*Candidatus* Blochmannia floridanus" in adults and larvae of the carpenter ant *Camponotus floridanus*. *Appl. Environ. Microbiol.* **68**:4187–4193.

Smith, D. C., and A. E. Douglas. 1987. *The Biology of Symbiosis*. Edward Arnold, London, United Kingdom.

Sprenger, W. W., M. C. van Belzen, J. Rosenberg, J. H. P. Hackstein, and J. T. Keltjens. 2000. *Methanomicrococcus blatticola* gen. nov., sp. nov., a methanol- and methylamine-reducing methanogen from the hindgut of the cockroach *Periplaneta americana*. *Int. J. Syst. Evol. Microbiol.* **50**:1989–1999.

Stackebrandt, E., and B. M. Goebel. 1994. Taxonomic note: a place for DNA-DNA reassociation and 16S rRNA sequence analysis in the present species definition in bacteriology. *Int. J. Syst. Bacteriol.* **44**:846–849.

Staley, J. T., and A. Konopka. 1985. Measurement of in situ activities of nonphotosynthetic microorganisms in aquatic and terrestrial habitats. *Annu. Rev. Microbiol.* 39:321–346.

Stouthamer, R., J. A. J. Breeuwer, and G. D. D. Hurst. 1999. *Wolbachia pipientis:* microbial manipulator of arthropod reproduction. *Annu. Rev. Microbiol.* 53:71–102.

Subandiyah, S., N. Nikoh, S. Tsuyumu, S. Somowiyarjo, and T. Fukatsu. 2000. Complex endosymbiotic microbiota of the citrus psyllid *Diaphorina citri* (Homoptera: Psylloidea). *Zool. Sci.* 17:983–989.

Sugimoto, A., D. E. Bignell, and J. A. MacDonald. 2000. Global impact of termites on the carbon cycle and atmospheric trace gases, p. 409–435. *In* T. Abe, D. E. Bignell, and M. Higashi (ed.), *Termites: Evolution, Sociality, Symbioses, Ecology.* Kluwer Academic Publishers, Dordrecht, The Netherlands.

Tamas, I., L. Klasson, B. Canbäck, A. K. Näslund, A.-S. Eriksson, J. J. Wernegreen, J. P. Sandström, N. A. Moran, and S. G. E. Andersson. 2002. 50 Million years of genomic stasis in endosymbiotic bacteria. *Science* 296:2376–2379.

Taprab, Y., M. Ohkuma, T. Johjima, Y. Maeda, S. Moriya, T. Inoue, P. Suwanarit, N. Noparatnaraporn, and T. Kudo. 2002. Molecular phylogeny of symbiotic basidiomycetes of fungus-growing termites in Thailand and their relationship with the host. *Biosci. Biotechnol. Biochem.* 66: 1159–1163.

Thao, M. L. L., N. A. Moran, P. Abbot, E. B. Brennan, D. H. Burckhardt, P. Baumann. 2000. Cospeciation of psyllids and their primary prokaryotic endosymbionts. *Appl. Environ. Microbiol.* 66:2898–2905.

Thao, M. L. L., P. J. Gullan, and P. Baumann. 2002. Secondary (γ-*Proteobacteria*) endosymbionts infect the primary (β-*Proteobacteria*) endosymbionts of mealybugs multiple times and coevolve with their hosts. *Appl. Environ. Microbiol.* 68: 3190–3197.

Tokuda, G., N. Lo, H. Watanabe, M. Slaytor, T. Matsumoto, and H. Noda. 1999. Metazoan cellulase genes from termites: intron/exon structures and sites of expression. *Biochim. Biophys. Acta* 1447:146–159.

Tokuda, G., I. Yamaoka, and H. Noda. 2000. Localization of symbiotic clostridia in the mixed segment of the termite *Nasutitermes takasagoensis* (Shiraki). *Appl. Environ. Microbiol.* 66:2199–2207.

Tsuchida, T., R. Koga, H. Shibao, T. Matsumoto, and T. Fukatsu. 2002. Diversity and geographic distribution of secondary enädosymbiotic bacteria in natural populations of the pea aphid, *Acyrthosiphon pisum. Mol. Ecol.* 11:2123–2135.

von Dohlen, C. D., S. Kohler, S. T. Alsop, and W. R. McManus. 2001. Mealybug β-proteobacterial endosymbionts contain γ-proteobacterial symbionts. *Nature* 412:433–436.

Wernegreen, J. J. 2002. Genome evolution in bacterial endosymbionts of insects. *Nat. Rev. Genet.* 3:850–861.

Werren, J., G. Hurst, W. Zhang, W. J. Breeuwer, R. Stouthamer, and M. Majerus. 1994. Rickettsial relative associated with male killing in the ladybird beetle (*Adalia bipunctata*). *J. Bacteriol.* 176:388–394.

Werren, J. H., D. Windsor, and L. Guo. 1995. Distribution of *Wolbachia* among neotropical arthropods. *Proc. R. Soc. Lond. Ser. B* 262:197–204.

Wilson, E. O. 1992. *The Diversity of Life.* W. W. Norton & Company, New York, N.Y.

Wren, H. N., and D. G. Cochran. 1987. Xanthine dehydrogenase activity in the cockroach endosymbiont *Blattabacterium cuenoti* (Mercier 1906) Hollande and Favre 1931 and in the cockroach fat body. *Comp. Biochem. Physiol.* 88:1023–1026.

Yamin, M. 1979. Flagellates of the orders Trichomonadida Kirby, Oxymonadida Grassé, and Hypermastigida Grassi & Foà reported from lower termites (Isoptera families *Mastotermitidae, Hodotermitidae, Termopsidae, Rhinotermitidae,* and *Serritermitidae*) and from the wood-feeding roach *Cryptocercus* (Dictyoptera: Cryptocercidae). *Sociobiology* 4:1–119.

Microbial Diversity and Bioprospecting
Edited by Alan T. Bull
© 2004 ASM Press, Washington, D.C.

Chapter 20

Microbial Symbioses with Plants

PETER JEFFRIES

WHAT ARE PLANT-MICROBE SYMBIOSES?

Plant-microbe symbioses are common, ubiquitous, and very varied. The health and vigor of plants is dependent on the multifarious relationships they have with symbiotic microbes. These interactions are a rich source of novel microbial taxa and have provided a variety of microbes with economic value in the pharmaceutical, food, agricultural, and biotechnology industries. Plant-microbe symbioses occur on the aerial tissues, within the plant (endophytic relationships), and on or around the roots (the rhizosphere). The relationships are often specific and complex and occur in most plant families both aquatic and terrestrial. Many of these symbioses are ancient and essential for plants to survive in natural ecosystems, where critical nutrients are often limiting or pathogen populations are highly active. Coevolution of plants and their associated microbial symbioses have been a key driver of the evolutionary process. For example, there is good evidence that the symbiosis of most land plants with arbuscular mycorrhizal fungi was a critical factor allowing the evolution of primitive aquatic forms into the predecessors of today's terrestrial flora (Remy et al., 1994). In the broadest sense, symbiosis can be considered in the context of the original concept of De Bary (1887) in which two organisms live together, but for the purpose of this chapter, I have chosen to restrict the term to a narrower definition that excludes antagonistic interactions. There is also a problem in deciding whether loose associations of beneficial microbes in the phyllosphere and rhizosphere are truly symbiotic or are casual relationships. The fact that the rhizosphere supports a much greater diversity and abundance of microbes than the root-free soil indicates that beneficial associations are significant. However, it is much clearer to identify symbioses when microbes enter into endophytic associations of mutual benefit. Recently it has become acknowledged that many plants form such relationships and harbor beneficial microbes within both their root systems and their aerial tissues. In some cases, these relationships result in the development of characteristic structures such as root nodules or mycorrhizas. In many cases, however, it now seems that endophytic microbes exist in many plants in symptomless ways. As with mycorrhizal fungi, the coevolution of many plants and their associated endosymbionts has resulted in an interdependence that now means that neither partner can flourish alone. Endophytes are thus the rule rather than the exception in most plants.

The term endophyte needs definition and there is a great deal of heterogeneity in the plant-microbe interactions included. The term is used here to include all microbes colonizing plant tissues without causing any immediate, overt negative effects. This definition is a broader modification of that of Hirsch and Braun (1992) who referred only to fungi. Stone et al. (2001) reviewed some of the more common concepts of endophytism and discussed the fact that mutualism is not necessarily an assumed feature of endophytic relationships. The discovery by Freeman and Rodriguez (1993) that the pathogen *Colletotrichum magna* could be converted into a symptomless endophyte through mutation of a single gene showed that endophytes are not unusual organisms. This same discovery opened the way for exploitation in biological control. The endosymbiotic relationship has become a focus for search and discovery in that there is a need for complex cell-cell signaling between partners in order to maintain a stable relationship. Endophytes need to interact biochemically with their hosts so as to initiate the interaction, neutralize host defense reactions, and initiate new developmental pathways necessary to establish the host-endophyte relationship. The mediators of this communication will include secondary metabolites that can regulate eukaryotic cell function in order to facilitate the association. Roth et al. (1986) hypothesized that cell regulatory mechanisms in mammals evolved from mi-

Peter Jeffries • Research School of Biosciences, University of Kent, Canterbury CT2 6NJ, United Kingdom.

crobial sources. There are a number of fungal hormones, for example, involved in sexual reproduction and these have structural analogues in mammalian sex hormones. Given this existing evolutionary link, the need for common regulatory molecules within a symbiotic relationship is even more obvious. Yet the products of microorganisms colonizing multicellular eukaryotes as mutualists are still underresearched, and only limited genomic sequences of pathogenic fungi are currently accessible through the internet (Soanes et al., 2002). The availability of the draft sequence of the rice blast fungus, *Magnaporthe grisea*, may aid the identification of specific targets for screening less-antagonistic symbionts.

Useful products have resulted from the investigation of plant-microbe symbioses. In particular, the numbers of fungi that grow with vascular plants are enormous and offer a rich pool of diversity. Detailed investigations of the fungi associated with specific plants always yield a high proportion of novel species. For example, 22% of the species isolated from a sedge community in the Alps were novel (Nograsek, 1990). Hawksworth (1991) suggests that similar figures in the range 20 to 49% would be typical. The classic example is the interaction of *Taxomyces andreanae* and *Pestalotiopsis microspora*, along with several other fungi, that occur within the bark of yew trees and are potential sources of the anticancer drug Taxol. There is also a wide range of secondary metabolites with eukaryotic cell activity that are produced by the well-researched grass endophytic fungi. Because of their symptomless nature, our knowledge of the diversity and frequency of many endophytes is limited, but there is now good evidence to suggest that the genetic diversity is very wide and a rich source of new species. An endophyte with more overt symptoms is *Claviceps purpurea*, the ergot fungus, a well-established industrial source of biopharmaceuticals. The alkaloids produced act as agonists of dopamine toward different receptor sites. The most obvious biotechnological applications of endophytes include their use as sources of new pharmaceuticals, hormones, and other cell regulators. However, they can also be exploited as biocontrol agents or as plant growth stimulators and as agents to protect plants from abiotic stress. During the past 10 years, over 80 biocontrol products have been marketed worldwide (Paulitz and Bélanger, 2001), and the increased environmental pressure for sustainable plant protection has consolidated this market for the future. Plant-microbe symbioses have also been exploited in programs of ecosystem restoration, particularly in desertified areas. Loss of plant communities through natural or anthropogenic activities leads to loss of associated beneficial microbes. Hence any programs designed to restore natural plant communi-

ties must also consider augmentation of the indigenous microbiota such that the necessary symbiotic inoculum potential is elevated to levels that can sustain the regenerating plant cover. Leguminous plants in particular are important pioneer plants in ecosystem restoration, and dual inoculation schemes using combinations of rhizobial bacteria and mycorrhizal fungi are being piloted (Requena et al., 2001). The most appropriate isolates to exploit in such schemes are to be found in similar natural ecosystems in the local geographical area as these isolates are adapted to prevailing edaphic conditions.

DIVERSITY WITHIN PROKARYOTIC SYMBIOSES

Plant Growth-Promoting Bacteria

The existence of plant growth-promoting rhizobacteria (PGPR) is well established (see Lynch, chapter 34, this volume). These bacteria stimulate plant growth directly or indirectly by suppressing plant pathogens or by increasing nutrient supplies, alleviating heavy metal toxicity, or affecting phytohormone balances. A variety of species are involved but most belong to the ubiquitous genera *Pseudomonas* and *Bacillus*. The former genus in particular has been subject to dramatic reorganization as a result of molecular phylogenetics, and plant-associated species are now allocated to at least four additional genera: *Acidovorax*, *Burkholderia*, *Herbaspirillum*, and *Ralstonia*. Rhizobia are dealt with separately as relationships are complex, but a number of other PGPR stimulate plant growth through N^2 fixation including *Azospirillum*.

Rhizobia

Rhizobia are the bacteria that associate with leguminous hosts to form nitrogen-fixing nodules. Molecular phylogenetic studies have supported the classic division of these bacteria into six genera: *Rhizobium*, *Bradyrhizobium*, *Mesorhizobium*, *Sinorhizobium*, *Allorhizobium*, and *Azorhizobium* (Coutinho et al., 2000). These include a total of 26 species, which contrasts sharply with the four species recognized earlier (Jordan, 1984) on the basis of nonmolecular approaches. The exploitation of this group has been mainly as promoters of the growth of important leguminous crop plants, particularly soybean, and a number of commercial products are available. Rhizobia have also been used in the restoration of degraded ecosystems (Requena et al., 2001) where soil erosion has resulted in a diminution of indigenous inoculum.

Endophytic Bacteria

Prokaryotic endophytes of plants are sometimes considered to be less common or less important than fungal endophytes (Stone et al., 2000). However, healthy-appearing plant tissues are often colonized by inter- or intracellular bacteria, and roots also form relationships with endophytic bacteria. The most well-known examples are the stem and root nodules formed by rhizobial N^2 fixing bacteria with their leguminous plant hosts.

There are, however, many other covert examples of such mutualistic relationships. Sugar cane in some parts of Brazil has apparently been grown for many years without the addition of nitrogen thanks to the presence of an N^2-fixing endophyte *Acetobacter diazotrophicus* in roots, stems, and leaves (Boddey et al., 1991).

Bacterial Biocontrol Agents

The rhizosphere provides a rich niche for active microbial populations to compete. This often involves antagonistic interactions. This can be exploited for the discovery of agents for biocontrol of root diseases. Multiple microbial interactions involving both fungi and bacteria have been demonstrated to show enhanced biocontrol activity in comparison with specific agents used singly (Whipps, 2001). Many of the bacteria involved produce antibiotics, and the production of phenazine-1-carboxylic acid by several *Pseudomonas* spp. has proved a model system for study (Keel and Défago, 1997). The regulation and signaling involved in production of this compound are well understood. Whipps (2001) provides a useful summary of other bacteria exploited for biocontrol of fungal plant pathogens.

DIVERSITY WITHIN EUKARYOTIC SYMBIOSES

Ectomycorrhizal Fungi

Many soil fungi form joint organs with plants called mycorrhizas. The easiest to observe are ectomycorrhizas formed on the short roots of most temperate forest trees (Fig. 1 and 2). Most ectomycorrhizal fungi (ECMF) (some 5,000 genera) come from the *Basidiomycota*, although several genera of *Ascomycota* are also involved. The molecular systematics of this group is rapidly advancing (Jeffries and Dodd, 2000) and thousands of species are involved. Several compounds with pharmacological activity have been established from ECMF. The antitumor properties of *Lentinus edodes* ("shiitake") have long

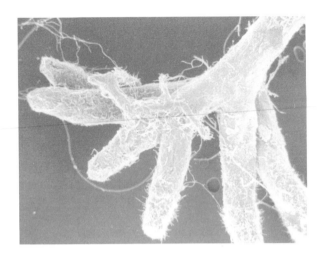

Figure 1. Ectomycorrhiza showing typical dichotomously branched short roots sheathed by fungal hyphae. Scanning electron micrograph; magnification; ×40. (S. Elphick and P. Jeffries, unpublished.)

been known, and extracts of the fruit bodies contain a range of pharmacologically active metabolites. However, the *Basidiomycota* are often challenging to grow in the laboratory and thus present a problem for pharmacological research. Field collection of large amounts of material is problematical and unpredictable. Nevertheless, the complexity of their cytodifferentiation and the uniqueness of the dikaryotic growth form suggest that they have enormous potential in drug discovery (Nisbet and Fox, 1991).

Some ECMF have evolved unique spore dispersal mechanisms whereby mammalian sex hormone analogues are used to attract wild rodents to dig up subterranean fruit bodies and carry them away. In contrast, others produce toxic metabolites to discour-

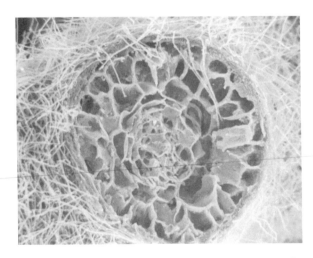

Figure 2. Cross section through ectomycorrhizal root showing fungal sheath surrounding root. Scanning electron micrograph; magnification, ×250. (S. Elphick and P. Jeffries, unpublished.)

age animal feeding, for example, the acetylcholine esterase inhibitors of *Amanita* spp.

Arbuscular Mycorrhizal Fungi

The most widespread symbiosis between soilborne fungi and plants is the arbuscular mycorrhizal symbiosis, with over 90% of land plants able to enter into this mutualsitic relationship. In contrast, the range of taxa of fungi involved is small, with a single order, the *Glomales*, comprising only six genera and around 150 species (and many of these are probably conspecific). However, these unique and ancient organisms display considerable interspecific genomic diversity at the level of the individual nuclei (Kuhn et al., 2001). Each mycelium or spore contains many thousands of nuclei that may differ from one another in key markers of phylogenetic diversity, e.g., the ribosomal genes (Rodriguez et al., 2001; Jansa et al., 2002). These fungi have been exploited as promoters of plant growth and health, and it is clear that phylogenetic markers do not give any clear guidance as to the potential benefits of a particular isolate in this context. Locally adapted strains of arbuscular mycorrhizal fungi (AMF) are usually the most appropriate source of isolates for use to stimulate plant growth in a particular ecosystem (Requena et al., 2001). The restoration of desertified Mediterranean ecosystems has been one area where AMF bring a range of benefits (Jeffries et al., 2002), including an improvement in nutrient uptake, alleviation of water stress, and protection against abiotic and biotic stress. These fungi have also been used to stimulate the development of natural plant communities on virgin soils. For example, inoculation with a selection of a mixture of indigenous strains of AMF was shown to improve survival rates and increase the fecundity of transplanted seedlings of *Elymus* on the reclamation platform created during the building of the Channel Tunnel between France and the United Kingdom (Dodd et al., 2002).

Other Mycorrhizal Fungi

The EMCF and AMF comprise the largest and most significant groups of mycorrhizal symbionts, but there are some other specific groups to note. Smith and Read (1997) give these groups a comprehensive treatment. All orchids are achlorophyllous in the early seedling stages and have an obligate mycorrhizal relationship with certain soil fungi from which they gain C-compounds. Many of these fungi belong to the form genus *Rhizoctonia*, and several have now been allied with other genera as a result of molecular analyses. In addition, several other hyphomycete gen-

era have been implicated as symbionts, along with a variety of basidiomycete species including the pathogen *Armillaria mellea*.

A second group of plants forming unique mycorrhizas is the Ericales. Considerable diversity of structures are formed, but in some cases the fungi are similar to those forming other types of mycorrhiza, which suggests that the plant plays an important role in regulation of development. Arbutoid mycorrhizas are formed between *Arbutus* and its relatives and several fungi which also form ectomycorrhizas, including *Hebeloma, Pisolithus, Rhizopogon,* and *Thelephora* species. The achlorophyllous genus *Monotropa* forms an unusual mycorrhizal type ("monotropoid") with a variety of fungal species, and molecular approaches have aided identification of *Suillus, Rhizopogon,* and *Russula* species. Finally there are also "ericoid" mycorrhizas, formed by members of the *Ericaceae* and *Epacridaceae,* in which the fungal partner is commonly *Hymenoscyphus ericae*. Other genera, such as *Oidiodendron,* have also been implicated. Again, molecular phylogenetics has helped elucidate the systematics of these symbionts.

Lichens

Lichens are formed from the stable mutualistic symbiosis of algae or cyanobacteria and certain fungi. There is a large range of such associations, and the fungal partners, or mycobionts, are most frequently from the *Ascomycota*. It is estimated that over 13,500 fungal species may take part in lichen symbioses, many of which are obligate (Hawksworth, 1988). They are prolific producers of secondary metabolites, with over 350 characterized (Nisbet and Fox, 1991). Numerous compounds isolated from lichen fungi have proven to have useful activities, such as polysaccharides with antitumor properties or orcinols and lactone derivatives with antibacterial action. The most widely used antibiotic from lichen sources is probably usmic acid (Katz, 2002). In addition, the fungi that grow on lichens (lichenicolous fungi) are also both numerous and diverse and a significant potential source of novel taxa (Hawksworth, 1991).

Plant Growth-Promoting Fungi

Many fungi that grow in the rhizosphere, such as PGPR, can stimulate plant growth in the absence of pathogen challenge. This phenomenon has been shown for *Trichoderma* spp., binucleate *Rhizoctonia* spp., and *Pythium* spp., but, in most cases, the mechanism is unknown and might provide an avenue for the discovery of novel products. In the case of *Trichoderma harzianum* 1295-27, the stimulatory activity was due to increased solubilization of the phosphate and micronutrients (see Lynch, chapter 34, this volume).

Endophytic Fungi of Leaves

Monaghan et al. (1995) estimated that at a rate of two to three unique endophytic fungi per plant species, there may be about 750,000 endophytic fungi that exist worldwide as sources of novel secondary metabolites. These authors concentrated on one genus, *Nodulisporium*, to determine the diversity of secondary metabolites produced by the endophytic isolates. Their study suggested that the major portion of shared metabolites produced by a particular taxon could be found by screening just a few cultures. However, the occurrence of novel metabolites was unpredictable, and even discarding morphologically similar isolates may lead to loss of novel products (Monaghan et al., 1995). This is a common phenomenon often ignored by microbiologists—each species comprises many individuals each of which may behave differently. A similar concentrated study of endophytes from a restricted site in Spain also found large differences among isolates from the same species with respect to their ability to produce metabolites with antimicrobial activity (Peláez et al., 1998). Strains isolated from the same habitat are not always similar and may show considerable differences (Arenal et al., 2002).

Phyllosphere Colonizers

The leaf surface is a harsh and changeable environment, and interactions between microbes are critical for survival. This aspect of ecology can be exploited in the search for compounds that inhibit the growth of other microbes or for the isolation of biocontrol agents of foliar plant disease. The production of antimicrobial compounds by *Trichoderma* spp., for example, is well documented, but more recently attention has been given to phyllosphere-inhabiting yeasts (Urquhart and Punja, 2002), particularly when sensitive methodology is used to detect production of antagonistic compounds. A similar novel methological approach was used by McCormack et al. (1994) to show that a wide variety of yeasts and yeastlike fungi produce antibacterial compounds, in contrast to their soilborne counterparts.

Fungal Biocontrol Agents

In addition to the bacterial examples of antagonistic interactions in the rhizosphere, there are also many of which involve soil fungi. These can be exploited for biocontrol through induced resistance, antibiosis and/or mycoparasitism (Jeffries and Young, 1996). A wide variety of fungal genera are involved, and a useful summary has been provided by Whipps (2001) and will not be discussed further here.

PROSPECTS FOR EXPLOITATION

World estimates of the numbers of undescribed species of microbes reinforce the potential for bioprospecting (see chapters 2 and 7). A high proportion of these undescribed species will form loose or strong symbiotic relationships with plants. Bioprospective strategies should include a component in which detailed inventory of plant-associated microbes in the target area is conducted. Many of these associations will be species specific, and thus each plant species will offer novelty in terms of the microbial spectrum revealed. Alternatively, some microbial products will occur across a wide variety of common microbial isolates. For example, synthesis of compactin or mevinolin is found in plant-associated genera *Aspergillus, Doratomyces, Eupenicilium, Gymnoascus, Hypomyces, Monascus, Paecilomyces, Penicillium, Phoma,* and *Trichoderma* (Monaghan and Tkacz, 1990). In all cases, screening of isolates should be based on infraspecific criteria to accomodate the diversity found within conventional species groups. Genotypic techniques tend to be more powerful for the discrimination and dereplication of strains in this respect than phenotypic ones (Arenal et al., 2002; Brandão et al., 2002). Subsequent laboratory screens for useful properties must then take into account the ecology of the interaction, and care must always be exercised in interpretation. For example, over 100 ectomycorrhizal fungi have been reported to produce antibiotics in culture, yet Rasanayagam and Jeffries (1992) showed that many of these effects could be due to pH changes in the medium induced by fungal growth rather than due to the production of secondary metabolites.

As mentioned above, the mutation of a pathogenic strain of *Colletotrichum magna* into a nonpathogenic, mutualistic endophyte has enabled this fungus to be used to control diseases of cucurbits and tomatoes (Freeman and Rodriguez, 1993; Redman et al., 2002). The mutant colonizes host tissues and is presumed to prime host defense mechanisms such that the plant becomes immune to further attack. A similar principle has been used in the exploitation of nonpathogenic strains of *Fusarium oxysporum* for the control of root diseases of several crops (Alabouvette et al., 1993). A commercial formulation, Fusclear, is available to farmers.

The diversity of plant-microbe interactions and the potential of novel products cannot be fully realized without a thorough understanding of systematics. The shift in phylogenetic approaches emphasized in this book has opened a myriad of opportunities for new discoveries. Loss of plant diversity is a potential concern for the future. In addition to the loss of macrobiota and fauna, the anthrogenic destruction of natural ecosystems will result in the loss of the species-specific microbial symbioses associated with particular hosts (see chapter 37, this volume). Thus conservation strategies are significant for their retention of plant-microbial symbioses as well as for their intrinsic protection of the plants themselves.

REFERENCES

Alabouvette, C., P. Lemanceau, and C. Steinberg. 1993. Recent advances in the biological control of *Fusarium* wilts. *Pesticide Sci.* 37:365–373.

Arenal, F., G. Platas, J. Martin, F. J. Asensio, O. Salazar, J. Collado, F. Vicente, A. Basilio, C. Ruibal, L. Royo, N. De Pedro, and F. Peláez. 2002. Comparison of genotypic and phenotypic techniques for assessing the variability of the fungus *Epicoccum nigrum. J. Appl. Microbiol.* 93:36–45.

Boddey, R. M., S. Urquiaga, and V. Reis. 1991. Biological nitrogen fixation associated with sugarcane. *Plant Soil* 137:111–117.

Brandão, P. F. B., M. Torimura, R. Kurane, and A. T. Bull. 2002. Dereplication for biotechnology screening: PyMS analysis and PCR-RFLP-SCCP (PRS) profiling of 16S rRNA genes of marine and terrestrial actinomycetes. *Appl. Microbiol. Biotechnol.* 58:77–83.

Coutinho, H. L. C., V. M. De Oliveira, and F. M. S. Moreira. 2000. Systematics of legume nodule nitrogen fixing bacteria, p. 107–134. *In* F. G. Priest and M. Goodfellow (ed.), *Applied Microbial Systematics.* Kluwer Academic Publishers, Dordrecht, The Netherlands.

De Bary, A. 1887. *Comparative Morphology and Biology of the Fungi, Mycetozoa and Bacteria.* Oxford University Press, Oxford, United Kingdom.

Dodd, J. C., T. A. Dougall, J. P. Clapp, and P. Jeffries. 2002. The role of arbuscular mycorrhizal fungi in plant community establishment at Samphire Hoe, Kent, UK—the reclamation platform created during the building of the Channel tunnel between France and the UK. *Biodiv. Consery* 11:39–58.

Freeman, S., and R. Rodriguez. 1993. Genetic conversion of a fungal plant pathogen to a nonpathogenic, endophytic mutualist. *Science* 260:75–78.

Hawksworth, D. L. 1988. Coevolution of fungi with algae and cyanobacteria in lichen symbioses, p. 125–149. *In* K. A. Pirozynski and D. L. Hawksworth (ed.), *Coevolution of Fungi with Plants and Animals.* Academic Press, London, United Kingdom.

Hawksworth, D. L. 1991. The fungal dimension of biodiversity: magnitude, significance, and conservation. *Mycol. Res.* 95:641–655.

Hirsch, G., and U. Braun. 1992. Communities of parasitic microfungi, p. 225–250. *In* W. Winterhoff (ed.), *Handbook of Vegetation Science*, vol. 19, *Fungi in Vegetation Science.* Kluwer Academic, Dordrecht, The Netherlands.

Jansa, J., A. Mozafar, S. Banke, B. A. McDonald, and E. Frossard. 2002. Intra- and intersporal diversity of ITS rDNA sequences in *Glomus intraradices* assessed by cloning and sequencing, and by SSCP analysis. *Mycol. Res.* 106:670–681.

Jeffries, P., and T. W. K. Young. 1996. *Interfungal Parasitic Relationships.* CAB Publishing, Wallingford, United Kingdom.

Jeffries, P., and J. C. Dodd. 2000. Molecular ecology of mycorrhizal fungi, p. 73–103. *In* F. G. Priest and M. Goodfellow (ed.), *Applied Microbial Systematics.* Kluwer Academic, Dordrecht, The Netherlands.

Jeffries, P., A. Craven-Griffiths, J. M. Barea, Y. Levy, and J. C. Dodd. 2002. Application of AMF in the revegetation of desertified ecosystems, p. 151–174. *In* S. Gianinazzi, H. Schüepp, J. M. Barea, and K. Haselwandter (ed.), *Mycorrhizal Technology in Agriculture: from Genes to Bioproducts.* Birkhäuser Verlag, Basel, Switzerland.

Jordan, D. C. 1984. *Rhizobiaceae*, 234–256. *In* N. R. Krieg and J. G. Holt (ed.), *Bergey's Manual of Systematic Bacteriology*, vol. 1. Williams & Wilkins, Baltimore, Md.

Katz, S. 2002. Beneficial uses of plant pathogens: anticancer and drug agents derived from plant pathogens. *Can. J. Plant Pathol.* 24:10–13.

Keel, C., and G. Défago. 1997. Interactions between beneficial soil bacteria and root pathogens: mechanisms and ecological impact, p. 27–47. *In* A. C. Gange and V. K. Brown (ed.), *Multitrophic Interactions in Terrestrial Ecosystems.* Blackwell Science, Oxford, United Kingdom.

Kuhn, G., M. Hijri, and I. R. Sanders. 2001. Evidence for the evolution of multiple genomes in arbuscular mycorrhizal fungi. *Nature* 414:745–748.

McCormack, P. J., H. G. Wildman, and P. Jeffries. 1994. Production of antibacterial compounds by phylloplane-inhabiting yeasts and yeastlike fungi. *Appl. Environ. Microbiol.* 60:927–931.

Monaghan, R. L., and J. S. Tkacz. 1990. Bioactive microbial products: focus on mechanism of action. *Annu. Rev. Microbiol.* 44:271–301.

Monaghan, R. L., J. D. Polishook, V. J. Pecore, G. F. Bills, M. Nallin-Omstead, and S. L. Streicher. 1995. Discovery of novel secondary metabolites from fungi—is it really a random walk through a random forest? *Can. J. Bot.* 73:S925–S931.

Nisbet, L. J., and F. M. Fox. 1991. The importance of microbial biodiversity to biotechnology, p. 229–244. *In* D. L. Hawksworth (ed.), *The Biodiversity of Microorganisms and Invertebrates: Its Role in Sustainable Agriculture.* CAB International, Wallingford, United Kingdom.

Nograsek, A. 1990. Ascomyceten auf Gefäßplanzen der Polsterseggenrasen in den Osteralpen. *Bibliotheca Lichenologica* 30:1–271. (Cited in Hawksworth [1991].)

Paulitz, T. C., and R. R. Bélanger. 2001. Biological control in greenhouse systems. *Annu. Rev. Phytopathol.* 31:103–133.

Peláez, F., J. Collado, F. Arenal, A. Basilio, A. Cabello, M. T. Diez Matas, J. B. Garcia, A. González Del Val, V. González, J. Gorrochategui, P. Hernández, I. Martin, G. Platas, and F. Vicente. 1998. Endophytic fungi from plants living on gypsum soils as a source of secondary metabolites with antimicrobial activity. *Mycol. Res.* 102:755–761.

Rasanayagam, S., and P. Jeffries. 1992. Production of acid is responsible for antibiosis by some ectomycorrhizal fungi. *Mycol. Res.* 96:971–976.

Redman, R. S., M. J. Roossinck, S. Maher, Q. C. Andrews, W. L. Schneider, and R. Rodriguez. 2002. Field performance of cucurbit and tomato plants colonized with a nonpathogenic, mutualistic mutant (path-1) of *Colletotrichum magna* (Teleomorph: *Glomerella magna*; Jenkins & Winstead). *Symbiosis* 32:55–70.

Remy, W., T. N. Taylor, H. Haas, and H. Kerp. 1994. Four hundred-million-year-old vesicular-arbuscular mycorrhizae. *Proc. Natl. Acad. Sci. USA* 91:11841–11843.

Requena, N., E. Perez-Solis, C. Azcón-Aguilar, P. Jeffries, and J. M. Barea. 2001. Management of indigenous plant-microbe symbioses aids restoration of desertified ecosystems. *Appl. Environ. Microbiol.* 67:495–498.

Rodriguez, A., T. Dougall, J. C. Dodd, and J. P. Clapp. 2001. The large subunit ribosomal RNA genes of *Entrophospora infrequens* comprise sequences related to two different glomalean families. *New Phytol.* **152:**159–167.

Roth, J., D. Leroith, E. S. Collier, A. Watkinson, and M. A. Lesnisk. 1986. The evolutionary origins of intercellular communication and the Maginot Lines of the mind. *Ann. N.Y. Acad. Sci.* **436:**1–11.

Smith, S. E., and D. J. Read. 1997. *Mycorrhizal Symbiosis*, 2nd ed. Academic Press, London, United Kingdom.

Soanes, D. M., W. Skinner, J. Keon, J. Hargreaves, and N. J. Talbot. 2002. Genomics of phytopathogenic fungi and the development of bioinformatic resources. *Mol. Plant Microbe Interact.* **15:**421–427.

Stone, J. K., C. W. Bacon, and J. F. White. 2000. An overview of endophytic microbes: endophytism defined, p. 3–29. *In* C. W. Bacon and J. F. White (ed.), *Microbial Endophytes.* Marcel Dekker, Inc., New York, N.Y.

Urquhart, E. J., and Z. K. Punja. 2002. Hydrolytic enzymes and antifungal compounds produced by *Tilletiopsis* species, phyllosphere yeasts that are antagonists of powdery mildew fungi. *Can. J. Microbiol.* **48:**219–229.

Whipps, J. M. 2001. Microbial interactions and biocontrol in the rhizosphere. *J. Exp. Bot.* **52:**487–511.

IV. BIOGEOGRAPHY AND MAPPING MICROBIAL DIVERSITY

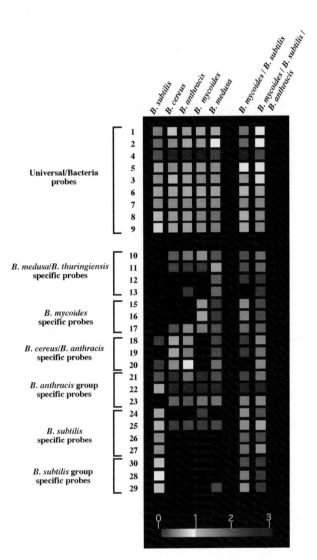

Color Plate 1 (Chapter 5). Low-density, DNA microchip fingerprints. 16S rRNA from single bacterial targets and as cocultures (each vertical column) were labeled with a fluorescent dye (Texas red) and hybridized with the oligonucleotide probles indicated by numbered rows. Melting curves were determined on the stage of a custom-made epifluorescence microscope equipped with appropriate fluorescence filters and a cooled charge-coupled device camera and controlled by a computer for image acquisition (for more details, see Liu et al., 2001). High levels of hybridization are indicated by red to pink to white as indicated by the scale at the bottom of the chip (from Liu et al., 2001, with permission).

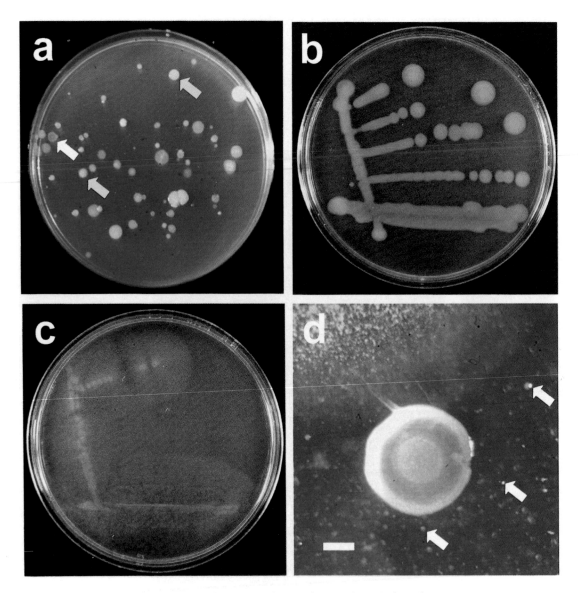

Color Plate 2 (Chapter 8). Isolation of aquatic bacteria from the *Bacteroidetes* phylum (previously *Cytophaga-Flexibacter-Bacteroides* [CFB]) from the River Taff in Cardiff, Wales, by plating onto agar. All plates were incubated for 7 days at 20°C. (a) Bacterial colonies on CPS agar (low organic carbon, 2.5 g of organic compounds per liter). Some potential CFB isolates that were identified as yellow, orange, red, or spreading colonies are indicated by the yellow arrows. (b) Isolate EP293 growing as large yellow colonies on plate count agar (high organic carbon, 8.5 g liter^{-1}), characterized by 16S rDNA sequencing as a possible new genus of the family *Flexibacteriaceae* most closely related phylogenetically to *Flectobacillus major* (93% sequence similarity). (c) Isolate EP233 growing as spreading, yellow colonies on CYT agar (low-organic carbon, 2 g liter^{-1}), characterized by 16S rDNA sequencing as a possible new species of *Flavobacterium* (family *Flavobacteriaceae*). (d) Very small colonies (≤0.1 mm in diameter) growing on CPS agar (some indicated by white arrows). A large doughnut-shaped colony (3.3 mm in diameter; similar to the one indicated in panel a by a white arrow) gives a size comparison; bar, 1 mm. Taxonomic names are given according to the latest Bergey's taxonomic outline (Garrity et al., 2002). Panels b and c are used with permission from Louise A. O'Sullivan, Cardiff University.

Color Plate 5 (Chapter 23). Composite maps can be constructed online in real time by clicking on the digital "layers" that one wants displayed. In this example, the base map, general locations, thermal features, major hydrography, lakes, and the boundary of YNP have been selected as the "visible" layers. Thermal features have been selected as the "active" layer from which simple queries can be performed.

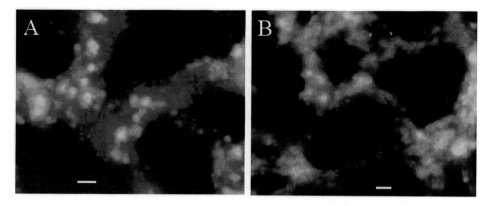

Color Plate 3 (Chapter 18). Epifluorescence micrograph of cryosections of *R. odorabile* visualized by FISH with Cy3-labeled probes specific for (A) all bacteria and (B) the *Cytophaga/Flavobacterium* group. Details are given by Webster et al. (2001). The scale bars are 10 μm.

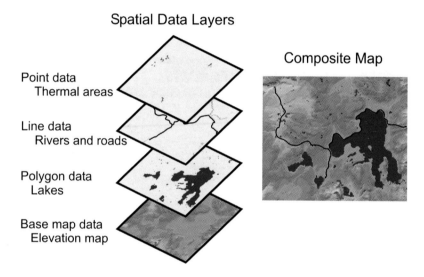

Color Plate 4 (Chapter 23). Maps are constructed from multiple digital layers. Base maps can be overlaid with polygon data, which represent the location of lakes and geothermal basins, line data indicating the location of rivers, roads, streams, etc., and point data, which indicate the location of springs or sampling areas. In this example, the topographical map is overlaid with the location of rivers, roads, and thermal areas.

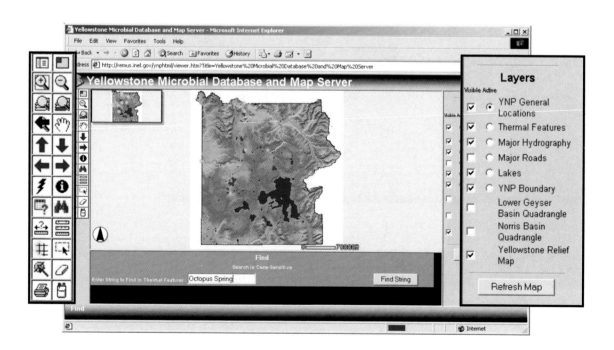

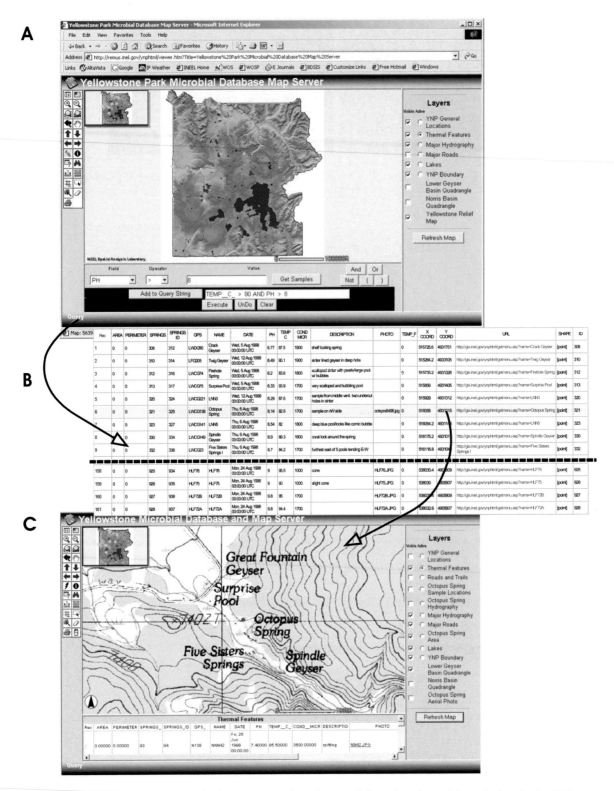

Color Plate 6 (Chapter 23). Multiple attributes can be used to search for springs that satisfy particular criteria. (A) A query was performed to search for springs that have a temperature greater than 80°C and a pH greater than 8. (B) The resulting data table lists all the springs that match these criteria. (C) A thermal spring (Octopus Spring) is selected from the list and the system zooms in on the location and launches a summary data table.

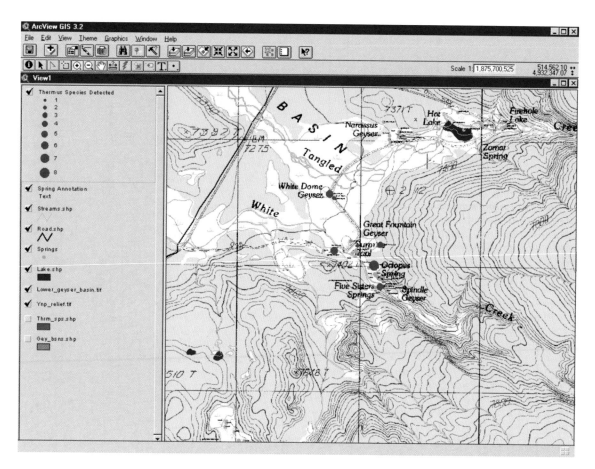

Color Plate 7 (Chapter 23). The results of a query, "Where have members of the genus *Thermus* been detected?" displayed in map format. Red dots indicate the locations where *Thermus* had been detected, and the size of the dot is proportional to the numbers of species or strains detected.

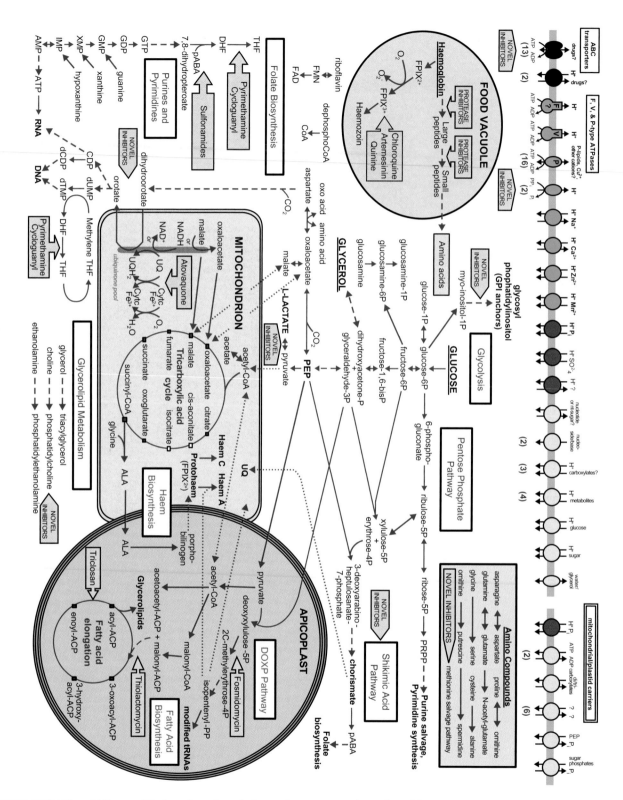

Color Plate 8 (Chapter 25). Overview of metabolism and transport in *Plasmodium falciparum*. (From Gardner et al. [2002] with permission.)

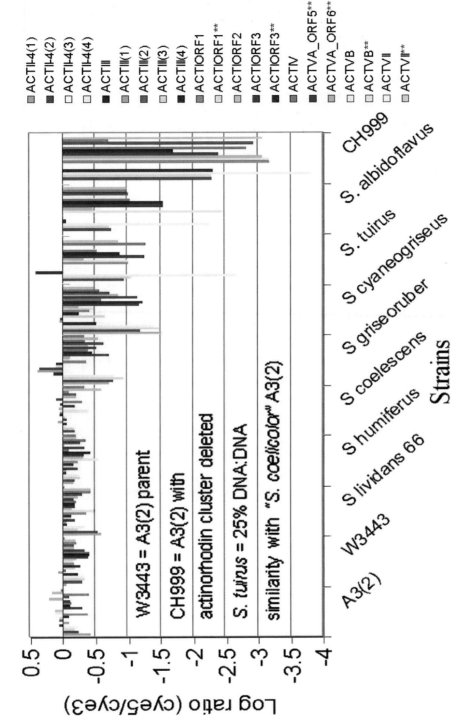

Color Plate 9 (Chapter 28). Presence of genes in the actinorhodin cluster of "*S. coelicolor*" A3(2) in the core genes of members of the *S. violaceoruber* genome species.

Color Plate 10 (Chapter 34). Increase in lettuce growth following treatment with *T. harzianum* at 0.1% (L) and 1% (H) (wt/wt) in the potting compost.

Color Plate 11 (Chapter 34). Increase in marigold flowering following treatment with *T. harzianum* at 1% (wt/wt) in potting compost.

PREAMBLE

The recognition of spatial scale as a determining factor in microbial ecology was discussed in chapter 7; at the global scale, the geographic distribution of organisms is known as biogeography. One of the questions we consider in this section is whether the concept of biogeographical distribution is apposite in the microbial world; in other words, how valid is the Beijerinck-Bass-Becking assertion that everything is everywhere? Bland Finlay and Genoveva Esteban (chapter 21) and Brian Hedlund and Jim Staley (chapter 22) present evidence and arguments that bear on a cosmopolitan or an endemic distribution of microorganisms. At the outset, Finlay and Esteban stress the need for a clear definition of terms: they refer to *ubiquitous dispersal* as the process of continuous worldwide dispersal of members of the same species, whereas *cosmopolitan* implies that a particular species can grow at a worldwide scale; an *endemic* organism, on the other hand, is restricted to a certain biogeographic region or locality within such a region. These authors conclude that ubiquitous dispersal appears to be common to most, if not all, microorganisms, and that there is little or no unequivocal evidence in support of microbial endemism. One corollary following from this conclusion is that microbial speciation due to geographic isolation will be rare and lead to low global species richness; the implication for bioprospecting is obvious.

Hedlund and Staley adopt a more circumspect position regarding possible biogeographic distribution of microorganisms. They argue that the everything is everywhere proposition has not been rigorously tested by molecular biological methods and that it would be premature to discount the possibility of endemic microorganisms. Accordingly, Staley and Gosink (1999) have proposed a set of postulates, based primarily on sequencing two or more genes, for evaluating bacterial biogeography. Both authors opine that absolute proof of endemism may not be possible because it requires proof of a negative—"Not finding an organism does not mean it is not there" (Staley and Gosink, 1999). However, should endemism be demonstrated convincingly among microorganisms, threats for their extinction (and actions for their conservation) become real and the implications for biotechnology prospecting again become obvious. The important outcome of proposing biogeographic criteria, as Staley (1999) urged, is to stimulate research in this area of microbial diversity.

So, is there anything more that is worth saying on the subject at this time? The relatively few publications that refer to biogeographic distribution issues in microbiology tend to do so only tangentially. Moreover, the designation of organisms as cosmopolitan or endemic will be strongly influenced by the procedures used to circumscribe them and, hence, the taxonomic rank at which the distinction is made. Staley and Gosink (1999) emphasized the need for phylogenetic analyses in which several appropriate genes are sequenced and noted that 16S rDNA sequences do not appear to provide sufficient resolving power. For example, an interesting study of actinomycete distribution (Rheims et al., 1996) revealed that three particular phylogroups were ubiquitously dispersed, but this conclusion rests only on short 16S rDNA clone sequences. More recent investigations have started to deploy combination molecular analyses in order to achieve greater taxonomic resolution. Laura Katz and her colleagues have characterized polymorphisms in the internally transcribed spacer regions (ITS-1, ITS-2) and sequenced α-tubulin and 5.8S rRNA genes to explore spatial and temporal diversity among planktonic ciliates. They conclude that, at least for some ciliate populations, biogeographic separation does exist, that ciliate diversification is an ongoing process, and that genetic heterogeneity underlies some very similar morphospecies (Snoeyenbos-West et al., 2001, 2002).

The most extensive and compeling studies to date, however, have focused on fluorescent *Pseudomonas* spp. (Cho and Tiedje, 2000) and *Sulfolobus* "*islandicus*" (Whitaker et al., 2000, cited in chapter 22) to test the cosmopolitan versus endemism hypothesis. Phylogenetic analyses were made on nearly 250 isolates from five continents, the methods being chosen to provide different degrees of resolution (16S rDNA restriction [ARDRA]), ITS-restriction fragment length polymorphism (ITS-RFLP), and repetitive extragenic palindromic fingerprinting with a BOX primer set (BOX-PCR). ARDRA profiles were very similar; ITS-RFLP data suggested weak endemicity at this level of resolution; BOX-PCR revealed that 35% of the isolates could be designated as strict site-endemic strains. It could be argued that the case for endemism here is being pushed to extremes by resorting to such a fine level of resolution. However, as we have reasoned earlier in this volume, circumscription at this intraspecific scale is fully justified—and necessary—for biotechnology prospecting. The Whitaker and Tiedje approaches should certainly encourage further, well-defined studies of biogeographic distribution.

Two final points might be made regarding microbial distribution. First, although the arguments have centered around free-living organisms, symbioses also need to be considered, some of which, such as mycorrhizas, are known to have evolved in the Devonian period (~400 BP). As Halling (2001) remarks, the existence of obligate fungus-plant specificities in mycorrhizal associations has a profound consequence on biogeographic distribution, whereas similar coevolutionary events have occurred in arthropod-bacteria endosymbiont interactions (for example, cockroaches [Clark et al., 2001]) with comparable biogeographic consequences. Turning to corals, 18 centers of reef endemism are recognized, and, as stated in chapter 7, they should be prime targets for testing the microbiological hot-spot hypothesis and for looking for endemic microbial symbionts. Second, we might prioritize refugia, i.e., those locations unaffected by large-scale environmental changes to the surrounding area in which previous biota are preserved, as targets for both microbial endemism studies and for microbial prospecting. Deep waters, especially isolated ocean trenches, polar areas including submarine lakes (see chapter 13), and mountains unaffected by Pleistocene glaciation, would be among the candidate refugia worth prospecting. Additionally, modern analogues of geological refugia, such as deep-sea microbial mats, which are widespread and known to harbor novel prokaryotic symbionts (Bernhard et al., 2000), also provide prime opportunities for biogeographic research.

Discussion of biogeography leads logically to the mapping of biodiversity. In essence a map is a representation of an area, and as such fauna and flora distribution maps are widely available. The recently published *World Atlas of Biodiversity* (Groombridge and Jenkins, 2002) provides an overview of the current state of global biodiversity, and it uses maps to chart biodiversity distribution, to highlight human impact on the environment, and to review planning and management intervention in response to such impact. The *World Atlas* makes no attempt to map the diversity, distribution, or ecosystem function of microorganisms, and this in turn prompts questions of the need for and the type of mapping that might be worth undertaking. Need could be justified in terms of biodiversity conservation and value for biotechnology prospecting; the type of mapping, on the other hand, is much more problematic and is concerned with how to turn given databases into a readable map and with issues of data quality. In a critical discussion of this topic, Geoffrey Bowker concluded that "there is no simple metadata solution to the problems of integrating information from multiple scientific disciplines in biodiversity and thus to problems of adopting GIS for producing maps of biodiversity" (Bowker, 2000). GIS (geographic information systems) is a computerized system for organizing and analyzing any spatially related collection of information and data sets; in the context of biological conservation it is a valuable tool for selecting locations for reserves by facilitating the mapping of species richness related to topology, geology, vegetation patterns, etc. (Meffe and Carroll, 1994). We will meet this problem of data integration again in other biotechnology contexts (see section V). Bowker contends that the current approach to the problem is largely to produce metadata standards, i.e., data about data, and argues that this is inadequate for dealing with the reconciliation of specific kinds of data necessary to develop useful biodiversity maps.

The American Academy of Microbiology (2002) recently called for the development of an integrated microbiology database that would include genomic and also habitat information (spatial sampling in four dimensions, activity assessments via DNA arrays, geology, and ecology), within a 6- to 10-year time frame. Chapter 23, by Daphne Stoner and her colleagues, describes the first such practical attempt to integrate microbiological, geochemical, geological, and habitat information (on Yellowstone National Park) into a geographical format and including an interactive website. Interestingly, in view of Bowker's comments, this project has adopted GIS as its research tool.

A final aspect of biodiversity mapping raises another integration issue, that of data that are collected and integrated into two distinct discourses—scientific and political (Bowker, 2001). Some facets of this issue are examined in section VIII.

REFERENCES

American Academy of Microbiology. 2002. *Microbial Ecology and Genomics: A Crossroads of Opportunity.* American Academy of Microbiology, Washington, D.C.

Bernhard, J. M., K. R. Buck, M. A. Farmer, and S. S. Bowser. 2000. The Santa Barbara Basin is a symbiosis oasis. *Nature* 403:77–80.

Bowker, G. C. 2000. Mapping biodiversity. *Int. J. Geogr. Infor. Sci.* 14:739–754.

Bowker, G. C. 2001. Biodiversity data diversity. *Soc. Stud. Sci.* 30:643–684.

Cho, J. C., and J. M. Tiedje. 2000. Biogeography and degree of endemicity of fluorescent *Pseudomonas* strains in soil. *Appl. Environ. Microbiol.* 66:5448–5456.

Clark, J. W., S. Hossain, C. A. Burnside, and S. Kambhampati. 2001. Coevolution between a cockroach and its bacterial endosymbiont: a biogeographical perspective. *Proc. R. Soc. Lond. Ser. B* 268:393–398.

Groombridge, B., and M. D. Jenkins. 2002. *World Atlas of Biodiversity. Earth's Living Resources in the 21st Century.* University of California Press, Berkeley, Calif.

Halling, R. E. 2001. Ectomycorrhizae: co-evolution, significance, and biogeography. *Ann. Missouri Bot. Gard.* 88:5–13.

Meffe, G. K., and C. R. Carroll. 1994. *Principles of Conservation Biology.* Sinauer Associates Inc., Sunderland, Mass.

Rheims, H., C. Spröer, F. A. Rainey, and E. Stackebrandt. 1996. Molecular biological evidence for the occurrence of uncultured members of the actinomycete line of descent in different environments and geographical locations. *Microbiology* (UK) 142:2863–2870.

Snoeyenbos-West, O. L. O., T. Salcedo, G. B. McManus, and I. A. Katz. 2001. Biogeography and level of endemism in ciliate populations from coastal environments. Abstract of the 148th Meeting, Soc. Gen. Microbiol. **SE 10:**2.

Snoeyenbos-West, O. L. O., T. Salcedo, G. B. McManus, and I. A. Katz. 2002. Insights into the diversity of choreotrich and logotrich ciliates (class: Spirotrichea) based on genealogical analyses of multiple loci. *Int. J. Syst. Evol. Microbiol.* **52:**1901–1913.

Staley, J. T. 1999. Bacterial biodiversity: a time for a place. *ASM News* **65:**681–687.

Staley, J. T., and J. J. Gosink. 1999. Poles apart: biodiversity and biogeography of sea ice bacteria. *Annu. Rev. Microbiol.* **53:**189–215.

Microbial Diversity and Bioprospecting
Edited by Alan T. Bull
© 2004 ASM Press, Washington, D.C.

Chapter 21

Ubiquitous Dispersal of Free-Living Microorganisms

BLAND J. FINLAY AND GENOVEVA F. ESTEBAN

THE KEY ROLE OF ABSOLUTE ABUNDANCE

It is difficult for the human brain to conceive of anything that varies by at least 12 orders of magnitude, but that is the case for the areal abundance of organisms that range in size from microbial eukaryotes to mammals (Finlay, 2002). The representatives of a protozoan species whose individuals have a mass of about 1 ng typically have areal abundances 12 orders of magnitude greater than that of an average-sized mammal. And with the abundance of free-living bacteria being roughly 1,000 times greater than that of protozoa (Berninger et al., 1991), the total range in areal abundance of individuals within a species approaches 15 orders of magnitude. For purely statistical reasons, the sheer weight of numbers of microorganisms may drive their large-scale, even ubiquitous, dispersal, and in this chapter we are concerned with exploring some of the evidence for this idea.

First it is necessary to clarify what is meant by ubiquity, because the terms "cosmopolitan" and "ubiquitous" are often used as if they had roughly the same meaning (i.e., found all over the world). We use the term ubiquitous dispersal to mean the process of continuous worldwide dispersal of individuals belonging to a species, whereas cosmopolitan refers to species that are capable of population growth in many different places worldwide. Representatives of a species may be continuously dispersed worldwide (i.e., ubiquitous dispersal), usually as cysts or as resting spores—but the species may only rarely find itself in a habitat that is suitable for population growth, in which case it is unlikely to be classed as cosmopolitan.

One consequence of ubiquitous dispersal is that it produces a local diversity of cryptic species that is significantly greater than the number of active species. When a small sample of pond sediment was examined microscopically, 20 active morphospecies of ciliated protozoa were detected, but after using enrichment techniques that provided a turnover of new niches for the seedbank of cryptic species, the cumulative number of active species had risen to 137 after 3 months (Finlay et al., 1996a; Fenchel et al., 1997). Within the prokaryotes too, it is estimated that the local diversity of cryptic species (where species are defined as an assemblage of strains sharing 70% or more DNA homology; but see chapters 3 and 4) in soil and sediment is very high—of the order of 10^4 species within a sample volume of 30 to 100 cm^3 (Torsuik et al., 2002). Even with the smaller metazoans, such as freshwater rotifers, the diversity of species represented by the egg bank within the sediment can be very large (Langley et al., 2001).

If ubiquitous dispersal is typical of free-living microbes, there are some important implications. With no effective barriers to dispersal, speciation that is due to geographic isolation will be rare, leading to low global species richness and high local:global species ratios. This of course is quite different from what happens in the case of macroscopic organisms, where geographic and physical barriers are more important, and local evolutionary diversification can take place. Kangaroos and cows, although adapted to similar niches, have evolved into quite different life forms in different regions of the world. But microbial morphospecies appear identical wherever they are collected because they are never restricted by geographical barriers and thus are never presented with the opportunity of local evolutionary diversification. The nature of biodiversity at the microbial level is obviously quite different from that of macroscopic organisms (Table 1).

EVIDENCE FOR UBIQUITOUS DISPERSAL

The evidence for ubiquitous dispersal is not always obvious because microorganisms are often

Bland J. Finlay and Genoveva F. Esteban • Centre for Ecology and Hydrology, Dorset, Winfrith Technology Centre, Winfrith Newburgh, Dorchester, Dorset DT2 8ZD, United Kingdom.

Table 1. Fundamental differences between macroscopic plants and animals and free-living microorganisms

Characteristic	Plants and animals	Microorganisms
Absolute abundance	Low	High
Rate of migration	Low	High
Rate of speciation	High	Low
Rate of species extinction	High	Low
Relative number of endemics	High	Low/None
Global number of species	High	Low
Local:global species richness	Low	High
Cryptic persistence of species	Variable	Yes

cryptic, or because of undersampling. There are some important examples, however, that present themselves as direct and indirect evidence. In the discussion that follows, the term microorganism includes the prokaryotes and the unicellular eukaryotes (mainly algae and protozoa).

Morphospecies

There is broad agreement that a large proportion of free-living flagellated protozoan morphospecies have worldwide distribution (Lee and Patterson, 2000). *Rhynchomonas nasuta,* for example, has been found in marine, freshwater, and soil sites at many locations in all biogeographic regions, including Europe and many islands in the Pacific Ocean (Larsen and Patterson, 1990). The resting cysts of the toxic dinoflagellate *Gymnodinium catenatum* have a cosmopolitan distribution in coastal marine sediments (Bolch and Reynolds, 2002); more than 75% of the global species richness of the chrysomonad *Paraphysomonas* was found in one small sample of sediment from a freshwater pond (Finlay and Clarke, 1999a,b); and the anaerobic flagellate *Postgaardi mariagerensis,* recorded for the first time in 1995 from an anoxic basin in a Danish inland fjord, was soon thereafter found in the sulfidic zone of a lake in Antarctica (Simpson et al., 1997). Atkins et al. (2000) found that flagellate species cultured from hydrothermal vents in the Pacific were typical of marine, freshwater, and terrestrial ecosystems worldwide, and Ekebom et al. (1996) concluded, following extensive examination of coral reef sediments, that the community of heterotrophic flagellates found there also had worldwide distribution. *Emiliana huxleyi*—the best known, most commonly encountered, and most abundant marine planktonic photosynthetic flagellate—has great phenotypic plasticity (Medlin et al., 1994). A single population can form blooms extending over an area greater than that of England, and it is generally agreed that the species is cosmopolitan throughout the world's oceans. Levinsen et al. (2000)

commented on the striking similarity of the heterotrophic dinoflagellate communities from an Arctic coastal ecosystem and from the Antarctic marginal ice zone, identical planktonic foraminiferan morphospecies have been found in Arctic and Antarctic waters (Darling et al., 2000), and some ciliate species live in sea ice in the Arctic and the Antarctic (Agatha et al., 1993). All 86 freshwater ciliate species identified from a volcanic crater lake in Australia in 1997 were already known from Northern Europe by the mid-1930s (Esteban et al., 2000; Finlay et al., 1999), and the ciliate community living in activated sludge plants is virtually identical worldwide (G.F. Esteban, unpublished).

Genotypes

Specific bacterial genotypes are now known to have large-scale geographical distribution in marine and freshwater environments (e.g., Kelly and Chistoserdov, 2001; Zwart et al., 2002). Glöckner et al. (2000) revealed 16 globally distributed sequence clusters within the *Actinobacteria* in freshwater ecosystems, and *Thiomicrospira* (sulfur-oxidizing chemolithotrophs) isolates from Atlantic and Pacific hydrothermal vent sites appear virtually identical genetically (Wirsen et al., 1998). Staley and Gosink (1999) showed that some bacterial genera occur in sea ice at both poles, an *Iceobacter* strain from the Arctic is more closely related to an Antarctic strain than to other strains from the Arctic, and hyperthermophilic archaea isolated from Alaskan oil reservoirs are identical to those from European thermal waters. A strain isolated from North Sea crude oil fields showed 100% similarity to a strain from an Italian hydrothermal system.

Lepère et al. (2000) analyzed molecular diversity (16S rDNA and intergenic spaces) within a large number of *Microcystis* strains from four continents and found them to be remarkably similar, if not identical genetically, and probably all members of a single, ubiquitous, genetic species. Different strains, however, still show morphological differences (e.g., in sheath structure) which are probably coded by parts of the genome that vary more rapidly than genes such as 16S rDNA.

The filamentous cyanobacterium *Arthrospira* (formerly *Spirulina*) is typically found in very high densities in the carbonate-bicarbonate-rich water bodies known as soda lakes. Baurain et al. (2002) collected 21 clonal strains from four continents and found a high degree of conservation in 16S rDNA and intergenic spaces sequence. They hypothesized that all strains belong to one, or at most two, cosmopolitan species. As the authors indicate, cosmopolitan distribution is easier to understand in the case of marine cyanobacteria, which may be transported long distances by seawater circulation. Soda lakes tend to be isolated ecosystems, often with restricted inflows and outflows.

Zwart et al. (2002) analyzed 689 bacterial (16S rDNA) sequences from the water column of lakes and rivers in North America, Europe, and Asia and dissected out a freshwater bacterial community that is characteristic of the diversity of freshwater ecosystems worldwide. They found identical bacterial sequences in aquatic sites separated by great distance and physical barriers, including two of the deepest lakes in the world—Lake Baikal in Siberia and Crater Lake in Oregon—the latter lying in a volcanic basin with no inlets or outlets and surrounded by rock walls 600 m high.

In the microbial eukaryotes, the same general picture is emerging—specific genotypes have very wide geographical distribution. The flagellate *Cafeteria roenbergensis,* originally isolated from shallow brackish waters in Denmark, was shown to be identical to *Cafeteria* isolates from hydrothermal vents in the Pacific (Atkins et al., 2000). Darling et al. (2000) sequenced the small-subunit (ssu) rDNA of three bipolar planktonic foraminiferal morphospecies and identified at least one genotype in all three morphospecies that was identical in both the Arctic and the Antarctic, indicating that transtropical gene flow must have occurred. Furthermore, following their collection of the foraminiferan *Orbulina universa* from both the Caribbean and the Coral seas, Darling et al. (1999) demonstrated complete sequence similarity throughout an amplified 980-bp fragment of the ssu rDNA.

Stoeck and Schmidt (1998) developed a method for the identification of species within the *Paramecium aurelia* complex using randomly amplified polymorphic DNA fingerprints and found a constant diagnostic band pattern for isolates of the same species, irrespective of geographic origin. The banding patterns for *P. primaurelia* were identical in isolates from the United States, Japan, and France. Ammermann et al. (1989) found that the mean genetic identity of European and the North American populations of *Stylonychia lemnae* was similar to that for different populations of the same species, indicating that American and European populations belong to the same species. Bowers and Pratt (1995), working with the soil ciliate *Colpoda inflata,* were unable to detect genetic differences between isolates from Chile, France, Malawi, Canada, Mexico, and Russia.

The cosmopolitan coccolithophorid *Emiliana huxleyi* can exist as one of two morphotypes, but these have identical base sequences in the ssu rDNA and in the spacer regions between the plastid-encoded large and small Rubisco subunits (Medlin et al., 1994). Similarly, Bolch and Reynolds (2002), after carrying out genetic studies (long-subunit rDNA sequencing) of marine *Gymnodinium* species, could find no detectable genetic variation among several globally distributed temperate and tropical populations. Toxic strains of the extraordinarily polymorphic dinoflagellate *Pfiesteria piscicida* have been found in the mid-Atlantic, in Gulf Coast estuaries, and in the waters of Northern Europe and New Zealand and are generally regarded as having worldwide distribution—a conclusion that was quickly reached after it was first suspected (in 1988) that the organism was economically important as the cause of large-scale fish kills (Burkholder et al., 2001).

Sibling Species and Physiological Species

Isolates of a single sibling species (a reproductively isolated gene pool) of the green alga *Gonium pectorale* collected from a very wide geographical area (e.g., California and Nepal) can interbreed, raising the possibility that the species has worldwide distribution (Coleman et al., 1994). The ciliate sibling species *Paramecium tredecaurelia* exists in culture as three stocks, originating from the River Seine in Paris, from Madagascar, and from Mexico, and it is possible to make crosses between all three (G.H. Beale, personal communication). Stoeck et al. (1998) could cross any of the three genotypes of *P. triaurelia,* even if they originated from geographically distant sites such as Florida and Northern Europe. Kusch et al. (2000) revealed that 18 isolates of the ciliate *Euplotes aediculatus* collected from ponds and streams in Europe and the United States were all genetically different, but the extent of similarity between genotypes from different sites was not significantly correlated with geographic distance. One would be as likely to find any particular genotype in neighboring or widely separated water bodies.

Prochlorococcus marinus is believed to be the smallest (0.6 μm) and the most abundant photosynthetic prokaryote, contributing anywhere up to 80% of local primary production in the ocean. The species consists of at least two different ecotypes that are adapted for growth at different light levels and therefore at different depths in the water column (Moore et al., 1998). At any particular site, high- and low-light-adapted ecotypes differ from each other by about 2.5% (16S rDNA), whereas each ecotype remains virtually identical genetically across sites. These distinctive physiological populations could be regarded as distinctive physiological species, and it is conceivable that there could be a large number of such species lying within the morphospecies. This, however, is unlikely if recent evidence from work on ciliates and flagellates is indicative of microbes in general, for these microbial eukaryotes demonstrate a remarkable ability to tolerate and adapt to wide ranges of ecologically important factors such as temperature and salinity. A marine *Paraphysomonas* was found to grow well in freshwater, and an isolate of the ciliate *Uronema* from the high Arctic could be adapted to grow at 37°C (Fenchel et al., in preparation), indicating that the number of discrete physiological species within morphospecies may not be very great.

Indirect Evidence

Because free-living microbes are extremely abundant, they are expected to have high rates of dispersal, and the combined effect of a large number of natural factors, some of which will operate against each other, will be to produce something approaching a random distribution over a range of spatial scales. For groups of organisms such as the testate amebae, which can be identified to species level even when they are in the encysted state, it has been shown that their fundamental spatial distribution following dispersal is, indeed, essentially random (Finlay, 2002; Finlay et al., 2001).

One likely consequence of large-scale random dispersal is that species that are locally rare or abundant will also be globally rare or abundant, and this has been observed for diatoms, testate amobae, ciliates, and chrysomonad flagellates (Finlay, 2002; Finlay and Clarke, 1999a, 1999b; Finlay et al., 2001, 2002). The species that are globally abundant tend to be the "weeds"—the generalist species with broad tolerance of factors such as salinity, temperature, and water pollution of various types (Finlay et al., 2002). Rare species, often the specialist species, are probably more limited in their tolerance of variation in environmental variables. Species with broad tolerance form populations in a broad range of environmental conditions, so they acquire high absolute abundance and colonize the globe with locally large populations. Species with more restrictive requirements produce smaller populations. The rate and scale of dispersal of a microbial species, therefore, is a function of the size of the global population.

There are some apparent exceptions among the microorganisms adapted for life in extreme habitat types that are relatively rare at the global scale (e.g., the filamentous cyanobacteria that thrive in African soda lakes). They can become extremely abundant locally, but as these specialized habitats are relatively rare worldwide, the global population size of these species is relatively low. Cyanobacteria are undoubtedly dispersed from soda lakes, but the absolute number of individuals finding suitable habitats elsewhere is probably very low.

If ubiquitous dispersal is a general characteristic of microbial species, the rate of allopatric speciation and the global species richness of microbes will be low. This applies at the level of morphospecies (Finlay and Esteban, 1998); for example, one of the best-studied groups, the free-living ciliates, is represented by only around 3,000 species (Finlay et al., 1996b), a small number when compared with macroscopic taxa that do have biogeographies (e.g., the insects, with ~5 million species).

A further consequence is that the local:global species ratio is significantly higher for microorganisms than it is for macroscopic organisms. The local species richness of marine meiofauna may represent <0.1% of the global species number (Fenchel, 1993), but the ratio is significantly higher for microbes (e.g., 10 to 20% for ciliates, ~50% for heliozoa, and roughly 100% for discrete groups of the smaller protists such as those within the genera *Paraphysomonas*, *Luffisphaera*, and *Cochliopodium* [Finlay 2002; K. J. Clarke, personal communication]).

Extraordinarily high local genetic diversities of prokaryotes are reported in numerous studies. Torsvik et al. (1990) deduced that the total number of bacteria per gram of dry soil was about 1.5×10^{10} and that this corresponded to about 4,000 bacterial genomes. But with ubiquitous dispersal as the relevant model of large-scale distribution, the global diversity of bacteria in soil would perhaps not be expected to be much greater than 4,000 genomes.

The question of the ubiquity of bacteria has rarely been explored in recent times, although it was accepted by Beijerinck and others more than 100 years ago (Van Iterson et al., 1983). The literature, however, is littered with circumstantial evidence. Consider the perchlorate-reducing bacteria that live anaerobically in petroleum-contaminated soils and sediments (Coates et al., 1998). They are capable of chlorite dismutation—a process that produces molecular oxygen, which facilitates hydrocarbon oxidation by indigenous aerobic hydrocarbon oxidizers. Interestingly, adding chlorite to "pristine" sites has the same effect, although there is a lag of about 24 h before the process is detectable, which may be consistent with adaptation and/or population growth of bacterial species that are typically rare in such sites (Coates et al., 1998).

A final, striking consequence of random dispersal is that species tend to thrive wherever a suitable habitat exists, even if the existence of such a habitat would not be readily predicted. Al-Rasheid (1996) found 21 species of typically marine ciliated protozoan species in the slightly brackish waters of an oasis located 50 km from the Arabian Gulf Coast.

Are There Any Endemic Microbial Species?

Fulthorpe et al. (1998) collected soil samples from pristine ecosystems on five continents. The samples were enriched with 3-chlorobenzoate, and 48 genotypes of 3-cba-degrading bacteria were discovered. Each genotype tended to be unique to the site from which it was isolated—implying, perhaps, some level of endemicity, but there was also evidence that this apparent correlation with geography was in fact a correlation with the different plant communities growing at the different sites. The quality of the habitat, rather than its geographical location, determined the likelihood of population growth of specific bacterial genotypes.

Antarctica, probably the most isolated region of the biosphere is a sensible place to look for endemic species of microbial eukaryotes. Tong et al. (1997) collected from a range of marine and freshwater sites and identified 35 species of heterotrophic flagellates, centrohelid heliozoa, and filose amebae. No evidence of endemism was found, and all species apart from a single new species had previously been observed at nonpolar sites. However, there are many curious reported examples of microbial endemics, including the diatom *Stephanodiscus yellowstonensis* that lives, apparently, only in one lake in Wyoming (Theriot, 1992), although it is 99.5 to 100% similar (in rDNA sequence) to the more widespread *Stephanodiscus niagarae* (Zechman et al., 1994).

The most famous microbial endemics are probably those reputed to live in Lake Baikal, the world's deepest lake. This lake is well known for its endemic fauna of larger organisms, but microbial endemics are also claimed (see Finlay et al., 2002). Of the 334 diatom species recorded from the lake, one-third are said to live in the lake and nowhere else, despite the fact that hundreds of river systems flow into the lake and it discharges into major rivers that flow for several thousands of kilometers. It is also claimed that Lake Baikal supports a fair number of endemic species of ciliated protozoa, but most if not all of these are dubious (Esteban et al., 2001).

Claims for endemic diatoms or any other microbial endemics probably cannot be supported because the absolute abundance of all free-living microbial eukaryote species is so great that ubiquitous dispersal is likely (Fenchel, 1993) and because it is effectively im-

possible to disprove the existence of endemics (i.e., including resting spores and other cryptic forms) elsewhere on the planet. The question of endemics, perhaps more than any other, highlights how different the character is of biodiversity and biogeography in micro- and macroorganisms. As an illustration, note that the algal flora of New Zealand does not contain a single endemic algal genus, yet 70% of New Zealand's indigenous seed plant species are endemic (see Lund, 2002).

Mechanisms of Dispersal

The scientific literature of the past 150 years contains a wealth of information referring to the mechanisms by which microorganisms are dispersed in the natural environment. Much of this early information can be found in Gislén's (1948) excellent review, where we find that, as early as the 1840s, Ehrenberg was examining dust deposited by the Sirocco wind, reportedly identifying 120 microbial forms therein and calculating that the weight of microorganisms transported and deposited by this mechanism was of the order of 1 ton per square mile. Gislén (1948) also records a duck with fresh mollusc spawn attached to its feet that was shot in the Sahara at least 100 miles from the nearest body of water; a tornado that lifted a church spire and transported it 28 km before coming to land; and the numerous records of spout formation in tornadoes with the power to suck up water and even small fishes (hence the biblical "rains of fishes"). When the airship *Graf Zeppelin* visited North America in 1929, 6 of the 20 insect species in bouquets decorating the cabins had not previously been reported from North America. One quickly realizes that any attempt at providing a precise answer to a question such as "How is microbial species x dispersed in the natural environment?" is not a simple task. There are innumerable, interacting forces and processes that drive the dispersal of microbes.

Transport by Wind and Water

It is inevitable that microorganisms are continuously being moved around the biosphere by flowing rivers and percolating groundwaters, advection via hydrothermal plume entrainment, and thermohaline circulation in the oceans and, in the air, by convection currents followed by precipitation. Schlichting (1964), for example, sampled air by drawing it through a bottle containing sterile soil-water extract and recorded 88 viable species of algae and protozoa that grew out.

But it is the extreme meteorological events such as hurricanes and dust storms that are more obviously associated with rapid and large-scale translocation of microbes. Campbell et al. (1999) recorded the transport of large amounts of pine and spruce pollen (particle sizes 30 to 55 μm) roughly 3,000 km into the high Arctic, the result of an unusually strong low-pressure system. Alternative causes of biological particles of this size injected into the high atmosphere include volcanic eruption, forest fires, and desert windstorms, so the rate of long-range transport of small biological particles (e.g., spores, cysts, pollen), although difficult to quantify, may be high. Marshall (1996) found large quantities of pollen of the southern beech (Nothofagus spp.) in spore traps in Antarctica. The material had originated in South America and was transported in the air roughly 1,500 km. It is well known that snow in the maritime Antarctic contains pollen from South American plants.

Aerial long-distance transport of fungi pathogenic on crop plants can be relatively rapid (Brown and Hovmøller, 2002). In June 1978, the sugar cane rust Puccinia melanocephala, carried by cyclonic winds, apparently made the single-step intercontinental jump from West Africa to the Dominican Republic.

The marine analogue of large-scale air transmission lies in the relatively slower but more predictable patterns of ocean circulation, and some of the best evidence comes from the planktonic foraminifera. Darling et al. (1999) obtained molecular data from Orbulina universa indicating gene flow between populations in the Pacific and the Atlantic, which is presumably facilitated by ocean surface circulation patterns. A more complex migration route, linking Arctic and Antarctic waters (which support identical foraminiferan genotypes), was suggested by Darling et al. (2000). The cool boundary currents off the coast of West Africa, together with deep submergence in the tropics, are believed to provide a corridor for the transport of viable organisms across the tropics.

Animal Vectors

There are numerous published examples of animals as vectors for the transport of micro-organisms, particularly algae and protozoa. Schlichting and Sides (1969) recorded passive transport on the external body surfaces of Hemiptera, and Milliger and Schlichting (1968) found a greater diversity of viable algae and protozoa in the guts and fecal deposits of migrating beetles than on the beetles' bod-

ies. Proctor (1966) and Atkinson (1980) recovered viable freshwater desmids and diatoms from the feces of ducks and other water birds up to 20 h after feeding. Maguire (1963) and Parsons et al. (1966) collected dragonflies, washed them with sterile soil-water extract, and subsequently recovered viable heliozoans, naked amebae, cyanobacteria, green algae, euglenoids, chrysomonads, and ciliates, especially soil ciliates. It is well known that many dragonfly species can travel great distances.

Field observations and experiments in a tropical forest carried out by Maguire and Belk (1967) confirmed that ciliates and other small aquatic organisms are transported between Heliconia flowers by terrestrial snails. One of the ciliates transported was Paramecium multimicronucleatum, in which cyst formation is unknown. Coesel et al. (1988) provided circumstantial evidence that the distributional patterns of neotropical desmid species are correlated with major routes of bird migration. Further remarkable examples of dispersal of aquatic organisms mediated by migratory (water) birds can be found in Clausen et al. (2002) and other articles in the same journal issue.

As microbes in some shape or form will always be associated with the internal or external body surfaces of animals, any dispersal event of larger animals (e.g., migrating birds with wet feet and feathers or any medium-to-large-sized mammal with wet fur carrying living microbes for many miles during a rainy night) is also an act of microbial dispersal. One particularly dramatic (and barely predictable) example was described by Censky et al. (1998), who recorded that, on October 1995, at least 15 individuals of the green iguana Iguana iguana made their first recorded appearance on Anguilla in the Caribbean. The animals were attached to a natural raft of logs and uprooted trees and had almost certainly come from the island of Guadeloupe, about 500 km distant. This route would have coincided with the tracks of two hurricanes that immediately preceded the iguanas' sea crossing.

Human Activities

With the continuing increase in size of the global human population, the growth in popularity of long-distance air travel, and the expansion of intercontinental trade in foodstuffs (particularly fresh fruit and vegetables) and ornamental plants, there can be no doubt that human activities contribute directly to the large-scale dispersal of microbes. Because of obvious difficulties, introductions of free-living microorganisms are rarely recorded, but there is a vast catalog of introductions of macroscopic organisms. Roughly 20 insect species become established on the Hawaiian Islands every year. In New Zealand, the 27 native species of freshwater fish are outnumbered by 30 nonnative species (and imported fish must also be associated with a diversity of aquatic microbes). The Asian tiger mosquito was brought into the United States in used automobile tires imported for retreading and resale (Vitousek et al., 1996). Larger-scale introductions are made when ships empty their ballast tanks. Galil and Hülsmann (1997) described the protist communities present in ballast tanks of cargo vessels arriving in Israeli Mediterranean ports and identified at least 198 species and Carlton and Geller (1993) counted 367 plankton taxa in ballast water released in Oregon.

CONCLUSION

Focusing on free-living bacterial and microbial eukaryote species, we have attempted to characterize the general pattern of large-scale distribution of microbial species. Ubiquitous dispersal driven by high absolute abundance appears to be common to most, if not all, species, with the rate and scale of dispersal of a species largely determined by its global population size. There is little if any convincing evidence supporting the existence of endemic microbial species. The factors and processes driving dispersal are numerous and varied (ducks' feet, ballast water of ocean-going ships, etc.), but their relative importance is poorly understood. One emerging view is that extreme or unusual meteorological events (hurricanes, dust storms, etc.) are particularly important in effecting rapid, long-distance dispersal, especially of bacteria (Roberts and Cohan, 1995) and spores and cysts (Brown and Hovmøller, 2002). For most microbial species, it is probably not possible to model or even fully describe the catalog of events and essentially random processes that bring about and sustain their global distribution. The ubiquitous distribution of microbial species may also point the way to alternative strategies of bioprospecting. Building on the pioneering work of Beijerinck and Winogradsky more that 100 years ago, it may be productive to design highly specific enrichment culture conditions that favor the growth of microbes with specific metabolic properties, perhaps using virtually any natural sample.

Acknowledgments. This work was carried out with financial support from the Natural Environment Research Council (U.K.) through the Marine and Freshwater Microbial Diversity Thematic Programme (grant no. NER/T/S/2000/1351).

REFERENCES

Agatha, S., M. Spindler, and N. Wilbert. 1993. Ciliated protozoa (Ciliophora) from Arctic Sea ice. *Acta Protozool.* **32:**261–268.

Al-Rasheid, K. A. S. 1996. Records of free-living ciliates in Saudi Arabia. II. Freshwater benthic ciliates of Al-Hassa Oasis, Eastern Region. *Arab Gulf J. Sci. Res.* **15:**187–205.

Ammermann, D., M. Schlegel, and K.-H. Hellmer. 1989. North American and Eurasian strains of *Stylonychia lemnae* (Ciliophora, Hypotrichida) have a high genetic identity, but differ in the nuclear apparatus and in their mating behavior. *Eur. J. Protistol.* **25:**67–74.

Atkins, M. S., A. P. Teske, and O. R. Anderson. 2000. A survey of flagellate diversity at four deep-sea hydrothermal vents in the Eastern Pacific Ocean using structural and molecular approaches. *J. Euk. Microbiol.* **47:**400–411.

Atkinson, K. M. 1980. Experiments in dispersal of phytoplankton by ducks. *Br. Phycol. J.* **15:**49–58.

Baurain, D., L. Renquin, S. Grubisic, P. Scheldeman, A. Belay, and A. Wilmotte. 2002. Remarkable conservation of internally transcribed spacer sequences of *Arthrospira* ("*Spirulina*") (Cyanophyceae, Cyanobacteria) strains from four continents and of recent and 30-year old dried samples from Africa. *J. Phycol.* **38:**384–393.

Bell, G. 2001. Neutral macroeclogy. *Science* **293:**2413–2418.

Berninger, U.-G., B. J. Finlay, and P. Kuuppo-Leinikki. 1991. Protozoan control of bacterial abundances in fresh water. *Limnol. Oceanogr.* **36:**139–147.

Bolch, C. J. S., and M. L. Reynolds. 2002. Species resolution and global distribution of microreticulate dinoflagellate cysts. *J. Plank. Res.* **24:**565–578.

Bowers, N. J., and J. R. Pratt. 1995. Estimation of genetic variation among soil ciliates of *Colpoda inflata* (Strokes) (Protozoa: Ciliophora) using the polymerase chain reaction and restriction fragment length polymorphism analysis. *Arch. Protistenkd.* **145:**29–36.

Brown, J. K. M., and M. S. Hovmøller. 2002. Aerial dispersal of pathogens on the global and continental scales and its impact on plant disease. *Science* **297:**537–541.

Burkholder, J. M., H. B. Glasgow, and N. Deamer-Melia. 2001. Overview and present status of the toxic *Pfiesteria complex* (Dinophyceae). *Phycologia* **40:**186–214.

Campbell, I. D., K. McDonald, M. D. Flannigan, and J. Kringayark. 1999. Long-distance transport of pollen into the Arctic. *Nature* **399:**29–30.

Carlton, J. T., and J. B. Geller. 1993. Ecological roulette: the global transport of nonindigenous marine organisms. *Science* **261:**78–82.

Censky, E. J., K. Hodge, and J. Dudley. 1998. Over-water dispersal of lizards due to hurricanes. *Nature* **395:**556.

Clausen, P., B. A. Nolet, A. D. Fox, and M. Klaassen. 2002. Long-distance endozoochorous dispersal of submerged macrophyte seeds by migratory waterbirds in northern Europe—a critical review of possibilities and limitations. *Acta Oecologica* **23:**191–203.

Coates, J. D., R. A. Bruce, and J. D. Haddock. 1998. Anoxic bioremediation of hydrocarbons. *Nature* **396:**730.

Coesel, P. F. M., S. R. Duque, and G. Arango. 1988. Distributional patterns in some neotropical desmid species (Algae, Chlorophyta) in relation to migratory bird routes. *Rev. Hydrobiol. Trop.* **21:**197–205.

Coleman, A. W., A. Suarez, and L. J. Goff. 1994. Molecular delineation of species and syngens in volvocacean green algae (Chlorophyta). *J. Phycol.* **30:**80–90.

Darling, K. F., C. M. Wade, D. Kroon, A. J. Leigh Brown, and J. Bijma. 1999. The diversity and distribution of modern planktonic foraminiferal subunit RNA genotypes and their potential as tracers of present and past ocean circulations. *Paleoceanography* 14:3–12.

Darling, K. F., C. M. Wade, I. A. Stewart, D. Kroon, R. Dingle, and A. J. Leigh Brown. 2000. Molecular evidence for genetic mixing of Arctic and Antarctic planktonic foraminifers. *Nature* 405: 43–47.

Ekebom, J., D. J. Patterson, and N. Vprs. 1996. Heterotrophic flagellates from coral reef sediments (Great Barrier Reef, Australia). *Arch. Protistenkd.* 146:251–272.

Esteban, G. F., B. J. Finlay, J. L. Olmo, and P. A. Tyler. 2000. Ciliated protozoa from a volcanic crater-lake in Victoria, Australia. *J. Nat. Hist.* 34:159–189.

Esteban, G. F., B. J. Finlay, N. Charubhun, and B. Charubhun. 2001. On the geographic distribution of *Loxodes rex* (Protozoa, Ciliophora) and other alleged endemic species of ciliates. *J. Zool.* 255:139–143.

Fenchel, T. 1993. There are more small than large species? *Oikos* 68:375–378.

Fenchel, T., G. F. Esteban, and B. J. Finlay. 1997. Local versus global diversity of microorganisms: cryptic diversity of ciliated protozoa. *Oikos* 80:220–225.

Finlay, B. J. 2002 Global dispersal of free-living microbial eukaryote species. *Science* 296:1061–1063.

Finlay, B. J. and K. J. Clarke. 1999a. Ubiquitous dispersal of microbial species. *Nature* 400:828.

Finlay, B. J., and K. J. Clarke. 1999b. Apparent global ubiquity of species in the protist genus *Paraphysomonas*. *Protist* 150:419–430.

Finlay, B. J. and G. F. Esteban. 1998. Freshwater protozoa: biodiversity and ecological function. *Biodiver. Conserv.* 7:1163–1186.

Finlay, B. J., G. F. Esteban, and T. Fenchel. 1996a. Global diversity and body size. *Nature* 383:132–133.

Finlay, B. J., J. O. Corliss, G. Esteban, and T. Fenchel. 1996b. Biodiversity at the microbial level: the number of free-living ciliates in the biosphere. *Q. Rev. Biol.* 71:221–237.

Finlay, B. J., G. F. Esteban, J. L. Olmo, and P. A. Tyler. 1999. Global distribution of free-living microbial species. *Ecography* 22:138–144.

Finlay, B. J., G. F. Esteban, K. J. Clarke, and J. L. Olmo. 2001. Biodiversity of terrestrial protozoa appears homogeneous across local and global spatial scales. *Protist* 152:355–366.

Finlay. B. J., E. B. Monaghan, and S. C. Maberly. 2002. Hypothesis: the rate and scale of dispersal of freshwater diatom species is a function of their global abundance. *Protist* 153:261–274.

Fulthorpe, R. R., A. N. Rhodes, and J. M. Tiedje. 1998. High levels of endemicity of 3-chlorobenzoate-degrading bacteria. *Appl. Environ. Microbiol.* 64:1620–1627.

Galil, B. S., and N. Hülsmann. 1997. Protist transport via ballast water—biological classification of ballast tanks by food web interactions. *Eur. J. Protistol.* 33:244–253.

Gislén, T. 1948. Aerial plankton and its conditions of life. *Biol. Rev.* 23:109–126.

Glöckner, F. O., E. Zaichikov, N. Belkova, L. Denissova, J. Pernthaler, A. Pernthaler, and R. Amann. 2000. Comparative 16S rRNA analysis of lake bacterioplankton reveals globally distributed phylogenetic clusters including an abundant group of Actinobacteria. *Appl. Environ. Microbiol.* 66:5053–5065.

Kelly, K. M., and A. Y. Chistoserdov. 2001. Phylogenetic analysis of the succession of bacterial communities in the Great South Bay (Long Island). *FEMS Microbiol. Ecol.* 35:85–95.

Kristiansen, J. 1996. Dispersal of freshwater algae—a review. *Hydrobiologia* 336:151–157.

Kusch J., H. Welter, M. Stremmel, and H. J. Schmidt. 2000. Genetic diversity in populations of a freshwater ciliate. *Hydrobiologia* 431:185–192.

Langley, J. M., R. J. Shiel, D. L. Nielsen, and J. D. Green. 2001. Hatching from the sediment egg-bank, or aerial dispersing?—the use of mesocosms in assessing rotifer biodiversity. *Hydrobiologia* 446/447:203–211.

Larsen, J., and D. J. Patterson. 1990. Some flagellates (Protista) from tropical marine sediments. *J. Nat. Hist.* 24:801–937.

Lee, W. J., and D. J. Patterson. 2000. Heterotrophic flagellates (Protista) from marine sediments of Botany Bay, Australia. *J. Nat. Hist.* 34:483–562.

Lepère, C., A. Wilmotte, and B. Meyer. 2000. Molecular diversity of *Microcystis* strains (Cyanophycease, Chroococcales). *Syst. Georg. Pants* 70:275–283.

Levinsen, H., T. G. Nielsen, and B. W. Hansen. 2000. Annual succession of marine pelagic protozoans in Disko Bay, West Greenland, with emphasis on winter dynamics. *Mar. Ecol. Prog. Ser.* 206:119–134.

Lund, J. W. G. 2002. *Seventieth Annual Report of the Freshwater Biological Association*, p. 43–46. Freshwater Biological Association, Ambleside, United Kingdom.

Maguire, B. 1963. The passive dispersal of small aquatic organisms and their colonization of isolated bodies of water. *Ecol. Monogr.* 33:161–185.

Maguire, B., and D. Belk. 1967. *Paramecium* transport by land snails. *J. Protozool.* 14:445–447.

Marshall, W. A. 1996. Biological particles over Antarctica. *Nature* 383:680.

Medlin, L. K., G. L. A. Barker, M. Baumann, P. K. Hayes, and M. Lange. 1994. Molecular biology and systematics, p. 393–411. *In* J. C. Green and B. S. C. Leadbeater (ed.), *The Haptophyte Algae*. Systematics Association, spec. vol. 51. Clarendon Press, Oxford, United Kingdom.

Milliger, L. E., and H. E. Schlichting. 1968. The passive dispersal of viable algae and protozoa by an aquatic beetle. *Trans. Am. Microsc. Soc.* 87:443–448.

Moore, L. R., G. Rocap, and S. W. Chisholm. 1998. Physiology and molecular phylogeny of coexisting *Prochlorococcus* ecotypes. *Nature* 393:464–467.

Parsons, W. M., H. E. Schlichting, and K. W. Stewart. 1966. In-flight transport of algae and protozoa by selected Odonata. *Trans. Am. Microsc. Soc.* 85:520–527.

Proctor, V. W. 1966. Dispersal of desmids by waterbirds. *Phycologia* 5:227–232.

Roberts, M. S., and F. M. Cohan. 1995. Recombination and migration rates in natural populations of *Bacillus subtilis* and *Bacillus mojavensis*. *Evolution* 49:1081–1094.

Schlichting, H. E. 1964. Meteorological conditions affecting the dispersal of airborne algae and protozoa. *Lloydia* 27:64–78.

Schlichting, H. E., and S. L. Sides. 1969. The passive transport of aquatic microorganisms by selected Hemiptera. *J. Ecol.* 57:759–764 (and references therein).

Simpson, A. G. B., J. van den Hoff, C. Bernard, H. R. Burtom, and D. J. Patterson. 1997. The ultrastructure and systematic position of the euglenozoon *Postgaardi mariagerensis*, Fenchel et al. *Arch. Protistenkd.* 147:213–225.

Staley, J. T., and J. J. Gosink. 1999. Poles apart: biodiversity and biogeography of sea ice bacteria. *Annu. Rev. Microbiol.* 53:189–215.

Stoeck, T., and H. J. Schmidt. 1998. Fast and accurate identification of European species of the *Paramecium aurelia* complex by RAPD-fingerprints. *Microb. Ecol.* 35:311–317.

Stoeck, T., E. Przybos, and H. J. Schmidt. 1998. A combination of genetics with inter- and intra-strain crosses and RAPD-fingerprints reveals different population structures within the *Paramecium aurelia* species complex. *Eur. J. Protistol.* 34:348–355.

Theriot, E. 1992. Clusters, species concepts, and morphological evolution of diatoms. *Syst. Biol.* 41:141–157.

Tong, S., N. Vprs, and D. J. Patterson. 1997. Heterotrophic flagellates, centrohelid heliozoa and filose amoebae from marine and freshwater sites in the Antarctic. *Polar Biol.* **18:**91–106.

Torsvik, V., J. Goksøyr, and F. I. Daae. 1990. High diversity in DNA of soil bacteria. *Appl. Environ. Microbiol.* **56:**782–787.

Torsvik, V., L. Øvreås, and T. F. Thingstad. 2002. Prokaryotic diversity—magnitude, dynamics, and controlling factors. *Science* **296:**1064–1066.

Van Iterson, G., L. E. Den Dooren de Jong, and A. J. Kluyver. 1983. *Martinus Willem Beijerinck, His Life and Work.* Science Tech, Inc. Madison, Wis.

Vitousek, P. M., C. M. D'Antonio, L. L. Loope, and R. Westbrooks. 1996. Biological invasions as global environmental change. *Am. Sci.* **84:**468–478.

Wirsen, C. O., T. Brinkhoff, J. Kuever, G. Muyzer, S. Molyneaux, and H. W. Jannasch. 1998. Comparison of a new *Thiomicrospira* strain from the Mid-Atlantic Ridge with known hydrothermal vent isolates. *Appl. Environ. Microbiol.* **64:**4057–4059.

Zechman, F. W., E. A. Zimmer, and E. C. Theriot. 1994. Use of ribosomal DNA internal transcribed spacers for phylogenetic studies of diatoms. *J. Phycol.* **30:**507–512.

Zwart, G., B. C. Crump, M. P. K. Agterveld, F. Hagen, and S.-K. Han. 2002. Typical freshwater bacteria: an analysis of available 16S rRNA gene sequences from plankton of lakes and rivers. *Aquat. Microb. Ecol.* **28:**141–155.

Microbial Diversity and Bioprospecting
Edited by Alan T. Bull
© 2004 ASM Press, Washington, D.C.

Chapter 22

Microbial Endemism and Biogeography

BRIAN P. HEDLUND AND JAMES T. STALEY

THE MICROBIAL BIOGEOGRAPHY DEBATE

The topic of microbial biogeography is almost 100 years old, however, when confronted with questions about the existence and extent of endemism in the microbial world, many microbiologists respond with opinions and theoretical arguments rather than examples of well-conducted studies. We begin this chapter with an overview of this debate as it applies to free-living prokaryotes in part because there are relatively few good microbial biogeography studies. Furthermore, the arguments help to frame microbial biogeography in the larger context of biodiversity in that if endemism is common, then many more species exist.

The Argument for Microbial Endemism

Arguments for endemism among free-living microbes generally draw from examples of endemism of other organisms. First, any international traveler knows that many pathogenic microbes, both viral and bacterial, have distinct biogeographies. For example, dogs and cats entering Australia are quarantined in order to prevent the introduction of the rabies virus onto the continent. And historians are quick to point out one of the great tragedies in human history, the re-peopling of the Americas by Europeans, who brought many new pathogens with them to the new world. This microbial invasion included the variola virus (smallpox), the measles virus, the influenza virus, the yellow fever virus, *Salmonella typhae* (typhoid fever), *Vibrio cholerae* (cholera), and *Yersinia pestis* (plague) and resulted in the infection and death of >95% of the pre-Columbian native population (Diamond, 1999). The fact that some pathogens have distinct biogeographies may suggest that some free-living microbes have biogeographies as well.

Another reason to suppose that endemism exists for free-living microbes is that most plant, animal, and some fungal species are endemic to one continent or another. Indeed, examples of plant and animal families and even orders are known to be endemic to isolated islands (Brown and Lomolino, 1998). Furthermore, a basic tenet of biodiversity is that species diversity increases with decreasing body size (May, 1988, Wilson, 1992). For terrestial animals, each 10-fold reduction in body length or 1,000-fold reduction in body weight corresponds to roughly a 100-fold increase in species diversity (May, 1988). If such a relationship held for microbes, the number of microbial species on our planet would be astronomical and it would seem that many microbes would have to be endemic to achieve this level of diversity. However, it is important to clarify that there is no theoretical ground for extrapolating a species richness-"body" size correlation to include microbes, and the current lack of a meaningful species definition for most microbes reduces any such relationship to a conceptual level rather than a truly quantitative level. Nevertheless, given that endemism is the norm, why should free-living microbes defy this pattern?

The Argument for Microbial Cosmopolitanism

The cosmopolitan view of the microbial world has a long history (Baas-Becking, 1934) and has withstood the test of time relatively well. Although some pathogens provide fuel for the endemism argument, others are clearly cosmopolitan, consisting of a limited number of globally distributed clonal lineages, including *Escherichia coli*, *Haemophilus influenzae*, *Neisseria meningitidis*, *Staphylococcus aureus*, and *Streptococcus pneumoniae*. Appropriately, Cho and Tiedje (2000) have pointed out that these bacteria are closely associated with humans and their biogeographies are inevitably affected by human activity, but it

Brian P. Hedlund • Department of Biological Sciences, University of Nevada, Las Vegas, 4505 Maryland Parkway, Las Vegas, NV 89154-4004. James T. Staley • Department of Microbiology, University of Washington, Seattle, WA 98195-7242.

could be argued that free-living microbes are also dispersed by human activities, leading to cosmopolitanism, as is also true of many plants and animals.

The cosmopolitan view has recently been staunchly advocated by the eukaryotic microbial ecologist Finlay, among others, who has published frequently in high-profile journals (Finlay, 2002; Finlay and Clarke, 1999; Finlay et al., 1996). The argument states that small body size and immense population size lead to constant dispersion of microbes by a variety of mechanisms (including human-mediated) and, consequently, cosmopolitanism. Some microbes form cysts or spores that are particularly well suited for long-term survival during passive dissemination, and even microbes that do not form specific survival stages are known to survive long periods in metabolically inactive states. These microbial seeds "bloom" when and where conditions favor their growth.

Based on these ideas and corroborating data on the cosmopolitanism of many protist morphospecies, Finlay (2002) has proposed that an ubiquity and biodiversity transition occurs at a body length of between 1 and 10 mm. According to this model, essentially all species with body lengths larger than this transition size have distinct biogeographies, whereas essentially all organisms smaller than the transition size are ubiquitous (i.e., cosmopolitan).

These arguments draw attention to the topic of microbial biogeography and help to set conceptual limits on microbial biogeography and diversity.

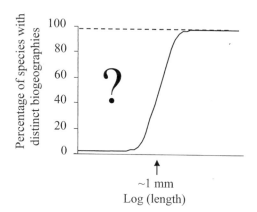

Figure 1. Conceptual figure showing alternative biogeography and body size models. Finlay (2002) has proposed that an ubiquity and biodiversity transition occurs at a body length of 1 to 10 mm, which suggests that many protozoa and nearly all prokaryotes are cosmopolitan (solid line). Alternatively, the apparent ubiquity and biodiversity transition may represent the size at which the morphospecies concept decays, and future molecular studies may show that most microbial species have distinct biogeographies, as do larger organisms (dotted line). It is likely that the percentage of microbes with distinct biogeographical patterns lies somewhere between the two extreme models (question mark).

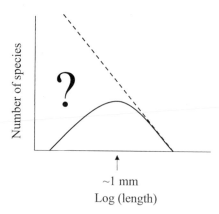

Figure 2. Conceptual figure showing the enormous effect of biogeography and body size models on biodiversity estimates. According to Finlay's model (2002), most of the species diversity on this planet would be in the size range of small insects (solid line). Alternatively, extrapolation of the body size and species diversity relationship (May, 1988) to include microbes would yield a vastly greater number of prokaryotic species (dotted line). In reality microbial diversity may lie between the two extremes (question mark).

They also clarify how little we know. Figure 1 shows Finlay's model (2002) of microbial biogeography (solid line), which invokes that essentially all prokaryotes are cosmopolitan. Although Finlay is dogmatic about this view, we believe that it is far from conclusive and present it here as a conceptual extreme in favor of microbial cosmopolitanism. The opposing extreme (dotted line) is that nearly all organisms have meaningful biogeographies at some taxonomic level, regardless of size. We believe it is important that microbiologists discuss, and eventually work toward resolving, the microbial biogeography debate because it has great implications for our understanding of biodiversity (Fig. 2). If Finlay is correct, then microbes are not particularly diverse (solid line). On the other hand, if microbial endemism is the norm, then the majority of diversity on our planet is microbial (dotted line). In the following sections we discuss some studies on microbial biogeography, which seem to suggest that Finlay's view may be correct on a certain taxonomic level—the protist morphospecies or the prokaryotic genus—but that some microbes have meaningful biogeographies below that level.

ARE MICROBIAL EUKARYOTES REALLY COSMOPOLITAN, OR DO WE NEED A BETTER MAGNIFYING GLASS?

The biogeography of free-living microbial eukaryotes has received a lot of attention; therefore it is significant that protozoologists seem to agree that most protist morphospecies are cosmopolitan. To illustrate this axiom, Finlay and Clarke (1999) used transmission electron microscopy to survey 25.2 μl of superficial sediment from Priest Pot, a small freshwater pond in the United Kingdom. In this tiny sample they observed siliceous scales representing 32 of the 41 known morphospecies of the ciliate *Paraphysomonas*. In addition, they found that the most numerous *Paraphysomonas* species in Priest Pot were also the most common *Paraphysomonas* species worldwide, as estimated from 73 published surveys of the genus. These data suggest that the more numerous *Paraphysomonas* species are more likely to be cosmopolitan because they produce larger numbers of cysts for dispersal.

More generally, Finlay et al. (1996) have pointed out that intensive surveying of benthic ciliates in several sites in Europe consistently recovered approximately 100 morphospecies, indicating that the diversity of ciliate morphospecies in that habitat is limited and well surveyed. However, it is unclear whether these studies recovered the same 100 morphospecies at all sites or whether some ciliates were particular to one site or another; therefore, it seems that the important details of biogeography were lost in the numbers. Finlay's group has also debunked several candidate endemics (e.g., Esteban et al., 2001), and these studies provide more definitive, if not universally applicable, arguments for the cosmopolitanism of protist morphospecies.

In a study designed to determine a relationship between protist size and cosmopolitanism, Wilkinson (2001) compiled data on polar surveys of testate amoebae. Contrary to the cosmopolitan morphospecies axiom, the survey of 13 Arctic and 13 Antarctic sites revealed that only 29 of the 127 (23%) morphospecies recorded in polar samples were found on both poles. It is important to note, however, that not finding an organism at one pole or another should not be regarded as evidence of its absence, particularly since sampling was limited. The author also noted that each of the 29 cosmopolitan species was small (<135 μm) and that a negative relationship exists between species size and the number of sites at which it was recorded. Based on this data, Wilkinson suggested that the maximum size for cosmopolitan testate amoebae is between 100 and 200 μm. However, an alternative explanation for this pattern would be that the apparent cosmopolitanism of small amoebae is an artifact of practical limitations of the morphospecies concept, as generally alluded to by Foissner (1999). Thus, it is important to ask: Is it species diversity (and therefore endemicity) that breaks down with decreasing size? Or is it our ability

to resolve morphologically similar (or identical) species that breaks down with decreasing size? It seems obvious that at some size or morphological complexity limit, the morphodiversity concept should give way to a more sensitive and less subjective molecular phylogenetic diversity concept. Perhaps Wilkinson has described this limitation rather than a true biogeographical pattern. Likewise, Finlay's ubiquity and biodiversity transition may instead be the morphodiversity-molecular phylogenetic diversity transition. In conclusion, it seems that molecular approaches would provide a better magnifying glass than the electron microscope alone for viewing the world of protist diversity and biogeography.

MANY PROKARYOTIC GENERA ARE COSMOPOLITAN

Bacteriologists have also focused on polar communities for studies of biogeography. Staley and Gosink (1999) hypothesized that gas vacuolate sea ice bacteria might be ideal candidates for endemism because they occupy specialized niches on opposite sides of the planet. Furthermore, these bacteria are the most psychrophilic organisms known, as many do not grow at temperatures of >10°C; thus, it was hypothesized that this extreme sensitivity may limit transit between the poles. In essence, their approach was to try to disprove their hypothesis by finding a species of gas vacuolate sea ice bacteria that inhabits both poles. Roughly 200 strains were grouped by fatty acid analyses and the unweighted pair group method with arithmetic mean (UPGMA) treeing and the most closely related representatives from each pole, representing three separate genera, were selected for 16S rDNA sequencing and pairwise DNA-DNA hybridization experiments. The end result was that the closest Arctic/Antarctic pairs differed by 11, 18, and 6 16S rDNA nucleotides for *Octadecabacter*, *Polaribacter*, and *Psychromonas* (formerly referred to as "*Iceobacter*"), respectively, and none exhibited more than 42% DNA-DNA hybridization. Thus, although the study recovered very close relatives from both poles, it failed to uncover a species that inhabits both poles (>70% DNA-DNA hybridization is the currently used criterion for delineating prokaryotic species [Wayne et al., 1987]). As with other studies, however, the absence of finding a cosmopolitan species in this study is not evidence that such species do not exist.

The study by Staley and Gosink (1999) exemplifies the general consensus among microbiologists that most prokaryotic genera are widely distributed in their respective habitats. Further examples of generic cosmopolitanism come from marine habitats at the opposite end of the thermal spectrum. Shallow hydrothermal fields, deep hot sediments, and hydrothermal vents are habitats for hyperthermophilic archaea and bacteria. A priori, it may seem that these habitats, too, are possible sites for endemic microbes since many are widely separated by oxygenated waters and most hyperthermophiles are extremely oxygen sensitive. However, the same genera of hyperthermophiles often inhabit distant thermal habitats. This pattern was first described in a pair of papers from Stetter's laboratory (Huber et al., 1990; Stetter et al., 1993). Enrichments from hydrothermal vent material from the Macdonald Seamount in Polynesia and from deep sediments from oil wells in the North Sea (off Scotland) and the North Slope of Alaska each showed high levels of DNA-DNA hybridization with probes derived from *Archaeoglobus*, *Thermococcus*, *Pyrococcus*, and *Pyrodictium* isolates from Vulcano, Italy. These studies also helped to decipher the mode of dissemination of marine hyperthermophiles. Using an extinction dilution procedure, Huber et al. (1990) found up to 10^6 viable anaerobic hyperthermophiles per liter of superficial seawater following eruptions of the Macdonald Seamount. Thus, it appears that hyperthermophiles are dispersed into the open ocean where they survive in metabolically inactive states and can be transported by currents.

More recently a very unusual hyperthermophile, "*Nanoarchaeum equitans*," was discovered from samples of hot rocks from the Kolbeinsey Ridge, a shallow hydrothermal system north of Iceland (Huber et al., 2002). "*N. equitans*" is among the smallest organisms known, a coccus with a diameter of 400 nm, and it seems to be an obligate extracellular symbiont of *Ignicoccus*. Phylogenetically, "*N. equitans*" represents a new phylum of *Archaea*, the "*Nanoarchaeota*," which was previously undetected in molecular ecology surveys because it has several base exchanges in the universal primer binding regions of the 16S rDNA. In an initial study of the distribution of the "*Nanoarchaeota*," Hohn et al. (2002) used primers specific for "*Nanoarchaeota*" 16S rRNA genes to amplify genes from DNA purified from a variety of hydrothermal habitats. One gene amplified from deep-sea vent material from the East Pacific Rise, off the west coast of Mexico, was identical to the "*N. equitans*" sequence. Thus, even marine hyperthermophiles that seem to require special symbioses can be cosmopolitan on the level of the 16S rRNA gene. But similarity in rRNA genes often belies differences throughout the rest of the genome and, ultimately, important ecological differences.

NARROWING THE FOCUS: THE GENOME AND ORGANISM SURROUNDING THE 16S RRNA GENE

The small-subunit RNA was chosen by Woese (1987) as the ideal molecule for creating the universal tree of life because it is critical for a major cell function, protein synthesis, and is therefore ubiquitous and highly constrained. The slow rate at which the gene evolves allowed comparison of even the most evolutionarily distant organisms. But it is precisely this slow rate of evolution that limits the gene's utility in subgenus-level biogeography studies. Examples of phenotypically and genetically (<70% DNA-DNA hybridization) distinguishable species of bacteria with zero to two 16S rDNA sequence differences are known (Stackebrandt and Goebel, 1994; Palys et al., 1997, 2000). Furthermore, phylogenetic analyses of protein-encoding genes from members of these closely related species have yielded clearly distinguishable groups with 6 to 10% nucleotide divergence between the species and 0 to 1.5% divergence among members of the same species (Palys et al., 2000).

A recent study of terrestrial thermophiles by Whitaker et al. (2002) is the most elegant example of microbial endemism to date and illustrates why the discovery of similar or identical small-subunit rRNA genes in distant locations is not necessarily evidence that the corresponding organism has no biogeography. Seventy-eight strains of *Sulfolobus* "*islandicus*" isolated from eight different sites in Iceland, the western United States, and Kamchatka, Russia, were used for the study. A 552-bp fragment of the 16S rRNA gene yielded only 10 polymorphic sites. However, sequencing of fragments of eight protein-encoding genes revealed that all but one was more variable than the 16S rDNA. When the nine gene fragments were concatenated and used in phylogenetic analyses, the Icelandic strains and the North American strains each formed monophyletic clades that were strongly supported. The North American isolates were further divided into groups from Lassen National Park, California, and Yellowstone National Park, Wyoming. Within the North American populations, even isolates from neighboring hot springs, 5.7 km distant, could be divided into monophyletic clades. The Russian strains were more diverse than the North American populations and formed two monophyletic clades that corresponded to the sampling locations from which the strains were isolated, the Mutnovskiy Volcano region and the Uzon Caldera and Geyser Valley region. Furthermore, in agreement with the conclusion that geographic isolation caused the phylogenetic pattern, and not an environmental factor, the authors noted a clear correlation between geographic distance and genetic distance for pairwise sequence comparisons, and the phylogenetic pattern could not be correlated with any environmental factor. Because *Sulfolobus "islandicus"* is composed of clearly distinguishable populations, this collection of strains is ideal for population ecology studies.

Cho and Tiedje (2000) examined the biogeography of fluorescent *Pseudomonas* strains and uncovered a more complex pattern of endemicity. The fluorescent pseudomonads comprise several closely related species. As a group, they are ubiquitous and typically abundant in soils, freshwater, and coastal marine sediments and water. Thus, superficially, it would seem that they are cosmopolitan bacteria with no meaningful biogeography. To dig deeper, the authors isolated 248 strains of fluorescent pseudomonads from 59 samples from Australia, South America (Chile), South Africa, and North America (California and Sakatchewan) and studied them on several taxonomic levels. On a coarse taxomonic level, the isolates were indeed cosmopolitan. Restriction analysis of PCR-amplified 16S rRNA genes with each of four different 4-bp recognition enzymes (amplified rDNA restriction analysis [ARDRA]) revealed only four types, two of which were found at every sampling region. To examine a finer taxonomic level, the authors amplified 16S-23S intergenic spacer (ITS) fragments and restricted them with each of three 4-bp recognition enzymes (ITS-restriction fragment length polymorphism [RFLP]), resulting in 39 types. Although most of the ITS-RFLP types were unique to a particular sample, suggesting endemicity, three types were found at more than one sampling region, indicating that the ITS-RFLP types are widely distributed. To focus even further, a PCR primer specific for repetitive extragenic palindromes was used in PCRs (BOX-PCR) and products were analyzed on polyacrylamide gels, resulting in 85 unique types. Each BOX-PCR type was unique to one site. Thus, a particular BOX-PCR type could be isolated from adjacent samples, between 5 and 175 m distant; however, it was never isolated from two locations, even sites with similar soil and vegetation within the same region. Similar to the analysis of *Sulfolobus "islandicus,"* the genetic distance between these BOX-PCR types correlated with the geographic distance, supporting the hypothesis that geographic separation allowed genetic differentiation. However, the relationships between distant BOX-PCR types were not easy to interpret because BOX-PCR's resolution saturates, preventing meaningful phylogenetic analyses between distantly related clades (Cho and Tiedje, 2000).

Roberts and Cohan (1995) studied the biogeography of desert-dwelling *Bacillus* species using an approach similar to that of Whitaker et al. (2002)

and concluded that two *Bacillus* species are cosmopolitan. Like *Pseudomonas, Bacillus* is a ubiquitous, abundant soil organism. Because it forms endospores, it could be thought of as the ideal candidate for cosmopolitanism because its hardiness enhances its survival during dispersion in comparison with other bacteria. Isolates from undisturbed soils in North Africa (Sahara), Asia (Gobi), and North America (Death Valley, Mojave, Sonora) were screened phenotypically, and approximately 100 *B. subtilis* and *B. mojavensis* strains were selected for detailed study. Three protein-encoding genes were PCR amplified from each strain, and nucleotide divergence was calculated from restriction patterns of PCR products digested with 4-bp recognition restriction enzymes. In contrast to the results for *Sulfolobus "islandicus,"* phylogenetic analyses of the *Bacillus* data did not produce clear biogeographic clades. Significantly, three genotypes were represented by isolates from separate continents, providing evidence for sub-16S rDNA level microbial cosmopolitanism. In addition, the sequence diversity within each *B. subtilis* local population (each desert) was similar to that of the world population (all deserts pooled), suggesting that each desert contains most or all of the world's diversity of these organisms, similar to the data Finlay and Clarke (1999) presented for *Paraphysomonas*. These data suggest that the rate of migration between the populations is sufficiently high to prohibit any divergence between populations on different continents. However, a closer look at the data suggests that the biogeography pattern is more complex. In particular, two well-separated clades were composed solely of Death Valley isolates (DV4-D-3 and DV2-D-1 groups) and could be regarded as candidate endemics.

CONCLUSIONS AND FUTURE DIRECTIONS

The existence and extent of endemism among free-living prokaryotes is frequently discussed and has important implications for our understanding of microbial evolution and diversity. Early in the past century it was stated that "everything is everywhere, the environment selects" (Baas-Becking, 1934), and this theory of microbial biogeography endures today in that microbiologists are still unaware of large taxonomic groups (genera, families, orders, etc.) that are restricted due to geographical barriers. However, recent investigations into microbial biogeography using molecular techniques have begun to elucidate a more complex pattern of prokaryotic biogeography below the genus level.

The emerging picture seems to be that some prokaryotes, for example, two species of *Bacillus* (Roberts and Cohan 1995), are cosmopolitan to the extent that isolates from opposite sides of the Earth are indistinguishable from each other by phylogenetic analyses of several protein-encoding genes, whereas others, for example, *S. "islandicus"* (Whitaker et al., 2002), consist of biogeographical groups that show strict endemism. The biogeographical pattern of a particular microbe may be related to its biology and its habitat, which together determine how easily it is dispersed, whether it is likely to remain viable during transport, and whether it is likely to encounter a favorable environment. From what we know of prokaryotic biogeography, cosmopolitanism does not necessarily follow global abundance, as has been suggested for microbial eukaryotes, since endemicity has been shown for fluorescent *Pseudomonas* isolates (Cho and Tiedje, 2000). Nevertheless, it seems reasonable to hypothesize that less-abundant microbes are more likely candidates for endemism. Although it is generally agreed that most protist morphospecies are cosmopolitan, protist biogeography has not yet been addressed by molecular approaches; therefore, details of the biogeography of microbial eukaryotes remain unresolved.

The studies by Whitaker et al. (2002) and Cho and Tiedje (2000) resoundingly dispel the myth that no microbes have biogeographies. Thus, future questions of microbial biogeography need not focus on whether microbial endemism exists. Rather, they should focus on how common endemism is and what a given biogeographical pattern means for the population ecology of a given microbe (e.g., migration rates, estimating times of microbial colonization events, whether biogeographic populations have diverged into separate ecotypes). Although we may be moving closer to understanding what a microbial species is (Cohan 2002), we believe that the use of different species definitions by different researchers, and even more so by researchers in different fields of biology, leads to unproductive arguments. For example, depending on what opinion a researcher is predisposed to, he or she might argue that the biogeographic *S. "islandicus"* clades described by Whitaker et al. (2002) are endemic species (an example of endemic species!) or that they are simply endemic populations (there are no endemic microbial species!). So, the challenge is this: can large numbers of closely related microbes be divided into biogeographic clusters? And do the genetic distances between clusters correlate with geographic distance or some obvious geographical demarcation? It seems that the approach of Whitaker et al. (2002), using phylogenetic analyses of concatenated gene fragments, provides the best model for biogeographical studies.

To get a handle on the extent and degree of endemism in the microbial world, we urge microbiologists to investigate microbes from different habitats and with different lifestyles. Marine plankton, for example, are ecologically important organisms that might be more likely candidates for cosmopolitanism due to constant mixing, yet the biogeography of planktonic bacteria has never been addressed.

REFERENCES

Baas-Becking, L. G. M. 1934. *Geobiologie of Inleiding Tot de Milieukunde*, p. 263. Van Stockkum & Zoon, The Hague: The Netherlands.

Brown, J. H., and M. V. Lomolino. 1998. *Biogeography*, 2nd ed. Sinauer Associates, Inc., Sunderland, Mass.

Cho, J.-C., and J. M. Tiedje. 2000. Biogeography and degree of endemism of fluorescent *Pseudomonas* strains in soil. *Appl. Environ. Microbiol.* 66:5448–5456.

Cohan, F. M. 2002. What are bacterial species? *Annu. Rev. Microbiol.* 56:457–487.

Diamond, J. 1999. *Guns, Germs, and Steel: the Fates of Human Societies*. W. W. Norton and Co., New York, N.Y.

Esteban, G. F., B. J. Finlay, N. Charubhun, and B. Charubhun. 2001. On the geographic distribution of *Loxodes rex* (Protozoa, Ciliophora) and other alleged endemic species of ciliates. *J. Zool. Lond.* 255:139–143.

Finlay, B. J. 2002. Global dispersal of free-living microbial eukaryote species. *Science* 296:1061–1063.

Finlay, B. J., and K. J. Clarke. 1999. Ubiquitous dispersal of microbial species. *Nature* 400:828.

Finlay, B. J., G. F. Esteban, and T. Fenchel. 1996. Global diversity and body size. *Nature* 383:132–133.

Foissner, W. 1999. Protist diversity: estimates of the near-imponderable. *Protist* 150:363–368.

Hohn, M. J., B. P. Hedlund, and H. Huber. 2002. Detection of 16S rDNA sequences representing the novel phylum "Nanoarchaeota": indication for a broad distribution in high temperature biotopes. *Syst. Appl. Microbiol.* 25:551–554.

Huber, H., M. J. Hohn, R. Rachel, T. Fuchs, V. C. Wimmer, and K. O. Stetter. 2002. A new phylum of Archaea represented by a nanosized hyperthermophilic symbiont. *Nature* 417:63–67.

Huber, R., P. Stoffers, J. L. Cheminee, H. H. Richnow, and K. O. Stetter. 1990. Hyperthermophilic archaebacteria within the crater and open-sea plume of erupting Macdonald Seamount. *Nature* 345:179–181.

May, R. M. 1988. How many species are there on earth? *Science* 241:1441–1449.

Palys, T., L. K. Nakamura, and F. M. Cohan. 1997. Discovery and classification of ecological diversity in the bacterial world: the role of DNA sequence data. *Int. J. Syst. Bacteriol.* 47:1145–1156.

Palys, T., E. Berger, I. Mitrica, L. K. Nakamura, and F. M. Cohan. 2000. Protein-coding genes as molecular markers for ecologically distinct populations. *Int. J. Syst. Environ. Microbiol.* 50:1021–1028.

Roberts, M. S., and F. M. Cohan. 1995. Recombination and migration rates in natural populations of *Bacillus subtilis* and *Bacillus mojavensis*. *Evolution* 49:1081–1094.

Stackebrandt, E., and B. M. Goebel. 1994. Taxonomic note: a

place for DNA-DNA reassociation and 16S rRNA sequence analysis in the present species definition in bacteriology. *Int. J. Syst. Bacteriol.* **44:**846–849.

Staley, J. T., and J. J. Gosink. 1999. Poles apart: biodiversity and biogeography of sea ice bacteria. *Annu. Rev. Microbiol.* **53:**189–215.

Stetter, K. O., R. Huber, E. Blöchl, M. Kurr, R. D. Eden, M. Fielder, H. Cash, and I. Vance. 1993. Hyperthermophilic archaea are thriving in deep North Sea and Alaskan oil reserves. *Nature* **365:**743–745.

Wayne, L. G., D. J. Brenner, R. R. Colwell, P. A. D. Grimont, O. Kandler, M. I. Krichevsky, L. H. Moore, W. E. C. Moore, E. Stackebrandt, M. P. Starr, and H. G. Trüper. 1987. Report of

the Ad Hoc Committee on reconciliation of approaches to bacterial systematics. *Int. J. Syst. Bacteriol.* **37:**463–464.

Whitaker, R., D. Grogan, and J. Taylor. 2002. Biogeographic patterns of divergence between populations of *Sulfolobus "islandicus."* *In* Abstracts of the 4th International Congress on Extremophiles 2002, Naples, Italy.

Wilkinson, D. M. 2001. What is the upper size limit for cosmopolitan distribution in free-living microorganisms? *J. Biogeog.* **28:**285–291.

Wilson, E. O. 1992. *The Diversity of Life.* Belknap Press, Cambridge, Mass.

Woese, C. R. 1987. Bacterial evolution. *Microbiol. Rev.* **51:**221–271.

Microbial Diversity and Bioprospecting
Edited by Alan T. Bull
© 2004 ASM Press, Washington, D.C.

Chapter 23

Mapping Microbial Biodiversity Case Study: The Yellowstone National Park Microbial Database and Map Server

DAPHNE L. STONER, RANDY LEE, LUKE WHITE, AND RON ROPE

We define biocartography as the analysis and display of microbiological data and associated geochemical, geological, and habitat information in a geographical format. Biocartography originates from the word cartography, which is the art and technique of making maps or charts. Although much of the handcrafted art of making maps is gone, we are now able to produce highly accurate maps using computers and geographic information system (GIS) software. A GIS is a computer-based tool for mapping and analyzing features and events that occur on Earth. GIS technology integrates common database operations, such as query and statistical analyses, with the unique visualization and geographic analysis benefits offered by maps. These abilities distinguish GIS from other information systems and make it valuable to a wide range of public and private enterprises for explaining events, predicting outcomes, and strategic planning. GIS saw its early development in the 1960s and is recognized as an essential tool in such diverse fields as natural resource management, environmental studies, ecology, urban response, and urban planning. The interactive capability of GIS allows the integration of data, the development of maps, and analysis of related information that satisfy the unique needs of each investigator.

THE YNP PROTOTYPE SYSTEM

A prototype system, YNP Microbial Database and Map Server (http://gis.inel.gov), represents a first application of an Internet-accessible GIS for microbial ecology and diversity (Stoner et al., 2001). The relational database was constructed using Microsoft Access (Microsoft, Inc., Redmond, Wash.) and linked to spatial data using ESRI Arc Internet Map Server software (ESRI Inc., Redlands, Calif.). The data compiled include base maps, thermal spring survey data from Yellowstone National Park (YNP), microbiological and geochemical data acquired from technical publications, and pictures. Octopus Spring, which is located in the Lower Geyser Basin, was used as an example to demonstrate several data access and display features.

GIS technologies are well suited for microbiological applications as the data system structure allows the integrations of microbiological, geochemical, and geographical data. GIS is a research tool that can be used for fundamental research of microbial ecology and diversity, biogeography, biogeochemistry, new product discovery, resource management, and environmental restoration activities. GIS has been demonstrated as a useful tool to track the dissemination of pathogens and to compile ancillary information needed to assess dispersal and infection rates (Kistemann et al., 2001; Strittholt et al., 1998; Zelicoff et al., 2001). GIS would also be useful to assess coevolution of pathogens with hosts and the transfer to new plant and animal hosts. The Rapid Syndrome Validation Project is an Internet-accessible GIS (http://rsvp.sandia.gov/) whose intended users are medical practitioners and state and federal health officials. The site is designed to act as a clearinghouse for the collection and rapid dissemination of information pertaining to emerging epidemics and diseases.

Biogeography, microbial ecology, and new product and species discovery are used to investigate the interaction of microbial species with their environment. Biogeography is considered an observational science for documenting and understanding the spa-

Daphne L. Stoner • Biotechnology Department, Idaho National Engineering and Environmental Laboratory, Idaho Falls, ID 83415-2203. **Randy Lee and Ron Rope** • Ecological and Cultural Resources, Idaho National Engineering and Environmental Laboratory, Idaho Falls, ID 83415-2213. **Luke White** • Programmatic Software Development, Idaho National Engineering and Environmental Laboratory, Idaho Falls, ID, 83415-3419.

tial distribution of organisms and the relationships between biodiversity and geochemical environments. Biogeography encompasses the study of the evolution, extinction, invasion, and dispersal of organisms and their evolution within specific locales. Microbial ecology examines the interactions among microbial species, microbial communities, the environment, and plant and animal species and the relationship of these interactions to diversity and function. The discovery of new biologically based products or detecting new microorganisms relies on understanding the relationships between habitat and microbial physiology. Resource management infers recognition of microbes as a natural resource and requires an inventory of species, their distribution, and habitat requirements so measures can be taken to address impacts from a variety of activities including research, recreation, and agriculture. Microbial cartography can address all these research needs.

Geographic positions alone provide minimal information to the biogeographer, microbial ecologist, or bioprospector. It is the integration of chemical, geological, geochemical, and physical data with the microbiological data that allows insight of relevance. GIS provides for such integration. GIS maps are compiled as layers (Color Plate 4) using spatial and tabulated data (Table 1). The spatial data provide the geographic context while the tabulated data provide

information that is linked to the geographic sites or areas. Spatial data include base maps, aerial and satellite imagery, photographs, and sample locations. In the YNP prototype system, spatial data include the relief map of Yellowstone National Park, topographic maps, roads, rivers, trails, aerial imagery, sample locations, and photographs for many of the thermal features. Maps can be created online by clicking on the attributes one wishes to include in the map (Color Plate 5).

Tabulated or attribute data consist of information that is conveyed in numerical or linguistic format. Tabulated data for the prototype system include dates and times, geographic position system (GPS) coordinates, feature or location names, microbiological data, geochemical data, physical data, methods, and source information, e.g., technical publications and unpublished sources. Ancillary information such as methods and sample collection facilitate the interpretation of the data. The data tables include the taxonomic, phylogenetic, and physiological data as well as the geochemical information required to understand the relationship between microbial diversity, activity, and habitat. Much of this information can be accessed through the search features that are available in the system (Color Plates 5 and 6). In general, GIS analysis results can be in map format (Color Plate 7), statistical summaries, or sorted data sets. However, it is the ability to conduct spatial analyses and display the results in the spatial context (a map) that is one of the greatest powers of GIS technology.

Tabulated data are compiled in a relational database where tables are linked to each other via a common attribute or field. All tables in the prototype system are linked either directly or indirectly to the General Field Data Table using sample, general field data, microbial, or reference identification numbers (Fig. 1). Any number of links can be formed among the tables in the database. Tables and links within the relational databases can be updated as the data requirements, concepts, and software capabilities change; however, developing a good database design from the onset is desirable to avoid the extra time required to change tables and associated links. In addition to its use in queries, the information in the data tables is compiled as a summary data report that is linked to each location (Fig. 2). The summary report includes the high, low, and average values of pH, temperature, and conductivity, as well as a list of the microbial species detected. Photographs are accessible via an active link to each site.

At present, the data tables related to microbial identification allow the inclusion of information pertaining to the sample, culture designation or name (e.g., genus, species, strain), culture bank identifica-

Table 1. Data types used in GIS

Data type	Definition	Examples
Spatial data		
Point	A single x-y coordinate	Sample sites, thermal springs
Line	Composed of a connected sequence of individual points	Rivers, streambed, roads, trails
Polygons	A closed area with shape and area characteristics	Wetlands, meadows, thermal basins, parking lots, areas with a specific soil type
Base maps	Maps or aerial photos showing basic features found on the Earth's surface	Digitized topographical maps, photographs, imagery
Attribute (tabulated) data		
Descriptive	Data that are incorporated in a primarily linguistic form	Investigator names, feature names, microbial identification information, methods, citations
Numerical	Data that are incorporated in a primarily numerical format	GPS locations, concentrations, temperature, pH, times and dates

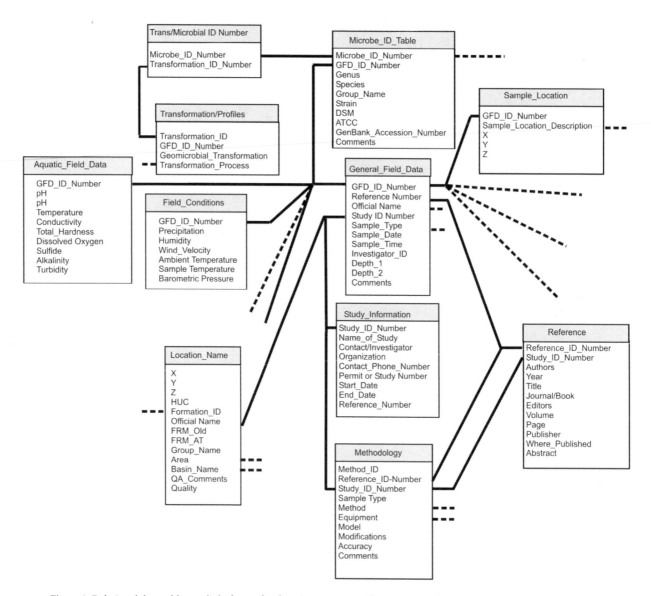

Figure 1. Relational data tables are linked to each other via common attributes. Depicted is a subset of the data tables from the prototype system, which shows the linking of various tables via common identification numbers.

tion numbers, and GenBank accession number. The data table architecture was designed to accommodate data for microorganisms that have been classified and formally named by molecular-based phylogenetic or phenotypic-based methods, as well as those whose presence are detected by metabolic activity or cellular constituents. Microorganisms can also be referred to by a strain, sample, or clone, e.g., "phylotype" designation, or grouped according to physiological, metabolic, morphological, or chemical attributes.

The knowledge and understanding that we now have about the microbial world has been gained from the integration of information acquired from a variety of techniques and approaches used. Unlike physical or chemical measurements such as tempera-

ture, pH, or an elemental analysis, there is not a standard method that is applied to the detection and identification of microorganisms or to the assessment of their function in the environment. To aid in data interpretation, the general approach used to detect, identify, or group microorganisms, e.g., cultivation-, molecular-, chemical analysis-, antigenic-, or spectral-based methods (Reysenbach et al., 1994; Shiea et al., 1991; Stahl et al., 1985; Stoner et al., 1994; Tayne et al., 1987; and Zeikus et al., 1979) has been noted in the database.

The linking of microbial data with geochemical and geophysical data is integral to the understanding of the interaction and function of microbiota within the environment. The YNP prototype system archi-

Octopus Spring Information

Overview conditions for Octopus Spring
225 records were found for this spring.

Average pH value: 8.236544
Maximum pH value: 9
Minimum pH value: 7.2

Average Temperature(C): 57.52648
Maximum Temperature: 91
Minimum Temperature: 23

Average Conductivity(uMho): 790.7808
Maximum Conductivity: 1700
Minimum Conductivity: 15.37

Microbe ID	Genus	Species	Strain
1	Synechoccus		
2	Chloroflexus		
4	Thermodesulfobacterium	commune	YSRB
5	Thermus	aquaticus	Y-VII-70
6	Thermus	aquaticus	Y-IV-69-1
7	Isocystis		
8	Bacillus		
9	Chloroflexus		
10	Methanobacterium		
11	Synechoccus		
12	Thermodesulfobacterium		
49	Chloroflexus		
50	Isocystis		
381	Thermus	sp	isolate ac-7
382	Thermus	sp	isolate ac-1
383	Thermus	sp	isolate ac-2
386	Clostridium	thermoautotrophicum	JW 701/3

Figure 2. Each site has a summary data table that can be accessed through a link associated with the site. Information included are the average, high, and low values for pH, temperature, and conductivity for the spring as well as the species and strains that have been detected at the site.

tecture includes geochemical, physical, general field data, and photos of the sampling areas. Geochemical data included in the data table represent the range of ionic and elemental species as well as organic and inorganic compounds that investigators have detected at the various locations throughout the park, as well as the general spring types that are present, e.g., acid sulfate, neutral chloride, calcium carbonate. General information and field data allow for the inclusion of time and date of sample collection or in situ measurement, the type of sample collected, ambient temperature, sample or in situ temperature, and specific comments pertaining to the acquisition or processing of the samples, e.g., pump sampling, grab sampling, or filtration.

FUTURE DEVELOPMENTS

Humans are limited in their ability to visualize a large number of dimensions or parameters, whereas computers are well suited for this type of activity. Even at its most rudimentary state of development, a GIS application, such as the prototype system described here, adds significant value to the data simply from the compilation of data, georeferencing the data, and the ability to access and display the information as a map. Currently, the prototype system can be used for selecting research sites and reviewing the limited amount of data and images that have been added to the system. Field sites can be identified and located using geographical coordinates or by searching for habitat characteristics such as pH and temperature, or sites where microorganisms have been detected can be located and data reviewed.

The prototype system is an example of what can be done with microbial diversity data and associated geochemical information. A fully developed system would require that the data table architecture be extended to include soil type, geological formation information, as well as host information such as plants or animals to examine their influences on microbial diversity. To examine spatial relationships below the submeter scales that are possible with current GPS systems would require a system of notation that would accommodate the finer resolution that would be of interest to microbiologists. The incorporation of time-dependent query capability would allow the generation of interactive maps that can be used to assess long-term trends in the distribution of microorganisms.

A great advance would occur through the development of software modules that could link GIS technology with phylogenetic tree capabilities or other molecular methods such as genome sequencing or microarray analysis. Phylogenetic trees linked with a GIS would allow the ready visualization of microbial diversity detected at a given location or area as well as the gaps in our knowledge of microbial diversity. Extending this capability to genomic sequences and microarray technologies would create yet another powerful research tool that could begin to address the fundamental questions regarding the interaction of microorganisms with the geochemical environment.

In summary, GIS is a research tool that can be used for fundamental research into microbial biodiversity and ecology, biogeography, biogeochemistry, new product discovery, and to support resource management and environmental restoration activities. Internet-accessible GIS applications provide a formal means to catalog microbiological data and facilitate acquisition and distribution of information. The continued compilation of information promotes and helps define gaps in our understanding of the distribution and diversity of microbiota and their functional role in the environment.

Acknowledgment. The authors are grateful for the support from the Laboratory Directed Research and Development Program under contract DE-AC07-99ID13727 from the U.S. Department of Energy to the Idaho National Engineering and Environmental Laboratory.

REFERENCES

Kistemann, T., F. Dangendorf, and M. Exner. 2001. A Geographical Information System (GIS) as a tool for microbial risk assessment in catchment areas of drinking water reservoirs. *Int. J. Hyg. Environ. Health* **203:**225–233.

Reysenbach, A.-L., G. S. Wickham, and N. R. Pace. 1994. Phylogenetic analysis of the hyperthermophilic pink filament community in Octopus Spring, Yellowstone National Park. *Appl. Environ. Microbiol.* **60:**2113–2119.

Shiea, J., S. C. Brassell, and D. M. Ward. 1991. Comparative analysis of free lipids in hot spring cyanobacterial and photosynthetic bacterial mats and their component photosynthetic bacteria. *Org. Geochem.* **17:**309–319.

Stahl, D. A., D. J. Lane, G. J. Olsen, and N. R. Pace. 1985. Characterization of a Yellowstone Hot Spring Microbial Community by 5S rRNA sequences. *Appl. Environ. Microbiol.* **49:**1379–1384.

Stoner, D. L., N. S. Burbank, and K. S. Miller. 1994. Anaerobic transformation of organosulfur compounds in microbial mats from Octopus Spring. *Geomicrobiol. J.* **12:**195–202.

Stoner, D. L., M. C. Geary, L. J. White, R. D. Lee, J. A. Brizzee, A. C. Rodman, and R. C. Rope. 2001. Mapping microbial biodiversity. *Appl. Environ. Microbiol.* **67:**4324–4328.

Strittholt, J. R., R. J. Garono, and P. A. Frost. 1998. Spatial Patterns in Land Use and Water Quality in the Tillamook Bay Watershed: a GIS Mapping Project. http://www.earthdesign.com/caforpt/report4b.html.

Tayne, T. A., J. E. Cutler, and D. M. Ward. 1987. Use of *Chloroflexus*-specific antiserum to evaluate filamentous bacteria of a hot spring microbial mat. *Appl. Environ. Microbiol.* **53:**1962–1964.

Zeikus, J. G., P. W. Hegge, and J. B. Andersen. 1979. *Thermoanaerobium brockii* gen. nov. and sp. nov., a new chemoorganotrophic, caldoactive, anaerobic bacterium. *Arch. Microbiol.* **122:**41–48.

Zelicoff, A. P., J. Brillman, D. W. Forslund, J. E. George, S. Zink, S. Koenig, T. Staab, G. Simpson, E. Umland, K. Bersell, R. M. Salerno, and G. Mann. 2001. The Rapid Syndrome Validation Project (RSVP) SAND No. 2001-2754J. Sandia National Laboratories, Livermore, Calif.

V. THE PARADIGM SHIFT: BIOINFORMATICS

PREAMBLE

The advent of the genomics era, and with it the burgeoning of bioinformatics, is transforming not only microbial prospecting but the whole of biology. In chapter 24, I argue the case for describing this transformation as a paradigm shift and justifying its status as a scientific revolution (sensu Kuhn). One consequence of genome sequencing and the wide span of functional genomics activities that has followed is a realization that the strong reductionist approach to biology in recent times has imposed severe limitations on our understanding of organismal complexity. The situation is posed succinctly by Nierman and Nelson (2002): "We have accumulated an impressive inventory of molecular mechanisms, components, knowledge of metabolic pathways and regulatory networks from reduction biology research, without making parallel advances towards a conceptual or theoretical framework of living systems." Interestingly, each of the contributors to this section sees the genomics revolution as leading to an holistic approach to biology aimed at understanding how genes and gene products integrate to produce a particular organism behavior or phenotype.

A substantial part of this section is devoted to the principal *"omic"* components of bioinformatics, namely, genomics, proteomics, and phenomics; Karen Nelson, Phil Cash, and Jennifer Reed and colleagues, respectively, guide us through the theory, practice, and critical evaluation of data and provide many examples of the ways in which these approaches are having an impact on contemporary biotechnology search and discovery activities. These contributors also reveal how rapidly the new technology is being adopted and how broad is its application, as strikingly illustrated by the penetration of microarray (DNA, protein) techniques. Microarrays now present exciting opportunities to tackle difficult problems in pathology and disease, microbial ecology, and bioprospecting, but also, returning to an earlier theme, they can catalyze "hypotheses, pointing the way to unexpected or unpredicted relationships between diverse sets of genes . . . and suggesting functional roles for proteins that have no known homologs" (Barry and Schroeder, 2000). Another feature of the paradigm shift is the reemergence of metabolic pathway engineering as a rational and applicable option for industrial process development (see chapters 25 and 27).

Accompanying these spectacular technological advances is the often-expressed view that a mere increase in the quantity of information does not per se lead to an increase in biological understanding. Moreover, as Attwood and Miller argue, "in the panic to automate the route from raw data to biological and medical insight, we are generating and propagating innumerable errors" (Attwood and Miller, 2001); this cautionary overview should be prescribed reading for all those espousing bioinformatics. In this context, I emphasize the problems of data and information integration and database interoperability (see chapter 24). This, in turn, forces us to define the boundaries of bioinformatics that from a biotechnology point of view, I would define in inclusive rather than in exclusive terms. Thus, although taxonomic databases, for example, are an important candidate for inclusion, there is an urgent need to establish unequivocal taxonomy-activity relationships. Bill Strohl in the next section (chapter 31) acknowledges that a primary use of taxonomic diversity in the pharmaceutical industry is as a surrogate for chemical diversity, but concludes that the approach is flawed. The contrary position is put by Alan Ward and Mike Goodfellow (chapter 28) who argue that improved taxonomic definition and more sophisticated screening procedures for chemical diversity will enable more reliable correlations to be made. Similarly, the integration of disparate information into biodiversity maps has been discussed in section IV. Mapping of a different kind has exercised Larry Wackett and his colleagues, who have established a Biocatalysis/Biodegradation Database that is founded on the principle of distinct organic functional groups, or inorganic elements, constituting unique compounds that can undergo transformations by specific enzyme types (http://umbbd.ahc.umn.edu/search/FuncGrps.html). The current database contains 50 functional groups, although natural product inventories reveal the existence of over 100. Wackett (2002) predicts that microbial catalysts will be found that will be able to metabolize many of these groups and natural inorganic compounds, and increasingly via metagenomic screening rather than from culture studies. Optimization of such a catalysis for defined biotechnology substrate targets then will be achieved by directed evolution (see chapters 24 and 33).

The traditional biological route and the bioinformatics route to discovery are discussed in chapter 24 (see Fig. 1 for a summary), and if the shift from the one to the other is indeed paradigmatic, we should be able to report initial achievements and identify deliverables. Kuhn (1970) asserted that paradigms achieve such status by being more successful than their competitors at problem solving. The early focus of genomics has been on medical problems such as infectious and metabolic diseases and on the search for drug and vaccine targets and new drugs. Probably in the long term the impact of pharmacogenomics (the interaction of genetic variation and drug intervention) will lead to the customizing of drug design and prescription. It has been estimated that the number of drug targets in humans could rise from the current 100 to as many as 10,000. Consequently, such large numbers of targets will necessitate the identity of protein domain families as targets for drug design. The recent article by Waldmann (2002) is very germane to this discussion. The large number of proteins in humans conform to 600 to 8,000 distinct domains, and Waldmann postulates that natural products bind to different proteins within a domain family; small-molecule scaffolds that bind to particular proteins are known as privileged structures. Thus, the discovery process might start with the bioinformatic identity of a defined protein domain for which ligands are already known and could serve as starting points for the design of novel drugs. Waldmann illustrates this strategy with reference to nakijiquinones (ex. marine sponge), the only known natural inhibitors of the Her-2/Neu receptor tyrosine kinase that is overexpressed in several carcinomers. Whereas none of the putative nakijiquinone-privileged structures was a significant inhibitor of the target kinase, a library of combinatorial analogues produced effective drug candidates, including the first known inhibitor of vascular endothelial growth factor 3. This result endorses the point made in chapter 1 about the relevance of basing combinatorial chemistry campaigns on natural product architechures.

Further evidence for the impact of the genomics revolution on biotechnology is provided in the excellent review of Nierman and Nelson (2002) and the following chapters in this section.

REFERENCES

Attwood, T. K., and C. J. Miller. 2001. Which craft is best in bioinformatics? *Comp. Chem.* 25:329–339.

Barry, C. E., III, and B. G. Schroeder. 2000. DNA microarrays: translational tools for understanding the biology of *Mycobacterium tuberculosis. Trends Microbiol.* 8:209–210.

Kuhn, T. S. 1970. *The Structure of Scientific Revolutions*, 2nd ed. University of Chicago Press, Chicago, Ill.

Nierman, W. C., and K. E. Nelson. 2002. Genomics for applied microbiology. *Adv. Appl. Microbiol.* 51:201–245.

Wackett, L. 2002. Expanding the map of microbial metabolism. *Environ. Microbiol.* 4:12–13.

Waldmann, H. 2002. Nature provides the answer. Natural product structures as guiding principle in combinatorial chemistry. *Screening* 6:46–48.

Microbial Diversity and Bioprospecting
Edited by Alan T. Bull
© 2004 ASM Press, Washington, D.C.

Chapter 24

The Paradigm Shift in Microbial Prospecting

ALAN T. BULL

In a manner reminiscent of the pioneering manned exploration of space, the general public and science community alike have become rather blasé about the advent of the genomics era, yet only 7 years have elapsed since the first complete genomic sequence—*Haemophilus influenzae*—was published (Fleischmann et al., 1995). This signal event, more than any other, was the portent of a major change in the way that biotechnology search and discovery could be conducted. This change is so profound that it truly embodies a paradigm shift. Paradigm, in Thomas Kuhn's defining terms (Kuhn, 1970), represents on the one hand "an entire constellation of beliefs, values, techniques and so on shared by members of a given community," and on the other hand "it denotes one sort of element in that constellation, the concrete puzzle-solutions which, employed as models or exemplars, can replace explicit rules as a basis for the solution of the remaining puzzles of normal science." Kuhn questioned what it was that the particular community shared and suggested "disciplinary matrix," "'disciplinary' because it refers to the common possession of the practitioners of a particular discipline; 'matrix' because it is composed of ordered elements of various sorts, each requiring further specification." The main sorts of elements in Kuhn's disciplinary matrix are (i) symbolic generalizations—expressions used without question by the community that can be set in a logical form; (ii) shared commitments—beliefs in particular models, including heuristic models; (iii) values, the most deeply held concerning predictions and hypothesis; and (iv) exemplars—the engagement with problem and solutions.

This brief reference to Kuhn's position is important, I believe, because while the term "paradigm" is increasingly used in a multiplicity of contexts, its use is often indiscriminate and inappropriate for changes that are incremental and predictable rather than those that are unpredictable and manifestly have the characteristics of scientific revolutions. Such issues should be kept in mind as we explore the posited paradigm shift from traditional microbiology to bioinformatics in the search for exploitable biology.

The traditional microbiology route is based on the isolation of cultivable organisms and the establishment of strain libraries that provide the raw material for the search strategy. Thereafter, targets are prescribed, specific assays developed, and collections of appropriate organisms screened for the desired property (Fig. 1). With the advent of PCR technology and the ability to recover amplifiable DNA from environmental samples, it has become possible to add as yet uncultured microorganisms to the screening pool in a variety of options including activity-based expression and metagenomic and biocombinatorial libraries (Fig. 1). However, as we have moved further into the postgenomics era, a whole raft of novel techniques have been developed and applied to exploit sequence information; these are the "omics" enterprises (Ward and White, 2002) based on comparative and functional versions of genomics, transcriptomics, proteomics, phenomics, metabolomics, etc. that are "coupled with a non-'omics' glue," bioinformatics (Fig. 1). In an excellent expositive article, Luscombe et al. (2001) define the aims of bioinformatics as (i) organizing data in a way that allows researchers to access extant information and to submit new data; (ii) developing tools and resources that aid data analysis; and (iii) using such tools for data analysis and interpretation in a biological meaningful manner. They propose the following definition: "Bioinformatics is conceptualizing biology in terms of macromolecules (in the sense of physical-chemistry) and then applying *'informatics techniques'* (derived from applied maths, computer science, and statistics) to *understand* and *organize* the *information* associated with these molecules, on a *large scale* . . . for . . . *practical applications*" (authors' key words are italicized). Essentially, bioinformatics refers to the computational working of omics data with the objective of solving

Alan T. Bull • Research School of Biosciences, University of Kent, Canterbury, Kent CT2 7NJ, United Kingdom.

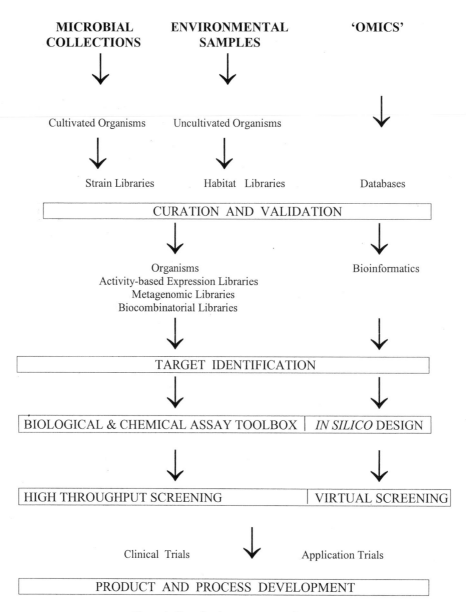

Figure 1. Biotechnology—routes to discovery.

biological problems. In the biotechnology search context (again after targets have been set), it enables in silico design and virtual screening that in turn can be translated into real chemical products. As Goodman (2002) observed, the precise boundaries of bioinformatics are elusive: thus the inclusion of taxonomic and phylogenetic databases is highly desirable in search and discovery operations. The sheer enormity of biological data being produced also is highlighted by Luscombe et al. (2001); for example, the GenBank entry of nucleic acid sequences that was 11,546,000 in April 2001 is 17,471,000 14 months later, while comparable entries for SWISS-PROT protein sequences were 95,320 and 111,820 (July 2002). Similarly, the number of complete genome sequences has

grown at an amazing rate: as of September 9, 2002, there were 115 microbial sequences (including two chromosomes) and another 584 genomes in progress of which 326 are *Bacteria* and 23 are *Archaea* (http://ergo.integratedgenomics.com/GOLD; see also Wheeler et al. [2002]).

The first step along the bioinformatics route is data collection but, because of the heterogeneity of the data and storage formats, the curation, retrieval, and analysis of information is an exacting process. Similarly, data quality and validation are essential prerequisites for the successful interrogation of databases. Note that entirely comparable problems are associated with curation and validation of microbial culture collections. The development of systems to store, analyze, and allow access to information is both complex and lagging behind the capacity to generate data (Valencia, 2002) and is further frustrated because most relevant data are not contained within databases but reside in the scientific and technical literature. Moreover, Attwood and Miller (2001) recently appraised the "unbidden troubles" presently contained in bioinformatics (troubles with counting genes, sequences, and structures; defining function prediction; homology; integrating tools and data) and concluded that none of the currently available tools is able to address biological complexity in an effective manner. These caveats are not highlighted in order to diminish the role of bioinformatics in biotechnology search and discovery, but rather to temper hasty and uncritical adoption of its tools.

One additional comment is necessary in light of the extraordinary pace of data acquisition and accumulation of data sets, namely, the concern that data-driven discovery (data mining) (see Brown and Botstein, 1999, for example) will displace traditional hypothesis-driven science. While some authors such as Brent (1999) have opined that the development of bioinformatics may presage a more observational phase of biology research, others have roundly condemned inductive reasoning and predict that "induction and data-mining, *uniformed by ideas* [my italics], can themselves produce neither knowledge nor understanding" (Allen, 2001). However, the view that informatics techniques can provide significant added value to traditional hypothesis-driven research if "investigators slightly alter their practice to take advantage of this opportunity" (Smalheiser, 2002) is the prevailing one in bioinformatics. The position is stated with admirable clarity by Kell and King (2000): scientific research deploys both deductive and inductive logic—in the former, given the truth of the axioms and observations, answers must be true; in the latter, even if all of the observations are true, inductive rules may be false. Both types of logic are needed to generate knowledge and understanding from the avalanche of data in this new postgenomic era: automated inductive reasoning to erect hypotheses or rules from the data, and deductions to confirm or refute the hypothesis and to open up new avenues of investigation.

INTEGRATION AND INTEROPERABILITY

I referred above to the extraordinary rate of biological data acquisition and deposition in the world's databases. The 2002 update of the Molecular Biology Database Collection alone lists 339 databases (Baxevanis, 2002), which represents a 20% increase from the previous year. These databases are classified by type (e.g., genomes, gene expression, proteins, RNA sequences, pathology), but the distinction between categories is often arbitrary and individual databases may provide more than one type of information. At the molecular level, research may be focused on the genome (DNA), transcriptome (mRNA), proteome (protein), or the metabolome (low-molecular-weight metabolites); at more holistic levels of biological organization, research focuses on the phenome and ecome, which can be defined in terms of phenotypes and ecological interactions. Apart from the genome, the other levels of investigation are context dependent and have generated their own appropriate techniques such as proteomics, metabolite measurements and flux analysis, and microarrays. Thus, in a major sense, the sequencing of the human genome simply heralds the even more demanding task of understanding biological function. Fuchs (2002) recently described the impact of bioinformatics on biological research: "The history of biology has been characterised by a continuous shift from the whole organism down to the molecular level . . . to today's molecular dissection of individual genes. It is likely that the sequencing of complete genomes will cause the pendulum to swing back in the other direction as [bioinformatics] forms a foundation on which we can gradually build a growing knowledge base of interactions, pathways, and increasingly more complex systems."

Various features of functional genomics are discussed in other chapters in this section of the book, and here brief reference is made to microarray technologies and their impact. High-density DNA microarrays comprising large gene sets in the form of oligonucleotides or PCR products were developed initially to monitor simultaneous changes in mRNA expression. The technology has developed to a state whereby (i) "conventional" glass substrate arrays containing 280,000 different nucleotides can be manufactured, thus enabling up to 10,000 genes to be monitored (Gingeras and Rosenow, 2000); (ii) fluidic arrays on microbeads for "massively parallel signature sequencing" can detect even rare mRNA species (Brenner et al., 2000); and (iii) protein microarrays can be used for quantitative measurements of mRNA translation, enzyme activity, and assessing protein-protein and protein-ligand interactions (Wilson and Nock, 2002). Other developments, including real-time analysis that can resolve binding kinetics, for example, are reviewed by Blohm and Guiseppie-Elie (2001). The possible applications of microarray technology are enormous (see "Applications" below) but, although the technology is at an early stage of development, several cautionary notes have been sounded.

Thus, the underlying assumption in functional genomics, that genes that appear to be expressed in similar patterns are related mechanistically, should be approached critically, and claims that similarity in expression corresponds to similarity in function warrant careful questioning Lucchini et al. (2001) make a similar point—the generation of reliable microarray data requires very rigorous experimental design—and they cite the transcriptional response of yeast to salt in this context. An exposure of 10 min to 0.4 M NaCl resulted in the induction of 1,300 genes, but only 170 induced genes were evident after 20 min (Posas et al., 2000).

Databases, irrespective of how many and how good, are but a component of the search and discovery armamentarium, and progress will best be made by integrating multiple sources of bioinformatics data. The situation is cogently stated by Edwards et al. (2002): "we are in a position, to a certain extent, to describe the details of *living*; the task that lies ahead is to integrate these details in such a way that we can better understand *life* as a whole." Integration of the molecular biology databases, although by no means perfect, has attracted serious attention and may be able to deliver useful information on issues such as a protein's structure, function, phylogenetic occurrence, expression, and protein-protein interactions (Gerstein, 2000). Similarly, the various bioinformatics research groups and commercial organizations are developing multidatabase query systems that enable interoperability of heterogeneous, formerly incompatible resources (see, for example, the ISYS platform [Siepel et al., 2001], www.negr.org/research/isys, and Valencia [2002]).

The problem of integration and interoperability becomes even more difficult when we wish to introduce microbiological and biodiversity data into the interrogation scenario. Despite the existence of large numbers of microbiology-related databases, it is difficult to secure answers to questions that rely on integrated information. Again the situation is further compromised by the variable quality and completeness of the data. An interoperable informatics platform that enabled microbial systematics, natural product, and ecological data to be interrogated would be an invaluable resource for biotechnology search activities. Such a resource is a distant goal, but current initiatives in biodiversity informatics are encouraging (Bisby, 2000). Some of these integrated databases have relatively restricted and specialist objectives; HICLAS (Hierachical Classification System, http://aims.cps.msu.edu/hiclas/), for example, seeks to compare and evaluate quantitative taxonomic analyses such as cladistics and classifications (Zhong et al., 1999), whereas the Taxonomy Workbench (http://mendel.imp.inivie.ac.at/taxonomy/) maps sequence data onto taxonomic trees (Wildpaner et al., 2001). This discussion is developed by Ward and Goodfellow (chapter 28), who examine the relationships between microbial taxonomy and natural products. The most ambitious biodiversity informatics project, the Global Biodiversity Information Facility (www.gbif.org), aims to coordinate and make interoperable worldwide biodiversity databases (Edwards et al., 2000) in order to provide a catalog of known organisms and a species bank compilation of facts about individual species. The Global Biodiversity Information Facility also intends to become interoperable with other biodiversity databases including sequence, geospatial, climate, ecological, and ecosystem data and therby enable questions to be posed that are germane to biotechnology search and discovery.

Ultimately, as Gerstein (2000) emphasizes, the most important issue in interactive database analysis and data mining lies in the drafting of scientifically apposite questions.

APPLICATIONS

"Paradigms gain their status because they are more successful that their competitors in solving . . . problems that the group of practitioners has come to recognise as acute" (Kuhn, 1970). Consequently, we need to examine the bioinformatics paradigm with respect to its success and potential in biotechnology and, particularly, in search and discovery. An earlier assessment of such impact was made by Bull et al. (2000). Deployment of bioinformatics in biotechnology and microbiology is now widespread and varied, and only a few pointers can be provided here to illustrate the trends. Although most attention has focused on medical applications, there is a growing appreciation of the benefits that bioinformatics techniques can bring to environmental and industrial microbiology.

Within the medical sector, bioinformatics objectives have included (i) the detection and identification of pathogens, (ii) the identity of drug targets, and (iii) the search for new drugs and vaccines. For example, a high-density microarray (53,660 probes) has been developed for a suite of pathogenic microorganisms (Wilson et al., 2002) that has good specificity and very high sensitivity (10 fg). The inability to culture the causal agent of syphilis, *Treponema pallidum*, highlights the benefits of genome sequencing in the study and fight against microbial pathogens. In their review of syphilis genomics, Norris and Weinstock (2001) emphasize that the genome sequence of *T. pallidum* will greatly facilitate research on its physiology, antigenesis, pathogenesis, and genetics that, in turn, can be expected to improve diagnosis and understanding of pathogenesis, give clues to the conditions for in vitro cultivation, and the development of effective vaccines. These authors also alert us to the limitations of genome sequencing and point to features of *T. pallidum* (and by extension all other organisms) that must be verified by experiment.

Drug Targets

Genomics provides an ever-increasing portfolio of genes across a wide range of organisms that can highlight potential targets that might be appropriate for pharmaceutical intervention; such targets include noninfectious and infectious diseases. Target identification uses strategies to identify taxon-specific genes and comparative genome studies of human (and other) disease states. Many human diseases involve multiple gene cascades, predisposition, and/or environmental interactions, a complexity that has led to the development of pharmacogenomics. Pharmacogenomics is concerned with identifying responders and nonresponders to candidate drugs and also adverse drug reactions (also termed toxicogenomics) (see Boorman et al., 2002). The discovery of single-nucleotide polymorphisms in the human genome opens up the possibility of devising drug regimes for the individual patient. Screening for taxon-specific genes can be made by transposon mutagenesis and PCR (Akerley et al., 1998), by various subtractive methods, and by differential genome analysis (e.g., *Helicobacter pylori* gene targeting [Huynen et al., 1998]). As the number of complete genome sequences of pathogenic microorganisms increases, so will the opportunities for bringing more such organisms into laboratory culture for improving the speed and accuracy of diagnostic tests, for the development of vaccines, and for an overall better understanding of pathogenic mechanisms. The use of DNA microarray analysis is well illustrated by recent work on *Candida albicans* (Hauser et al., 2002), the switch to the hyphal morphotype of which is a major virulence factor. Arrays containing most of the complete genome were used to analyze *C. albicans* grown under hyphae-inducing and noninducing conditions and hence to identify genes associated with transcription factors known to be implicated in the dimorphic switch.

Malaria remains a major human pathogen and killer for which genomics is providing new insights and options for drug and vaccine research and development. The genome of the malaria parasite (sequenced by the *Plasmodium falciparum* Genome Consortium [http://PlasmoDB.org; Gardner et al., 2002]) comprises three classes of DNA: a nucleus (about 5,300 genes) and two circular elements. One of the latter (35 kb in size) is contained within a discrete, recently identified organelle, the apicoplast (see Roos et al., 2002). The phylogenetic origin of the apicoplast DNA appears to be complex given its affinity with dinoflagellates and ciliates and with cyanobacteria. Roos et al. (2002) conclude that "a common ancestor of all apicoplexan parasites 'ate' a eukaryotic alga, whose ancestor had previously engulfed a cyanobacterium, and progeny parasites retained the genome of the algal plastid." The apicoplast genome has the distinction of having the lowest information content of any organellar genome, but the fact that apicoplexans have maintained the organelle suggests that it may have a crucial role, and the encoding of some enzymes involved in isoprenoid and fatty acid synthesis is emerging as one such role. Thus, the prokaryotic character of the apicoplast and the identity of some of its functions recommend it as a primary target for the development of new parasitocidal drugs.

Searching for New Drugs

The discovery of small-molecule drugs remains a primary activity in the pharmaceutical industry (Dean et al., 2001), and high-throughput screening (HTS) is the procedure used to screen natural and synthetic compound libraries. Dean and colleagues opine that the results of HTS have been less attractive than hoped; they offer two reasons for this situation: (i) the theoretical coverage of molecular diversity within the screening library is limited (cf. natural product diversity, see chapter 1) and (ii) the numbers of compounds that can be screened economically are small compared with the available chemical space. The latter limitation is being tackled with innovative in silico screening: "Simply generate all possible chemical combinations against a target (ligand), filter these and select out best candidates for synthesis and testing" (Scott, 2002). Scott discusses how the astronomical number of small organic molecules ($\sim 10^{200}$ for molecular weight ≤ 500) can be reduced logically to a representative set of in silico compounds and the success of the approach. Worst- and best-case hit rates using this technology are of the order of 1 and 6 to 7%, respectively (cf. random screening rates of 0.1 to 1%). However, the attractions of virtual screening notwithstanding, Dean et al. (2001) believe that "true *de novo* design of drug candidates is in a class of its own" and that new design algorithms can deliver large numbers of structural types for a given protein target site. They conclude that the coupling of such algorithms to large-scale structural genomics and HTS will deliver novel and patentable compounds within a fast time frame.

Rosamond and Allsop (2000) have reviewed the shift that has been made in the search for new antibiotics and other drugs from direct screening to rational target-based strategies, in particular transcript profiling using DNA microarrays. Referring to work on the immunosuppressive drug FK506 and on sterol biosynthesis in yeast, they demonstrate the efficacy of transcript profiling in revealing secondary drug targets, confirming mechanisms of drug action, and discriminating between drug candidates acting at different steps in a metabolic pathway. These authors believe that if microarrays can identify reproducible and statistically significant changes in global gene expression, the application of transcript profiling would streamline most phases of drug discovery and development.

Members of the genus *Streptomyces* have an unparalleled record as sources of bioactive natural products, and thus the sequencing of the genomes of *S. avermitilis* (Omura et al., 2001) and *S. coelicolor* A3(2) (formally a synonym of *S. violaceoruber* [Bentley et al., 2002]) are landmark events. The linear chromosome genomes of these species are about 8.7×10^6 bp in size and represent the largest bacterial genomes sequenced to date; that of *S. coelicolor* is predicted to contain 7,825 genes. Secondary metabolite gene clusters comprise a significant proportion of the genome of both species (at least 25 and >20 in *S. avermitilis* and *S. coelicolor*, respectively, which is equivalent to 6.4% of the genome in the former species), facts that resonate strongly with the predictions of Watve et al. (2001) on the predicted antibiotic potential of the genus (see chapter 1). In our opinion (Bull et al., 2000), functional genomics of *Streptomyces* species will be a powerful aid in discovering novel organisms and new bioactive natural products, identifying their roles in ecosystem function, and enabling improvements in bioprocess control. Given the diversity and significance of the streptomycete clade in bioactive metabolite production, sequencing selective members of the major clusters is justified to determine how representative are those of *S. avermitilis* and *S. coelicolor*.

Functional genomics also is being brought to bear on vaccine development, a good example being research on *Haemophilus influenzae* (Moxon et al., 2002). The lipopolysaccharide (LPS) of the outer membrane of this bacterium is implicated at each stage of the pathogenicity and has been the focal point of vaccine development work. The complete genome sequence enabled the identification of 25 candidate LPS genes, about 60% of which could not have been identified reliably from DNA sequence data alone. Such information together with a range of LPS mutants constructed by Moxon's group has allowed correlation to be made between LPS structure and virulence. Thus, the comparatively conserved heptose-containing inner core region of the molecule appears to be a good candidate for a lower respiratory tract vaccine. Genomic approaches also are being used in attempts to develop vaccines for *Leishmania*; the genome is a similar size to that of *Plasmodium falciparum* (35 Mb, ca. 8,500 genes) and is expected to be sequenced by the end of 2002 (http://www.ebi.ac.uk/parasites/leish.html). The approach adopted by Almeida et al. (2002) is to identify genes expressed in those forms of the parasite that invade and survive in the host; of the 1,094 unique genes that are specifically upregulated in amastigotes, 100 new vaccine candidates have been identified and are being tested.

Environmental and Industrial Microbiology

Bioinformatics techniques are starting to be introduced within an ever-increasing range of basic and applied microbiology. For example, Cho and Tiedje (2001) have provided proof-of-concept support for DNA:DNA hybridization using genome fragment and DNA microarrays as a replacement for the conventional DNA reassociation procedure (see chapter 3), and microarray technology also can be deployed for direct detection of microorganisms in environmental samples. The genomes of a large number of bacteria and archaea used in, or having potential as, bioremediation agents currently are being sequenced (http://www.sc.doe.gov). They include degraders of polychlorinated biphenyls and polyaromatic hydrocarbons (*Burkholderia* sp. LB400; *Pseudomonas fluorescens* PF0-1), metal-sequestering organisms (*Acidithiobacillus ferroxidans*, *Ralstonia metallidurans* CH34), and metal and sulfate reducers (*Desulfovibrio vulgaris*). Comparative genomics also is providing new insights into the mechanisms of adaptation to extreme conditions such as alkalinity (Takami et al., 2000), radiation (Makarova et al., 2001; Earl et al., 2002), and high and low temperatures. Metabolic engineering is another area in which theoretical tools and informatics are required to engage with the great complexity of biosynthetic networks. Recently Van Dien and Lidstrom (2002) have modeled the central metabolic network of *Methylobacterium extorquens* AM1 and used genome data to compensate for enzymes not previously reported for this bacterium. Modeled values for biomass yields, and theoretical behavior of null mutants for central pathway enzymes, agreed very well with experimentally determined values and provided greater understanding of the metabolic capabilities of biotechnologically relevant organisms.

A final illustration of the ambitious use to which bioinformatics is being explored for industrial processing is through the construction of in silico strains that can be used for modeling process optimization and for producing novel real organisms by genome engineering customized for specific applications. Palsson and colleagues (Schilling et al. [1999] and chapter 26) developed an algorithm for constructing genome-specific stoichiometric matrices that enable metabolic capabilities to be determined and metabolic phenotypes to be predicted. The algorithm was tested by constructing an in silico strain of *Escherichia coli* K-12 comprising 587 open reading frames that accounted for the metabolic properties of the real bacterium. Schilling et al. (1999) showed how in silico strains can be used for a variety of virtual investigations that are highly relevant to industrial processes, including the prediction of metabolic shift following gene deletions and multiple gene knockouts and metabolic phenotypes under different growth conditions. The E-CELL project launched by a group at Keio University (Tomita, 2001; http://www.e-cell.org) also has the objective of constructing virtual whole-cell models. This group initially constructed a hypothetical cell largely based on the genes of *Mycoplasma genitalium*. This virtual "self-surviving cell" model comprised 127 genes that accommodated sufficient information for transcription, translation, energy production, and phospholipid synthesis (Tomita, 1999). The Keio group point out that many of the genes of natural microorganisms are unnecessary for application in industrial environments, which, as opposed to ecosystem conditions, are closely controlled; indeed their expression represents an energy and materials drain and thereby compromises the yield of desired products. Consequently, genome trimming is being prioritized as a means of optimizing the productivity of industrial strains.

DIRECTED EVOLUTION

Introduction

The final stratagem to be considered here for producing novel exploitable biology is directed evolution, which, as genome sequencing projects continue to grow, promises to become a principal route for search and discovery. Evolution via natural and artificial (sensu traditional animal and plant breeding) selection is dependent on genetic variation, inheritance, and differential reproduction. Directed evolution in vitro manipulates these same factors but the technology—genomics, molecular structure determination, and monitoring expression levels—is able

to provide unparalleled levels of evolution (Bull and Wichman, 2001). As Schmidt-Dannert (2001) has pointed out, although directed evolution aims to mimic natural evolution, the two are fundamentally different. Thus, in vitro evolution is a process guided toward a predetermined goal resulting largely in the accumulation of adaptive mutations, whereas natural evolution accumulates adaptive and neutral mutations; and properties targeted in in vitro evolution "often go beyond requirements that would make biological sense." Genetic variation in vitro can be generated by random-point mutation and the improved variants passed to iterative rounds of mutation and selection in order to accumulate further improved variants.

Random-point mutagenesis is not suitable for introducing simultaneous amino acid changes, and for this purpose the complementary recombination technique of gene shuffling is used. Thus, "family shuffling" consists of recombining homologous genes in vitro or in vivo using a range of methods. Recently work in Arnold's laboratory (Joern et al., 2002) on the composition of shuffled gene libraries has produced valuable insights into recombination efficiency and guides for optimization protocols. Libraries constructed from several dioxygenase genes revealed biases both in the points of crossover and in which parent genes were involved. Such information is important in devising strategies for avoiding or minimizing duplicate chimeras, the reassembly of full-length wild-type genes, and for identifying key sequence-function relationships. Family shuffling can accommodate large numbers of homologous genes as illustrated by work by Stemmer's group; over 20 human alpha interferon genes were shuffled that eventually led to a chimeric cytokine with a 185-fold improved activity over the best parent (Chang et al., 1999).

The pace of technology development and of the applications (see below) of directed evolution have been spectacular. Some indication of the technology innovations is given by the following examples. (i) Directed evolution can be accelerated in mutator strains that have defects in DNA repair mechanisms, whereas subsequent restoration of normal low mutation rates to prevent the accumulation of undesirable spontaneous mutations can be achieved by curing the bacteria of mutator plasmids (Selifonova et al., 2001). (ii) Pairwise deletions or repeats in fragments of the same gene can be obtained via exonuclease treatment; by controlling the truncation, mutagenesis can be limited to a target sequence (Pikkemaat and Janssen, 2002). (iii) Nonhomologous recombination methods have been developed for exploring larger sequence spaces.

Applications

The uses to which directed evolution is being put are truly enormous and certain to grow in scale and variety. The targets for directed evolution span single genes, metabolic pathways, and whole genomes (Table 1). For further details of these and other applications, the reader is referred to Bull and Wichman (2001), Orencia et al. (2001), Powell et al. (2001), Schmidt-Dannert (2001), and Zhao et al. (2002).

Rational Design?

The irrational design approach embodied in directed evolution has its counterpart in the rational design of proteins, whereby specific amino acids are manipulated on the basis of knowledge of protein structure and function. Zhao et al. (2002) opine that recent advances in structure determination and increased computing power are improving the prospects for rational design and that, when combined with directed evolution, will provide the most powerful approach to biocatalyst development. An illustration of this combined technology is found in the engineering of indole-3-glycerol phosphate synthase to very low-activity phosphoribosylanthranilate isomerase; subsequent DNA shuffling produced an engineered variant that had a sixfold higher isomerase activity than the wild type and no synthase activity (Altamirano et al., 2000).

Finally, the consensus approach for stabilizing proteins should be mentioned as another bioinformatics-based route to the design of improved or novel molecules. Comparison of amino acid sequences of homologous proteins is made and a consensus sequence calculated using a sequence analysis software package. The consensus approach does not require three-dimensional structures, and a given residue is replaced only by an amino acid that has already proven its "evolutionary fitness" at the corresponding position of at least one other homologous wild-type protein. Lehmann et al. (2002) have used the technique to increase the thermostability of wild-type

fungal phytase (see chapter 1) from 48 to 90°C in a consensus phytase.

REFERENCES

Akerley, B. J., E. J. Rubin, A. Camilli, D. J. Lampe, H. M. Robertson, and J. J. Mekalanos. 1998. Systematic identification of essential genes by *in vitro* mariner mutagenesis. *Proc. Natl. Acad. Sci. USA* **95**:8927–8932.

Allen, J. F. 2001. Bioinformatics and discovery: induction beckons again. *BioEssays* **23**:104–107.

Almeida, R., A. Norrish, M. Levick, D. Vetrie, T. Freeman, et al. 2002. From genomes to vaccines: *Leishmania* as a model. *Phil. Trans. R. Soc. Lond. Ser. B* **357**:5–11.

Altamirano, M. M., J. M. Blackburn, C. Aguayo, and A. R. Fersht. 2000. Directed evolution of new catalytic activity using the alpha/beta-barrel scaffold. *Nature* **403**:617–622.

Attwood, T. K., and C. J. Miller. 2001. Which craft is best in bioinformatics? *Comp. Chem.* **25**:329–339.

Baxevanis, A. D. 2002. The molecular biology database collection: 2002 update. *Nucleic Acids Res.* **30**:1–12.

Bentley, S. D., K. F. Chater, A. M. Cerdeno-tarraga, G. L. Chalis, N. R. Thomson, et al. 2002. Complete genome sequence of the model actinomycete *Streptomyces coelicolor* A3(2). *Nature* **417**:141–147.

Bisby, F. A. 2000. The quiet revolution: biodiversity informatics and the Internet. *Science* **289**:2309–2312.

Blohm, D. H., and A. Guiseppie-Elie. 2001. New developments in microarray technology. *Curr. Opin. Biotechnol.* **12**:41–47.

Boorman, G. A., S. P. Anderson, W. M. Casey, R. H. Brown, L. M. Crosby, et al. 2002. Toxicogenomics, drug discovery, and the pathologist. *Toxicol. Pathol.* **30**:15–27.

Brenner, S., M. Johnson, J. Bridgham, G. Golda, D. H. Lloyd, et al. 2000. Gene expression analysis by massively parallel signature sequencing (MPSS) on microbead arrays. *Nat. Biotechnol.* **18**:630–634.

Brent, R. 1999. Functional genomics: learning to think about gene expression data. *Curr. Biol.* **9**:R338–R341.

Brown, P. O., and D. Botstein. 1999. Exploring the new world of the genome with DNA microarrays. *Nat. Genet. Suppl.* **21**:33–37.

Bull, A. T., A. C. Ward, and M. Goodfellow. 2000. Search and discovery strategies for biotechnology: the paradigm shift. *Microbiol. Mol. Biol. Rev.* **64**:573–606.

Bull, J. J., and H. A. Wichman. 2001. Applied evolution. *Annu. Rev. Ecol. Syst.* **32**:183–217.

Chang, C. C. J., T. T. Chen, B. W. Cox, G. N. Dawes, W. P. C. Stemmer, et al. 1999. Evolution of a cytokine using DNA family shuffling. *Nat. Biotechnol.* **17**:793–797.

Cho, J.-C., and J. M. Tiedje. 2001. Biogeography and degree of endemicity of fluorescent *Pseudomonas* strains in soil. *Appl. Environ. Microbiol.* **66**:5448–5456.

Dean, P. M., E. D. Zanders, and D. S. Bailey. 2001. Industrial-scale genomics-based drug design and discovery. *Trends Biotechnol.* **19**:288–292.

Earl, A. M., S. K. Rankin, K. P. Kim, O. N. Lamendola, and J. R. Battista. 2002. Genetic evidence that the *usvE* gene product of *Deinococcus radiodurans* R1 is a UV damage endonuclease. *J. Bacteriol.* **184**:1003–1009.

Edwards, J. L., M. A. Lane, and E. B. Nielsen. 2000. Interoperability of biodiversity databases: biodiversity information on every desktop. *Science* **289**:2312–2314.

Edwards, J. S., M. Covert, and B. Palsson. 2002. Metabolic modelling of microbes: the flux-balance approach. *Environ. Microbiol.* **4**:133–140.

Table 1. Impact of directed evolution on biotechnology

Category	Target
Enzymes	Activity, specificity, enantioselectivity, stability
Proteins	Antibodies, receptors, fluorescence signal
Pathways	Novel chemicals (e.g., carotenoids), improved product yields (e.g., antibiotics), pH tolerance, detoxification (e.g., arsenate)
Other	Viruses (gene therapy, vaccines), control of crystal growth (nanotechnology application), predicting emergence of antibiotic resistance

Fleischmann, R. D., M. D. Adams, O. White, R. A. Clayton, E. F. Kirkness, et al. 1995. Whole-genome random sequencing and assembly of *Haemophilus influenzae* Rd. *Science* **269**:496–512.

Fuchs, R. 2002. From sequence to biology: the impact on bioinformatics. *Bioinformatics* **18**:505–506.

Gardner, M. J., N. Hall, E. Fung, O. White, M. Berriman et al. 2002. Genome sequence of the human malaria parasite *Plasmodium falciparum*. *Nature* **419**:498–511.

Gerstein, M. 2000. Integrative database analysis in structural genomics. *Nat. Struct. Biol.* **7**:960–963.

Gingeras, T. R., and C. Rosenow. 2000. Studying microbial genomics with high-density oligonucleotide arrays. *ASM News* **66**:463–469.

Goodman, N. 2002. Biological data becomes computer literate: new advances in bioinformatics. *Curr. Opin. Biotechnol.* **13**:68–71.

Hauser, N. C., K. Fellenberg, and S. Rupp. 2002. How to discover pathogenic mechanisms. *Screening* **4**:28–31.

Huynen, M., T. Dandekar, and P. Bork. 1998. Differential genome analysis applied to the species-specific features of *Helicobacter pylori*. *FEBS Lett.* **426**:1–5.

Joern, J. M., P. Meinhold, and F. H. Arnold. 2002. Analysis of shuffled gene libraries. *J. Mol. Biol.* **316**:643–656.

Kell, D. B., and R. D. King. 2000. On the optimization of classes for the assignment of unidentified reading frames in functional genomics programmes: the need for machine learning. *Trends Biotechnol.* **18**:93–98.

Kuhn, T. S. 1970. *The Structure of Scientific Revolutions*, 2nd ed. University of Chicago Press, Chicago, Ill.

Lehmann, M., C. Loch, A. Middendorf, D. Studer, S. F. Lassen, et al. 2002. The consensus concept for thermostability engineering of proteins: further proof of concept. *Prot. Eng.* **15**:403–411.

Lucchini, S., A. Thompson, and J. C. D. Hinton. 2001. Microarrays for microbiologists. *Microbiology (UK)* **147**:1403–1414.

Luscombe, N. M., D. Greenbaum, and M. Gerstein. 2001. What is bioinformatics? a proposed definition and overview of the field. *Methods Inform. Med.* **40**:346–358.

Makarova, K. S., L. Ararvind, Y. I. Wolf, R. L. Tatusov, K. W. Minton, E. V. Koonin, and M. J. Daly. 2001. Genome of the extremely radiation-resistant bacterium *Deinococcus radiodurans* viewed from the perspective of comparative genomics. *Microbiol. Mol. Biol. Rev.* **65**:44–79.

Moxon, E. R., D. W. Hood, N. J. Saunders, E. K. H. Schweda, and J. C. Richards. 2002. Functional genomics of pathogenic bacteria. *Phil. Trans. R. Soc. Lond. Ser. B* **357**:109–116.

Norris, S. J., and G. M. Weinstock. 2001. The genome sequence of *Treponema pallidum*, the syphilis spirochaete: will clinicians benefit? *Curr. Opin. Infect. Dis.* **13**:29–36.

Omura, S., H. Ikeda, J. Ishikawa, A. Hanamoto, C. Takahashi, et al. 2001. Genome sequence of an industrial microorganism *Streptomyces avermitilis*: deducing the ability of producing secondary metabolites. *Proc. Natl. Acad. Sci. USA* **98**:12215–12220.

Orencia, M. C., J. S. Yoon, J. E. Ness, W. P. C. Stemmer, and R. C. Stevens. 2001. Predicting the emergence of antibiotic resistance by directed evolution and structural analysis. *Nat. Struct. Biol.* **8**:238–242.

Pikkemaat, M. G., and D. B. Janssen. 2002. Generating segmental mutations in haloalkane dehalogenase: a novel part in the directed evolution toolbox. *Nucleic Acids. Res.* **30**:e35.

Posas, F., J. R. Chambers, J. A. Heyman, J. P. Hoeffler, E. de Nadal, and J. Arino. 2000. The transcriptional response of yeast to saline stress. *J. Biol. Chem.* **275**:17249–17255.

Powell, K. A., S. W. Ramer, S. B. del Cardayré, W. P. C. Stemmer,

M. B. Tobin, et al. 2001. Directed evolution and biocatalysis. *Angew. Chem. Int. Ed.* **40**:3948–3959.

Roos, D. S., M. J. Crawford, R. G. K. Donald, M. Fraunholz, O. S. Harb, et al. 2002. Mining the *Plasmodium* genome database to define organellar function: what does the apicoplast do? *Phil. Trans. R. Soc. Lond. Ser. B* **357**:e1–e12.

Rosamond, J., and A. Allsop. 2000. Harnessing the power of the genome in the search for new antibiotics. *Science* **287**:1973–1976.

Schilling, C. H., J. S. Edwards, and B. O. Palsson. 1999. Towards metabolic phenomics: analysis of genomic data using flux balances. *Biotechnol. Prog.* **15**:288–295.

Schmidt-Dannert, C. 2001. Directed evolution of single proteins, metabolic pathways, and viruses. *Biochemistry* **40**:13125–13134.

Scott, R. K. 2002. Chemical space in *in silico* screening. *Screening* **4**:32–34.

Selifonova, O., F. Valle, and V. Schellenberger. 2001. Rapid evolution of novel traits in microorganisms. *Appl. Environ. Microbiol.* **67**:3645–3649.

Siepel, A., A. Farmer, A. Toopko, M. Zhuang, P. Mendes, W. Beavis, and B. Sobral. 2001. ISYS: a decentralized, component-based approach to the integration of heterogenous bioinformatics resources. *Bioinformatics* **17**:83–94.

Smalheiser, N. R. 2002. Informatics and hypothesis-driven research. *EMBO Rep.* **3**:702–703.

Takami, H., K. Nakasone, Y. Takaki, G. Maeno, R. Sasaki, et al. 2000. Complete genome sequence of the alkaliphilic *Bacillus halodurans* and genomic sequence comparison with *Bacillus subtilis*. *Nucleic Acids Res.* **28**:4317–4331.

Tomita, M. 1999. E-CELL: software environment for whole cell simulation. *Bioinformatics* **15**:72–84.

Tomita, M. 2001. Towards computer aided design (CAD) of useful microorganisms. *Bioinformatics* **17**:1091–1092.

Valencia, A. 2002. Search and retrieve. Large-scale generation is becoming increasingly important. *EMBO Rep.* **3**:396–405.

Van Dien, S. J., and M. E. Lidstrom. 2002. Stoichiometric model for evaluating the metabolic capabilities of the facultative methylotroph *Methylobacterium extorquens* AM1, with application to reconstruction of C_3 and C_4 metabolism. *Biotechnol. Bioeng.* **78**:296–312.

Ward, D. C., and D. C. White. 2002. The new omics era. *Curr. Opin. Biotechnol.* **13**:11–13.

Watve, M. G., R. Tickoo, M. M. Jog, and B. D. Bhole. 2001. How many antibiotics are produced by the genus *Streptomyces*? *Arch. Microbiol.* **176**:386–390.

Wheeler, D. L., D. M. Church, A. E. Lash, D. D. Leipe, T. L. Madden, et al. 2002. Databases resources of the National Center for Biotechnology Information: 2002 update. *Nucleic Acids Res.* **30**:13–16.

Wildpaner, M., G. Schneider, A. Schleiffer, and F. Eisenhaber. 2001. Taxonomy workbench. *Bioinformatics* **17**:1179–1182.

Wilson, D. S., and S. Nock. 2002. Functional protein microarrays. *Curr. Opin. Chem. Biol.* **6**:81–85.

Wilson, W. J., C. L. Strout, T. Z. DeSantis, J. L. Stilwell, A. V. Carrano, and G. L. Andersen. 2002. Sequence-specific identification of 18 pathogenic microorganisms using microarray technology. *Mol. Cell. Probes* **16**:119–127.

Zhao, H., K. Chockalingam, and Z. Chen. 2002. Directed evolution of enzymes and pathways for industrial biocatalysts. *Curr. Opin. Biotechnol.* **13**:104–110.

Zhong, Y., Y. Luo, S. Pramanik, and J. H. Beaman. 1999. HI-CLAS: a taxonomic database system for displaying and comparing biological classification and phylogenetic trees. *Bioinformatics* **15**:149–156.

Microbial Diversity and Bioprospecting
Edited by Alan T. Bull
© 2004 ASM Press, Washington, D.C.

Chapter 25

Genomics

Karen E. Nelson

The availability of the complete genome sequence of the free-living bacterium *Haemophilus influenzae* (Fleischmann et al., 1995) in 1995 opened the field of microbial genomics. Since then, more than 100 microbial genomes have been completely sequenced and published, and another 200 are estimated to be in progress worldwide. In the early stages of the development of this field, microbial choices for whole-genome sequencing were clearly geared toward organisms of medical importance such as *H. influenzae* (Fleischmann, et al., 1995) and *Mycoplasma genitalium* (Fraser et al., 1995), undoubtedly because the characterization of the major pathogens was anticipated to increase our understanding of the basic biology of these species and increase the opportunities for the identification of antimicrobial targets (Hoffman et al., 1998). This initial focus on pathogenic species later changed to include the sequencing of microbes of agricultural environmental, evolutionary, and biotechnological importance (for a review, see Nelson et al., [2000]). In addition to being able to gain insights into lateral gene transfer (Nelson et al., 1999), environmental applications (Nelson et al., 2002) and virulence mechanisms in many of these species (Tettelin et al., 2002, 2001, 2000), the technique of genome sequencing has allowed us to enter other avenues where genomic information previously could not be derived from microbial populations.

For example, the high level of success associated with the complete sequencing of the genomes of cultivated microbial species has increased possibilities for deciphering the genetic information contained in organisms that have not been cultured, as well as to organisms that are not in pure culture. Deciphering the genetic information of uncultured species can currently be achieved by sequencing genomic libraries that are created directly from environmental DNA (Beja et al., 2002a). This development has taken advantage of the successful construction of large insert libraries up to 120 kb in size, as well as our ability to

generate a vast number of sequences at a relatively low cost from any environment of choice. Initial surveys have demonstrated an unanticipated level of microbial diversity that remains to be explored (Beja et al., 2002a and 2000b).

Beyond sequencing, there have been major advances in the field of functional genomics where whole genomes are being characterized in more detail using proteomic and microarray technologies. DNA microarrays that allow for the determinatin of genes that are turned on or off under different environmental conditions on a genome-wide scale and comparative genome hybridization (CGH) studies that employ DNA microarrays are also revealing the extent of diversity across related and unrelated microbial species.

WHOLE GENOME SEQUENCING AND ANALYSIS

Sequencing Methodology

The sequencing and assembly of the genetic information of an organism (both prokaryotes and eukaryotes) establishes the unambiguous order of the As, Ts, Gs, and Cs in the DNA molecules, the absolute size, and the structure (linear versus circular). Although a variety of sequencing technologies have been used for genome sequencing projects, the random shotgun sequencing strategy has demonstrated itself to be the most successful and efficient and has become the preferred method for whole-genome sequencing. In the random shotgun strategy, genomic libraries that vary in insert size are constructed from the DNA of the organism of choice. The DNA is fragmented into pieces of 2,000 to 10,000 base pairs (bp) in length, depending on the desired final insert size. These DNA fragments are then ligated to a plasmid vector to create DNA constructs (plasmids) that can

Karen E. Nelson • The Institute for Genomic Research, 9712 Medical Center Drive, Rockville, MD 20850.

be inserted into *Escherichia coli,* one plasmid per cell, for propagation and amplification. Larger-sized insert libraries such as lambda, fosmid, or bacterial artificial chromosome (BAC) libraries are also often constructed to act as a scaffold on the assembled sequences.

Following on the growth of the *E. coli* cells containing the plasmid and purification of the plasmid DNA, sequence is obtained from both ends of the plasmid insert fragment. Generally, for the organism being sequenced, 15,000 sequences are obtained for each million base pairs of DNA representing approximately an eightfold redundancy of the genome. This number varies depending on the success rate of sequencing and the average read length of the sequences that can vary significantly by organism and the kind of technology that is being used in the project. A variety of assembly algorithms can be used to generate large contiguous pieces of genomic sequence that are invariably separated by sequencing and physical gaps. Physical and sequencing gaps occur because of the nonrandomness in the clone libraries that are initially constructed, or as a result of unclonable regions that are present in the genome. Sequencing gaps are spanned by small segments of large insert clones, but do not have sufficient sequence coverage to successfully close the gaps. Walking the spanning clones usually resolves these gaps. Physical gaps represent areas of the genome where there is no apparent linking information, and these gaps are resolved by a variety of techniques including multiplex and combinatorial PCR and the sequencing of the newly generated PCR products. Generally, the final genome sequence is edited to resolve any remaining ambiguities, low-coverage areas, or highly repetitive regions of the genome that may have been incorrectly assembled.

Bioinformatics Analysis of the Genome Sequence

Following on the closure and assembly of the DNA replicons, a complete analysis of the genome necessitates the identification of all DNA-encoded open reading frames (ORFs) or candidate genes in the DNA sequence and the assignment of gene names and associated function to these ORFs. Various ORF finding softwares (e.g., GLIMMER [Salzberg et al., 1998a, 1999]) in combination with a comparison of the DNA and amino acid sequences encoded in the ORFs to that of all known genes (e.g., BLAST [Altschul et al., 1990, 1997]) enables gene identification. Initial searches of the predicted coding regions can be performed with BLASTP (Fleischmann et al., 1995). The protein-protein matches can be aligned with a modified Smith-Waterman algorithm that maximally extends regions of similarity across protein coding regions in the DNA

sequence (Waterman, 1988). Gene assignment can be facilitated by searching all the ORFs against a database of nonredundant proteins (*nraa*) as has been developed at the Institute for Genomic Research (TIGR) and curated from the public archives GenBank (http://www.ncbi.nlm.nih.gov/), Genpept (http://helix.nih.gov/apps/bioinfo/), PIR (http://pir.georgetown.edu/pirwww/dbinfo/pirpsd.html), and Swiss-Prot (http://www.expasy.ch/sprot/sprot-top.html). These searches result in a gene assignment that has a corresponding role, common name, percent identity and similarity of match, the pairwise sequence alignment, and taxonomy associated with the match assigned to the predicted coding region and stored in a database. Regions of the genome without predicted coding regions and GLIMMER predictions with no database match can be reevaluated using BLASTX as the initial search, and newly identified genes can then be extrapolated from regions of alignment. The Gene Ontology system is now being adopted by a number of sequencing centers so that "a structured, precisely defined, common controlled vocabulary for describing the roles of genes and gene products in any organism" (Ashburner et al., 2000) can be developed.

Tools based on multiple sequence alignment and family building (Eddy, 1998; Sonnhammer et al., 1998a, 1998b) can be used to enhance gene identifications. Paralogous gene families can be created from multiple sequence alignments made with the sequenced genome's predicted amino acid sequences and built with the MKDOM software (Gouzy et al., 1999). The multiple sequence alignments group similar proteins into families for verification of annotation and identification of family members not recognized by simple pairwise alignment. The ORFs can also be aligned with a variety of tools against a growing database of hidden Markov models (HMMs) built on protein family and superfamily multiple sequence alignments (Bateman et al., 2002, 2000; Haft et al., 2001).

In addition to ORF analysis and gene identification, a number of other features of the genome can be identified using a variety of available tools. TopPred enables the identification of potential membrane-spanning domains in proteins (Claros and von Heijne, 1994). Signal peptides and the probable position of a cleavage site in secreted proteins can be detected with SignalP (Nielsen et al., 1997). Gene coding for untranslated RNAs can be identified by database searches at the nucleotide level, and searches for tRNAs can be performed using tRNAScan-SE (Lowe and Eddy, 1997). Repetitive sequences can be identified by various repeat finding programs, as well as by using an algorithm based on suffix trees (Delcher et al., 1999; also, see below). Statistical analysis of nucleotide frequency across the genome can be conducted to locate such features as the origin of replication (Salzberg et al., 1998b), regions of atypical composition, as well as regions of the genome that may have resulted from gene transfer from closely as well as distantly related species (Nelson et al., 1999). Finally, the reconstruction of biochemical pathways and transporter profiles associated with the organism of interest allows for an overview of the metabolic capacity of the cell and often reveals new aspects of the basic biochemistry of the species (see below and Color Plate 8).

Following the completion of the annotation phase of the sequencing project, the sequence allows for detailed comparative and functional genomics. In contrast to comparative studies that rely heavily on bioinformatic tools to interpret the genomic data, functional genomic studies aid in the assignment of a function to each gene in the genome, as well as in developing an understanding of the regulatory circuits that control the metabolic and other activities of the organism. Available tools of functional genomics include expression profiling, identification and analysis of protein-protein interactions, deletion phenotype analysis, and proteomics.

TOOLS FOR COMPARATIVE GENOME ANALYSIS

Current estimates from the analysis of competed microbial genomes (http://www.tigr.org/tigr-scripts/ CMR2/CMRHomePage.spl) are that more than 247,314 genes are available in public databases. This includes genes from 16 archaeal species and 72 bacterial species. As a result, bioinformatics tools have had to be developed that can handle these quantities of information, as well as that are capable of doing comparisons of genome information derived from closely as well as distantly related strains and species.

MUMmer (Delcher et al., 1999), for example, is a system that allows the rapid alignment of whole-genome sequences. The algorithm is based on a suffix tree data structure, which can be built and searched in linear time and which occupies only linear space. With the currently available version of MUMmer (http://www.tigr.org/software/mummer/), two complete genomes that range in size from 3 to 4 Mbp can be aligned in less than 30 s. The NUCmer utility that is also included with the system can align sequences from genomes that have not been closed, being capable of aligning thousands of smaller assemblies to another sequence data set. The PROmer utility generates alignments based on the six-frame translations of both input sequences and permits the alignment of

genomes for which the proteins are similar but the DNA sequence is too divergent to detect similarity.

The Comprehensive Microbial Resource (CMR) (http://www.tigr.org/tigr-scripts/CMR2/CMRHomePage.spl) is one of the few publicly available tools that allows for access to all the prokaryotic genomes or any subset of prokaryotic genomes that have been completed to date. The CMR was introduced primarily to reduce annotation inconsistency across completed genomes (Peterson et al., 2001). In situations where the genome was not sequenced at TIGR, the CMR displays two kinds of annotation: the primary annotation taken from the genome sequencing center where the organism was completed and the TIGR annotation generated by an automated annotation process at TIGR. Complex queries based on role assignments, database matches, protein families, membrane topology, and other features are feasible. The CMR also provides access to web-based tools that allow data mining using pre-run homology searches, whole-genome dot plots, batch downloading, and traversal across genomes using a variety of datatypes.

The phylome refers to the phylogenetic trees that represent a reconstruction of all the genes in a genome (Sicheritz-Ponten and Andersson, 2001). Tools have recently been developed for the large-scale automatic reconstruction of phylogenetic relationships (Sicheritz-Ponten and Andersson, 2001). This reconstruction is based on a set of python scripts and modules for automatic, large-scale reconstruction of phylogenetic relationships. PyPhy consists of AutoTree that automatically generates phylogenetic trees for each amino acid sequence in a FASTA file, and *Xphylome*, which generates and visualizes the Phylome maps for a microbial genome (http://www.cbs.dtu.dk/ staff/thomas/pyphy/). Currently with this tool, generating all phylogenetic trees for a typical bacterial genome can be accomplished in 1 to 2 hours.

The periodicity atlas was developed by David Ussery at the Technical University of Denmark (http://www.cbs.dtu.dk/services/GenomeAtlas/). This tool is a method of visualizing structural features within large regions of DNA and can map various properties of the DNA sequence along the chromosome. It includes a combination of structural parameters plus information about global repeats and base composition. It was originally designed for analysis of complete genomes, but can also be used for smaller regions of DNA. This tool allows for visualization of DNA structures within bacterial chromosomes that includes various types of repeats; DNA helix families, which is caused by certain stretches of purines (or pyrimidines) for A-DNA; and certain stretches of alternating pyrimidine and purines for Z-DNA. The various conformations of these different sequences have putative biological functions, based in part on these structures.

Insights into Bacterial Metabolic Diversity

In addition to the above listed features that can be identified through bioinformatics analysis, insights can also be gained into previously unidentified biochemical pathways and transporter systems (Nelson et al., 2000; Paulsen et al., 1998). Some environmental species such as *Pseudomonas putida*, for example (Nelson et al., 2002), have revealed a higher number of metabolic pathways for the conversion of atypical compounds than have been previously identified. Some organisms such as *Caulobacter crescentus* that have been sequenced for insights into biological processes such as cell cycle control have revealed the presence of unsuspected pathways such as the β-ketoadipate pathway for the metabolism of atypical compounds. Considering that on average 40% of each microbial genome is considered to be hypothetical or conserved hypothetical proteins, it is obvious that a significant amount remains to be elucidated about the biology of microbial species. The magnitude of the possible diversity that exists is evident when we consider that more than 99% of microbial species remain to be identified.

In the absence of appropriate tools for conducting comparative genomic studies, careful human analysis of the predicted coding regions allows for the identification of pathways, as well as for a detailed prediction of the possible overall metabolic profile and basic biology of the species in question. These types of analyses and conclusions rely heavily on having an accurate manual curation of the genome. Automated annotation programs that generate results without human intervention often miss predicted coding regions, make overcalls and overpredictions, and generally are not able to recognize operonic structures. At TIGR, for example, we have successfully reconstructed metabolic profiles for a number of bacterial and archaeal species (Eisen et al., 2002; Nelson et al., 1999, 2002; Tettelin, et al., 2001) and more recently for the eukaryotic species *Plasmodium falciparum* (Gardner et al., 2002) (Plate 8). Again, it should be highlighted that, although tremendous insight is gained into the metabolic diversity of the species that is being analyzed, many other pathways are likely missed because of the limited characterization of many of these species that is reflected in the high number of conserved hypothetical and hypothetical proteins that remain at the end of the average genome sequencing project.

Other interesting findings have resulted from biochemical reconstruction of pathways present in many prokaryotic genomes. The *Thermotoga maritima* genome (Nelson, et al., 1999), for example, revealed a number of pathways for the metabolism of plant compounds including cellulose and xylan, as well for the metabolism of sugars. The bacterium also has a significantly high number of transporter systems that are devoted to the import of polysaccharides and oligopeptides and that appear to be a reflection of the environmental niche that this bacterium occupies. Currently, based on the predictions that have been made from the genome sequence, this bacterium (and some close relatives such as *T. neopolitana*) is being investigated for its potential to produce hydrogen gas as a renewal source of energy using a range of carbohydrates (S. E. Van Otengham, personal communication). Similar analysis of the *Streptococcus pneumoniae* genome (Tettelin et al., 2001) highlighted the ability of this organism to import and metabolize a range of sugars, many of which had not been previously identified as potential substrates. Extracellular enzymes that enable degradation of host polymers apparently also have a secondary role in making substrates available to the bacterium. *Streptomyces coelicolor* is a soil inhabitant that produces most of the natural antibiotics that are used in human and veterinary medicine (Bentley et al., 2002). The analysis of the genome reveals a lot about the adaptations of this bacterium to the soil environment. The large size of the genome may in part be a reflection of the genes that have allowed the organism to exploit a range of nutrients and for the production of secondary metabolites including antibiotics, polyketides, and siderophores. The production of natural metabolites by this bacterium is one aspect for which there is tremendous potential for future exploitation of this species. This includes the development and production of new antimicrobials.

Porphyromonas gingivalis is one of the many bacterial species that reside in the oral cavity, where microorganisms in the supragingival plaque are exposed to the host's dietary intake and many ferment carbohydrates to acidic end products. Anaerobic species such as *P. gingivalis* that occupy the subgingival region are not exposed to the dietary fraction, but are exposed to the host tissue proteins and metabolic end products from other microbial species (Shah and Williams, 1987). In an attempt to further characterize the biology of this organism, a whole-genome-sequencing project was initiated at TIGR and completed in 2002 (Nelson et al., in press). Some of the highlights from analysis of the physiology of this bacterium are detailed below.

Glucose utilization by *P. gingivalis* is known to be very poor, and glucose and other carbon sources do not appear to support growth (Shah and Williams,

1987). The sequenced strain, however, does contain putative ORFs for all enzymes of the glycolytic pathway. In addition, there are ORFs for a putative glucose and galactose transporter and glucose kinase. Four putative ORFs for the pentose phosphate pathway were identified, and it is possible that this pathway is used to generate precursor metabolites during anaerobic growth. Aspartate, asparagine, and glutamine are readily utilized by this bacterium (Shah and Williams, 1987; Takahashi et al., 2000), and pathways have been identified from the genome sequence that suggest that additional amino acids can be utilized. In total, 44 peptidases could be identified. In addition, *P. gingivalis* possesses enzymes for the degradation of complex amino sugars, and the bacterium encodes glucose and galactose aminidase activities. It is not known whether these complex sugars are metabolized, but it is possible that the removal of amino sugars from glycoproteins renders the glycoproteins more susceptible to degradation by proteinases. The results from the whole-genome analysis also suggest that the major fermentation end products of *P. gingivalis* include propionate, butyrate, isobutyrate, isovalerate, actetate, ethanol, and butanol. Some of these fermentation end products have been shown to be toxic to host tissues and may result in cell death, disruption of immunocyte activity, and cytokine networks (Niederman et al., 1990, 1997). These end products may therefore act as virulence agents against the human host.

Examples from Eukaryotic Species: *Plasmodium falciparum*

As another example of recent analysis of completed genome sequences, reconstruction of biochemical pathways of *P. falciparum* (Gardner et al., 2002) has given tremendous insight into the biology of this organism and also for the identification of novel putative drug targets. For example, all of the enzymes necessary for a functional glycolytic pathway could be identified, but a fructose bisphosphatase that would be necessary for gluconeogenesis suggests that this pathway is absent. Similarly, candidate genes for all but one of those necessary for the pentose phosphate pathway were identified. The presence of a phosphoenolpyruvate carboxylase implies that *P. falciparum* may cope with a drain of intermediates from the tricarboxylic acid cycle by using this enzyme to replenish oxaloacetate from cytosolic phosphoenolpyruvate and bicarbonate. Biochemical, genetic, and chemotherapeutic data suggest that malaria and other apicomplexan parasites possess the ability to synthesize chorismate from erythrose 4-phosphate and phosphoenol pyruvate via the shikimate pathway

(Roberts et al., 2002; Roberts et al., 1998). Apart from chorismate synthase, the genes for the enzymes in the pathway could not be identified with any certainty. Finally, the malaria parasite utilizes hemoglobin from the host cytoplasm as a food source, hydrolyzing globin and releasing heme that is detoxified in the form of haemazoin. It was unclear whether de novo synthesis occurs using imported host enzymes or using the parasite's own enzymes. We could identify orthologs for every enzyme in the pathway except for the uroporphyrinogen-III synthase.

MICROBIAL GENOMICS: INCREASING OUR UNDERSTANDING OF EVOLUTION

The ability to generate complete genome sequences has significantly increased our awareness of lateral gene transfer and has demonstrated the inadequacies and misleading information that can occur through the analysis of single genes to describe the evolution of a species. Before the availability of complete genome sequences, there was a high dependence on rRNA sequences to demonstrate evolutionary relationships, the analysis of this sequence being used to generate the "universal tree of life". Many examples exist of other genes and protein phylogenies that do not agree with the universal tree. With the advent of complete genome sequencing, it has become evident that gene transfer is rampant and is one of the major shaping forces in the evolution of microbial species.

Analysis of the genome of *T. maritima* for example (Nelson et al., 1999) showed that close to 25% of the genes in this bacterium were more similar to genes from archaeal species. Based on the presence of atypical regions in the genome, genes that had a higher level of significance to genes from archaeal species, genes that had similarities to genes found only in archaeal species, and the conservation of gene order that was shared with archaeal species, it was concluded that there had been extensive gene transfer between *Thermotoga* and various archaeal species. Subsequent biochemical studies that have employed techniques including gene amplification with degenerate PCR primers and subtractive hybridization have also demonstrated extensive genomic diversity, as well as gene exchange in this genus (Nesbø, et al., 2001, 2002).

Nesbø and coworkers (Nesbø et al., 2001) conducted a study with 16 strains of *Thermotoga* and other related members of the *Thermotogales* where they investigated the distribution of two of the many predicted "archaeal-like" genes based on the analysis of the complete genome sequence. The genes that were investigated were those that encoded the large subunit of glutamate synthetase and myo-inositol 1P synthase. The distribution patterns of these two genes showed that they had been acquired from multiple archaeal lineages during the divergence of the *Thermotogales*, to the exclusion of other bacterial species. In a subsequent study, Nesbø and colleagues (2002) used suppressive subtractive hybridization (SSH) techniques to identify genes that are present in different *Thermotoga* strains that do not have homologs in the sequenced *Thermotoga* genome. Their studies focused on *Thermotoga* sp. strain RQ2 that differs from the sequenced *T. maritima* MSB8 by only 0.3% in the 16 rRNA gene, and numerous differences between these two strains could be identified.

At TIGR, we are currently using CGH on a range of *Thermotoga* strains that have been provided by Karl Stetter and Robert Huber (University of Regensburg, Germany) to further elucidate the extent of gene transfer in this genus. Preliminary results suggest that *T. maritima* MSB8 shares the highest level of genome conservation with *Thermotoga* sp. RQ2, with which it shares 99.7% identity in the small-subunit rRNA sequence. At least 7% (129 ORFs) in the MSB8 genome do not have homologous sequences in the RQ2 genome. These include 45 hypothetical proteins, 13 conserved hypothetical proteins, and 23 (18% of total) that are devoted to transport. Of these 129, only 18 occur as single ORFs, and the remaining correspond to islands that range in size from 2 to 38 kb. The largest island (TM0616 to TM0651 on the MSB8 genome) corresponds to a region that is predicted to be of atypical composition and includes ORFs that are involved in the biosynthesis of lipopolysaccharides. Further details on the methodology of CGH are discussed below.

Finally, Anderssen and colleagues (Sicheritz-Ponten and Andersson, 2001) have been able to reconstruct all possible gene phylogenies for the genes in the *T. maritima* genome using their phylome reconstruction method (see above). Similar to the findings from the main genome paper published in 1999 (Nelson et al., 1999), they have shown that close to a quarter of the *T. maritima* genome is archaeal-like in nature based on phylogenetic placings. Although the mechanisms of gene transfer in the *Thermotogales* remain to be elucidated, this genome sequence remains the clearest examples and is the most significant example to date of gene transfer across the prokaryotic domains.

The availability of genome sequences from related strains and species is also giving insight into the role that lateral gene transfer is playing in the acquisition of virulence factors, and ultimately the role of this mechanism in the emergence of pathogenic species. The bacterium *E. coli* O157:H7, for example,

is a worldwide threat to public health and has been implicated in many outbreaks of haemorrhagic colitis. Comparison of *E. coli* O157:H7 to the genome of the nonpathogenic laboratory strain *E. coli* K-12 allowed for the identification of genes that may be responsible for pathogenesis (Perna et al., 2001). A total of 1,387 new genes encoded in strain-specific clusters of diverse sizes were found in O157:H7. These include candidate virulence factors, alternative metabolic capacities, and several prophages. These studies along with others that have conducted comparative analyses of other pathogenic species have demonstrated that bioinformatics tools can be used to identify regions of gene transfer as well as to increase knowledge on genes that are involved in virulence (Tettelin et al., 2002). These types of studies will ultimately result in more rational drug design for antimicrobial targets. This is particularly significant with regard to the rise in antibiotic resistance among a variety of bacterial pathogens during the past two decades.

In other examples that have looked at genome rearrangements, comparative genome analysis across three sequenced *Pyrococcus* species (*Pyrococcus abysii*, *P. horokoshii*, and *P. furiosus*) highlights differential conservation of genes, reorganization of DNA by genetic elements, and the rearrangements that tend to occur at the replication terminus (Lecompte et al., 2001). Genetic elements involved in the reorganization of a genome can be seen in the *Sulfolobus solfataricus* genome (She et al., 2001) where the presence of 200 diverse insertion sequence elements (that accounted for close to 10% of genome), long tandem repeats, and evidence of integrase-mediated insertion events can be observed. It is evident that the hyperthermophilic species that have been sequenced show high levels of genetic rearrangements and examples of gene transfer.

FUNCTIONAL GENOME ANALYSIS OF MICROBIAL SPECIES

In the postgenomic era, the discipline of functional genomics is facing the challenge of associating function to the thousands of genes of unknown function that remain at the end of each genome project. DNA microarrays are rapidly becoming standard laboratory tools for investigating gene expression under different conditions as well as for looking at the presence and absence of genes in different stains or species that are related to a reference genome. Their applicability extends to many areas of microbial research including microbial physiology, pathogenesis, epidemiology, ecology, phylogeny, and pathway engineering.

DNA microarray technology allows a parallel analysis of RNA abundance and DNA homology for thousands of genes in a single experiment. Microarray expression analysis, for example, is ideal for profiling mRNA expression (Schena et al., 1995). DNA segments that represent the genes to be profiled, or the entire genome sequence, are amplified by PCR and mechanically spotted at high density on glass microscope slides using relatively simple *x-y-z* stage robotics systems. With such a system, microarrays containing the entire set of genes from a microbial genome can be easily constructed. The microarrays are queried in a cohybridization assay using two or more fluorescently labeled probes prepared from mRNA from the cellular conditions of interest (Shalon et al., 1996). The kinetics of hybridization allows the determination of relative expression levels based on the ratio with which each probe hybridizes to the individual array element. Hybridization is assayed with a confocal laser scanner to measure fluorescence intensities, which allows the simultaneous determination of the relative expression levels of all the genes represented in the array. Whole-genome microarray expression studies have been initiated for several pathogenic and environmental microorganisms, and an extensive informatics infrastructure has been developed and implemented to support these studies.

Neisseria meningitidis serogroup B gene regulation during interaction with human epithelial cells has been followed using microarrays (Grifantini et al., 2002). In this study, it was found that contact between host and cell induced expression of 347 genes. Genes that were upregulated included transporters of iron, chloride, amino acids, and sulfate, many virulence factors, and the entire pathway of sulfur-containing amino acids. This study also showed that microarray technology is valid for identifying new vaccine candidates. Currently we are conducting expression studies at TIGR that look at a number of pathogenic species including *Staphylococcus aureus*, *S. epidermidis*, and *S. pneumoniae*.

DNA-DNA microarrays can also be used to do CGH experiments that allow for the identification of genomic differences across closely related strains and species. CGH experiments have successfully been conducted for a number of microbial species including *T. maritima*, *P. furiosus*, *Bacillus anthracis*, *S. pneumoniae*, and *Deinococcus radiodurans*. Recently, comparisons between the sequenced *S. agalactiae* serotype V strain 2603 V/R and 19 *S. agalactiae* strains (Tettelin et al., 2002) have shown the genetic heterogeneity among *S. agalactiae* strains, even of the same serotype, and provide evidence that gene acquisition, duplication, and reassortment have produced the genetic diversity within the species that has permitted *S. agalactiae* to adapt to new environmental niches and to emerge as a major human pathogen.

In other situations where a complete genome sequence is available, and a microarray has not been prepared to conduct CGH experiments, the complete genome sequencing of a closely related strain becomes an impractical and expensive alternative to identifying differences across strains. CGH also allows only for the identification of genes that are shared across species, and the sequenced genome always becomes the reference genetic information. CGH, as valuable a technique as it is, does not give any direct sequence information on the new (test) strains, but tells only what is shared between the test and the reference species. The technique of SSH allows for the identification of differences across closely related strains by allowing strain-specific sequences to be generated. SSH studies have been conducted on a number of microbial species that have been completely sequenced including *T. maritima* (Nesbp et al., 2002), *Helicobacter pylori* (Akopyants et al., 1998), and *E. coli* (Miyazaki et al., 2002). The latter study employed this technique to identify the virulence-related genes in uropathogenic *E. coli* that show invasiveness to T-24 bladder cancer cells.

Because microarray analysis can demonstrate gene expression, and give insights into genes that are turned on and off in a pathway, predictions from the genome can be tested using mass spectrometry analysis. Substrate disappearance from the medium and end product formation from the bacterial species under test can be determined. This is about to be tested on the complete genome sequence of *Epulopiscium* sp., an organism that to date has not been grown in pure culture. It is anticipated that the ability to reconstruct the growth medium of a species based on its sequenced genome will have applications to successfully growing uncultured species based on information from their genome.

GENOMICS AS A TOOL TO UNDERSTAND UNCULTURABLE SPECIES

The ability to utilize the cultivation-independent technique of sequencing rRNA genes has allowed for the identification of novel microbial lineages, many of which have no cultured representatives. These molecular methods have revealed that traditional culturing methods have failed to represent the true level of microbial diversity in nature. The ability to analyze large sections of genomic DNA that have been created from BAC libraries that are created directly from environmental samples has become very promising as a technique to address the issue of uncultivated species. DeLong and coworkers (Beja et al., 2000) have successfully used this technique to construct libraries and sequence DNA generated from planktonic marine samples. In that study, a significant portion of the libraries was derived from previously uncultivated microbial species, and the results verified the utility of BAC libraries for providing access to the genomes of as yet uncultivated microbial species. It is anticipated that similar analysis of BAC libraries can be applied to other environments such as soils, the oral cavity, and the gastrointestinal tract, such that significant insight into the genomic potential of these natural populations can be gained.

TIGR has recently been funded to generate the complete genome sequence of the uncultured bacterial species *Epulopiscium fishelonii*. It is expected that the analysis and metabolic reconstruction of the complete *Epulopiscium* genome will enable the development of the appropriate culture medium that will allow for growth of this unusual bacterial species in the laboratory. Ultimately, these kinds of approaches will have to be expanded in an attempt to identify and cultivate the many unculturable species that remain to be identified in nature.

REFERENCES

Akopyants, N. S., A. Fradkov, L. Diatchenko, J. E. Hill, P. D. Siebert, S. A. Lukyanov, E. D. Sverdlov, and D. E. Berg. 1998. PCR-based subtractive hybridization and differences in gene content among strains of *Helicobacter pylori*. *Proc. Natl. Acad. Sci. USA* 95:13108–13113.

Altschul, S. F., W. Gish, W. Miller, E. W. Myers, and D. J. Lipman. 1990. Basic local alignment search tool. *J. Mol. Biol.* 215:403–410.

Altschul, S. F., T. L. Madden, A. A. Schaffer, J. Zhang, Z. Zhang, W. Miller, and D. J. Lipman. 1997. Gapped BLAST and PSI-BLAST: a new generation of protein database search programs. *Nucleic Acids Res.* 25:3389–3402.

Ashburner, M., C. A. Ball, J. A. Blake, D. Botstein, H. Butler, J. M. Cherry, A. P. Davis, K. Dolinski, S. S. Dwight, J. T. Eppig, M. A. Harris, D. P. Hill, L. Issel-Tarver, A. Kasarskis, S. Lewis, J. C. Matese, J. E. Richardson, M. Ringwald, G. M. Rubin, and G. Sherlock. 2000. Gene ontology: tool for the unification of biology. The Gene Ontology Consortium. *Nat. Genet.* 25:25–29.

Bateman, A., E. Birney, R. Durbin, S. R. Eddy, K. L. Howe, and E. L. Sonnhammer. 2000. The Pfam protein families database. *Nucleic Acids Res.* 28:263–266.

Bateman, A., E. Birney, L. Cerruti, R. Durbin, L. Etwiller, S. R. Eddy, S. Griffiths-Jones, K. L. Howe, M. Marshall, and E. L. Sonnhammer. 2002. The Pfam protein families database. *Nucleic Acids Res.* 30:276–280.

Beja, O., M. T. Suzuki, E. V. Koonin, L. Aravind, A. Hadd, L. P. Nguyen, R. Villacorta, M. Amjadi, C. Garrigues, S. B. Jovanovich, R. A. Feldman, and E. F. DeLong. 2000. Construction and analysis of bacterial artificial chromosome libraries from a marine microbial assemblage. *Environ. Microbiol.* 2:516–529.

Beja, O., E. V. Koonin, L. Aravind, L. T. Taylor, H. Seitz, J. L. Stein, D. C. Bensen, R. A. Feldman, R. V. Swanson, and E. F. DeLong. 2002a. Comparative genomic analysis of archaeal genotypic variants in a single population and in two different oceanic provinces. *Appl. Environ. Microbiol.* 68:335–345.

Beja, O., M. T. Suzuki, J. F. Heidelberg, W. C. Nelson, C. M. Preston, T. Hamada, J. A. Eisen, C. M. Fraser, and E. F. DeLong. 2002b. Unsuspected diversity among marine aerobic anoxygenic phototrophs. *Nature* 415:630–633.

Bentley, S. D., K. F. Chater, A. M. Cerdeno-Tarraga, G. L. Challis, N. R. Thomson, K. D. James, D. E. Harris, M. A. Quail, H. Kieser, D. Harper, A. Bateman, S. Brown, G. Chandra, C. W. Chen, M. Collins, A. Cronin, A. Fraser, A. Goble, J. Hidalgo, T. Hornsby, S. Howarth, C. H. Huang, T. Kieser, L. Larke, L. Murphy, K. Oliver, S. O'Neil, E. Rabbinowitsch, M. A. Rajandream, K. Rutherford, S. Rutter, D. Saunders, S. Sharp, R. Squares, S. Squares, K. Taylor, T. Warren, A. Wietzorrek, J. Woodward, B. G. Barrell, J. Parkhill, and D. A. Hopwood. 2002. Complete genome sequence of the model actinomycete *Streptomyces coelicolor* A3(2). *Nature* 417:141–147.

Claros, M. G., and G. von Heijne. 1994. TopPred II: an improved software for membrane protein structure predictions. *Comput. Appl. Biosci.* 10:685–686.

Delcher, A. L., D. Harmon, S. Kasif, O. White, and S. L. Salzberg. 1999. Improved microbial gene identification with GLIMMER. *Nucleic Acids Res.* 27:4636–4641.

Eddy, S. R. 1998. Profile hidden Markov models. *Bioinformatics* 14:755–763.

Eisen, J. A., K. E. Nelson, I. T. Paulsen, J. F. Heidelberg, M. Wu, R. J. Dodson, R. Deboy, M. L. Gwinn, W. C. Nelson, D. H. Haft, E. K. Hickey, J. D. Peterson, A. S. Durkin, J. L. Kolonay, F. Yang, I. Holt, L. A. Umayam, T. Mason, M. Brenner, T. P. Shea, D. Parksey, W. C. Nierman, T. V. Feldblyum, C. L. Hansen, M. B. Craven, D. Radune, J. Vamathevan, H. Khouri, O. White, T. M. Gruber, K. A. Ketchum, J. C. Venter, H. Tettelin, D. A. Bryant, and C. M. Fraser. 2002. The complete genome sequence of *Chlorobium tepidum* TLS, a photosynthetic, anaerobic, green-sulfur bacterium. *Proc. Natl. Acad. Sci. USA* 99:9509–9514.

Fleischmann, R. D., M. D. Adams, O. White, R. A. Clayton, E. F. Kirkness, A. R. Kerlavage, C. J. Bult, J. F. Tomb, B. A. Dougherty, J. M. Merrick, et al. 1995. Whole-genome random sequencing and assembly of *Haemophilus influenzae* Rd. *Science* 269:496–512.

Fraser, C. M., J. D. Gocayne, O. White, M. D. Adams, R. A. Clayton, R. D. Fleischmann, C. J. Bult, A. R. Kerlavage, G. Sutton, J. M. Kelley, et al. 1995. The minimal gene complement of *Mycoplasma genitalium*. *Science* 270:397–403.

Gardner, M. J., N. Hall, E. Fung, O. White, M. Berriman, R. W. Hyman, J. M. Carlton, A. Pain, K. E. Nelson, S. Bowman, I. T. Paulsen, K. James, J. A. Eisen, K. Rutherford, S. L. Salzberg, A. Craig, S. Kyes, M. S. Chan, V. Nene, S. J. Shallom, B. Suh, J. Peterson, S. Angiuoli, M. Pertea, J. Allen, J. Selengut, D. Haft, M. W. Mather, A. B. Vaidya, D. M. Martin, A. H. Fairlamb, M. J. Fraunholz, D. S. Roos, S. A. Ralph, G. I. McFadden, L. M. Cummings, G. M. Subramanian, C. Mungall, J. C. Venter, D. J. Carucci, S. L. Hoffman, C. Newbold, R. W. Davis, C. M. Fraser, and B. Barrell. 2002. Genome sequence of the human malaria parasite *Plasmodium falciparum*. *Nature* 419:498–511.

Gouzy, J., F. Corpet, and D. Kahn. 1999. Whole genome protein domain analysis using a new method for domain clustering. *Comput. Chem.* 23:333–340.

Grifantini, R., E. Bartolini, A. Muzzi, M. Draghi, E. Frigimelica, J. Berger, G. Ratti, R. Petracca, G. Galli, M. Agnusdei, M. Monica Giuliani, L. Santini, B. Brunelli, H. Tettelin, R. Rappuoli, F. Randazzo, and G. Grandi. 2002. Previously unrecognized vaccine candidates against group B meningococcus identified by DNA microarrays. *Nat. Biotechnol.* 20:914–921.

Haft, D. H., B. J. Loftus, D. L. Richardson, F. Yang, J. A. Eisen, I. T. Paulsen, and O. White. 2001. TIGRFAMs: a protein family resource for the functional identification of proteins. *Nucleic Acids Res.* **29:**41–43.

Hoffman, S. L., W. O. Rogers, D. J. Carucci, and J. C. Venter. 1998. From genomics to vaccines: malaria as a model system. *Nat. Med.* **4:**1351–1353.

Lecompte, O., R. Ripp, V. Puzos-Barbe, S. Duprat, R. Heilig, J. Dietrich, J. C. Thierry, and O. Poch. 2001. Genome evolution at the genus level: comparison of three complete genomes of hyperthermophilic archaea. *Genome Res* **11:**981–993.

Lowe, T. M., and S. R. Eddy. 1997. tRNAscan-SE: a program for improved detection of transfer RNA genes in genomic sequence. *Nucleic Acids Res.* **25:**955–964.

Miyazaki, J., W. Ba-Thein, T. Kumao, H. Akaza, and H. Hayashi. 2002. Identification of a type III secretion system in uropathogenic *Escherichia coli*. *FEMS Microbiol. Lett.* **212:** 221–228.

Nelson, K. E., R. A. Clayton, S. R. Gill, M. L. Gwinn, R. J. Dodson, D. H. Haft, E. K. Hickey, J. D. Peterson, W. C. Nelson, K. A. Ketchum, L. McDonald, T. R. Utterback, J. A. Malek, K. D. Linher, M. M. Garrett, A. M. Stewart, M. D. Cotton, M. S. Pratt, C. A. Phillips, D. Richardson, J. Heidelberg, G. G. Sutton, R. D. Fleischmann, J. A. Eisen, C. M. Fraser, et al. 1999. Evidence for lateral gene transfer between Archaea and bacteria from genome sequence of *Thermotoga maritima*. *Nature* **399:**323–329.

Nelson, K. E., I. T. Paulsen, J. F. Heidelberg, and C. M. Fraser. 2000. Status of genome projects for nonpathogenic bacteria and archaea. *Nat. Biotechnol.* **18:**1049–1054.

Nelson, K. E., R. D. Fleischmann, R. T. DeBoy, I. T. Paulsen, D. E. Fonts, J. A. Eisen, S. Daugherty, R. J. Dodson, S. Durkin, M. Gwinn, D. Haft, J. Kolonay, W. Nelson, T. Mason, L. Tallon, J. Gray, D. Granger, H. Tettelin, H. Dong, J. L. Galvin, M. J. Duncan, F. E. Dewhirst, and C. M. Fraser. The complete genome sequence of the oral pathogenic bacterium *Porphyromonas gingivalis* strain W83. *J. Bacteriol*, in press.

Nelson, K. E., C. Weinel, I. T. Paulsen, R. J. Dodson, H. Hilbert, D. Fouts, S. R. Gill, M. Pop, V. Martins Dos Santos, M. Holmes, L. Brinkac, M. Beanan, R. Deboy, S. Daugherty, J. Kolonay, R. Madupu, W. Nelson, and O. White. 2002b. Complete genome sequence and comparative analysis of the metabolically versatile *Pseudomonas putida* KT2440. *Environ. Microbiol.* **4:**799–808.

Nesbø, C. L., S. L'Haridon, K. O. Stetter, and W. F. Doolittle. 2001. Phylogenetic analyses of two "archaeal" genes in *Thermotoga maritima* reveal multiple transfers between archaea and bacteria. *Mol. Biol. Evol.* **18:**362–375.

Nesbø, C. L., K. E. Nelson, and W. F. Doolittle. 2002. Suppressive subtractive hybridization detects extensive genomic diversity in *Thermotoga maritima*. *J. Bacteriol.* **184:**4475–4488.

Niederman, R., B. Brunkhorst, S. Smith, R. N. Weinreb, and M. I. Ryder. 1990. Ammonia as a potential mediator of adult human periodontal infection: inhibition of neutrophil function. *Arch. Oral Biol.* **35:**205S–209S.

Niederman, R., J. Zhang, and S. Kashket. 1997. Short-chain carboxylic-acid-stimulated, PMN-mediated gingival inflammation. *Crit. Rev. Oral Biol. Med.* **8:**269–290.

Nielsen, H., J. Engelbrecht, S. Brunak, and G. von Heijne. 1997. A neural network method for identification of prokaryotic and eukaryotic signal peptides and prediction of their cleavage sites. *Int. J. Neural Syst.* **8:**581–599.

Paulsen, I. T., M. K. Sliwinski, and M. H. Saier, Jr. 1998. Microbial genome analyses: global comparisons of transport capabilities based on phylogenies, bioenergetics and substrate specificities. *J. Mol. Biol.* **277:**573–579.

Perna, N. T., G. Plunkett, III, V. Burland, B. Mau, J. D. Glasner, D. J. Rose, G. F. Mayhew, P. S. Evans, J. Gregor, H. A.

Kirkpatrick, G. Posfai, J. Hackett, S. Klink, A. Boutin, Y. Shao, L. Miller, E. J. Grotbeck, N. W. Davis, A. Lim, E. T. Dimalanta, K. D. Potamousis, J. Apodaca, T. S. Anantharaman, J. Lin, G. Yen, D. C. Schwartz, R. A. Welch, and F. R. Blattner. 2001. Genome sequence of enterohaemorrhagic *Escherichia coli* O157:H7. *Nature* **409:**529–533.

Peterson, J. D., L. A. Umayam, T. Dickinson, E. K. Hickey, and O. White. 2001. The Comprehensive Microbial Resource. *Nucleic Acids Res.* **29:**123–125.

Roberts, C. W., F. Roberts, R. E. Lyons, M. J. Kirisits, E. J. Mui, J. Finnerty, J. J. Johnson, D. J. Ferguson, J. R. Coggins, T. Krell, G. H. Coombs, W. K. Milhous, D. E. Kyle, S. Tzipori, J. Barnwell, J. B. Dame, J. Carlton, and R. McLeod. 2002. The shikimate pathway and its branches in apicomplexan parasites. *J. Infect. Dis.* **185:**S25–S36.

Roberts, F., C. W. Roberts, J. J. Johnson, D. E. Kyle, T. Krell, J. R. Coggins, G. H. Coombs, W. K. Milhous, S. Tzipori, D. J. Ferguson, D. Chakrabarti, and R. McLeod. 1998. Evidence for the shikimate pathway in apicomplexan parasites. *Nature* **393:**801–805.

Salzberg, S. L., A. L. Delcher, S. Kasif, and O. White. 1998a. Microbial gene identification using interpolated Markov models. *Nucleic Acids Res.* **26:**544–548.

Salzberg, S. L., A. J. Salzberg, A. R. Kerlavage, and J. F. Tomb. 1998b. Skewed oligomers and origins of replication. *Gene* **217:** 57–67.

Salzberg, S. L., M. Pertca, A. L. Delcher, M. J. Gardner, and H. Tettelin. 1999. Interpolated Markov models for eukaryotic gene finding. *Genomics* **59:**24–31.

Schena, M., D. Shalon, R. W. Davis, and P. O. Brown. 1995. Quantitative monitoring of gene expression patterns with a complementary DNA microarray. *Science* **270:**467–470.

Shah, H. N., and R. A. D. Williams. 1987. Utilization of glucose and amino acids by *Bacteroides intermedius* and *Bacteroides gingivalis*. *Curr. Microbiol.* **15:**241–246.

Shalon, D., S. J. Smith, and P. O. Brown. 1996. A DNA microarray system for analyzing complex DNA samples using two-color fluorescent probe hybridization. *Genome Res.* **6:**639–645.

She, Q., R. K. Singh, F. Confalonieri, Y. Zivanovic, G. Allard, M. J. Awayez, C. C. Chan-Weiher, I. G. Clausen, B. A. Curtis, A. De Moors, G. Erauso, C. Fletcher, P. M. Gordon, I. Heikamp-de Jong, A. C. Jeffries, C. J. Kozera, N. Medina, X. Peng, H. P. Thi-Ngoc, P. Redder, M. E. Schenk, C. Theriault, N. Tolstrup, R. L. Charlebois, W. F. Doolittle, M. Duguet, T. Gaasterland, R. A. Garrett, M. A. Ragan, C. W. Sensen, and J. Van der Oost. 2001. The complete genome of the crenarchaeon *Sulfolobus solfataricus* P2. *Proc. Natl. Acad. Sci. USA* **98:**7835–7840.

Sicheritz-Ponten, T., and S. G. Andersson. 2001. A phylogenomic approach to microbial evolution. *Nucleic Acids Res.* **29:**545–552.

Sonnhammer, E. L., S. R. Eddy, E. Birney, A. Bateman, and R. Durbin. 1998a. Pfam: multiple sequence alignments and HMM-profiles of protein domains. *Nucleic Acids Res.* **26:**320–322.

Sonnhammer, E. L., G. von Heijne, and A. Krogh. 1998b. A hidden Markov model for predicting transmembrane helices in protein sequences. *Proc. Int. Conf. Intell. Syst. Mol. Biol.* **6:**175–182.

Takahashi, N., T. Sato, and T. Yamada. 2000. Metabolic pathways for cytotoxic end product formation from glutamate- and aspartate-containing peptides by *Porphyromonas gingivalis*. *J. Bacteriol.* **182:**4704–4710.

Tettelin, H., N. J. Saunders, J. Heidelberg, A. C. Jeffries, K. E. Nelson, J. A. Eisen, K. A. Ketchum, D. W. Hood, J. F. Peden, R. J. Dodson, W. C. Nelson, M. L. Gwinn, R. DeBoy, J. D. Peterson, E. K. Hickey, D. H. Haft, S. L. Salzberg, O. White, R. D. Fleischmann, B. A. Dougherty, T. Mason, A. Ciecko, D. S. Parksey, E. Blair, H. Cittone, E. B. Clark, M. D. Cotton, T. R. Utterback, H. Khouri, H. Qin, J. Vamathevan, J. Gill, V. Scarlato,

V. Masignani, M. Pizza, G. Grandi, L. Sun, H. O. Smith, C. M. Fraser, E. R. Moxon, R. Rappuoli, and J. C. Venter. 2000. Complete genome sequence of *Neisseria meningitidis* serogroup B strain MC58. *Science* **287:**1809–1815.

Tettelin, H., K. E. Nelson, I. T. Paulsen, J. A. Eisen, T. D. Read, S. Peterson, J. Heidelberg, R. T. DeBoy, D. H. Haft, R. J. Dodson, A. S. Durkin, M. Gwinn, J. F. Kolonay, W. C. Nelson, J. D. Peterson, L. A. Umayam, O. White, S. L. Salzberg, M. R. Lewis, D. Radune, E. Holtzapple, H. Khouri, A. M. Wolf, T. R. Utterback, C. L. Hansen, L. A. McDonald, T. V. Feldblyum, S. Angiuoli, T. Dickinson, E. K. Hickey, I. E. Holt, B. J. Loftus, F. Yang, H. O. Smith, J. C. Venter, B. A. Dougherty, D. A. Morrison, S. K. Hollingshead, and C. M. Fraser. 2001. Complete genome sequence of a virulent isolate of *Streptococcus pneumoniae*. *Science* **293:**498–506.

Tettelin, H., V. Masignani, M. J. Cieslewicz, J. A. Eisen, S. Peterson, M. R. Wessels, I. T. Paulsen, K. E. Nelson, I. Margarit, T. D. Read, L. C. Madoff, A. M. Wolf, M. J. Beanan, L. M. Brinkae, S. C. Daugherty, R. T. DeBoy, A. S. Durkin, J. F. Kolonay, R. Madupu, M. R. Lewis, D. Radune, N. B. Fedorova, D. Scanlan, H. Khouri, S. Mulligan, H. A. Carty, R. T. Cline, S. E. Van Aken, J. Gill, M. Scarselli, M. Mora, E. T. Iacobini, C. Brettoni, G. Galli, M. Mariani, F. Vegni, D. Maione, D. Rinaudo, R. Rappuoli, J. L. Telford, D. L. Kasper, G. Grandi, and C. M. Fraser. 2002. Complete genome sequence and comparative genomic analysis of an emerging human pathogen, serotype V *Streptococcus agalactiae*. *Proc. Natl. Acad. Sci. USA* **99:**12391–12396.

Waterman, M. S. 1988. Computer analysis of nucleic acid sequences. *Methods Enzymol.* **164:**765–793.

Microbial Diversity and Bioprospecting
Edited by Alan T. Bull
© 2004 ASM Press, Washington, D.C.

Chapter 26

Bacterial Proteomics

PHILLIP CASH

The proteome is defined as the total protein complement expressed by a cell (Wasinger et al., 1995; Wilkins et al., 1996). In the case of a multicellular organism each functional cell type expresses a distinctive proteome that can also vary depending on the activation state of the individual cell. For single-celled microbes the proteome will be for the intact organism, but as described below the proteome changes in response to environmental stimuli. Thus, the proteome is a highly dynamic entity that changes in response to the external cellular environment as well as to structural changes in the genome of the cell or organism. Although an organism has but a single genome, which is relatively stable, it will exhibit many distinctive proteomes. The complexity of the proteome represents a challenge to many different disciplines ranging from the wet bench, specifically the technologies required to display the proteome, to the virtual bench, with the establishment of bioinformatic methods for interpreting the large amounts of data that even modest proteomics-based experiments generate.

Over the past few years there has been an exponential growth in studies aimed at defining the proteomes of a range of prokaryotic and eukaryotic organisms fueled largely by the increasing amount of complete genome sequence data now available. Despite major advances in the determination of the genome sequences for many organisms, including humans, there is an awareness that these data alone do not provide a complete picture of the functions of a cell or microorganism. To achieve this objective one must move beyond the genome to look at the proteins. In silico analyses of the raw genome sequence data can predict the coding capacities of that genome as well as provide predictions on the likely functions for the proteins encoded by the open reading frames (ORFs). Nevertheless, a significant proportion of the predicted ORFs represent hypothetical proteins unique to a single genome or with homologies to the hypothetical proteins identified from other genome sequences (Fleischmann et al., 1995; Fraser et al., 1995; Tomb et al., 1997). Even for those ORFs encoding previously characterized proteins, the genome sequence alone is insufficient to reliably predict the occurrence or extent of posttranslational modifications to the protein products. It has been extensively documented that a single ORF can give rise to more than one protein product through posttranslational modification. For example, it has been estimated that in *Escherichia coli* each ORF leads on average to 1.4 proteins (Tonella et al., 2001) and 2.7 protein products for *Helicobacter pylori* (Lock et al., 2001). The determination of the cellular proteome therefore provides added value to the basic genome sequence data.

Beyond confirming the genome coding sequences, proteomics plays a key role in the rapidly expanding field of functional genomics, specifically in being one of a number of approaches used to provide a holistic view of global cellular gene expression (Humphery-Smith et al., 1997). Along with methods that examine gene expression at the level of the transcriptome, the techniques of proteomics can be used to investigate gene expression on a global scale with the simultaneous analysis of a significant proportion of the genome coding capacity. One of the aims of these global approaches to the analysis of gene expression is to provide valuable insights into novel genetic pathways and networks. These may not otherwise have been demonstrated through the more traditional reductionist approach of analyzing the functions of individual genes frequently in isolation from the rest of the genetic information. Compared with studies of the transcriptome and metabolome, investigations at the level of the proteome can provide significant data on the interaction of the cell and environment because proteins function at the interface of the cell and its environment frequently in bidirectional processes. It is not possible to predict the biosynthetic levels of the encoded proteins solely

---

Phillip Cash • Department of Medical Microbiology, University of Aberdeen, Foresterhill, Aberdeen AB32 6QX, Scotland.

from the gene sequence, and the measurement of specific mRNA synthesis is also a poor indicator of the level of synthesis of the corresponding protein product (Anderson and Seilhamer, 1997; Gygi et al., 1999b). Consequently, the only option for investigating the cellular proteome accurately is to target directly the proteins themselves. Although, as discussed below, there has been significant progress in developing technologies for elucidating the proteome, it remains a formidable task to understand, in its entirety, the proteome of even a simple unicellular organism. Proteomics encompasses many areas of protein analysis and includes the determination and classification of protein structure. Much of the following discussion looks at "expression proteomics," i.e., the determination of which proteins are expressed under a specific experimental stimulus along with their level of synthesis.

When using two-dimensional gel electrophoresis (2-DGE) to monitor protein synthesis, the initial analyses can be carried out without prior knowledge of the protein identities or their function. Thus, in many of the studies described below, 2-DGE is used to identify putative targets for identification that can be further investigated using alternative but complementary methods. This strategy is illustrated in Fig. 1 for the analysis of an external stimulus. The same approach is used for the comparison of, for example, virulent and avirulent bacterial isolates. A number of investigators have proposed the existence of so-called "proteomic signatures," which are characteristic patterns of protein synthesis at both a qualitative and a quantitative level that are indicative of a specific metabolic status of the cell. Evers and Gray (2001) made a similar proposal in the study of drug action in eukaryotic cells, i.e., drugs with similar modes of action are expected to interact with the proteome in the same way and induce characteristic changes in protein synthesis. In the field of microbiology, Singh et al. (2001) presented a similar proposal for antibiotic action with antibiotics inducing a characteristic pattern of protein changes in bacteria related to the antibiotic's mode of action. VanBogelen et al. (1999) reviewed the use of proteomics for defining the physiological status of *E. coli* and considered a variety of physiological conditions and their response on the bacterial proteome. It was concluded that the proteomic signature represents a subset of proteins whose altered expression pattern is characteristic for a specific response. The proteins affected under these conditions are often related to either a specific pathway or function. The characteristic changes in protein synthesis may be based on either qualitative (presence or absence) or quantitative changes in abundance. Because several proteomic signatures within a cell can

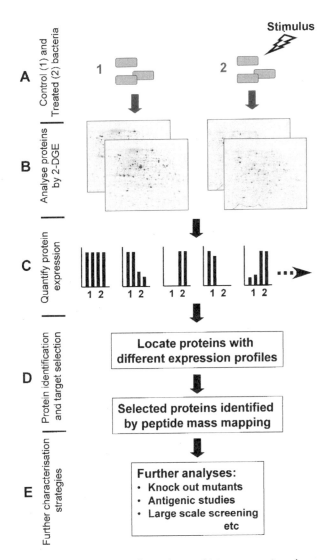

Figure 1. Typical strategy for studies combining proteomic and genomic analysis. This example illustrates the process for investigating bacterial response to a defined stimulus. (A) Bacteria are exposed to the stimulus under study and grown in parallel with control bacterial cultures. (B) The bacterial proteins are extracted from the cells and analyzed by 2-DGE; all assays are replicated to account for the inherent variability associated with 2-DGE. (C) Computer analysis is used to compare the protein profiles from different 2-D gels and to quantify the individual proteins across the gel series. Each histogram shows the amount of individual proteins across the series of four gels. (D) Proteins showing differential levels of expression are located and identified by peptide mass mapping. (E) Once target proteins are identified, complementary technologies (e.g., recombinant DNA techniques) can be used to further define the response of the bacteria.

interact at any one time, it is not possible to recognize the signature simply by comparing protein profiles; instead a catalog has to be created to reliably identify the specific response. Although not stated specifically, this concept of proteomics signatures is a recurring theme throughout many of the investigations discussed below for bacterial systems. The same view of the proteomic signature is used in the analysis of the proteomes of higher eukaryotes in the identification of disease markers (Petricoin et al., 2002).

The characterization of microbial proteomes is associated with many different areas of investigation of microbial gene expression and physiology (reviewed by Cash [1998]). The current review draws on experiments primarily from work carried out on pathogenic bacterial systems.

EXPERIMENTAL APPROACHES TO PROTEOME CHARACTERIZATION

No single technology is used throughout proteomics because of the wide variety of areas that come under the proteomics umbrella. Some of the principal techniques for characterizing eukaryotic and prokaryotic proteomes have been used for many years. For example, one of the key methods for protein separation, high-resolution 2-DGE, was first documented over 27 years ago (Klose, 1975; O'Farrell, 1975). Although the underlying separation strategy of 2-DGE has remained largely unchanged since that time, there have been technical refinements to improve the reliability and reproducibility of the method over the years.

What might now be considered as "classical proteomics" uses 2-DGE to resolve the individual proteins present in the complex mixture of the cell and then peptide mass fingerprinting with mass spectrometry (MS) to identify the individual proteins that have been resolved. This experimental strategy has been widely used over the years, and many of the studies detailed below have used this approach. The basic methods of 2-DGE and peptide mass mapping have been widely reviewed elsewhere (Dunn, 1987; Jensen et al., 1999). The combination of 2-DGE and peptide mass fingerprinting are fairly robust, and with care and attention to the techniques, useful data can be achieved with material from a wide range of sources. However, 2-DGE in particular has severe limitations over the classes of proteins that are amenable to analysis. In the majority of protein separations utilizing 2-DGE, hydrophobic and high-molecular-weight proteins are underrepresented. In addition, without the use of prefractionation techniques to the sample prior to analysis, the low-abundance proteins present in the cell are not included in the data. Within a typical cell, proteins are synthesized over a wide dynamic range. For example, in *E. coli* ribosomal proteins may exist at approximately 70,000 copies per cell, whereas LacI can be present in as few as 20 copies per cell. In the case of peptide mass fingerprinting, this technique performs best when applied to microbes with known

genome sequences. Nevertheless, the combined approach of 2-DGE and peptide mass mapping has been successfully applied to looking at the proteomes of microbes, which have been poorly characterized at the molecular level (Cordwell et al., 1997)

Complementary techniques to 2-DGE have been applied in proteomics, and these include multidimensional chromatographic separations to improve the coverage of the proteome for those protein classes not resolved by 2-DGE (Opiteck et al., 1998). More specialized methods for resolving proteins combined with their identification have been developed around advanced MS techniques.

Isotope-coded affinity tagging is one such method to provide comparative quantitative data in proteomics (Griffin et al., 2001; Gygi et al., 1999a). Essentially, cellular proteins are tagged on cysteine residues, and following trypsin digestion the cysteine-containing peptides are selected and resolved by capillary electrophoresis coupled to a tandem mass spectrometer (MS/MS). The use of tags that can be distinguished by mass spectrometry provides a means of carrying out dual labeling for quantitative comparisons of paired samples (Gygi, et al., 1999a). More recently, protein chips, analogous nucleic acid chips for mRNA analysis, have been documented. The protein chips can be prepared with specific binding characteristics or specific antibody coatings to allow the selective binding of proteins from complex mixtures (Cahill, 2001; Merchant et al., 2000) The bound proteins can then be detected, for example, by mass spectrometry. At present many of the high-resolution MS techniques and the use of protein chips remain in the realm of specialized laboratories.

PROTEOME DATABASES

As genome sequence databases become more extensive, there has been a movement toward the development of comprehensive proteome databases. These may represent a direct catalog of the proteins predicted to be encoded by a specific genome sequence. Alternatively, the proteome database can be developed from the analysis of protein expression under defined conditions. The latter form of database includes accurate identifications on only a proportion (perhaps just a few hundred) of the proteins expressed and currently do not represent the complete proteome. However, these can be of major importance for extended proteomic analyses containing data on protein expression under different physiological conditions. The basic idea of establishing a comprehensive protein database through the analysis of protein synthesis by 2-DGE is not new. Anderson and

Anderson (1982), in fact, proposed the development of a human protein index for studying human disease in the early 1980s. In a later review (Anderson and Anderson, 1996) the same authors commented that this proposal was unfortunately overtaken by technological developments in nucleotide sequencing, which provided faster and more comprehensive data on genome coding than could be achieved by the protein analytical techniques available at that time. Although current proteome databases still make extensive use of 2-DGE for resolving cellular proteins, the availability of peptide mass mapping for the spot identification step coupled with the extensive gene sequence data now available makes this a more viable experimental approach. However, these databases remain limited in their coverage of specific protein classes because of the technical limitations of the methods mentioned above.

A number of proteome databases have been developed for bacterial systems and are available either as print versions or dedicated websites (Table 1). In principal, proteomic databases developed through the analysis of protein expression provide a better representation of the highly dynamic proteome compared with those databases derived from the predicted coding of the genome. Buttner et al. (2001) highlighted the importance of incorporating data on bacterial physiology to increase the coverage achieved by the proteome database. In their work they commented that, although technological developments will improve our ability to resolve, detect, and identify more of the bacterial proteome, it is only by investigating the proteome under different environmental and stress conditions that a complete picture can be established. This approach has been used in the development of the *Bacillus subtilis* proteome database (Buttner et al., 2001). As discussed below for *Mycobacterium* and *Helicobacter*, other proteome databases have used a comparative approach by comparing the proteomes of different bacterial isolates with distinctive biological properties.

Proteome databases based on the proteins resolved by 2-DGE are being developed for a number of model organisms including *E. coli* (Tonella et al., 2001), *B. subtilis* (Buttner et al., 2001), and the cyanobacterium *Synechocystis* sp. strain PCC6803 (Sazuka et al., 1999), each of which can be accessed via the Internet. There are also a limited number of databases developed for clinically relevant bacteria. Some of the key outcomes of representative proteome databases are discussed below.

E. coli

The *E. coli* genome is approximately 4,639 kbp in length and is predicted to contain 4,288 ORFs capable of encoding proteins of which 38% have no known function (Blattner et al., 1997). Work on the *E. coli* Gene-Protein Index was initiated by Neidhardt and his colleagues and is currently based on *E. coli* K-12 W3110 (VanBogelen et al., 1996), although *E. coli* B/r (derivative NC3) was used in some of the earlier investigations (Bloch et al., 1980; Phillips et al., 1980). Over the years that this protein index has been

Table 1. Representative bacterial proteome databases

Organism	Database name	Reference	URL
E. coli		VanBogelen et al. (1996)	
	SWISS-2DPAGE	Tonella et al. (2001)	www.expasy.ch/ch2d/ch2d-top.html
H. influenzae		Langen et al. (2000)	
		Link et al. (1997)	
		Cash et al. (1997)	www.abdn.ac.uk/~mmb023/2dhome.htm
Synechocystis sp. PCC6803	Cyano2Dbase	Sazuka et al. (1999)	www.kazusa.or.jp/cyano/Synechocystis/cyano2D/
B. subtilis	Sub2D	Schmid et al. (1997)	http://microbio2.biologie.unigreifswald.de:8880/index.htm
H. pylori		Jungblut et al. (2000)	www.mpiib-berlin.mpg.de/2D-PAGE/
		Lock et al. (2001)	
		Cho et al. (2002)	
	Protein-protein interaction map	Rain et al. (2001)	
M. tuberculosis		Jungblut et al. (1999b)	www.mpiib-berlin.mpg.de/2D-PAGE/
	SSI-2DPAGE	Rosenkrands et al. (2000a)	www.ssi.dk/en/forskning/tbimmun/tbhjemme.htm
Borrelia garinii		Jungblut et al. (1999a)	www.mpiib-berlin.mpg.de/2D-PAGE/
Mycoplasma pneumoniae			www.mpiib-berlin.mpg.de/2D-PAGE/
Brucella melitensis		Wagner et al. (2002)	
Xylella fastidiosa			proteome.ibi.unicamp.br/databases/index.html

developed, the proteins have been identified by a variety of methods in addition to modern methods of peptide mass mapping (Bloch et al., 1980; Neidhardt et al., 1983; Phillips et al., 1980). In 1996 the *E. coli* Gene-Protein Index developed into two complementary areas, a Genome Expression Map, cataloging the products expressed by each ORF, and a Response/Regulation Map that details the expression pattern of the proteins resolved by 2-DGE under different growth conditions (VanBogelen et al., 1996). At present the *E. coli* Gene-Protein Index contained the identities of approximately 25% of the 1,550 proteins sports resolved by 2-DGE (VanBogelen et al., 1997, 1996).

A second *E. coli* two-dimensional (2-D) protein database has been established as part of SWISS-2DPAGE on the ExPASy WWW server (Pasquali et al., 1996). This database contrasts with the *E. coli* Gene-Protein Index described above in two areas. First, immobilized pH gradient (IPG) gels are used in 2-DGE for the first-dimension charge separation of the proteins instead of carrier ampholytes. Second, the proteins resolved by 2-DGE were identified using biochemical analysis of the protein spots (Pasquali et al., 1996; Wilkins et al., 1996) rather than the more traditional mapping methods used by Neidhardt's group. The *E. coli* reference gel accessed as part of SWISS-2DPAGE (Table 1) resolves a total of 1900 spots, detected by silver staining, within a pH range of 3.5 to 10 and an M_r range of 8 to 250 kDa (Pasquali et al., 1996). A total of 153 proteins were identified on this reference map by comparison with the 2-D protein profiles produced by VanBogelen et al., (1992). The identities of 56 of these proteins were confirmed by amino acid composition analysis and N-terminal microsequencing either alone or in combination (Pasquali et al., 1996). Tonella et al. (2001) extended the coverage of the *E. coli* proteome for SWISS-2DPAGE by using a series of narrow overlapping pH range IPG gels covering a total pH range of 4 to 11. The authors estimate that this approach displays approximately 71% of the entire *E. coli* proteome. However, of the 4,950 protein spots resolved by these gels, only 313 have been identified and these belong to 222 different ORFs. It is likely that the non-identified proteins are either hydrophobic proteins or encoded by nonexpressed genes (Tonella et al., 2001).

B. subtilis

B. subtilis is the type species for the genus *Bacillus,* which are gram-positive, spore-forming bacilli. The genus has a number of important features of interest to bacteriologists. First, the bacteria form endospores, making them a useful model to investigate molecular details of a simple developmental system, and second, some members of the genus produce products (for example, enzymes, antibiotics, and toxins) of economic interest to the biotechnology industry (Devine, 1995). The *B. subtilis* genome is 4,214 kbp in length, and the complete nucleotide sequence for *B. subtilis* 168 has been determined (Kunst et al., 1997). The genome is predicted to encode 4,100 genes, of which 25% correspond to gene families that have been expanded through gene duplication.

The data derived from the *B. subtilis* proteome forms the "Sub2D" protein database (Table 1) (Buttner et al., 2001). The type isolate used as the reference strain for the proteome database is *B. subtilis* 168. Proteins have been primarily analyzed using IPG gels covering the range of pH 4 to 7 (Schmid et al., 1997), which includes two-thirds of the proteins from the theoretical *B. subtilis* proteome (Buttner et al., 2001). In addition, basic proteins (pI > 7) have been separated using alkaline IPG gels (Ohlmeier et al., 2000). The current proteome database includes identifications on over 420 proteins expressed in bacterial cells or detected in extracellular compartments. Although the majority of the proteins identified could be assigned to known metabolic pathways, a significant proportion were gene products with unknown functions whose existence was first shown experimentally through the development of the *B. subtilis* database.

H. influenzae

The *Haemophilus influenzae* proteome has been investigated by at least three groups (Cash et al., 1995, 1997; Langen et al., 1997, 2000; Link et al., 1997). However, the group of Fountoulakis and Langen has provided the most detailed picture of the proteome to date. The series of papers published by this group provides a clear strategy with which to dissect out the components of a bacterial proteome. Moreover, the study highlights the importance of using multiple methods of sample preparation to achieve the maximum coverage of the proteome. The analyses have centered on the use of the Rd strain of *H. influenzae,* which is the avirulent laboratory strain used for the determination of the complete genome sequence (Fleischmann et al., 1995). Direct analysis of cellular proteins by 2-DGE, without prior fractionation of the cell lysate, resolved proteins in the pH range of 4 to 9.5 and mass range of 10 to 150 kDa (Langen et al., 1997). This first pass of the proteome assigned identifications to 119 protein spots through the use of peptide mass mapping, amino acid composition analysis, and N-terminal sequencing. As with all analyses based on the use of 2-DGE, these data led to an underrepresentation of low-abundance and hydrophobic proteins. Subsequent studies employed enrichment protocols prior to 2-DGE to look for these "problem" proteins. Heparin chromatography (Fountoulakis et al., 1997) and chromatofocusing (Fountoulakis et al., 1998b) were used to enrich for low-abundance proteins. Fractionation of bacterial cell lysates on heparin-actigel led to the enrichment of 160 proteins, of which 110 were not detectable in nonfractionated cell lysates (Fountoulakis et al., 1997). Similarly, fractionation of the cell lysates by chromatofocusing led to the enrichment of 125 proteins of which 75 were novel (Fountoulakis et al., 1998b).

In the case of *H. influenzae*, the chromatofocusing provided the direct purification to homogeneity, or near homogeneity, of three bacterial proteins: major ferric iron-binding protein (HI0097), a hypothetical protein (HI0052), and 5'-nucleotidase (HI0206). There was no clear separation of the proteins solely on the basis of their pI by chromatofocusing with some pools collected from the columns containing proteins with pI's ranging between 5 and 9. The authors considered that coelution of these proteins may have been due to the presence of protein complexes (Fountoulakis et al., 1998b). Neither of these chromatographic approaches enriched solely for the low-copy-number proteins, and no single class of protein was selected by either method. Fountoulakis and Takacs (1998b) proposed the detailed cataloging of the proteins enriched by specific chromatographic steps, for example, heparin-actigel, which might be useful for the detailed analysis of specific protein classes.

A similar spectrum of proteins is enriched by the same protocols from a range of different bacteria (Fountoulakis and Takacs, 1998; Fountoulakis et al., 1999), suggesting that data obtained with one microorganism can be extrapolated to others. Other classes of *H. influenzae* proteins have been specifically analyzed with the inclusion of basic (pI 6 to 11) (Fountoulakis et al., 1998) and low-molecular-mass (5 to 20 kDa) (Fountoulakis et al., 1998a) proteins into the database. Both of these approaches modified the 2-DGE separation technology itself to improve the analytical window available for protein separation. The combined data from these analyses have provided a detailed proteome map for *H. influenzae*, with the assignment of 502 of the 1,742 predicted proteins to the map, and this map currently represents one of the most detailed microbial databases now available. This study also highlights the fact that no single method can reveal the entire proteome simultaneously and that complementary procedures are required to reveal low-abundance proteins as well as proteins with extremes of pI and molecular mass.

A second catalog of *H. influenzae* proteins has been produced for a nontypeable strain of *H. influenzae*, designated NCTC 8143 (Link et al., 1997). Proteins were separated by either one-dimensional or 2-D electrophoresis to optimize the number of bacterial proteins amenable to analysis. Out of approximately 400 unique protein spots detected, 303 of the most abundant proteins were characterized using coupled microcolumn liquid chromatography and MS/MS. Of these 303 proteins, 263 were identified as unique proteins. Forty-two protein spots were not identified because of the absence of complete MS data for various reasons. One of the 263 proteins was identified as tryptophanase on the basis of its homology to *tnaA* of *E. coli*; this gene was absent from the Rd strain genome sequence (Link et al., 1997). This observation emphasizes the benefits of characterizing multiple isolates for a bacterial group in order to fully define the proteome. This particular study examined the most abundant proteins expressed by *H. influenzae* based on their detection by Coomassie brilliant blue staining. It was observed that, among these abundant proteins, the products of 19 hypothetical ORFs were readily identified, indicating the potential complementary benefits arising from combining protein and nucleic acid analyses in defining bacterial genome structure and function. Both Link et al. (1997) and Langen et al. (1997) (Fountoulakis et al., 1997) made significant progress in linking the *H. influenzae* proteins resolved by 2-DGE to the genome sequence. However, because of the use of different bacterial strains in these two reports, it is difficult to compare the published 2-D protein profiles to determine the extent of overlap between the studies.

H. pylori

H. pylori is a major cause of gastrointestinal infections and is of significant clinical interest because it is implicated as a predisposing cause of gastric cancer. The genome is predicted to encode 1,590 proteins (Tomb et al., 1997), which is slightly larger than the *H. influenzae* genome but still within the scope of the current proteome technologies of 2-DGE and peptide mass fingerprinting (Bumann et al., 2001). A number of groups have investigated the proteome of *H. pylori*.

Jungblut et al. (2000) and Bumann et al. (2001) have undertaken a detailed analysis of the *H. pylori* proteome and characterized the two sequenced *H. pylori* strains as well as a type strain used for animal studies. Up to 1,800 protein species were resolved by 2-DGE when silver staining was used to locate the proteins following electrophoresis. Peptide mass fingerprinting was used to identify over 200 of the bacterial proteins for the 26695 strain, which included 50 proteins not previously described at the protein level. A number of the proteins identified in the database are likely to play important roles in metabolic regulation and host pathogen interactions. Preliminary data also revealed the pH-dependent expression of five protein spots (Jungblut et al., 2000). Because of the site of infection in the stomach, the bacteria must survive extremes of pH, and the detailed characterization of proteins under acid conditions is certain to be a key area of investigation. As with previous comparative studies (Dunn et al., 1989; Enroth et al., 2000) of *Helicobacter* by 2-DGE, Jungblut et al. (2000) also found extensive variation at the level of the proteome. Other groups have also presented data on the proteome of *H. pylori,* which largely provide catalogs of the proteins that can be resolved by 2-DGE under defined conditions. Cho et al. (2002), working on the 26695 strain, described alternative approaches to sample preparation to increase the coverage of the proteome open to analysis. Lock et al., (2001) identified the abundant proteins resolved by 2-DGE for the *H. pylori* type strain NTC 11637.

The study of the *H. pylori* proteome has expanded beyond the purely gel-based approach for its characterization. Rain et al. (2001) have presented a protein-protein interaction map for *H. pylori* which provides connections between 46.6% of the proteins encoded by the proteome. Inspection of the protein interactions revealed by this approach is bringing to light specific biological pathways and contributes toward the prediction of protein function.

M. tuberculosis

The proteome of *Mycobacterium tuberculosis* has been extensively characterized to look for virulence markers as well as specific changes induced during the interaction between bacteria and eukaryotic cells, and these aspects of the proteome are considered below. The size of the genome of *M. tuberculosis* is 4,441 kbp, with a predicted coding capacity of 3,924 genes (Cole et al., 1998).

Urquhart et al. (1997, 1998) compared the predicted proteome map with that obtained experimentally from a series of six overlapping zoom 2-D gels, which yielded 493 unique proteins, equivalent to approximately 12% of the expected protein coding. The predicted proteome map displays a bimodal distribution for the predicted pI values of the predicted proteins (Mattow et al., 2001; Urquhart et al., 1998). A similar bimodal distribution has also been observed for plots of the theoretical proteomes of, for example, *E. coli* (VanBogelen et al., 1997), *H. influenzae* (Link

et al., 1997), and *B. subtilis* (Buttner et al., 2001). Comparisons were made between the predicted and experimental determined proteome maps for *M. tuberculosis.* When the predicted and observed maps were superimposed, outlier protein spots were observed in the observed map that were not present in the predicted proteome. These included 13 proteins with pI values of <3.3, i.e., the lowest pI predicted from the genome sequence, as well as a cluster of low-molecular-mass (<10 kDa) proteins with pI's between 5 and 8. None of these outlier proteins was identified, although it was suggested that they might have been derived through fragmentation or post-translational modification of the products of predicted ORFs.

Two detailed studies of the *M. tuberculosis* proteome are under way (Jungblut et al., 1999b; Mollenkopf et al., 1999; Rosenkrands et al., 2000), both of which can be accessed via the Internet. Jungblut and his colleagues (1999b; Mollenkopf et al., 1999) have compared two virulent and two vaccine strains of *M. tuberculosis.* Up to 1,800 and 800 proteins were resolved from either cell lysates or culture media, respectively. A total of 263 protein spots were identified by peptide mass fingerprinting. This initial study has been expanded to look specifically at low-molecular-weight acidic proteins as part of a systematic analysis of the *M. tuberculosis* proteome (Mattow et al., 2001). Seventy-six proteins that had molecular masses between 6 and 15 kDa and migrated in the pH range 4 to 6 were excised and processed for peptide mass fingerprinting. Seventy-two of the proteins were identified and found to include about 50 structural proteins. In the second *M. tuberculosis* proteome database currently available, Rosenkrands et al. (2000a) analyzed the H37Rv strain of *M. tuberculosis* and identified 288 proteins either in culture filtrates or cell lysates.

Jungblut et al. (2001) have highlighted the capacity of proteomics to complement genome sequencing and in silico determination of genome coding for *M. tuberculosis.* Six genes were identified by proteomics (2-DGE and peptide mass fingerprinting) that were not predicted from the genome sequence of the H37Rv strain. The identified proteins had molecular masses of <11.5 kDa and migrated in the pH range of 4.5 to 5.9. Partial sequencing by MS/MS of one of the proteins showed that the predicted DNA sequence derived from the peptide was present in the genome but that it had not been assigned in the original determination of predicted ORFs. Similar observations have been reported by Rosenkrands et al. (2000b), who identified a 9-kDa (pI 4.9) protein that was not predicted from the known genome sequence.

MICROBIAL TYPING AT THE LEVEL OF THE PROTEOME

Many studies in microbiology at the organismal level require, at some stage, the analysis of genetically variable organisms. This can either be a positive benefit or a problem for data interpretation depending on the research objectives. These advantages and disadvantages are not necessarily specific for proteomic applications. For example, difficulties can arise when using well-characterized laboratory strains of bacteria as a reference against which to compare recently derived clinical isolates of the same bacterial genus or species. Novel ORFs may be present in one bacterial isolate but not in the corresponding sequenced strain, or vice versa. As discussed above, tryptophanase is detected by 2-DGE in the *H. influenzae* NCTC 8143 strain, but the enzyme is absent in the sequenced *H. influenzae* Rd strain. Protein heterogeneity between bacterial isolates can make the cross-matching of 2-D protein gels difficult and unreliable unless coelectrophoresis of protein samples is used (Cash et al., 1997). Some of the problems of strain variability in studies of microbial pathogenesis and antibiotic resistance are discussed below. In contrast, the use of proteomics in epidemiology and taxonomy takes advantage of the observed genetic variation to differentiate related bacterial isolates.

Molecular techniques have been used extensively to investigate the epidemiology and taxonomy of microorganisms. Preeminent among the technologies used have been those in which the genomic DNA is analyzed (for example, Leaves and Jordens [1994]). Proteins can also be used as molecular markers based on differences in their electrophoretic mobilities. Comparative studies of the unfractionated cellular proteins resolved by 2-DGE to differentiate clinical isolates have been described for bacterial including *Neisseria* sp. (Klimpel and Clark, 1988), *Campylobacter* sp. (Dunn et al., 1989), *Haemophilus* sp. (Cash et al., 1995), and *Mycoplasma* sp. (Watson et al., 1987). Early applications of 2-DGE used the method simply as a sensitive technique to differentiate closely related bacterial isolates on the basis of protein charge and molecular weight. It is generally assumed for these applications that comigrating protein spots are functionally equivalent proteins with amino acid homology. It must also be considered that functionally equivalent proteins differing by only a single charged amino acid may appear as unique protein spots when analyzed by 2-DGE. In fact, bacterial strains can sometimes show a greater degree of variability by 2-DGE than observed by DNA-DNA hybridization (Rodwell and Rodwell, 1978). Jackson et al. (1984) compared, by 2-DGE, the proteins of multiple isolates of *Neisseria gonorrhoeae*, *Neisseria meningitidis*, and *Branhamella catarrhalis*. Quantitative and qualitative comparisons of the approximately 200 proteins resolved by 2-DGE showed that the differences in the 2-D protein profiles observed for these bacteria followed their generally recognized taxonomic classification (Jackson et al., 1984). Similarly, six *Mycoplasma* strains, representing four different species, showed the equivalent degree of similarity by 2-DGE as was obtained when they were compared by DNA-DNA hybridization (Andersen et al., 1984).

An extensive study of *Listeria* sp. classified 29 isolates, representing six different species, on the basis of the cellular proteins resolved by 2-DGE (Gormon and Phan-Thanh, 1995). Genetic differences between the bacteria were determined on the basis of the comigration of proteins resolved by 2-DGE. This approach readily distinguished the *Listeria* species as well as separating the 19 strains of *Listeria monocytogenes* present in the collection into two distinct clusters. The clustering of the isolates determined by 2-DGE agreed with that obtained using other molecular methods. The authors stressed the need for highly controlled growth conditions for the bacteria as well as the need for care and attention to the technical aspects of the 2-DGE analyses to achieve high-quality data (Gormon and Phan-Thanh, 1995). Comparative analyses of *H. influenzae* isolates in which comigrating proteins were scored showed that genetic differences indicated by 2-DGE followed trends similar to those previously obtained using molecular methods to define *H. influenzae* clonal relationships (Cash et al., 1995). Coelectrophoresis of the proteins prepared from the *H. influenzae* isolates as well as other representatives of the *Haemophilus* genus was required to accurately compare the protein profiles (Cash et al., 1997).

The extent of protein conservation (specifically conservation of electrophoretic mobility) has been reported for bacteria that were compared using 2-DGE. In the taxonomic studies of *Listeria* described above, up to 600 protein spots were resolved for each of the species analyzed but relatively few of these protein spots were conserved between the isolates. Fourteen protein spots were conserved among all six *Listeria* species examined, and 75 spots were conserved for all of the 19 strains of *L. monocytogenes* (Gormon and Phan-Thanh, 1995).

A similar classification of bacterial strains using 2-DGE has been described for *H. pylori* isolates collected from patients with different disease symptoms (Enroth et al., 2000). In agreement with the extensive heterogeneity observed at the genome level (Taylor et al., 1992), there was a high level of protein variation revealed by 2-DGE. Classification of the data suggested that there might be some clustering of the isolates according to the disease with which the specific bacteria were associated (i.e., duodenal ulcer, gastric cancer, and gastritis). The authors suggest that there may be some as yet unidentified disease-specific proteins contributing to the clustering of the bacterial isolates. However, the significance of this clustering must remain in doubt since Hazell et al. (1996) failed to demonstrate any disease-specific clustering among *H. pylori* isolates using genome mapping by restriction enzyme digestion.

These comparative studies demonstrate the potential of proteomics to distinguish related bacterial isolates. Whether the differences in the proteome are due to changes in the coding regions of the variable proteins or whether they indicate a fundamental difference in how the bacteria respond to the same environment awaits to be seen. Identification of the variable proteins is sure to provide valuable insights into bacterial metabolism and, where appropriate, bacterial pathogenesis.

INVESTIGATIONS OF BACTERIAL PATHOGENESIS AT THE LEVEL OF THE PROTEOME

Analysis of Virulent and Avirulent Strains In Vitro

The ability of pathogenic bacteria to cause disease in a susceptible host is determined by multiple factors acting individually or together at different stages of infection. For example, two key virulence determinants, *vacA* and *cagA*, are involved in distinct aspects of *H. pylori* pathogenesis (Atherton, 1998); the *cagA* gene is itself a marker for a pathogenicity island containing approximately 29 ORFs (Tomb et al., 1997). The investigation of such complex phenomena on a gene-by-gene basis where the role of each gene is investigated in isolation is unreasonable. The availability of extensive gene sequence data and in some cases complete genome sequences for many bacterial pathogens opens up strategies for analyzing global gene expression that are more appropriate for investigating polygenic phenomena such as pathogenesis. The capacity of proteomics to analyze global protein synthesis and, by extension, gene expression in a nonspecific manner makes this approach a powerful tool for identifying and characterizing the expression of bacterial pathogenic determinants. A simple approach to investigate pathogenesis is to compare the proteins synthesized by virulent and avirulent bacterial strains, grown under standard conditions, in order to identify proteins that correlate with virulence. However, this approach has major

limitations. First, when comparing naturally occurring bacterial variants, protein differences occur that are unrelated to virulence. Second, bacteria grown in vitro on defined culture media do not necessarily express all of the proteins encoded by the genome at levels characteristic of in vivo growth in the organism's natural host. This is apparent when comparing the proteins synthesized by facultative intracellular bacterial pathogens grown in defined culture media with the same bacteria growing in association with eukaryotic cells.

Despite these restrictions, high-resolution 2-DGE is a popular method for identifying virulence determinants at the level of protein synthesis. Early studies comparing virulent and avirulent *Mycoplasma pneumoniae* isolates identified three novel proteins expressed by the virulent isolates that were absent from avirulent strains (Hansen et al., 1979). Comparisons of a virulent parental strain of *M. pneumoniae* with two derived avirulent mutant strains revealed both quantitative and qualitative differences in the protein profiles when analyzed by 2-DGE (Hansen et al., 1981). No data were provided on the identities of these proteins. A similar approach was used to compare virulent and avirulent vaccine strains of *Brucella abortus* together with lipopolysaccharide deficient strains derived from each of the parental isolates (Sowa et al., 1992). Up to 935 proteins were resolved by 2-DGE for the four strains. This was fewer than the expected 2,129 proteins predicted based on the size of the 3.13×10^6-bp *B. abortus* genome (Allardet-Servent et al., 1991). The virulent and vaccine strains showed 98.4 to 99.3% homology at the DNA level. The amount of DNA equivalent to this difference in homology has a potential coding capacity of up to 34 proteins. The comparison of the 2-D protein profiles of the virulent and avirulent strains identified 86 qualitative and 6 quantitative protein differences. Although the accumulation of point mutations in coding and regulatory sequences could have occurred, this alone would be insufficient to account for the observed difference in the DNA homology. An alternative proposal was that a genome rearrangement or deletion occurred that indirectly altered the expression of additional genes required to maintain cellular homeostasis (Sowa et al., 1992).

Proteomics has been used to identify the virulence determinants of *M. tuberculosis* by comparing virulent and vaccine BCG strains (Jungblut et al., 1999b; Mahairas et al., 1996; Mollenkopf et al., 1999; Sonnenberg and Belisle, 1997; Urquhart et al., 1997, 1998). Relatively few differences are detected at the proteome level for these bacteria (Jungblut et al., 1999b), consistent with the limited genetic variability found from the sequence comparison of 26 genes among 842 *M. tuberculosis* complex isolates (Sreevatsan et al., 1997). Urquhart et al. (1997) used a high-resolution multigel system to compare the virulent *M. tuberculosis* H37Rv and *Mycobacterium bovis* BCG (Pasteur ATCC 35734) strains. Up to 772 protein spots were identified for *M. bovis* BCG and virulent *M. tuberculosis* over a pH range of 2.3 to 11 with apparent molecular masses between 10 and 216 kDa. Some differences were observed between the two strains under these analytical conditions, but their significance remains to be established. A detailed comparison of virulent *M. tuberculosis* (H37Rv and Erdman) and vaccine (BCG Chicago and BCG Copenhagen) strains has been reported by Jungblut et al., (1999b). The virulent and avirulent strains were grown in defined culture media, and the bacterial proteins were extracted from both the cell lysates and the culture supernatants. The majority of the proteins associated with the bacterial cells, as well as those released into the culture media, were common for all four isolates. The most extensive variation was found between H37Rv and BCG Chicago with 31 variant spots; 21 were qualitative and 10 were significant quantitative differences. There were also 18 and 3 protein spot variants observed in comparisons of the two virulent and two vaccine strains, respectively. Although no novel virulence determinants were found for *M. tuberculosis* by using proteomics, some virulence determinants previously identified by other means were assigned to the proteome (Jungblut et al., 1999b). The authors commented on the fact that amino acid substitutions leading to electrophoretic mobility variants might be useful for vaccine development if the substitution(s) occurred in T-cell epitopes. Betts et al. (2000) compared the proteomes of two virulent strains of *M. tuberculosis*, the laboratory-adapted H37Rv strain and CDC1551, a recent clinical isolate (Valway et al., 1998) that has been partially sequenced. The analysis demonstrated that the classic virulent H37Rv strain used as the basis for many proteomic analyses had retained the features of the more recent virulent *M. tuberculosis* isolates. A total of 1,750 intracellular proteins were resolved by 2-DGE over a range of pH of 3 to 10 for each of the two strains examined when grown in vitro. Comparative studies of the proteomes of the bacteria assayed at various times during their growth revealed just 13 consistent spot differences between the isolates. Seven and three spots were specific for the CDC 1551 and H37Rv strains, respectively. A further two spots were increased in abundance for H37Rv compared with CDC 1551. A single protein, identified as ribosome recycling factor, showed a mobility difference between the two isolates. Peptide mass mapping and matrix-assisted laser desorption ionization–time-of-

flight (MALDI-TOF) MS were used to identify nine of the proteins exhibiting differences, and identities were obtained. Four of the protein differences corresponded to mobility variants of MoxR (Rv1479); two variants were specific for each of the two bacterial strains. One of the CDC 1551-specific proteins identified was a probable alcohol dehydrogenase (Rv0927c). One H37Rv-specific protein was identified as HisA, and the two H37Rv-induced spots were electrophoretic variants of alkyl hydroperoxide reductase chain C. The difference in the mobility of the MoxR protein was consistent with a nucleotide change observed in the gene for this protein. Consistent proteomes were observed for both isolates over the 12-day growth curve despite the fact that the CDC 1551 strain entered stationary phase in advance of H37Rv.

Comparisons of the genome sequences for the virulent and avirulent strains of *M. tuberculosis* demonstrated that the majority of the genome differences were single-nucleotide changes of which approximately 1,000 have been identified (Betts et al., 2000). Apart from the MoxR protein, none of the protein differences correlated with genome differences. Many of the predicted differences in the genomes of virulent and avirulent strains lie within the acidic, glycine-rich proteins of the PE (Pro-Glu) and PPE (Pro-Pro-Glu) families. The PE and PPE multigene families consist of 167 members and account for approximately 10% of the bacterial genome coding capacity. The function of these proteins is unknown, but they may be involved in antigenic variation. None of the 167 members of the PE and PPE families has been identified to date in the *M. tuberculosis* proteome databases (Jungblut et al., 1999b; Mollenkopf et al., 1999; Rosenkrands et al., 2000a). This is despite the fact that 70% of the PE and PPE families are within the mass and pI range typically covered by 2-DGE and, if expressed in sufficient amounts, ought to be detectable (Rosenkrands et al., 2000a). Rosenkrands et al. (2000a) suggested that the proteins might have an unusual amino acid composition that prevents them from being detected by 2-DGE.

One difficulty in interpreting the data obtained from the comparative studies described above is in the selection of the specific virulent and avirulent strains used as the basis for the comparison. In many cases, the virulent and avirulent strains are genetically distinct, and differences in their proteomes may be present that are unrelated to virulence. Mahairas et al. (1996) used subtractive genomic hybridization to locate genetic differences between *M. bovis* BCG vaccine and virulent isolates of *M. bovis* and *M. tuberculosis*. Three regions were deleted in the BCG vaccine compared with the virulent strains. One 9.5-kb segment (designated RD1) was absent from six BCG substrains but present in virulent *M. tuberculosis* and *M. bovis* strains as well as in 62 clinical *M. tuberculosis* isolates analyzed. Based on its sequence, RD1 contains at least eight ORFs. The virulent *M. bovis* and *M. tuberculosis* showed indistinguishable protein profiles when compared by 2-DGE. In contrast, the BCG vaccine strains expressed at least 10 additional proteins and induced expression levels for a number of other proteins. When the RD1 region was introduced into the BCG genome to generate BCG::RD1, the protein profile of BCG::RD1 was indistinguishable from virulent *M. bovis*. This suggested that parts of RD1 caused a specific suppression of protein synthesis in virulent mycobacteria. Some low-molecular-weight proteins were identified for BCG::RD1 that were consistent in size with short ORFs encoded in the RD1 sequence. More data are required to link these observations to the protein differences reported by Jungblut et al. (1999b) and Urquhart et al. (1997), but once available a more complete picture of the mycobacterial virulence determinants is sure to emerge.

In Vivo-Induced Protein Synthesis

There are many limitations to identifying the determinants of bacterial pathogenesis by simply comparing bacterial isolates grown under laboratory conditions on defined culture media. A large number of studies have demonstrated that facultative bacterial pathogens express novel genes when they infect eukaryotic cells. Extensive efforts are being made by many research groups to identify these specific gene sets since they may represent novel pathogenic determinants or serve as potential targets for new therapeutic drugs. Recombinant DNA technologies, which take advantage of the existing genome sequence data now available, play a key role in identifying the bacterial genes specifically expressed in vivo, i.e., when the bacteria are in association with the eukaryotic cell. Reporter genes (for example, green fluorescent protein, β-galactosidase, and luciferase) have been linked to bacterial promoters to monitor their in vivo induction (reviewed by Hautefort and Hinton, 2001). An important experimental strategy that has been developed for looking at the expression of genes in vivo is that of in vivo expression technology (IVET). The use of IVET can rapidly identify those bacterial genes that are expressed specifically in vivo either in cell culture systems or the intact animal (Slauch et al., 1994). Proteomics provides a valuable complement to these DNA-based technologies to examine in vivo gene expression. However, there are severe restrictions in the use of proteomics for this field of research. Using current 2-D gel-based technologies, proteomics is limited to analyzing in vitro cell systems as described below. Using in vitro grown cell lines actually limits the type of data that can be derived. Although producing valuable and important data, in vitro-grown cells do not mimic entirely the intact animal, where the bacteria are interacting at a number of levels with functionally distinct cell types present in an intact organism or tissue. At present there are no means to investigate protein synthesis of bacterial pathogens when grown in vivo with the equivalent sensitivity and power of IVET. The low recovery of bacterial cells from the host and an absence of a protein amplification method, analogous to the PCR for nucleic acids, present a number of technical difficulties to the investigator. Growth of the bacteria in artificial culture media, even for a limited time, would be required, which would invalidate the identification of in vivo-specific protein synthesis. This may very well change in the future once protein array technologies become generally available.

Nevertheless, progress can be made with in vitro model systems, and proteomics has been widely used to examine the interaction of bacteria with eukaryotic cells. The use of in vitro cell lines has the advantage that they can be infected under reproducible controlled conditions, and radioactive amino acids can be used to increase the sensitivity of protein detection. Typically, a combination of antibiotics and radioactive amino acids is used to selectively radiolabel the proteins synthesized by intracellular bacteria (Abshire and Neidhardt, 1993; Hanawa et al., 1995). Briefly, eukaryotic cells are infected with bacteria and incubated with gentamicin to kill extracellular bacteria; the intracellular bacteria retain their viability. After a predetermined time interval, radioactive amino acids are added in the presence of cycloheximide, which inhibits cellular but not bacterial protein synthesis. The radiolabeled proteins synthesized by the intracellular bacteria are compared with radiolabeled proteins prepared from bacteria grown in defined culture media. The data obtained by this approach should be considered in the light of differences in the growth phase of the two bacterial populations as well as possible nonspecific effects of the antibiotics used to inhibit cellular protein synthesis. The use of a cycloheximide block to inhibit cellular protein synthesis has been shown to be unnecessary for the analysis of *M. bovis* protein synthesis during infection of macrophage cells (Monahan et al., 2001). It was demonstrated that, after radiolabeling in the absence of cycloheximide, the macrophage cells could be lysed with sodium dodecyl sulfate and the bacteria could then be collected by centrifugation and washed with Tween-80. Under these conditions, control studies showed that there was minimal carry-over of cellular proteins with the bacterial pellet. It was suggested that the ability to omit the cycloheximide allowed the cross signaling between the cell and the bacteria that might otherwise be inhibited in the presence of cycloheximide and so influence the pattern of bacterial growth within the cell (Monahan et al., 2001).

Cocultivation of bacteria with eukaryotic cells alone can be sufficient to induce the synthesis of proteins not expressed by bacteria grown in defined media. During the cocultivation of *Campylobacter jejuni* with INT 407 cells, an epithelial cell line, at least 14 proteins showed increased biosynthetic levels. These changes in protein synthesis were revealed using 2-DGE combined with either metabolic radiolabeling or immunoblotting (Konkel and Cieplak, 1992; Konkel et al., 1993). The induction of a subset of the 14 proteins also occurred following exposure of the bacteria to either cell culture medium alone or INT 407 cell-conditioned media (Konkel and Cieplak, 1992). It was suggested that the de novo synthesis of these proteins was required for the subsequent internalization of the bacteria into the epithelial cells.

Intracellular bacteria growing in the cell's phagosome are exposed to a variety of stress conditions, including extremes of acidity, oxygen, and nutrients (Kwaik and Harb, 1999). In contrast, those intracellular bacterial pathogens that migrate out of the intracellular vacuoles into the cytoplasm are exposed to a reduced level of stress. During the infection of macrophage cells, *Legionella pneumophila*, *B. abortus*, and *Salmonella enterica* serovar Typhimurium remain associated with the phagosome and may interfere with its maturation. A consistent observation for these intracellular bacterial pathogens is that the synthesis of specific bacterial proteins is either induced or repressed during the intracellular growth phase compared with bacteria growing in artificial culture media. Moreover, a number of the bacterial proteins induced during intracellular growth also show altered biosynthesis under in vitro stress conditions. The synthesis of the bacterial heat shock proteins GroEL and DnaK are induced during *B. abortus* infection of bovine macrophages (Rafie-Kolpin et al., 1996); the induced synthesis of GroEL has also been demonstrated in *B. abortus*-infected murine macrophages (Lin and Ficht, 1995). The same two heat shock proteins also form part of the spectrum of proteins induced during *S. enterica* serovar Typhimurium infection of macrophages (Buchmeier and Heffron, 1990). Two major bacterial proteins homologous to the stress proteins DnaK and cross-reacting protein antigen (also known as Hsp60) are induced during *Yersinia enterocolitica* infection of J774 cells,

a murine macrophage cell line; these proteins are also induced in the bacteria by heat shock and oxidative stress in vitro (Yamamoto et al., 1994). Of 67 bacterial proteins induced during *L. pneumophila* infection of the U937 macrophage cell line, 32 are also induced by in vitro stress conditions. These include the heat shock proteins GroEL and GroES (Abu Kwaik et al., 1993). A protein, global stress protein (GspA), is expressed by *L. pneumophila* in response to all stress conditions examined to date as well as during intracellular replication. GspA is induced at higher levels in intracellular bacteria, suggesting that the bacteria in this particular environment may be exposed to multiple simultaneous stress conditions (Abu Kwaik et al., 1997; Kwait and Harb, 1999). Although the spectrum of proteins exhibiting altered synthesis by intracellular bacteria and bacteria stressed in vitro have many similarities, the changes observed for the former are not simply a summation of the in vitro stress responses (Abshire and Neidhardt, 1993), suggesting that there are specific responses induced during the intracellular replication phase.

Contrasting data to those described above are found for *L. monocytogenes* infection of J774 cells. A range of in vitro stress conditions, including heat shock and oxidative stress which induce the synthesis of GroEL and DnaK, induce none of the 32 proteins observed for intracellular bacteria (Hanawa et al., 1995). The absence of known stress-induced proteins expressed by the intracellular bacteria is believed to be due to the rapid migration of *Listeria* from the phagosome to the cytoplasm during intracellular growth (Hanawa et al., 1995).

M. bovis BCG infection of the THP-1 macrophage cells results in the induced synthesis of bacterial proteins not expressed by bacteria grown in artificial growth media (Monahan et al., 2001). These proteins were demonstrated using radiolabeling as well as by immunoblotting against human *M. tuberculosis*-infected sera. The induced immunogenic proteins may serve as future immunoprotective antigens. Under the conditions used, at least 20 proteins, differentially expressed in BCG-infected macrophages, either were specific for the infected macrophage cells or exhibited significantly induced levels of synthesis. Six of the proteins that were induced at 24 h postinfection were identified using a combination of MALDI-TOF and nanoES-MS. These proteins were the GroEL homologues, GroEL-1/GroEL-2, InhA, 16-kDa antigen (α-crystallin Hsp-X), EF-Tu, and a 31-kDa hypothetical protein. As Monahan et al. (2001) commented, these data are significant because they represent one of the earliest successful identifications of proteins recovered from intracellular bacteria using MS techniques.

As might be expected, the source of the eukaryotic cell influences the outcome of the bacterial infection. This has been shown for *M. tuberculosis* and *Mycobacterium smegmatis*, in which the type and origin of the eukaryotic cell influence both the initial interaction between the bacterium and the eukaryotic cell as well as the growth kinetics of the bacteria (Barker et al., 1996; Mehta et al., 1996). The response of the bacterial proteome to bacterial growth in host cells of differing origin has been followed in *Mycobacterium avium* infection of bone marrow macrophages and J774 cells (Sturgill-Koszychki et al., 1997). *M. avium* infection of J774 cells results in the specific induced synthesis of bacterial proteins. The induced bacterial protein synthesis commenced by 6 h postinfection and continued until at least 96 h postinfection. None of the induced proteins showed altered synthesis when bacteria were exposed to in vitro stress conditions. In contrast to J774 cells, *M. avium* infection of primary bone macrophages followed different kinetics of bacterial replication and protein synthesis. Intracellular bacteria radiolabeled at 5 and 12 days postinfection showed significant differences in their protein synthesis. These data were consistent with the bacteria initially entering a stasis phase early in infection before commencing a normal replication cycle between 5 and 12 days postinfection. At 12 days postinfection the bacteria synthesized a similar range of proteins as *M. avium* grown in J774 cells. These observed variations in the replication cycle and protein synthesis of the bacteria using macrophage cells of different sources demonstrate the care required in selecting the host cell used to investigate intracellular bacterial replication at the level of the proteome.

The above discussion considered the response of bacterial gene expression to the intracellular environment of the host. Similar approaches can, in principle, be used to investigate the host cell response to the bacterial infection. Phagocytosis of intracellular pathogens by macrophages initiates a cascade of processes involving both the phagosomal and the intracellular components of the endocytic pathway (Kovarova et al., 2002). The intracellular organisms sense these changes and adapt to them to promote their survival. Kovarova et al. (1992) described the global changes in macrophage protein synthesis following *Francisella tularensis* infection of mice. Principal component analysis was used to identify significant changes in protein synthesis following infection with *F. tularensis* between infected and uninfected cells. It was possible to distinguish uninfected and infected macrophage cells on the basis of their overall protein patterns obtained using 2-DGE. The infected macrophage cells were further subdivided into

macrophages collected 3 to 7 days postinfection and cells collected at 10 days postinfection (Kovarova et al., 1992). The analysis has been further refined to look at the phagosomal compartments from infected macrophages (Kovarova et al., 2002). Several proteins were identified that were specific for the *F. tularensis* live vaccine strain containing phagosomes prepared from infected B10R macrophages. The proteins that were identified include an *F. tularensis* hypothetical 23-kDa protein, chaparonin GroEL, and host-putative proteins that appear to be ATP synthase β-chain and NADH-ubiquinone oxidoreductase. The expression of the 23-kDa protein appears to correlate with the pathogenesis of *F. tularensis* with high levels of expression occurring in the phagosomes prepared from infected permissive macrophages. In addition, the 23-kDa protein is also induced in *F. tularensis* exposed to hydrogen peroxide, suggesting that the protein helps to protect the bacteria against hostile intracellular environments including that found intracellularly (Golovliov et al., 1997).

This same bacterial-eukaryotic cell system has been used to define the role of the *Nramp1* gene in defining susceptibility to intracellular bacterial growth (Govoni et al., 1996). *Nramp1* has a number of pleiotropic effects, and 2-DGE combined with principal component analysis has been used to identify those genes regulated by *Nramp1* (Kovarova et al., 1998). Comparisons of protein synthesis in macrophages from mice carrying either the resistant or sensitive allele of *Nramp1* showed at least four proteins whose synthesis was influenced by *Nramp1*. Two of the proteins with induced synthesis in macrophages carrying the resistant allele were provisionally identified as Mn-superoxide dismutase and bcl-2 (Kovarova et al., 1998). The expression of the resistance allele of *Nramp1* leads to modifications in the expression of a number of macrophage signal transduction pathways, which may be involved in providing resistance to microbial infection (Kovarova et al., 2001).

The Bcg locus/*Nramp1* gene controls the natural resistance of mice to infection by *F. tularensis* live vaccine strain (LVS) and other intracellular bacterial pathogens. However, in the case of *F. tularensis* the effect of *Nramp1* is opposite of that described for other Bcg/*Nramp1*-controlled pathogens, for example, several *Leishmania donovani* and *S. enterica* serovar Typhimurium strains (Bradley, 1977; Plant and Glynn, 1976). The mutant Bcg/*Nramp1* allele confers natural resistance to *F. tularensis* LVS infection, whereas the wild-type allele makes the cells susceptible to infection. To determine if the differential expression of the Bcg/*Nramp1* gene modifies the composition of the *F. tularensis* LVS-containing phagosomes (FCP), FCP were isolated from infected Bcg congenic B10R and B10S macrophages of susceptible and resistant phenotypes, respectively. Comparisons of the proteomes of these two phagosomal compartments revealed several proteins that were specific for FCP prepared from B10R macrophages including the 23-kDs a protein described above. *Nramp1* is a member of the divalent cation transporter proteins, with the typical structure of an integral membrane protein (Cellier et al., 1995). Thus, it was proposed that *F. tularensis* might exploit the ion transport function of Bcg/*Nramp1* to aid its intracellular survival. In this model it is proposed that the transport function of *Nramp1* allows *F. tularensis* to utilize ions (for example, nutrient iron) to overcome the host immune response of the B10 macrophages. The differential allelic expression of the host Bcg/*Nramp1* affects the phagosome composition, which then contributes toward the permissiveness of B10R macrophages for *F. tularensis* infection. This might occur through the acquisition of an essential nutrient iron via the ion-transport function of Bcg/*Nramp1* (Kovarova et al., 2002).

The effect of infecting bacteria on the host cell proteome has also been examined for *M. tuberculosis*. The interaction of *M. tuberculosis* with the functions of the host cell phagosome has been extensively investigated using a number of criteria including the response of the cellular proteome. During the intracellular replication of *M. tuberculosis* in macrophages, the *M. tuberculosis*-containing phagosomal compartments fail to fuse with the lysosomes, the normal fate of intracellular phagosomes. Essentially, the bacteria arrest the development and processing of the phagosome. Comparisons of the phagosomal compartments from mycobacterial-infected cells with the same structures from uninfected cells show a number of differences in the cellular proteomes. Fratti et al. (2000) proposed that one of the key features of the *M. tuberculosis* phagosome structure was the exclusion of EEA-1 (early endosomal autoantigen). This protein plays a role in vesicle tethering and also in endosomal fusion. At present, the question remains of determining the identity of the protein or proteins expressed by *M. tuberculosis* which lead(s) to this exclusion of EEA-1.

DEVELOPMENT OF THERAPEUTIC STRATEGIES THROUGH PROTEOMICS

In parallel with genome-based technologies, proteomics has played a role in the development of therapeutic strategies via the identification of novel vaccine and antibiotic targets (Nilsson et al., 2000;

Rosamond and Allsop, 2000). Potential vaccine candidate proteins can be identified through the identification of in vivo immunogenic proteins. Antigenic proteins can be identified by characterizing the immune response following either natural or laboratory infection with the bacterium of interest. Proteins that are recognized by either the humoral or cellular immunes responses can be readily identified; for example, *Borrelia burgdoferi* (Jungblut et al. 1999c), *Streptococcus pyogenes* (Lemos et al., 1998), *Brucella ovis* (Teixeira-Gomes et al., 1997), and *M. tuberculosis* (Andersen, 1994; Pal and Horwitz, 1992; Roberts et al., 1995). The global analysis of antigenic proteins using proteomics is a potentially useful approach to identify novel antigenic determinants for inclusion in future vaccines. This is illustrated in studies of the immunogenic proteins of *H. pylori*. *H. pylori* cellular proteins were analyzed by 2-DGE, and the antigenic proteins were identified by immunoblotting against pooled human sera. Over 30 immunogenic protein spots were identified by this approach (McAtee et al., 1998a, b). The antigenic proteins were characterized further by N-terminal sequencing and peptide mass mapping. Twenty nine of the antigenic protein spots were identified and 15 were assigned to known ORFs in the *H. pylori* genome sequence; the remaining protein spots had homologies with bacterial proteins not present in the current *H. pylori* genome sequence. One 30-kDa protein spot contained two protein species both derived by posttranslational processing from the *H. pylori* ORF, HP0175 (McAtee et al., 1998c), which had not previously been shown to encode an immunogenic protein. Comparisons of the proteins expressed by different isolates of *H. pylori* demonstrated that the majority of the antigenic proteins were expressed by all of isolates, although differences in the expression levels of the flagellin protein and catalase were observed between some isolates (McAtee et al., 1998b).

To optimize the identification of vaccine candidates, specific protein classes can be selectively analyzed. For example, it is reasonable to expect that the membrane proteins are major targets for the host immune response. Nilsson et al. (2000) used a liquid-phase charge separation to resolve putative membrane proteins of *H. pylori*. Bacterial cells were solubilized with *n*-octylglucoside, and the proteins were analyzed with no further enrichment or fractionation. The bacterial cell proteins were fractionated by isoelectric focusing in liquid phase for the first dimension and by continuous elution of isoelectric focusing fractions by sodium dodecyl sulfate-polyacrylamide gel electrophoresis. Fifteen of the 40 proteins identified following this separation were membrane or membrane-associated proteins, and many of these had not been previously identified using standard protocols of 2-DGE. Chakravarti et al. (2000) described a bioinformatics approach to the identification of vaccine candidate proteins which takes advantage of the extensive gene sequence data that are now available. The *H. influenzae* Rd genome sequence was searched in silico for potential vaccine candidate proteins. The identification of the candidate proteins was based on the detection of specific characteristics of outer membrane proteins because these are the most likely to be immunogenic and so most useful in vaccines. A complementary proteomic approach to identify the proteins from soluble outer membrane fractions of *H. pylori* was also described. Although this is a potentially valuable approach, the authors did not confirm that the proteins identified by these criteria were in fact immunogenic under natural conditions. The secreted proteins from *H. pylori* are a further category of proteins that may contain potential therapeutic targets. Bumann et al. (2002) have initiated a study of the proteins secreted by *H. pylori* during growth in liquid media. Twenty six of the 33 reproducibly detected proteins were identified. These contained the previously documented virulence determinants of VacA and urease as well as proteins of an unknown function.

The spread of antibiotic and drug resistance among microbial pathogens is a major problem for the control of infection. An understanding of the mechanism(s) by which drug resistance develops will lead to improvements in extending the efficacy of current antimicrobial agents. Proteomics can contribute toward determining antimicrobial resistance mechanisms through the capacity to analyze global changes in microbial proteins. Qualitative and quantitative changes can be identified in a nonspecific manner without making preconceived judgments on the potential importance of different components. Resistance to β-lactam antibiotics has been investigated for *Pseudomonas aeruginosa*. A diminished expression of a 47-kDa (pI 5.2) outer membrane protein has been found for imipenem-resistant *P. aeruginosa* (Vurma-Rapp et al., 1990). Michea-Hamzehpour et al. (1993) reported similar data with the loss of an outer membrane protein (pI 5.2) in imipenem-resistant *P. aeruginosa*. N-terminal sequencing showed that the protein was homologous to the porin outer membrane protein D (Michea-Hamzehpour et al., 1993). Changes in outer membrane proteins have also been shown among ceftazidime-resistant *P. aeruginosa* isolates with the expression of a basic protein homologous to the *ampC* gene product (Michea-Hamzehpour et al., 1993).

Penicillin resistance is increasing among clinical isolates of *Streptococcus pneumoniae*, and resistance to erythromycin, used as an alternative antibiotic to penicillin, has now emerged as a potential problem (Johnson et al., 1996). Two erythromycin-resistant phenotypes are recognized for *S. pneumoniae*, specifically the MLS (macrolide, lincosamide, and streptogramin antibiotic-resistant) and M phenotypes (Johnson et al., 1996). Erythromycin-resistant *S. pneumoniae* possessing the MLS phenotype owe their resistance to the methylation of rRNA by the product of the *erm* gene located on the transposon Tn*1545* (Clewell et al., 1995; Trieu-Cuot et al., 1990). *S. pneumoniae* isolates with the M phenotype are less well characterized, although resistance appears to be linked to the expression of a gene called *mefE* that is believed to encode a membrane transporter protein that reduces the intracellular levels of erythromycin (Sutcliffe et al., 1996). Proteomics has been used to further investigate the method of erythromycin resistance in M phenotype *S. pneumoniae* isolates (Cash et al., 1999). Cellular proteins were prepared from erythromycin-resistant (M phenotype) and susceptible isolates of *S. pneumoniae* and analyzed using 2-DGE. *S. pneumoniae* isolates with the M-phenotype erythromycin resistance showed a characteristic induced synthesis of a 38.5-kDa protein (Cash et al., 1999). None of the erythromycin-sensitive *S. pneumoniae* strains showed this induced protein synthesis. The 38.5-kDa protein was identified as glyceraldehyde phosphate dehydrogenase (GAPDH) using peptide mass mapping and its peptide homology to GAPDH encoded by *Streptococcus equisimilis* and *S. pyogenes*. Three electrophoretic variants of GAPDH that differed in their pI were resolved for M-phenotype resistant isolates, with the induced protein having the most basic isoelectric point. Thus, the abnormal synthesis of GAPDH in the M-phenotype isolates may be related to an altered pattern of posttranslational modification. The disrupted GAPDH synthesis was not related to erythromycin resistance per se since the MLS phenotype was indistinguishable from sensitive *S. pneumoniae* isolates. Using the *S. equisimilis* and *S. pyogenes gap* gene sequences, DNA primers were designed to analyze the GAPDH gene for the erythromycin-sensitive and -resistant *S. pneumoniae* isolates analyzed by 2-DGE (Amezaga, Carter, Cash and McKenzie [2000], The 2nd International Symposium on Pneumococcal Diseases, South Africa). A characteristic base mutation causing an amino acid substitution was shown to correlate completely with the protein profile demonstrated by 2-DGE. This combination of proteomics and genomics to investigate erythromycin resistance clearly illustrates the power of proteomics to locate putative gene targets for later study using genomic technologies.

The mechanism of antibiotic action can also be investigated by locating proteins that show differential expression patterns when bacteria are grown in the presence or absence of antibiotics. This experimental approach has been used to investigate isoniazid-induced gene expression in *M. tuberculosis* using nucleic acid microarrays (Wilson et al., 1999). The response of the bacterial proteome following their exposure to antibiotics has been investigated for *Staphylococcus aureus* (Singh et al., 2001). *S. aureus* grown in the presence of inhibitory concentrations of oxacilin, a cell-wall active antibiotic, resulted in elevated expression for at least nine proteins, as shown by 2-DGE. Five of the induced proteins were identified by N-terminal sequencing as methionine sulfoxide reductase, enzyme IIA component of the phosphotransferase system, signal transduction protein, GroES, and GreA. A similar pattern of induced protein synthesis was found with other antibiotics that act on the bacterial cell wall but not for antibiotics acting on other targets, thus suggesting that the observed induced protein expression might represent a "proteomic signature" for this response (Singh et al., 2001). A similar approach has been used to examine metronidazole-induced gene expression in *H. pylori* (McAtee et al., 2001). Metronidazole resistance generally depends on the mutation of the RDXA (NADH reductase) gene with higher levels of resistance arising through mutations leading to the loss of function in additional reductase genes. When functional, the reductase genes convert metronidazole from a harmless drug to mutagenic and bactericidal products and, in the process, may generate reactive oxygen metabolites. Metronidazole-resistant *H. pylori* were grown in sublethal concentrations of metronidazole, and the differential protein expression was determined by 2-DGE. In the presence of metronidazole, 19 protein spots exhibited differential expression; 12 spots showed a greater than twofold decrease in expression, 3 spots showed a greater than twofold increased expression, and no quantitative data were presented for the remaining 4 spots. Proteins that showed an increased expression level in the presence of metronidazole were identified by peptide mass mapping as alkylhydroperoxide reductase (AHP) (two protein spots) and aconitase B. AHP is known to protect against oxygen toxicity, and it was proposed that the increased expression of AHP in metronidazole-resistant *H. pylori* is important in the generation of the resistance phenotype.

CONCLUDING COMMENTS

Data derived through analyzing the proteome represent a crucial step toward understanding cellular metabolism at a global level. It is only by studying the proteome that many aspects of protein structure and function can be correctly determined; for example, the specific details of posttranslational modification and protein turnover cannot be established from the gene sequence alone. As illustrated above, there are a number of advanced studies on microbial proteomes. In many cases an experimentally described protein has now been assigned to what was originally a hypothetical protein predicted from the genome sequence. Microbial systems are relatively simple, with limited numbers of genes encoded by the genome and only restricted levels of posttranslational modification, suggesting that a complete picture of cellular metabolism for a free-living organism will soon emerge and that this will be based on a bacterium.

Although there are many advantages to investigations at the level of the proteome, there are limitations in reaching the desired total picture of cellular metabolism. The widely used methods of 2-DGE and peptide mass mapping are both labor intensive and limited for analyzing some protein classes. There are limitations due to the biology of the system under investigation that are particularly apparent when investigating the proteomes of pathogenic bacteria in association with their hosts in vivo. Current technologies provide only a limited view of the microbial proteome under these conditions. Recent developments in the field of MS and in the preparation of protein arrays are likely to overcome some if not all of these problems. It is certain that there will be continued expansion in the data on microbial proteomes with the identification of further novel genetic interactions and therapeutic targets.

REFERENCES

Abshire, K. Z., and F. C. Neidhardt. 1993. Analysis of proteins synthesized by *Salmonella typhimurium* during growth within a host macrophage. *J. Bacteriol.* 175:3734–3743.

Abu Kwaik, Y., B. I. Eisenstein, and N. C. Engleberg. 1993. Phenotypic modulation by *Legionella pneumophila* upon infection of macrophages. *Infect. Immun.* 61:1320–1329.

Abu Kwaik, Y., L. Y. Gao, O. S. Harb, and B. J. Stone. 1997. Transcriptional regulation of the macrophage-induced gene (*gspA*) of *Legionella pneumophila* and phenotypic characterization of a null mutant. *Mol. Microbiol.* 24:629–642.

Allardet-Servent, A., M. J. Carles-Nurit, G. Bourg, S. Michaux, and M. Ramuz. 1991. Physical map of the *Brucella melitensis* 16 M chromosome. *J. Bacteriol.* 173:2219–2224.

Andersen, H., G. Christiansen, and C. Christiansen. 1984. Electrophoretic analysis of proteins from *Mycoplasma capricolum* and related serotypes using extracts from intact cells and from minicells containing cloned mycoplasma DNA. *J. Gen. Microbiol.* 130:1409–1418.

Andersen, P. 1994. Effective vaccination of mice against *Mycobacterium tuberculosis* infection with a soluble mixture of secreted mycobacterial proteins. *Infect. Immun.* 62:2536–2544.

Anderson, L., and J. Seilhamer. 1997. A comparison of selected mRNA and protein abundances in human liver. *Electrophoresis* 18:533–537.

Anderson, N. G., and L. Anderson. 1982. The human protein index. *Clin. Chem.* 28:739–748.

Anderson, N. G., and N. L. Anderson. 1996. Twenty years of two-dimensional electrophoresis: past, present and future. *Electrophoresis* 17:443–453.

Atherton, J. C. 1998. *H. pylori* virulence factors. *Brit. Med. Bull.* 54:105–120.

Barker, K., H. Fan, C. Carroll, G. Kaplan, J. Barker, W. Hellmann, and Z. A. Cohn. 1996. Nonadherent cultures of human monocytes kill *Mycobacterium smegmatis*, but adherent cultures do not. *Infect. Immun.* 64:428–433.

Betts, J. C., P. Dodson, S. Quan, A. P. Lewis, P. J. Thomas, K. Duncan, and R. A. McAdam. 2000. Comparison of the proteome of *Mycobacterium tuberculosis* strain H37Rv with clinical isolate CDC 1551. *Microbiology* 146:3205–3216.

Blattner, F. R., G. Plunkett, III, C. A. Bloch, N. T. Perna, V. Burland, M. Riley, J. Collado-Vides, J. D. Glasner, C. K. Rode, G. F. Mayhew, J. Gregor, N. W. Davis, H. A. Kirkpatrick, M. A. Goeden, D. J. Rose, B. Mau, and Y. Shao. 1997. The complete genome sequence of *Escherichia coli* K-12. *Science* 277:1453–1474.

Bloch, P. L., T. A. Phillips, and F. C. Neidhardt. 1980. Protein identifications of O'Farrell two-dimensional gels: locations of 81 *Escherichia coli* proteins. *J. Bacteriol.* 141:1409–1420.

Bradley, D. J. 1977. Regulation of *Leishmania* populations within the host. II. genetic control of acute susceptibility of mice to *Leishmania donovani* infection. *Clin. Exp. Immunol.* 30:130–140.

Buchmeier, N. A., and F. Heffron. 1990. Induction of *Salmonella* stress proteins upon infection of macrophages. *Science* 248:730–732.

Bumann, D., T. F. Meyer, and P. R. Jungblut. 2001. Proteome analysis of the common human pathogen *Helicobacter pylori*. *Proteomics* 1:473–479.

Bumann, D., S. Aksu, M. Wendland, K. Janek, U. Zimny-Arndt, N. Sabarth, T. F. Meyer, and P. R. Jungblut. 2002. Proteome analysis of secreted proteins of the gastric pathogen *Helicobacter pylori*. *Infect. Immun.* 70:3396–3403.

Buttner, K., J. Bernhardt, C. Scharf, R. Schmid, U. Mader, C. Eymann, H. Antelmann, A. Volker, U. Volker, and M. Hecker. 2001. A comprehensive two-dimensional map of cytosolic proteins of *Bacillus subtilis*. *Electrophoresis* 22:2908–2935.

Cahill, D. J. 2001. Protein and antibody arrays and their medical applications. *J. Immunol. Methods* 250:81–91.

Cash, P. 1998. Characterisation of bacterial proteomes by two-dimensional electrophoresis. *Anal. Chim. Acta* 372:121–146.

Cash, P., E. Argo, and K. D. Bruce. 1995. Characterisation of *Haemophilus influenzae* proteins by 2-dimensional gel electrophoresis. *Electrophoresis* 16:135–148.

Cash, P., E. Argo, P. R. Langford, and J. S. Kroll. 1997. Development of an *Haemophilus* two-dimensional protein database. *Electrophoresis* 18:1472–1482.

Cash, P., E. Argo, L. Ford, L. Lawrie, and H. McKenzie. 1999. A proteomic analysis of erythromycin resistance in *Streptococcus pneumoniae*. *Electrophoresis* 20:2259–2268.

Cellier, M., G. Prive, A. Belouchi, T. Kwan, V. Rodrigues, W. Chia, and P. Gros. 1995. Nramp defines a family of membrane proteins. *Proc. Natl. Acad. Sci. USA* 92:10089–10093.

Chakravarti, D. N., M. J. Fiske, L. D. Fletcher, and R. J. Zagursky. 2000. Application of genomics and proteomics for identification of bacterial gene products as potential vaccine candidates. *Vaccine* 19:601–612.

Cho, M. J., B. S. Jeon, J. W. Park, T. S. Jung, J. Y. Song, W. K. Lee, Y. J. Choi, S. H. Choi, S. G. Park, J. U. Park, M. Y. Choe, S. A.

Jung, E. Y. Byun, S. C. Baik, H. S. Youn, G. H. Ko, D. Lim, and K. H. Rhee. 2002. Identifying the major proteome components of *Helicobacter pylori* strain 26695. *Electrophoresis* 23:1161–1173.

Clewell, D. B., S. E. Flannagan, and D. D. Jaworski. 1995. Unconstrained bacterial promiscuity: the Tn916-Tn1545 family of conjugative transposons. *Trends Microbiol.* 3:229–236.

Cole, S. T., R. Brosch, J. Parkhill, T. Garnier, C. Churcher, D. Harris, S. V. Gordon, K. Eiglmeter, S. Gas, C. E. Barry III, et al. 1998. Deciphering the biology of *Mycobacterium tuberculosis* from the complete genome sequence. *Nature* 393:537–544. (Erratum, 396:190.)

Cordwell, S. J., D. J. Basseal, B. Bjellqvist, D. C. Shaw, and I. Humphery-Smith. 1997. Characterisation of basic proteins from *Spiroplasma melliferum* using novel immobilised pH gradients. *Electrophoresis* 18:1393–1398.

Devine, K. M. 1995. The *Bacillus subtilis* genome project: aims and progress. *Trends Biotechnol.* 13:210–216.

Dunn, B. E., G. I. Perez-Perez, and M. J. Blaser. 1989. Two-dimensional gel electrophoresis and immunoblotting of *Campylobacter pylori* proteins. *Infect. Immun.* 57:1825–1833.

Dunn, M. J. 1987. Two-dimensional polyacrylamide gel electrophoresis. *Adv. Electrophor.* 1:1–109.

Enroth, H., T. Akerlund, A. Sille, and L. Engstrand. 2000. Clustering of clinical strains of *Helicobacter pylori* analyzed by two-dimensional gel electrophoresis. *Clin. Diagn. Lab. Immunol.* 7:301–306.

Evers, S., and C. P. Gray. 2001. Application of proteome analysis to drug development and toxicology, p. 225–236. *In* S. R. Pennington and M. J. Dunn (ed.), *Proteomics from Protein Sequence to Function*. BIOS, Oxford, United Kingdom.

Fleischmann, R. D., M. D. Adams, O. White, R. A. Clayton, E. F. Kirkness, A. R. Kerlavage, C. J. Bult, J. F. Tomb, B. A. Dougherty, J. M. Merrick, K. McKenney, G. Sutton, W. FitzHugh, C. Fields, J. D. Gocayne, J. Scott, R. Shirley, L. I. Liu, A. Glodek, J. M. Kelley, J. F. Weidman, C. A. Phillips, T. Spriggs, E. Hedblom, M. D. Cotton, T. R. Utterback, M. C. Hanna, D. T. Nguyen, D. M. Saudek, R. C. Brandon, L. D. Fine, J. L. Fritchman, J. L. Fuhrmann, N. S. M. Geoghagen, C. L. Gnehm, L. A. McDonald, K. V. Small, C. M. Fraser, H. O. Smith, and J. C. Venter. 1995. Whole-genome random sequencing and assembly of *Haemophilus influenzae* Rd. *Science* 269: 496–511.

Fountoulakis, M., and B. Takacs. 1998. Design of protein purification pathways: application to the proteome of *Haemophilus influenzae* using heparin chromatography. *Protein Expr. Purif.* 14:113–119.

Fountoulakis, M., H. Langen, S. Evers, C. Gray, and B. Takacs. 1997. Two-dimensional map of *Haemophilus influenzae* following protein enrichment by heparin chromatography. *Electrophoresis* 18:1193–1202.

Fountoulakis, M., J. F. Juranville, D. Roder, S. Evers, P. Berndt, and H. Langen. 1998a. Reference map of the low molecular mass proteins of *Haemophilus influenzae*. *Electrophoresis* 19: 1819–1827.

Fountoulakis, M., H. Langen, C. Gray, and B. Takacs. 1998b. Enrichment and purification of proteins of *Haemophilus influenzae* by chromatofocusing. *J. Chromatogr. A* 806:279–291.

Fountoulakis, M., B. Takacs, and H. Langen. 1998c. Two-dimensional map of basic proteins of *Haemophilus influenzae*. *Electrophoresis* 19:761–766.

Fountoulakis, M., M. F. Takacs, P. Berndt, H. Langen, and B. Takacs. 1999. Enrichment of low abundance proteins of *Escherichia coli* by hydroxyapatite chromatography. *Electrophoresis* 20:2181–2195.

Fraser, C. M., J. D. Gocayne, O. White, M. D. Adams, R. A. Clayton, R. D. Fleischmann, C. J. Bult, A. R. Kerlavage, G. Sutton, J.

M. Kelley, J. L. Fritchman, J. F. Weidman, K. V. Small, M. Sandusky, J. Fuhrmann, D. Nguyen, T. R. Utterback, D. M. Saudek, C. A. Phillips, J. M. Merrick, J. F. Tomb, B. A. Dougherty, K. F. Bott, P. C. Hu, T. S. Lucier, S. N. Peterson, H. O. Smith, C. A. Hutchison, and J. C. Venter. 1995. The minimal gene complement of *Mycoplasma genitalium*. *Science* 270:397–403.

Fratti, R. A., I. Vergne, J. Chua, J. Skidmore, and V. Deretic. 2000. Regulators of membrane trafficking and *Mycobacterium tuberculosis* phagosome maturation block. *Electrophoresis* 21:3378–3385.

Golovliov, I., M. Ericsson, G. Sandstrom, A. Tarnvik, and A. Sjostedt. 1997. Identification of proteins of *Francisella tularensis* induced during growth in macrophages and cloning of the gene encoding a prominently induced 23-kilodalton protein. *Infect. Immun.* 65:2183–2189.

Gormon, T., and L. Phan-Thanh. 1995. Identification and classification of *Listeria* by two-dimensional protein mapping. *Res. Microbiol.* 146:143–154.

Govoni, G., S. Vidal, S. Gauthier, E. Skamene, D. Malo, and P. Gros. 1996. The Bcg/Ity/Lsh locus: genetic transfer of resistance to infections in C57BL/6J mice transgenic for the Nramp1 Gly169 allele. *Infect. Immun.* 64:2923–2929.

Griffin, T. J., D. K. Han, S. P. Gygi, B. Rist, H. Lee, R. Aebersold, and K. C. Parker. 2001. Toward a high-throughput approach to quantitative proteomic analysis: expression-dependent protein identification by mass spectrometry. *J. Am. Soc. Mass Spectrom.* 12:1238–1246.

Gygi, S. P., B. Rist, S. A. Gerber, F. Turecek, M. H. Gelb, and R. Aebersold. 1999a. Quantitative analysis of complex protein mixtures using isotope-coded affinity tags. *Nat. Biotechnol.* 17:994–999.

Gygi, S. P., Y. Rochon, B. R. Franza, and R. Aebersold. 1999b. Correlation between protein and mRNA abundance in yeast. *Mol. Cell. Biol.* 19:1720–1730.

Hanawa, T., T. Yamamoto, and S. Kamiya. 1995. *Listeria monocytogenes* can grow in macrophages without the aid of proteins induced by environmental stresses. *Infect. Immun.* 63:4595–4599.

Hansen, E. J., R. M. Wilson, and J. B. Baseman. 1979. Two-dimensional gel electrophoretic comparison of proteins from virulent and avirulent strains of *Mycoplasma pneumoniae*. *Infect. Immun.* 24:468–475.

Hansen, E. J., R. M. Wilson, W. A. Clyde, Jr., and J. B. Baseman. 1981. Characterization of hemadsorption-negative mutants of *Mycoplasma pneumoniae*. *Infect. Immun.* 32:127–136.

Hautefort, I., and J. C. Hinton. 2001. Measurement of bacterial gene expression in vivo. *Phil. Trans. R. Soc. London Ser. B* 355:601–611.

Hazell, S. L., R. H. Andrews, H. M. Mitchell, G. Daskalopoulous, Q. Jiang, K. Hiratsuka, and D. E. Taylor. 1996. Variability of gene order in different *Helicobacter pylori* strains contributes to genome diversity. *FEMS Microbiol. Lett.* 20:833–842.

Humphery-Smith, I., S. J. Cordwell, and W. P. Blackstock. 1997. Proteome research: complementarity and limitations with respect to the RNA and DNA worlds. *Electrophoresis* 18:1217–1242.

Jackson, P., M. J. Thornley, and R. J. Thompson. 1984. A study by high resolution two-dimensional polyacrylamide gel electrophoresis of relationships between *Neisseria gonorrhoeae* and other bacteria. *J. Gen. Microbiol.* 130:3189–3201.

Jensen, O. N., M. Wilm, A. Shevchenko, and M. Mann. 1999. Sample preparation methods for mass spectrometric peptide mapping directly from 2-DE gels, p. 513–530. *In* A. J. Link (ed.), *2-D Proteome Analysis Protocols*. Humana Press, Totowa, N.J.

Johnson, A. P., D. C. Speller, R. C. George, M. Warner, G. Domingue, and A. Efstratiou. 1996. Prevalence of antibiotic resistance and serotypes in pneumococci in England and Wales: results of observational surveys in 1990 and 1995. *Br. Med. J.* 312:1454–1456.

Jungblut, P. R., G. Grabher, and G. Stoffler. 1999a. Comprehensive detection of immunorelevant *Borrelia garinii* antigens by two-dimensional electrophoresis. *Electrophoresis* 20:3611–3622.

Jungblut, P. R., U. E. Schaible, H. J. Mollenkopf, U. Zimny-Arndt, B. Raupach, J. Mattow, P. Halada, S. Lamer, K. Hagens, and S. H. Kaufmann. 1999b. Comparative proteome analysis of *Mycobacterium tuberculosis* and *Mycobacterium bovis* BCG strains: towards functional genomics of microbial pathogens. *Mol. Microbiol.* 33:1103–1117.

Jungblut, P. R., U. Zimny-Arndt, E. Zeindl-Eberhart, J. Stulik, K. Koupilova, K. P. Pleissner, A. Otto, E. C. Muller, W. Sokolowska-Kohler, G. Grabher, and G. Stoffler. 1999c. Proteomics in human disease: cancer, heart and infectious diseases. *Electrophoresis* 20:2100–2110.

Jungblut, P. R., D. Bumann, G. Haas, U. Zimny-Arndt, P. Holland, S. Lamer, F. Siejak, A. Aebischer, and T. F. Meyer. 2000. Comparative proteome analysis of *Helicobacter pylori*. *Mol. Microbiol.* 36:710–725.

Jungblut, P. R., E. C. Muller, J. Mattow, and S. H. Kaufmann. 2001. Proteomics reveals open reading frames in *Mycobacterium tuberculosis* H37Rv not predicted by genomics. *Infect. Immun.* 69:5905–5907.

Klimpel, K. W., and V. L. Clark. 1988. Multiple protein differences exist between *Neisseria gonorrhoeae* type 1 and type 4. *Infect. Immun.* 56:808–814.

Klose, J. 1975. Protein mapping by combined isoelectric focusing and electrophoresis of mouse tissues. *Humangenetik* 26:231–243.

Konkel, M. E., and W. Cieplak. 1992. Altered synthetic response of *Campylobacter jejuni* to cocultivation with human epithelial cells is associated with enhanced internalization. *Infect. Immun.* 60:4945–4949.

Konkel, M. E., D. J. Mead, and W. Cieplak. 1993. Kinetic and antigenic characterization of altered protein synthesis by *Campylobacter jejuni* during cultivation with human epithelial cells. *J. Infect. Dis.* 168:948–954.

Kovarova, H., J. Stulik, A. Macela, I. Lefkovits, and Z. Skrabkova. 1992. Using two-dimensional gel electrophoresis to study immune response against intracellular bacterial infection. *Electrophoresis* 13:741–742.

Kovarova, H., D. Radzioch, M. Hajduch, M. Sirova, V. Blaha, A. Macela, J. Stulik, and L. Hernychova. 1998. Natural resistance to intracellular parasites: a study by two-dimensional gel electrophoresis coupled with multivariate analysis. *Electrophoresis* 19:1325–1331.

Kovarova, H., R. Necasova, S. Porkertova, D. Radzioch, and A. Marcela. 2001. Natural resistance to intracellular pathogens: modulation of macrophage signal transduction related to the expression of the Bcg locus. *Proteomics* 1:587–596.

Kovarova, H., P. Halada, P. Man, I. Golovliov, Z. Krocova, J. Spacek, S. Porkertova, and Necasova. 2002. Proteome study of *Francisella tularensis* live vaccine strain-containing phagosome in Bcg/Nramp1 congenic macrophages: resistant allele contributes to permissive environment and susceptibility to infection. *Proteomics* 2:85–93.

Kunst, F., N. Ogasawara, I. Moszer, A. M. Albertini, G. Alloni, V. Azevedo, M. G. Bertero, P. Bessieres, A. Bolotin, S. Borchert, R. Borriss, L. Boursier, A. Brans, M. Braun, S. C. Brignell, S. Bron, S. Brouillet, C. V. Bruschi, B. Caldwell, V. Capuano, N. M. Carter, S. K. Chol, J. J. Codani, I. F. Connerton, and A. Danchin. 1997. The complete genome sequence of the gram positive bacterium *Bacillus subtilis*. *Nature* 390:249–256.

Kwaik, Y. A., and O. S. Harb. 1999. Phenotypic modulation by intracellular bacterial pathogens. *Electrophoresis* 20:2248–2258.

Langen, H., C. Gray, D. Roder, J. F. Juranville, B. Takacs, and M. Fountoulakis. 1997. From genome to proteome: protein map of *Haemophilus influenzae*. *Electrophoresis* 18:1184–1192.

Langen, H., B. Takacs, S. Evers, P. Berndt, H. W. Lahm, B. Wipf, C. Gray, and M. Fountoulakis. 2000. Two-dimensional map of the proteome of *Haemophilus influenzae*. *Electrophoresis* 21: 411–429.

Leaves, N. I., and J. Z. Jordens. 1994. Development of a ribotyping scheme for *Haemophilus influenzae* type b. *Eur. J. Clin. Microbiol. Infect. Dis.* 13:1038–1045.

Lemos, J. A., M. Giambiagi-Demarval, and A. C. Castro. 1998. Expression of heat-shock proteins in *Streptococcus pyogenes* and their immunoreactivity with sera from patients with streptococcal diseases. *J. Med. Microbiol.* 47:711–715.

Lin, J., and T. A. Ficht. 1995. Protein synthesis in *Brucella abortus* induced during macrophage infection. *Infect. Immun.* 63:1409–1414.

Link, A. J., L. G. Hays, E. B. Carmack, and J. R. Yates. 1997. Identifying the major proteome components of *Haemophilus influenzae* type-strain NCTC 8143. *Electrophoresis* 18:1314–1334.

Lock, R. A., S. J. Cordwell, G. W. Coombs, B. J. Walsh, and G. M. Forbes. 2001. Proteome analysis of *Helicobacter pylori*: major proteins of type strain NCTC 11637. *Pathology* 33:365–374.

Mahairas, G. G., P. J. Sabo, M. J. Hickey, D. C. Singh, and C. K. Stover. 1996. Molecular analysis of genetic differences between *Mycobacterium bovis* BCG and virulent *M. bovis*. *J. Bacteriol.* 178:1274–1282.

Mattow, J., P. R. Jungblut, E. C. Muller, and S. H. Kaufmann. 2001. Identification of acidic, low molecular mass proteins of *Mycobacterium tuberculosis* strain H37Rv by matrix-assisted laser desorption/ionization and electrospray ionization mass spectrometry. *Proteomics* 1:494–507.

McAtee, C. P., K. E. Fry, and D. E. Berg. 1998a. Identification of potential diagnostic and vaccine candidates of *Helicobacter pylori* by "proteome" technologies. *Helicobacter* 3:163–169.

McAtee, C. P., M. Y. Lim, K. Fung, M. Velligan, K. Fry, T. Chow, and D. E. Berg. 1998b. Identification of potential diagnostic and vaccine candidates of *Helicobacter pylori* by two-dimensional gel electrophoresis, sequence analysis, and serum profiling. *Clin. Diagn. Lab. Immunol.* 5:537–542.

McAtee, C. P., M. Y. Lim, K. Fung, M. Velligan, K. Fry, T. P. Chow, and D. E. Berg. 1998c. Characterization of a *Helicobacter pylori* vaccine candidate by proteome techniques. *J. Chromatogr.* 714:325–333.

McAtee, C. P., P. S. Hoffman, and D. E. Berg. 2001. Identification of differentially regulated proteins in metronidozole resistant *Helicobacter pylori* by proteome techniques. *Proteomics* 1:516–521.

Mehta, P. K., C. H. King, E. H. White, J. J. Murtagh, and F. D. Quinn. 1996. Comparison of in vitro models for the study of *Mycobacterium tuberculosis* invasion and intracellular replication. *Infect. Immun.* 64:2673–2679.

Merchant, M., S. R. Weinberger, R. W. Nelson, D. Nedelkov, and K. A. Tubbs. 2000. Recent advancements in surface-enhanced laser desorption/ionization-time of flight-mass spectrometry. *Electrophoresis* 21:1164–1177.

Michea-Hamzehpour, M., J. C. Sanchez, S. F. Epp, N. Paquet, G. J. Hughes, D. Hochstrasser, and J. C. Pechere. 1993. Two-dimensional polyacrylamide gel electrophoresis isolation and microsequencing of *Pseudomonas aeruginosa* proteins. *Enzyme Protein* 47:1–8.

Mollenkopf, H. J., P. R. Jungblut, B. Raupach, J. Mattow, S. Lamer, U. Zimny-Arndt, U. E. Schaible, and S. H. Kaufmann.

1999. A dynamic two-dimensional polyacrylamide gel electrophoresis database: the mycobacterial proteome via Internet. *Electrophoresis* 20:2172–2180.

Monahan, I. M., J. Betts, D. K. Banerjee, and P. D. Butcher. 2001. Differential expression of mycobacterial proteins following phagocytosis by macrophages. *Microbiology* 147:459–471.

Neidhardt, F. C., V. Vaughn, T. A. Phillips, and P. L. Bloch. 1983. Gene-protein index of *Escherichia coli* K-12. *Microbiol. Rev.* 47:231–284.

Nilsson, C. L., T. Larsson, E. Gustafsson, K. A. Karlsson, and P. Davidsson. 2000. Identification of protein vaccine candidates from *Helicobacter pylori* using a preparative two-dimensional electrophoretic procedure and mass spectrometry. *Anal. Chem.* 72:2148–2153.

O'Farrell, P. H. 1975. High resolution two-dimensional electrophoresis of proteins. *J. Biol. Chem.* 250:4007–4021.

Ohlmeier, S., C. Scharf, and M. Hecker. 2000. Alkaline proteins of *Bacillus subtilis*: first steps towards a two-dimensional alkaline master gel. *Electrophoresis* 21:3701–3709.

Opiteck, G. J., S. M. Ramirez, J. W. Jorgenson, and M. A. Moseley, III. 1998. Comprehensive two-dimensional high-performance liquid chromatography for the isolation of overexpressed proteins and proteome mapping. *Anal. Biochem.* 258:349–361.

Pal, P. G., and M. A. Horwitz. 1992. Immunization with extracellular proteins of *Mycobacterium tuberculosis* induces cell-mediated immune responses and substantial protective immunity in a guinea pig model of pulmonary tuberculosis. *Infect. Immun.* 60: 4781–4792.

Pasquali, C., S. Frutiger, M. R. Wilkins, G. J. Hughes, R. D. Appel, and A. Bairoch. 1996. 2-dimensional gel electrophoresis of *Escherichia coli* homogenates—the *Escherichia coli* SWISS-2DPAGE database. *Electrophoresis* 17:547–555.

Petricoin, E. F., A. M. Ardekani, B. A. Hitt, P. J. Levine, V. A. Fusaro, S. M. Steinberg, G. B. Mills, C. Simone, D. A. Fishman, E. C. Kohn, and L. A. Liotta. 2002. Use of proteomic patterns in serum to identify ovarian cancer. *Lancet* 359:572–577.

Phillips, T. A., P. L. Bloch, and F. C. Neidhardt. 1980. Protein identifications on O'Farrell two-dimensional gels: locations of 55 additional *Escherichia coli* proteins. *J. Bacteriol.* 144:1024–1033.

Plant, J., and A. A. Glynn. 1976. Genetics of resistance to infection with *Salmonella typhimurium* in mice. *J. Infect. Dis.* 133:72–78.

Rafie-Kolpin, M., R. C. Essenberg, and J. H. Wyckoff, III. 1996. Identification and comparison of macrophage-induced proteins and proteins induced under various stress conditions in *Brucella abortus*. *Infect. Immun.* 64:5274–5283.

Rain, J. C., L. Selig, H. De Reuse, V. Battaglia, C. Reverdy, S. Simon, G. Lenzen, F. Petel, J. Wojcik, V. Schachter, Y. Chemama, A. Labigne, and P. Legrain. 2001. The protein-protein interaction map of *Helicobacter pylori*. *Nature* 409:211–215.

Roberts, A. D., M. G. Sonnenberg, D. J. Ordway, S. K. Furney, P. J. Brennan, J. T. Belisle, and I. M. Orme. 1995. Characteristics of protective immunity engendered by vaccination of mice with purified culture filtrate protein antigens of *Mycobacterium tuberculosis*. *Immunology* 85:502–508.

Rodwell, A. W., and E. S. Rodwell. 1978. Relationships between strains of *Mycoplasma mycoides* subspp *mycoides* and *capri* studied by two-dimensional gel electrophoresis of cell proteins. *J. Gen. Microbiol.* 109:259–263.

Rosamond, J., and A. Alisop. 2000. Harnessing the power of the genome in the search for new antibiotics. *Science* 287:1973–1976.

Rosenkrands, I., A. King, K. Weldingh, M. Moniatte, E. Moertz, and P. Andersen. 2000a. Towards the proteome of *Mycobacterium tuberculosis*. *Electrophoresis* 21:3740–3756.

Rosenkrands, I., K. Weldingh, S. Jacobsen, C. V. Hansen, W. Florio, I. Gianetri, and P. Andersen. 2000b. Mapping and identification of *Mycobacterium tuberculosis* proteins by two-dimensional gel electrophoresis, microsequencing and immunodetection. *Electrophoresis* 21:935–948.

Sazuka, T., M. Yamaguchi, and O. Ohara. 1999. Cyano2Dbase updated: linkage of 234 protein spots to corresponding genes through N-terminal microsequencing. *Electrophoresis* 20:2160–2171.

Schmid, R., J. Bernhardt, H. Antelmann, A. Volker, H. Mach, U. Volker, and M. Hecker. 1997. Identification of vegetative proteins for a two-dimensional protein index of *Bacillus subtilis*. *Microbiology* 143:991–998.

Singh, V. K., R. K. Jayaswal, and B. J. Wilkinson. 2001. Cell wall-active antibiotic induced proteins of *Staphylococcus aureus* identified using a proteomic approach. *FEMS Microbiol. Lett.* 199:79–84.

Slauch, J. M., M. J. Mahan, and J. J. Mekalanos. 1994. In vivo expression technology for selection of bacterial genes specifically induced in host tissues. *Methods Enzymol.* 235:481–492.

Sonnenberg, M. G., and J. T. Belisle. 1997. Definition of *Mycobacterium tuberculosis* culture filtrate proteins by two-dimensional polyacrylamide gel electrophoresis, N-terminal amino acid sequencing, and electrospray mass spectrometry. *Infect. Immun.* 65:4515–4524.

Sowa, B. A., K. A. Kelly, T. A. Ficht, and L. G. Adams. 1992. Virulence associated proteins of *Brucella abortus* identified by paired two-dimensional gel electrophoretic comparisons of virulent, vaccine and LPS deficient strains. *Appl. Theor. Electrophor.* 3:33–40.

Sreevatsan, S., X. Pan, K. E. Stockbauer, N. D. Connell, B. N. Kreiswirth, T. S. Whittam, and J. M. Musser. 1997. Restricted structural gene polymorphism in the *Mycobacterium tuberculosis* complex indicates evolutionary recent global dissemination. *Proc. Natl. Acad. Sci. USA* 94:9869–9874.

Sturgill-Koszycki, S., P. L. Haddix, and D. G. Russell. 1997. The interaction between Mycobacterium and the macrophage analyzed by two-dimensional polyacrylamide gel electrophoresis. *Electrophoresis* 18:2558–2565.

Sutcliffe, J., A. Tait-Kamradt, and L. Wondrack. 1996. *Streptococcus pneumoniae* and *Streptococcus pyogenes* resistant to macrolides but sensitive to clindamycin: a common resistance pattern mediated by an efflux system. *Antimicrob. Agents Chemother.* 40:1817–1824.

Taylor, D. E., M. Eaton, N. Chang, and S. M. Salama. 1992. Construction of a *Helicobacter pylori* genome map and demonstration of diversity at the genome level. *J. Bacteriol.* 174:6800–6806.

Teixeira-Gomes, A. P., A. Cloeckaert, G. Bezard, R. A. Bowden, G. Dubray, and M. S. Zygmunt. 1997. Identification and characterization of *Brucella ovis* immunogenic proteins using two-dimensional electrophoresis and immunoblotting. *Electrophoresis* 18:1491–1497.

Tomb, J. F., O. White, A. R. Kerlavage, R. A. Clayton, G. G. Sutton, R. D. Fleischmann, K. A. Ketchum, H. P. Klenk, S. Gill, B. A. Dougherty, K. Nelson, J. Quackenbush, L. Zhou, E. F. Kirkness, S. Peterson, B. Loftus, D. Richardson, R. Dodson, H. G. Khalak, A. Glodek, K. McKenney, L. M. Fitzgerald, N. Lee, M. D. Adams, E. K. Hickey, D. E. Berg, J. D. Gocayne, T. R. Utterback, J. D. Peterson, J. M. Kelley, M. D. Cotton, J. M. Weidman, C. Fujii, C. Bowman, L. Watthey, E. Wallin, W. S. Hayes, M. Borodovsky, P. D. Karp, H. O. Smith, C. M. Fraser, and J. C. Venter. 1997. The complete genome sequence of the gastric pathogen *Helicobacter pylori*. *Nature* 388:539–547.

Tonella, L., C. Hoogland, P. A. Binz, R. D. Appel, D. F. Hochstrasser, and J. C. Sanchez. 2001. New perspectives in the *Escherichia coli* proteome investigation. *Proteomics* 1:409–423.

Trieu-Cuot, P., C. Poyart-Salmeron, C. Carlier, and P. Courvalin. 1990. Nucleotide sequence of the erythromycin resistance gene of the conjugative transposon Tn1545. *Nucleic Acids. Res.* 18:3660.

Urquhart, B. L., T. E. Atsalos, D. Roach, D. J. Basseal, B. Bjellqvist, W. L. Britton, and I. Humphery-Smith. 1997. "Proteomic contigs" of *Mycobacterium tuberculosis* and *Mycobacterium bovis* (BCG) using novel immobilised pH gradients. *Electrophoresis* 18:1384–1392.

Urquhart, B. L., S. J. Cordwell, and I. Humphery-Smith. 1998. Comparison of predicted and observed properties of proteins encoded in the genome of *Mycobacterium tuberculosis* H37Rv. *Biochem. Biophys. Res. Commun.* 253:70–79.

Valway, S. E., M. P. Sanchez, T. F. Shinnick, I. Orme, T. Agerton, D. Hoy, J. S. Jones, H. Westmoreland, and I. M. Onorato. 1998. An outbreak involving extensive transmission of a virulent strain of *Mycobacterium tuberculosis*. *N. Engl. J. Med.* 338:633–639.

VanBogelen, R. A., P. Sankar, R. L. Clark, J. A. Bogan, and F. C. Neidhardt. 1992. The gene-protein database of *Escherichia coli*: edition 5. *Electrophoresis* 13:1014–1054.

VanBogelen, R. A., K. Z. Abshire, A. Pertsemlidis, R. L. Clark, and F. C. Neidhardt. 1996. Gene-protein database of *Escherichia coli* K-12, Edition 6, p. 2067–2117. *In* F. C. Neidhardt, R. Curtiss, J. L. Ingraham, E. C. C. Lin, K. B. Low, B. Magasanik, W. S. Reznikoff, M. Riley, M. Schaechter, and H. E. Umbarger (ed.), *Escherichia coli and Salmonella: Cellular and Molecular Biology*. ASM Press, Washington, D.C.

VanBogelen, R. A., K. Z. Abshire, B. Moldover, E. R. Olson, and F. C. Neidhardt. 1997. *Escherichia coli* proteome analysis using the gene-protein database. *Electrophoresis* 18:1243–1251.

VanBogelen, R. A., E. E. Schiller, J. D. Thomas, and F. C. Neidhardt. 1999. Diagnosis of cellular states of microbial organisms using proteomics. *Electrophoresis* 20:2149–2159.

Vurma-Rapp, U., F. H. Kayser, K. Hadorn, and F. Wiederkehr. 1990. Mechanism of imipenem resistance acquired by three *Pseudomonas aeruginosa* strains during imipenem therapy. *Eur. J. Clin. Microbiol. Infect. Dis.* 9:580–587.

Wagner, M. A., M. Eschenbrenner, T. A. Horn, J. A. Kraycer, C. V. Hagius, P. Elzer, and V. G. DelVecchio. 2002. Global analysis of the *Brucella melitensis* proteome: identification of proteins expressed in laboratory grown culture. *Proteomics* 2:1047–1060.

Wasinger, V. C., S. J. Cordwell, A. Cerpa-Poljak, J. X. Yan, A. A. Gooley, M. R. Wilkins, M. W. Duncan, R. Harris, K. L. Williams, and I. Humphery-Smith. 1995. Progress with gene-product mapping of the Mollicules: *Mycoplasma genitalium*. *Electrophoresis* 16:1090–1094.

Watson, H. L., M. K. Davidson, N. R. Cox, J. K. Davis, K. Dybvig, and G. H. Cassell. 1987. Protein variability among strains of *Mycoplasma pulmonis*. *Infect. Immun.* 55:2838–2840.

Wilkins, M. R., C. Pasquali, R. D. Appel, K. Ou, O. Golaz, J. C. Sanchez, J. X. Yan, A. A. Gooley, G. Hughes, I. Humphery-Smith, K. L. Williams, and D. F. Hochstrasser. 1996. From proteins to proteomes: large scale protein identification by two-dimensional electrophoresis and amino acid analysis. *Biotechnology* 14:61–65.

Wilson, M., J. DeRisi, H. Kristensen, P. Imboden, S. Rane, P. O. Brown, and G. K. Schoolnik. 1999. Exploring drug-induced alterations in gene expression in Mycobacterium tuberculosis by microarray hybridization. *Proc. Natl. Acad. Sci. USA* 96:12833–12838.

Yamamoto, T., T. Hanawa, and S. Ogata. 1994. Induction of *Yersinia enterocolitica* stress proteins by phagocytosis with macrophage. *Microbiol. Immunol.* 38:295–300.

Microbial Diversity and Bioprospecting
Edited by Alan T. Bull
© 2004 ASM Press, Washington, D.C.

Chapter 27

Phenomics

Jennifer L. Reed, Stephen S. Fong, and Bernhard Ø. Palsson

The genomics era has arrived (see Fig. 1). To date, the genomes of over 100 organisms have been sequenced and annotated (Drell, 2002; see also chapters 24 and 25 in this book). Genomics, including functional genomics, leads to an organism-specific "parts catalog." This parts catalog does not indicate when or how the parts are used together in a coordinated manner. Although biological studies over the past century have taken a reductionist approach to characterize these components, the field of systems biology takes an integrative approach, aiming to understand how these gene products work together as a system to produce a cellular behavior or a phenotype (Palsson, 2000).

In bacteria, a one-to-one relationship exists between a gene and its protein; however, complicated interactions among gene products lead to highly complex genotype-phenotype relationships. This is evident by the fact that multiple phenotypes can be observed for a defined genotype. Understanding the genotype-phenotype relationship is crucial in predicting the phenotypic outcome that is due to genome alterations, ranging from single-nucleotide polymorphisms to entire gene additions or deletions, or due to responses to environmental parameters. The ability to predict phenotypic behavior will guide the development of industrially favorable microorganisms and help understand the mechanisms of disease.

Phenomics is the study of a cell's phenotypic behavior, not only the expressed phenotypes but also how a defined genotype produces an observed phenotype (Schilling et al., 1999). Phenomic studies are often done with simple, single-cell organisms whose genomes have been sequenced and are capable of being genetically modified. Organisms with annotated genomes can be tested experimentally to better understand their phenotypic characteristics, thus yielding a pool of phenomic information that can be used to correlate the genotype-phenotype relationship. Increases in the amount of phenomic data will enable better predictions of an organism's phenotype from its known genotype. In this chapter we focus on types of phenotypic measurements, technological advances that enable high-throughput phenotypic measurements, computational tools that are used to interpret and predict phenotypic behavior, and the impact that the field of phenomics will have on biotechnology.

PHENOTYPIC MEASUREMENTS

Phenomics looks to elucidate the genotype-phenotype relationship; as such, current work is primarily limited to studying organisms with a well-characterized genotype. The current state of phenomics is focused on studying behaviors that are typical of simpler organisms. Phenotypic behaviors of interest are related to the basic growth and metabolic functions of the cell. These include the specific growth rate of the cell, the substrate uptake rate, the oxygen uptake rate, and the amounts and types of by-products secreted (see Fig. 2). The challenge is to devise methods for measuring these quantities accurately and reproducibly in a high-throughput manner.

The growth properties of a microorganism can now be assessed in a high-throughput manner. For bacteria, the specific growth rate is a characteristic trait of a particular strain under given growth conditions and thus is a good measurement for identifying phenotypic differences. The specific growth rate of a microorganism can be easily and reproducibly measured by growing a culture and determining the change in optical density (OD) over time. Because the OD is proportional to the cell density (for low cell densities,) the OD can be plotted on a log scale over time, and the specific growth rate is given as the slope of the plotted line. Most bacteria have a doubling time (calculated as ln[2]/growth rate) that is on the

Jennifer L. Reed, Stephen S. Fong, and Bernhard Ø. Palsson • Department of Bioengineering, University of California, San Diego, La Jolla, CA 92093-0419.

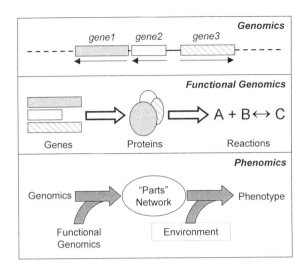

Figure 1. The classification of "omics" data. Genomics involves the study of an organism's genome, including sequencing of its open reading frame, identification, and annotation. Functional genomics is an area of genomics that involves the functional assignment of genes. As diagrammed, the three proteins encoded by genes 1, 2, and 3 are subunits of a functional enzyme that catalyzes the conversion of metabolites A and B into C. Phenomics is the study of phenotypes, and this information is used to understand the genotype-phenotype relationship, which is also dependent on environmental conditions.

order of hours and can be determined with a high degree of confidence.

The substrate uptake rates, oxygen uptake rate, and by-product secretions of a microorganism can be more difficult to determine, but can be measured reproducibly if well-controlled protocols are used. These parameters are often expressed in units of millimoles per gram dry weight per hour. The substrate

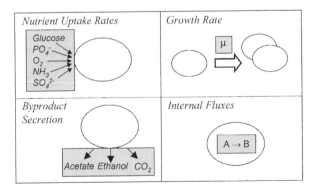

Figure 2. Phenotypic measurements. There are four basic types (boxed in gray) of phenotypic measurements that can be made. Nutrient uptake rates can be measured, which would include carbon source (in this case glucose), oxygen, nitrogen, sulfur, and phosphate uptake rates. Growth rate (μ) can be measured with high-throughput technology. By-product secretion, including identification of the compounds and measurement of the secretion rates, can be evaluated. By using radioactively labeled substrates, internal fluxes can also be measured.

uptake rate(s) can be determined by growing a culture in medium of known substrate concentration(s) and then sampling the culture over time for depletion of the substrate(s), while taking into account changes in biomass as the culture grows. In a similar manner, the oxygen uptake rate can be easily determined by placing a sample of a growing culture (with a known quantity of cells) into a respirometer and using a dissolved oxygen probe to measure the rate of oxygen depletion, again accounting for changes in biomass. Secretion of by-products by microorganisms is of interest to many fields and, again, can be a distinct phenotypic characteristic of a particular strain. Samples of medium can be taken from a growing culture and sterile filtered to remove cells. These samples can then be analyzed by various analytical chemistry methods (high-pressure liquid chromatography, mass spectrometry, nuclear magnetic resonance, enzymatic assay) to identify secreted by-products and to determine their concentrations.

Metabolic measurements are not limited to the outward or external characteristics of the cell; measurements of the internal state of the cell can also be made. Internal flux measurements can be made using ^{14}C or ^{13}C tracing experiments, which has recently been reviewed (Wiechert, 2001), or rapidly developing metabolomic approaches (Buchholz et al., 2001; Fiehn, 2001).

PHENOMIC MEASUREMENT TOOLS

Largely because of the newness of the field, the number of high-throughput technologies useful for phenomics is limited. Most tools that are particularly useful for phenomics are focused on measuring growth capabilities of microorganisms in a high-throughput manner.

One system that has been developed in this area is the Phenotype Microarrays developed by Biolog (Hayward, Calif.). The Phenotype Microarrays are composed of sets of 96-well plates that are used to test for growth under different substrate conditions. Each well contains a different substrate. Cellular respiration is monitored in each well by a colorimetric change (www.biolog.com). Cell growth is indicated by a positive colorimetric change; the absence of growth results in no colorimetric change. The degree of cell growth is measured by the amount of colorimetric change. In this manner, a strain of cells can be quickly and simultaneously tested for its ability to grow under thousands of different substrate conditions.

It is also possible to directly measure specific growth rate by using a plate reader. One such system is the Bioscreen C system by Thermo Finnigan (San

Jose, Calif.). This combination plate reader and incubator is designed to measure growth rates (www.allmicrobiology.com). The system has been designed to precisely control the temperature while measurements are being made, and also includes the option to shake the culture plates, allowing for aeration. This system produces a quantitative measurement of the growth rate of the microorganism and allows for media formulations that are not premade (unlike the Phenotype Microarrays).

A third method of determining growth on particular substrates was developed by Becton, Dickinson and Co. (Franklin Lakes, N.J.). They developed an insert that fluoresces in the absence of oxygen (www.bdbiosciences.com). The insert is placed in the bottom of a well to which cells and media are added. For a dilute culture (when the culture is first inoculated), the cells will not utilize all of the available oxygen dissolved in the medium, so oxygen will reach the bottom of the well and quench the fluorescence of the insert. As the culture grows, the cells will begin to utilize all of the available oxygen, thus not allowing any oxygen to reach the insert at the bottom of the well, causing the insert to fluoresce.

The systems described above are some of the available tools for use in the study of phenomics. Useful information can be obtained from studies utilizing these high-throughout technologies. For example, the function of an unknown gene could be tested by creating a knockout of that gene and measuring the mutant's growth on a variety of substrates. The observed changes in growth properties may lead to some insight into the function of the unknown gene.

In the near future, one should expect the development of new technologies that will help propel the field of phenomics forward. Numerous cell-based assays are being formulated to test for various functions. Time-lapse microscopy and the associated image processing can be used to observe certain cellular characteristics such as morphology, motility, and differentiation. New techniques are being developed to facilitate genetic manipulation in various organisms to allow for more comprehensive mutagenesis studies (Badarinarayana et al., 2001). These advances will enable the further characterization of cellular phenotypes and, in combination with in silico modeling approaches, will lead to a better understanding of the genotype-phenotype relationship.

PREDICTING AND ANALYZING PHENOMIC DATA

Metabolic reactions are responsible for producing the biochemical precursors and energy needed to carry out cellular functions, such as growth, tran-

scription, and translation. Metabolism has been studied in molecular terms for over 70 years and is well characterized biochemically and genetically. The field of metabolic engineering, whose principle goal is to design more useful organisms, has proven to be a driving force for the development of different modeling methodologies.

The most obvious method for identifying metabolic engineering targets is to look at the metabolic network and generate an educated hypothesis. For example, the secretion of acetate by *Escherichia coli* reduces the production of recombinant proteins as well as the cellular growth rate (Aristidou et al., 1999). A prediction can be made that deleting the genes that encode enzymes that produce acetate would eliminate acetate production and increase the production of recombinant proteins. However, deletion of these genes might have other deleterious effects that are not apparent. Instead, more robust and comprehensive analytical methods have been developed and used to analyze metabolic networks; they include cybernetic modeling (Dhurjati et al., 1985; Kompala et al., 1986; Varner and Ramkrishna, 1999), metabolic control analysis (Cascante et al., 2002), flux-balance analysis (FBA) (Varma and Palsson 1994a; Edwards et al., 2002a), inverse flux analysis (Delgado and Liao, 1997), biochemical systems theory (Savageau et al., 1987a, 1987b), elementary mode analysis (Schuster and Hilgetag, 1994), and extreme pathway analysis (Schilling et al., 2000).

Genome-scale models of metabolic networks are needed to interpret and predict phenomic data. So far, these genome-scale models have only been analyzed by using FBA (Edwards and Palsson, 1999, 2000; Schilling et al., 2002) and extreme pathway analysis (Schilling and Palsson, 2000; Papin et al., 2002; Price et al., 2002; Schilling et al., 2002), mainly due to a lack of measured enzymatic parameters that are needed in some of the other modeling methodologies (Bailey, 2001). For this reason the principles behind FBA and its uses are discussed here, and readers are encouraged to access the given references for descriptions of the other modeling methodologies.

Flux-Balance Analysis

Metabolic phenotypes can be defined by the flux values through the reactions in a metabolic network; FBA is just one method that can be used to predict these flux values. This analysis technique is diagrammed in Fig. 3. Once all of the reactions in a metabolic network are specified, steady-state mass balances for each metabolite can be written, where the sum of the fluxes that produce that metabolite must be equal to the sum of the fluxes that consume

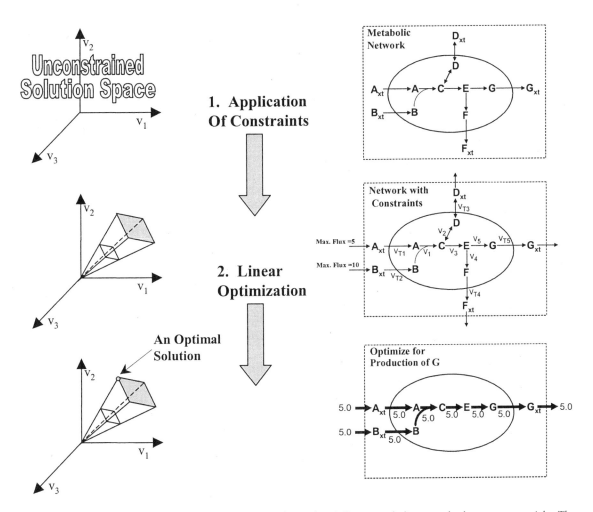

Figure 3. Steps involved in FBA. A metabolic genotype can be used to define a metabolic network, shown at upper right. The range of possible flux values through the individual reactions in the network lie in a high-dimensional flux space. A schematic representation of a three-dimensional projection of this high-dimensional space is shown on the left. Note that the solution spaces on the left are not specific for the actual network depicted on the right. The individual flux values in this network without any applied constraints can be any real number, and this is represented by an unconstrained solution space, shown in the upper left. Application of constraints (shown in the middle right diagram), including mass-balance, flux capacity, and thermodynamic constraints, leads to a convex cone that is the allowable solution space (shown in the middle left diagram) and contains all phenotypes that obey the applied constraints. The second step in FBA involves using linear optimization to select one solution (or phenotype) from the allowable solution space that optimizes a selected objective. In this case, the production of G was used as the objective. Optimality is illustrated by a point in the allowable solution space (shown in the lower left) or as a set of flux values through the network (shown in the lower right).

that metabolite. Other constraints can be applied, such as thermodynamic constraints regarding the reversibility of an individual reaction or enzyme capacity constraints that limit the maximum (or minimum for the case of a reversible reaction) allowable fluxes through a reaction. These constraints limit the allowable cellular behavior or the allowable metabolic phenotypes achievable by the defined metabolic genotype (Edwards et al., 2002a).

If an objective is chosen, exploration of the allowable solution space can lead to identification of points or solutions that maximize the stated objective. Linear optimization is used to find a solution that maximizes a given objective; however, multiple equiv-

alent maxima can exist within the solution space. One objective that might be used is the production of ATP. Another objective that has proven useful in predicting *E. coli* phenotypic behavior is a combination of energy and biomass constituents (including the energy required to sustain growth and the metabolic products that make up the cell biomass) (Varma and Palsson, 1994b; Edwards et al., 2001). The combination of applied constraints and linear optimization allows for the calculation of all the fluxes in a metabolic network, and thus a metabolic phenotype. This modeling method has been used to predict the phenotypic changes that result from gene deletions and that arise from different nutritional conditions (Varma and

Palsson, 1994b; Edwards and Palsson, 2000; Edwards et al., 2001; Schilling et al., 2002).

Gene Deletions

FBA has been used to predict growth capabilities of mutants that arise from the deletion of metabolic and regulatory genes. Deletion of genes can be simulated in silico by constraining the flux(es) through the reaction(s) encoded by the deleted gene to zero. The gene can be classified as essential or nonessential by comparing the calculated optimal growth rates before and after gene the deletion (essential indicating that the resulting in silico metabolic genotype cannot support growth and nonessential indicating that the cell can still grow in silico). The predicted effects on growth rate can be compared with experimental data gathered from literature.

In silico gene deletion studies using FBA have been performed for *E. coli* (Edwards and Palsson, 2000; Burgard and Maranas, 2002; Covert and Palsson, 2002), *Helicobacter pylori,* and *Saccharomyces cerevisiae* (J. Forster et al., unpublished data). For *E. coli,* the effects on growth caused by gene deletions under various environmental conditions were predicted correctly for 97 out of 116 cases; when regulatory effects were taken into account, the predictions were more accurate (106 out of 116) (Covert and Palsson, 2002). The *S. cerevisiae* model was able to predict over 526 out of 599 cases examined (Forster

et al., unpublished data). For *H. pylori* the accuracy of the model was significantly lower, with only 10 of 17 gene deletions being in agreement with known growth capabilities (Schilling et al., 2002).

PhPP Analysis

Phenotypic phase plane (PhPP) analysis has been developed to understand the phenotypic behavior for a defined metabolic genotype under a variety of environmental conditions (Edwards et al., 2002b). Variations in carbon uptake rates or oxygen uptake rates are experimentally measurable parameters that often result in different observable phenotypes. For example, the metabolic by-products that are secreted during aerobic versus anaerobic growth can be quite different. PhPP analysis is a computational tool that allows for the identification of conditions where the metabolic network is qualitatively used in the same manner; these appear as phases in the PhPP.

A PhPP is generated by calculating optimal solutions for a range of environmental conditions, often simulated by varying the carbon uptake rate and the oxygen uptake rate. Lines that appear on the phase plane demarcate the different regions. The PhPP calculated for succinate minimal media conditions using a genome-scale metabolic model of *E. coli* is shown in Fig. 4, where succinate uptake and oxygen uptake rates are incrementally varied. Four regions exist, each with its own characteristics. Region 1 corre-

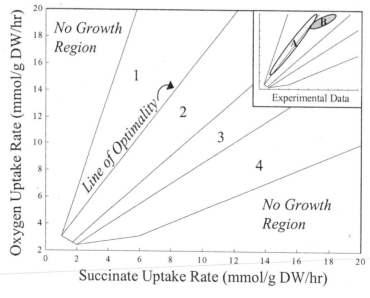

Figure 4. *E. coli* PhPP varying succinate and oxygen uptake rates. Four regions emerge from the PhPP analysis, labeled 1 through 4. The areas to the left of region 1 and below region 4 are conditions in which the model does not predict growth, and they are labeled as such. The line of optimality separates regions 1 and 2. It corresponds to conditions that result in the highest biomass yield (with respect to succinate) and can easily be identified by fixing one flux and allowing the other to vary and then calculating the optimal solution. Experimental studies (inset) found that fully aerobic growth operates around the line of optimality (A) whereas deviations from the line led to acetate secretion (B).

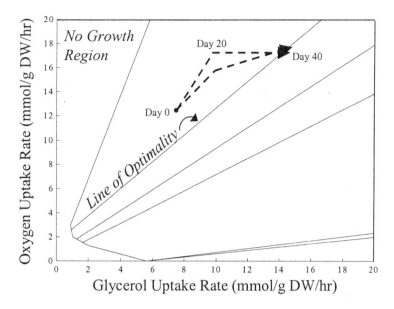

Figure 5. Tracking adaptive evolution of *E. coli* on the glycerol-oxygen PhPP. The glycerol-oxygen PhPP for *E. coli* has six distinct regions. Initial experiments with *E. coli* K-12 MG1655 show that the strain does not operate optimally, the experimental points lie off the line of optimality. Using growth rate as a selective pressure, *E. coli* was grown in serial batch culture for 40 days. Two independent evolutionary trajectories are shown as dashed lines. In both cases, the growth rates doubled as compared to the initial starting point (day 0). Further testing of the endpoint strains (day 40) showed that they remained on the line of optimality, and the growth rate stabilized.

sponds to an abundance of oxygen, and a futile cycle is used to deplete ATP. Region 2 is dual-substrate limited, where an increase in oxygen or succinate uptake rates would increase growth rates. Acetate is expected to be secreted if the cell is operating in this region. Region 3 is also a dual-substrate-limited region; however, the growth rate is mainly limited by the oxygen uptake rate. In contrast, an increase in succinate uptake rate in region 4 results in a decrease in growth rate (Edwards et al., 2001).

The line of optimality separates regions 1 and 2; points on this line correspond to optimal solutions if the carbon uptake rate is fixed and the oxygen uptake rate is not constrained. For example, if the succinate uptake rate is fixed at 10 mmol/g dry weight per h and the oxygen uptake rate is unconstrained, then the maximum growth rate would occur when the oxygen uptake rate was around 17 mmol/g dry weight per h. No by-products are predicted to be secreted if the cell is operating along the line of optimality.

Experimental studies have been performed to test the accuracy of these phase planes toward predicting growth behavior. *E. coli* K-12 MG1655 has been grown with acetate, succinate, malate, and glycerol as the carbon source (Edwards et al., 2001; R. U. Ibarra et al., 2002). With the exception of glycerol, *E. coli* operated along the line of optimality for the different substrates tested. Cells grown on glycerol, however, did not initially exhibit optimal behavior

but evolved toward optimality (Fig. 5) (Ibarra et al., 2002). These results illustrate that not only can the phenotypic behavior be predicted, but also that the metabolic network is optimized by the cell to produce biomass.

HOW WILL PHENOMICS IMPACT BIOTECHNOLOGY?

Understanding how a particular phenotype is established from a defined genotype will enable scientists to design genotypes that express a desired phenotype. This capability is expected to have dramatic impacts on the field of biotechnology, where microorganisms are utilized for the production of biochemicals (such as ethanol for use as an alternative fuel or lactate as a precursor for polymers), recombinant proteins (such as insulin), drugs (such as penicillin), and for the consumption of chemical waste. Strains with higher product yields, for example, can be designed in silico based on phenomic, genomic, and functional genomic information.

Examples exist where computational studies have provided insights into the design of "better" strains (Delgado and Liao, 1997; et al., 1996; Carlson et al., 2002; Burgard and Maranas, 2001). As mentioned previously, acetate secretion by *E. coli* reduces its cellular growth rate. Inverse flux analysis was applied to a central metabolic network of *E. coli* to evaluate the effects that overexpression of *ppc* or upregulating the glyoxylate shunt has on the secretion of acetate (Delgado and Liao, 1997). The *ppc* gene encodes phosphoenolpyruvate decarboxylase (PPC), which catalyzes the conversion of phosphoenolpyruvate to oxaloacetate (OAA). The glyoxylate shunt bypasses parts of the tricarboxylic acid cycle, converting acetyl-coenzyme A and isocitrate to succinate and OAA. Experimental studies showed that when *E. coli* was transformed with a *ppc*-containing plasmid (leading to the overexpression of the PPC enzyme), acetate secretion was dramatically reduced (Farmer and Liao, 1997). Deletion of *fadR*, which normally downregulates the glyoxylate shunt, also reduced acetate secretion but to a lesser extent (Farmer and Liao, 1997).

Recent work with *S. cerevisiae* has also shown how in silico calculations can lead to the design of organism strains (Carlson et al., 2002). The central metabolic network was constructed for *S. cerevisiae*, and elementary mode analysis was used to explore the capabilities of the network for producing poly-β-hydroxybutyrate (PHB). Elementary mode analysis was used to identify which gene additions would increase the predicted maximum yield of PHB. Because acetate is a metabolic precursor of PHB, the study predicts that addition of ATP citrate-lyase or transhydrogenase to the metabolic network will increase the theoretical production of PHB.

The cases discussed here are just two examples illustrating how the field of phenomics will affect biotechnology. Phenomics will lead to an understanding of the genotype-phenotype relationship, resulting in the ability to make predictions about the effect of genome alterations on a cell's phenotype.

SUMMARY

Phenomics is the study of phenotypes for a known genotype and thus is focused on the genotype-phenotype relationship. High-throughput phenotypic measurements can be made involving the growth capabilities of different strains. However, phenomics is not limited to these types of measurements; other measurements can involve substrate uptake rates, oxygen uptake rates, by-product secretion rates, and internal flux measurements.

The genotype-phenotype relationship is intricate and highly complex. Phenomics seeks to analyze, interpret, and predict a phenotype from a defined genotype. Combined with genomics and functional genomics, computational methods can be developed and utilized to predict phenomic properties of novel genotypes, thus greatly enriching the rapidly developing field of biotechnology.

REFERENCES

Aristidou, A. A., K. Y. San, and G. N. Bennett. 1999. Metabolic flux analysis of *Escherichia coli* expressing the *Bacillus subtilis* acetolactate synthase in batch and continuous cultures. *Biotechnol. Bioeng.* **63:**737–749.

Badarinarayana, V., P. W. Estep III, J. Shendure, J. Edwards, S. Tavazoie, F. Lam, and G. M. Church. 2001. Selection analyses of insertional mutants using subgenic-resolution arrays. *Nat. Biotechnol.* **19:**1060–1065.

Bailey, J. E. 2001. Complex biology with no parameters. *Nat. Biotechnol.* **19:**503–504.

Buchholz, A., R. Takors, and C. Wandrey. 2001. Quantification of intracellular metabolites in *Escherichia coli* K12 using liquid chromatographic-electrospray ionization tandem mass spectrometric techniques. *Anal. Biochem.* **295:**129–137.

Burgard, A. P., and C. D. Maranas. 2001. Probing the performance limits of the *Escherichia coli* metabolic network subject to gene additions or deletions. *Biotechnol. Bioeng.* **74:**364–375.

Carlson, R., D. Fell, and F. Srienc. 2002. Metabolic pathway analysis of a recombinant yeast for rational strain development. *Biotechnol. Bioeng.* **79:**121–134.

Cascante, M., L. G. Boros, B. Comin-Anduix, P. de Atauri, J. J. Centelles, and P. W. Lee. 2002. Metabolic control analysis in drug discovery and disease. *Nat. Biotechnol.* **20:**243–249.

Covert, M. W., and B. O. Palsson. 2002. Transcriptional regulation in constraints-based metabolic models of *Escherichia coli*. *J. Biol. Chem.* **277:**28058–28064.

Delgado, J., and J. C. Liao. 1997. Inverse flux analysis for reduction of acetate excretion in *Escherichia coli*. *Biotechnol. Prog.* **13:**361–367.

Dhurjati, P. D., D. Ramkrishna, M. C. Flickinger, and G. T. Tsao. 1985. A cybernetic view of microbial growth: modeling of cells as optimal strategists. *Biotechnol. Bioeng.* **27B:**1–9.

Drell, D. 2002. The Department of Energy microbial cell project: a 180° paradigm shift for biology. *Omics* **6:**3–9.

Edwards, J. S., and B. O. Palsson. 1999. Systems properties of the Haemophilus influenzae Rd metabolic genotype. *J. Biol. Chem.* **274:**17410–17416.

Edwards, J. S., and B. O. Palsson. 2000. The *Escherichia coli* MG1655 in silico metabolic genotype: its definition, characteristics, and capabilities. *Proc. Natl. Acad. Sci. USA* **97:**5528–5533.

Edwards, J. S., R. U. Ibarra, and B. O. Palsson. 2001. In silico predictions of *Escherichia coli* metabolic capabilities are consistent with experimental data. *Nat. Biotechnol.* **19:**125–130.

Edwards, J. S., M. Covert, and B. O. Palsson. 2002a. Metabolic modelling of microbes: the flux-balance approach. *Environ. Microbiol.* **4:**133–140.

Edwards, J. S., R. Ramakrishna, and B. O. Palsson. 2002b. Characterizing the metabolic phenotype: a phenotype phase plane analysis. *Biotechnol. Bioeng.* **77:**27–36.

Farmer, W. R., and J. C. Liao. 1997. Reduction of aerobic acetate production by *Escherichia coli*. *Appl. Environ. Microbiol.* **63:**3205–3210.

Fell, D. 1996. *Understanding the Control of Metabolism*, p. 1–301. Portland Press, London, United Kingdom.

Fiehn, O. 2001. Combining genomics, metabolome analysis, and biochemical modelling to understand metabolic networks. *Comp. Funct. Genom.* **2:**155–168.

Förster, J., I Famili, B. O. Palsson, and J. Nielsen. 2003. Large-scale evaluation of in silico gene deletions in *Saccharomyces cerevisiae*. *OMICS* 7:193–202.

Ibarra, R. U., J. S. Edwards, and B. O. Palsson. 2002. *Escherichia coli* K-12 undergoes adaptive evolution to achieve in silico predicted optimal growth. *Nature* 420:186–189.

Kompala, D. S., D. Ramkrishna, N. B. Jansen, and G. T. Tsao. 1986. Investigation of bacterial growth on mixed substrates. Experimental evaluation of cybernetic models. *Biotechnol. Bioeng.* 28:1044–1055.

Liao J. C., S. Y. Hou, and Y. P. Chao. 1996. Pathway analysis, engineering, and physiological considerations for redirecting central metabolism. *Biotechnol. Bioeng.* 52:129–140.

Palsson, B. O. 2000. The challenges of *in silico* biology. *Nat. Biotechnol.* 18:1147–1150.

Papin, J. A., N. D. Price, J. S. Edwards, and B. O. Palsson. 2002. The genome-scale metabolic extreme pathway structure in *Haemophilus influenzae* shows significant network redundancy. *J. Theor. Biol.* 215:67–82.

Price, N. D., J. A. Papin, and B. O. Palsson. 2002. Determination of redundancy and systems properties of *Helicobacter pylori*'s metabolic network using genome-scale extreme pathway analysis. *Genome Res.* 12:760–769.

Savageau, M. A., E. O. Voit, and D. H. Irvine. 1987a. Biochemical systems theory and metabolic control theory. I. Fundamental similarities and differences. *Math. Biosci.* 86:127–145.

Savageau, M. A., E. O. Voit, and D. H. Irvine. 1987b. Biochemical systems theory and metabolic control theory. II. The role of summation and connectivity relationships. *Math. Biosci.* 86:147–169.

Schilling, C. H., J. S. Edwards, and B. O. Palsson. 1999. Towards metabolic phenomics: analysis of genomic data using flux balances. *Biotechnol. Prog.* 15:288–295.

Schilling, C. H., D. Letscher, and B. O. Palsson. 2000. Theory for the systemic definition of metabolic pathways and their use in interpreting metabolic function from a pathway-oriented perspective. *J. Theor. Biol.* 203:229–248.

Schilling, C. H., and B. O. Palsson. 2000. Assessment of the metabolic capabilities of *Haemophilus influenzae* Rd through a genome-scale pathway analysis. *J. Theor. Biol.* 203:249–283.

Schilling, C. H., M. W. Covert, I. Famili, G. M. Church, J. S. Edwards, and B. O. Palsson. 2002. Genome-scale metabolic model of *Helicobacter pylori* 26695. *J. Bacteriol.* 184:4582–4593.

Schuster, S., and C. Hilgetag. 1994. On elementary flux modes in biochemical reaction systems at steady state. *J. Biol. Syst.* 2:165–182.

Varma, A., and B. O. Palsson. 1994a. Metabolic flux balancing: basic concepts, scientific and practical use. *Bio/Technology* 12:994–998.

Varma, A., and B. O. Palsson. 1994b. Stoichiometric flux balance models quantitatively predict growth and metabolic by-product secretion in wild-type *Escherichia coli* W3110. *Appl. Environ. Microbiol.* 60:3724–3731.

Varner, J., and D. Ramkrishna. 1999. Metabolic engineering from a cybernetic perspective. 1. Theoretical preliminaries. *Biotechnol. Prog.* 15:407–425.

Weichert, W. 2001. C-13 metabolic flux analysis. *Metab. Eng,* 3:195–206.

Microbial Diversity and Bioprospecting
Edited by Alan T. Bull
© 2004 ASM Press, Washington, D.C.

Chapter 28

Phylogeny and Functionality: Taxonomy as a Roadmap to Genes

ALAN C. WARD AND MICHAEL GOODFELLOW

For microbiologists, the confounding complexity which makes search and discovery for new natural product drugs so difficult arises from the following: microbial diversity, which is very large (Bull et al., 2000); biogeography and ecology, which mean that diversity is heterogeneously distributed over environments, space (Staley and Gosink, 1999; Hedlund and Gosink, chapter 22, this book), and time; and the encapsulation of the chemical diversity, which is sought within this biological diversity. Historically, the diversity has been sampled, and individual organisms have been screened for their ability to express bioactive natural products. However, rediscovery of known metabolites has exceeded the discovery of novelty, and major efforts have been directed away from natural product screening (Watve et al., 2001). Although new paradigms for generating natural product diversity, including combinatorial biosynthesis (Tsoi and Khosla, 1995; Walsh, 2002) and sampling the environmental metagenome (Krsek and Wellington, 1999; Rondon et al., 2000; Brady et al., 2001; Sosio et al., 2001), may provide novel routes to new products, they are still dependent on sampling microbial diversity. Meanwhile, strategies independent of natural products—"driven by chemistry, guided by pharmacology"—have not improved overall productivity, and "it is difficult to judge the success of the new paradigm" (J. Drews, 2000). Therefore, dereplication of biological diversity still has an important role to play in selecting the input to new high-throughput, target-based (Chalker and Lunsford, 2002), and high-technology screens (Hill et al., 1998; Harvey, 2000). The challenge is to understand how evolutionary and environmental forces have shaped the distribution of bioactive natural products across biological diversity and the implications for search and discovery strategies.

The potential confounding complexities are:

- Microbial diversity—the microbial diversity, encapsulating the chemical diversity, is very large, but log-normally distributed (Curtis et al., 2002) so that most diversity is present in rare organisms. More effort is required to isolate, characterize, and identify rare strains (Lazzarini et al., 2001; Monciardini et al., 2002).
- Biogeography—the microbial diversity is very large and heterogeneously distributed over environments, space, and time. More effort is required to sample new environments (Colquhoun et al., 2000; Ivanova et al., 2001; Mincer et al., 2002).
- Viable but nonculturable (Roszak and Colwell, 1987)—the microbial diversity is, in practice, mostly noncultured. This situation requires isolation, cloning, and expression of the environmental metagenome (Krsek and Wellington, 1999; Rondon et al., 2000; Sosio et al., 2001) as well as better methods for cultivation (Sait et al., 2002).
- Gene expression—the chemical diversity detected by contemporary screening is low, and organisms have the potential to express much more chemical diversity (Challis and Ravel, 2000; Sosio et al., 2000; Ōmura et al., 2001). Improved methods for eliciting gene expression are needed (Huang et al., 2001; Zahn et al., 2001).
- Lateral gene transfer—although microbial and chemical diversity are both large, they are also uncoupled by lateral gene transfer, so that the chemical diversity is, more or less, randomly distributed (Monaghan et al., 1995). High-throughput screening and special-purpose dereplication are required.

Alan C. Ward and Michael Goodfellow • School of Biology, University of Newcastle upon Tyne, Newcastle upon Tyne, NE1 7RU, United Kingdom.

Only domination by the last of these factors can make understanding biological diversity an ineffective roadmap to metabolic potential. Taxonomy, the principles and practice of classification of the organisms that make up this diversity, enables their detection and identification. The extent to which a general, phylogenetically based taxonomy can be predictive depends on the relative contributions of horizontal and vertical transmission of genes. To exploit this predictive power will depend on our understanding of biological diversity, the distribution of genes within that diversity, and the ecology and biogeography of the organisms. Taxonomy is a key prerequisite to this understanding.

TAXONOMY

The beginning of wisdom is calling things by their right names. (Chinese proverb)

History

The taxonomy of prokaryotes is relatively recent. Carl von Linné (1707–1778) classified microscopic organisms in the genus "chaos" (see G. Drews, 2000), and in 1874 Theodor Billroth (1829–1894) believed that there was only one bacterial species, *Coccobacteria septica*, which could occur in any form depending on growth conditions (Bullock, 1938), although Ferdinand Cohn (1828–1898) was convinced that the species concept should also be applied to bacteria (see G. Drews, 2000). The primacy of morphological criteria was seen as an article of faith by early systematists (see Murray, 1962; Kandler, 1985). Indeed, morphological properties were instrumental in establishing phylogenetic arrangements in plants and animals. Initial studies on the taxonomy of prokaryotes and protozoa were modeled on higher eukaryote systematics, as exemplified by Ehrenberg (1795–1876) who classified them into several genera, and Ferdinand Cohn, a trained botanist and adherent of Ehrenberg, who was one of the first to hold that bacteria could be arranged into genera and species that had a high degree of constancy (see Bullock, 1938), although his views did not begin to be accepted until the isolation of pure cultures (e.g., *Bacillus anthracis* [Koch, 1877; G. Drews, 2000]). The most widely accepted classification, strictly based on morphological criteria and summarizing all the species described by the end of the 19th century, was that of Migula (1897).

The Concept of a Bacterium

This emphasis on morphology represented the beginning of a trend, as microbiologists struggled with inadequate concepts and techniques to capture the fundamental nature of prokaryotic taxa (Stanier and van Niel, 1962; Woese, 2002). Morphology, a basis for phylogenetic classifications for many higher organisms, has often led to misleading hypotheses and misclassification in the prokaryotes. Morphology was responsible for the actinomycetes being considered as boundary organisms between bacteria and fungi (Waksman, 1961). The reliance on morphology became increasingly difficult to justify as the 20th century progressed and taxa such as the aerobic, endospore-forming bacilli (Stackebrandt and Swiderski, 2002), aerobic, gram-positive cocci (Goodfellow, 1985), coryneform bacteria (Bousfield and Callely, 1978), and nocardioform actinomycetes (Goodfellow et al., 1999) were recognized as catchalls for a multiplicity of taxonomically diverse bacteria. Morphologically based classifications are artificial, derived from a narrow database, and consequently unreliable.

The Modern Era

Prokaryotic classification is markedly data dependent (Goodfellow and O'Donnell, 1993), is still relatively data poor, and, with each influx of new technology and new data, is in a constant state of flux. Technological developments such as the development and application of chemotaxonomy (Goodfellow and O'Donnell, 1994), numerical taxonomy (Sneath and Sokal, 1973), small-subunit (SSU) rDNA (SSU rRNA = 16S rRNA in prokaryotes) sequencing (Woese, 1987), DNA:DNA pairing (Grimont, 1988), and molecular fingerprinting techniques (Stackebrandt and Goodfellow, 1991; Mougel et al., 2002) have led to fundamental reappraisals. The integrated application of many techniques, applying the concept of polyphasic taxonomy (Colwell, 1970), and as described by Wayne et al. (1987), has underpinned prokaryote systematics for the past 20 years (Vandamme et al., 1996; Priest and Goodfellow, 2000). Although the concept is universally accepted, there is often no well-defined practice.

Phylogeny

A parallel development has been the generation of a phylogenetic framework based on the molecular record (Zuckerkandl and Pauling, 1965) and implemented with SSU rRNA sequencing (Ludwick and Schleifer, 1994; Ludwig and Klenk, 2001). Using the SSU rDNA as a universal semantide has revealed fundamental taxonomic relationships from Domains (Woese, 1987) to the diversity of prokaryotes in ecological niches (Hugenholtz et al., 1998; Ward, 1998). SSU rRNA sequencing, at least in prokaryotes, is an essential component of modern polyphasic taxonomy (Stackebrandt et al., 2002), but polyphasic taxonomy is not seen as essential for sequencing and phylogenetic analysis.

There is a tension between sequencing—which is universally applicable, cumulative, rapid, and cheap for characterizing prokaryotes (Stackebrandt and Goebel, 1994), but lacks resolution at the species level—and polyphasic, natural classifications, which often include taxon-specific procedures and are demanding of resources, time, and expertise (see "Taxonomy now," below). The latter have been the basis of prokaryote taxonomy for most of the modern period of systematics, reflected in the continuing dominance of DNA:DNA reassociation as the "gold standard" for defining bacterial species (Wayne et al., 1987) right through the 16S rRNA sequencing era (see Fig. 1). As the genomic era begins, DNA:DNA reassociation still claims a place in any prokaryotic species definition (Rosselló-Mora and Amann, 2001; Stackebrandt et al., 2002).

Molecular Ecology and Biodiversity

The application of SSU rRNA sequencing for cultivation-free molecular ecology (Stahl et al., 1985) has challenged the conservative nature of the practical species concept (Goodfellow et al., 1997) vis-à-vis a natural species concept (Ward, 1998), but at the same time, it has not been able to satisfactorily resolve species even at the present conservative levels (Fig. 1).

These are all fundamental problems. Woese (2002) has argued that the "radical insight" of Pace (Stahl et al., 1985) to use rRNA for cultivation-independent characterization of prokaryotes in the environment "freed microbiologists to explore the microbial world in its entirety" and "made irrelevant whether organisms existed in pre-culture." Although molecular ecology has undoubtedly revealed the extent of our ignorance of prokaryotic diversity (Bull et al., 2000; Whitman et al., 1998), it has still not provided the ability to detect and identify organisms and define ecological niches in organismal terms with which Woese (2002) distinguished the classical ecology of higher organisms from the "pseudo-ecology" possible for microbiologists. The reasons for this are evident (Stackebrandt and Goebel, 1994): 16S rRNA sequence data are noisy (Fig. 1), and the SSU rRNA molecule is too conserved to discriminate between prokaryotic species.

Taxonomy Now

A recent history of prokaryotic taxonomy, reliance on morphology, difficulty in defining the basic unit of diversity (i.e., the species [Lan and Reeves, 2000; Lawrence, 2002]), and funding have all resulted in 5,988 validly described prokaryotic species (J.P. Euzéby, http://www.bacterio.cict.fr) and 62,700 prokaryotic 16S rRNA sequences in the Ribosomal Database Project, the RDP II database (Maidak et al., 2001, http://rdp.cme.msu.edu/html/). However,

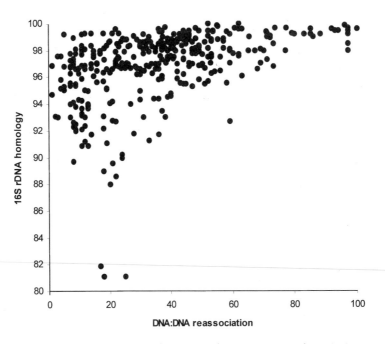

Figure 1. DNA:DNA homology versus 16S rRNA similarity. Data for actinomycetes, from the literature. Adapted from J. Chun, Ph.D. thesis, University of Newcastle upon Tyne, updated by J. E. M. Stach and L. A. Maldonado.

among the 880 streptomycete sequences in the RDP database, only 118 of the 501 validly described streptomycete species are present, and of the 880 sequences, 336 are partial sequences, though full sequences (>1,300 bases, <0.5% ambiguities [Stackebrandt et al., 2002]) are needed.

Systematics and taxonomy, like the rest of biology, are currently in the throes of a new technological revolution. Whole-genome sequencing and the commensurate developments in high-throughput sequencing technology are providing unprecedented amounts of data. Prokaryotic systematics is undoubtably wrestling with the imbalance between high-throughput sequencing and the concept of polyphasic taxonomy (Stackebrandt et al., 2002). Currently, taxonomy is reliable only at the level of broad phylogenetic groups (well delineated by even partial 16S sequences) and at the species level within certain well-studied taxa such as the genus *Mycobacterium* (Goodfellow and Magee, 1998). For many genera, identification of species remains a major problem, as exemplified by the genera *Nocardia* (Goodfellow et al., 1999) and *Rhodococcus* (Goodfellow et al., 1998).

GENES

Genes, as the units of inheritance and the genetic basis for specific phenotypic properties, have been studied since Mendel (Dobzhansky, 1967) and Beadle and Tatum (1941). But only indirectly, genetics has been the study of phenotypic properties often in single-organism model systems and transmitted by Darwinian inheritance. Few studies were complemented by extensive gene sequencing (Dykhuizen and Green, 1991). Consequently, the relationship between taxonomy and genes has been indirect, and views on the distribution of genes across microbial diversity based on observations of phenotypic properties and concepts of inheritance which began to be challenged only with the recognition of widespread transfer of antibiotic resistance genes (see Davies, 1996).

Historically, classification of prokaryotes has also been based on phenotypic properties, and large-scale evolutionary lineages such as actinomycetes, archaea, cyanobacteria, proteobacteria, spirochetes, and green sulfur bacteria can all be recognized by their shared phenotypic properties (Krieg and Holt, 1984; Williams et al., 1989; Boone et al., 2001).

Chemotaxonomy—Inferring Genes from Biomarkers

Chemical data from the analysis of whole-organism and cell components, using methods such as gas, thin-layer, and high-performance liquid chromatography, have been used extensively to classify microorganisms according to the discontinuous distribution of specific compounds (Goodfellow and Minnikin, 1985; Goodfellow and O'Donnell, 1994). Chemotaxonomic analyses of macromolecules, especially amino acids and peptides (from peptidoglycan and pseudomurien), isoprenoid quinones (e.g., menaquinones and ubiquinones), lipids (lipopolysaccharides and fatty acids, including mycolic acids), polysaccharides and related polymers (e.g., methanochondroitin and wall sugars), proteins (e.g., bacteriochlorophylls and whole-organism protein patterns), and enzymes (e.g., hydrolases, lyases, and multilocus enzyme electrophoretic patterns) were used to classify innumerable prokaryotic taxa prior to the introduction of 16S rDNA sequencing. Indeed, chemotaxonomy remains central to prokaryotic systematics, as witnessed by the minimal standards recommended for the description of *Mycobacterium* (Vincent Lévy-Frébault and Portaels, 1992), *Staphylococcus* (Freney et al., 1999), and the genera and cultivable species of the family *Flavobacteriaceae* (Bernardet et al., 2002). Chemotaxonomic data proved to be of particular value in the classification of the actinomycetes and coryneform bacteria, which initially was essentially morphological in concept (Goodfellow and Cross, 1983; Suzuki et al., 1984). Data from wall amino acid and sugar analyses promoted an extensive reappraisal of the classification of these taxa (Lechevalier, 1976; Keddie and Bousfield, 1980), with the actinomycetes being assigned to eight wall chemotypes based on the discontinuous distribution of a limited number of major wall components (Becker et al., 1965; Lechevalier and Lechevalier, 1970). The diamino acid of the wall peptidoglycan was, and remains, a particularly reliable chemical marker. Wall and whole-organism analyses provided the first unambiguous evidence that the genus *Corynebacterium* was closely related to the genera *Mycobacterium, Nocardia,* and *Rhodococcus;* members of all these taxa contain major amounts of *meso*-diaminopimelic acid (*meso*-A$_2$pm), arabinose, and galactose, that is, they have a wall chemotype IV. In contrast, members of the genus *Streptomyces* are characterized by the presence of LL-A$_2$pm and glycine in the wall (wall chemotype I sensu Lechevalier and Lechevalier [1970]).

More precise data for the classification of gram-positive bacteria came from analyses of the primary structure of the wall peptidoglycan, notably the mode of cross linkage and the composition of interpeptide bridges (Schleifer and Kandler, 1972). The variation in peptidoglycan structure provided the framework for reclassification of coryneform bacteria (Keddie and Cure, 1978; Minnikin et al., 1978). Structural variation in the peptidoglycan of wall chemotype IV actinomycetes shows that the muramic acid moieties of *Gordonia, Mycobacterium, Nocardia, Rhodococcus, Skermania,* and *Tsukamurella* are N-glycolated whereas those of *Corynebacterium, Dietzia,* and most other actinomycetes are N-acetylated (Uchida and Aida, 1979; Goodfellow et al., 1998; Society for Actinomycetes Japan, 2000). The variations among the various lipid classes have been extensively used in bacterial classification (Goldfine, 1972; Embley and Wait, 1994) and, in particular, have led to marked improvements in the classification of actinomycetes and coryneform bacteria (Minnikin et al., 1978; Minnikin and O'Donnell, 1984). Long-chain 2-alkyl-branched 3-hydroxy acids (mycolic acids), for example, are found only in some wall chemotype IV taxa; details of their structure allow the separation of these bacteria into the genera *Corynebacterium, Dietzia, Gordonia, Mycobacterium, Nocardia, Rhodococcus, Skermania,* and *Tsukamurella* (Goodfellow et al., 1998, 1999). Similarly, analyses of other lipids, notably long-chain fatty acids, isoprenoid quinones, pigments, and polar lipids, provide valuable information for the classification of prokaryotes, including actinomycetes and coryneform bacteria (Lechevalier et al., 1977; Minnikin and Goodfellow, 1980; Collins, 1994).

Chemotaxonomy, part of the disparate, labor-intensive character of polyphasic taxonomic characterization, will continue to be important with the availability of high-throughput chemical fingerprinting methods for characterization and identification, such as Fourier-transform infrared spectroscopy (Helm et al., 1991; Oberreuter et al., 2002), pyrolysis mass spectrometry (Goodfellow et al., 1994; Colquhoun et al., 2000), matrix-assisted laser desorption–ionization with time of flight (Claydon et al., 1996; Conway et al., 2001), and spray-ionization mass spectrometry (Vaidyanathan et al., 2001). These high-throughput chemical fingerprinting methods offer the possibility of integration between genomic and phenotypic characterization of organisms, which are important to understand much of our current data and to exploit technology to solve the major problem of rapid and reliable identification. In general, good congruence has been found between the discontinuous distribution of chemical markers and the positions of the corresponding taxa in the phylogenetic tree (Goodfellow and O'Donnell, 1994), as clearly shown with respect to the actinomycetes (Embley and Stackebrandt, 1997; Goodfellow et al., 1998, 1999). Indeed, chemotaxonomic markers can be used to evaluate the structure of phylogenetic trees and to adjudicate between phylogenies. The concept that specific molecular species act as biomarkers is not just exploited for taxonomy (Lechevalier and Lechevalier, 1988), but has been widely employed as a molecular tool to investigate microbial community structure (Guezennec and Fiala-Medioni, 1996). Biomarkers indirectly reflect the distribution of the genes responsible for their biosynthesis, although additional complexity is often revealed when the specific genes responsible for metabolic processes and products are identified, as illustrated below.

PUFA Biosynthesis

Polyunsaturated fatty acids (PUFAs) are an example in which the distribution of the biochemical marker, the taxonomy of the producing organisms, and the genes of the biosynthetic pathways involved have all been studied. The ability to biosynthesize PUFAs is distributed throughout animals, plants, and prokaryotes, with the greatest diversity of PUFAs among microorganisms (algae, bacteria, and fungi). Typical PUFAs include linoleic acid (LA, $18:2\omega6$), α-linolenic acid (ALA, $18:3\omega3$), γ-linolenic acid (GLA, $18:3\omega6$), eicosapentaenoic acid (EPA, $20:5\omega3$), docasahexaenoic acid (DHA, $22:6\ \omega3$), and arachidonic acid (AA, $20:4\ \omega6$). These are generated from C_{18} fatty acids by an array of desaturase and elongase enzymes (Napier et al., 1999a) that enable de novo synthesis or elaboration from precursor PUFAs. Algae, bacteria, fungi, and insects can synthesize various PUFAs de novo and are primary producers. Higher plants do not generally contain complex PUFAs, as they lack the necessary biosynthetic enzymes, and there is a restricted taxonomic distribution of C_{18} PUFAs such as GLA and ALA.

PUFAs are essential for higher animals that possess a limited repertoire of desaturases, typically the Δ^4-, Δ^5-, Δ^6-, Δ^9-desaturases, although mammals also lack the Δ^4-desaturase and rely on a dietary supply of essential fatty acids (simple PUFAs) and subsequent downstream desaturation and elongation (Napier et al., 1999a,b) to produce more complex PUFAs. Complex PUFAs are essential dietary components for many marine organisms (Kanazawa et al., 1979), and fish are an important source for human diet and health (Gill and Valivety, 1997; Broun et al., 1999). The supply of PUFAs in the marine environment has long been ascribed to phototrophic algae; their presence in prokaryotes was first reported in *Flexibacter polymorphus* (Johns and Perry, 1977) and highlighted by DeLong and Yayanos (1986) and Yazawa (1996), and their prevalence was reviewed by Nichols and McMeekin (2002). The ability of prokaryotes to produce PUFAs is restricted to members of two taxa (Nichols and McMeekin, 2002), the *Cytophagales* and the *Gammaproteobacteria*, but the genes for EPA biosynthesis appear to be present in the whole genome sequence of "*Streptomyces coelicolor*" M145 (Bentley et al., 2002). The cloning and sequencing of a 3-kb DNA fragment from a *Shewanella* sp. was postulated to encode aerobic desaturases and elongases (Watanabe et al., 1997), but the *Shewanella* open reading frames (ORFs) have been shown to encode a novel polyketide synthase (PKS)-like biosynthetic pathway for C_{20} PUFAs (Metz et al., 2001).

The presence of PUFAs has been found in just two strains in the *Cytophagales*, *F. polymorphus* and *Psychroflexus torquis*. Each has its own characteristic fatty acid composition, including PUFA—EPA in *F. polymorphus*, and EPA + AA in *P. torquis*—but the *Cytophagales* are a heterogeneous group in need of taxonomic revision (Nakagawa and Yamasato, 1993).

The genera *Colwellia*, *Moritella*, and *Shewanella* of the *Gammaproteobacteria* include strains that contain PUFAs, namely, EPA in some *Shewanella* or DHA in some *Colwellia* and all *Moritella* strains. There is a discontinuous distribution of PUFAs in members of these genera (Fig. 2). Nichols and McMeekin (2002) correlate the possession of PUFAs in these strains with psychrophilic, marine-adapted organisms. In deep-sea bacteria, PUFA content was found to increase with hydrostatic pressure (DeLong and Yayanos, 1986). When mapped onto the 16S rDNA phylogenetic tree, this pattern is consistent with Darwinian inheritance and multiple gene loss events, a pattern that will require DNA sequencing to confirm unambiguously. The interaction between phylogeny and environment is reflected in this pattern. It reflects the potential gene pool available from Darwinian inheritance and the selective pressure from the environment leading to gene loss and a discontinuous distribution of genes.

In the lower eukaryotes, a marine protist, a *Thraustochrytrium* sp., from the *Thraustochytriidae*, was found to contain a Δ^4-desaturase that reinforced the expectation that the major input of PUFAs into the marine food web from lower eukaryotes was through the sequential aerobic desaturation and elongation reactions. However, the marine protist characterized by Metz et al. (2001), *Schizochytrium*, is also a member of the *Thraustochytriidae*, contains *Shewanella* PKS-like ORFs, and seems likely to synthesize PUFAs by this system. Drawing conclusions about the extent and role of lateral gene transfer in this distribution is difficult on the basis of such a limited taxonomic sampling.

Isoprenoid Biosynthesis

Isopentenyl diphosphate, or its isomer dimethylallyl diphosphate, is the precursor of all isoprenoids, of which there are more than 22,000 (Connolly and Hill, 1992), including those essential for the isoprenyl glycerol ether lipids of archaea, the sterols of eukaryotes, and ubiquinones and menaquinones in electron transport chains in bacteria. These precursors are synthesized either via the mevalonate pathway or the recently discovered 1-deoxy-D-xylulose 5-phosphate pathway (Lange et al., 2000). The distribution of genes from these two pathways is only partially consistent with vertical transmission and acquisition by eukaryotes from the genomes of chloroplast and mitochondrial endosymbionts.

Nonphotosynthetic eukaryotes use the mevalonate pathway, whereas most algae and higher plants have the mevalonate pathway in the cytoplasm and the 1-deoxy-D-xylulose 5-phosphate pathway in the chloroplast. However, unicellular algae only have the 1-deoxy-D-xylulose 5-phosphate pathway genes in the nuclear DNA, although the proteins are still targeted to the chloroplast, consistent with acquisition of the 1-deoxy-D-xylulose 5-phosphate pathway, present in all cyanobacteria (Disch et al., 1998), by photosynthetic eukaryotes from the chloroplast endosymbiont and loss of the mevalonate pathway genes in unicellular algae. The archaea only have the mevalonate pathway, but DNA matching the genes for the last two enzymes of the pathway, found in fungi, mammals, and plants (phosphomevalonate kinase and diphosphomevalonate decarboxylase), has not been found, suggesting novel archaeal enzymes.

In the bacteria, the spirochete *Borrelia burgdorferi* and some low-G-C Gram-positive organisms have some or all the genes for the mevalonate pathway and/or have been shown to use the pathway without the 1-deoxy-D-xylulose 5-phosphate pathway. Most bacteria use the 1-deoxy-D-xylulose 5-phosphate pathway with either some or all the genes of the mevalonate pathway (Boucher and Doolittle, 2000).

Against this background of gene distribution, largely congruent with taxonomy, are several examples of probable lateral gene transfer (Boucher and Doolittle, 2000; Katz, 2002), as follows:

- The deep-sea hyperthermophilic archeon *Archaeoglobus fulgidus* possesses a 3-hydroxy-3-methylglutaryl coenzyme A (CoA) reductase of class 2 (found in bacteria), not class 1 (found in archaea and eukaryotes). Phylogenetic analysis places the *A. fulgidus* sequence in the bacterial clade (Doolittle and Logsdon, 1998). Sequencing of 3-hydroxy-3-methylglutaryl CoA reductase genes in relatives of *A. fulgidus*, and representatives of each of the main euryarchaeal groups, found that all cultured members of the *Archeoglobales* and two species of the *Thermoplasmatales* contain a bacterial enzyme matching that from *Pseudomonas mevalonii* (Boucher et al., 2001).

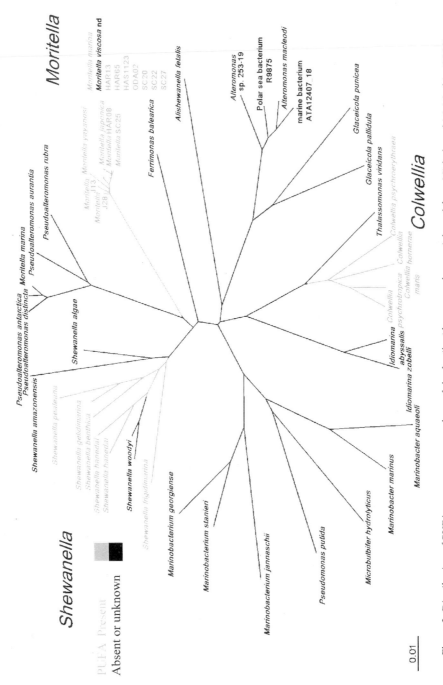

Figure 2. Distribution of PUFAs among some members of the family *Alteromonadaceae*. Adapted from Nichols and McMeekin (2002).

- Conversely, genome sequence data show that *Vibrio cholerae* has a 3-hydroxy-3-methylglutaryl CoA reductase that falls into the archaeal clade (Boucher and Dolittle, 2000).
- Most streptomycetes use the 1-deoxy-D-xylulose 5-phosphate pathway (Orihara et al., 1998); "*Streptomyces aerouvifer*" CL190 also synthesizes menaquinone via the 1-deoxy-D-xylulose 5-phosphate pathway, but synthesizes isoprenoid-containing secondary metabolites such as naphterpin via the mevalonate pathway (Seto et al., 1996). The 3-hydroxy-3-methylglutaryl CoA reductase of "*S. aerouvifer*", *Streptomyces griseolosporeus,* and *Streptomyces* sp. strain KO-3880 (Dairi et al., 2000) are class 1 and fall into the archaeal clade on phylogenetic analysis (Boucher and Doolittle, 2000). Other isoprenoid antibiotics, such as novobiocin, from *Streptomyces niveus* (Orihara et al., 1998) and *Streptomyces spheroides* (Li et al., 1998) are synthesized using the 1-deoxy-D-xylulose 5-phosphate pathway.
- Eukaryotic 3-hydroxy-3-methylglutaryl CoA reductases fall into the archaeal clade (Boucher and Doolittle, 2000), but the sequence of the enzyme from *Giardia lamblia,* a parasitic, amitochondriate, diplomonad protozoan, falls into the bacterial clade.

These potential lateral gene transfers are all reinforced by rigorous phylogenetic analysis, which is a minimum requirement (Boucher and Doolittle, 2000; Gribaldo and Philippe, 2002; Katz, 2002).

Nitrogenase

The discontinuous distribution of nitrogen fixation was one of the earliest indicators of a more extensive role for lateral gene transfer in prokaryotic evolution (Postgate, 1982). Nitrogen fixation is a prokaryotic function carried out by members of the *Archaea* and *Bacteria*. In the *Archaea,* the ability to fix N_2 occurs in the methanogens, whereas it occurs in the *Bacteria* among gram positive bacteria, cyanobacteria, green sulfur bacteria, and the proteobacteria. The mechanism of N fixation is strongly conserved as are the enzyme structure and protein sequence (Postgate, 1982). The discontinuous distribution of N fixation could be explained as an ancient, highly conserved, ancestral molecule with extensive lineage-specific gene loss (Normand and Bousquet, 1989) and/or lateral gene transfer (Postgate, 1982).

Studies of the structural genes for N fixation largely indicate that they, in particular the *nifH* gene, reflect the phylogeny of the bacterial groups in which they occur (Young, 1992), for example, the concurrent evolution of nitrogenase and 16S rRNA in *Rhizobium* species (Hennecke et al., 1985) and the classification of all nitrogen-fixing endospore-forming bacilli in the evolutionary radiation of the *Paenibacillus* clade (Achouak et al., 1999). The phylogenetic trees generated from *nifH* sequences (Hurek et al., 1997) show that gene clusters are congruent with the major bacterial groups of their host bacteria: *Alpha-, Beta-,* and *Gammaproteobacteria,* cyanobacteria, gram-positive bacteria, and archaea. Evidence for lateral gene transfer is most clear cut in the genus *Azoarcus* (Hurek et al., 1997); the sequence data for two strains fall, anomalously, within the *Alphaproteobacteria* whereas the *nifH* sequences from the other *Azoarcus* strains fall, as expected, within the *Betaproteobacteria.* The position of cyanobacteria and frankiae close to the *Alphaproteobacteria* clade could be the result of lateral gene transfer (Normand and Bousquet, 1989), but analysis artifacts are also a possibility (Gribaldo and Philippe, 2002). The archaeal *nifD* gene in *Methanosarcina barkeri* clusters with that of *Clostridium pasteurianum* (Chien and Zinder, 1994), but there is limited taxonomic sampling.

The simple picture of nitrogenase as an $\alpha_2\beta_2$ tetramer encoded by *nifD* and *nifK* and containing FeMoCo (Yates, 1992), in which the phylogenetic distribution of N fixation can be correlated with the distribution of genes encoding homologous proteins, has been made more complex by the discovery of molybdenum-independent nitrogenases, either V-nitrogenases with vanadium encoded by *vnfD* and *vnfK* or Fe-nitrogenases encoded by *anfD* and *anfK* (Bishop and Premakumar, 1992). Similarly, the nitrogenase reductase present in the N-fixing protein complex is encoded, classically, by *nifH* but in the alternative systems by *vnfH* or *anfH,* and a δ-subunit, encoded by *vnfG* or *anfG,* is also found associated with Mo-independent nitrogenases. N-fixing organisms such as *Azotobacter chroococcum* (Robson et al., 1989) and *M. barkeri* (Chien and Zinder, 1996) possess multiple nitrogen-fixing systems. The *vnfH* and *anfH* sequences are found in system-related clades.

Nitrate Reductase

Nitrate reductase, like nitrogenase, always catalyzes the same chemical reaction-conversion, that of nitrate to nitrite—but is a widely distributed step in the biogeochemical N cycle. Nitrate reduction plays different assimilatory and dissimilarity roles in the cellular metabolism of organisms, and nitrate reductases belong to two different protein families, the sulfite oxidase and the dimethyl sulfoxide (DMSO) reductase protein families (Hille, 1996), which contain forms of the pyranopterin cofactor. There is continuing interest in the nitrate reductases, only heightened by the availability of new sequence data from genome projects. Stolz and Basu (2002) cite the availability of more than 67 prokaryotes with sequence data for one or more nitrate reductases, but point out that less than half are amenable to phylogenetic analysis. Many sequence studies, carried out, for example, for structure and function determination, have generated only partial sequences, whereas data from whole-genome annotations often come with no experimental verification.

Phylogenetic analysis (Stolz and Basu, 2002) clusters nitrate reductases into groups that can be distinguished on the basis of taxonomy, specifically eukaryotic and prokaryotic nitrate reductases, metabolic function, and cellular location. Eukaryotic nitrate reductases (Euk-NRs) are found in algae, fungi, and plants and belong to the sulfite oxidase protein family. All prokaryotic nitrate reductases belong to the DMSO reductase protein family and may be cytoplasmic assimilatory nitrate reductases (Nas), periplasmic facing membrane-associated reductases (Nap), or cytoplasmic facing membrane-associated (Nar) reductases. The phylogenetic tree forms three major clades consisting of Euk-NR, Nar, and Nas/Nap.

The Euk-NR clade mirrors the taxonomy with fungal and plant clades that can be further subdivided into algae, monocotyledons, and dicotyledons within the plant clade and ascomycetes, basidiomycetes, yeasts, and an oomycete in the fungal clade. All algae and higher plants have Euk-NR present in the cytoplasm, but not Nas (the nitrate reductase present in cyanobacteria), implying loss of the endosymbiotic nitrate reductase gene during the evolution of the chloroplast, unlike the plant assimilatory nitrite reductase (Tischner, 2000), that is related to the cyanobacterial protein sequence and is located in the chloroplast. The prokaryotic Nar clade shows some evidence of taxonomic congruence; the sequence from *Haloarcula marismortui* (a halophilic *Euryarchaeota*) is the deepest branch, but *Aeropyrum pernix* (the representative of the other archaeal order, the *Crenarchaeota*) clusters with the thermophilic eubacterium *Thermus thermophilus*. The sequences of high-G-C gram-positive bacteria, as exemplified by *Mycobacterium tuberculosis*, *Mycobacterium bovis*, and "*Streptomyces coelicolor*" A3(2), cluster together, as do those of the low-G-C gram-positive bacteria *Bacillus subtilis*, *Staphylococcus aureus*, and *S. carnosus*, with the sequences from gram-negative bacteria, including *Escherichia coli*, *Pseudomonas aeruginosa*, and *P. fluorescens*. The denitrifying sequences from the pseudomonads cluster with the dissimilatory nitrate reductase from *E. coli*, which has

been used to argue that the high degree of sequence similarity (69%) indicates a common ancient lineage (Castresana and Moreira, 1999; Petri and Imhoff, 2000) or suggests horizontal gene transfer (Lawrence, 1999; Stolz and Basu, 2002).

Prokaryotic Nap and Nas nitrate reductases form a third clade of clearly related sequences (Stolz and Basu, 2002). The Nap sequences are highly conserved, can be involved in different metabolic functions, and are found in diverse taxa from *E. coli* to *Desulfovibrio desulfuricans*, *nap* genes have been found on plasmids in *Paracoccus denitrificans*, *Ralstonia eutrophus*, and *Rhodobacter* strains (Moreno-Vivian et al., 1999), so the clustering together of these strains in the Nap phylogeny and the high degree of conservation across the 16S rRNA phylogenetic tree can be interpreted as strong evidence that the *nap* gene is a candidate for horizontal gene transfer. The Nas clade, on the other hand, seems to possess the least-conserved sequence similarity of all the nitrate reductases, with several deep branches in the phylogenetic tree. Cyanobacterial (*Oscillatoria chalybea*, *Synechococcus* sp. strans PCC 7002 and PCC 7942, *Synechocystis* sp. stran PCC 6803), gram-negative (*Caulobacter crescentus*, *Klebsiella oxytoca*, *Klebsiella pneumoniae*, *Pseudomonas putida*), high-G-C (*Amycolatopsis mediterranei*), and low-G-C (*B. subtilis*, *Bacillus halodurans*) gram-positive bacteria cluster together. The *nas* gene seems to be a very rapidly evolving gene sequence.

Different rates of evolution and limited taxonomic sampling are complicating factors in phylogenetic reconstruction, which make deductions less reliable.

As well as these classic molybdenum-containing enzymes, alternative nitrate reduction systems are suggested by the isolation of metal-free, vanadium-containing, and tungsten-dependent enzymes from *Geobacter metallireducens*, *Pseudomonas isachenhovii*, and *Pyrobaculum aerophilum*.

Ammonia Monooxygenase

Autotrophic ammonia-oxidizing bacteria (AOB) convert ammonia to hydroxylamine using the enzyme ammonia monooxygenase. The *amo* operon contains the three genes that encode the three subunits of the enzyme and is present in AOB in both the *Betaproteobacteria* and *Gammaproteobacteria* (Teske et al., 1994; Norton et al., 2002). Ammonia monooxygenase shows evolutionary relationships to the particulate methane monooxygenase found in methane-oxidizing members of the *Gammaproteobacteria* (Rotthauwe et al., 1997). The *Betaproteobacteria* are mainly terrestrial, and freshwater

AOB, whereas AOB in the *Gammaproteobacteria* are marine. The copy number of the *amo* operon is 2 or 3 near-identical operons in the members of the β subdivision, whereas the γ-subdivision bacteria contain a single operon (Norton et al., 2002). Phylogenetic analysis (Rotthauwe et al., 1997) of *amo* protein sequences and 16S rDNA gave congruent trees and support the concept of vertical inheritance rather than lateral gene transfer. Sequences of 16S RNA, 23S RNA, ATPase β-subunit, and elongation factor Tu support the phylogenetic relationship of the *Betaproteobacteria* as a subgroup of the *Gammaproteobacteria* (Schleifer, 1998). The suggestion that the multiple copies of the *amo* operon in the *Betaproteobacteria* were acquired by lateral gene transfer was not supported by studies on codon substitution patterns (Klotz and Norton, 1998), and gene duplication and gene loss events are preferred to account for the phylogenetic distribution of *amoABC* and *pmoABC* operons in the *Beta/Gamma* proteobacterial clade.

Sulfite Reductase

Dissimilarity sulfite reductases catalyze the reduction of sulfite to sulfide in anaerobic sulfate respiration in members of the Domain *Archaea* (the *Euryarchaeota*), in the eubacterial divisions *Nitrospira* and *Thermodesulfobacterium*, and in low-G-C gram-positive bacteria and *Deltaproteobacteria*. The *dsrAB* genes are conserved across these divergent evolutionary lines, and the *dsrAB* phylogenetic tree is very largely congruent with the 16S rRNA phylogeny (Wagner et al., 1998), supporting the early origin of sulfate respiration and vertical transmission from a single ancestor. However, *dsrAB* sequences from *Desulfotomaculum thermocisternum* and *Desulfotomaculum ruminis* (Larsen et al., 1999) were not congruent and suggested the possibility of lateral gene transfer.

A more extensive phylogenetic analysis of *dsrAB* gene sequences from sulfate-reducing prokaryotes (Klein et al., 2001) found evidence for several lateral gene transfer events between major divisions in the *Bacteria* and between *Bacteria* and *Archaea*. The phylogeny derived from the *dsrAB* gene sequences is still largely congruent with 16S rRNA phylogeny except for members of the genera *Archaeoglobus*, *Desulfobacula*, *Desulfotomaculum*, and *Thermodesulfobacterium*. The 13 species of the genus *Desulfotomaculum* form a single monophyletic clade with the low-G-C gram-positive taxa in the 16S rRNA tree, but seven of the *dsrAB* gene sequences from this genus fall into the *Deltaproteobacteria*, including *D. thermocisternum* as well as the single strain of *Desulfobacula toluolica* and the two strains of *Thermodesulfobacterium*.

Sulfate reduction is limited to the genus *Archaeoglobus* in the *Archaea*. The evolutionary distance between the *dsrAB* genes of *Archaeoglobus* and the *Bacteria* is much smaller than for 16S rRNA, suggesting possible lateral gene transfer (Klein et al., 2001), a possibility that will be resolved only with improved taxonomic coverage.

O-Antigens

Within gram-negative pathogens, the surface polysaccharide is antigenic, and antigenic diversity is a widely used mechanism to evade the immune system. In *Salmonella enterica* there are 46 variations; only 5 variants are present in a single subspecies, and their distribution shows evidence of extensive lateral gene transfer (Reeves, 1995). The *gnd* gene is adjacent to the O-antigen gene locus, and subspecies-specific gene sequences are transferred with the O-antigen genes, providing additional evidence from strain-specific signature sequences (Thampapillai et al., 1994). The antigenic cell surface of pathogens is under intense selective pressure, and within-species variation by frequent homologous recombination events is a feature of these organisms.

Multilocus sequence typing of pathogens (Maiden et al., 1998) has provided the data to calculate the relative frequencies of mutation and recombination in generating allelic variation in these organisms (Feil et al., 2001). The estimated relative probability that a nucleotide will change by recombination or mutation ranges from 24:1 for *S. aureus* to 100:1 for *Neisseria meningitidis*.

These data, together with whole-genome data (see "Whole-Genome Sequence Comparisons," below), raise the possibility that lateral gene transfer is so extensive as to completely erase any organism-based phylogenetic signal. This largely intraspecific, homologous recombination, however, is different in mechanism and effect from lateral gene transfer across diverse phylogenetic lineages. Its main effect may be at the level of defining a "genome species" concept for prokaryotes (Lan and Reeves, 2000; Boucher and Doolittle, 2000), leading to a fuzzy species boundary (Lawrence, 2002).

Whole-Genome Sequence Comparisons

Whole-genome sequence data from across phylogenetic diversity reflects the 3.8×10^9 years (Gyr) of evolution with genome sizes from <0.5 Mb, e.g., obligate endosymbionts such as *Buchnera* spp. (Gil et al., 2002), to 3 Gb at the upper size for eukaryotes (International Human Genome Sequencing Consortium, 2001). Searches for conserved, universal genes find mostly ribosomal proteins and only 76 genes common to all 54 of the genomes available (Karlin et al., 2002), and the homology between proteins with common structure and function can be unrecognizable. Comparisons between available whole-genome sequences reveal extensive gene transfer (Doolittle, 1999), gene duplication (Henikoff et al., 1997), and gene deletion (Lawrence and Roth, 1999). Nevertheless, phylogenies derived from rRNA, ribosomal proteins (Hansmann and Martin, 2000), and translational proteins (Brochier et al., 2002), gene content (Bansal and Meyer, 2002), and genome tree approaches (Wolf et al., 2001) are largely congruent (Brochier and Philippe, 2002) or not congruent (Pennisi, 1998), according to different points of view.

E. coli

Prior to whole-genome sequence data, many examples, such as those above, indicated that lateral gene transfer played a role in evolution, but quantification of its contribution and effects was more difficult. With much greater access to direct gene sequence data, the sequence data provide evidence of lateral gene transfer. Anomalous base composition and patterns of codon usage are widely used to determine the extent of lateral gene transfer. Accordingly, about 24% of the genome of *Thermotoga maritima* is proposed to have originated from *Archaea* (Nelson et al., 1999) associated with *T. maritima* in its hyperthermophilic environmental niche.

E. coli is still probably the most completely understood organism. Following the availability of the complete genome sequence of strain MG1655 (K-12), Lawrence and Ochman (1998) estimated that 755 (totaling 548 kb) out of 4,288 ORFs (and 4,639 kb) had been introduced into the *E. coli* genome since it diverged from *Salmonella enterica* about 100 million years (Myr) ago.

Genes recently transferred from organisms with different patterns of DNA base composition match the donor patterns; with time these compositional biases are ameliorated by a mutationally biased change in the host. By applying specific rates of substitution at synonomous and nonsynonomous sites determined for *E. coli* and starting from the anomalous patterns of base composition and codon usage, genes can be reverse ameliorated (Lawrence and Ochman, 1997) until they match an extant bacterial group. On this basis the age of the transferred gene can be estimated; in *E. coli* the average age of the transferred DNA is estimated at 6.7 Myr. Much of this recently acquired DNA is insertion sequence elements and fragments of prophages: remnants of the transfer mechanisms, which are rapidly lost. Once this DNA is excluded,

the remaining potentially beneficial genes have an average age of 14.4 Myr, giving an estimated rate of accumulation corresponding to ~1,600 kb of such DNA acquired since diverging from *S. enterica*. Of this, 548 kb is left; greater than half is recently acquired insertion sequence and prophage DNA, and the rest is mostly recently acquired. Thus, most recently transferred genes are short-term acquisitions.

Nevertheless, all phenotypic characters observed to separate *E. coli* and *S. enterica* arise from laterally transferred genes and none arise from the estimated 22 kb of changes introduced by mutation (Lawrence and Ochman, 1998). Lawrence (2002) discusses the complex interactions leading to speciation and the generation of fuzzy species.

Human Genome

A culmination of the role of lateral gene transfer in confounding the ability to deduce whole-organism phylogenies was the announcement of several hundred recently acquired (within vertebrate evolution) bacterial genes in the human genome (International Human Genome Sequencing Consortium, 2001). This dramatic news prompted reanalysis of the data and methods of evaluating lateral gene transfer (Andersson et al., 2001; Ponting, 2001; Salzberg et al., 2001). The more conservative estimate of ~100 genes from the consortium was a more realistic initial estimate. The potential candidates decreased as more nonvertebrate genomes were included in the analysis. Allowing for differences in rates of evolution also decreased the number of candidates, leaving less than half as candidates for lateral gene transfer. Finally, removing potential annotation errors, deleting mitochondrial genes, and trimming left about 40 candidates (Salzberg et al., 2001).

Katz (2002) identifies discordant gene genealogies as the most direct evidence for lateral gene transfer, but data on most of these candidate gene phylogenetic trees have not been published yet. The example of hyaluronan synthase in Salzberg et al. (2001) is consistent with their statement that most candidate gene phylogenetic trees did not support lateral gene transfer. Similarly, Andersson et al. (2001) found only a single example in seven phylogenies that they constructed, N-acetylneuraminate lyase, in which the phylogeny was not inconsistent with lateral gene transfer. Unlike the case of N-acetylneuraminate lyase in *Trichomonas vaginalis* (De Koning et al., 2000), the phylogenetic tree lacks sufficient vertebrate, nonvertebrate, and bacterial sequence data to confirm the placement of the vertebrate (human, mouse, and pig) or the bacterial (*Yersinia pestis* and *V. cholerae*) sequences that they most closely resemble in a vertebrate, eukaryote, or eubacterial clade.

TAXONOMY AS A ROADMAP TO GENES FOR BIOPROSPECTING

Prokaryotic 16S rDNA

Bacterial genomes contain between 1 and 10 *rrn* operons. There are relatively few organisms in which these individual genes have been sequenced, and usually the different genes have been found to be the same or to differ by only a few nucleotides. However, specific examples may differ significantly. In *Haloarcula marismortui*, two rRNA sequences differ by 74 of 1,472 bases (Mylvaganam and Dennis, 1992); in *Thermobispora bispora* it is 98 out of 1,530 (Wang et al., 1997); in a *Mycobacterium terrae* isolate, a slow-growing mycobacterium but with two 16S rDNA genes, they differ by 18 out of 1,500 bases (Ninet et al., 1996); and in *Rhodococcus koreensis*, two genes differ by 10 nucleotides (Yoon et al., 2000). However, a comparison of two out of the six operons in "*S. coelicolor*" A3(2), *rrnA* and *rrnD*, showed only 99% similarity, equivalent to 8 of 1,529 differences (van Wezel et al., 1991). In the whole-genome sequence there are only 1 base and 2 base differences in two of the six operons (Bentley et al., 2002). About half the of whole-genome sequences that contain more than one *rrn* operon show similar small numbers of nucleotide differences. In a study based on a 120-bp region of the 16S sequence (positions 1 to 100 in the *E. coli* numbering system) determined for 475 streptomycetes, 33 strains showed heterogeneity in the variable region (Ueda et al., 1999).

To explain the lack of congruence between protein gene phylogenies and SSU rRNA phylogeny, Jain et al. (1999) proposed the complexity hypothesis based on the proposition that genes whose products were involved in networks of interactions were less likely to be subject to lateral gene transfer than those operational genes involved in functions such as biosynthesis or catabolism of a specific substrate. The SSU rRNA, at the core of the protein translation machinery, exemplifies such a gene, interacting with more than 100 other molecules (Green and Noller, 1997). Nevertheless, Yap et al. (1999) conclude that the *rrnB* operon of *Thermomonospora chromogena* was acquired from *Thermobispora bispora* via lateral gene transfer. Wang and Zhang (2000) proposed that anomalies in the 16S and 23S gene sequences observed in members of the genus *Nonomuraea* were examples of a homologous recombination of small segments of 20 to 70 bp between strains, although this analysis applies to only three species of a genus with 18 representatives. Thus, despite the probably optimal choice of SSU rRNA as a universal semantide, there is accumulating evidence that it may not be infallible.

Taxonomic Distribution of Natural Product Biosynthesis

Actinomycetes have produced two-thirds of natural products (Bérdy, 1995), with streptomycetes responsible for more than 80% of these. At the higher taxonomic levels, despite the confounding effect of screening efforts (see Watve et al., 2001; Strohl, chapter 31, this volume), for example, in exploring natural products from myxobacteria (Dawid, 2000; Reichenbach, 2001; Altaha et al., 2002), it seems clear that particular taxonomic groups are prolific natural product producers. This conclusion is confirmed by the whole-genome sequencing to date. The "*S. coelicolor*" A3(2) (Bentley et al., 2002) and *Streptomyces avermitilis* (Ōmura et al., 2001) genomes both contain more than 20 natural product gene clusters, far in excess of other genomes (for example, *B. subtilis* with three [Kunst et al., 1997], four in *P. aeruginosa* PAO1 [Stover et al., 2000], and two in *Ralstonia solanacearum* [Salanoubat et al., 2002]), but most genomes lack any detected natural product gene clusters (Bentley et al., 2002).

Polyketide Keto Synthases

Polyketides are prime examples of bioactive secondary metabolites whose search and discovery epitomizes past success and from which lessons might be learned for future strategies. Macrolides synthesized by prokaryotic type I PKSs (polyketide synthases) (Hopwood, 1997) are produced by actinomycetes, bacilli, and myxobacteria. Aromatic polyketides from type II iterative PKSs are produced primarily by actinomycetes (Hopwood, 1997). Type I iterative PKSs are found exclusively in fungi (Nicholson et al., 2001). Polyenes are produced from PKSs in many different microorganisms, including actinomycetes, bacilli, myxobacteria, proteobacteria, and eukaryotic algae. Among a restricted range of prokaryotic PUFA producers, PKSs are also involved in their biosynthesis (Napier, 2002).

Metsä-Ketelä et al. (2002) have compared the phylogeny derived from the sequences of fragments from PKS (a 613-bp fragment) and the γ-variable region of the 16S rDNA, a 194-bp fragment that includes the 120-bp fragment of Ueda et al. (1999). A set of 12 reference *Streptomyces* strains and 87 isolates were compared, and Metsä-Ketelä and colleagues concluded that the two trees are incongruent. The streptomycetes belong to a taxonomically complex group, and complete 16S rDNA sequences (>1,300 nucleotides and <0.5% ambiguity [Stackebrandt et al., 2002]) are recommended as a minimal requirement for phylogenetic analysis. As described in "Taxonomy," above), even full 16S sequence data do not necessarily resolve taxa at the species level. It is also well known that phylogenetic trees can be sensitive to treeing algorithms and choice of strains—isolated sequences without neighbors can move around in the tree. Thus, the inclusion of 12 type strains does not provide a satisfactory phylogenetic framework for 87 uncharacterized isolates in an rRNA phylogenetic tree.

The whole-genome sequence data for "*S. coelicolor*" A3(2) and *S. avermitilis* show multiple polyketide synthetase gene clusters, but only single instances were cloned and sequenced from each isolate. Thus, the clade of organisms, strains TA11, TA19, and TA32, that cluster with "*S. coelicolor*" A3(2) in both the 120-bp 16S tree and with the *actI-orf1* PKS gene cannot be compared with the "*S. coelicolor*" *whiE* gene (in a clade of spore pigment-related genes) because only one sequence from each isolate was included and no potential homologue of *whiE* is available. Another group of closely related sequences from isolates AUSB4, AUSA89, JHA37, and AUSA67 cluster closely to *whiE* and are closely grouped together in the 120-bp 16S tree in the same part of the tree as *S. coelicolor* A3(2), but no other PKS sequences are available for these strains nor are related sequences from *whiE* homologues from other strains in that part of the 120-bp tree available. The spore pigment-related sequences form a closely related clade of sequences in the overall tree. In contrast, Metsä-Ketelä et al. (2002) cite counterexamples. There seems little doubt that lateral gene transfer has played a part in the evolution of natural product biosynthesis in streptomycetes, but the conclusion that lateral gene transfer has left no correlation between organism and gene phylogenies is premature.

Streptomycin

Antibiotic resistance genes, part of antibiotic gene clusters, can be subjected to high selective pressure and are extensively transferred (Mazel and Davies, 1999). The acquisition of such resistance genes has been documented in actinomycetes (Ravel et al., 1998; Tolba et al., 2002), and Wellington's group (Huddleston et al., 1997; Egan et al., 2001) has provided evidence for the transfer of streptomycin resistance, but also, in some instances, biosynthetic genes. Streptomycin biosynthesis is distributed extensively among streptomycetes, as is the biosynthesis of the anthracycline daunorubicin (Strohl et al., 1997). These examples suggest that lateral gene transfer may have played a significant role in the current distribu-

tion of natural product biosynthetic capacity across microbial biodiversity.

The *S. violaceusniger* Clade

The *Streptomyces violaceusniger* clade contains streptomycetes with distinctive morphological properties that make them relatively easy to select from isolation plates and confirm them as putative members of the clade. They are found in soil and rhizosphere samples from many sites around the world, and they form a distinct clade in the 16S rDNA *Streptomyces* phylogenetic tree (Fig. 3).

Streptomyces hygroscopicus subsp. *hygroscopicus* (NRRL 5491) and some other isolates identified as *S. hygroscopicus* produce various immunosuppressants, including rapamycin (Vezina et al., 1975), FK506 (tacrolimus), and FK520 (ascomycin, an active FK506 derivative in which the C21 allyl group is replaced by an ethyl group). "*Streptomyces geldamyceticus*," another member of the *S. violaceusniger* 16S rRNA clade (Sembiring et al., 2000), produces geldanamycin (De Boer et al., 1970; Allen and Ritchie, 1994). Rapamycin, FK506, and FK520 are important new immunosuppressive drugs for transplant therapy, and geldanamycin has potential application as an antitumor agent. They are also potent antifungal antibiotics; although this is not their potential field of application, it is probably their natural role, and some strains are important biocontrol agents (Trejo-Estrada et al., 1998). Other secondary metabolites found in members of this clade (elaiophylin, nigericin, and an unnamed polyene) are also antifungal, for example, elaiophylin has only weak antifungal activity itself but potentiates the antifungal action of rapamycin (Fang et al., 2000).

Members of the *S. violaceusniger* clade are characterized by their ability to form a gray aerial spore mass and a grayish-yellow substrate mycelium on oatmeal agar and to form spiral chains of rugose, ornamented spores (Sembiring et al., 2000). The taxon currently encompasses 10 validly described species that include *S. hygroscopicus* (Jensen, 1931, emended in Labeda and Lyons, 1991), *S. violaceusniger* (Waksman and Curtis, 1916, emended in Labeda and Lyons, 1991), *Streptomyces malaysiensis* Al-Tai et al., 1998, and *Streptomyces melanosporofaciens* (Arcamone et al., 1959). The remaining species, *Streptomyces asiaticus, Streptomyces cangkringensis, Streptomyces indonesiensis, Streptomyces javensis, Streptomyces rhizosphaericus,* and *Streptomyces yogyakartensis,* were described following an extensive

Figure 3. *S. violaceusniger* clade in the streptomycete phylogenetic tree.

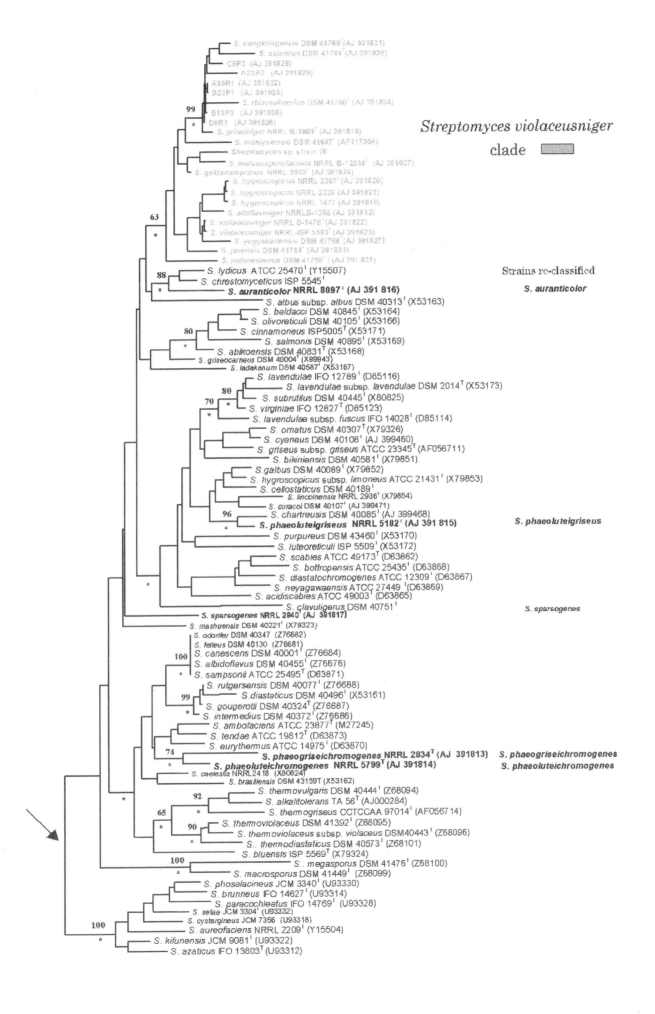

Streptomyces violaceusniger
clade

Strains re-classified
S. auranticolor

S. phaeoluteigriseus

S. sparsogenes

S. phaeogriseichromogenes
S. phaeoluteichromogenes

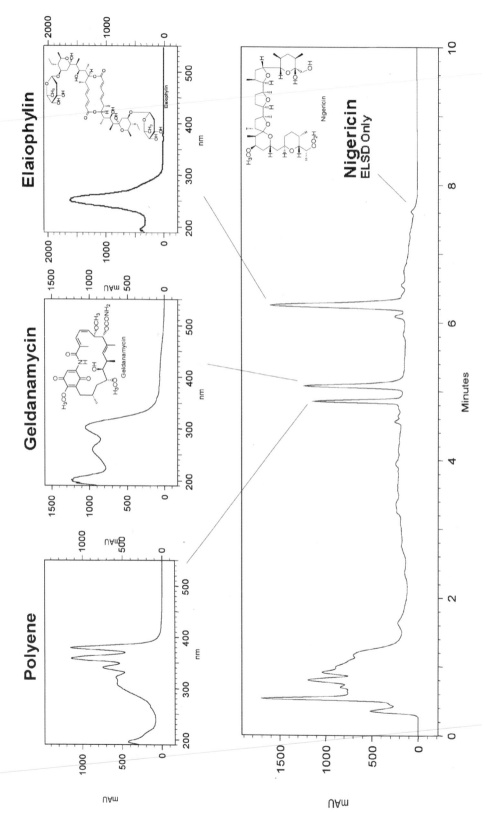

Figure 4. HPLC profile of secondary metabolites from a representative strain of the *S. violaceoruber* clade. ELSD, evaporative light-scattering detector.

polyphasic taxonomic study of *S. violaceusniger*-related strains and isolates from the rhizosphere of *Paraserianthes falcateria* (Sembiring, 2000; Sembiring et al., in press), a tropical leguminous tree used extensively for growth on poor soils, soil improvement, and the supply of wood pulp for paper.

Among the 12 reference strains classified as *S. hygroscopicus* or *S. violaceusniger* and included in this study, 5 did not possess rugose spores and did not fall in the *S. violaceusniger* 16S rRNA clade (Fig. 3), namely, *S. auranticolor* NRRL 8097[T], *S. phaeoluteichromogenes* NRRL B-5799[T], *S. phaeogriseichromogenes* NRRL 2834[T], *S. phaeoluteigriseus* NRRL 5182[T], and *S. sparsogenes* NRRL 2940[T]. In contrast, three additional species were added to the *S. violaceusniger* clade, namely, isolates described as *S. albiflaviniger* and *S. griseiniger* and the marker strain *S. geldanamycinus* of the previously illegimately described "*S. geldamyceticus*."

The metabolic profiles of well-characterized members of the *S. violaceusniger* clade have been determined by growth under conditions to optimize production of secondary metabolites and by high-pressure liquid chromatography (HPLC) analysis (Ward et al., 2000). All the members of the *S. violaceusniger* clade tested show the same pattern of HPLC-detected secondary metabolites, consisting of geldanamycin, eliaophylin, nigericin, and a polyene (Fig. 4), consistent with those found by Allen and Ritchie (1994) and Fang et al. (2000).

S. hygroscopicus subsp. *hygroscopicus* (NRRL 5491) is a member of the *S. violaceusniger* clade and produces rapamycin. Fang et al. (2000) have shown that the same secondary metabolites detected by HPLC (see above) are also produced and play a role in potentiating the activity of rapamycin. Biological screening of nine representatives of the *S. violaceusniger* clade using isogenic *Saccharomyces cerevisiae* strains that are sensitive and resistant to rapamycin (Cardenas et al., 1999) indicates that they all produce rapamycin (Khodobakshian, 2002). However, there are suggestions from base sequence anomalies that the rapamycin gene cluster in *S. hygroscopicus* has been acquired by lateral gene transfer (Molnar et al., 1996).

The *S. violaceoruber* Clade

"*S. coelicolor*" A3(2) is a member of the *S. violaceoruber* clade. Many of the members of this clade have virtually identical 16S rRNA sequences and DNA:DNA similarities that are >90% (Fig. 5). Com-

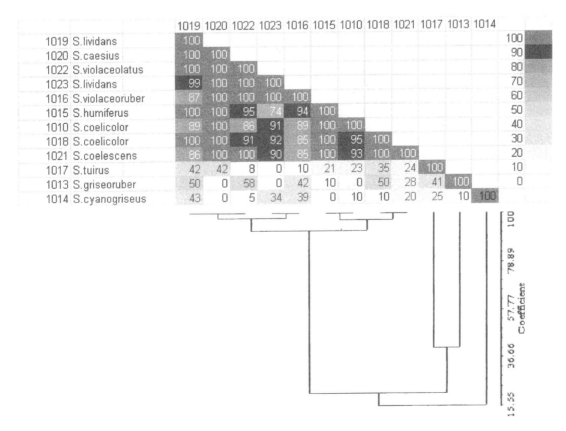

Figure 5. DNA:DNA homology of members of the *S. violaceoruber* clade.

petitive hybridization of cye-5-labeled genomic DNA from these strains against cye-3-labeled genomic DNA from "*S. coelicolor*" M145 [the strain of "*S. coelicolor*" A3(2) sequenced] on glass slide microarrays of PCR fragments from the whole-genome ORFs shows a pattern of "core" genes (normalized $\log_2$ ratios around zero) with a shoulder of several hundred missing genes for each strain (Fig. 6) in the putative species genome of *S. violaceoruber*.

The actinorhodin gene cluster is one of four secondary metabolite gene clusters detected by the expression of the bioactive natural products (actinorhodin, prodigiosin, calcium-dependent antibiotic, and methylenomycin). The presence or absence of the genes of the actinorhodin gene cluster in the species genome of *S. violaceoruber* is shown in Color Plate 9. All strains in the *S. violaceoruber* species group contain actinorhodin as a member of the core genes.

TAXONOMY IS NOT PHYLOGENY

We have argued that taxonomy is not a luxury (Bull et al., 2000). In the context of bioprospecting, taxonomy has three roles: enabling the classification and then the detection and identification of organisms; predicting metabolic potential; and dereplicating isolates for screening. The threats to this strategy are that oversimplistic identification of the organisms and partial characterization of the metabolites leave a confused and incongruent pattern, which may be rationalized as being the result of extensive lateral gene transfer.

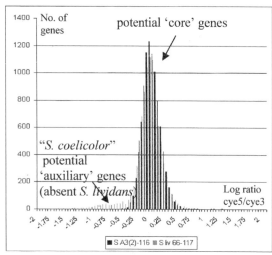

S. coelicolor A3(2) vs *S. coelicolor* M145 log ratio compared with *S. lividans* 66 vs *S. coelicolor* M145 log ratio

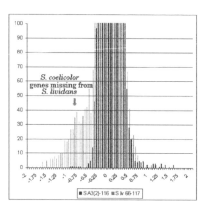

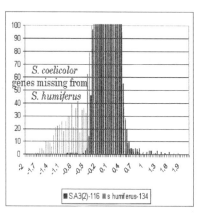

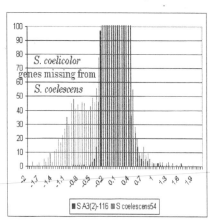

Figure 6. Comparative genomics of some members of the *S. violaceoruber* clade. Genes present and absent detected by "*S. coelicolor*" microarrays.

The Effect of Lateral Gene Transfer

In this overview of the relationship between taxonomy and genes we have tried to reconcile the emerging evidence for extensive lateral gene transfer with the view that a general-purpose hierarchical classification of microorganisms is both possible and predictive. The opposing view is that search and discovery must employ a special-purpose, artificial classification to detect and identify diversity for bioprospecting.

Much of the recent evidence that lateral gene transfer is extensive has come from statistical comparisons of DNA base sequence data. Gribaldo and Philippe (2002) discuss some of the issues with this approach. Many conclusions based on anomalous sequence patterns are not subjected to the scrutiny that followed the results of such an analysis in the human genome (see "Human Genome," above). One of the major factors highlighted in that reanalysis was the importance of taxonomic coverage, which is a special problem for data derived from whole-genome sequence data analysis.

Whole-genome sequencing has, commendably, sampled genomes from across the full phylogeny. This has led to initial comparative genomic studies to address the issues of large-scale evolution—the origin of life and the last common ancestor, the three Domains, and the branching order of major lineages. These are difficult questions, and extensive lateral gene transfer over evolutionary time scales means that new questions are asked and new hypotheses are proposed. But taxonomy is not phylogeny (Sneath, 1988); indeed the International Code of Nomenclature of Bacteria (Sneath, 1992) does not even address categories above the rank of class.

After a long period of accepting the unavailability of phylogenetic information, phylogeny is now at the heart of microbial classification. But although phylogenetic information and relationships provide understanding and guide interpretation, the classification of organisms into related groups does not depend on phylogeny and can be useful and predictive despite phylogenetic uncertainty.

Nevertheless, the major evolutionary lineages are, in the main, well defined, and it is their relationships at the root of the phylogenetic tree that are less well determined. Comparative genomics of whole-organism sequences (Brochier et al., 2002) is generating new hypotheses on relationships, and, with new computational methods and ongoing analyses, a consensus of congruent trees is emerging from core genes (Brochier et al., 2002; Gribaldo and Philippe, 2002), supporting the idea that a phylogenetic signal and a Darwinian tree can represent evolutionary relatedness right back to the roots of evolution.

Major Lineages as Prolific Producers

Even without success in deciphering such ancient phylogenetic relationships, the broad shape of the major evolutionary lineages is clear, and the chemical diversity of organisms can be superimposed on these phylogenetically congruent groups in such a way that chemotaxonomy can be used to identify organisms and predict aspects of their chemical composition (see "Chemotaxonomy—Inferring Genes from Biomarkers," above). In this book, Strohl (chapter 31) is able to identify confidently the taxonomic groups in which bioactive antimicrobial agents will be found. The debate about whether *Streptomyces* are really prolific producers of secondary metabolites or merely the beneficiaries of extensive screening seems to have been decisively answered even by the limited whole-genome sequences available (see "Taxonomic Distribution of Natural Product Biosynthesis," above). Taxonomy is already predictive to the extent that the discovery of a novel drug class, epothilones (Bollag et al., 1995) in myxobacteria, and other indications of their ability to produce secondary metabolites, will lead to an extensive search and discovery effort among myxobacteria and related organisms. The question is whether there are more novel bioactive compounds to be discovered, as estimated by Watve et al. (2001), or if streptomycetes have already been exhaustively screened. This depends on the extent to which streptomycete diversity has been sampled and the distribution of secondary metabolite genes across that diversity.

Dereplication at the Species Level

Comparison of closely related strains—*Helicobacter pylori* (Alm and Trust, 1999); *Chlamydia pneumoniae* and *Chlamydia trachomatis* (Read et al., 2000); *Neisseria menigitidis* (Lan and Reeves, 2000); *Bacillus halodurans* and *B. subtilis* (Takami et al., 2000); *Pyrococcus furiosus* with *Pyrococcus horikoshi* (Maeder et al., 1999); *E. coli, S. enterica,* and *K. pneumoniae* (McClelland et al., 2000)—all shows similar patterns, with significant differences between closely related strains. These comparisons indicate the need to define a species genome in which a species is characterized by a core of genes, present in most strains of the species, and an auxiliary set of genes found only in some of the strains (Lan and Reeves, 2000; Boucher et al., 2001). These ideas were expressed some years ago by Ruth Gordon (Gordon, 1968, 1978), when the presence, absence, and relatedness of genes could be inferred only from studying phenotypic properties of multiple strains. Because the full set of auxiliary genes can be determined only from such a set of strains, Boucher et al. (2001) conclude that we have not actually finished sequencing the genome of even one prokaryotic species.

So, once a target taxon has been identified, can taxonomy act as a roadmap to bioprospecting within it? Strategies employed to sample and characterize strains (Lazzarini et al., 2001; Monciardini et al., 2002), to sample new environments (Colquhoun et al., 2000; Ivanova et al., 2001; Mincer et al., 2002), to devise better methods for cultivation (Sait et al., 2002), and to elicit gene expression (Huang et al., 2001; Zahn et al., 2001) can representatively sample biodiversity, whether as isolates or environmental DNA, and screen isolates by means of biological, chemical, and molecular methods under optimized conditions if taxonomic diversity is reasonably congruent with chemical diversity.

The genome species concept depends on the idea of core and auxiliary genes. Lan and Reeves (2000) suggest that core genes should be those found in >95% of strains in the species whereas auxiliary genes should be found in 1 to 95% of strains. Secondary metabolite gene clusters need not be core genes for the above strategies to achieve results. In the example of O-antigen genes (see "O-Antigens," above), these genes are under intense selective pressure, and antigenic variation is extensive within species in some pathogens (Lan and Reeves, 2000). If antibiotic genes were being exchanged as freely, then taxonomy at the species level would be no guide to chemical diversity. Streptomycin might be an example of such extensive exchange, although neither its distribution nor that of the other streptomycin-related antibiotics nor the phylogenetic relationships of many producing strains, are well defined. Rapamycin, eliaophylin, nigericin, and geldanamycin may be a counterexample in which members of a large clade of related streptomycetes produce a congruent set of related metabolites. But if *S. violaceusniger* has diversified to occupy a nonrhizosphere environmental niche, then gene loss might leave a pattern of secondary metabolite production like that of PUFAs in *Colwellia* and *Shewanella* spp. (see "PUFA Biosynthesis," above), in which biosynthesis is dependent on the environment as well as phylogeny.

This pattern may be emerging at the species level in the *S. violaceoruber* clade. The microarray data for strains closely related to "*S. coelicolor*" A3(2) closely match the genome species concept, as do similar studies in *S. enterica* (Chan et al., 2000) and *S. aureus* (Fitzgerald et al., 2001), for example, with several hundred genes falling into the auxiliary gene set. The actinorhodin genes are present in all the strains assigned to the *S. violaceoruber* clade (Color Plate 9),

including "*Streptomyces lividans*" 66, but though these genes are readily expressed in "*S. coelicolor*" A3(2), conditions for their expression in "*S. lividans*" have only recently been established (Kim et al., 2001), with differences in their environmental niche perhaps reflected at the regulatory level.

The End of "Grab and Go" Search and Discovery?

It is significant that an estimate that there are >100,000 bioactive metabolites still to be discovered from members of the *Streptomyces* genus (Watve et al., 2001) is difficult to evaluate. This is a reflection of the exploitation of a strategy that, in a computing environment, would be described as a greedy algorithm. It seems unlikely that a strategy of random sampling, genus-level dereplication, and screening only for commercial significance will have exhausted natural product diversity. But how much more is there to discover and how much effort will it take? It is clear that brute force screening of more of the same will be successful only by dint of ever-decreasing chance.

Rational search and discovery strategies will need much more effort to characterize and map bio- and chemical diversity, but equally it is not clear that screening combinatorial numbers of simple molecules, or clone libraries generated from environmental metagenomes of unknown size, will require less effort. Secondary metabolite gene clusters, such as PKS, have evolved to generate diversity and are more complex than primary metabolism. Mapping that extensive biosynthetic diversity into a taxonomic unit as small as a genus requires much finer phylogenetic and taxonomic discrimination, just the level at which current methods, such as SSU rRNA sequencing, run out of resolution. However, a research agenda to define prokaryotic species and to determine their diversity and their role in the environment resonates well with the technologies and challenges of the genomic era. We expect that new high-throughput techniques for phenotypic characterization, whole-genome and multilocus sequencing, microarrays, and subtractive hybridization will define the unit of diversity for prokaryotes—the genome species concept—and provide a roadmap to help guide the new search and discovery strategies.

Acknowledgments. We thank colleagues and students for discussions and data, in particular L. A. Maldonado, L. Sembiring, and J. S. Stach for 16S sequence data; K. Duangmal for microarray data and E. Lacey and colleagues at Microbial Screening Technologies, Yarramdo, NSW, Australia, for metabolite profiles of strains in the *S. violaceusniger* clade. "*S. coelicolor*" microarrays were from Colin Smith at University of Manchester Science and Technology, U.K., and Camilla Kao, Stanford University, Stanford, Calif., who also generously gave help with microarray techniques.

REFERENCES

Achouak, W., P. Normand, and T. Heulin. 1999. Comparative phylogeny of *rrs* and *nifH* genes in the *Bacillaceae*. *Int. J. Syst. Bacteriol.* **49**:961–967.

Allen, I. W., and D. A. Ritchie. 1994. Cloning and analysis of DNA-sequences from *Streptomyces hygroscopicus* encoding geldanamycin biosynthesis. *Mol. Gen. Genet.* **243**:593–599.

Alm, R. A., and T. J. Trust. 1999. Analysis of the genetic diversity of *Helicobacter pylori*: the tale of two genomes. *J. Mol. Med.* **77**:834–846.

Altaha, R., T. Fojo, E. Reed, and J. Abraham. 2002. Epothilones: a novel class of nontaxane microtubule-stabilizing agents. *Curr. Pharm. Des.* **8**:1707–1712.

Al-Tai, A., B. Kim, S. B. Kim, G. P. Manfio, and M. Goodfellow. 1998. *Streptomyces malaysiensis* sp. nov., a new streptomycete species with rugose, ornamented spores. *Int. J. Syst. Bacteriol.* **49**:1395–1402.

Andersson, J. O., W. F. Doolittle, and C. L. Nesbs. 2001. Genomics—are there bugs in our genome? *Science* **292**:1848–1850.

Arcamone, F. C., C. Bertazzoli, M. Ghione, and T. Scotti. 1959. Melanosporin and elaiophylin, a new antibiotic from *Streptomyces melanosporus* (sive melanosporofaciens) n. sp. *Giorn. Microbiol.* **7**:207–216.

Bansal, A. K., and T. E. Meyer. 2002. Evolutionary analysis by whole genome comparisons. *J. Bacteriol.* **184**:2260–2272.

Beadle, G. W., and E. L. Tatum. 1941. Genetic control of biochemical reactions in *Neurospora*. *Proc. Natl. Acad. Sci. USA* **27**:499–506.

Becker, B., M. P. Lechevalier, and H. A. Lechevalier. 1965. Chemical composition of cell wall preparations from strains of various form genera of aerobic actinomycetes. *Appl. Microbiol.* **13**:236–243.

Bentley, S. D., K. F. Chater, A.-M. Cerdeño-Tárraga, G. L. Challis, N. R. Thompson, D. James, D. E. Harris, M. A. Quail, H. Kieser, D. Harper, A. Bateman, S. Brown, G. Chandra, C. W. Chen, M. Collins, A. Cronin, A. Fraser, A. Goble, J. Hildago, T. Hornsby, S. Howarth, C.-H. Huang, T. Kieser, L. Larke, L. Murphy, K. Oliver, S. O'Neil, E. Rabbinowitsch, M. A. Rajandream, K. Rutherford, S. Rutter, K. Seeger, D. Saunders, S. Sharp, R. Squares, S. Squares, K. Taylor, T. Warren, A. Wietzorrek, J. Woodwark, B. G. Barrell, J. Parkhill, and D. A. Hopwood. 2002. Complete genome sequence of model actinomycete *Streptomyces coelicolor* A3(2). *Nature* **417**:141–147.

Bérdy, J. 1995. Are actinomycetes exhausted as a source of secondary metabolites? *Biotechnologia* **7–8**:13–14.

Bernardet, J.-F., Y. Nakagawa, and B. Holmes. 2002. Proposed minimal standards for describing new taxa of the family *Flavobacteriaceae* and emended description of the family. *Int. J. Syst. Evol. Microbiol.* **52**:1049–1070.

Bishop, P. E., and R. Premakumar. 1992. Alternative nitrogen fixation systems, p. 736–762. *In* R. H. Burris and H. J. Evans (ed.), *Biological Nitrogen Fixation*. Chapman & Hall, New York, N.Y.

Bollag, D. M., P. A. McQueney, J. Zhu, O. Hensens, L. Koupal, J. Liesch, M. Goetz, E. Lazarides, and C. M. Woods. 1995. Epothilones, a new class of microtubule-stabilizing agents with a taxol-like mechanism of action. *Cancer Res.* **55**:2325–2333.

Boone, D. R., R. W. Casterholtz, and G. M. Garrity (ed.). 2001. *Bergey's Manual of Systematic Bacteriology*, 2nd ed., vol. 1. *The Archaea and the Deeply Branching and Phototrophic Bacteria*. Springer Verlag, New York, N.Y.

Boucher, Y., and W. F. Doolittle. 2000. The role of lateral gene transfer in the evolution of isoprenoid biosynthesis pathways. *Mol. Microbiol.* **37**:703–716.

Boucher, Y., H. Huber, S. L'Haridon, K. O. Stetter, and W. F. Doolittle. 2001. Bacterial origin for the isoprenoid biosynthesis

enzyme HMG-CoA reductase of the archaeal orders Thermoplasmatales and Archaeoglobales. *Mol. Biol. Evol.* **18**:1378–1388.

Bousfield, I. J., and A. G. Callely. 1978. *Coryneform Bacteria*. Academic Press, London, United Kingdom.

Brady, S. F., C. J. Chao, J. Handelsman, and J. Clardy. 2001. Cloning and heterologous expression of a natural product biosynthetic gene cluster from eDNA. *Org. Lett.* **3**:1981–1984.

Brochier, C., and H. Philippe. 2002. Phylogeny: a non-hyperthermophilic ancestor for bacteria. *Nature* **417**:244.

Brochier, C., E. Bapteste, D. Moreira, and H. Philippe. 2002. Eubacterial phylogeny based on translational apparatus proteins. *Trends Genet.* **18**:1–5.

Broun, P., S. Gettner, and C. Somerville. 1999. Genetic engineering of plant lipids. *Annu. Rev. Nutr.* **19**:197–216.

Bull, A. T., A. C. Ward, and M. Goodfellow. 2000. Search and discovery strategies for biotechnology: the paradigm shift. *Microbiol. Mol. Biol. Rev.* **64**:573–606.

Bullock, W. 1938. *The History of Bacteriology*. Oxford University Press, London, United Kingdom.

Cardenas, M. E., M. C. Cruz, M. del Poeta, N. J. Chung, J. R. Perfect, and J. Heitman. 1999. Antifungal activities of antineoplastic agents: *Saccharomyces cerevisiae* as a model system to study drug action. *Clin. Microbiol. Rev.* **12**:583–611.

Castresana, J., and D. Moreira. 1999. Respiratory chains in the last common ancestor of living organisms. *J. Mol. Evol.* **49**:453–460.

Chalker, A. F., and R. D. Lunsford. 2002. Rational identification of new antibacterial drug targets that are essential for viability using a genomics-based approach. *Pharmacol. Therapeut.* **95**:1–20.

Challis, G. L., and J. Ravel. 2000. Coelichelin, a new peptide siderophore encoded by the *Streptomyces coelicolor* genome: structure prediction from the sequence of its non-ribosomal peptide synthetase *FEMS Microbiol. Lett.* **187**:111–114.

Chan, K., S. Baker, C. C. Kim, C. S. Detweiler, G. Dougan, and S. Falkow. 2000. Genomic comparison of *Salmonella enterica* serovars and *Salmonella bongori* by use of an *S. enterica* Typhimurium DNA microarray. *J. Bacteriol.* **185**:553–563.

Chien, Y.-T., and S. H. Zinder. 1994. Cloning, functional organization, transcript studies and phylogenetic analysis of the complete nitrogenase structural genes (*nifHDK2*) and associated genes in the archeon *Methanosarcina barkeri* 227. *J. Bacteriol.* **178**:143–148.

Chien, Y.-T., and S. H. Zinder. 1996. Cloning, DNA-sequencing and characterisation of a *nifD*-homologous gene from the archeon *Methanosarcina barkeri* 227 which resembles *nifD1* from the eubacterium *Clostridum pasteurianum*. *J. Bacteriol.* **176**:6509–6598.

Claydon, M. A., S. N. Davey, V. Edwards-Jones, and D. B. Gordon. 1996. The rapid identification of intact micro-organisms using mass spectrometry. *Nat. Biotech.* **14**:1584–1586.

Collins, M. D. 1994. Isoprenoid quinones, p. 265–309. *In* M. Goodfellow and A. G. O'Donnell (ed.), *Chemical Methods in Prokaryotic Systematics*. John Wiley & Sons, Chichester, United Kingdom.

Colquhoun, J. A., J. Zulu, M. Goodfellow, K. Horikoshi, A. C. Ward, and A. T. Bull. 2000. Rapid characterisation of deep-sea actinomycetes for biotechnology screening programmes. *Antonie Leeuwenhoek* **77**:359–367.

Colwell, R. R. 1970. Polyphasic taxonomy of the genus *Vibrio*: numerical taxonomy of *Vibrio cholerae*, *Vibrio parahaemolyticus*, and related *Vibrio* species. *J. Bacteriol.* **104**:410–433.

Connolly, J. D., and R. A. Hill. 1992. *Dictionary of Terpenoids*. Chapman & Hall, New York, N.Y.

Conway, G. C., S. C. Smole, D. A. Sarracino, R. D. Arbeit, and P. E. Leopold. 2001. Phyloproteomics: species identification of Enterobacteriaceae using matrix-assisted laser desorption/ionization time-of-flight mass spectrometry. *J. Mol. Microbiol. Biotech.* 3:103–112.

Curtis, T. P., W. T. Sloan, and J. W. Scannell. 2002. Estimating prokaryotic diversity and its limits. *Proc. Natl. Acad. Sci. USA* 99:10494–10499.

Dairi, T., Y. Motohira, T. Kuzuyama, S. Takahashi, N. Itoh, and H. Seto. 2000. Cloning of the gene encoding 3-hydroxy-3-methylglutaryl CoA reductase from terpenoid antibiotic-producing *Streptomyces* strains. *Mol. Gen. Genet.* 262:957–964.

Davies, J. 1996. Origins and evolution of antibiotic resistance. *Microbiologia* 12:9–16.

Dawid, W. 2000. Biology and global distribution of myxobacteria in soils. *FEMS Microbiol. Rev.* 24:403–427.

De Boer, C., P. A. Meulman, R. J. Wnuk, and D. H. Peterson. 1970. Geldanamycin, a new antibiotic. *J. Antibiot.* 23:442–447.

De Koning, A. P., F. S. L. Brinkman, S. J. M. Jones, and P. J. Keeling. 2000. Lateral gene transfer and metabolic adaptation in the human parasite *Trichomonas vaginalis*. *Mol. Biol. Evol.* 17:1769–1773.

DeLong, E. F., and A. A. Yayanos. 1986. Biochemical functional and ecological significance of novel bacterial lipids in deep-sea prokaryotes. *Appl. Environ. Microbiol.* 51:730–737.

Disch, A., J. Schwender, C. Muller, H. K. Lichtenthaler, and M. Rohmer. 1998. Distribution of the mevalonate and glyceraldehydes phosphate/pyruvate pathways for isoprenoid biosynthesis in unicellular algae and the cyanobacterium *Synechococcus* PCC 6714. *Biochem. J.* 333:381–388.

Dobzhansky, T. 1967. Looking back at Mendel's discovery. *Science* 156:1588–1589.

Doolittle, W. F. 1999. Phylogenetic classification and the universal tree. *Science* 284:2124–2129.

Doolittle, W. F., and J. M. Logsdon, Jr. 1998. Archaeal genomics: do archaea have a mixed heritage? *Curr. Biol.* 8:R209–R211.

Drews, J. 2000. Drug discovery: a historical perspective. *Science* 287:1960–1964.

Drews, G. 2000. The roots of microbiology and the influence of Ferdinand Cohn on microbiology of the 19th century. *FEMS Microbiol. Rev.* 24:225–249.

Dykhuizen, D. E., and L. Green. 1991. Recombination in *Escherichia coli* and the definition of biological species. *J. Bacteriol.* 173:7257–7268.

Egan, S., P. Wiener, D. Kallifidas, and E. M. H. Wellington. 2001. Phylogeny of *Streptomyces* species and evidence for horizontal transfer of entire and partial antibiotic gene clusters. *Antonie Leeuwenhoek* 79:127–133.

Embley, T. M., and E. Stackebrandt. 1997. Species in practice: exploring uncultured prokaryotic diversity in natural samples, p. 61–81. *In* M. F. Claridge, H. A. Dawah, and M. R. Wilson (ed.), *Species: the Units of Diversity*. Chapman & Hall, London, United Kingdom.

Embley, T. M., and R. Wait. 1994. Structural lipids of eubacteria, p. 121–161. *In* M. Goodfellow and A. G. O'Donnell (ed.), *Chemical Methods in Prokaryotic Systematics*. John Wiley & Sons, Chichester, United Kingdom.

Fang, A. Q., G. K. Wong, and A. L. Demain. 2000. Enhancement of the antifungal activity of rapamycin by the coproduced elaiophylin and nigericin. *J. Antibiot.* 53:158–162.

Feil, E. J., E. C. Holmes, D. E. Bessen, M.-S. Chan, N. P. J. Day, M. C. Enright, R. Goldstein, D. W. Hood, A. Kalia, C. E. Moore, J. Zhou, and B. G. Spratt. 2001. Recombination within natural populations of pathogenic bacteria: short term empirical estimates and long term phylogenetic consequences. *Proc. Natl. Acad. Sci. USA* 98:182–187.

Fitzgerald, J. R., D. E. Sturdevant, S. M. Mackie, S. R. Gill, and J. M. Musser. 2001. Evolutionary genomics of *Staphylococcus au-*

reus: insights into the origin of methicillin-resistant strains and the toxic shock syndrome epidemic. *Proc. Natl. Acad. Sci. USA* 98:8821–8826.

Freney, J., W. E. Kloos, V. Hajek, J. A. Webster, M. Bes, Y. Brun, and C. Vernozy-Rozand. 1999. Recommended minimal standards for description of new staphylococcal species. *Int. J. Systeriol. Bact.* 49:489–502.

Gil, R., B. Sabater-Munoz, A. Latorre, F. J. Silva, and A. Moya. 2002. Extreme genome reduction in *Buchnera* spp.: toward the minimal genome needed for symbiotic life. *Proc. Natl. Acad. Sci. USA* 99:4454–4458.

Gill, I., and R. Valivety. 1997. Polyunsaturated fatty acids. Part I. Occurrence, biological activities and applications. *Trends Biotechnol.* 15:401–409.

Goldfine, H. 1972. Comparative aspects of bacterial lipids. *Adv. Microbiol. Physiol.* 8:1–58.

Goodfellow, M. 1985. Staphylococcal systematics: past, present and future. *Zbl. Bakt. Suppl.* 14:69–81.

Goodfellow, M., and T. Cross. 1983. Classifications, p. 7–164. *In* M. Goodfellow, M. Mordarski, and S. T. Williams (ed.), *The Biology of Actinomycetes*. Academic Press, London, United Kingdom.

Goodfellow, M., and J. G. Magee. 1998. Taxonomy of myxobacteria, p. 1–71. *In* P. R. J. Gangadharan and P. A. Jenkins (ed.), *Myxobacteria I Basic Aspects*. Chapman & Hall, London, United Kingdom.

Goodfellow, M., and D. E. Minnikin (ed.). 1985. *Chemical Methods in Bacterial Systematics*. Academic Press, London, United Kingdom.

Goodfellow, M., and A. G. O'Donnell. 1993. Roots of bacterial systematics, p. 3–54. *In* M. Goodfellow and A. G. O'Donnell (ed.), *Handbook of New Bacterial Systematics*. Academic Press, London, United Kingdom.

Goodfellow, M., and A. G. O'Donnell (ed.). 1994. *Chemical Methods in Prokaryotic Systematics*. Wiley & Sons, Chichester, United Kingdom.

Goodfellow, M., J. Chun, E. Atalan, and J. J. Sanglier. 1994. Curie point pyrolysis mass spectrometry and its application to bacterial systematics, p. 87–104. *In* F. G. Priest, A. Ramos-Cormenzana, and B. J. Tindall (ed.), *Bacterial Diversity and Systematics*. Plenum Press, New York, N.Y.

Goodfellow, M., G. P. Manfio, and J. Chun. 1997. Towards a practical species concept for cultivable bacteria, p. 25–60. *In* M. F. Claridge, H. A. Dawah, and M. R. Wilson (ed.), *Species: the Units of Biodiversity*. Chapman & Hall, London, United Kingdom.

Goodfellow, M., G. Alderson, and J. Chun. 1998. Rhodococcal systematics: problems and developments. *Antonie Leeuwenhoek* 74:3–20.

Goodfellow, M., K. Isik, and E. Yates. 1999. Actinomycetes systematics: an unfinished synthesis. *Nova Acta Leopold.* 312:47–82.

Gordon, R. E. 1968. The taxonomy of soil bacteria, p. 293–321. *In* T. R. G. Gray and D. Parkinson (ed.), *The Ecology of Soil Bacteria*. Liverpool University Press, Liverpool, United Kingdom.

Gordon, R. E. 1978. A species definition. *Int. J. Syst. Bacteriol.* 28:605–607.

Green, R., and H. F. Noller. 1997. Ribosomes and translation. *Annu. Rev. Biochem.* 66:679–716.

Gribaldo, S., and H. Philippe. 2002. Ancient phylogenetic relationships. *Theor. Pop. Biol.* 61:391–408.

Grimont, P. A. D. 1988. Use of DNA re-association in bacterial classification. *Can. J. Microbiol.* 34:541–546.

Guezennec, J., and A. Fiala-Medioni. 1996. Bacterial abundance and diversity in the Barbados Trench determined by phospholipids analysis. *FEMS Microbiol. Ecol.* 19:83–93.

Hansmann, S., and W. Martin. 2000. Phylogeny of 33 ribosomal and six other proteins encoded in an ancient gene cluster that is conserved across prokaryotic genomes: influence of excluding poorly alignable sites from analysis. *Int. J. Syst. Evol. Microbiol.* **50:**1655–1663.

Harvey, A. 2000. Strategies for discovering drugs from previously unexplored natural products. *Drug Discovery Today* **5:**294–300.

Helm, D., H. Labischinski, G. Schallehn, and D. Naumann. 1991. Classification and identification of bacteria by Fourier-transform infrared spectroscopy. *J. Gen. Microbiol.* **137:**69–79.

Henikoff, S., E. A. Greene, S. Pietrokovski, P. Bork, T. K. Attwood, and L. Hood. 1997. Gene families: the taxonomy of protein paralogs and chimeras. *Science* **278:**609–614.

Hennecke, H., K. Kaluza, B. Thöny, M. Fuhrmann, W. Ludwig, and E. Stackebrandt. 1985. Concurrent evolution of nitrogenase genes and 16S rRNA in *Rhizobium* species and other nitrogen-fixing bacteria. *Arch. Mikrobiol.* **142:**342–348.

Hill, D. C., S. K. Wrigley, and L. J. Nisbet. 1998. Novel screen methodologies for identification of new microbial metabolites with pharmacological activity. *Adv. Biochem. Eng. Biotechnol.* **59:**75–121.

Hille, R. 1996. The mononuclear molybdenum enzymes. *Chem. Rev.* **96:**2757–2816.

Hopwood, D. A. 1997. Genetic contributions to understanding polyketide synthases. *Chem. Rev.* **97:**2465–2498.

Huang, J., C. J. Lih, K. H. Pan, and S. N. Cohen. 2001. Global analysis of growth phase responsive gene expression and regulation of antibiotic biosynthetic pathways in *Streptomyces coelicolor* using DNA microarrays. *Genes Dev.* **15:**3183–3192.

Huddleston, A. S., N. Cresswell, M. C. Neves, J. E. Beringer, S. Baumberg, D. I. Thomas, and E. M. H. Wellington. 1997. Molecular detection of streptomycin-producing streptomycetes in Brazilian soils. *Appl. Environ. Microbiol.* **63:**1288–1297.

Hugenholtz, P., C. Pitulle, K. L. Hershberger, and N. R. Pace. 1998. Novel division level bacterial diversity in a Yellowstone hot spring. *J. Bacteriol.* **180:**366–376.

Hurek, T., T. Egener, and B. Reinhold-Hurek. 1997. Divergence in nitrogenases of *Azoarcus* spp., *Proteobacteria* of the β subclass. *J. Bacteriol.* **179:**4172–4178.

International Human Genome Sequencing Consortium. 2001. Initial sequencing and analysis of the human genome. *Nature* **409:**860–921.

Ivanova, V., M. Oriol, M. J. Montes, A. Garcia and J. Guinea. 2001. Secondary metabolites from a *Streptomyces* strain isolated from Livingston Island, Antartica. *Z. Naturforsch* **56:**1–5.

Jain, R., M. C. Rivera, and J. A. Lake. 1999. Horizontal gene transfer among genomes: the complexity hypothesis. *Proc. Natl. Acad. Sci. USA* **96:**3801–3806.

Jensen, H. L. 1931. Contribution to our knowledge of the *Actinomycetales* II. The definition and subdivision of the genus *Actinomyces* with a preliminary account of Australian soil actinomycetes. *Proc. Linnean Soc. NSW* **51:**364–376.

Johns, R. B., and G. J. Perry. 1977. Lipids of the marine bacterium *Flexibacter polymorphus*. *Arch. Microbiol.* **114:**267–271.

Kanazawa, A., S. I. Teshima, and O. Kazuo. 1979. Relationship between essential fatty acid requirements of aquatic animals and the capacity for bioconversion of linolenic acid to highly unsaturated fatty acids. *Comp. Biochem. Physiol.* **63B:**295–298.

Kandler, O. 1985. Evolution of systematics of bacteria, p. 335–361. *In* K. H. Schleifer and E. Stackebrandt (ed.), *Evolution of Prokaryotes*. Academic Press, London, United Kingdom.

Karlin, S., L. Brocchieri, J. Trent, B. E. Blaisdell, and J. Mrázek. 2002. Heterogeneity of genome and proteome content in *Bacteria*, *Archaea*, and Eukaryotes. *Theor. Pop. Biol.* **61:**367–390.

Katz, L. A. 2002. Lateral gene transfers and the evolution of eukaryotes: theories and data. *Int. J. Syst. Evol. Microbiol.* **52:**1893–1900.

Keddie, R. M., and I. J. Bousfield. 1980. Cell wall composition in the classification and identification of coryneform bacteria, p. 167–188. *In* M. Goodfellow and R. G. Board (ed.), *Microbial Classification and Identification*. Williams & Wilkins, Baltimore, Md.

Keddie, R. M., and G. I. Cure. 1978. Cell wall composition of coryneform bacteria, p. 47–83. *In* I. J. Bousfield and A. G. Callely (ed.), *Coryneform Bacteria*. Academic Press, London, United Kingdom.

Khodobakshian, M. 2002. Biodiversity and natural product synthesis in members of the Streptomyces violaceusniger. M.Sc. thesis. University of Newcastle upon Tyne, Newcastle upon Tyne, United Kingdom.

Kim, E. S., H. J. Hong, C. Y. Choi, and S. N. Cohen. 2001. Modulation of actinorhodin biosynthesis in *Streptomyces lividans* by glucose repression of *afsR2* gene transcription. *J. Bacteriol.* **183:**2198–2203.

Klein, M., M. Friedrich, A. J. Roger, P. Hugenholtz, S. Fishbain, H. Abicht, L. L. Blackall, D. A. Stahl, and M. Wagner. 2001. Multiple lateral transfers of dissimilatory sulfite reductase genes between major lineages of sulfate-reducing prokaryotes. *J. Bacteriol.* **183:**6028–6035.

Klotz, M. G., and J. M. Norton. 1998. Multiple copies of ammonia mono-oxygenase (*amo*) operons have evolved under biased AT/GC mutational pressure in ammonia-oxidising autotrophic bacteria. *FEMS Microbiol. Lett.* **168:**303–311.

Koch, R. 1877. Die Ätiologie der Milzbrandkrankheit, begrüdet auf die Entwicklungsgeschüchte des *Bacillus anthracts*. *Beitr. Biol. Pflanz.* **2:**277–310.

Krieg, N. R., and J. G. Holt (ed.). 1984. *Bergey's Manual of Systematic Bacteriology*, vol. I. Williams & Wilkins, Baltimore, Md.

Krsek, M., and E. M. H. Wellington. 1999. Comparison of different methods for the isolation and purification of total community DNA from soil. *J. Microbiol. Methods* **39:**1–16.

Kunst, F., N. Ogasawara, I. Moszer, A. M. Albertini, G. Alloni, V. Azevedo, M. G. Bertero, P. Bessières, A. Bolotin, et al. 1997. The complete genome sequence of the Gram-positive bacterium *Bacillus subtilis*. *Nature* **390:**249–256.

Labeda, D. P., and A. J. Lyons. 1991. The *Streptomyces violaceusniger* cluster is heterogeneous in DNA relatedness among strains: emendation of the description of *Streptomyces violaceusniger* and *Streptomyces hygroscopicus*. *Int. J. Syst. Bacteriol.* **41:**398–401.

Lan, R., and P. R. Reeves. 2000. Intraspecies variation in bacterial genomes: the need for a species concept. *Trends Microbiol.* **8:**396–401.

Lan, R., and P. R. Reeves. 2001. When does a clone deserve a name? A perspective on bacterial species based on population genetics. *Trends Microbiol.* **9:**419–424.

Lange, B. M., T. Rujan, W. Martin, and R. Croteau. 2000. Isoprenoid biosynthesis: the evolution of two ancient and distinct pathways across genomes. *Proc. Natl. Acad. Sci. USA* **97:**13172–13177.

Larsen, O., T. Lien, and N. K. Birkeland. 1999. Dissimilatory sulfite reductase from *Archaeoglobus profundus* and *Desulfotomaculum thermocisternum*: phylogenetic and structural implications from gene sequences. *Extremophiles* **3:**63–70.

Lawrence, J. G. 1999. Selfish operons: the evolutionary impact of gene clustering in prokaryotes and eukaryotes. *Curr. Opin. Genet. Dev.* **9:**642–648.

Lawrence, J. G. 2002. Gene transfer in bacteria: speciation without species? *Theor. Pop. Biol.* **61:**449–460.

Lawrence, J. G., and H. Ochman. 1997. Amelioration of bacterial genomes: rates of change and exchange. *J. Mol. Evol.* **44:**383–397.

Lawrence, J. G., and H. Ochman. 1998. Molecular archaeology of the *Escherichia coli* genome. *Proc. Natl. Acad. Sci. USA* **95:** 9413–9417.

Lawrence, J. G., and J. R. Roth. 1999. Genomic flux: genome evolution by gene loss and acquisition, p. 263–289. *In* R. L. Charlebois (ed.), *Organization of the Prokaryotic Genome.* ASM Press, Washington, D.C.

Lazzarini, A., G. Cavaletti, G. Toppo, and F. Marinelli. 2001. Rare genera of actinomycetes as potential producers of new antibiotics. *Antonie Leeuwenhoek* **79:**399–405.

Lechevalier, M. P. 1976. The taxonomy of the genus *Nocardia:* some light at the end of the tunnel? p. 1–38. *In* M. Goodfellow, G. H. Brownell, and J. A. Serrano (ed.), *The Biology of the Nocardiae.* Academic Press, London, United Kingdom.

Lechevalier, M. P., and H. A. Lechevalier. 1970. Chemical composition as a criterion in the classification of aerobic actinomycetes. *Int. J. Syst. Bacteriol.* **20:**435–443.

Lechevalier, H. A., and M. P. Lechevalier. 1988. Chemotaxonomic use of lipids—an overview, p. 869–902. *In* C. Ratledge, and S. G. Wilkinson (ed.), *Microbial Lipids.* Academic Press, London, United Kingdom.

Lechevalier, M. P., C. DeBière, and H. Lechevalier. 1977. Chemotaxonomy of aerobic actinomycetes: phospholipid composition. *Biochem. Syst. Ecol.* **5:**249–260.

Li, S. M., S. Hennig, and L. Heide. 1998. Biosynthesis of the dimethylallyl moiety of novobiocin *via* the non-mevalonate pathway. *Tetrahedron Lett.* **39:**2717–2720.

Ludwick, W., and K. H. Schleifer. 1994. Bacterial phylogeny based on 16S and 23S rRNA sequence analysis. *FEMS Microbiol. Rev.* **15:**155–173.

Ludwick, W., and H.-P. Klenk. 2001. Overview: a phylogenetic backbone and taxonomic framework for prokaryotic systematics, p. 49–65. *In* D. R. Boone, R. W. Castenholtz, and G. M. Garrity (ed.), *Bergey's Manual of Systematic Bacteriology,* 2nd ed. Springer, New York, N.Y.

Maeder, D. L., R. B. Weiss, D. M. Dunn, J. L. Cherry, J. M. Gonzalez, J. DiRuggiero, and F. T. Robb. 1999. Divergence of the hyperthermophilic archaea *Pyrococcus furiosus* and *Pyrococcus horikoshi* inferred from complete genome sequences. *Genetics* **152:**1299–1305.

Maidak, B. L., J. R. Cole, T. G. Lilburn, C. T. Parker, P. R. Saxman, R. J. Farris, G. M. Garrity, G. J. Olsen, T. M. Schmidt, and J. M. Tiedje. 2001. The RDP-II (Ribosomal Database Project). *Nucleic Acids Res.* **29:**173–174.

Maiden, M. C. J., J. A. Bygraves, E. Feil, G. Morelli, J. E. Russell, R. Unwin, Q. Zhang, J. Zhou, K. Zurth, D. A. Caugant, I. M. Feavers, M. Achtman, and B. G. Spratt. 1998. Multilocus sequence typing: a portable approach to the identification of clones within populations of pathogenic microorganisms. *Proc. Natl. Acad. Sci. USA* **95:**3140–3145.

Mazel, D., and J. Davies. 1999. Antibiotic resistance in microbes. *Cell Mol. Life Sci.* **56:**742–754.

McClelland, M., L. Florea, K. Sanderson, S. W. Clifton, J. Parkhill, C. Churcher, G. Dougan, R. K. Wilson, and W. Miller. 2000. Comparison of the *Escherichia coli* K-12 genome with sampled genomes of a *Klebsiella pneumoniae* and three *Salmonella enterica* serovars Typhimurium, Typhi and Paratyphi. *Nucleic Acids Res.* **28:**4974–4986.

Metsä-Ketelä, M., L. Halo, E. Munukka, J. Hakala, P. Mäntsälä, and K. Ylihonko. 2002. Molecular evolution of aromatic polyketides and comparative sequence analysis of polyketide ketosynthase and 16S ribosomal DNA genes from various *Streptomyces* species. *Appl. Environ. Microbiol.* **68:**4472–4479.

Metz, J. G., P. Roessler, D. Facciotti, C. Levering, F. Dittrich, M. Lassner, R. Valentine, K. Lardizabal, F. Domergue, A. Yamada, K. Yazawa, V. Knauf, and J. Browse. 2001. Production of polyunsaturated fatty acids by polyketide synthases in both prokaryotes and eukaryotes. *Science* **293:**290–293.

Migula, W. 1897. *System der Bakterien.* Gustav Fischer Verlag, Jena, Germany.

Mincer, T. J., P. R. Jensen, C. A. Kaufman and W. Fenical. 2002. Widespread and persistent populations of a major new marine actinomycete taxon in ocean sediments *Appl. Environ. Microbiol.* **68:**5005–5011.

Minnikin, D. E., and M. Goodfellow. 1980. Lipid composition in the classification and identification of acid fast bacteria, p. 189–256. *In* M. Goodfellow and R. G. Board (ed.), *Microbiological Classification and Identification.* Academic Press, London, United Kingdom.

Minnikin, D. E., and A. G. O'Donnell. 1984. Actinomycete envelope lipid and peptidoglycan composition, p. 337–388. *In* M. Goodfellow, M. Mordarski, and S. T. Williams (ed.), *The Biology of the Actinomycetes.* Academic Press, London, United Kingdom.

Minnikin, D. E., M. Goodfellow, and M. D. Collins. 1978. Lipid composition in the classification and identification of coryneform and related bacteria, p. 85–160. *In* I. J. Bousfield and A. G. Callely (ed.), *Coryneform Bacteria.* Academic Press, London, United Kingdom.

Molnar, I., J. F. Aparicio, S. F. Haydock, L. E. Khaw, T. Schwecke, A. Konig, J. Staunton, and P. F. Leadlay. 1996. Organisation of the biosynthetic gene cluster for rapamycin in *Streptomyces hygroscopicus:* analysis of genes flanking the polyketide synthase. *Gene* **169:**1–7.

Monaghan, R. L., J. D. Polshook, V. J. Pecore, G. F. Bills, M. Nallin-Omstead, and S. L. Streicher. 1995. Discovery of novel secondary metabolites from fungi—is it really a random walk through a random forest? *Can. J. Bot.* **73:**S925–S931.

Monciardini, P., M. Sosio, L. Cavaletti, C. Chiocchini, and S. Donadio. 2002. New PCR primers for the selective amplification of 16S rDNA from different groups of actinomycetes. *FEMS Microbiol. Ecol.* **42:**419–429.

Moreno-Vivian, C., P. Cabello, M. Martinez-Luque, R. Blasco, and F. Castillo. 1999. Prokaryotic nitrate reduction: molecular properties and functional distinction among bacteriol nitrate reductases. *J. Bacteriol.* **181:**6573–6584.

Mougel, C., J. Thioulouse, G. Perrière, and X. Nesme. 2002. A mathematical method for determining genome divergence and species delineation using AFLP. *Int. J. Syst. Evol. Microbiol.* **52:**573–586.

Murray, R. G. E. 1962. Time structure and taxonomy of bacteria, p. 119–144. *In* G. C. Ainsworth and P. H. A. Sneath (ed.), *Microbial Classification.* Cambridge University Press, Cambridge, United Kingdom.

Mylvaganam, S., and P. P. Dennis. 1992. Sequence heterogeneity between the two genes encoding 16S rRNA from the halophilic archaebacterium *Haloarcula marismortui. Genetics* **130:**399–410.

Nakagawa, Y., and K. Yamasato. 1993. Phylogenetic diversity of the genus *Cytophaga* revealed by 16S rRNA sequencing and menaquinone analysis. *J. Gen. Microbiol.* **139:**1155–1161.

Napier, J. A. 2002. Plumbing the depths of PUFA biosynthesis: a novel polyketide synthase-like pathway from marine organisms. *Trends Plant Sci.* **7:**51–54.

Napier, J. A., L. V. Michaelson, and A. K. Stobart. 1999a. Plant desaturases: harvesting the fat of the land. *Curr. Opin. Plant Biol.* **2:**123–127.

Napier, J. A., O. Sayanova, P. Sperling, and E. Heinz. 1999b. A growing family of cytochrome b5-doman fusion proteins. *Trends Plant Sci.* **4:**2–4.

Nelson, K. E., R. A. Clayton, S. R. Gill, M. L. Gwinn, R. J. Dodson, D. H. Haft, E. K. Hickey, J. D. Peterson, W. C. Nelson, K. A. Ketchum, L. McDonald, T. R. Utterback, J. A. Malek, K. D. Linher, M. M. Garrett, A. M. Stewart, M. D. Cotton, M. S. Pratt, C. A. Phillips, D. Richardson, J. Heidelberg, G. G. Sutton, R. D. Fleischmann, J. A. Eisen, and C. M. Fraser. 1999. Evidence for lateral gene transfer between *Archaea* and bacteria from genome sequence of *Thermotoga maritima*. *Nature* **399**:323–329.

Nichols, D. S., and T. A. McMeekin. 2002. Biomarker techniques to screen for bacteria that produce polyunsaturated fatty acids. *J. Microbiol. Methods* **48**:161–170.

Nicholson, T. P., B. A. M. Rudd, M. Dawson, C. M. Lazarus, T. J. Simpson, and R. J. Cox. 2001. Design and utility of oligonucleotide gene probes for fungal polyketide synthases. *Chem. Biol.* **8**:157–178.

Ninet, B., M. Monod, S. Emler, J. Pawlowski, C. Metral, P. Rohner, R. Auckenthaler, and B. Herschel. 1996. Two different 16S rRNA genes in a mycobacterial strain. *J. Clin. Microbiol.* **34**:2531–2536.

Normand, P., and J. Bousquet. 1989. Phylogeny of nitrogenase sequences in *Frankia* and other nitrogen-fixing organisms. *J. Mol. Biol.* **29**:1599–1602.

Norton, J. M., J. J. Alzerreca, Y. Suwa, and M. G. Klotz. 2002. Diversity of ammonia mono-oxygenase operon in autotrophic ammonia-oxidising bacteria. *Arch. Microbiol.* **177**:139–149.

Oberreuter, H., H. Seiler, and S. Scherer. 2002. Identification of coryneform bacteria and related taxa by Fourier-transform infrared (FT-IR) spectroscopy. *Int. J. Syst. Evol. Microbiol.* **52**:91–100.

Ōmura, S., H. Ikeda, J. Ishikawa, A. Hanamotto, C. Takahashi, M. Shinosa, Y. Takahashi, H. Horikawa, H. Nakazawa, T. Osonoe, H. Kikuchi, T. Shiba, Y. Sakaki, and M. Hattori. 2001. Genome sequence of an industrial microorganism *Streptomyces avermitilis*: deducing the ability of producing secondary metabolites. *Proc. Natl. Acad. Sci. USA* **98**:12215–12220.

Orihara, N., T. Kuzuyama, S. Takahashi, K. Furihata, and H. Seto. 1998. Studies on the biosynthesis of terpenoid compounds produced by actinomycetes. 3. Biosynthesis of the isoprenoid side chain of novobiocin *via* the non-mevalonate pathway in *Streptomyces niveus*. *J. Antibiot.* (Tokyo) **51**:676–678.

Pennisi, E. 1998. Genome data shake the tree of life. *Science* **280**:672–674.

Petri, R., and J. F. Imhoff. 2000. The relationship of nitrate reducing bacteria on the basis of narH gene sequences and comparison of narH and 16S rDNA based phylogeny. *Syst. Appl. Microbiol.* **23**:47–57.

Ponting, C. P. 2001. Plagiarized bacterial genes in the human book of life. *Trends Genet.* **17**:235–237.

Postgate, J. R. 1982. *The Fundamentals of Nitrogen Fixation*. Cambridge University Press, Cambridge, United Kingdom.

Priest, F. G., and M. Goodfellow (ed.) 2000. *Applied Microbial Systematics*. Kluwer Academic Publishers, Dordrecht, The Netherlands.

Ravel, J., M. J. Amoroso, R. R. Colwell, and R. T. Hill. 1998. Mercury-resistant actinomycetes from Chesapeake Bay. *FEMS Microbiol. Lett.* **162**:177–184.

Read, T. D., R. C. Brunham, C. Shen, S. R. Gill, J. F. Heidelberg, O. White, E. K. Hickey, J. Peterson, T. Utterback, K. Berry, S. Bass, K. Linher, J. Weidman, H. Khouri, B. Craven, C. Bowman, R. Dodson, M. Gwinn, W. Nelson, R. DeBoy, J. Kolonay, G. McClarty, S. L. Salzberg, J. Eisen, and C. M. Fraser. 2000. Genome sequences of *Chlamydia trachomatis* MoPn and *Chlamydia pneumoniae* AR39. *Nucleic Acids Res.* **28**:1397–1406.

Reeves, P. R. 1995. Role of O-antigen variation in the immune response. *Trends Microbiol.* **3**:381–386.

Reichenbach, H. 2001. Myxobacteria, producers of novel bioactive substances. *J. Ind. Microbiol. Biotech.* **27**:149–156.

Robson, R. L., P. R. Woodey, R. N. Pau, and R. R. Eady. 1989. Structural genes for the vanadium nitrogenase from *Azotobacter chroococcum*. *EMBO J.* **8**:1217–1224.

Rondon, M. R., P. R. August, A. D. Betterman, S. F. Brady, T. H. Grossman, M. R. Liles, K. A. Loiacono, B. A. Lynch, I. A. McNeil, C. Minor, C. L. Tiong, M. Gilman, M. S. Osburne, J. Clardy, J. Handelsman, and R. M. Goodman. 2000. Cloning the soil metagenome: a strategy for accessing the genetic and functional diversity of uncultured microorganisms. *Appl. Environ. Microbiol.* **66**:2541–2547.

Rosselló-Mora, R., and R. Amann. 2001. The species concept for prokaryotes. *FEMS Microbiol. Rev.* **25**:39–67.

Roszak, D. B., and R. R. Colwell. 1987. Survival strategies of bacteria in the natural environment. *Microbiol. Rev.* **51**:365–379.

Rotthauwe, J.-H., K.-P. Witzel, and W. Liesack. 1997. The ammonia mono-oxygenase structural gene amoA as a functional marker: molecular fine-scale analysis of natural ammonia-oxidising populations. *Appl. Environ. Microbiol.* **63**:4704–4712.

Sait, M., P. Hugenholtz, and P. H. Janssen. 2002. Cultivation of globally distributed soil bacteria from phylogenetic lineages previously only detected in cultivation-independent surveys. *Environ. Microbiol.* **4**:654–666.

Salanoubat, M., S. Genin, F. Artiguenave, J. Gouzy, S. Mangenot, M. Ariat, A. Billault, P. Brottier, J. C. Camus, L. Cattolico, M. Chandler, N. Choisne, C. Claudel-Renard, S. Cunnae, N. Demange, C. Gaspin, M. Lavie, A. Molsan, C. Robert, W. Saurin, T. Schlex, P. Siguier, P. Thébault, M. Whalen, P. Wincker, M. Levy, J. Weissenbach, and C. A. Boucher. 2002. Genome sequence of the plant pathogen *Ralstonia solanacearum*. *Nature* **415**:497–502.

Salzberg, S. L., O. White, J. Peterson, and J. A. Eisen. 2001. Microbial genes in the human genome: lateral transfer or gene loss. *Science* **292**:1903–1906.

Schleifer, K. H. 1998. Phylogeny of prokaryotes beyond 16S rRNA sequencing, p. 30. *In Abstracts of the 98th General Meeting*. American Society for Microbiology, Washington, D.C.

Schleifer, K. H., and O. Kandler. 1972. Peptidoglycan types of bacterial cell walls and their taxonomic implications. *Bacteriol. Rev.* **36**:407–477.

Sembiring, L. 2000. The selective isolation and characterization of the *Streptomyces violaceusniger* clade associated with the roots of *paraserianthes falcataria* (L) Nielson. Ph.D. thesis. University of Newcastle upon Tyne, Newcastle upon Tyne, United Kingdom.

Sembiring, L., M. Goodfellow, and A. C. Ward. 2000. The selective isolation and characterization of the *Streptomyces violaceusniger* clade associated with the roots of *Paraserianthes falcataria*. *Antonie Leeuwenhoek* **78**:353–366.

Sembiring, L., M. Goodfellow, and A. C. Ward. The *Streptomyces violaceusniger* clade: a home for streptomycetes with rugose ornamented spores. *Int. J. Syst. Evol. Microbiol.*, in press.

Seto, H., H. Watanabe, and K. Funihata. 1996. Simultaneous operation of the mevalonate and non-mevalonate pathways in the biosynthesis of isopentenyl diphosphate in *Streptomyces aerouvifer*. *Tetrahedron Lett.* **37**:7979–7982.

Sneath, P. H. A. 1988. The phenetic and cladistic approaches, p. 252–273. *In* D. L. Hawksworth (ed.), *Propsects in Systematics*. Clarendon Press, Oxford, United Kingdom.

Sneath, P. H. A. 1992. *International Code of Nomenclature of Bacteria*. American Society for Microbiology, Washington, D.C.

Sneath, P. H. A., and R. R. Sokal. 1973. *Numerical Taxonomy*. W. H. Freeman & Company, San Francisco, Calif.

Society for Actinomycetes Japan. 2000. *Identification Manual for Actinomycetes*. Japan Basineis Center for Academic Societies, Tokyo, Japan.

Sosio, M., E. Bossi, A. Bianchi, and S. Donadio. 2000. Multiple peptide synthetase gene clusters in Actinomycetes. *Mol. Gen. Genet.* **264:**213–221.

Sosio, M., E. Bossi, and S. Donadio. 2001. Assembly of large genomic segments in artificial chromosomes by homologous recombination in *Escherichia coli. Nucleic Acids Res.* **29:**1–8.

Stackebrandt, E., and B. M. Goebel. 1994. Taxonomic note: a place for DNA:DNA reassociation and 16S rRNA sequence analysis in the present species definition in bacteriology. *Int. J. Syst. Bacteriol.* **44:**846–849.

Stackebrandt, E., and M. Goodfellow. 1991. *Nucleic Acid Techniques in Bacterial Systematics.* John Wiley & Sons, Chichester, United Kingdom.

Stackebrandt, E., and J. Swiderski. 2002. From phylogeny to systematics: the dissection of the genus *Bacillus*, p. 8–22. *In* R. Berkeley, M. Heyndrickx, N. Logan, and P. de Vos (ed.), *Applications and Systematics of Bacillus and Relatives.* Blackwell, London, United Kingdom.

Stackebrandt, E., W. Frederiksen, G. M. Garrity, P. A. D. Grimont, P. Kämpfer, M. C. J. Maiden, X. Nesme, R. Rosselló-Mora, J. Swings, H. G. Trüper, L. Vauterin, A. C. Ward, and W. B. Whitman. 2002. Report of the *ad hoc* committee for the re-evaluation of the species definition in bacteriology. *Int. J. Syst. Evol. Microbiol.* **52:**1043–1047.

Stahl, D. A., D. J. Lane, G. T. Olsen, and N. R. Pace. 1985. Characterization of a Yellowstone hot spring microbial community by 5S rRNA sequences. *Appl. Environ. Microbiol.* **45:**1379–1384.

Staley, J. T., and J. J. Gosink. 1999. Poles apart: biodiversity and biogeography of sea ice bacteria. *Annu. Rev. Microbiol.* **53:**189–215.

Stanier, R. Y., and C. B. van Niel. 1962. The concept of a bacterium. *Arch. Mikrobiol.* **42:**17–35.

Stoiz, J. F., and P. Basu. 2002. Evolution of nitrate reductase: molecular and structural variations on a common function. *ChemBioChem* **3:**198–206.

Stover, C. K., X. Q. Pham, A. L. Erwin, S. D. Mizoguchi, P. Warrener, M. J. Hickey, F. S. L. Brinkman, W. O. Hufnagle, D. J. Kowalik, M. Lagrou, R. L. Garber, L. Goltry, E. Tolentino, S. Westbrock-Wadman, Y. Yuan, L. L. Brody, S. N. Coulter, K. R. Folger, A. Kas, K. Larbig, R. Lim, K. Smith, D. Spencer, G. K.-S. Wong, Z. Wu, I. T. Paulsen, J. Reizer, M. H. Saier, R. E. W. Hancock, S. Lory, and M. V. Olson. 2000. Complete genome sequence of *Pseudomonas aeruginosa* PAO1, an opportunistic pathogen. *Nature* **406:**959–964.

Strohl, W. R., M. L. Dickens, V. B. Rajgarhia, A. J. Woo, and N. D. Priestley. 1997. Anthracyclines, p. 577–657. *In* W. R. Strohl (ed.), *Biotechnology of Antibiotics*, 2nd ed. Marcel Dekker Publishers, New York, N.Y.

Suzuki, K., M. Goodfellow, and A. G. O'Donnell. 1983. Cell envelopes and classification, p. 195–250. *In* M. Goodfellow and A. G. O'Donnell (ed.), *Handbook of New Bacterial Systematics.* Academic Press, London, United Kingdom.

Takami, H., K. Nakasone, Y. Takaki, G. Maeno, R. Sasaki, N. Masui, F. Fuji, C. Hirama, Y. Nakamura, N. Ogasawara, S. Kuhara, and K. Horikoshi. 2000. Complete genome sequence of the alkaliphilic bacterium *Bacillus halodurans* and genomic sequence comparisons with *Bacillus subtilis. Nucleic Acids Res.* **28:**4317–4331.

Teske, A., E. Alm, J. M. Regan, S. Toze, B. E. Ritmann, and D. A. Stahl. 1994. Evolutionary relationships among ammonia- and nitrite-oxidizing bacteria. *J. Bacteriol.* **176:**6623–6630.

Thampapillai, G., R. T. Lan, and P. R. Reeves. 1994. Molecular evolution in the *gnd* locus of *Salmonella enterica. Mol. Biol. Evol.* **11:**813–828.

Tischner, R. 2000. Nitrate uptake and reduction in higher and lower plants. *Plant Cell Environ.* **23:**1005–1024.

Tolba, S., S. Egan, D. Kallifidas, and E. M. H. Wellington. 2002. Distribution of streptomycin resistance and biosynthesis genes in streptomycetes recovered from different soil sites. *FEMS Microbiol. Ecol.* **42:**269–276.

Trejo-Estrada, S. R., A. Paszczynski, and D. L. Crawford. 1998. Antibiotics and enzymes produced by the biocontrol agent *Streptomyces violaceusniger* YCED-9. *J. Ind. Microbiol. Biotech.* **21:**81–90.

Tsoi, C. J., and C. Khosla. 1995. Combinatorial biosynthesis of unnatural natural products—the polyketide example. *Chem. Biol.* **2:**355–362.

Uchida, K., and K. Aida. 1979. Taxonomic significance of cell wall acyl type in *Corynebacterium-Mycobacterium-Nocardia* group by glycolate test. *J. Gen. Appl. Microbiol.* **25:**169–183.

Ueda, K., T. Seki, T. Kudo, T. Yoshida, and M. Kataoka. 1999. Two distinct mechanisms cause heterogeneity of 16S rRNA. *J. Bacteriol.* **181:**78–82.

Vaidyanathan, S., J. J. Rowland, D. B. Kell, and R. Goodacre. 2001. Discrimination of aerobic endospore-forming bacteria *via* electrospray mass spectrometry of whole cell suspensions. *Anal. Chem.* **73:**4134–4144.

Vandamme, P., B. Pot, M. Gillis, P. de Vos, K. Kersters, and J. Swings. 1996. Polyphasic taxonomy: a consensus approach to bacterial systematics. *Microbiol. Rev.* **60:**407–443.

van Wezel, G. P., E. Vijgenboom, and L. Bosch. 1991. A comparative study of the ribosomal RNA operons of *Streptomyces coelicolor* A3(2) and sequence analysis of *rrnA. Nucleic Acids Res.* **19:**4399–4403.

Vezina, C., A. Kudelski, and S. Sehgal. 1975. Rapamycin (AY-22,989), a new antifungal antibiotic. I. Taxonomy of the producing streptomycete and isolation of the active principle. *J. Antibiot.* **28:**721–726.

Vincent Lévy-Frébault, V., and F. Portaels. 1992. Proposed minimal standards for the genus *Mycobacterium* and for the description of new slowly growing *Mycobacterium* species. *Int. J. Syst. Bacteriol.* **42:**315–323.

Wagner, M., A. J. Roger, J. L. Flax, G. A. Brusseau, and D. A. Stahl. 1998. Phylogeny of dissimilatory sulfite reductases supports an early origin of sulfate respiration. *J. Bacteriol.* **180:**2975–2982.

Waksman, S. A. 1961. *The Actinomycetes*, vol. II: *Classification, Identification and Descriptions of Genera and Species.* Bailliére, Tindall and Cox Ltd., London, United Kingdom.

Waksman, S. A., and R. E. Curtis. 1916. The *Actinomyces* of the soil. *Soil Sci.* **1:**99–134.

Walsh, C. T. 2002. Combinatorial biosynthesis of antibiotics: challenges and opportunities. *ChemBioChem* **3:**124–134.

Wang, Y., and Z. Zhang. 2000. Comparative sequence analyses reveal frequent occurrence of short segments containing an abnormally high number of non-random base variations in bacterial rRNA genes. *Microbiology* (UK) **146:**2845–2854.

Wang, Y., Z. S. Zhang, and N. Ramanan. 1997. The actinomycete *Thermobispora bispora* contains two distinct types of transcriptionally active 16S rRNA genes. *J. Bacteriol.* **179:**3270–3276.

Ward, A. C., E. Rashidian, L. Sembiring, E. Lacey, S. W. Robinson, J. H. Gill, J. H. Tennont, B. A. Boyleson, and M. Goodfellow. 2001. Mapping metabolites to streptomycete taxonomy. *In 12th International Symposium Biological Actinomycetes*, Vancouver, Canada.

Ward, D. M. 1998. A natural species concept for prokaryotes. *Curr. Opin. Microbiol.* **1:**271–277.

Watanabe, K., K. Yazawa, K. Kondo, and A. Kawaguchi. 1997. Fatty acid synthesis of an eicosapentaenoic acid-producing bacterium: *de novo* synthesis, chain elongation and desaturation systems. *J. Biochem.* **122:**467–473.

Watve, M. G., R. Tickoo, M. M. Jog, and B. D. Bhole. 2001. How many antibiotics are produced by the genus *Streptomyces? Arch. Microbiol.* **176:**386–390.

Wayne L. G., D. J. Brenner, R. R. Colwell, P. A. D. Grimont, O. Kandler, M. L. Krichevsky, L. H. Moore, W. E. C. Moore, R. G. E. Murray, E. Stackebrandt, M. P. Starr, and H. G. Trüper. 1987. Report of the ad hoc committee on reconciliation of approaches to bacterial systematics. *Int. J. Syst. Bacteriol.* **37:**463–464.

Williams, S. T., M. E. Sharpe, and J. G. Holt (ed.) 1989. *Bergey's Manual of Systematic Bacteriology*, vol. 4. Williams & Wilkins, Baltimore, Md.

Whitman, W. B., D. C. Coleman, and W. T. Wiese. 1998. Prokaryotes: the unseen majority. *Proc. Natl. Acad. Sci. USA* **95:**6578–6583.

Wolf, Y. I., I. B. Rogozin, N. V. Grishin, R. L. Tatusov, and E. V. Koonin. 2001. Genome trees constructed using five different approaches suggest new major bacterial clades. *BMC Evol. Biol.* **1:**8.

Woese, C. R. 1987. Bacterial evolution. *Microbiol. Rev.* **51:**221–271.

Woese, C. R. 2002. Perspective: microbiology in transition, p. xvi–xxxii. *In* J. T. Staley and A.-L. Reysenbach (ed.), *Biodiversity of Microbial Life: Foundation of Earth's Biosphere.* John Wiley & Sons, Inc., New York, N. Y.

Yap, W. H., Z. S. Zhang, and Y. Wang. 1999. Distinct types of rRNA operons exist in the genome of the actinomycete *Thermomonospora chromogena* and evidence for horizontal transfer of an entire rRNA operon. *J. Bacteriol.* **181:**5201–5209.

Yates, M. G. 1992. The enzymology of molybdenum-dependent nitrogen fixation, p. 685–735. *In* R. H. Burris and H. J. Evans (ed.), *Biological Nitrogen Fixation.* Chapman & Hall, New York, N. Y.

Yazawa, K. 1996. Production of eicosapentaenoic acid from marine bacteria. *Lipids* **31:**S297–S300.

Yoon, J.-H., Y. G. Cho, S. S. Kang, S. B. Kim, S. T. Lee, and Y. H. Park. 2000. *Rhodococcus koreensis* sp. nov., a 2,4-dinitrophenol-degrading bacterium. *Int. J. Syst. Evol. Microbiol.* **50:**1193–1201.

Young, J. P. W. 1992. Phylogenetic classification of nitrogen-fixing organisms, p. 43–86. *In* G. Stacey, R. H. Burris, and H. J. Evans (ed.), *Biological Nitrogen Fixation.* Chapman & Hall, New York, N. Y.

Zahn, J. A., R. E. Higgs, and M. D. Hilton. 2001. Use of direct-infusion electro-spray mass spectrometry to guide empirical development of improved conditions for expression of secondary metabolites from actinomycetes. *Appl. Environ. Microbiol.* **67:**377–386.

Zuckerlandl, E., and L. Pauling. 1965. Molecules as documents of evolutionary history. *J. Theor. Biol.* **8:**357–366.

VI. PROSPECTING: THE TARGETS

PREAMBLE

A number of industrial sectors where biotechnology has made an impact are considered in this section, and these include examples from the manufacturing and processing, service, and agricultural industries (see Table 1, chapter 29). It is worth reiterating that biotechnology-based industries and industries benefiting from biotechnology inputs are becoming progressively diverse, and, although pharmaceuticals exert a major influence over much of biotechnology, it is not an all-inclusive one. In an interesting analysis of technological innovation in the chemicals industry, Walsh and Lodorfos (2002) opine that increasing costs, globalization of production, markets, and research and development, for example, have effected a significant diversification into high-value-added areas including agrochemicals and specialty chemicals, drugs, advanced materials, and catalysts (including biocatalysts). Much of this diversification was concomitant with or catalyzed by the appearance of biotechnology and an increased focus of research and development on life sciences.

In a book of this type it is not possible to make a comprehensive survey of the biotechnology industries; instead I have selected cases of established and emerging biotechnology industries that illustrate technological and market trends. Notable omissions from these discussions are the food and beverage, paper and pulp, textiles, and polymer sectors, some of which already have significant biotechnology-related market shares. Brief reference to some of these latter sectors is made in chapter 1, and further information is readily available in the review literature.

Regarding the emerging commercial biotechnology, plant growth-promoting and antifouling agents are included as exemplars of industrial developments whose success is dependent on careful dissection of biological interactions and a preparedness to pursue exploitable biological behavior. Other emerging biotechnology can be found in such diverse activities as nutriceuticals, mineral processing, and biomaterials. Nutriceuticals are a new range of food ingredients that have beneficial effects on human health; they include low-calorie sugars, probiotics (nondigestible oligo- and polysaccharides that selectively stimulate intestinal bacteria), and B-vitamins produced by food-grade bacteria (Hugenholtz and Smid, 2002). Industrial-scale mineral processing in bioreactors was established in the early 1990s, and today several plants worldwide are operating at a rate of 100 to 1,000 metric tons per day to liberate gold from refractory gold concentrates; similarly, nickel and copper processes have been developed to pilot-plant scale (Norris et al., 2000). In this field, current attention is focused on thermoacidophilic archaea and their promise for high-temperature release of copper from chalcopyrite. Current understanding of the microbiology, however, is one of the chief bottlenecks in process optimization. Mineral dissolution is usually achieved with uncharacterized mixed populations. Cultured thermoacidophiles have been obtained from some of these communities and appear to be species of *Metallosphera*. In other cases, phylotypes such as those recovered from hot springs in Monserrat cluster most closely with *Acidianus* but with low bootstrap values. The characterization and cultivation of this group of extremophiles is a prime requirement for the further understanding and optimization of mineral extraction. Finally, under the head of emerging technology, it is interesting to consider developments in the field of biomaterials and biomimetics. Some products here, such as biopolymers, are well established; others, including biomimetics and the exciting new developments coming from biosilicification, still require considerable development but offer very promising routes to novel, high-performance materials (see chapter 29 for a brief review).

Current and projected market sizes and sales are referred to briefly in individual chapters in this section, and pointers are also provided to potential markets for new products. A good example here is biofouling, which, as Carola Holmström and colleagues (chapter 36) remark, is a major problem for marine and aquatic industries. The annual cost to marine shipping alone is estimated to be about $5 billion. Hence, the incentive to develop new antifoulants from marine microorganisms is a powerful one both in economic terms and also as environmentally clean replacements for the present nonselective, heavy metal-based formulations. The contributions that biotechnology is able to make to the promotion of clean, sustainable industrial products and processes are discussed in chapter 1, where it was also emphasized that economic considerations continue to be the main driving force for innovation. Further, in order to generate commercial innovation it is increasingly necessary for companies to have, or to have access to, integrated technology platforms. This point is made very clearly by Thomas Schäfer and Torban Borchert in their account of bioprospecting for industrial enzymes (chapter 33). Thus, although traditional enzyme discovery is based on methods such as enrichment isolation culture incorporating target substrates and the screening of culture collections, these are increasingly being complemented by molecular screening and metagenomics (see chapter 31) strategies. Thereafter, catalyst optimization and customization for specific industrial applications may be achieved via protein engineering or, more commonly, by one of the methods of artificial evolution.

Effective discovery programs ultimately are dependent on the quality, stringency, and inventiveness of the screening methods employed at each stage of the search and discovery campaign. The so-called classical screening procedures tended to be dependent on large numbers, randomness, and empiricism and were gradually supplanted by "intelligent" screening developments (Bull et al., 1992), epitomized by the enzyme inhibition strategem introduced by Umezawa in the 1960s. Subsequently, great strides have been made in assay technology, particularly for novel drug discovery, among which are receptor-ligand binding, reporter gene, adhesion, proteosome inhibition, signal transduction, and cell-cell communication-based assays (Hill et al., 1998; Wrigley, chapter 32). In many search and discovery programs where novel natural product chemistry is sought, a hierarchy or sequence of screening operations may be required to ensure effective dereplication. The starting point may be the dereplication of collections of target organisms, followed by chemical dereplication (of candidate organisms) of the sort developed by Hans-Peter Fiedler (see chapters 30 and 32). Bill Strohl concludes his chapter with a thought-provoking epilogue: "what you seek is what you find," and, with reference to the search for natural antimicrobial and antitumor compounds, he suggests a number of approaches that might enhance our ability to make effective discoveries. The latter range from the development of better, specific in vitro functional assays and cancer-specific mechanism assays to the prospecting of new environments for exploitable microbial diversity.

Douglas Hofstadter (1980), in discussing mechanized intelligence, included in his essential abilities for intelligence: flexible response to situations, taking advantage of fortuitous circumstances, making sense out of ambiguous messages, finding similarities between situations despite apparent differences, distinguishing between situation despite apparent links, synthesizing new concepts by reassessing old ones, and coming up with novel ideas. The contributions in this section demonstrate clearly how contemporary biotechnology innovations and applications are the products of intelligence, sagacity, and lateral thinking, and interestingly, as Vincent and Mann (2002) would have us consider, inventive problem solving as a conduit to natural design solutions.

REFERENCES

Bull, A. T., M. Goodfellow, and J. H. Slater. 1992. Biodiversity as a source of innovation in biotechnology. *Annu. Rev. Microbiol.* 42:219–257.

Hill, D. C., S. K. Wrigley, and L. J. Nisbet. 1998. Novel screen methodologies for identification of new microbial metabolites with pharmacological activity. *Adv. Biochem. Eng. Biotechnol.* 59:75–121.

Hofstadter, D. R. 1980. *Gödel, Escher and Bach: An Eternal Golden Braid. A Metaphysical Fugue on Minds and Machines in the Spirit of Lewis Carroll.* Vintage Books Edition, Random House, Inc., New York, N.Y.

Hugenholtz, J., and E. J. Smid. 2002. Nutriceutical production with food-grade microorganisms. *Curr. Opin. Biotechnol.* 13:497–507.

Norris, P. R., N. P. Burton, and N. A. M. Foulis. 2000. Acidophiles in bioreactor mineral processing. *Extremophiles* 4:71–76.

Vincent, J. F. V., and D. L. Mann. 2002. Systematic technology transfer from biology to engineering. *Phil. Trans. R. Soc. London Sev.* 360:159–173.

Walsh, V., and G. Lodorfos. 2002. Technological and organizational innovation in chemicals and related products. *Technol. Anal. Strat. Manage.* 14:273–298.

Microbial Diversity and Bioprospecting
Edited by Alan T. Bull
© 2004 ASM Press, Washington, D.C.

Chapter 29

Sectors and Markets

ALAN T. BULL

SECTORS

The adoption of modern biotechnology typifies the development of any new technology, namely, a slow initial phase followed by a period of substantial growth, which, in the case of biotechnology, has been mainly focused on health care, then the entry into a mature phase of consolidation and progressive penetration into the manufacturing, service, and agricultural industry sectors (Table 1).

The range of techniques defined by biotechnology is very large and varied, and none of them is likely to be applicable across the whole range of sectors shown in Table 1. However, the versatility of biotechnology is such that many industries that have not previously considered using biological systems are now exploring these options.

But what factors might encourage industry to adopt biotechnology? The combination of factors driving biotechnology will vary according to the particular company, country, market, environmental regulations, social climate, and so on, but high on the agenda will be profitability, a positive investment climate, novelty and the opportunity to develop new markets, the competitive replacement of extant goods and services (including product differentiation), and environmental performance or sustainability. At the company level, answers to the following types of question also will be crucial in deciding whether to adopt biotechnology. Can biotechnology improve my, or my competitor's, process? Does the entire process have to be changed, or can biotechnology be integrated as one or more unit stages? Can natural organisms or biocatalysts be used, or will the process be dependent on genetically manipulated organisms or reagents? If genetically manipulated technology is involved, will the process and/or product have public acceptability? Is the chosen process environmentally more sustainable than any alternative? (Bull, 2001). A recent Organisation for Economic Cooperation and Development

(OECD) survey (OECD, 2001) concluded that "Process innovations which 'only' improve environmental performance do not give companies sufficient incentive to modify their operations . . . Only economic advantages convince decision makers in companies to apply ecologically advantageous, innovative processes." Such arguments notwithstanding, every case study investigated by the OECD demonstrated an improvement in sustainability following the introduction of a biotechnology process or unit stage (a small selection of these case studies was reviewed in chapter 1 of this volume). The environmental benefits reported in these examples were realized as the sparing of energy and raw materials and as reduced waste arisings to air and water; economic benefits were realized in reduced operating costs, as illustrated in Table 2.

Focus on enzymes and whole-cell biocatalysts reveals quite dramatically how biotechnology has penetrated a very wide spectrum of commercial activities. A number of industries, notably food, feed, starch processing, leather, textile, and cleaning, have a well-established history of using biocatalysts, probably, as pointed out by van Beilen and Li (2002), because the raw materials and products of these industries are biomaterials that can be produced, modified, and degraded enzymatically. In the past, problems for deploying biocatalysts in the chemicals industry, for example, have stemmed from their fragility under operating conditions, their cost, and requirements for large concentrations of water. However, the developments that have occurred in recent years in biochemical engineering and biocatalyst design have opened the way for biocatalyst intervention in a much wider range of industries (Table 3).

Huisman and Gray (2002) remind us that the "metabolic space" of an organism defines the window of physicochemical parameters within which natural enzymes function optimally; in an analogous way, "functional space" limits the activity of enzyme activity per se. The physicochemical environments that are

Alan T. Bull •   Research School of Biosciences, University of Kent, Canterbury, Kent CT2 7NJ, United Kingdom

Table 1. Penetration of biotechnology by industrial sector

Manufacturing and processing industries	Service industries	Agriculture
Pharmaceuticals	Bioenergy	Animal feed and feed additives
Fine chemicals	Environment	Animal health
Laboratory, diagnostic, and forensic reagents	Bioremediation Ecosystem restoration	Soil fertility Pest control
Enzymes	Clean technology	Plant-growth-promoting agents
Biosensors	Waste treatment Oil recovery Antifouling agents	
Commodity chemicals	Health care	
Plastics and polymers	Personal care Cosmetics	
Food and food additives Nutriceuticals		
Beverages		
Textiles		
Leather		
Pulp and paper		
Minerals recovery and processing		
Biomaterials and biomimetics		

Table 3. Examples of industrial applications of biocatalysts[a]

Industry	Application
Food	Fruit products; milk clotting; cheese flavoring
Baking	Dough strength and stability
Starch	Liquefaction and saccharification; cyclodextrin production
Beverages	Juice clarification; mashing
Fats and oils	Transesterification; degumming
Agriculture	Animal fodder digestibility; ensilage
Pulp and paper	Pitch control; drainage improvement; de-inking
Textiles	Depiling; desizing; "biostoning" of denim
Leather	Bating
Detergents	Protein, lipid, and starch stain removal
Chemicals	Acrylamide; nicotinamide; amino acids; polymers
Pharmaceuticals	Semisynthetic antibiotics; chiral compounds
Bioprocessing	Enantioselective production and resolution
Energy	Fuel alcohol
Personal care	Antimicrobials; bleaching; soaps

[a] Sources: van Beilin and Li (2002), Kirk et al. (2002), Huisman and Gray (2002), Schmid et al. (2002).

typical of the chemical industry bear little resemblance to natural environments, ipso facto most natural enzymes function suboptimally under such circumstances, and thereby stimulating enzyme design and engineering. Biocatalyst design, whether by directed evolution or rational approach, has enabled enzymes to be customized with respect to specificity, efficiency, and stability (Bull et al., 1999; see also chapters 24 and 33). Similarly, biochemical engineering innovations including protocols for enzyme preparation and pre-

Table 2. Cost benefits resulting from biotechnology innovation[a]

Product, process	Reduction in operating costs (%)
Riboflavin (Hoffmann-LaRoche)	50
Amino acids (Tanabe)	43
Acrylic acid (Ciba)	54 (raw materials)
7-Amino-cephalosporanic acid (Biochemie)	90 (environment related)
Cephalexin (Dutch State Mines)	Considerable
Vegetable oil processing (Cereol)	40
Vegetable processing (Pasfrost)	30 (groundwater recovery)
Removal of bleach residues (Windel)	9

[a] For full details of these and other case studies, see OECD (2001).

sentation, manipulation of the reaction environment (e.g., supercritical fluids), and stabilization (e.g., immobilization, imprinting, biocatalytic plastics, and cross-linked enzyme crystals) (Bull, 2001) have greatly extended the range of process conditions under which biocatalysts can be used. And, of course, to these inventions can be added the discovery of novel catalysts, including extremozymes, and the evaluation of reactions such as oxidations (effective means for cofactor regeneration) and carbon-carbon bond synthesis.

The early euphoria of biotechnology 20 to 30 years ago was strongly science driven, and its products were largely expected in the health care sector. Maturity in the intervening years has directed attention increasingly toward bringing products to market and to widening the range of sectors within which biotechnology innovation can be applied. Moses and Moses (1995) opined that marketing was "likely to be the main driving force in the biotechnology of the future" and would be based on (i) recognizing what the market wants, (ii) ensuring that companies would be able to provide products and services to meet such needs, (iii) providing appropriate information to clients and customers about products and services before and as they become available, and (iv) determining appropriate pricing policies. They conclude, rightly in my opinion, that there are rules about spotting winners in biotechnology. These rules are the same as in other businesses: identify the market niches, get the products right, and sell. Consequently this section of the book is intended to illustrate only a few current trends in commercial biotechnology, without laying claim to being comprehensive. One of the many omissions in this context—biomaterials and biomimetics—is treated briefly as a comment on how efforts to mimic nature (biodiversity) are leading to biotechnology innovations.

MARKETS

Biotechnology is now established as a robust, reliable, and relatively low-risk technology (current debates on genetically modified organisms notwithstanding) and is capable of being implemented on a large scale and across the full range of industrial sectors (Bull, 2001; Berry, 2002). At this point it is worth reiterating the main drivers of biotechnology: economic competitiveness, government policy, and public interest (Bull et al., 1998). Market contributions of biotechnology products can be categorized as (i) sales of new products directly attributable to the use of modern biotechnology, (ii) sales of products manufactured by improved processes that make direct use of modern biotechnology (direct impact, e.g., recombinant products), and (iii) sales of products manufactured by improved processes using the products of biotechnology arising in other industries (indirect impact, e.g., biocatalysts). These categories represent total biotechnology-related sales (BRS). Recent estimates of biotechnology markets, expressed as the shares of worldwide and forecasts for 2005 are shown in Table 4 for seven major industrial sectors.

The impact of biotechnology to date has been most pronounced in the pharmaceuticals sector (see chapters 31 and 32), but it is clear that enormous potential exists in all of the other sectors for biotechnology penetration even though short-term forecasts show no change in those sectors in which the market share is very low.

BIOMATERIALS AND BIOMIMETICS

In a stimulating essay on systematic technology transfer, Vincent and Mann (2002) comment that

during the process of innovation we commonly ignore the solutions and practices developed in other sciences and technologies and, in particular, "fail to tap in to the four billion years' worth of 'R&D' in the natural world." Of course, biologists, including microbiologists, as a result of careful observation and serendipity, have enabled a portion of this natural resource to be exploited. However, Vincent and Mann stress two critical points. Their first point is that the key to integrating biological knowledge has been functionality, and they illustrate the potential of biology-based technology (biomimetics) with a number of cases that have resulted in patents or production—for example, underwater glues based on mussel adhesive, Velcro conceptualized from plant burs, tough composites based on wood fibers, extrusion technology inspired by spider spinerets, and self-cleaning surfaces based on the structure of the lotus leaf. Second, although there is a growing interest in biomaterials and biomimetics, the process by which engineers (and technologists in general) are made aware of nature's solutions is largely ineffective. Vincent and Mann advocate the theory of inventive problem solving (Altshuller, 1999) as an objective framework for providing access to natural design solutions, but this approach notwithstanding, microbiologists could do much more to advertise the idiosyncratic potential of their organisms. A few pointers in this direction are given in the following paragraphs.

A useful starting point for this discussion is biopolymers and their derivatives, where the microbial world is a rich source of chemical structures and functionality. Microbial polysaccharides and polyesters are particularly well known and exploited for a wide range of uses that depend on their hydrocolloid, adhesive, gel, ionic, plastic, and other properties. Microbial peptides, proteins and polyamides provide similar opportunities for innovative products and activities. For example, cyanophycin, an arginyl–L-aspartic acid polymer that is produced by most cyanobacteria, in a chemically reduced arginine form can be used as a biodegradable substitute for polyacrylates. Biopolymers have diverse applications in medicine, bioengineering, food processing, agriculture, wastewater treatment, and so on, an excellent comprehensive account of which can be found in Steinbüchel (2002).

Increased understanding of the behavior of microbial proteins can provide unexpected options for technological development. Take the TlpA protein of *Salmonella,* for example. This coiled-coil protein is encoded on the virulence plasmid and functions as a temperature-sensing gene regulator in vivo, and Stone and colleagues (Naik et al., 2001) are exploring its potential in sensor applications. TlpA is unique in the

Table 4. Worldwide market share of biotechnology (BRS) for selected sectors[a]

Sectors	1996 (%)	Forecast 2005 (%)
Chemical products	<1	<1
Pharmaceuticals/fine chemicals	5–11	10–22
Pulp and paper	5	35
Food	1–2	2–4
Textiles	<1	<1
Leather	<1	<1
Energy	<1	<1

[a] Source: Bull et al. (1998).

extreme rapidity of its temperature-stimulated un-folding and refolding, a property that is apposite for a number of biosensor developments.

A biomimetic development resulting from the discovery of eukaryotic defense peptides has been described recently by Tew et al. (2002). Peptides such as the magainins are facially amphiphilic in structure and exert an antibacterial effect by causing pore formation without affecting the host cell. Unfortunately the production and cost of the natural peptides and peptidomimetics synthesized from β-amino acids are difficult and expensive. Tew et al. have designed various facially amphiphilic acrylamide polymers that mimic the physical and bioactive properties of the natural agents, the deployment of which as antibacterial chemicals and surfaces is very attractive.

A potent example of the Vincent and Mann (2002) philosophy is found in inorganic materials, where the fields of biological assembly and inorganic chemistry have become increasingly linked. Mann and Ozin (1996) were early advocates for considering biological concepts such as morphogenesis, replication, and self-organization as constructs for developing novel ways of synthesizing complex inorganic materials over a wide range of scale (nanoscopic [<1.5 nm] to macroscopic [>100 μm]). Their focus was on a conceptional framework that could lead to systems with enhanced or highly adjustable properties, engineered biomaterials, bioactive ceramic-matrix composites, environmentally responsive systems, organized catalyst supports, efficient membranes for liquid-gas separations, and so on. During the intervening years considerable progress has been made in using biomolecules to direct the self-assembly of inorganic nanostructures. For example, phage display has been used to select peptides that can recognize and control the growth of inorganic particles such as semiconductor nanocrystals (Seeman and Belcher, 2002). Probably the greatest microbiological inspiration for inorganic materials research and development has come from diatoms and sponges (Vrieling et al., 1999; Morse, 1999). The growing demand for new materials and the increased understanding of diatom silicon biomineralization (Vrieling et al., 2000) is providing alternatives to existing silicas and zeolites, for example, and completely novel applications; thus, silica formation at the nanoscale appears to be a uniform process while pH and salinity are controlling factors for polymerization and the grade of silicification. As Morse (1999) observed, such biologically produced silicas "exhibit a genetically controlled precision of nanoscale architecture that, in many cases, exceeds the capabilities of present-day human engineering," a production level moreover that amounts to gigatons per annum in the world's oceans.

Morse and his colleagues have researched biosilicification in the marine sponge *Tethya aurantia* (Morse, 1999) and have characterized and cloned proteins (silicateins) that control the process. Silicatein-α has hydrolase activity in vitro and converts silicon alkoxides to silanols that, in turn, condense spontaneously to polysiloxanes. The in vivo precursors for silica synthesis remain to be identified. Nevertheless, such in vitro catalysis presents opportunities for developing clean technology routes to high-performance materials having novel functionality.

Subsequently, Stone's group have used combinatorial phage display technology to isolate selective silica-binding peptides and thence to identify amino acid sequences that promote silica nucleation and precipitation (Naik et al., 2002); the latter had the same overall characteristics of previously described silica-binding and -precipitating biological and synthetic polyamines. A further breakthrough in the understanding of biosilification has come recently from studies on the freshwater diatom *Navicula pelliculosa*, which provide, for the first time, direct evidence of an organosilicon compound formed in vivo (Kinrade et al., 2002). Such understanding eventually promises to furnish low-cost routes to the production of high-performance silicon materials. Other innovative developments are based on molecular imprinting (Mosbach, 2001) and in situ "click chemistry," in which drug-target receptors or enzymes are involved in the direct selection of building blocks in the synthesis of their own inhibitors (Lewis et al., 2002).

REFERENCES

Altshuller, G. 1999. The innovation algorithm, TRIZ, systematic innovation and technical activity. Technical Innovation Center, Inc., Worcester, Mass.

Berry, S. 2002. Biotech meets the investors. *Trends Biotechnol.* 20:370–371.

Bull, A. T. 2001. Biotechnology for industrial sustainability. *Korean J. Chem. Eng.* 18:137–148.

Bull, A. T., B. L. Marrs, and R. Kurane. 1998. *Biotechnology for Clean Industrial Products and Processes. Towards Industrial Sustainability*, p. 1–200. Organisation for Economic Cooperation and Development, Paris, France.

Bull, A. T., A. W. Bunch, and G. K. Robinson. 1999. Biocatalysts for clean industrial products and processes. *Curr. Opin. Microbiol.* 2:246–251.

Huisman, G. W., and D. Gray. 2002. Towards novel processes for the fine-chemical and pharmaceutical industries. *Curr. Opin. Biotechnol.* 13:352–358.

Kinrade, S. D., A.-M. E. Gillson, and C. T. G. Knight. 2002. Silicon-29 NMR evidence of a transient hexavalent silicon complex in the diatom *Navicula pelliculosa*. *J. Chem. Soc. Dalton Trans.* 3:307–309.

Kirk, O., T. V. Borchert, and C. C. Fuglsang. 2002. Industrial enzyme applications. *Curr. Opin. Biotechnol.* 13:345–351.

Lewis, W. G., L. G. Green, F. Grynszpan, Z. Radic, P. R. Carlier, P. Taylor, M. G. Finn, and K. B. Sharpless. 2002. Click chemistry in situ: acetylcholinesterase as a reaction vessel for the selective assembly of a femtomolar inhibitor from an array of building blocks. *Angew. Chem. Int. Ed.* **41**:1053–107.

Mann, S., and G. A. Ozin. 1996. Synthesis of inorganic materials with complex form. *Nature* **382**:313–318.

Morse, D. E. 1999. Silicon biotechnology: harnessing biological silica production to construct new materials. *Trends Biotechnol.* **17**:230–232.

Mosbach, K. 2001. Towards the next generation of molecular imprinting with emphasis on the formation, by direct molding, of compounds with biological activities (biomimetics). *Anal. Chim. Acta* **435**:3–8.

Moses, V., and S. Moses. 1995. *Exploiting Biotechnology.* Harwood Academic Publishers GmbH, Chur, Switzerland.

Naik, R. R., S. M. Kirkpatrick, and M. O. Stone. 2001. The thermostability of an α-helical coiled-coil protein and its potential use in sensor applications. *Biosens. Bioelectron.* **16**:1051–1057.

Naik, R. R., L. L. Brott, S. J. Clarson, and M. O. Stone. 2002. Silica-precipitating peptides isolated from a combinatorial phage display peptide library. *J. Nanosci. Nanotechnol.* **2**:95–100.

Organisation for Economic Cooperation and Development. 2001. *The Application of Biotechnology to Industrial Sustainability.* Organisation for Economic Cooperation and Development Publications, Paris, France.

Schmid, A., F. Hollmann, J. B. Park, and B. Bühler. 2002. The use of enzymes in the chemical industry in Europe. *Curr. Opin. Biotechnol.* **13**:359–366.

Seeman, N. C., and A. M. Belcher. 2002. Emulating biology: building nanostructures from the bottom up. *Proc. Natl. Acad. Sci. USA* **99**:6451–6455.

Steinbüchel, A. (ed.) 2002. Biopolymers, vol. 1 to 10. Wiley-VCH Verlag GmbH, Weinheim, Germany.

Tew, G. N., D, Liu, B. Chen. R. J. Doerksen, J. Kaplan, P. J. Carroll, M. L. Klein, and W. F. DeGrado. 2002. De novo design of biomimetic antimicrobial polymers. *Proc. Natl. Acad. Sci. USA* **99**:5110–5114.

van Beilen, J. B., and Z. Li. 2002. Enzyme technology: an overview. *Curr. Opin. Biotechnol.* **13**:338–344.

Vincent, J. F. V., and D. L. Mann. 2002. Systematic technology transfer from biology to engineering. *Phil. Trans. R. Soc. London Ser. A* **360**:159–173.

Vrieling, E. G., T. P. M. Beelen, R. A. van Santen, and W. W. C. Gieskes. 1999. Diatom silicon biomineralization as an inspirational source of new approaches to silica production. *J. Biotechnol.* **70**:39–51.

Vrieling, E. G., T. P. M. Beelen, R. A. van Santen, and W. W. C. Gieskes. 2000. Nanoscale uniformity of pore architecture in diatomaceous silica: a combined small and wide angle X-ray scattering study. *J. Phycol.* **36**:146–159.

Microbial Diversity and Bioprospecting
Edited by Alan T. Bull
© 2004 ASM Press, Washington, D.C.

Chapter 30

Screening for Bioactivity

HANS-PETER FIEDLER

The present situation in the screening for bioactive compounds is determined by the nearly unlimited capacity in the throughput of assays. In the latter half of the 1990s, high-throughput screening was established by automated robot systems using 96- and 384-well microtiter plates having a test capacity up to 100,000 samples per month. At the beginning of this millennium, ultra-high-throughput screening using 1536-well microtiter plates was developed, with a capacity of the same sample throughput within a few days. The problem for natural compounds is based on the availability of the amount of samples, not only within old laboratory stocks, but also for new metabolites obtained from newly isolated strains by fermentation and preparative isolation. Therefore, chemical libraries obtained from combinatorial chemistry seemed to be favorable in the sight of some managers in the pharmaceutical industry, because the capacity of combinatorial chemistry can be increased nearly in the same way as assay capacity. Nevertheless, it must be considered that the chemical diversity of chemical libraries differs from that of natural compounds (Henkel et al., 1999) and that the chemical diversity of natural compounds cannot be mimicked by organic chemistry. That is one fact. Another fact is that nature is a nearly inexhaustible source for both microorganisms as potential producers of natural compounds and the multitude of novel natural compounds themselves (Bérdy, 1995; also see chapter 1, this book). A further argument that speaks for natural products is their huge diversity in biological action. Members of the same group of natural products can act as selective antibacterial, antifungal, insecticidal, or immunosuppressive agents, as in the case of macrolide antibiotics.

To discover new bioactive metabolites it will be helpful to look back at the "old" antibiotics and the screening methods by which most of the drugs based on natural products were discovered and placed in the top 100 of the pharmaceutical market and by which nearly all antibiotics were found. In my opinion it is worthwhile to reflect on new modifications of the traditional screening methods to detect drugs that are active at new and as yet nonconsidered targets, as well as for new lead structures. It is a necessity that we feed the pipeline in the near future with new drugs that act upon new targets to overcome the increasing problems of multiresistant pathogens whose resistance is based on target modifications (Zähner and Fiedler, 1995). Chemically or genetically engineered modifications of well-introduced block buster drugs cannot solve the imminent disaster of resistant bacterial or fungal pathogens in medical therapy.

CLASSICAL ANTIBACTERIAL ASSAYS

Agar Plate Diffusion Assay

Alexander Fleming was one of the first scientists who used the agar plate diffusion assay to detect antibacterial activity. At the beginning of the 1920s, Fleming was strongly involved in the action of lysozyms as an antibacterial agent (Fleming, 1922). Some years later, it was a mere accident that one of his glass plates filled with blood agar on which he cultivated staphylococci was contaminated by a fungus. This contamination led to a complete suppression of the staphylococci surrounding the fungus. He determined the contamination to be *Penicillium notatum.* Fleming realized that a diffusible compound had caused this effect, and he named it penicillin (Fleming, 1929). He determined the antimicrobial spectrum of the diffusible compound by challenging various pathogenic bacteria with the fungus. This "good stroke of fortune," as Fleming later called his discovery, was the beginning of one of the most successful eras in the fields of microbiology and medicine.

During the 1940s, various groups modified the efficiency and reproducibility of the test. One modifi-

Hans-Peter Fiedler • Mikrobiologisches Institut, Universität Tübingen, Auf der Morgenstelle 28, D-72076 Tübingen, Germany.

cation was to place agar pieces with actinomycetes on the surface of agar plates that were seeded with the bacterial test organisms. Such an assay is shown in Fig. 1. It allowed the screening of hundreds of actinomycetes for the production of antibiotic activity by scaling up the rectangular agar plates.

This assay using whole cells and mycelia was modified by testing only culture filtrates and organic extracts from filtrate and mycelium. The fluids were applied either on cylinders or on holes that were punched with a cork borer. The cylinder plate assay was performed with open cyclinders manufactured from glass, porcelain, or stainless steel. They were standardized in their dimensions: 8-mm outside diameter, 6-mm inside diameter, and 10-mm length. The hole assay is more sensitive than the cylinder assay because of the increased surface for diffusion. Both assay modifications realized the preconditions needed for the screening of a high number of strains: (i) they were very simple techniques, (ii) they were relatively insensitive against contaminations, (iii) they permitted testing nonsterile solutions, (iv) they permited interpolation of single data, and (v) smaller volumes of test solutions were needed.

Today most laboratories prefer the more easily executable filter disk plate diffusion assay instead of the cylinder and hole assays. The test solutions are placed on filter disks of 6 mm in diameter and applied to the agar plate, as shown in Fig. 2. The sensitivity of the agar plate diffusion assay can be increased by (i) prediffusion, (ii) the density of the bacterial inoculum, and (iii) using a thin top agar layer containing the test organism and ground agar without the test

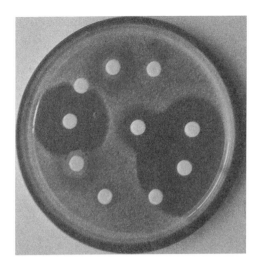

Figure 2. Agar plate diffusion assay using *Streptomyces viridochromogenes* as the test organism. Test solutions are applied on filter disks.

organism. All these methods, especially their evaluation, were reviewed by Grove and Randell (1950) and Cooper (1963).

Important Antibiotics Detected by Agar Plate Diffusion Assays

In 1939 René Dubos, a former student of Selman Waksman, discovered a microbial complex named tyrothricin (tyrocidin) (Dubos, 1939), which was produced by the soil isolate *Bacillus brevis*. Industrial production of the complex was set back, however, because of the affairs of the Second World War, during which time penicillin gained greater significance, and being developed at the U.S. Department of Agriculture's Northern Regional Research Laboratories in Peoria, Illinois, together with the U.S. pharmaceutical companies Merck, Pfizer, Squibb, and Lederle.

In the meantime, Waksman turned his research interests to the discovery of new broad-spectrum antibacterial agents from the actinomycetes. Woodruff, a graduate student in Waksman's lab, continued Dubos's enrichment techniques for the isolation of actinomycetes from soils and testing against gram-negative bacteria, which were unaffected by penicillin. This approach led to the discovery of actinomycin, a lipophilic, extractable compound produced by *Streptomyces antibioticus* (Waksman and Woodruff, 1940). Waksman proposed the name "antibiotic" for this new class of antimicrobial substance of microbial origin. The toxicity associated with actinomycin dropped it from consideration as an antibacterial drug. Years later, however, when it was found to be effective in the treatment of cancer, its significance increased again.

Figure 1. Agar plate diffusion assay using *Bacillus subtilis* as the test organism and agar pieces from actinomycetes cultures. Reprinted from Zähner (1965) by permission of Springer-Verlag.

A short time later, streptothricin, a broad-spectrum antibacterial compound, was isolated from a streptomycete (Waksman and Woodruff, 1942), but its severe kidney toxicity made it unacceptable as a commercial product. Subsequently another antibiotic produced by a streptomycete was isolated by Waksman's graduate student, Albert Schatz; it had a similar antibacterial spectrum, but was different in structure from streptothricin. This antibiotic was called streptomycin (Schatz et al., 1944). Shortly after its efficacy against tuberculosis was detected, streptomycin became the first commercial natural product produced at Merck by fermentation.

The detection of antibiotically active metabolites in streptomycete cultures by the agar plate diffusion assay increased dramatically after Waksman's successful application of this revolutionary technique and led to an often simultaneous but independent publication of the same antibiotics mainly by research groups of pharmaceutical companies in the United States, Europe, and Japan. It was the beginning of an era that can be called a "new gold rush." Thus, chloramphenicol (chloromycetin) was discovered in *Streptomyces venezuelae* independently by Ehrlich and coworkers in 1947 at Parke Davis (Ehrlich et al., 1947), by Gottlieb and coworkers in 1948 at the University of Illinois (Gottlieb et al., 1948), and by Umezawa and coworkers in 1948 in Japan (Umezawa et al., 1948). Chloramphenicol was the first broad-spectrum antibiotic and was in clinical use over some decades. Today chloramphenicol belongs to the category of "historic" antibiotics because of its insufficient antibacterial activity and toxic side effects, such as aplastic anemia.

Tetracyclines were discovered between 1947 and 1950. Chlorotetracycline (aureomycin) was isolated from *Streptomyces aureofaciens* by Finlay and coworkers in 1947 at Lederle and oxytetracycline (terramycin) was isolated from *Streptomyces rimosus* by Duggar and coworkers in 1950 at Pfizer. Tetracycline production is widely distributed within the genus *Streptomyces*. The history of the discovery of tetracyclines has been summarized by Dürkheimer (1975).

Erythromycin, a 14-membered macrolide antibiotic, was isolated by McGuire and coworkers in 1952 from *Streptomyces erythraea* (McGuire et al., 1952) (which was later reclassified as *Saccharopolyspora erythraea*). In clinical application, erythromycin is one of the safest anti-infective drugs; its antimicrobial spectrum includes gram-positive bacteria, such as streptococci, pneumococci, clostridia, and *Listeria*, and some gram-negative pathogens such as *Haemophilus neisseriae, Bordetella pertussis*, and mycoplasmas. However, erythromycin was not the first

discovered macrolide antibiotic. Picromycin, produced by *Streptomyces felleus*, was discovered in 1950 (Brockmann and Henkel, 1950) but, because of its lower activity and greater toxicity compared with erythromycin, it gained no importance for therapeutic use.

The discovery of cephalosporin C goes back to 1945, when Giuseppe Brotzu at the Hospital of Cagliari, Italy, found that culture extracts from *Cephalosporium acremonium* (later reclassified as *Acremonium chrysogenum*) exhibited activities against gram-positive and gram-negative bacteria (Brotzu, 1948). Cephalosporin C was isolated in 1953 (Newton and Abraham, 1953), and its structure was elucidated by the same group in 1959. The second wave of resistance against antibiotics during the 1960s—seen in gram-negative bacteria such as *Shigella, Salmonella*, and *Klebsiella*, which showed multiple resistance against tetracyclines, chloramphenicol, and aminoglycosides—intensified efforts to overcome this resistance. An answer was found in the chemical modification of cephalosporin C (Flynn, 1972). This led to a large number of semisynthetic cephalosporins, which showed differences in their antibacterial spectrum, oral and parenteral applications, and resistance to beta-lactamase (Bryskier, 2000).

Cycloserine (oxamycin, seromycin, PA-94) was discovered and published simultaneously by three research groups working with *Streptomyces orchidaceus* (Harned et al., 1955), *Streptomyces lavendulae* (Shull and Sardinas, 1955), and *Streptomyces garyphalus* (Harris et al., 1955). Cycloserine had prominent activity against *Mycobacterium tuberculosis*, but it was associated with severe physiological disturbances in animal studies, which prevented its broad application.

Novobiocin was reported in 1955 to 1956 by several pharmaceutical companies under the names cathomycin, cathocin, streptonivicin, albamycin, PA-93, and cardelmycin (Wallick et al., 1955), as produced by *Streptomyces sphaeroides* and *Streptomyces niveus*. It was proved to be of clinical importance because it was active against penicillin-resistant gram-positive pathogens. In 1956 an agreement was reached to give all six of these antibiotics the generic name novobiocin.

Rifamycin, a new ansamycin-type antibiotic produced by *Streptomyces mediterranei*, was first announced by Lepetit in Italy (Sensi et al., 1959). The producing strain was later reclassified as *Nocardia mediterranei* and ultimately as *Amycolatopsis mediterranei*. An intensive cooperative effort between Lepetit and Ciba-Geigy discovered the mode of action, a specific inhibition of bacterial RNA

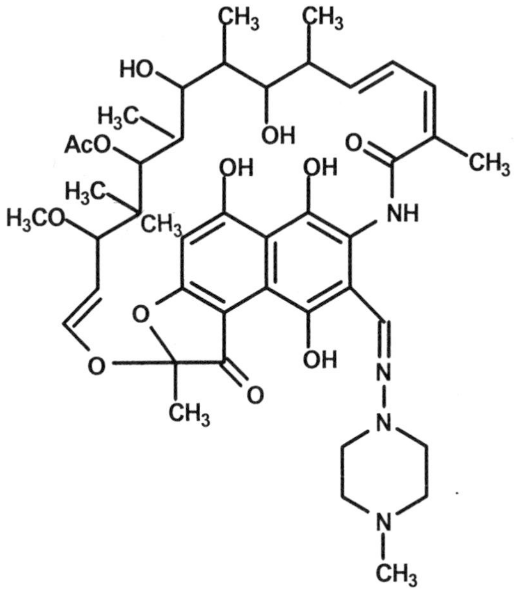

Figure 3. Structure of rifampin, a semisynthetic therapeutic agent against tuberculosis.

polymerase, and a new semisynthetic oral rifamycin named rifampicin, or, more recently, rifampin, was introduced. Rifampin (Fig. 3) showed a stronger antibacterial activity, mainly against gram-positive pathogens, a better tissue penetration, and oral resorption. It was highly active against the problematic pathogen *M. tuberculosis*.

Modifications of the Classical Agar Plate Diffusion Assay

Screen based on the formation of spheroplasts

Eugene Dulaney developed a screening assay based on the formation of spheroplasts in the presence of actinomycete culture filtrates. The test required the microscopic observation of a gram-negative cell suspension shortly after the culture filtrate from actinomycetes was added. Some years later, this assay led to the detection of fosfomycin (phosphonomycin), a wide spectral antibiotic produced by *Streptomyces fradiae* (Hendlin et al., 1969), that showed activity against pathogenic *Pseudomonas* strains. Fosfomycin

gained importance in medicine because of its safety, though it could only be used in infusion therapy.

Edward Stapley developed a screening protocol that highlighted both novel cell wall activities and resistance against beta-lactamases. The modified assay led to the discovery of cephamycin C, an antibiotic produced by *Nocardia lactamdurans* that was effective against most penicillin- and cephalosporin-resistant bacteria (Stapley et al., 1979). A semisynthetic derivative, cefoxitin, was introduced successfully to clinical use.

Thienamycin, isolated from *Streptomyces cattleya*, possessed a beta-lactam resistance and astonishing broad-spectrum activity against gram-positive and gram-negative bacteria (Kahan et al., 1979). After serious problems were overcome—such as low fermentation titer, purification, instability, and degradation in the urinary tract, causing persistence of pathogens—thienamycin was brought on the market as Primaxin (imipenem), a high-active, broad-spectrum, parenteral antibiotic that possessed resistance to most beta-lactamases. The stuctures of thienamycin and imipenem are shown in Fig 4.

The narrow-spectrum screening concept

A new approach developed by Hans Zähner and coworkers (Zähner et al., 1982) based on the idea that if, in addition to the usual test bacteria, unusual test organisms were used, such as clostridia, streptomycetes, *Halobacterium*, etc., and selection was done only for those activities that were effective against the unusual test organisms but ineffective against the unusual ones, there was a good chance that novel antibiotics would be obtained. With this screen modification, a group at the University of Tübingen discovered the tetracenomycins, kirromycin (mocimycin) and kirrothricin, brevienomycins, haloquinone, imacidin C, stenothricin, and tirandamycin B (Kuhn and Fiedler, 1995).

Screen against the permeation barrier

The same group at Tübingen developed another screening model that considered antibiotics that

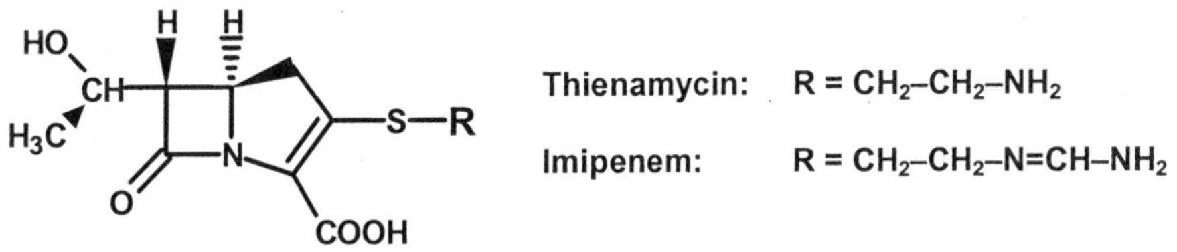

Thienamycin: R = CH₂–CH₂–NH₂

Imipenem: R = CH₂–CH₂–N=CH–NH₂

Figure 4. Thienamycin, a broad-spectrum antibacterial beta-lactam antibiotic produced by *Streptomyces cattleya*.

were unable to penetrate whole cells (Zähner and Fiedler, 1995). The addition of EDTA increased cell permeability to many compounds. Only those antibiotics were of interest that were active only in the presence of EDTA. This method discovered the new diphenylethers ethericin A and B, which show ionophoric properties for bivalent cations (König et al., 1978).

CLASSICAL ANTIFUNGAL ASSAYS

Agar Plate Diffusion Assay

All the old antifungal antibiotics such as nystatin, amphotericin B, pyrrolnitrin, and echinocandin B were discovered by the classical agar plate diffusion assay. The most prominent antifungal antibiotics belong to the group of macrolide polyenes, having a tetraene, pentaene, hexaene, and heptaene structure. The first member of a polyene macrolide antifungal antibiotic was nystatin (fungicidin), which has a tetraene structure and was isolated from the soil actinomycete *Streptomyces noursei* (Hazen and Brown, 1950). Nystatin showed a strong and broad antifugal action and interacted with sterol in the fungal cell membrane. Nystatin was introduced into clinical use together with the pentaene macrolide filipin (Whitfield et al., 1955) and amphotericin B, which was isolated from *Streptomyces nodosus* (Sternberg et al., 1956) and became the most prominent antifungal polyene antibiotic in medical use.

Two antibiotics that were first detected in classical antifungal agar plate diffusion assays were rediscovered some years later as very potent immunosuppressive agents. Cyclosporine A, a cyclic polypeptide antibiotic, was discovered at Sandoz in *Trichoderma polysporum* (Dreyfuss et al., 1976); rapamycin (sirolimus), a triene macrolide antibiotic produced by *S. hygroscopicus*, was detected by Vézina and coworkers at Wyeth-Ayerst (Vézina et al., 1975). Cyclosporin was introduced to clinical use in 1982 as Sandimmun and ranks among the top 20 on the pharmaceutical market. Rapamycin was introduced to clinical as use Rapamun in 1999.

Screening for Antifungal Antibiotics Causing Morphological Changes of Fungal Hyphae

Another assay is based on growth inhibition of filamentous fungi, such as *Botrytis* species, together with morphological changes of its hyphae which cause a so-called bulging or curling effect. This effect was first described by Brian and coworkers in 1946

(Brian et al., 1946). Brian et al. rediscovered the antifungal antibiotic griseofulvin, which was first detected by Oxford and coworkers in 1939 in a culture of *Penicillium griseofulvum* (Oxford et al., 1939). The assay gave rise to the detection of inhibitors of fungal cell wall synthesis in the presence of polyene antibiotics, which do not cause such morphological changes. The abnormal cell morphology can be observed under the microscope around the inhibition zone on the agar plate, as shown in Fig. 5. Other prominent antifungal antibiotics that cause the same effect are the nikkomycins, detected in the culture filtrates of *Streptomyces tendae* (Dähn et al., 1976). Nikkomycins are potent inhibitors of chitin synthase, as are the structurally related polyoxins, and show strong fungicidal, insectial, and acaricidal efficacy (Fiedler et al., 1993).

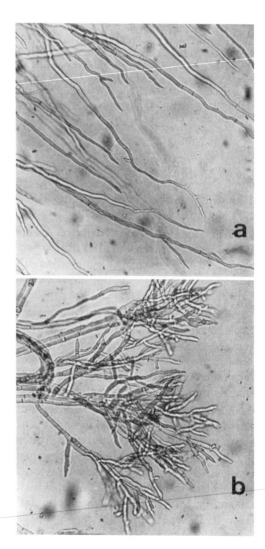

Figure 5. Bulging or curling effect caused by inhibitors of the fungal cell wall biosynthesis on the hyphae of *Botrytis cinerea*; (a) normal hyphae, (b) abnormal hyphae.

Screening for Inhibition of Fungal Cell Wall Synthesis

The growing need for safe and effective antifungal agents in clinical use stemmed from the rapidly increasing population of immunosuppressed patients and led to the development of specific targets. Most assays are focused on the inhibition of the polysaccharide network of the cell wall of many fungi, which consists of β-1,3-glucan, chitin, and mannan. An assay of chitin synthesis, based on the inhibition of chitin synthase isolated from *Pyricularia oryzae* using radioactive-labeled UDP-*N*-[^{14}C]acetylglucosamine as substrate, was developed by Hori and coworkers in 1974 (Hori et al., 1974). Assays of β-1,3-glucan synthesis and mannan synthesis have been based on the inhibition of particulate fractions of *Saccharomyces cerevisiae* as enzyme sources and radioactively labeled UDP-[^{14}C] glucose and GDP-[^{14}C] mannose as substrates, respectively (Satomi et al., 1982). These targets were reviewed by Debono and Gordee (1994).

One of the most promising new antifungals that were detected as glucan synthase inhibitors is the echinocandin lipopeptide group (Benz et al., 1974). Cilofungin (LY121019), a semisynthetic analogue of echinocandin B (Debono et al., 1988), is especially useful for treatment of candidiasis, caused by *Candida albicans*.

ANTITUMOR ASSAYS

In Vitro Antitumor Assays

In 1951 Hamao Umezawa, at the University of Tokyo, initiated the study of antitumor antibiotics by testing culture filtrates of actinomycetes against experimental animal tumor cell lines in proliferation assays. This pioneering study resulted within a few years in the discovery of caryomycin, sarkomycin, actinoleukin, ractinomycins, pluramycins, raromycin, and peptimycin (Takeuchi, 1987). This group isolated phleomycins in 1956 (Maeda et al., 1956), and the structurally related bleomycins in 1966 (Umezawa et al., 1966). Phleomycin produced a strong inhibition against Ehrlich carcinoma, but caused an irreversible renal toxicity in dogs. Therefore, bleomycins were developed as a therapeutic agent (Fig. 6). More than 400 derivatives have been obtained by fermentation and chemical derivatization of bleomycinic acid. Peplomycin, one of the semisynthetic bleomycins, was developed as a second-generation bleomycin, showing less pulmonary toxicity and a broader antitumor spectrum. One candidate from the third bleomycin generation, liblomycin, showed a stronger antitumor activity and no pulmonary toxicity. Bleomycin-type antitumor antibiotics are used worldwide for the

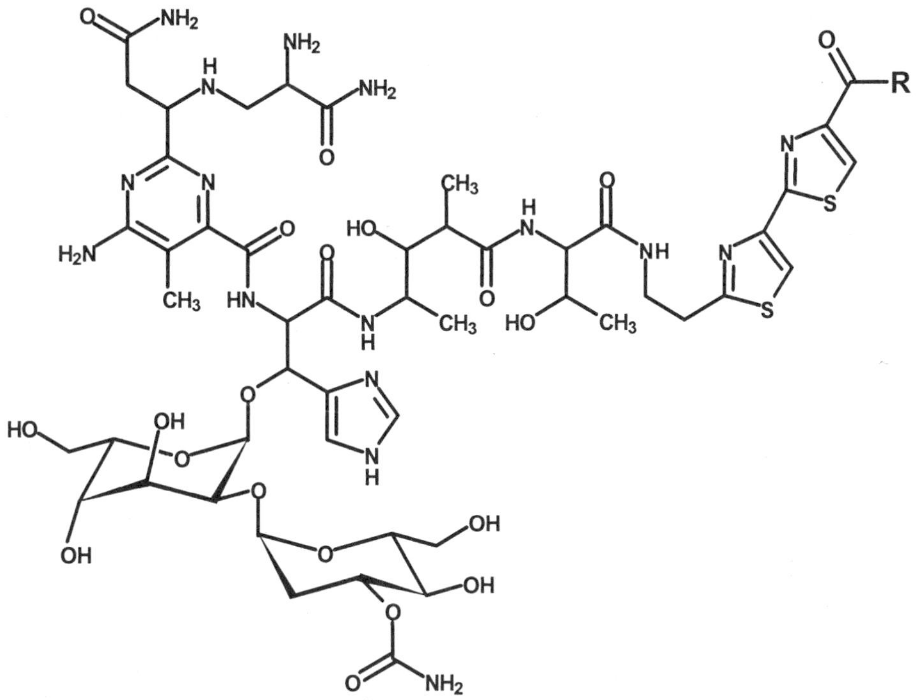

Figure 6. Structure of bleomycin, an antitumor antibiotic produced by *Streptomyces verticillatus*.

treatment of squamous cell carcinoma, Hodgkin's disease, testis tumors, and prostatic cancer.

The most prominent members of the group of anthracycline antibiotics showing antitumor activities are daunorubicin and adriamycin. The first anthracycline to be described was β-rhodomycin (Brockmann and Bauer, 1950). Daunorubicin was discovered independently in 1963 by groups working at Farmitalia in Italy (Grein et al., 1963) and at Rhône-Poulenc in France (Dubost et al., 1963). The group at Farmitalia isolated daunomycin from *Streptomyces peucetius*, the group at Rhône-Poulene isolated rubidomycin from *Streptomyces cueruleorubidus* and, because both compounds were identical, they were named daunorubicin. In addition to antibacterial and antifungal activity, daunorubicin showed an interesting activity against several human tumor cell lines and has progressed to clinical trials as an anticancer antibiotic. Later, adriamycin was discovered during a mutation program to find an improved daunorubicin (Arcamone et al., 1969). Adriamycin showed not only an improved therapeutic index but also a wider antitumoral spectrum than daunorubicin and gained enormous importance in the therapy of leukemia, lymphoma, and breast cancer. In 1971 Umezawa's group discovered aclacinomycin in the culture filtrate of *Streptomyces galilaeus*; aclacinomycin is a further member of the anthracycline group of antibiotics and shows fewer side effects than adriamycin (Oki et al., 1975). The compounds were detected in an agar plate diffusion assay using *Bacillus subtilis* as the test organism. Aclacinomycin A was successfully introduced into clinical use as an alternative to adriamycin.

Antitumor assays for high-throughput screening are mainly based on the specific inhibition of factors belonging to regulatory cascades. Such assays include the inhibition of protein-tyrosin phosphatases, which control the cell cycle (Galaktionov et al., 1996), or the inhibition of protein-tyrosin kinases (Levitzki and Gazit, 1995). A summary of screening systems for antitumor compounds of microbial origin is given by Komiyama and Funayama (1992).

In Vivo Antitumor Assays

Activities against transplantable tumors, especially slow-proliferating tumors, are tested on nude mice, which lack an immune system. Up to 12 solid tumors can be tested on one nude mouse. Test samples are applied in general as intraperitoneal injections, and the evaluation is performed by observing the survival days and measuring the tumor diameter.

ENZYME INHIBITITORY ASSAYS

Screening for low-molecular-weight enzyme inhibitors in micoroorganisms started independently in the latter half of the 1960s in the groups of Umezawa at the University of Tokyo and Sawao Murao at the Kumamoto Institute of Technology. To succeed in screening for enzyme inhibitors, it is necessary to use purified enzymes, or specific substrates if the enzymes cannot be purified enough. The screening was done with culture filtrates and extracts for inhibitory activities against endopeptidases; exopeptidases; enzymes related to the sugar, lipid, and nucleic acid pathways; phosphatases and kinases; enzymes related to neurotransmitters; and many other enzyme activities. A number of reviews have been published concerning the principles and techniques (Hill et al., 1998; Tanaka et al., 1992; Umezawa, 1982; and chapter 32 this book). Two examples are considered here because of their importance on the pharmaceutical market.

HMG-CoA Reductase Inhibitors

Hydroxymethylglutaryl (HMG)-coenzyme A (CoA) reductase (EC 1.1.1.88) is a key enzyme of cholesterol biosynthesis in the mevalonate pathway. In case of hypercholesterolemia, cholesterol is overexpressed, causing arteriosclerosis. A potent inhibitor of HMG-CoA reductase, compactin, was discovered independently in 1976 by Brown and coworkers at Beecham in the United Kingdom and by Endo and coworkers at Sankyo in Japan in cultures of *Penicillium* strains (Brown et al., 1976; Endo et al., 1976). A natural derivative of compactin from a *Monascus* species, monacolin K, was described by Endo in 1979 (Endo, 1979). One year later the same compound (mevinolin, lovastatin) was isolated by Alberts and coworkers at Merck Sharp & Dohme (Alberts et al., 1980). The statins, which is the generic name of this compound type, were introduced in 1987 to clinical use and remain one of the most profitable drugs on the pharmaceutical market, with total sales of US$15 billion in 2001. The structures of statins are shown in Fig. 7.

Glucosidase Inhibitors

S-AI, a potent inhibitor of α-amylase (EC 3.2.1.1), was discovered in *Streptomyces diastaticus* (Murao and Ohyama, 1975) and in *Actinoplanes* strain (BAYg5421) (Schmidt et al., 1977). BAYg5421 was introduced into clinics in 1990 as Acarbose, for treatment of obesity, and ranks within the top 20 most widely used in the pharmaceutical market.

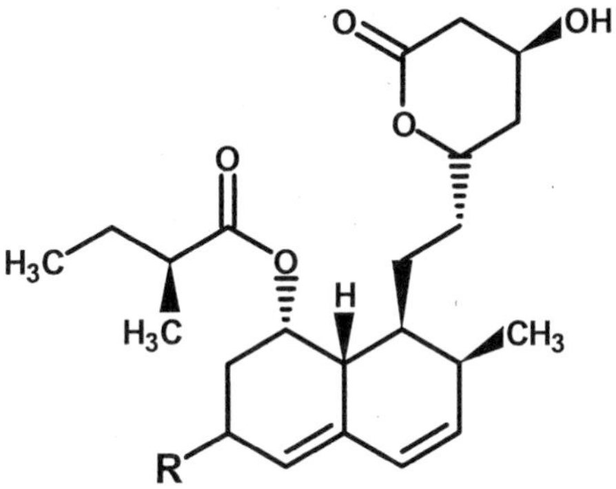

Figure 7. HMG-CoA reductase inhibitors of the compactin type (statins), isolated from various fungi.

ANTIPARASITIC ASSAYS

A huge percentage of the human population is affected with parasitic diseases caused by protozoa and helminths, especially in the tropical and subtropical areas. Despite the need for effective and less-toxic drugs for therapy of malaria (300 million people infected), schistosomiasis (200 million), filiariasis (130 million), leishmaniasis (12 million), and trypanosomasiasis (24 million), fewer efforts are undertaken by pharmaceutical companies to develop compounds for the market because of the smaller profits that can be gained in developing countries.

In Vitro Antiparasitic Screening

An examination of the literature shows a great variety of screening assays that are focused on antiparasitic action. Most of them are in vitro assays because of their easy handling in multiwell plates and reduced costs compared with in vivo animal assays. The free-living nematode *Caenorhabditis elegans* (Simpkin and Coles, 1981) and the plant pathogenic nematodes *Meloidogyne incognita* (Anke and Sterner, 1997) and *Bursaphelenchus lignicolus* (Otoguro, et al., 1988) can be successfully used in screening for anthelmintic antibiotics, although the efforts for growth and maintenance of the cultures are considerable. A simple alternative bioassay using the brine shrimp (*Artemia salina*) (Blizzard et al., 1989; Michael et al., 1956) can be used for detection of general toxicity of organic compounds and is highly sensitive to avermectins.

In Vivo Antiparasitic Screening

The most successful antiparasitic compound introduced to the market during the past decades is avermectin. This macrolide antibiotic was discovered in *Streptomyces avermitilis* by Ōmura and coworkers at the Kitasato Institute in close collaboration with Merck Sharp & Dohme (Burg et al., 1979). Avermectins were detected in an in vivo animal screening model showing a potent nematocidal activity against *Nematospiroides dubius* grown in mice. The most active components, avermectins B_{Ia} and B_{Ib} (Fig. 8), were chemically modified by hydratization at C_{22} and C_{23}, yielding ivermectin, which was introduced to the market as a broad-spectrum antiparastic agent for animals and humans. Ivermectin is used against gastrointestinal nematodes in domestic animals and as an insecticide and acaricide agent in agriculture. In human medicine, ivermectin is used for onchocerciasis (river blindness) caused by the nematode *Onchocerca volvulus* and for bancroftosis (elephantiasis) caused by *Wuchereria bancrofti*. Approximately 50 million people are infected by these nematodes, mainly in Africa. Successful therapy can be achieved by a single dose of ivermectin.

A simple in vivo screening model for detecting antiparasitic action, developed by Zähner (Tejmar-Kolar and Zähner, 1984), consisted of *Dinophilus gyrociliatus* as the host worm and *Eucoccidium dinophili* as the parasite. The entire development of the parasite, a regular change of gametogony and sporogony, occurs extracellularly in the body of the worm and can be observed microscopically.

HERBICIDAL ASSAYS

Herbicidal activity can be determined using *Lemna minor* L. (duckweed) or *Lepidium sativum* L. (garden cress) (Maier et al., 1999). The test is performed in multiwell plates, e.g., 24-well plates, using a chemically defined medium to which the test solution is added. After cooling of the agar, either one young plant of *Lemna minor* having two leaves is de-

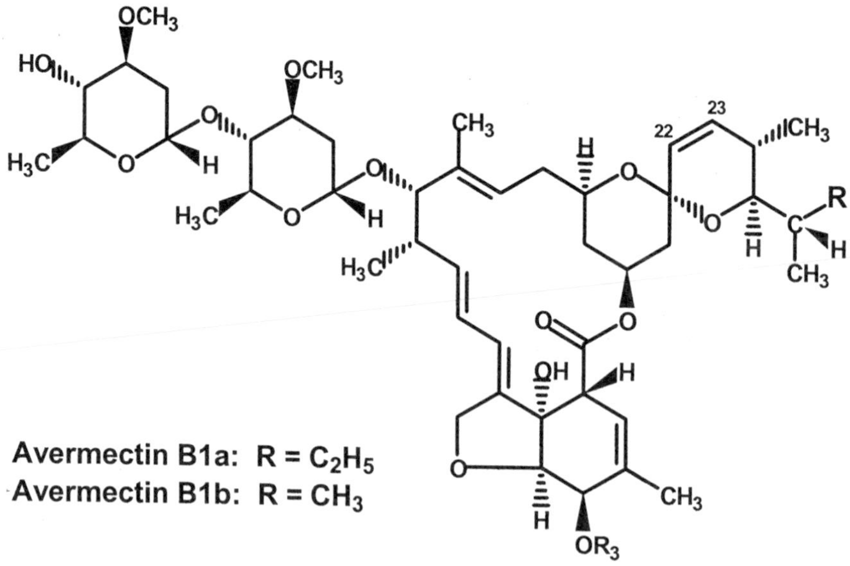

Avermectin B1a: R = C₂H₅
Avermectin B1b: R = CH₃

Figure 8. Structures of anthelminthic avermectins produced by *Streptomyces avermitilis.*

posited in each well on the surface of the medium or a *Lepidium sativum* seed is placed on the agar. Incubation times are 3 and 5 days in a climatic chamber with a 16/8 h light/dark photoperiod at 24°C and 60% humidity. Activity is evaluated by rating the ratio of bleaching and growth in the case of *Lemna minor* and as inhibition of germination in the case of *Lepidium sativum.*

ALGICIDAL ASSAYS

An algicidal assay based on the classical agar plate diffusion assay has been developed using the green alga *Chlorella fusca* (Schulz et al., 1995). The test is performed with a chemically defined medium, in which the alga is seeded in the soft top agar. Test solutions are applied on filter disk and, after incubation for 3 to 4 days with a 16/8 light/dark photoperiod, inhibition zones can be measured.

CHEMICAL SCREENING

Chemical screening is focused, in the first stage only, on the chemical diversity produced by microorganisms and on the assumption that each secondary metabolite produced has or had a biological function in the producing organism. The screen, which was originally based on thin-layer chromatography and staining reagents, was introduced by Umezawa (Umezawa et al., 1970) and was continued a few years

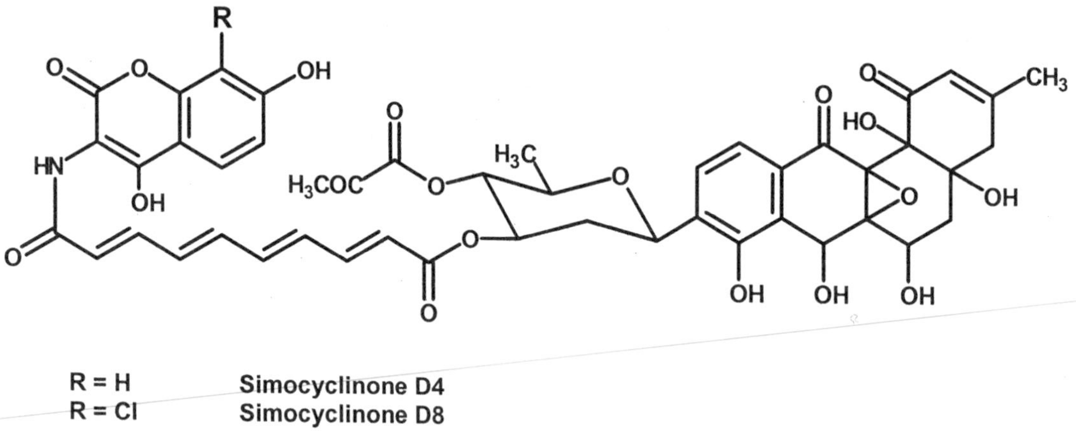

R = H   Simocyclinone D4
R = Cl   Simocyclinone D8

Figure 9. Structure of simocyclinones (antibacterial and antitumor antibiotics), inhibitors of gyrase and protein kinase, produced by *Streptomyces antibioticus* Tü 6040.

later by Ōmura, who detected staurosporin by this assay (Ōmura et al., 1977). Zähner and coworkers modified the method with regard to staining reagents, sample preparation, and variation of culture conditions of the microorganisms, and his group described numerous secondary metabolites having antibacterial, antifungal, antitumor, enzyme inhibitory, or insecticidal activities (Kuhn and Fiedler, 1995).

In 1984 Fiedler's group introduced high-performance liquid chromatography with computer-assisted diode-array detection and database technology to the chemical screening strategy. Coupling with electrospray mass spectrometry led to additional information of the molecular mass and permitted the identification of known and unknown metabolites in raw extracts without their isolation. The resulting secondary metabolites were summarized by Fiedler and Zähner (2001). Examples of prominent antibiotics that were detected by this screening method are the simocyclinones. These are novel "natural hybrid" polyketide antibiotics consisting of an unusual angucyclinone ring with a tetraene side chain and a coumarin ring (Schimana et al., 2000). The main congeners, simocyclinones D4 and D8 (Fig. 9), showed activity against gram-positive bacteria and several human tumor cell lines, and they are gyrase and protein kinase inhibitors.

REFERENCES

Alberts, A. W., J. Chen, G. Kuron, V. Hunt, J. Huff, C. Hoffman, J. Rothrock, M. Lopez, H. Joshua, E. Harris, A. Patchett, R. Monaghan, S. Currie, E. Stapley, G. Albers-Schonberg, O. Hensens, J. Hirschfield, K. Hoogsteen, L. Liesch, and J. Springer. 1980. Mevilonin: a potent competitive inhibitor of hydroxymethylglutaryl-coenzyme A reductase and a cholesterol-lowering agent. *Proc. Natl. Acad. Sci. USA* **77**:3957–3961.

Anke, H., and O. Sterner. 1997. Nematicidal metabolites from higher fungi. *Curr. Org. Chem.* **1**:361–374.

Arcamone, F., G. Cassinelli, G. Fantini, A. Grein, P. Orezzi, C. Pol, and C. Spalla. 1969. Adriamycin, 14-hydroxydaunomycin, a new antitumor compound from *S. peucetius* var. *caesius*. *Biotechnol. Bioeng.* **11**:1101–1110.

Benz, F., F. Knüsel, J. Nüesch, H. Treichler, and W. Voser. 1974. Echinocandin B. Ein neuartiges Polypeptid-Antibiotikum aus *Aspergillus nidulans* var. *echinulatus*: Isolierung und Bausteine. *Helv. Chim. Acta* **57**:2459–2477.

Bérdy, J. 1995. Are actinomycetes exhausted as a source of secondary metabolites? *Biotekhnologiya* **7**:3–23.

Blizzard, T. A., C. L. Ruby, H. Mrozik, F. A. Preiser, and M. H. Fisher. 1989. Brine shrimp (*Artemia salina*) as a convenient bioassay for avermectin analogs. *J. Antibiot.* **42**:1304–1307.

Brian, P. W., P. J. Curtis, and H. Hemming. 1946. A substance causing abnormal development of fungal produced by *Penicillium janczewskii*. I. Biological assay, production and isolation of "curling factor." *Trans. Br. Mycol. Soc.* **29**:173–187.

Brockmann, H., and W. Henkel. 1950. Pikromycin, ein neues Antibiotikum aus Actinomyceten. *Naturwissensch.* **37**:138–139.

Brockmann, H., and K. Bauer. 1950. Rhodomycin, ein rotes Antibiotikum aus Actinomyceten. *Naturwissensch.* **37**:492–493.

Brotzu, G. 1948. Richerche su di un nuovo antibiotico. *Lavori Dell Instituto d'Igiene di Cagliari* **1948**:m10–11.

Brown, A. G., T. C. Smale, T. J. King, R. Hasenkamp, and R. H. Thompson. 1976. Crystal and molecular structure of compactin, a new antifungal metabolite from *Penicillium brevicompactum*. *J. Chem. Soc. Perkin Trans.* **1**:1165–1170.

Bryskier, A. 2000. Cephems: fifty years of continuous research. *J. Antibiot.* **53**:1028–1037.

Burg, R. W., B. M. Miller, E. E. Baker, J. Birnbaum, S. A. Currie, R. Hartman, Y. L. Kong, R. L. Monaghan, G. Olson, I. Putter, J. B. Tunac, H. Wallick, E. O. Stapley, R. Oiwa, and S. Ōmura. 1979. Avermectins, new family of potent anthelmintic agents: producing organism and fermentation. *Antimicrob. Agents Chemother.* **15**:361–367.

Cooper, K. E. 1963. The theory of antibiotic inhibition, p. 1–86. *In* D. Kavanagh (ed.), *Analytical Microbiology*. Academic Press, New York, N.Y.

Dähn, U., H. Hagenmaier, H. Höhne, W. A. König, G. Wolf, and H. Zähner. 1976. Nikkomycin, ein neuer Hemmstoff der Chitinsynthese bei Pilzen. *Arch. Microbiol.* **107**:143–160.

Debono, M., and R. S. Gordee. 1994. Antibiotics that inhibit fungal cell wall development. *An. Rev. Microbiol.* **48**:471–497.

Debono, M., B. J. Abbott, J. R. Turner, L. C. Howard, and R. S. Gordee. 1988. Synthesis and evaluation of LY121019, a member of a series of semi-synthetic analogues of the antifungal lipopeptide echinocandin B. *Ann. N. Y. Acad. Sci.* **544**:152–167.

Dreyfuss, M., E. Haerri, H. Hofmann, H. Kobel, W. Pache, and H. Tscherter. 1976. Cyclosporin A and C. New metabolites from *Trichoderma polysporum* (Link ex Pers.) Rifai. *Eur. J. Appl. Microbiol.* **3**:125–133.

Dubos, R. L. 1939. Bactericidal agent extracted from a soil bacillus. I. Preparation of the agent. Its activity in vitro. *J. Exp. Med.* **70**:1–10.

Dubost, M., P. Ganter, R. Maral, L. Ninet, S. Pinnert, J. Preud'Homme, and G. H. Werner. 1963. Un nouvel antibiotique a propriétés cytostatiques: la rubidomycine. *C.R. Acad. Sci. Paris* **257**:1813–1815.

Dürkheimer, W. 1975. Tetracyclines: chemistry, biochemistry, and structure-activity relations. *Angew. Chem.* **14**:721–734.

Ehrlich, J., Q. R. Bartz, R. M. Smith, D. A. Joslyn, and P. R. Burkholder. 1947. Chloromycetin, a new antibiotic from a soil actinomycete. *Science* **106**:417.

Endo, A. 1979. Monacolin K, a new hypocholesterolemic agent produced by a Monascus species. *J. Antibiot.* **32**:852–854.

Endo, A., M. Kuroda, and Y. Tsujita. 1976. ML-236a, ML-236b, and ML-236c, new inhibitors of cholesterogenesis produced by *Penicillium citrum*. *J. Antibiot.* **29**:1346–1348.

Fiedler, H.-P., and H. Zähner. 2001. Screening for new secondary metabolites from microorganisms, p. 16–51. *In* V. Braun and F. Götz (ed.), *Microbial Fundamentals of Biotechnology*. Wiley-VCH, Weinheim, Germany.

Fiedler, H.-P., T. Schüz, and H. Decker. 1993. An overview of nikkomycins: history, biochemistry, and applications, p. 325–352. *In* J. W. Rippon and R. A. Fromtling (ed.), *Cutaneous Antifungal Agents: Compounds in Clinical Practice and Development*. Marcel Dekker, New York, N.Y.

Fleming, A. 1922. Remarkable bacteriolytic element found in tissues and secretions. *Proc. R. Soc.* **93B**:306–317.

Fleming, A. 1929. The antibacterial action of cultures of a *Penicillium* with special reference to their use in the isolation of *B. influenzae*. *Br. J. Exp. Pathol.* **10**:226–236.

Flynn, E. H. 1972. *Cephalosporins and Penicillins. Chemistry and Biology*. Academic Press, New York, N.Y.

Galaktionov, K., X. Chen, and D. Beach. 1996. CdC25 cell-cycle phosphatase as a target of c-*myc*. *Nature* **382**:511–517.

Gottlieb, D., F. K. Bhattacharyya, H. W. Anderson, and H. E. Carter. 1948. Some properties of an antibiotic obtained from a species of *Streptomyces*. *J. Bacteriol.* 55:409–417.

Grein, A., C. Spalla, A. Di Marco, and G. Canevazzi. 1963. Descrizione e classificazione di un attinomicete (*Streptomyces peuticus* sp. nova) productore di una sostanza ad attivita antitumorale: la daunomicina. *G. Microbiol.* 11:109–118.

Grove, D. C., and W. A. Randell. 1950. *Assay Methods of Antibiotics*. Med. Encyclopaedia, New York, N.Y.

Harned, R. L., P. H. Hidy, and E. K. La Baw. 1955. Cycloserine. I. A preliminary report. *Antibiot. Chemother.* 5:204–205.

Harris, D. A., M. Ruger, M. A. Reagan, F. J. Wolf, R. L. Peck, H. Wallick, and H. B. Woodruff. 1955. Discovery, development and antimicrobial properties of D-4-amino-3-isoxazolidone (oxamycin), a new antibiotic produced by *S. garyphalus* n. sp. *Antibiot. Chemother.* 5:183.

Hazen, E. L., and R. Brown. 1950. Two antifungal agents produced by a soil actinomycete. *Science* 112:423.

Hendlin, D., E. O. Stapley, M. Jackson, H. Wallick, A. K. Miller, F. J. Wolf, T. W. Miller, L. Chaiet, F. M. Kahan, E. L. Foltz, H. B. Woodruff, J. M. Mata, S. Hernandez, and S. Mochales. 1969. Phosphonomycin, a new antibiotic produced by strains of streptomycetes. *Science* 166:122–123.

Henkel, T., R. M. Brunne, H. Müller, and F. Reichel. 1999. Statistische Untersuchung zur Strukturkomplexität von Naturstoffen und synthetischen Substanzen. *Angew. Chem.* 111:688–691.

Hill, D. C., S. K. Wrigley, and L. J. Nisbet. 1998. Novel screen methodologies for identification of new microbial metabolites with pharmaceutical activity. *Adv. Biochem. Eng. Biotechnol.* 59:73–121.

Hori, M., K. Kakiki, and T. Misato. 1974. Further study on the relation of polyoxin structure to chitin synthetase inhibition. *Agric. Biol. Chem.* 38:691–698.

Kahan, J. S., F. M. Kahan, R. Goegelmann, S. A. Currie, M. Jackson, E. O. Stapley, T. W. Mille, A. K. Miller, D. Hendlin, S. Mochales, S. Hernandez, H. B. Woodruff, and J. Birnbaum. 1979. Thienamycin, a new beta-lactam antibiotic. I. Discovery, taxonomy, isolation and physical properties. *J. Antibiot.* 32:1–12.

Komiyama, K., and S. Funayama. 1992. Antitumor agents, p. 79–103. *In* S. Ōmura (ed.), *The Search for Bioactive Compounds from Microorganisms*. Springer-Verlag, New York, N.Y.

König, W. A., K.-P. Pfaff, W. Loeffler, D. Schanz, and H. Zähner. 1978. Ethericin A: Isolierung, Charakterisierung und Strukturaufklärung eines neuen, antibiotisch wirksamen Diphenylethers. *Liebigs. Ann. Chem.* 1978:1289–1296.

Kuhn, W., and H.-P. Fiedler. 1995. *Sekundärmetabolismus bei Mikroorganismen*. Attempto Verlag, Tübingen, Germany.

Levitzki, A., and A. Gazit. 1995. Tyrosine kinase inhibition: an approach to drug development. *Science* 267:1782–1788.

Maeda, K., H. Kosaka, K. Yagishita, and H. Umezawa. 1956. A new antibiotic, phleomycin. *J. Antibiot.* 9:82–85.

Maier, A., J. Müller, P. Schneider, H.-P. Fiedler, I. Groth, F. S. K. Tayman, F. Teltschik, C. Günther, and G. Bringmann. 1999. (2E, 4Z)-Decadienoic acid and (2E, 4Z,7Z)-decatrienoic acid, two herbicidal metabolites from *Streptomyces viridochromogenes* Tü 6105. *Pestic. Sci.* 55:733–739.

McGuire, J. M., R. L. Bunch, R. C. Anderson, H. E. Boaz, E. H. Flyan, H. M. Powell, and J. W. Smith. 1952. "Ilotycin," a new antibiotic. *Antibiot. Chemother.* 2:281–283.

Michael, A. S., C. G. Thompson, and M. Abramovitz. 1956. *Artemia salina* as a test organism for bioassay. *Science* 123:464.

Murao, S., and K. Ohyama. 1975. New amylase inhibitor (S-AI) from *Streptomyces diastaticus* var. *amylostaticus* No. 2476. *Agric. Biol. Chem.* 39:2271–2273.

Newton, G. G. F., and E. P. Abraham. 1953. Cephalosporin C, a new antibiotic containing sulfur and D-α-aminoadipinic acid. *Nature* 175:548–556.

Oki, T., Y. Matsuzawa, A. Yoshimoto, K. Numata, I. Kitamura, S. Hori, A. Takamatsu, H. Umezawa, M. Ishizuka, H. Naganawa, H. Suda, M. Hamada, and T. Takeuchi. 1975. New antitumor antibiotics, aclacinomycins A and B. *J. Antibiot.* 28:830–834.

Ōmura, S., Y. Iwai, A. Hirano, A. Nakagawa, J. Awaya, H. Tsuchiya, Y. Takahashi, and R. Masuma. 1977. A new alkaloid AM-2282 of *Streptomyces* origin, taxonomy, fermentation, isolation and preliminary characterization. *J. Antibiot.* 30:275–282.

Otoguro, K., Z.-H. Liu, K. Fukuda, Y. Li, Y. Iwai, H. Tanaka, and S. Ōmura. 1988. Screening for new nematocidal substances of microbial origin by a new method using the pine wood nematode. *J. Antibiot.* 41:573–575.

Oxford, A. E., H. Raistrick, and P. Simonart. 1939. Studies on the biochemistry of microorganisms. 60. Griseofulvin, $C_{17}H_{17}O_6Cl$, a metabolic product of *Penicillium* Dierekx. *Biochem. J.* 33:240–248.

Satomi, T., H. Kusakabe, G. Nakamura, T. Nishio, M. Uramoto, and K. Isono. 1982. Neopeptins A and B, new antifungal antibiotics. *Agric. Biol. Chem.* 46:2621–2623.

Schatz, A., A. Bugle, and S. A. Waksman. 1944. Streptomycin, a substance exhibiting antibiotic activity against Gram-positive and Gram-negative bacteria. *Proc. Soc. Exp. Biol. Med.* 55:66–69.

Schimana, J., H.-P. Fiedler, I. Groth, R. Süssmuth, W. Beil, M. Walker, and A. Zeeck. 2000. Simocyclinones, novel cytostatic angucyclinone antibiotics produced by *Streptomyces antibioticus* Tü 6040. *J. Antibiot.* 53:779–787.

Schmidt, D. D., W. Frommer, B. Junge, L. Müller, W. Wingender, and E. Truscheit. 1977. Alpha-glucosidase inhibitors, new complex oligosaccharides of microbial origin. *Naturwissensch.* 64:535–536.

Schulz, B., J. Sucker, H. J. Aust, K. Krohn, K. Ludewig, P. G. Jones, and D. Döring. 1995. Biologically active secondary metabolites of endophytic *Pezicula* species. *Mycol. Res.* 99:1007–1015.

Sensi, P., P. Margalith, and M. T. Timbal. 1959. Rifomycin, a new antibiotic. Preliminary report. *Farmaco Ed. Sci.* 14:146–147.

Shull, G. M., and J. L. Sardinas. 1955. PA-94, an antibiotic identical with D-4-amino-3-isoxazolidone (cycloserine, oxamycin). *Antibiot. Chemother.* 5:398–399.

Simpkin, K. G., and G. C. Coles. 1981. The use of *Caenorhabditis elegans* for anthelmintic screening. *J. Chem. Technol. Biotechnol.* 31:66–69.

Stapley, E. O., J. Birnbaum, A. K. Miller, H. Wallick, D. Hendlin, and H. B. Woodruff. 1979. Cefoxitin and cephamycins: microbiological studies. *Rev. Infect. Dis.* 1:73–89.

Sternberg, T. H., E. T. Wright, and M. Oura. 1956. A new antifungal antibiotic amphotericin B. *Antibiot. Annu.* 1955/56:566–573.

Takeuchi, T. 1987. *Institute of Microbial Chemistry 1962–1987*. Microbial Chemistry Research Foundation, Tokyo, Japan.

Tanaka, H., K. Kawakita, N. Imamura, K. Tsuzuki, and K. Shiomi. 1992. General screening of enzyme inhibitors, p. 117–160. *In* S. Ōmura (ed.), *The Search for Bioactive Compounds from Microorganisms*. Springer-Verlag, New York, N.Y.

Tejmar-Kolar, L., and H. Zähner. 1984. Search for effective substances against parasitic protozoa: an attempt to develop a new screening model. *FEMS Microbiol. Lett.* 24:21–24.

Umezawa, H. 1982. Low-molecular-weight enzyme inhibitors of microbial origin. *Annu. Rev. Microbiol.* 36:75–99.

Umezawa, H., T. Tazaki, H. Kanari, Y. Okami, and S. Fukuyama. 1948. Isolation of crystalline antibiotic substance from a strain of *Streptomyces* and its identity with chloromycetin. *Jpn. Med. J.* 1:358–363.

Umezawa, H., K. Maeda, T. Takeuchi, and Y. Okami. 1966. New antibiotics, bleomycin A and B. *J. Antibiot.* **19:**200–209.

Umezawa, H., T. Tsuchiya, K. Tatsuta, Y. Horiuchi, T. Usui, H. Umezawa, M. Hamada, and A. Yagi. 1970. A new antibiotic, dienomycin. I. Screening method, isolation and chemical studies. *J. Antibiot.* **23:**20–27.

Vézina, C., A. Kudelski, and S. N. Sehgal. 1975. Rapamycin (AY-22,989), a new antifungal antibiotic. I. Taxonomy of the producing streptomycete and isolation of the active principle. *J. Antibiot.* **28:**721–726.

Waksman, S. A., and H. B. Woodruff. 1940. Bacteriostatic and bactericidal substances produced by soil actinomycetes. *Proc. Soc. Exp. Biol. Med.* **45:**609–614.

Waksman, S. A., and H. B. Woodruff. 1942. Streptothricin, a new selective bacteriostatic and bactericidal agent particularly against Gram-negative bacteria. *Proc. Soc. Exp. Biol. Med.* **49:**207–209.

Wallick, H., D. A. Harris, M. A. Reagan, and H. B. Woodruff. 1955. Discovery and antimicrobial properties of cathomycin, a new antibiotic produced by *Streptomyces spheroides* n. sp. *Antibiot. Annu.* **1955–1956:**909–917.

Whitfield, G. B., T. D. Brock, A. Ammann, D. Gottlieb, and H. E. Carter. 1955. Filipin, an antifungal antibiotic: isolation and properties. *J. Am. Chem. Soc.* **77:**4799–4801.

Zähner, H. 1965. *Biologie der Antibiotica.* Springer, Berlin, Germany.

Zähner, H., and H.-P. Fiedler. 1995. The need for new antibiotics: possible ways forward, p. 67–84. *In* G. K. Darby, P. Hunter, and D. Russell (ed.), *Fifty Years of Antimicrobials.* Cambridge University Press, Cambridge, United Kingdom.

Zähner, H., H. Drautz, and W. Weber. 1982. Novel approaches to metabolite screening, p. 51–70. *In* J. D. Bu'Lock, I. J. Nisbet, and D. J. Winstanley (ed.), *Bioactive Metabolite Products: Search and Discovery.* Academic Press, London, United Kingdom.

Microbial Diversity and Bioprospecting
Edited by Alan T. Bull
© 2004 ASM Press, Washington, D.C.

Chapter 31

Antimicrobials

WILLIAM R. STROHL

To date, the vast majority of clinically relevant antibiotics, including antibacterials, antifungals, and antitumor drugs, have been either natural products or derived from natural products (Strohl, 1997; Craig et al., 1997; Shu, 1998). As of 1997, 26% of the marketed antibacterial, antifungal, and antitumor drugs (combined) were unmodified natural products, 38% were semisynthetic derivatives of natural products, 4% were chemically synthesized but patterned directly from bioactive natural products, and 31% were chemically synthesized drugs (Strohl, 1997). Of these, 80% of the antibacterial drugs owed their core scaffold to a bioactive natural product. In a separate analysis, of the 520 new pharmaceuticals approved for all therapeutic categories between 1983 and 1994, 5.8% were found to be unmodified natural products, 24.4% were semisynthetic natural product-based compounds, 8.8% were chemically synthesized drugs patterned after natural products, 55.6% were chemically synthesized, and 5.4% were "biologics" (e.g., recombinant proteins) (Craig et al., 1997). Thus, approximately 40% of the new drugs approved in that period were derived from natural product leads. Natural products dominated the antibacterial (63% were derived from natural product leads) and the anticancer (61% of the new drugs were derived from natural product leads) markets during that period (Craig et al., 1997). Thus, when one considers the discovery of new antibiotics, it seems clear that, at least from an historical perspective, natural products must comprise a significant component of the search.

Table 1 lists the major antibacterial scaffolds that have either resulted in marketed drugs or in promising antibacterials and the year that the discovery of the compound or class of compounds was first reported. After the discovery of penicillin by Alexander Fleming in 1928 and its subsequent development as the first major antibacterial drug during World War II, the pharmaceutical industry and well-known academic scientists, such as Selman Waksman of Rutgers University, worked feverishly to discover new antibacterials that would eliminate bacterial infections of all kinds. This enormous effort paid off in the discovery by 1962 of a wide variety of potent natural product antibacterial compounds, including streptomycin, the tetracyclines, chloramphenicol, cephalosporin C, neomycin, erythromycin, vancomycin, the rifamycins, lincomycin, and the streptogramins (Table 1). In fact, with the exceptions of the carbapenems (e.g., imipenem) and monobactams (e.g., Aztreonam), both classes of which were modeled after natural products during the 1970s, most current natural product antibacterial scaffolds were known by about 1970. The targets for these drugs, including cell wall, protein, and nucleic acid biosynthesis, all have been targeted many times over by analogues, as well as by other compounds, both synthetic and naturally derived. As documented below, there is a critical need for new, potent antibacterials to which resistance is not easily gained.

The natural product polyene antifungal drugs nystatin and amphotericin B also were discovered in the 1950s, making them a half-century old. One of the newest antifungal entries onto the market, Cancidas, is a semisynthetic echinocandin, the first of which were discovered in 1974 (Turner and Current, 1997). Thus, although new natural product antibacterial and antifungal scaffolds are discovered each year by academic and pharmaceutical company researchers, no new scaffold discovered within the past two decades has been developed and licensed. Nevertheless, the discovery of new natural products in general still continues at a rapid pace. Based on an analysis of papers published in the *Journal of Antibiotics,* Bérdy (1995) indicated that about 11,900 antibiotics and bioactive natural products had been discovered through 1994, with increasing rates of

William R. Strohl • Department of Biologics Research, Merck Research Laboratories, P.O. Box 2000, Mail-drop RY80Y-215, Rahway, N.J. 07065.

Table 1. Major groups of antibacterial, antifungal, and bioactive antibiotics; their source; and the year their discovery was publicly disclosed in the literature or the patent literature

Compound	Activity	Scaffold	Original microorganism producing the scaffold	Year of discovery
Penicillins	AB	β-Lactam (penam)	*Penicillium notatum*	1928
Sulfanilamide	AB	Sulfa	[Chemically synthesized]	1934
Tyrothricin	AB	Peptides mixture	*Bacillus* sp.	1936
Gramicidin	AB	Cyclic peptide	*Bacillus brevis*	1941
Streptomycin	AB	Aminoglycoside	*Streptomyces griseus*	1944
Nisin	AB	Lantibiotic	*Lactococcus lactis*	1944
Bacitracin	AB	Peptide	*Bacillus licheniformis*	1945
Chloramphenicol	AB	Phenylpropanoid	*Streptomyces venezuelae*	1948
Cephalosporins	AB	β-Lactam (ceph-3-em)	*Acremonium chrysogenum*	1948
Polymyxin B	AB	Cyclic peptide	*Bacillus polymyxa*	1948
Chlortetracycline	AB	Tetracycline	*Streptomyces aureofaciens*	1948
Neomycin	AB	Aminoglycoside	*Streptomyces* sp.	1949
Oxytetracycline	AB	Tetracycline	*Streptomyces rimosus*	1950
Nystatin	AF	Polyene	*Streptomyces noursei*	1950
Erythromycin	AB	14-Membered macrolide	*Saccharopolyspora erythrea*	1953
Synercid[b]	AB	Streptogramins	*Streptomyces pristinaespiralis*	1953
Tetracycline	AB	Tetracycline	*Streptomyces viridofaciens*	1953
Virginiamycin	FA	Streptogramins	*Streptomyces virginiaensis*	1955
Vancomycin	AB	Glycopeptide	*Streptomyces orientalis*	1955
Novobiocin	AB	Coumarin	*Streptomyces spheroides*	1955
Cycloserine	AB	Amino acid analogue	*Streptomyces orchidaceus*	1955
Amphotericin B	AF	Polyene	*Streptomyces nodosus*	1955
Rifampin	AB	Rifamycin	*Streptomyces mediterranei*	1959
Kanamycin	AB	Aminocyclitol	*Streptomyces kanamycetus*	1960
Tylosin	FA	16-Membered macrolide	*Streptomyces fradiae*	1961
Avilamycin	FA	Orthosomycin	*Streptomyces viridochromogenes*	1961
Spectinomycin	AB	Aminocyclitol	*Streptomyces spectabilis*	1961
Lincomycin	AB	Lincomycin	*Streptomyces lincolnensis*	1962
Ciprofloxacin	AB	Nalidixic acid (quinolone)	[Chemically synthesized]	1962
Fusidic acid	AB	Sterol	*Fusidium coccineum*	1962
Everninomycin	AB	Orthosomyen	*Micromonospora carbonacea*	1964
Monensin	FA	Polyesther	*Streptomyces cinnamonensis*	1967
Fosfomycin	AB	Phosphonic acid	*Streptomyces* sp.	1969
Miconazole	AF	Azole	[Chemically synthesized]	1969
Cefoxitin	AB	β-Lactam (cephamycin)	*Nocardia lactamdurans*	1971
Mupirocin	AB	Pseudomonic acid	*Pseudomonas fluorescens*	1971
Cuspofungin	AF	Echinocandin	*Glarea lozoyensis*	1974
Clavulanic acid	AB	β-Lactam (clavam)	*Streptomyces clavuligerus*	1975
Imipenem	AB	β-Lactam (carbapenem)	*Streptomyces cattleya*	1976
Aztreonam	AB	β-Lactam (monobactam)	*Chromobacterium violaceum*	1981
Terbinafine	AF	Allylamine	[Chemically synthesized]	1981
Linezolid	AB	Oxazolidinone	[Chemically synthesized]	1996

[a] AB, antibacterial; AF, antifungal; FA, feed additive. Not all clinically relevant compounds are listed (e.g., only the first azole is listed).
[b] For semisynthetic examples, the date is listed for the discovery of the scaffold rather than the development of the actual semisynthetic (e.g., whereas Synercid was first described in the literature as a potential development candidate in 1989, the first streptogramins from which it was derived were discovered in 1953).

discovery every year through 1990. He projected that the total of new, novel natural product structures discovered by the year 2000 would top 16,000 (Bérdy, 1995). The annual rate of new natural product descriptions published is now more than 500 per year, up from 200 to 300 per year 20 years ago (Bérdy, 1995; Strohl, 1997). However, sheer numbers obviously are not enough to put new natural products on the market. There is a significant qualitative difference, for example, between the discov-ery of a totally new bioactive scaffold versus the discovery of new analogues within known scaffold series. Unfortunately, new bioactive natural product scaffolds have not been advanced to the clinic since the mid-to-late 1970s (Table 1). Hence, one of the key points raised in this chapter is that screening for bioactive natural products in the present and future should include new approaches to find new natural chemical scaffolds on which to build future antibiotics.

NEED FOR NEW NATURAL PRODUCT ANTIBIOTICS

The total world sales for prescription pharmaceuticals was estimated in a September 1998 news release to be $308 billion. The total world market for antiinfective agents in 1996 was reportedly $23 billion (Strohl, 1999). The 1995 U.S. antibacterials market alone was greater than $8 billion, with cephalosporins (45%), penicillins (15%), fluoroquinolones (11%), tetracyclines (6%), and macrolides (5%) comprising the majority of sales. Of these top-selling anti-infective agents, only the fluoroquinolones are not derived from a natural product scaffold. Thus, in deference to their origins, most antibacterial drugs are still natural products, semisynthetic compounds derived from fermentation products or patterned after natural products. The world antifungal market in 1995 was approximately $3 billion. Although this represents a much smaller market than the antibacterial market, the antifungal market has been projected to grow at 20% annually. This rapid growth in antifungals unfortunately is driven by the dramatic increase in life-threatening fungal diseases, mostly as a result of opportunistic fungal pathogens infecting immuno-compromised patients.

The Pharmaceutical Research and Manufacturers of America organization announced that, as of September 1998, 136 new anti-infective drugs were in advanced stages of development by 78 different companies. Of these, 27 compounds were antibacterials, 31 were antivirals, 12 were antifungals, 42 were vaccines, and 24 were in the "other" category, but still relating to anti-infection indications. As described throughout this chapter, even with the development of promising new antibiotics, several unmet needs continue to exist in virtually all fields of antibiotics.

There are three important needs in the antibacterial market. First, new parenteral drugs that can successfully treat life-threatening, broadly antibiotic-resistant systemic infections are the most critical need. The incidence of methicillin-resistant *Staphylococcus aureus* (MRSA), vancomycin-resistant enterococci (VRE) (particularly by *Enterococcus faecium* and *Enterococcus faecalis*), and β-lactam-resistant *Streptococcus pneumoniae* are among the most serious problems, but are by no means the only ones (Lederberg et al., 1992). The incidence of nosocomial (i.e., hospital-acquired) infections is still on the rise, and of those, an increasingly greater percentage of the infectious agents (typically MRSA and VRE) are drug resistant (Weinstein, 1998). Of the 35 to 40 million hospitalizations in the United States each year, as many as 5% result in nosocomial infections, of which some 88,000 result in deaths annually (Weinstein,

1998). Additionally, new pathogens are evolving at an alarming rate. For example, today there are more than 30 new infectious diseases (e.g., AIDS, Ebola, Legionnaires disease, hantavirus, Lyme disease, West Nile virus, and foodborne *E. coli* O157:H7 infections) that were virtually unknown 20 years ago (Strohl, 1997, 1999; World Health Organizaiton, 1996). Finally, even with the development of remarkable anti-infective drugs with outstanding biological activities, there are still certain pathogens that naturally defy even the best efforts to defeat them. Perhaps the best example of this is the difficulty in treating successfully *Pseudomonas aeruginosa* infections resulting from cystic fibrosis (Strohl, 1997). Also, there are strains of *Acinetobacter* spp. (mostly *A. baumannii*) that are resistant to virtually every antibiotic tested (Mahgoub et al., 2002). These persistent organisms can be lethal nosocomial pathogens and are beginning to pose a serious threat, particularly in intensive care units (Wisplinghoff et al., 2000).

Vancomycin has long been considered the last line of defense against antibiotic-resistant bacteria, particularly MRSAs and β-lactam- and aminoglycoside-resistant enterococci. Just a little over a decade ago, it was thought that microorganisms would never become resistant to vancomycin due to its unusual mechanism of action (Rowe, 1996). VRE, however, were found in England and France in 1987 (Uttley et al., 1988), making experts rethink their approaches. VREs now are being discovered at an ever-increasing rate, with no new "last line of defense" substitute firmly entrenched on the market (French, 1998). The greatest concern of infectious disease experts, however, has been the emergence of vancomycin-resistant MRSAs (methicillin-resistant *S. aureus*) (Walsh and Howe, 2002). This spectre was first raised a few years ago, when vancomycin-intermediate (resistance) *S. aureus* (VISA; also known as GISA; "glycopeptide-intermediate resistant *S. aureus*") were described (Ploy et al., 1998; Walsh and Howe, 2002). Unfortunately, highly disturbing reports of truly vancomycin-resistant bacteria have very recently emerged. An *S. aureus* strain was isolated from a patient in Illinois that developed an MIC to vancomycin of 12 µg/ml via a resistance mechanism that appears to be different from that observed in VISA strains (Boyle-Vavra et al., 2001). Sievert et al. (2002a, 2002b) very recently described an MRSA isolate from a patient in Michigan that also was resistant to 128 µg of vancomycin per ml. Perhaps most disturbing was the presence in this isolate of the *vanA* locus on a 60-kbp plasmid. Furthermore, a VRE strain was isolated from the same infection, suggesting the that *S. aureus* *vanA* had been acquired in vivo from *vanA* enterococci (Sievert et al., 2002a, 2002b). The potential for this type of interspecies transfer had been shown experimentally a decade ago (Noble et al., 1992), but this was the first clinical *S. aureus* isolate obtained with the *vanA* locus (Woodford, 2001; Walsh and Howe, 2002). This raises the very real possibility that potentially lethal bacterial pathogens may become resistant to every available antibacterial drug. Although the incidence of a true vancomycin-resistant *S. aureus* is apparently very low today, it is not unrealistic to suppose that, due to the potentially high-frequency nature of *vanA* interspecies transfer, it may exist broadly in a decade. The fact that it typically takes more than a decade and >$800 million to discover and develop new antibiotics implies that we must make significant breakthroughs now to stave off potentially serious threats of resistant infectious bacteria in the near future.

Second, even with a seemingly crowded pediatric market, there is still need and a market for an orally absorbed, pediatric-safe antibacterial that can successfully treat upper respiratory tract infections (primarily otitis media caused by β-lactam-resistant streptococci, moraxellae, and nontypeable *Haemophilus influenzae*) and not be compromised by β-lactam, aminoglycoside, or macrolide-lincosamide-streptogramin (MLS) resistance or degraded by extended spectrum β-lactamases (Sunekawa et al., 1995). The pediatric antibacterial market, which has grown sharply since the early 1980s, has become one of the most important anti-infective markets today. Of the six top-selling antibacterials in 2001, four were major components of this upper respiratory pediatric market (2001 sales: Augmentin [amoxicillin/clavulanic acid], $2.06 billion; Zithromax [azithromycin], $1.51 billion; Biaxin [clarithromycin], $1.16 billion; and Rocephin [ceftriaxone], $1.07 billion). The other two top sellers were Cipro (ciprofloxacin; $1.76 billion) and Levaquin/Floxin (levofloxacin/ofloxacin; $1.07 billion).

The final critical antibacterial need is for a tuberculocidal drug. Tuberculosis is one of the most devastating infectious diseases in the world today, with approximately 2 billion people currently infected with *Mycobacterium tuberculosis* (Kamholz, 2002). Moreover, an additional ca. 9 million new cases are diagnosed annually, and 2.6 million deaths worldwide are attributable to tuberculosis yearly (Kamholz, 2002). Additionally, antibiotic resistance in *M. tuberculosis* is a significant and rapidly growing problem. Currently, combination treatment with isoniazid and rifamycin (both tuberculostats) is the method of choice, but approximately 14% of all tuberculosis cases reported today involve *M. tuberculosis* strains that are resistant to one or more antibiotics (Bloom and Murray, 1992; Kamholz, 2002). The

problem facing doctors today is that there is yet no proven effective alternative on the market to treat isoniazid- and rifamycin-resistant *M. tuberculosis* (Bloom and Murray, 1992).

Even with several new recent entries on the market, there is always a need for new antifungal drugs. Fungal infections can either be localized, dermal infections such as ringworm, athlete's foot, and nail infections (onychomycoses), or those that are disseminated and life threatening. Life-threatening, disseminated fungal diseases have risen dramatically over the past several years, making this a critical area for new drug development. Typically, life-threatening fungal infections are associated with conditions of immunosuppression and neutropenia that occur in patients with acquired immunodeficiency syndrome (AIDS), under treatment with immunosuppressive drugs (usually after organ transplants, to prevent tissue rejection), with chemotherapy treatment, or with severe burns. Additionally, with the potentially broader use of immunosuppressive drugs for other indications such as rheumatoid arthritis, fungal infections may be expected to continue to rise.

The most important pathogenic fungi are *Candida albicans* (disseminated candidiasis), *Candida* spp., *Aspergillus fumigatus* (systemic aspergillosis), *Cryptococcus neoformans* (disseminated cryptococcosis leading to meningoencephalitis), *Pneumocystis carinii* (*P. carinii* pneumonia), *Histoplasma capsulatum* (histoplasmosis), and *Coccidioides immitis* (D. Ellis, 2002; M. Ellis, 2002). The most important fungal pathogens associated with immunocompromised patients, such as those with AIDS or receiving immunosuppressive therapy, are *P. carinii*, *C. neoformans*, and *Candida* spp.

Some of the historically used antifungal drugs suffer from problems that make them less than optimal for clinical use. Polyenes such as amphotericin B, nystatin, candicidin, or faerifungin target fungal cell membranes by forming transmembrane pores and disrupting membrane function (Gil and Martin, 1997). Amphotericin B, the historical antifungal drug of choice in life-threatening cases, suffers from acute toxicity (D. Ellis, 2002).

Azoles (e.g., clotrimazole, fluconazole, ketoconazole), which are widely popular due to their oral administration, target lanosterol 14-demethylase, a P450 enzyme involved in the synthesis of ergosterol (Tkacz and DiDomenico, 2001), a sterol found in fungal cell membranes but not mammalian cell membranes. Azoles are generally considered to be static in vivo, i.e., they stop fungi from growing but do not kill them, relying on the immune system to clear the existing fungal infection; a few studies, however, have demonstrated in animal models the ability of azoles to prevent outgrowth even after cessation of antibiotic treatment (Tkacz and DiDomenico, 2001). Certainly, however, resistance to azoles, especially during long treatment regimens, is considered a growing problem, particularly among *Candida* spp. (Tkacz and DiDomenico, 2001).

The newest antifungal drug on the market is Cancidas (caspofungin), a semisynthetic echinocandin that inhibits the synthesis of β-1,3-glucans, a major component of the fungal cell wall (Powles et al., 1998; Tkacz and DiDomenico, 2001). Cancidas specifically targets the Rho1p dissociable subunit of the heterodimeric transmembrane 1,3-β-(D)-glucan synthase (Turner and Current, 1997; Powles et al., 1998; Tkacz and DiDomenico, 2001). Cancidas, currently approved for use against disseminated aspergillosis, has been shown to possess in vivo efficacy against *A. fumigatus*, *C. albicans*, *H. capsulatum*, and *P. carinii* (but not against *C. neoformans*), suggesting that this would make an outstanding potential broad-spectrum candidate for disseminated, life-threatening fungal diseases, particularly in immunocompromised patients (Powles et al., 1998). Other echinocandins such as anidulafungin and micafungin also are being developed as antifungal agents targeting glucan synthesis (Tkacz and DiDomenico, 2001).

Genomics, particularly the sequencing of the yeast genome, has helped considerably to identify new fungal targets. Of some 6,000 genes analyzed from the *Saccharomyces cerevisiae* genome sequencing project, only a dozen or so potential unique targets were identified, including chitin synthetase, β-1,4-glucan synthetase, tubulin, elongation factor 2, N-myristoyl transferase, acetyl-CoA carboxylase, inositol phosphoryl ceramide synthase, membrane ATPase, mannosyl transferase, tRNA synthetases, lanosterol dehydrogenase, lanosterol synthase, and squalene epoxidase (Koltin, 1998). There are several natural products that already have been shown to possess significant activities against some of these targets. The leading antifungal chitin biosynthesis-inhibitor drug candidate is nikkomycin Z, a nucleoside dipeptide substrate analog of UDP-N-acetylglucosamine that competitively inhibits chitin synthases (Graybill et al., 1998). Rustmicin (also called galbonolide A), a 14-membered macrolide produced by *Micromonospora chalcea*, inhibits inositol phosphoceramide synthase at pM levels, resulting in the loss of all complex sphingolipids and the accumulation of ceramide (Mandala et al., 1998). The pradamycins are D-amino acid-substituted benzonaphthacene quinone glycosides that appear to act by binding to the saccharide moieties of the mannoproteins in fungal cell walls, leading to a loss of intracellular potassium and resulting in gross morphological changes

(Shu, 1998). Another unusual antifungal compound is sordarin, which exerts antifungal activity via inhibition of the elongation step in protein synthesis (Justice et al., 1998). Glaxo-Wellcome has developed a semisynthetic sordarin, GM237354, which possesses a broad spectrum of activity and high in vivo potency (Shu, 1998). Significantly, resistance against single enzyme targets in yeasts and fungi does not develop as quickly as observed with bacteria. Thus, additional new natural product antifungal entries that target cell wall synthesis, that yield fungicidal activity in vivo, or that block key gene products required for growth of fungal and yeast pathogens are attractive possibilities.

CHEMICAL DIVERSITY VERSUS BIOLOGICAL DIVERSITY

The first issue to be addressed is what constitutes microbial diversity in the context of natural products. According to the experts, only a few percent of the bacteria and filamentous fungi have supposedly been isolated, cultivated, and tested for the production of antimetabolites. Does this statistic presuppose that the remaining 95-plus percent of untested microorganisms will bring us a bounty of completely new structures, new scaffolds, and new paradigms to the antimetabolite field? In reality, only a small fraction of microorganisms are genetically predisposed to producing natural products of interest, and a wide cross section of these have been mined significantly for their antibacterial, antifungal, and anticancer products over the past 50 years (Tables 2 and 3). Bérdy (1995) showed that of the 11,900 natural products that had been discovered through 1994, approximately 6,600 (55%) were produced by *Streptomyces* spp. The remainder of producing organisms included the filamentous fungi, which produced 2,600 (22%); nonactinomycete bacteria, which produced 1,400 (12%); and non-*Streptomyces* strains of actinomycetes, which produced 1,300 (11%) (Bérdy, 1995). When broken down by pharmacological activity, however, the streptomycetes and other actinomyctes produced 66% of the compounds which demonstrated antibacterial, antifungal, and antitumor activities (Bérdy, 1995). Filamentous fungi and nonactinomycetous bacteria produced only 22% and 12%, respectively, of the natural products with "antibiotic" activities. On the other hand, 63% of the compounds isolated from extracts of filamentous fungi were bioactive for pharmacological targets other than antibiotics. Thus, although penicillin, the first natural product, was a fungal-derived antibacterial agent, it seems that fungal production of devel-

Table 2. Biogenic classes of natural product antibiotics and examples of producing microorganisms

Biogenic class and subclass	Producing microbes[a]	Example compound, activity, and producer
Peptide synthetases		
Diketopiperazines	F, P	Tryprostatin (antimitotic; *Aspergillus* sp.)
Cyclic peptides	A, B, C, E, F, M, P	Gramicidin (antibacterial; *Bacillus* sp.)
Depsipeptide	A, B, C, E, F, M, P	Valinomycin (ionophore; *Streptomyces* sp.)
Lipopeptides	A, B, C, F, M, P	Pneumocandin (antifungal; *Glaria* sp.)
Glycopeptides	A, E, M	Vancomycin (antibacterial; *Actinomadura* sp.)
β-Lactams	A, F, P	Penicillin (antibacterial; *Penicillium* sp.)
Lantibiotic	B, L	Nisin (antibacterial: *Lactococcus* sp.)
Peptidolactones	A	Actinomycin D (antitumor; *Streptomyces* sp.)
Phosphinopeptide	A, B	Bialaphos (phytotoxic; *Streptomyces* sp.)
Polyketides		
Type I modular		
Macrolides	A, B, C, M	Erythromycin (antibacterial; *Saccharopolyspora* sp.)
Pentacyclic lactones	A	Avermectin (insecticidal; *Streptomyces* sp.)
Ascomycins	A	Tacrolimus (immunosuppressive; *Streptomyces* sp.)
Ansamycins	A	Rifamycin (tuberculostatic; *Streptomyces* sp.)
Polyenes	A, B, E, M, P	Amphotericin B (antifungal; *Streptomyces* sp.)
Polyethers	A, E	Monensin (growth promotant; *Streptomyces* sp.)
Type I iterative	A	Lovastatin (cholesterol-lowering; *Aspergillus* sp.)
Type II iterative	A	Oxytetracycline (antibacterial; *Streptomyces* sp.)
Type I (*Pseudomonas*)	P	Mupirocin (antibacterial)
Aminoglycosides		
Aminoglycosides	A, B, P	Streptomycin (antibacterial; *Streptomyces* sp.)
Aminocyclitols	A	Spectinomycin (antibacterial; *Streptomyces* sp.)
Cyclitol	A	Kasugamycin (antibacterial; *Streptomyces* sp.)
Isoprenoids		
Sesquiterpene	A, E, F, M	Trichothecene (toxin; *Fusarium* sp.)
Diterpene	A, C, E, F	Nodulisporic acid (insecticide; *Nodulisporium* sp.)
Triterpene	A, B, E, F, P	Fusidic acid (antibacterial; *Fusidium* sp.)
Indole alkaloid	E, F	Hypaphorine (plant auxin antagonist; *Pisolithus* sp.)
Other		
Amino acis; dipeptides	A, B	Cycloserine (antibacterial; *Streptomyces* sp.)
Chorismate derived	A	Chloramphenicol (antibacterial; *Streptomyces* sp.)
Lincosamides	A	Lincomycin (antibacterial; *Streptomyces* sp.)
Mitosanes	A	Mitomycin C (antitumor; *Streptomyces* sp.)
Nucleosides	A, B, C, P	Nikkomycin (antifungal; *Streptomyces* sp.)
Oligosaccharides	A	Avilamycin (antibacterial; *Streptomyces* sp.)
Tetramic acid derivatives	A, F, P	Maltophilin (antifungal toxin; *Stenotrophomonas* sp.)

[a] A, actinomycetes (including the genus *Streptomyces*); B, *Bacillus* spp.; C, cyanobacteria; E, eukaryotic algae (red, green, or brown); F, filamentous fungi; M, myxobacteria; P, proteobacteria.

opable antibacterial compounds may be more of an exception than the rule.

Table 2 gives a broad view of the types of chemistry produced by each of the major groups of microorganisms normally considered as producers of natural products. Although the chemistry represented in Table 2 is far from all-inclusive, it does give an indication of the types of natural product molecules that have been found from the various groups of microorganisms. Henkel et al. (1999) recently carried out a brilliant study demonstrating that natural products in general offer a large pool of diverse structures that differs significantly from those found in most synthetic chemical libraries.

The largest family of natural products is probably the terpenoids, with a reported 22,000 known examples (Connolly and Hill, 1992). Although ter-

penoids are produced by virtually all of the major groups of microorganisms (Table 2), most of them are produced by plants and filamentous fungi. Also, there is considerable bias in production of terpenoid subgroups (e.g., monoterpenes, sesquiterpenes, diterpenes, triterpenes, indole diterpenes) by groups of microorganisms. For example, indole diterpenes are produced primarily by filamentous fungi such as *Aspergillus* spp., *Penicillium* spp., and *Claviceps* spp., as well as by some eukaryotic algae. Important to the tenet of this chapter is the fact that a new diterpene was just recently discovered from *Streptomyces griseolosporeus* (Dairi et al., 2001), demonstrating that even after years of screening actinomycetes, new scaffolds can be found from these organisms. In a screen of actinomycetes for genes associated with terpenoid biosynthesis, Sigmund et al. (2003) very recently

Table 3. Prolificity and structural diversity of natural products by microorganisms

Microorganism[a]	Prolificity score	Structure diversity score	Ease of manipulation	No. of compounds isolated[b]	Comments
Streptomyces spp.	++++	+++	+++	>7,000	Most prolific commercial antibiotic producers; antibacterials; antitumors; antifungals
Filamentous fungi	++++	++++	++	>3,000	Most prolific natural product producers; antifungals; β-lactams; insecticides; toxins
Actinomycetes	+++	+++	+++	>2,000	Prolific antibacterial and antitumor producers
Myxobacteria	++	++++	±	~80	Prolific producers of unusual scaffolds but difficult to isolate and grow
Bacillus spp.	+	+	++++	~50	Producers of cyclic peptides, lipopeptides, and polyenes, but with limited overall structural diversity
Gammaproteobacteria	+	+	++++	~40	β-lactams, peptides, indoles, polyenes, macrolides, and amides
Pseudomonads	+	+	++++	~35	Producers of a few interesting antifungal compounds
Eukaryotic algae	+++	++	±	~35	Terpenoids, sterols, cyclic peptides
Cyanobacteria	±	±	+	~25	Producers of cyclic and linear peptides; mostly cytotoxins
Betaproteobacteria	±	±	++++	~10	A few β-lactams and antifungals
Dinoflagellates	±	±	±	~10	Polyenes, macrolides, terpenoids
Cytophaga group	±	±	++++	~5	A few β-lactams and peptides
Alphaproteobacteria	±	±	++++	~5	A few β-lactams and polyenes

[a] *Gammaproteobacteria* producing natural products include members of *Lysobacter, Serratia, Vibrio, alteromonas, Erwinia, Xanthomonas, Stenotrophomonas, Xenorhabdus, Photorhabdus, Chromatium; Betaproteobacteria* include *Burkholderia, Chromobacterium; Alphaproteobacteria* include *Gluconobacter, Methylobacterium, Agrobacterium.* The category Actinomycetes represents non-*Streptomyces* actinomycetes.
[b] Numbers are approximations that are derived from various literature sources plus a survey of references cited in PubMed from 1980 to the present.

demonstrated that these pathways were only present in less than 1% of isolates tested and that they were also associated with marine microorganisms.

In addition to the terpenoids, the various non-ribosomally synthesized peptides and the polyketides are the most commonly produced natural products. The cyclic-, depsi-, and lipopeptides as a group are produced by the broadest range of microorganisms. These groups of natural products, produced by non-ribosomal peptide synthetase mechanisms, are found in virtually every taxonomic group (i.e., proteobacteria, bacilli, actinomycetes, cyanobacteria, myxobacteria, eukaryotic algae, filamentous fungi) of natural product producers. The polyketides are interesting in that, although they are produced by several of the taxonomic groups, the mechanisms for their biosynthesis are more taxonomically restricted. Macrolides, for example, which are synthesized via a prokaryotic type I polyketide synthase (PKS) (Hopwood, 1997), are produced nearly exclusively by actinomycetes, bacilli, and myxobacteria. Aromatic polyketides derived from type II iterative PKSs are produced primarily by actinomycetes (Hopwood, 1997), and type I iterative PKSs are exclusively found in filamentous fungi (Nicholson et al., 2001). Polyenes, which are produced by a wide variety of prokaryotic microorganisms including actinomycetes, bacilli, myxobacteria, and proteobacteria (as well as by eukaryotic algae), are probably the most broadly distributed polyketides produced by microorganisms (Table 2).

BACTERIAL DIVERSITY WITH RESPECT TO NATURAL PRODUCT ANTIBIOTICS

Although it is nearly impossible to predict the true number of bacterial species (Aman et al., 1995; Pace, 1997; Hugenholz et al., 1998), it is generally accepted that the ca. 4,500 to 5,000 bacteria that have been isolated, characterized, and described taxonomically probably represent less than 1% of the potential diversity available. This suggests either that a great number of bacteria are not easily, or at all, capable of being isolated, or that conditions have not been found to isolate them. This has led some researchers to suggest that an incredible diversity of natural products is yet to be discovered from this enormous pool of as-yet-undiscovered bacteria. Although there is clearly new chemical diversity still to be found from bacterial natural products, it is more likely that a relatively broad sampling of this diversity has already occurred. There are multiple reasons for this supposition. First, the best producers of natural product antibiotics are members of the actinomycetes, and according to Hugenholz et al. (1998), approximately 30% or more of these have been discovered at one time or another. This has been the result of some 50 years of intense antibiotic screening programs conducted by pharmaceutical industry natural products scientists.

Based on environmental DNA sampling, a large fraction of unisolated bacteria, in fact, are now thought to be members of the *OP11*, *Verrumicrobia*, and *Holophaga-Acidobacterium* divisions, few members of which have been isolated and described (Hugenholz et al., 1998; Rondon et al 1999). Considering that most of the prolific bacterial producers (e.g., actinomycetes, bacilli, pseudomonads, *Gammaproteobacteria*) of natural products are fairly well-studied microorganisms, it is not likely that there are significant, new unexplored division-level groups of natural product-forming bacteria yet to be found. On the other hand, Rheims et al. (1996) showed 16S rDNA data suggesting that several new actinomycete-related phylogenetic lines of descent could be found in DNA isolated from soil. Additionally, Rheims and Stackebrandt (1999) showed clear rDNA evidence for the existence of a deep-rooted actinobacterial class for which member strains have not yet been isolated. This is supported by the fact that seven new genera within the *Actinobacteria* group have been named just since 1999, suggesting that additional undiscovered or uncharacterized diversity of this natural product-rich group is still bountiful. Moreover, a recent study by Lazzarini et al. (2001) suggests that many of the so-called "rare actinomycetes" can be cultured readily using selective isolation procedures, making them not so rare after all. Lazzarini et al. (2001) also made a strong case for some of these organisms, particularly those of the family *Streptosporangiaceae*, being rich in chemical diversity. Finally, a new phylogenetically distinct group of obligately marine actinomycetes, collectively named *MAR1*, was recently isolated from marine sediments, suggesting that undiscovered diversity within the *Actinobacteria* still exists in underexploited marine environments (Mincer et al., 2002). Moreover, the authors showed that isolates from this novel taxonomic group possessed significant antibacterial, antifungal, and cytotoxicity activities (Mincer et al., 2002). From these various lines of evidence, it seems likely that more selective isolation techniques and (or) searches for actinomycetes in unusual environments, combined with greater understanding of the physiological requirements for culturing these new organisms, may lead to the discovery of additional diversity within this group.

Additionally, myxobacteria, which are gram-negative (*Deltaproteobacteria*), fruiting body-producing, gliding bacteria (Reichenbach, 2001), are clearly underrepresented in terms of numbers of natural product structures elucidated (Table 3). Myxobacteria are well documented to produce highly complex and chemically diverse natural products with very interesting chemistries and biological functions (Reichenbach, 2001). The major obstacle to mining these organisms in a more high-throughput manner, however, is that it takes weeks to months to purify each strain, they are relatively difficult to grow in fermentors, and they generally produce minute quantities of the natural products. However, it has been shown that myxobacteria are very widely distributed and taxonomically diverse (Dawid, 2000), so if new methods can be devised to isolate them more rapidly, these organisms could potentially play a much more significant role in natural products drug discovery (Dawid, 2000; Reichenbach, 2001). Nevertheless, one myxobacterial natural product, epothilone, already is of significant clinical interest as an antitumor drug with microtubule-stabilizing activity (Bollag et al., 1995; Altaha et al., 2002), and another, the highly adhesive antibacterial compound "TA," has been partially developed as an oral antigingivitis antibiotic (Manor et al., 1989).

The second reason for believing that a broad sampling of prokaryotic natural product antibiotics has already been accomplished is that no significant new bacterially derived scaffolds leading to clinical antibiotics have been discovered in the past 20 to 25 years or so (Table 1). The last two major new prokaryotic-derived scaffolds to be developed into a clinically relevant antibiotic were the carbapenems and monobactams (Table 1).

Depending on the study cited, approximately 65 to 75% of all antibiotics have been isolated from actinomycetes. The most prolific genus within this group has long been considered to be *Streptomyces*, members of which produce an incredibly wide array of natural product metabolites (Tables 2 and 3; Bérdy, 1995; Strohl, 1997; Watve et al., 2001). In fact, the single greatest problem with isolating new natural product antibiotics from actinomycetes is the reisolation of known compounds and dereplication of those extracts versus extracts that possess potentially new activities, especially if whole-cell killing assays are used as the primary screening strategy (Watve et al., 2001). This has resulted in more recent efforts to focus on more obscure or rarely found species of newly isolated actinomycetes (Lazzarini et al 2001), or on actinomycetes that are isolated from unusual locations, such as Antarctica (Ivanova et al., 2001) and marine sediments (Mincer et al., 2002).

One of the primary uses of taxonomic analyses in the pharmaceutical industry is the use of taxonomic diversity as a surrogate for chemical diversity, i.e., the concept that finding new strains, the more novel the better, of actinomycetes should lead to discovery of new natural products. Although this approach has had some historical merit, it is also basically flawed for several reasons. First, it is well known that a variety of different strains of a "species" of actinomycete is linked with the production of a wide diversity of natural products. For example, a survey of the literature linked the "species" names *Streptomyces antibioticus*, *Streptomyces griseus*, and *Streptomyces hygroscopicus* with the production of 13, 32, and 46 different structural types of natural products, respectively. Examples of compounds produced by strains characterized as *S. hygroscopicus* include rapamycin, spectinomycin, bialophos, hygromycin, and a variety of polyethers. Likewise, strains identified as *S. griseus* produce streptomycin, candicidin, valinomycin, nonactin, daunorubicin, as well as macrolides, polyenes, and carbapenems. A caveat to this, of course, is that the taxonomy of actinomycetes has undergone significant changes over the years (Anderson and Wellington, 2001), and strains originally identified as belonging to *S. griseus*, for example, may now be identified through genetic or other means to other species. Nevertheless, genomic evidence supports the contention that individual strains or species may produce dozens of different natural products. The genomes of *Streptomyces coelicolor* and *Streptomyces avermitilis* each contain more than 20 clusters of potential natural product biosynthesis genes (Ōmura et al., 2001; Bentley et al., 2002). Of all the eubacterial genomes sequenced to date, only *Bacillus subtilis*, with three natural products gene clusters, comes close (Kunst et al., 1997). Finally, recent studies by Anderson et al. (2002) and Sigmund et al. (2002) also demonstrate clearly that taxonomy is not necessarily a good indicator of bioactive potential.

Second, even though a given strain of actinomycete may possess the genetic capability to produce 20 or more different natural products, it likely produces them under different growth conditions. It is now well documented within the industry that changing growth conditions in a natural product screening program changes the spectrum of metabolites produced (Monaghan et al., 1995; Burkhardt et al., 1996). The problem, of course, is that each strain is unique and therefore responds uniquely to changes in growth conditions (Monaghan et al., 1995). The generation of more sophisticated chemical analyses for measuring product diversity (Julian et al., 1998; Zahn et al., 2001; Gräfe et al., 2001; An et al., 2003) and genetic probes for determining whether certain types of gene clusters are present (Huddleston et al., 1997; Bingle et al., 1999; Sosio et al., 2000; Nicholson et al., 2001) and/or are being transcribed (Ye et al., 2000; Huang et al., 2001) is an advance in this field that should help to delineate optimal growth conditions for expression of different natural product biosynthesis pathways.

Third, especially if an organism produces large quantities of known compounds and minor quantities of potentially undiscovered novel compounds, it is difficult to find those minor activities in the extracts, much like panning for gold in a stream full of pyrite. Chemical methods to single out the potential poorly produced gems can assist this process (Julian et al., 1998; Zahn et al., 2001; Gräfe et al., 2001; An et al., 2003), but ultimately it remains a game of chance.

Finally, it is well known that a wide diversity of actinomycetes can produce the same natural product (or, in some cases, very close analogues). One well-documented case of this is exemplied by anthracycline-producing actinomycetes. Members of five different genera within the actinomycetes, including at least 60 different strains or species, produce compounds of the anthracycline group. Of these, at least a dozen different strains or species produce daunorubicin as one of the metabolites (Strohl et al., 1997). Thus, in this case taxonomy is not a reasonable surrogate for structural diversity. Additionally, it has been shown recently that actinomycete strains can transfer both large, linear plasmids (Ravel et al., 2000) as well as antibiotic resistance and biosynthesis genes from one strain to taxonomically dissimilar strains (Huddleston et al 1997). These results provide at least a partial explanation for the diversity of actinomycetes producing similar compounds, as well as the prevalence of individual actinomycete strains or species producing a broad range of compounds.

Other prokaryotes also produce natural product antibiotics, although none of them are likely to be as prolific as, or to produce the chemical diversity of, the actinomycetes (Tables 2 and 3). Some caution, however, needs to be taken in interpreting Tables 2 and 3, because considerably more effort has been spent searching for natural products in actinomycetes than from any other prokaryotes. Moreover, the numbers for Table 3 come from a survey of the abstracted literature from approximately 1980 to the present. Nevertheless, with the possible exception of the myxobacteria (known producers of diverse natural product structures, but with a very low throughput), industrial experience has shown that most nonactinomycete bacteria produce significantly lower numbers and diversity of natural products than those from the actinomycete group. Of the nonactinomycete bacteria, it is clear that various members of the bacilli and the *Gammaproteobacteria* are reasonably good natural product producers (Table 3). On the other hand, members of the *Betaproteobacteria* and *Alphaproteobacteria* have not, at least historically, been producers of many natural products. Moreover, the natural products from the nonpseudomonad *Gammaproteobacteria* as well as all of the *Alpha-* and *Betaproteobacteria*, proteobacters, have been mostly limited in diversity to a narrow range of β-lactams, peptides, polyenes, and a few other relatively simple compounds.

These data are supported by the available genomic sequences. As mentioned above, both actinomycete genomes sequenced contained at least 20 natural product gene clusters each. The *B. subtilis* genome yielded three potential natural product gene clusters, *srf*, *pps*, and *pks*, which might be employed for the production of surfactin, fengycin, and difficidin, respectively (Kunst et al., 1997). The genome of *P. aeruginosa* PAO1 appeared to contain clusters of genes encoding phenazine, pyoverdine, and the siderophore pyochelin (Stover et al., 2000). The genome of *M. tuberculosis* contained genes encoding four unique types of PKSs (Cole et al., 1998), whereas the genome of *Xylella fastidiosa* yielded a PKS gene, a pteridine-dependent deoxygenase, and a few other genes that were not apparently associated with any large, identifiable natural product biosynthesis gene clusters (Simpson et al., 2000). PKS and nonribosomal peptide synthase genes also were found in the genome of the plant pathogen *Ralstonia solanacearum*; the peptide synthase genes were hypothesized to code for syringomycin biosynthesis (Salanoubat et al., 2002). Finally, a few potential natural products biosynthesis genes were found scattered in the genomes of two sequenced *Xanthomonas* strains, but these were not related to any large, obvious natural product biosynthesis gene clusters (de Silva et al., 2002). As far as can be deciphered, all other prokaryotes sequenced thus far lacked natural product biosynthesis gene clusters, further supporting the long-held notion that such pathways are indeed restricted taxonomically.

FUNGAL DIVERSITY WITH RESPECT TO NATURAL PRODUCT ANTIBIOTICS

There are estimated to be more than 1.5 million different fungal species, of which only approximately 70,000 have been characterized to any significant degree (Jiang and An, 2000). The fungi are classified in four phyla, *Basidiomycota* (containing the class *Basidiomycetes*; e.g., mushrooms, puffballs, rusts,

smuts), *Ascomycota* (containing the classes *Ascomycetes* and *Deuteromycetes*), *Zygomycota* (containing the classes *Zygomycetes* and *Trichomycetes*), and *Chytridiomycota*, containing the single class *Chytridiomycetes* (the chytrids). Of these groups, a large fraction of the natural product-forming fungi have thus far belonged to the *Ascomycota*, and a smaller portion belong to the *Basidiomycota* (Jiang and An, 2000). In our experience, the genera within the *Ascomycota* that provide the most diverse and prolific extracts are (by order) *Eurotiales* (*Aspergillus, Penicillium, Monascus*), *Xylariales* (*Hypoxylon, Nodulisporium, Xylaria*), *Hypocreales* (*Trichoderma, Fusarium, Paecilomyces, Claviceps, Tolypocladium*), and *Pleosporales* (*Phoma, Alternaria*).

Examples of some significant natural products derived from filamentous fungi include penicillin (antibacterial), cyclosporin (immunosuppressant), lovastatin (antihypercholesterolemic), caspofungin (antifungal), fusidic acid (antibacterial), mycophenolic acid (immunosuppressant), ergotamine (pain), and strobilurins (agricultural fungicides). Other than the β-lactams, which are produced by several different species of fungi, and fusidic acid, very few fungal-derived compounds have gone forward into development as therapeutic antibacterial antibiotics. Additionally, other than a handful of widely distributed mycotoxins that demonstrate broad prokaryotic and eukaryotic cytotoxicity, relatively few classes of antibacterial compounds have been found in decades of screening fungal extracts. Thus, screening fungal extracts for antibacterials is likely to cause more problems (i.e., dereplication for known penicillins and nuisance compounds) than are worth the effort, considering the relatively few non-β-lactams that have been found thus far.

As stated above for the myxobacteria, filamentous fungi are well documented to produce highly complex and chemically diverse natural products with very interesting chemistries and biological functions. These organisms also take weeks to months to purify, they can be relatively difficult to grow in fermentors, and they often produce minute quantities of the natural products. Yet broad taxonomic groups of filamentous fungi are very widely distributed, so if new methods can be devised to isolate them more rapidly, then these organisms could potentially play even a greater role in natural products drug discovery than they currently do.

Filamentous fungi make a variety of compounds that exhibit both broad- and narrow-spectrum activities against other fungi, making them an excellent source of antifungal compounds. One recent example of multiple compound classes yielding activities on the same target is in β-1,3-glucan synthesis, for which both echinocandins (e.g., caspofungin [Tkacz and DiDomenico, 2001]) and enfumafungins (Onishi et al., 2000) are bioactive. As mentioned above, fungi produce a wide variety of cytotoxic compounds, including trichothecenes, zearalenones, wortmannin, cytochalasins, beauvericin, aflatoxins, and other mycotoxins (e.g., isocoumarins, substituted anthraquinones), although these generally are considered as nuisance compounds rather than potential scaffolds for cytotoxic antitumor agents. Thus, at least based on historical data, filamentous fungi as a group are excellent sources of antifungal compounds and for other nonantibiotic pharmacophores (Bérdy, 1995), but not for non-β-lactam antibacterials.

OTHER MICROORGANISM SOURCES OF NATURAL PRODUCTS

It is becoming more apparent with each year that marine microorganisms, perhaps especially those isolated from marine invertebrates, offer potentially new sources of bioactive compounds (Faulkner, 2001; Kelecom, 2002). It is clear that marine microorganisms across a wide range of diversity, from *Lyngbia* (cyanobacterium), to *Alteromonas* (*Gammaproteobacteria*), to red and brown algae, to marine actinomycetes and fungi, produce a diverse set of natural product antibiotics (Faulkner, 2001). Although efforts have increased dramatically over the past decade to isolate and characterize these natural products, this type of effort needs to be carried out in a more systematic and high-throughput fashion. It also has been surmised that a subset of natural products thought to derive from tunicates and similar marine macroorganisms are probably produced by the microbial flora they harbor (Stierle et al., 1988; Webster et al., 2001; Kelecom, 2002). If this premise turns out to be true, even in part, it offers additional motivation to search for natural products from marine microorganisms. One caveat to the broader search for marine natural product antibiotics is that these compounds in general tend to be more polar than those found in extracts from terrestrial microbes, which may make isolation more challenging. On the other hand, the more hydrophobic marine compounds often tend to be polyhalogenated (J. Liesch, personal communication).

ACCESSING DNA DIVERSITY

In recent years, a significant level of effort has gone toward developing methods by which natural products might be discovered from environmental DNA (Seow et al., 1997; Watts et al., 1998; Handelsman et al., 1998; Rondon et al., 1999, 2000; MacNeil et al., 2001) or from the recombination of DNA from environmental isolates with standard lab strains (An et al., 2003). Methods for isolation of large DNA fragments from natural environments (Krsek and Wellington, 1999; Rondon et al., 2000; Short and Mathur, 2002), vectors generating libraries of such fragments in natural product-friendly hosts (Rondon et al., 2000; Sosio et al., 2001), and probes to identify the presence of potential natural product biosynthesis pathways (Seow et al., 1997; Bingle et al., 1999; Sosio et al., 2000; Nicholson et al., 2001; Metsä-Ketelä et al., 2002) all have been developed for these purposes. In reality, however, it is likely that in most cases, the output of these intensive efforts will not be chemically diverse or complex enough, or biologically relevant enough, to offset the cost and efforts required to make them work.

There are at least two separate documented successes in efforts to make "natural" libraries for the discovery of new natural products. A recent paper by Jo Handelsman and her colleagues demonstrated that large BAC-sized libraries of soil DNA expressed in *Escherichia coli* could result in the production of the indolic compounds turbomycins A and B (Gillespie et al., 2002; see also Chapter 11 in this book). A similar indole also was shown to be a product of the same library in an earlier publication (MacNeil et al., 2001). Although these compounds do exhibit some weak antibiotic activities, they are probably not of the quality required for development into novel therapeutic antibacterial antibiotics.

Similarly, several cosmid libraries made of cDNA from difficult-to-grow filamentous fungi were introduced into and expressed in *Aspergillus nidulans* (An et al., 2003). From over 7,700 transformants analyzed, approximately 400 extracts were demonstrated by LC-MS to be "outliers" statistically, of which several produced antibacterial activity (compared to a no activity background). The output of these libraries included several extracts that inhibit *S. aureus* and other bacteria. Three compounds, although not bioactive as antibiotics, were isolated and characterized as asperugins A and B and violaceol I (An et al., 2003). These products were not made by the parental strain under a wide variety of growth conditions, but because they are known products of some aspergilli, we cannot be certain what the contribution of the recombinant DNA was to the formation or expression of these compounds (An et al., 2003).

Thus far, it is significant that all of the compounds reported to be isolated from either soil DNA libraries or libraries from difficult-to-grow organisms are rather simple, noncomplex structures. Thus, based on the enormous effort required to make such libraries, one has to be skeptical of this approach as a practical avenue for future antibiotics. There are, however, a few potential exceptions to this notion: (i) the highly industrialized environmental DNA approach being taken by Diversa (Short, 2002a, 2002b; Short and Mathur, 2002), the company with the most significant experience in generating products from environmental DNA; (ii) the whole-organism gene shuffling approach (and obvious "discovery-based" variations therefrom) recently shown by Maxygen for the improvement of tylosin production by an actinomycetes (Zhang et al., 2002); and (iii) the high-throughput natural product gene cluster isolation approach being taken by Ecopia Biosciences (Farnet, 2000; www.ecopiabio.com).

For the high-throughput method of obtaining actinomycete natural product (and antibiotic) biosynthesis gene clusters, Ecopia Biosciences (Montreal) generates both small and large insert libraries of actinomycete DNA, followed by end-forward and reverse-primer sequencing of the small inserts to search for conserved regions of known natural product biosynthesis genes (e.g., PKS, nonribosomal peptide synthetase, P450 hydroxylase, methyltransferase, etc.). They also have developed a sophisticated software program that helps to discern primary from secondary metabolism genes to rapidly identify genes of interest. Because all actinomycete natural product biosynthesis genes are found in clusters, the discovery of a single such gene implies that a cluster has been found, allowing Ecopia to rapidly obtain the cluster from this sequence by using the small insert of interest to probe the large insert library. In this manner, Ecopia has generated a database and natural product gene library consisting of more than 300 genetic "pathways" containing more than 14,000 individual genes, which can be used for a variety of recombinant approaches to generate new natural product antibiotics. The output of the Ecopia approach likely will be minor to significant modifications of known scaffolds. Likewise, the Maxygen genomic shuffling approach for discovery purposes will likely result in modifications of known scaffolds. It is likely, however, that the Diversa approach of large-scale, high-throughput, soil DNA expressed in an actinomycete host has the best potential for discovering novel scaffold diversity.

SOURCES FOR FUTURE ANTIBACTERIAL ANTIBIOTICS

Notwithstanding the fact that the first and most famous antibacterial, the penicillins, were discovered as products of filamentous fungi, very few of today's classes of antibacterials are products of fungal fermentations. Thus, from an historical context, filamentous fungi may not make the best source of the next generation of antibacterial antibiotics. One possibly significant caveat to this notion, however, is that a large number of taxa within poorly understood orders of fungi have never been screened for antibacterials due to the relative difficulty of isolating, fermenting, and extracting the strains for high-throughput analysis.

Actinomycetes still appear to remain the best potential source of antibacterial compounds (Watve et al., 2001; Lazzarini et al., 2001). Watve et al. (2001) have gone so far as to suggest that only a few percent of the predicted 150,000 or more potential antibiotics produced by members of the genus *Streptomyces* have been discovered to date. Although, in principle, their analysis may have merit, it also leaves several issues unanswered. For example, they conclude that fewer natural products from actinomycetes have been isolated in recent years due to the downturn in screening. This is certainly true, but the downturn in screening is a result of the vastly diminishing returns obtained from the screens. At least for assays involving traditional targets or whole-cell screens, dereplication has become the most significant issue and the most prevalent deterrent to additional screening (Zähner and Fiedler, 1995; Gräfe et al., 2001; Watve et al., 2001). Moreover, Watve et al. (2001) did not distinguish between the discovery of new minor analogues of known structures versus the discovery of significantly different scaffolds. As stated above, there is a qualitative difference between the two types of new natural product discoveries. Thus, although the actinomycetes are still attractive producers of future antibacterials, they need to be screened using assays that will reduce the required dereplication of old standbys such as tetracycline, chloramphenicol, rifamycin, and similar compounds.

Several approaches are being taken to address the issues of MRSA, VRE, other antibiotic-resistant bacteria, newly emerging infectious diseases, and related problems. The targets for most existing natural-product-based antibacterial drugs, including cell wall biosynthesis, protein biosynthesis, and nucleic acid biosynthesis, all have been targeted many times over by analogues of them, as well as by other compounds, both synthetic and naturally derived. There is obviously a critical need for new, potent antibacterials to which resistance is not easily developed. Given the caveats discussed above, the best way to search for new natural product antibacterials from any source is to develop screens that discriminate between new functions and already discovered functions. It has long been hoped that the genomics efforts on bacterial pathogens would result in the discovery of many new targets for antibacterial antibiotics. The problem with this approach, however, is that many of the targets from genomics efforts are single enzyme targets. First, any single-site bacterial target for which single mutations can render resistance cannot be seriously considered due to the high frequency of mutations (10^{-8}) that occur in bacteria (Silver and Bostian, 1993). Second, many resistance mechanisms are passed horizontally among bacterial populations by plasmid and transposon-mediated transfer, which renders many antibiotics useless. Even the fluoroquinolones, which for years were thought to target only DNA gyrase, are now known to target at least two separate enzymes (Pan and Fisher, 1997). This is critical, too, because single mutations in bacterial gyrases have been shown recently to yield resistance to several fluoroquinolones (Waters and Davies, 1997).

What is the best answer for new antibacterial strategies? Although there is probably no single answer that addresses all issues, the best approach to obtain a successful, bactericidal antibiotic likely is to inhibit cell wall biosynthesis with a non-β-lactam compound, preferably with a novel mechanism of action or at several sites (Woodruff et al., 1977; Zafriri et al., 1981; DeCenzo et al., 2002). One particular assay that in the past produced a bounty of cell wall-active antibacterials at Merck was the "Sphero assay," the use of which resulted in the discovery of both fosfomycin and cephamycin C (Woodruff et al., 1977; Stapley et al., 1979; Strohl et al., 2001). Alternatively, inhibition of systems or processes, such as protein or nucleic acid synthesis, again at multiple sites, is still attractive, as long as the mechanism falls outside of existing resistance patterns such as MLS, as exemplified by the oxazolidinones (Aoki et al., 2002). The conundrum, however, is that protein synthesis inhibitors are by most definitions bacteriostatic agents, which are likely not the best mechanisms for microorganisms that cause life-threatening diseases, especially if the patients are unable to mount a robust immune response.

Finally, there are several very attractive approaches for new antibacterial natural products that are based generally on the success of clavulanic acid as the β-lactamase inhibitor component of the two-component antibacterial drug, Augmentin. In the first instance, broader spectrum β-lactamase inhibitors, capable of inhibiting even the broadest-spectrum enzymes available today (Sandanayaka and Prashad, 2002), offer the hope of extending the overall life of the β-lactams, of which there are many excellent, potent bactericidal examples on the market. Alternatively, drugs that target specific resistance mecha-

nisms, in particular efflux-mediated resistance mechanisms, may be reasonable alternatives within the same overall concept (Wright, 2000; Poole et al., 2001; Markham and Neyfakh, 2001). Most of the major resistance mechanisms are known today, and it should be possible to engineer specific and sensitive screens to address such resistance mechanisms.

Additionally, recent studies have shown a great potential for synergism between known classes of antibiotics, and therefore by extension, to new potential antibiotics as well. For example, Synercid is the combination of the streptogramin antibiotics quinupristin and dalfopristin, the combination of which works significantly better than either substance alone. It was recently shown that several different combinations of antibiotics plus Synercid (dalfopristin-quinupristin) on VREs in a classical checkerboard type of experiment demonstrated that gentamicin synergized with Synercid to provide a significantly greater killing effect than obtained with gentamicin alone (Eliopoulos and Wennersten, 2002). On the other hand, a significant antagonism was observed in using Synercid in combinations with certain other antibiotics (Fuchs et al., 2001; Eliopoulos and Wennersten, 2002), highlighting the caution needed in taking an approach to discovering potential synergists. Other investigators have shown that addition of other synergists can reduce the MICs of important antibiotics. For example, one of the key interests in the lantibiotics (e.g., natural cationic peptide antibiotics such as magainins, defensins, cecropins) has been the potential for them to act as synergists or sensitizers for existing antibiotics (Hancock and Falla, 1997). Moreover, Zhao et al. (2001) found that the combination of the cell wall-binding compound, epigallocatechin gallate, with various β-lactams, resulted in synergistic killing of staphylococci; and Liu et al. (2000) found that addition of the flavone, baicalin, a component of the traditional Chinese herb Xi-nan Haungqin, to β-lactams drastically reduced MICs on MRSAs. These types of studies mirror what may occur in the body, i.e., that multiple natural antimicrobial defense factors (e.g., defensins, lysozyme, lactoferrin) may be synergistic in killing microorganisms in vivo (Singh et al., 2000).

SOURCES FOR FUTURE ANTIFUNGAL ANTIBIOTICS

Historically, actinomycetes have produced a variety of antifungal compounds, including the various polyenes (e.g., nystatin, amphotericin B), as well as a variety of other types of compounds (e.g., rustmicin, nikkomycin, pamamycin, pradamycin, natamycin, pimaricin, bafilomycin, rapamycin, resormycin, dihydromaltophilin, blasticidin S, sinefungin, and oligomycin). A wide diversity of actinomycetes also have been shown to exhibit significant, unisolated antifungal activities against plant-pathogenic fungi (Lee and Hwang, 2002), demonstrating the breadth of the potential untapped antifungal activities from this group of organisms. Likewise, filamentous fungi also have produced a variety of antifungal compounds, including the various echinocandins, enfumafungin, sordarin, illicicolin, ergokinin A, sphingofungin, peptaibol, and several other compounds with a diversity of core structures. A very high proportion of the compounds isolated from myxobacteria have antifungal activity (Reichenbach, 2001), including sorangicin, epothilone, ambruticin, haliangicin, myxin, stigmatellin, myxalamide, cystathiazole, glidobactins, and ratjadon. Finally, it should not be overlooked that a variety of pseudomonads have been shown to produce seed- and crop-protecting antifungal compounds (e.g., pyrrolnitrin, pyoluteorin, 2,4-diacetylphloroglucinol, syringomycin, syringopeptin, tubermycin, tensin, cepacidine A, and viscosinamide), a group of which certainly could be scrutinized more closely as a source for potential therapeutic antifungal drugs (Bender et al., 1999). Although marine sources are a prolific source of new antifungal compounds (Mohapatra et al., 2002), many marine compounds with antifungal activity may be nonspecific cytotoxins. Nevertheless, extracts derived from marine microorganisms are likely to provide a new source of antifungal scaffolds.

SOURCES FOR FUTURE ANTITUMOR ANTIBIOTICS

The search for natural product anticancer agents that began in the late 1950s resulted in the discovery in 1964 of daunorubicin, a cytotoxic, antitumor drug of the anthracycline class with efficacy against a wide variety of solid tumors and leukemias (Strohl et al., 1997). Doxorubicin (14-hydroxy-analogue of daunorubicin), discovered at Farmitalia in 1969, showed a wider range of efficacy for treatment of various cancers and has been a mainstay in chemotherapy for over 25 years now. It was shown recently that doxorubicin as well as many other cytotoxic antitumor drugs (e.g., etoposide) act by trapping topoisomerase II with DNA in what has been referred to as the "cleavable complex," resulting in multiple DNA strand breaks (Strohl et al., 1997). Another class of anthracyclines, exemplified by aclarubicin (aclacinomycin A), were recently shown to inhibit by a different mechanism of action, i.e., by inhibition of topoisomerae II-DNA interaction (reviewed by Strohl et al., 1997).

The major natural product antitumor drugs are listed in Table 4. Other natural product antitumor chemotherapy agents used for several years, but more restrictively than doxorubicin, include mithramycin and the mitomycins, first described in the 1950s; bleomycin, described in 1965; and the vinca alkaloids, vinblastine and vincristine, which were first described around 1960. In all of these cases, the mode of action is cytotoxicity through drug-DNA interactions. With the ultimate desire to get away from cy-

Table 4. Clinical and lead experimental natural product-derived anticancer drugs

Anticancer drug	Type	Comments
Aclacinomycin A	Anthracycline	Produced by *Streptomyces galilaeus*; used in Japan and France
Bleomycin	Glycopeptide	Produced by *Streptomyces verticillus*
Carminomycin	Anthracycline	Produced by *Actinomadura carminata*; used in Russia
Dactinomycin	Acylpeptidolactone	Produced by *Streptomyces parvulus*
Daunorubicin	Anthracycline	Produced by *Streptomyces peucetius*
Docetaxel	Taxane	Taxotere; approved for treatment of ovarian cancer
Doxorubicin	Anthracycline	Produced by *S. peucetius* subsp. *caesius*
Epothilones B, D	Substituted macrolides	Produced by *Sorangium cellulosum*; both in clinical trials
Idarubicin	Anthracycline	Synthetic (mimetic), orally available analog of doxorubicin
Irinotecan	Camptothecin	Approved for metastatic colon and rectal cancers
Mitomycin C	Cytotoxic drug	Produced by *Streptomyces caespitosus*
Mitoxantrone	Anthracenedione	Biomimetic of two different natural product structures
Paclitaxel	Taxane	Produced by *Taxus* spp. (yew trees); approved for use in 1994
Pentostatin	Nucleoside analog	Produced by *Streptomyces antibioticus*
Plicamycin	Polyketide	Mithramycin; produced by *Streptomyces argillaceus*
Tenipocide	Podophyllotoxin	Cytotoxic antitumor drug
Topotecan	Camptothecin	Semisynthetic; from *Camptotheca acuminata* (Chinese tree)
Vinblastine sulfate	Vinca alkaloid	Produced by *Catharanthus roseus* (periwinkle)
Vincristine sulfate	Vinca alkaloid	Produced by *Catharanthus roseus* (periwinkle)
Vinorelbine tartrate	Vinca alkaloid	Semisynthetic vinblastine analog

totoxic drugs that are detrimental to all cells and not just cancerous cells, these cytotoxic natural product antitumor drugs are generally considered to be drugs of the past rather than the future.

Interest in natural product antitumor drugs was renewed, however, with the discovery of taxols from the Pacific yew tree. Taxol, a complex diterpenoid alkaloid, was discovered as a potential antitumor agent in 1971 (Wani et al., 1971), and it took over two decades to bring it to the market. One of the unique features that helped move the taxanes from experimental drugs to clinical candidates was the finding that they acted as tubulin stabilizers, a novel mode of action (Schiff et al., 1979). Taxotere (Docetaxel) and paclitaxel (Taxol) have excellent activity against ovarian and breast cancers. Additionally, taxol and semisynthetic taxanes may be efficacious against a wide range of tumors. Thus far, the greatest impediment to the development of these drugs on a large scale has been that they are only found at levels of about 0.01% (wt/wt) in the bark of several related yew trees, *Taxus brevifolia* (Pacific yew, an endangered species), *Taxus baccata*, *Taxus cuspidata* (Japanese yew), and others (Yukimune et al., 1996). Moreover, the chemical synthesis of paclitaxel is very difficult, requiring more than 30 steps. Thus, paclitaxel is currently made semisynthetically from 10-deacetylbaccatin III, produced by the needles of the Himalayan yew (van Rozendaal et al., 2000). Other novel antitumor drugs developed in recent years include irinotecan (Camptosar; Pharmacia-Upjohn) and topotecan (Hycamptin), both which are inhibitors of topoisomerase I.

The most recent natural product cytotoxic agent of significant interest is epothilone, which was first discovered as a weak antifungal agent of rust fungi (Gerth et al., 1996), and subsequently shown to stabilize microtubules much in the same manner as taxol (Bollag et al., 1995; Giannakakou et al., 2000). The various epothilones in development (e.g., epothilones B and D have entered phase II and phase I clinical trials for various forms of solid tumors, respectively) have several potential advantages over the taxanes, such as much high potency (Altaha et al., 2002), production in fermentation vats, and potential production by faster-growing and overproducing recombinant hosts (Tang et al., 2000). The discovery of tubulin inhibitors and stabilizers as potential cancer drugs has touched off a new cycle of searches for novel, higher-potency but lower-toxicity antitumor drugs with similar mechanisms of action (Mooberry et al., 1999; Giannakakou et al., 2000; Roberge et al., 2000; Altaha et al., 2002). One of the most recent of these to spark significant interest is the microtubule stabilizing agent from a marine sponge, laulimalide (Mooberry et al., 1999), which has been shown to bind tubulin at a site different from that of paclitaxel (Pryor et al., 2002). Another tubulin-stabilizing agent recently isolated from a marine sponge is peloruside A, which possesses a 16-membered macrocycle structure not unlike that of epothilone (Hood et al., 2002). Additionally, a wide range of marine actinomycetes, in particular those in the genus *Micromonospora*, were found to produce antitumor activities (Zheng et al., 2000). These recent findings exemplify the potentially rich source of largely untapped cytotoxic agents that both marine sponges and marine microorganisms offer (Fenical, 1997).

EPILOGUE: "WHAT YOU SEEK IS WHAT YOU FIND"

There is no question that the future of antimicrobial and antitumor chemotherapy depends on the discovery of new agents to remain one step ahead of the ever-expanding resistance problems. There are three approaches that should help natural products to make an impact in these areas. First, it is increasingly difficult to dereplicate known bioactive natural products in extracts assayed in whole-cell killing assays from potential novel activities within and among them. Thus, to be able to find and identify novel antibiotics in the future, it becomes even more critical that specific in vitro functional assays be developed that can sort out those new nuggets, potentially produced in very small quantities, from the large quantities of pyrite with which they are found.

What is the best antibacterial target? Although there is probably no single best answer for all resistance problems, the clear-cut best approach still seems to be the inhibition of cell wall biosynthesis, preferably with a novel mechanism of action or at several sites. Alternatively, inhibition of systems or processes, such as protein or nucleic acid synthesis, again at multiple sites, is a reasonable approach, with the caveat that protein synthesis inhibitors are bacteriostatic agents, limiting their usefulness against organisms like *S. aureus*. Nevertheless, the newer macrolides, dirithromycin, clarithromycin, and azithromycin, all protein synthesis-inhibiting bacteriostats, are effective drugs, especially in the lucrative pediatric upper-respiratory market. Additionally, linezolid, a protein synthesis inhibitor with a new mechanism of action (Thompson et al., 2002), has been a recent success and offers up hope that additional new mechanisms of action can be found and addressed with new scaffolds.

The second suggested approach is to more extensively mine new environments or niches for known antibiotic-producing microorganisms (Harvey, 2000). Examples include actinomycetes from Antarctica (Ivanova et al., 2001), tropical fungi (Bills et al., 2002), and marine microorganisms (both fungal and bacterial), particularly those isolated from marine invertebrates (Faulkner, 2001; Kelekom, 2002). Although no existing antibiotics have yet been developed from these novel sources or environments, it is likely that the chemical diversity will be present to support screening programs.

The third suggestion here is that actinomycetes, myxobacteria, and filamentous fungi will continue to be excellent sources for natural products, especially as genomic and proteomic assays for their production of different metabolites become more readily useful to guide fermentation conditions. For antibacterials, focus should be on mining the actinomycetes, myxobacteria, and marine microbes, but with the caveat mentioned above, i.e., use of novel in vitro target activity-driven isolations. For antifungals, all four groups should be screened vigorously, using targets such as those mentioned above.

Finally, the issue of new antitumor drugs from microorganisms brings forth both a reality check and a rhetorical question. The reality check is that all natural product antitumor drugs thus far taken to the market are cytotoxic drugs with cell-killing mechanisms of action, either through interactions with DNA, topoisomerases, microtubulin, or similar such cell-requiring functions. The question is, how many more new and different cytotoxic agents do we need for the treatment of various forms of cancer? If resistance is the major issue for the limited efficacy of existing antitumor drugs, then perhaps the answer is yes. On the other hand, if the cytotoxic nature of the mechanism of action is the limiting factor, then perhaps as pharmaceutical scientists, we are better off focusing on more cancer-specific mechanisms of action (e.g., tyrosine kinase bcr-abl, the target of Gleevec [Fabro et al., 2002]), which may include natural products in the mix. One ray of hope for this approach is that screening on the potential specific cancer target, farnesyl-protein transferase, resulted in the discovery of a wide diversity of potent, selective, and nontoxic bioactive natural products (Singh and Lingham, 2002). Although none of these was taken forward into the clinic, it does suggest that cancer target-specific, nontoxic natural products can be found when put to the test.

Natural products discovery programs in industry have been a critical part of the drug discovery process since the end of World War II. In general, there have been two major periods, the antibiotic period from 1945 to 1970, during which virtually all of the scaffolds for today's antibiotics and antitumor dugs were discovered (see Tables 1 and 4), and the pharmacology period, from 1970 to the present, during which the discovery of natural product pharmacophores has expanded the role that natural products play in drug discovery. These nonantibiotic pharmacophores include lovastatin and pravistatin (hypercholesterolemia), cyclosporin, rapamycin, tacrolimus (immunomodulatory), avermectin (antiparasitic), spinosyn (insecticide), and xenical (cholesterol absorbant). The primary issues faced by natural products drug discovery programs today include their role vis-á-vis high-throughput screening, the ability to find new gems out of extracts quickly and efficiently, and developing assays that will differentiate between known natural products and potentially new, exciting natural product leads (Strohl, 2000; Harvey, 2000).

Acknowledgments. I thank Paula Heck for information on current market sizes for antibacterials and Zhiqiang An, Gerald Bills, Annaliesa Anderson, and Jerrold Liesch for discussions.

REFERENCES

Altaha, R. T. Fojo, E. Reed, and J. Abraham. 2002. Epothilones: a novel class of non-taxane microtubule-stabilizing agents. *Curr. Pharm. Des.* 8:1707–1712.

Amann, R. K., W. Ludwig, and K. H. Schleifer. 1995. Phylogenetic identification and in situ detection of individual microbial cells without cultivation. *Microbiol. Rev.* 59:143–169.

An, Z., G. Harris, D. Zink, R. Giacobbe, R. Sangari, P. Lu, J. Greene, G. Bills, C. Meyers, S. Smith, V. Svetnik, B. Gunter, A. Liaw, P. Masurekar, J. Liesch, S. Gould, and W. Strohl. Expression of cosmid-size DNA of slow-growing fungi in *Aspergillus nidulans* for secondary metabolite screening. *In* Z. An (ed.), *Handbook of Industrial Mycology*, in press. Marcel-Dekker, New York, N.Y.

Anderson, A. S., and E. M. H. Wellington. 2001. The taxonomy of *Streptomyces* and related genera. *Int. J. Syst. Evol. Microbiol.* 51:797–814.

Anderson, A. S., D. Clark, P. Gibbons, and J. Sigmund. 2002. The detection of diverse aminoglycoside phosphotransferases within natural populations of actinomycetes. *J. Ind. Microbiol. Biotechnol.* 29:60–69.

Aoki, H., L. Ke, S. M. Poppe, T. J. Poel, E. A. Weaver, R. C. Gadwood, R. C. Thomas, D. L. Shinabarger, and M. C. Ganoza. 2002. Oxazolidinone antibiotics target the P site on *Escherichia coli* ribosomes. *Antimicrob. Agents Chemother.* 46:1080–1085.

Bender, C., V. Rangaswamy, and J. Loper. 1999. Polyketide production by plant-associated pseudomonads. *Annu. Rev. Phytopathol.* 37:175–196.

Bentley, S. D., K. F. Chater, A.-M. Cerdeño-Tárraga, G. L. Challis, N. R. Thomson, K. D. James, D. E. Harris, M. A. Quail, H. Kieser, D. Harper, A. Bateman, S. Brown, G. Chandra, C. W. Chen, M. Collins, A. Cronin, A. Fraser, A. Goble, J. Hidalgo, T. Hornsby, S. Howarth, C.-H. Huang, T. Kieser, L. Larke, L. Murphy, K. Oliver, S. O'Neil, E. Rabbinowitsch, M.-A. Rajandream, K. Rutherford, S. Rutter, K. Seeger, D. Saunders, S. Sharp, R. Squares, S. Squares, K. Taylor, T. Warren, A. Wietzorrek, J. Woodward, B. G. Barrell, J. Parkhill, and D. A. Hopwood. 2002. Complete genome sequence of model actinomycete *Streptomyces coelicolor* A3(2). *Nature* 417:141–147.

Bérdy, J. 1995. Are actinomycetes exhausted as a source of secondary metabolites?, p. 13–34. *In* V. G. Debabov, Y. V. Dudnik, V. N. Danilenko (ed.), *Proceedings of the Ninth International Symposium on the Biology of the Actinomycetes.* All-Russia Scientific Research Institute for Genetics and Selection of Industrial Microorganisms, Moscow, Russia.

Bills, G., A. Dombrowski, F. Pelaez, J. Polishook, and Z. An. 2002. Recent and future discoveries of pharmacologically active metabolites from tropical fungi, p. 165–193. *In* R. Watling, J. C. Frankland, M. Ainsworth, S. Isaac, and C. H. Robinson (ed.), *Tropical Mycology, Vol. 2. Micromycetes.* Oxford University Press, Inc., Oxford, United Kingdom.

Bingle, L. E., T. J. Simpson, and C. M. Lazarus. 1999. Ketosynthase domain probes identify two subclasses of fungal polyketide synthase genes. *Fungal Genet. Biol.* **26:**209–223.

Bloom, B. R., and C. J. Murray. 1992. Tuberculosis: commentary on a reemergent killer. *Science* **257:**1055–1064.

Bollag, D. M., P. A. McQueney, J. Zhu, O. Hensens, L. Koupal, J. Liesch, M. Goetz, E. Lazarides, and C. M. Woods. Epothilones, a new class of microtubule-stabilizing agents with a taxol-like mechanism of action. *Cancer Res.* **55:**2325–2333.

Boyle-Vavra, S., R. B. Carey, and R. S. Daum. 2001. Development of vancomycin and lysostaphin resistance in a methicillin-resistant *Staphylococcus aureus* isolate. *J. Antimicrob. Chemother.* **48:**617–625.

Burkhardt, K., H. P. Fiedler, S. Grabley, R. Thiericke, and A. Zeeck. 1996. New cineromycins and musacins obtained by metabolite pattern analysis of *Streptomyces griseoviridis* (FH-S 1832). I. Taxonomy, fermentation, isolation, and biological activity. *J. Antibiot.* **49:**432–437.

Cole, S. T., R. Brosch, J. Parkhill, T. Garnier, C. Churcher, D. Harris, S. V. Gordon, K. Eiglmeir, S. Gas, C. E. Barry III, F. Tekala, K. Badcock, D. Basham, D. Brown, T. Chillingworth, R. Connor, R. Davies, K. Devlin, T. Feltwell, S. Gentles, N. Hamlin, S. Holroyd, T. Hornsby, K. Jagels, A. Krogh, J. McLean, S. Moule, L. Murphy, K. Oliver, J. Osborne, M. A. Quail, M.-A. Rajandream, J. Rogers, S. Rutter, K. Seeger, J. Skelton, R. Squares, S. Squares, J. E. Sulston, K. Taylor, S. Whitehead, and B. G. Barrell. 1998. Deciphering the biology of *Mycobacterium tuberculosis* from the complete genome sequence. *Nature* **393:**537–544.

Connolly, J. D., and R. A. Hill. 1992. *Dictionary of Terpenoids.* Chapman & Hall, New York, N.Y.

Craig, G. M., D. J. Newman, and K. M. Snader. 1997. Natural products in drug discovery and development. *J. Nat. Prod.* **60:**52–60.

Dairi, T., Y. Hamano, T. Kuzuyama, N. Itoh, K. Furihata, and H. Seto. 2001. Eubacterial diterpene cyclase genes essential for production of the isoprenoid antibiotic terpentecin. *J. Bacteriol.* **183:**6085–6094.

da Silva, A. C. R., J. A. Ferro, F. C. Reinach, C. S. Farah, L. R. Furlan, R. B. Quaggio, C. B. Monteiro-Vitorello, M. A. Van Sluys, N. F. Almeida, L. M. C. Alves, et al. 2002. Comparison of the genomes of two *Xanthomonas* pathogens with differing host specificities. *Nature* **417:**459–463.

Dawid, W. 2000. Biology and global distribution of myxobacteria in soils. *FEMS Microbiol. Rev.* **24:**403–427.

DeCenzo, M., M. Kuranda, S. Cohen, J. Babiak, Z. D. Jiang, D. Su, M. Hickey, P. Sancheti, P. A. Bradford, P. Youngman, S. Projan, and D. M. Rothstein. 2002. Identification of compounds that inhibit late steps of peptidoglycan synthesis in bacteria. *J. Antibiot.* **55:**288–295.

Eliopoulos, G. M., and C. B. Wennersten. 2002. Antimicrobial activity of quinupristin-dalfopristin combined with other antibiotics against vancomycin-resistant enterococci. *Antimicrob. Agents Chemother.* **46:**1319–1324.

Ellis, D. 2002. Amphotericin B: spectrum and resistance. *J. Antimicrob. Chemother.* **49**(Suppl.):7–10.

Ellis, M. 2002. Invasive fungal infections: evolving challenges for diagnosis and therapeutics. *Mol. Immunol.* **38:**947–957.

Fabro, D., S. Ruetz, E. Buchdunger, S. Cowan-Jacob, G. Fendrich, J. Liebetanz, J. Mestan, T. O'Reilly, P. Traxler, B. Chaudhuri, H. Fretz, J. Zimmermann, T. Meyer, G. Caravatti, P. Furet, and P. Manley. 2002. Protein kinases as targets for anticancer agents: from inhibitors to useful drugs. *Pharmacol. Ther.* **93:**79–98.

Farnet, C. M. 2000. From genes to molecules: genomics in natural product discovery and development. *Abstr. Annu. Meet. Soc. Ind. Microbiol.* **561:**63.

Faulkner, D. J. 2001. Marine natural products. *Nat. Prod. Rep.* **19:**1–48.

Fenical, W. 1997. New pharmaceuticals from marine organisms. *Trends Biotechnol.* **15:**339–341.

French, G. L. 1998. Enterococci and vancomycin resistance. *Clin. Infect. Dis.* **27**(Suppl.):S75–S83.

Fuchs, P. C., A. L. Barry, and S. D. Brown. 2001. Interactions of quinupristin-dalfopristin with eight other antibiotics as measured by time-kill studies with 10 strains of *Staphylococcus aureus* for which quinupristin-dalfopristin alone was not bactericidal. *Antimicrob. Agents Chemother.* **45:**2662–2665.

Gerth, K., N. Bedorf, G. Hofle, H. Irschik, and H. Reichenbach. 1996. Epothilones A and B: antifungal and cytotoxic compounds from *Sorangium cellulosum* (Myxobacteria). Production, physico-chemical and biological properties. *J. Antibiot.* **49:**560–563.

Giannakakou, P., R. Gussio, E. Nogales, K. H. Downing, D. Zaharevitz, B. Bollbuck, G. Poy, D. Sackett, K. C. Nicolaou, and T. Fojo. 2000. A common pharmacophore for epothilone and taxanes: molecular basis for drug resistance conferred by tubulin mutations in human cancer cells. *Proc. Natl. Acad. Sci. USA* **97:**2904–2909.

Gil, J. A., and J. F. Martin. 1997. Polyene antibiotics, p. 551–575. *In* W. R. Strohl (ed.), *Biotechnology of Antibiotics,* 2nd ed. Marcel Dekker, Inc., New York, N. Y.

Gillespie, D. E., S. F. Brady, A. D. Bettermann, N. P. Cianciotto, M. R. Liles, M. R. Rondon, J. Clardy, R. M. Goodman, and J. Handelsman. 2002. Isolation of antibiotics turbomycin A and B from a metagenomic library of soil microbial DNA. *Appl. Environ. Microbiol.* **68:**4301–4306.

Gräfe, U., S. Heinze, B. Schlegel, and A. Hartl. 2001. Disclosure of new and recurrent microbial metabolites by mass spectrometric methods. *J. Ind. Microbiol. Biotechnol.* **27:**136–143.

Graybill, J. R., L. K. Najvar, R. Bocanegra, R. F. Hector, and M. F. Luther. 1998. Efficacy of nikkomycin Z in the treatment of murine histoplasmosis. *Antimicrob. Agents Chemother.* **42:**2371–2374.

Hancock, R. E. W., and T. J. Falla. 1997. Cationic peptides, p. 471–496. *In* W. R. Strohl (ed.), *Biotechnology of Antibiotics,* 2nd ed. Marcel Dekker, Inc., New York, N. Y.

Handelsman, J., M. R. Rondon, S. F. Brady, J. Clardy, and R. M. Goodman. 1998. Molecular biological access to the chemistry of unknown soil microbes: a new frontier for natural products. *Chem. Biol.* **5:**R245–R249.

Harvey, A. 2000. Strategies for discovering drugs from previously unexplored natural products. *Drug Discovery Today* **5:**294–300.

Henkel, T., R. M. Brunne, H. Mueller, and F. Reichel. 1999. Statistical investigation into the structural complementarity of natural products and synthetic compounds. *Angew. Chem. Int. Ed.* **38:**643–647.

Hood, K. A., L. M. West, B. Rouwe, P. T. Northcote, M. V. Berridge, S. J. Wakefield, and J. H. Miller. 2002. Peloruside A, a novel antimitotic agent with paclitaxel-like microtubule-stabilizing activity. *Cancer Res.* **62:**3356–3360.

Hopwood, D. A. 1997. Genetic contributions to understanding polyketide synthases. *Chem. Rev.* **97:**2465–2498.

Huang, J., C. J. Lih, K. H. Pan, and S. N. Cohen. 2001. Global analysis of growth phase responsive gene expression and regulation of antibiotic biosynthetic pathways in *Streptomyces coelicolor* using DNA microarrays. *Genes Dev.* **15:**3183–3192.

Huddleston, A. S., N. Cresswell, M. C. Neves, J. F. Beringer, S. Baumberg, D. I. Thomas, and E. M. Wellington. 1997. Molecular dissection of streptomycin-producing streptomycetes in Brazilian soils. *Appl. Environ. Microbiol.* **63:**1288–1297.

Hugenholz, P., B. M. Goebel, and N. Pace. 1998. Impact of culture-independent studies on the emerging phylogenetic view of bacterial diversity. *J. Bacteriol.* **180:**4765–4774.

Ivanova, V., M. Oriol, M. J. Montes, A. Garcia, and J. Guinea. 2001. Secondary metabolites from a *Streptomyces* strain isolated from Livingston Island, Antarctica. *Z. Naturforsch.* **56:**1–5.

Jiang, Z.-D., and Z. An. 2000. Bioactive fungal natural products through classic and biocombinatorial approaches. *Stud. Nat. Prod. Chem.* **22:**245–272.

Julian, R. K., R. E. Higgs, J. D. Gygi, and M. D. Hilton. 1998. A method for quantitatively differentiating crude natural extracts using high-performance liquid chromatography-electrospray mass spectroscopy. *Anal. Chem.* **70:**3249–3254.

Justice, M. C., M.-J. Hsu, B. Tse, T. Ku, J. Balkovec, D. Schmatz, and J. Nielsen. 1998. Elongation factor 2 as a novel target for selective inhibition of fungal protein synthesis. *J. Biol. Chem.* **273:**3148–3151.

Kamholz, S. L. 2002. Drug resistant tuberculosis. *J. Assoc. Acad. Minor Phys.* **13:**53–56.

Kelecom, A. 2002. Secondary metabolites from marine microorganisms. *Ann. Acad. Bras. Ciencias* **74:**151–170.

Koltin, Y. 1998. Drug discovery—changing the paradigm, Abstr. 8th Intl. Symp. Genet. Industr. Microorg., Jerusalem, Israel.

Krsek, M., and E. M. Wellington. 1999. Comparison of different methods for the isolation and purification of total community DNA from soil. *J. Microbiol. Methods* **39:**1–16.

Kunst, F., N. Ogasawara, I. Moszer, A. M. Albertini, G. Alloni, V. Azevedo, M. G. Bertero, P. Bessières, A. Bolotin, S. Borchert, et al. 1997. The complete genome sequence of the Gram-positive bacterium *Bacillus subtilis*. *Nature* **390:**249–256.

Lazzarini, A., Cavaletti, G. Toppo, and F. Marinelli. 2001. Rare genera of actinomycetes as potential producers of new antibiotics. *Antonie Leeuwenhoek* **79:**399–405.

Lederberg, J., R. E. Shope, and S. C. Oaks (ed). 1992. *Emerging Infections: Microbial Threats to Health in the United States.* Institute of Medicine, National Academy of Sciences Press, Washington, D.C.

Lee, J. Y., and B. K. Hwang. 2002. Diversity of antifungal actinomycetes in various vegetative soils of Korea. *Can. J. Microbiol.* **48:**407–417.

Liu, I. X., D. G. Durham, and R. M. E. Richards. 2000. Baicalin synergy with beta-lactam antibiotics against methicillin-resistant *Staphylococcus aureus* and other beta-lactam-resistant strains of *S. aureus*. *J. Pharm. Pharmacol.* **52:**361–366.

MacNeil, I. A., C. L. Tiong, C. Minor, P. R. August, T. H. Grossman, K. A. Loiacono, B. A. Lynch, T. Phillips, S. Narula, R. Sundaramoorthi, A. Tyler, T. Aldredge, H. Long, M. Gilman, D. Holt, M. S. Osburne. 2001. Expression and isolation of antimicrobial small molecules from soil DNA libraries. *J. Mol. Microbiol. Biotechnol.* **3:**301–308.

Mahgoub, S., J. Ahmed, and A. E. Glatt. 2002. Completely resistant *Acinetobacter baumannii* isolates. *Infect. Control Hosp. Epidemiol.* **23:**477–479.

Mandala, S. M., R. A. Thornton, J. Milligan, M. Rosenbach, M. Garcia-Calvo, H. G. Bull, G. Harris, G. K. Abruzzo, A. M. Flattery, C. J. Gill, K. Bartizal, S. Dreikorn, and M. B. Kurtz. 1998.

Rustmicin, a potent antifungal agent, inhibits sphingolipid synthesis at inositol phosphoceramide synthase. *J. Biol. Chem.* **273:**14942–14949.

Manor, A., I. Eli, M. Varon, H. Judes, and E. Rosenberg. 1989. Effect of adhesive antibiotic TA on plaque and gingivitis in man. *J. Clin. Periodontol.* **16:**621–624.

Markham, P. N., and A. A. Neyfakh. 2001. Efflux-mediated drug resistance in Gram-positive bacteria. *Curr. Opin. Microbiol.* **4:**509–514.

Metsä-Ketelä, M., L. Halo, E. Munukka, J. Hakala, P. Mäntsälä, and K. Ylihonko. 2002. Molecular evolution of aromatic polyketides and comparative sequence analysis of polyketide ketosynthase and 16S ribosomal DNA genes from various *Streptomyces* species. *Appl. Environ. Microbiol.* **68:**4472–4479.

Mincer, T. J., P. R. Jensen, C. A. Kauffman, and W. Fenical. 2002. Widespread and persistent populations of a major new marine actinomycete taxon in ocean sediments. *Appl. Environ. Microbiol.* **68:**5005–5011.

Mohapatra, B. R., M. Bapuji, and A. Sree. 2002. Antifungal efficacy of bacteria isolated from marine sedentary organisms. *Folia Microbiol.* **47:**51–55.

Monaghan, R. L., J. D. Polishook, V. J. Pecore, G. F. Bills, M. Nallin-Omstead, and S. L. Streicher. 1995. Discovery of novel secondary metabolites from fungi—is it really a random walk through a random forest? *Can. J. Bot.* **73:**S925–S931.

Mooberry, S. L., G. Tien, A. H. Hernandez, A. Plubrukarn, and B. S. Davidson. 1999. Laulimalide and isolaulimalide, new paclitaxel-like microtubule-stabilizing agents. *Cancer Res.* **59:**653–660.

Nicholson, T. P., B. A. M. Rudd, M. Dawson, C. M. Lazarus, T. J. Simpson, and R. J. Cox. 2001. Design and utility of oligonucleotide gene probes for fungal polyketide synthases. *Chem. Biol.* **8:**157–178.

Noble, W. C., Z. Virnai, and R. G. Cree. 1992. Co-transfer of vancomycin and other resistance genes from *Enterococcus faecalis* NCTC 12201 to *Staphylococcus aureus*. *FEMS Microbiol. Lett.* **93:**195–198.

Omura, S., H. Ikeda, J. Ishikawa, A. Hanamoto, C. Tkahashi, M. Shinose, Y. Takahashi, H. Horikawa, H. Nakazawa, T. Osonoe, H. Kikuchi, T. Shiba, Y. Sakaki, and M. Hattori. 2001. Genome sequence of an industrial microorganism *Streptomyces avermitilis*: deducing the ability of producing secondary metabolites. *Proc. Natl. Acad. Sci. USA* **98:**12215–12220.

Onishi, J., M. Meinz, J. Thompson, J. Curotto, S. Dreikorn, M. Rosenbach, C. Douglas, G. Abruzzo, A. Flattery, L. Kong, A. Cabello, F. Vincente, F. Pelaez, M. T. Diez, I. Martin, G. Bills, R. Giacobbe, A. Dombrowski, R. Schwartz, S. Morris, G. Harris, A. Tsipouras, K. Wilson, and M. B. Kurtz. 2000. Discovery of novel antifungal (1,3)-β-D-glucan synthase inhibitors. *Antimicrob. Agents Chemother.* **44:**368–377.

Pace, N. 1997. A molecular view of microbial diversity and the biosphere. *Science* **276:**734–740.

Pan, X.-S., and L. Fisher. 1997. Targeting of DNA gyrase in *Streptococcus pneumoniae* by sparfloxacin: selective targeting of gyrase or topoisomerase IV by quinolones. *Antimicrob. Agents Chemother.* **41:**471–474.

Ploy, M. C., C. Grelaud, C. Martin, L. de Lumley, and F. Denis. 1998. First clinical isolate of vancomycin-intermediate *Staphylococcus aureus* in a French hospital. *Lancet* **351:**1212.

Poole, K. 2001. Multidrug resistance in Gram-negative bacteria. *Curr. Opin. Microbiol.* **4:**500–508.

Powles, M. A., P. Liberator, J. Anderson, Y. Karkhanis, J. F. Dropinski, F. A. Bouffard, J. M. Balkovec, H. Fujioka, M. Aikawa, D. McFadden, and D. Schmatz. 1998. Efficacy of MK-991 (L-743,872), a semisynthetic pneumocandin, in murine models of *Pneumocystis carinii*. *Antimicrob. Agents Chemother.* 42:1985–1989.

Pryor, D. E., A. O'Brate, G. Bilcer, J. F. Diaz, Y. Wang, Y. Wang, M. Kabaki, M. K. Jung, J. M. Andreu, A. K. Ghosh, P. Giannakakou, and E. Hamel. 2002. The microtubule stabilizing agent laulimalide does not bind in the taxoid site, kills cells resistant to paclitaxel and epothilones, and may not require its epoxide moiety for activity. *Biochemistry* 41:9109–9115.

Ravel, J., E. M. Wellington, and R. T. Hill. 2000. Interspecific transfer of *Streptomyces* giant linear plasmids in sterile amended soil microcosms. *Appl. Environ. Microbiol.* 66:529–534.

Reichenbach, H. 2001. Myxobacteria, producers of novel bioactive substances. *J. Ind. Microbiol. Biotechnol.* 27:149–156.

Rheims, H., and E. Stackebrandt. 1999. Application of nested polymerase chain reaction for the detection of as yet uncultured organisms of the class *Actinobacteria* in environmental samples. *Environ. Microbiol.* 1:137–143.

Rheims, H., C. Sproer, F. A. Rainey, and E. Stackebrandt. 1996. Molecular biological evidence for the occurrence of uncultured members of the actinomycete line of descent in different environments and geographical locations. *Microbiology* 142:2863–2870.

Roberge, M., B. Cinel, H. J. Anderson, L. Lim, X. Jiang, L. Xu, C. M. Bigg, M. T. Kelly, and R. J. Anderson. 2000. Cell-based screen for antimitotic agents and identification of analogues of rhizoxin, eleutherobin, and paclitaxel in natural extracts. *Cancer Res.* 60:5052–5058.

Rondon, M. R., R. M. Goodman, and J. Handelsman. 1999. The earth's bounty: assessing and accessing soil microbial diversity. *Trends Biotechnol.* 17:403–409.

Rondon, M. R., P. R. August, A. D. Bettermann, S. F. Brady, T. H. Grossman, M. R. Liles, K. A. Loiacono, B. A. Lynch, L. A. MacNeil, C. Minor, C. L. Tiong, M. Gilman, M. S. Osburne, J. Clardy, J. Handelsman, and R. M. Goodman. 2000. Cloning the soil metagenome: a strategy for accessing the genetic and functional diversity of uncultured microorganism. *Appl. Environ. Microbiol.* 66:2541–2547.

Rowe, P. M. 1996. Preparing for the battle against vancomycin resistance. *Lancet* 347:252.

Salanoubat, M., S. Genin, F. Artiguenave, J. Gouzy, S. Mangenot, M. Ariat, A. Billault, P. Brottler, J. C. Camus, L. Cattolico, M. Chandler, N. Choisne, C. Claudel-Renard, S. Cunnac, N. Demange, C. Gaspin, M. Lavie, A. Molsan, C. Robert, W. Saurin, T. Schlex, P. Siguier, P. Thébault, M. Whalen, P. Wincker, M. Levy, J. Weissenbach, and C. A. Boucher. 2002. Genome sequence of the plant pathogen *Ralstonia solanacearum*. *Nature* 415:497–502.

Sandanayaka, V. P., and A. S. Prashad. 2002. Resistance to β-lactam antibiotics: structure and mechanism based design of β-lactamase inhibitors. *Curr. Med. Chem.* 9:1145–1165.

Schiff, P. B., J. Fant, and S. B. Horowitz. 1979. Promotion of microtubule assembly in vitro by taxol. *Nature* 277:665–667.

Seow, K.-T., G. Meurer, M. Gerlitz, E. Wendt-Pienkowski, C. R. Hutchinson, and J. Davies. 1997. A study of iterative type II polyketide synthases, using bacterial genes cloned from soil DNA: a means to access and use genes from uncultured microorganisms. *J. Bacteriol.* 179:7360–7368.

Short, J. M. April 2002a. Screening for novel bioactivities. U.S. patent 6,368,798 B1.

Short, J. M. September 2002b. Sequence based screening. U.S. patent 6,455,254 B1.

Short, J. M., and E. J. Mathur. September 2002. Production and use of normalized DNA libraries. U.S. patent 6,444,426 B1.

Shu, Y.-Z. 1998. Recent natural products based drug development: a pharmaceutical industry perspective. *J. Nat. Prod.* 61:1053–1071.

Sievert, D. M., G. Boulton, D. Stoltman, M. G. Johnson, F. P. Stobierski, P. A. Downes, J. T. Somsel, E. Rudrik, W. Brown, T. Hafeez, E. Lundstrom, R. Flansgan, J. Jonson, J. Mitchell, and S. Chang. 2002a. *Staphylococcus aureus* resistant to vancomycin—United States. *Morb. Mortal. Wkly. Rep.* 51:565–567.

Sievert, D. M., S. Chang, J. Hageman, S. K. Fridkin, and VRSA Investigation Team. 2002b. Investigation of a *vanA*-positive vancomycin-resistant *Staphylococcus aureus* infection. Prog. Abstr. 42nd Intersci. Conf. Antimicrob. Agents Chemother. Abstract LB-6. American Society for Microbiology, Washington, D.C.

Sigmund, J. M., D. J. Clark, F. A. Rainey and A. S. Anderson. 2003. Detection of eubacterial 3-hydroxy-3-methylglutaryl coenzyme A reductases from natural populations of actinomycetes. *Microb. Ecol.* 46:106–112.

Silver, L. L., and K. A. Bostian. 1993. Discovery and development of new antibiotics: the problem of antibiotic resistance. *Antimicrob. Agents Chemother.* 37:377–383.

Simpson, A. J. G., F. C. Reinach, P. Arruda, F. A. Abreu, M. Acencio, R. Alvarenga, L. M. C. Alves, J. E. Araya, G. S. Bala, C. S. Baptista, et al. 2000. The genome sequence of the plant pathogen *Xylella fastidiosa*. *Nature* 406:151–159.

Singh, P. K., B. F. Tack, P. B. McCray, Jr., and M. J. Welsh. 2000. Synergistic and additive killing by antimicrobial factors found in human airway surface liquid. *Am. J. Physiol. Lung Cell Mol. Physiol.* 5:L799–L805.

Sosio, M., E. Bossi, A. Bianchi, and S. Donadio. 2000. Multiple peptide synthetase gene clusters in actinomycetes. *Mol. Gen. Genet.* 264:213–221.

Sosio, M., E. Bossi, and S. Donadio. 2001. Assembly of large genomic segments in artificial chromosomes by homologous recombination in *Escherichia coli*. *Nucleic Acids Res.* 29:1–8.

Stapley, E. O., J. Birnbaum, A. K. Miller, H. Wallick, D. Hendlin, and H. B. Woodruff. 1979. Cefoxitin and cephamycins: microbiological studies. *Rev. Infect. Dis.* 1:73–89.

Stierle, A. C., J. H. Cardellina II, and F. L. Singleton. 1988. A marine *Micrococcus* produces metabolites ascribed to the sponge *Tedania ignis*. *Experientia* 44:1021.

Stover, C. K., X. Q. Pham, A. L. Erwin, S. D. Mizoguchi, P. Warrener, M. J. Hickey, F. S. L. Brinkman, W. O. Hufnagle, D. J. Kowalik, M. Lagrou, R. L. Garber, L. Goltry, E. Tolentino, S. Westbrock-Wadman, Y. Yuan, L. L. Brody, S. N. Coulter, K. R. Folger, A. Kas, K. Larbig, R. Lim, K. Smith, D. Spencer, G. K.-S. Wong, Z. Wu, I. T. Paulsen, J. Reizer, M. H. Saler, R. E. W. Hancock, S. Lory, and M. V. Olson. 2000. Complete genome sequence of *Pseudomonas aeruginosa* PA01, an opportunistic pathogen. *Nature* 406:959–964.

Strohl, W. R. 1997. Industrial antibiotics: today and the future, p. 1–47. *In* W. R. Strohl (ed.), *Biotechnology of Antibiotics*, 2nd ed. Marcel Dekker Publishers, New York, N.Y.

Strohl, W. R. 1999. Natural product antibiotics, p. 2348–2365. *In* M. C. Flickinger and S. W. Drew (ed.), *The Encyclopedia of Bioprocess Technology: Fermentation, Biocatalysis, and Bioseparation*. Wiley & Sons, Inc., New York, N.Y.

Strohl, W. R. 2000. The role of natural products in a modern drug discovery program. *Drug Discovery Today* 5:39–41.

Strohl, W. R., M. L. Dickens, V. B. Rajgarhia, A. J. Woo, and N. D. Priestley. 1997. Anthracyclines, p. 577–657. *In* W. R. Strohl (ed.), *Biotechnology of Antibiotics*, 2nd ed. Marcel Dekker Publishers, New York, N.Y.

Strohl, W. R., H. B. Woodruff, R. L. Monaghan, D. Hendlin, S. Mochales, A. L. Demain, and J. Liesch. 2001. The history of natural products research at Merck & Co., Inc. *Soc. Ind. Microbiol. News* 51:5–19.

Sunekawa, K., H. Akita, S. Iwata, Y. Sato, and R. Fujii. 1995. Rational use of oral antibiotics for pediatric infections. *Infection* 23(Suppl.):S74–S78.

Tang, L., S. Shah, L. Chung, J. Carney, L. Katz, C. Khosla, and B. Julien. 2000. Cloning and heterologous expression of the epothilone gene cluster. *Science* **287**:640–642.

Thompson, J., M. O'Connor, J. A. Mills, and A. E. Dahlberg. 2002. The protein inhibitors oxozal idinones and chloromphanicol cause extensive translational inaccuracy *in vivo. J. Mol. Biol.* **322**:273–279.

Tkacz, J., and B. DiDomenico. 2001. Antifungals: what's in the pipeline. *Curr. Opin. Microbiol.* **4**:540–546.

Turner, W. W., and W. L. Current. 1997. Echinocandin antifungal agents, p. 315–334. *In* W. R. Strohl (ed.), *Biotechnology of Antibiotics*, 2nd ed. Marcel Dekker, Inc., New York, N.Y.

Uttley, A. H. C., C. H. Collins, J. Naidoo, and R. C. George. 1988. Vancomycin-resistant enterococci. *Lancet* **1**:57–58.

van Rozendaal, E. L., G. P. Lelyveld, and T. A. van Beek. 2000. Screening of the needles of different yew species and cultivars for paclitaxel and related taxoids. *Phytochemistry* **53**:383–389.

Walsh, T. R., and R. A. Howe. 2002. The prevalence and mechanisms of vancomycin resistance in *Staphylococcus aureus. Annu. Rev. Microbiol.* **56**:657–675.

Wani, M. C., H. L. Taylor, M. E. Wall, P. Coggen, and A. T. McPhail. 1971. Plant antitumor agents. VI. The isolation and structure of taxol, a novel antileukemic and antitumor agent from *Taxus brevifolia. J. Am. Chem. Soc.* **93**:2325–2327.

Waters, B., and J. Davies. 1997. Amino acid variation in the GyrA subunit of bacteria potentially associated with natural resistance to fluoroquinolone antibiotics. *Antimicrob. Agents Chemother.* **41**:2766–2769.

Watts, J. E. M., A. S. Huddleston-Anderson, and E. M. H. Wellington. 1998. Bioprospecting, p. 631–641. *In* A. L. Demain and J. E. Davies (ed.), *Manual of Industrial Microbiology and Biotechnology.* ASM Press, Washington, D. C.

Watve, M. G., R. Tickoo, M. M. Jog, and B. D. Bhole. 2001. How many antibiotics are produced by the genus *Streptomyces? Arch. Microbiol.* **176**:386–390.

Webster, N. S., K. J. Wilson, L. L. Blackall, and R. T. Hill. 2001. Phylogenetic diversity of bacteria associated with the marine sponge *Rhopaloeides odorabile. Appl. Environ. Microbiol.* **67**:434–444.

Weinstein, R. A. 1998. Nosocomial infection update. *Emerging Infect. Dis.* **4**:416–420.

Wisplinghoff, H., M. B. Edmond, M. A. Pfaller, R. N. Jones, R. P. Wenzel, and H. Seifert. 2000. Nosocomial bloodstream infections caused by *Acinetobacter* species in United States hospitals: clinical features, molecular epidemiology, and antimicrobial susceptibility. *Clin. Infect. Dis.* **31**:690–697.

Woodford, N. 2001. Epidemiology of the genetic elements responsible for acquired glycopeptide resistance in enterococci. *Microb. Drug Resist.* **7**:229–236.

Woodruff, H. B., J. M. Mata, S. Hernandez, S. Mochales, A. Rodriguez, E. O. Stapley, H. Wallick, A. K. Miller, and D. Hendlin. 1977. Fosfomycin: laboratory studies. *Chemotherapy* **23**(Suppl.):1–22.

World Health Organization. 1996. *World Health Report.* Office of World Health Reporting, World Health Organization, Geneva, Switzerland.

Wright, G. D. 2000. Resisting resistance: new chemical strategies for battling superbugs. *Chem. Biol.* **7**:R127–R132.

Ye. R. W., W. Tao, L. Bedzyk, T. Young, M. Chen, and L. Li. 2000. Global gene expression profiles of *Bacillus subtilis* grown under anaerobic conditions. *J. Bacteriol.* **182**:4458–4465.

Yukimune, Y., H. Tabata, Y. Higashi, and Y. Hara. 1996. Methyl jasmonate-induced overproduction of paclitaxel and baccatin III in Taxus cell suspension cultures. *Nat. Biotechnol.* **14**:1129–1132.

Zafriri, D., E. Rosenberg, and D. Mirelman. 1981. Mode of action of *Myxococcus xanthus* antibiotic TA. *Antimicrob. Agents Chemother.* **19**:349–351.

Zahn, J. A., R. E. Higgs, and M. D. Hilton. 2001. Use of direct-infusion electrospray mass spectrometry to guide empirical development of improved conditions for expression of secondary metabolites from actinomycetes. *Appl. Environ. Microbiol.* **67**:377–386.

Zähner, H., and H. Fiedler. 1995. The need for new antibiotics: possible ways forward, p. 67–85. *In* P. A. Hunter, G. K. Darby, and N. J. Russell (ed.), *Fifty Years of Antimicrobials: Past Perspective and Future Trends.* SGM Symp. 53, Cambridge University Press, Cambridge, United Kingdom.

Zhang, Y.-X., K. Perry, V. A. Vinci, K. Powell, W. P. C. Stemmer, and S. B. del Cardayré. 2002. Genome shuffling leads to rapid phenotypic improvement in bacteria. *Nature* **415**:644–646.

Zhao, W. H., Z. Q. Hu, S. Okubo, Y. Hara, and T. Shimamura. 2001. Mechanism of synergy between epigallocetechin gallate and β-lactams against methicillin-resistant *Staphylococcus aureus. Antimicrob. Agents Chemother.* **45**:1737–1742.

Zheng, Z, W. Zeng, Y. Huang, Z. Yang, J. Li, H. Cai, and W. Su. 2000. Detection of antitumor and antimicrobial activities in marine organism associated actinomycetes isolated from the Taiwan Strait, China. *FEMS Microbiol. Lett.* **188**:87–91.

Microbial Diversity and Bioprospecting
Edited by Alan T. Bull
© 2004 ASM Press, Washington, D.C.

Chapter 32

Pharmacologically Active Agents of Microbial Origin

STEPHEN K. WRIGLEY

PHARMACOLOGICAL AGENTS AND IMPORTANCE OF MICROBIAL PRODUCTS

Pharmacologically active natural products are generally classified as those useful for treating human diseases and disorders including hyperlipidemia, cancer, immunoregulatory disorders, inflammation, and neurological and metabolic diseases. Microbial products have a rich heritage in this area and notable successes include the indole alkaloids from ergot (*Claviceps purpurea*) and their derivatives, agonists and antagonists of adrenergic, dopaminergic, and serotoninergic receptors with utility in the treatment of migraine, Parkinson's disease, and postpartum haemorrhage (Shu, 1998); antitumor agents such as the anthracyclines daunomycin and doxorubicin, bleomycins, and mitomycins from various streptomycetes (Newman et al., 2000); cyclosporin A from *Tolypocladium inflatum,* the first of a new class of immunosuppressants that is used widely for the prevention of organ transplant rejection (Borel et al., 1976; Mann, 2001); and the hydroxymethylglutaryl coenzyme A reductase inhibitor mevinolin from *Aspergillus terreus,* the first in a very successful class of drugs lowering the levels of cholesterol and low-density lipoprotein cholesterol in patients with hypercholesterolemia (Alberts et al., 1980; Endo and Hasumi, 1993). Approaches to the discovery of pharmacologically active natural products and the development status of leads arising from such discovery programs have been reviewed regularly over the past decade or so (see, for example, Franco and Coutinho, 1991; Hill et al. 1998; Newman et al., 2000; Nisbet and Porter, 1989; Shu, 1998).

The past decade has seen a great deal of screening of natural product-derived libraries against pharmacological targets, some of the outputs of which are discussed in "Pharmacologically Active Microbial Products," below. It has also seen much change in the pharmaceutical industry, notably great progress in

the area of genomics and the expectancy of a large increase in the number of new targets that will be available for drug discovery, the much-increased availability of synthetic chemical diversity for screening in the form of commercially available synthetic and combinatorial chemical libraries, and the consequent "industrialization" of high-throughput drug discovery screening designed to significantly reduce project timelines (Drews, 2000). Synthetic chemical diversity may well be limited in comparison with natural product structural diversity, but it is viewed as being more convenient as compounds in synthetic libraries should be pure and structurally characterized at the point of screening and should be easy to generate in larger quantities. These are factors that are currently limiting the deployment of natural product screening against pharmacological targets. In this chapter I review the discovery of pharmacologically active microbial metabolites, with an emphasis on recent developments, and consider the specific challenges involved in screening samples derived from microbial fermentations.

HIGH-THROUGHPUT SCREENING ASSAYS AND MICROBIAL PRODUCT DIVERSITY

High-Throughput Screening Assays and Strategy

High-throughput screening assays used for the discovery of pharmacologically active agents are discussed only briefly here, as screening was discussed in detail in chapter 30. Specific examples of assays that have resulted in the discovery of new pharmacological agents of microbial origin are given in "Pharmacologically Active Microbial Products," below. These assays fall into two broad categories: molecular interaction assays, either enzyme or receptor based, or cell-based assays, mostly using eukaryotic cells. The use of isolated enzymes to discover pharmacologically active enzyme inhibitors of microbial origin was

Stephen K. Wrigley • Cubist Pharmaceuticals (UK) Ltd., 545 Ipswich Road, Slough, Berkshire SL1 4EP, United Kingdom.

pioneered by Umezawa in the 1960s (Umezawa, 1982). The use of receptor binding assays based on a target of interest in a purified system for the discovery of pharmacologically active microbial metabolites had become widespread by the late 1980s; an early example of a significant discovery was the potent and selective cholecystokinin antagonist asperlicin from *Aspergillus alliaceus* (Chang et al., 1985). The design of high-throughput enzyme- or receptor-based screens often involves compromises such as the use of nonphysiological enzyme substrates or receptor ligands for economic reasons. Both enzyme- and receptor-based assays can be susceptible to high false-positive rates due to compounds in microbial samples with nonspecific effects on enzymes, receptors or their environments (see "Microbial Product Diversity, Reactivity, and Toxicity," below). Screening strategies should ideally involve the use of secondary assays, or comparative screening data from related and unrelated targets, to assess selectivity of action and more physiologically relevant functional assays.

In their simplest form, cell-based assays using tumor cell lines continue to be used for the discovery of new cytotoxic agents. Increased understanding of cell biology has enabled more sophisticated mechanism-based screens to be employed in the search for molecules inhibitory to a particular cellular function. Compounds detected in such screens may be active at any of a number of relevant targets ranging from ligand-receptor binding, through intracellular signal transduction, to effector protein synthesis and secretion. Such mechanism-based cellular screens are used to search for inhibitors of signal transduction cascades as potential immunosuppressants or antiproliferative agents (Cardenas et al., 1998). Screens using engineered cells, for example, can be used to search for transcriptional activation inhibitors (Hertzberg, 1993). Controls for false positives caused by toxic substances are essential and are usually performed as part of the assay protocol (but see also "Microbial Product Diversity, Reactivity, and Toxicity" below). As there are a number of targets with which compounds identified in cell-based screens can interact, various secondary assays may be required to determine the precise target and mechanism of action. The use of such screens does, however, offer advantages for the discovery of pharmacologically active substances: the target of the compound is an environment that more closely resembles that in vivo, leading to better prospects for activity in pharmacological models; the need for cellular penetration to reach intracellular targets will eliminate many of the nonspecific false positives detected in screens using purified enzymes or receptors; cellular penetration is also required for pharmacological activity against many tar-

gets (Hertzberg, 1993). This last point is also a disadvantage in that potential leads that inhibit the target, and would be active on purified enzymes or receptors but are unable to cross the cell membrane, will be missed. Such compounds can still form useful starting points for chemical modification.

Sample Presentation and Chemical Dereplication

Samples produced by microbial fermentation have traditionally been tested in drug discovery screens as clarified broths or extracts presented in water-miscible organic solvents such as methanol or dimethyl sulfoxide. Broth samples are not compatible with screens involving isolated enzymes or receptors as they contain high-molecular-weight species such as proteinases, which can attack the protein constituents of the assay and lead to high false-positive rates. The effects of such species can be reduced by heat treatment or ultrafiltration, but neither method is ideal. Solvent extracts are therefore the sample of choice for testing in pharmacological discovery assays, but solvent tolerance may be a problem, particularly in cell-based assays. A concentration step in the extraction procedure minimizes the volume of solvent required for sample addition.

Extracts consist of mixtures of metabolites present at unknown concentrations. At the stage when a number of extracts with selective activity against a target have been found, assay-guided purification is required to separate the components of each extract and determine whether activity is caused by the presence of a high concentration of a component with low to moderate potency or a highly active minor component. At the same time, chemical dereplication by high-pressure liquid chromatography (HPLC) with photodiode array (PDA) and mass spectrometric (MS) detection can begin. The aims of chemical dereplication are the identification of samples containing compounds known to be active against the target, the identification of known microbial metabolites, and the grouping of samples containing similar, or the same, active components. Identification of novel active components nearly always involves refermentation at a scale of several liters. This and the subsequent purification and structure elucidation are resource intensive, and these resources need to be focused on the most promising actives in terms of biological activity and potential novelty. The time taken from active extract detection through dereplication, refermentation, purification, and structure elucidation is usually 3 months at a minimum and significantly longer if multiple refermentations are required. By contrast, the structures of synthetic library compounds are usually available at the point of screening, enabling much quicker progression decisions.

An alternative to extract library screening is to screen libraries of purified natural products (Ainsworth et al., 2001; Bindseil et al., 2001; Grabley and Thiericke, 1999). Such libraries are designed to be submitted to high-throughput screening drug discovery operations under the same conditions as synthetic libraries, minimizing the time needed for identification of lead compounds. One such library was assembled by first screening microbial extracts physicochemically by HPLC-PDA-MS; organisms producing rare or potentially novel metabolites were then selected for scale-up fermentation. Compounds were then purified on a multimilligram scale to permit extensive screening at known concentrations, to facilitate rapid prioritization by potency, and to permit structure elucidation without the need for refermentation (Ainsworth et al., 2001). Purity checking also involves physicochemical characterization of library compounds, ensuring dereplication and avoidance of duplication. The drawbacks of such libraries of purified compounds are that they are expensive to produce and they may not capture effectively all the metabolites produced by an organism, necessarily focusing on those produced above a certain concentration threshold and missing potentially highly active minor components.

Microbial Product Diversity, Reactivity, and Toxicity

When considering sources of chemical diversity for drug discovery programs, strengths of natural products are generally regarded as being their structural diversity and complexity and their high incidence of biological activity (see chapter 1, this book). A recent statistical comparison of natural product and synthetic compound structural features found that natural products were sterically more complex and that there were distinct differences between natural and synthetic compounds, with around 40% of natural compounds having no synthetic equivalents (Henkel et al., 1999). The same study found that natural products produced by different sources including bacteria, fungi, rare actinomycetes, and plants were complementary and concluded that natural products remain an important source for lead finding. The belief in the high incidence in biological activity follows from the proposal that all secondary metabolites serve their producing organisms by improving their survival fitness. Complex structures that are programmed by many kilobases of DNA were not evolved, and are not produced, lightly when considered in terms of metabolic energy expenditure (Williams et al., 1989).

This expectation is apparently generally confirmed when inspecting hit rates obtained on screening microbial product extract libraries. High hit rates are, however, best approached with caution. Many nonspecific biologically active substances are present in microbial fermentation extracts. There are different classes of problematic compounds for molecular interaction and cell-based assays.

For assays involving purified protein targets, compounds that are chemically reactive toward protein bind irreversibly and cause nonspecific hits. Such compounds are also present in synthetic libraries but are easily recognizable on inspecting their chemical structures (Rishton, 1997). Extracts active against a target cannot be sorted in this way. One approach to reducing this problem is to look for the presence of known reactive compounds by HPLC-PDA-MS, but this will not eliminate novel reactive compounds. Examples of reactive compounds in microbial samples are polyphenolics, epoxides, and α,β-unsaturated carbonyl-containing compounds. Compounds such as the sesquiterpenyl 2-hydroxymethylbenzaldehyde L-671,776 (Fig. 1) from *Memnoniella echinata* (Lam et al., 1992) and other fungal metabolites such as the azaphilones are reactive toward amines and are possibly useful for their producing organisms for nitrogen sequestration purposes; unfortunately this makes them reactive toward free amine groups in proteins, explaining the numerous weak activities toward pharmacological targets reported for these classes (Vertesy et al., 2001). The problem of these nonspecifically reactive metabolites is best managed by looking for selectivity of activity across a number of related targets and confirming activity in more complex, physiologically relevant secondary assays. Selectivity is key: selectively reactive microbial metabolites, which bind irreversibly to their targets with high affinity can provide useful leads; some examples of such compounds are discussed in "Pharmacologically Active Microbial Products," below.

For cell-based assays, false positives may be caused by subtly toxic molecules not detected in controls and by compounds having effects unrelated to the target mechanism (Hertzberg, 1993). Three examples of such compounds are shown in Fig. 1. β-Methoxyacrylates such as 9-methoxystrobilurin E from *Favolaschia pustulosa* (Wood et al., 1996) bind to cytochrome *b* and inhibit mitochondrial respiration. 9-Methoxystrobilurin E was isolated as a potent inhibitor of interleukin-4 (IL-4)-dependent sCD23 production in Jijoye cells with no apparent toxicity. When the cytotoxicity assay was extended from the normal 2 to 5 days, the toxic effects of 9-methoxystrobilurin E became apparent. Geldanamycin and monorden (or radicicol) are fairly common metabo-

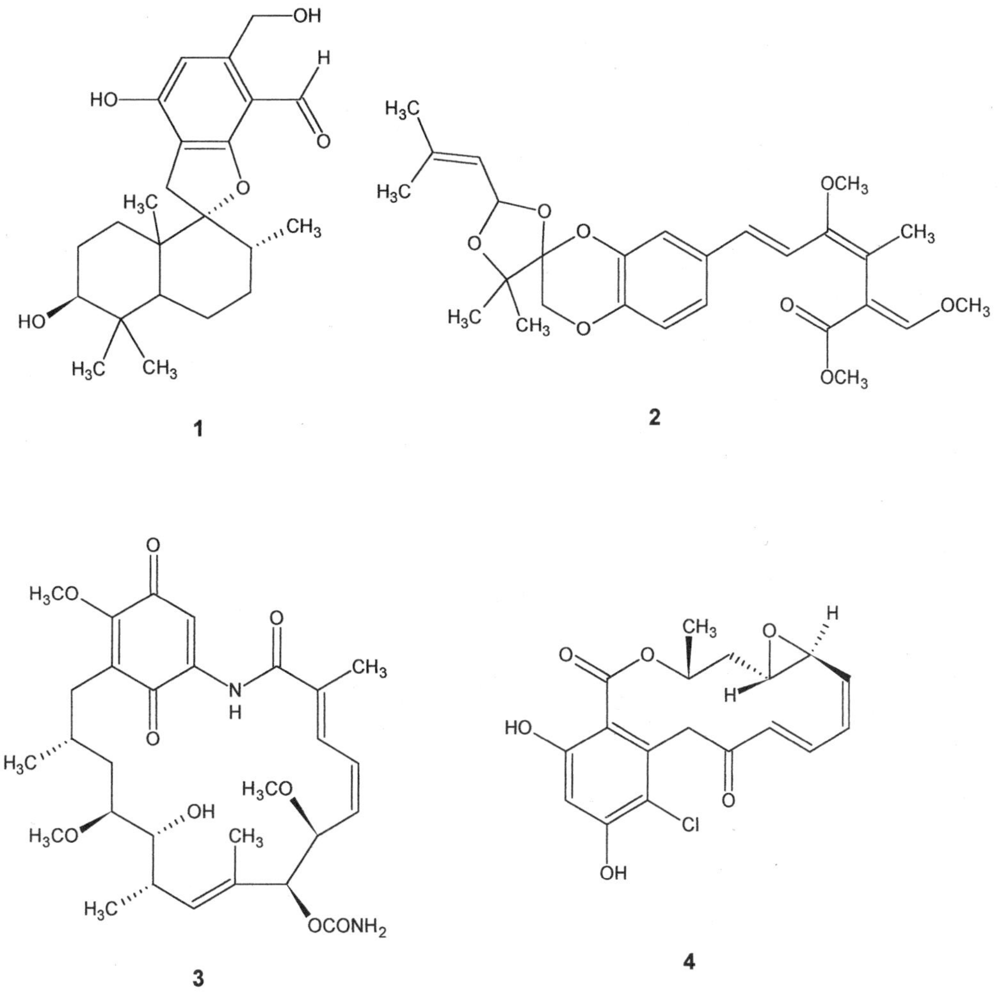

Figure 1. Structures of the *Memnoniella echinata* metabolite L-671,776, (structure 1), the *Favolaschia* sp. metabolite 9-methoxystrobilurin E (structure 2), the actinomycete metabolite geldanamycin (structure 3), and the fungal metabolite monorden (radicicol) (structure 4).

lites of actinomycetes and fungi, respectively, that cause potent inhibition in cellular assays for signal transduction cascades. For many years they were thought to be protein kinase inhibitors, but they are now both known to be inhibitors of the ATPase activity of the heat shock protein 90 molecular chaperone (Roe et al., 1999). Both compounds are in fact toxic, and a geldanamycin derivative, 17-allylamino,17-demethoxygeldanamycin, is currently being assessed for antitumor potential in the clinic (Kelland et al., 1999). Despite being produced in low concentrations, geldanamyin, monorden, and strobilurins are straightforward to detect by HPLC-PDA-MS. Novel variants will obviously not be characterized as easily. Varying the conditions of cytotoxicity control assays and looking at cytotoxicity against more than one cell line best manage the problem of nonobviously toxic molecules such as these.

PHARMACOLOGICALLY ACTIVE MICROBIAL PRODUCTS

The following is a review of selected pharmacologically active agents of microbial origin with utility or potential utility in the areas of immunosuppression, cancer, and cardiovascular and metabolic disease.

Immunosuppressive Agents

The cyclic undecapeptide cyclosporin A (Fig. 2) was originally isolated from *Tolypocladium inflatum* after detection of its antifungal activity. The isolated but still structurally uncharacterized product was subsequently shown to be a potent inhibitor of antibody production in mice, effective solubilization of the lipophilic compound proving to be key in obtain-

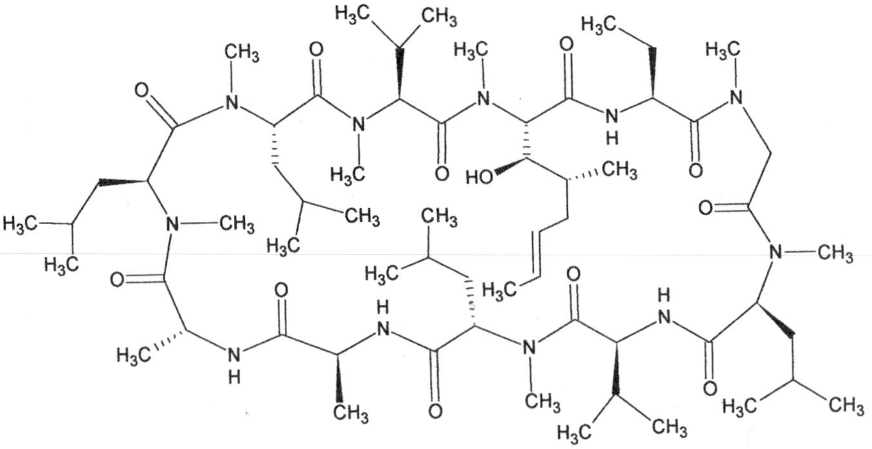

Figure 2. Structures of the immunosuppressants, cyclosporin A (structure 1), from *Tolypocladium inflatum*; FK506 (structure 2) from *Streptomyces tsukubaensis*; and rapamycin (structure 3) from *Streptomyces hygroscopicus*.

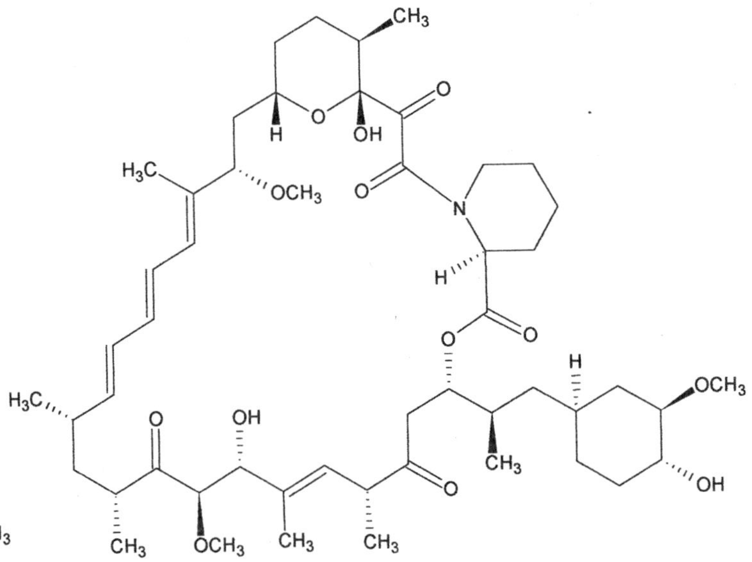

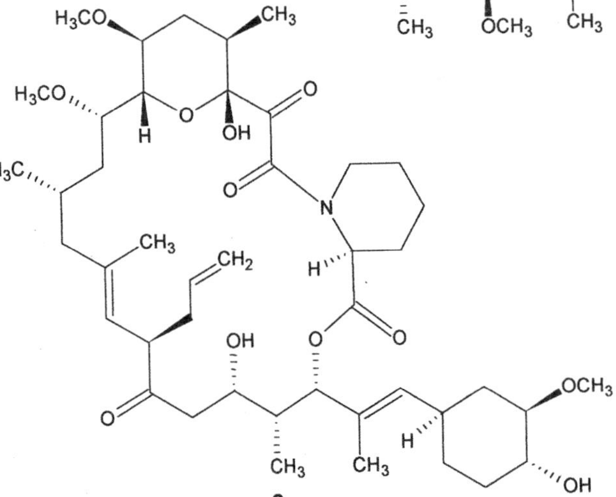

ing a positive in vivo result (Stahelin, 1996; Borel et al., 1976). Cyclosporin A was then developed as an immunosuppressant selectively inhibiting cytotoxic T cells and has achieved widespread use in the prevention of organ transplant rejection; it may well prove to have wider utility against autoimmune diseases including rheumatoid arthritis, systemic lupus erythematosus, and Crohn's disease (Mann, 2001). The immunosuppressive action of cyclosporin A is characterized by the inhibition of production of T-cell-derived cytokines such as IL-2 and the inhibition of cytotoxic T-cell proliferation. The macrolide immunosuppressant FK-506 (Fig. 2) was discovered as a product of *Streptomyces tsukubaensis* following a program of screening for inhibitors of the mouse mixed lymphocyte reaction with selective effects on IL-2 production (Kino et al., 1987a, 1987b). The in vitro immunosuppressive effects of FK-506 were 100-fold more potent than those of cyclosporin A. FK-506 was subsequently approved for clinical use. A third significant microbial metabolite with immunosuppressive properties was also originally discovered as a result of its antifungal properties: rapamycin, isolated from a strain of *Streptomyces hygroscopicus* (Vezina, et al., 1975). Rapamycin was later found to have in vivo immunosuppressive properties, prolonging the survival of highly histoincompatible heart grafts in mouse and rat transplant recipients without causing any cytotoxicity (Morris and Meiser, 1989). It has recently been approved for clinical use as an immunosuppressant (Mann, 2001).

Cyclosporin A, FK-506, and rapamycin are the three microbial metabolites that have had the most impact to date as immunosuppressants. Mechanism of action studies on the three compounds have proved to be very fruitful in elucidating the steps involved in T-cell activation and signaling (Mann, 2001). FK-506 and cyclosporin A bind to and cause potent inhibition of two different proteins with peptidyl-prolyl *cis-trans* isomerase (rotamase) activity: FKBP12 (FK-506 binding protein 12) and cyclophilin, respectively. This was originally thought to be the key to the mechanism of immunosuppression of these molecules. Rapamycin, however, also binds to and causes potent inhibition of FKBP12 but functionally inhibits a separate T-cell signaling pathway. The actual mechanisms of immunosuppression are related, but proved to be more complex. The natural products complexed to their binding proteins (collectively termed immunophilins) cause immunosuppression by binding to secondary protein targets. The cyclosporin A-cyclophilin and FK-506-FKBP12 complexes both bind to the phosphatase calcineurin, thereby inhibiting IL-2 production and T-cell proliferation. The rapamycin-FKBP12 complex binds to and inhibits a protein termed mTOR

(mammalian target of rapamycin, or FRAP—FK-506-binding protein), which is involved in the phosphatidylinositol-3-kinase signaling pathway, resulting in cell cycle arrest at the G1 phase. The actions of cyclosporin A, FK-506, and rapamycin are covered in more detail in the review on natural product immunosuppressants by Mann (2001). The mechanism of action of rapamycin also lends itself to the design of selective antitumor agents, and a number of semisynthetic derivatives of rapamycin are now undergoing clinical evaluation (Huang and Houghten, 2002).

Two other notable examples of compounds of microbial origin that have reached clinical use were also previously characterized compounds that were subsequently found to have immunosuppressive properties. The fungal metabolite mycophenolic acid inhibits the enzymes inosine monophosphate dehydrogenase and guanylate synthetase, causing selective inhibition of DNA synthesis in lymphocytes and is administered in conjunction with other drugs for the prevention of renal graft rejection as the prodrug mycophenolate mofetil (Sievers et al., 1997). Spergualin was originally isolated from *Bacillus lateosporus* as an antitumor agent (Takeuchi et al., 1981). A synthetic derivative, 15-deoxyspergualin, is being used in the prevention of organ transplant rejection (Mann, 2001).

The myriocins, cyclophilin A-binding compounds, and cytokine production inhibitors provide more recent examples of interesting microbial immunosuppressants and discovery approaches. The myriocins, originally discovered as antifungal agents, were rediscovered as metabolites of the fungus *Isaria sinclairii*, an imperfect stage of *Cordyceps sinclairii*, following a screening program to find inhibitors of the mouse allogeneic mixed lymphocyte reaction (Fujita et al., 1994). Myriocin A, a sphingosine-like compound (Fig. 3), was equipotent with FK-506 and 10- to 100-fold more potent than cyclosporin A in in vitro and in vivo immunosuppressive assays. Myriocin A is now known to be a highly potent (effective at picomolar concentrations) inhibitor of serine palmitoyl transferase, which constitutes the rate-limiting step in the biosynthesis of sphingoloipids, important mediators of cellular proliferation and programmed cell death (Chen et al., 1999). The sanglifehrins, from a *Streptomyces* sp. (Sanglier et al., 1999), and cymbimicins A and B, from a *Micromonospora* sp. (Fehr et al., 1997), were discovered in a screening program to identify novel cyclophilin-binding compounds that might interfere with T-cell signaling pathways not involving calcineurin. The screening assay comprised an enzyme-linked immunosorbent assay between microbial extracts and biotinylated cyclophilin A competing for binding to a cyclosporin derivative. After incuba-

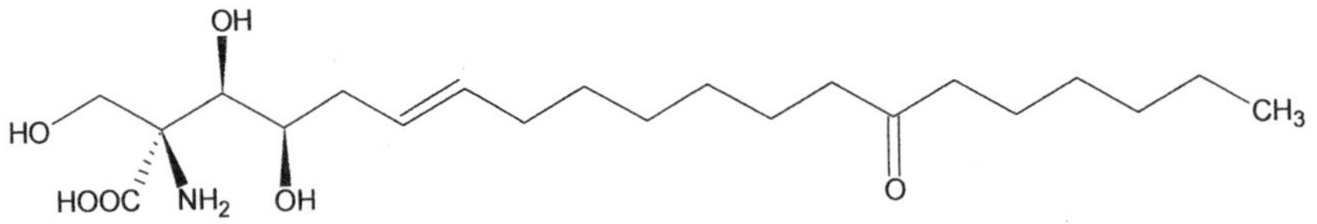

Figure 3. Structures of the immunosuppressive lead compounds myriocin A (structure 1) from *Isaria sinclairii;* sanglifehrin A (structure 2) from a *Streptomyces* sp.; XR774 (structure 3) from *Cladosporium* cf. *cladosporioides;* and CJ-14,897 (structure 4) from a *Marasmiellus* sp.

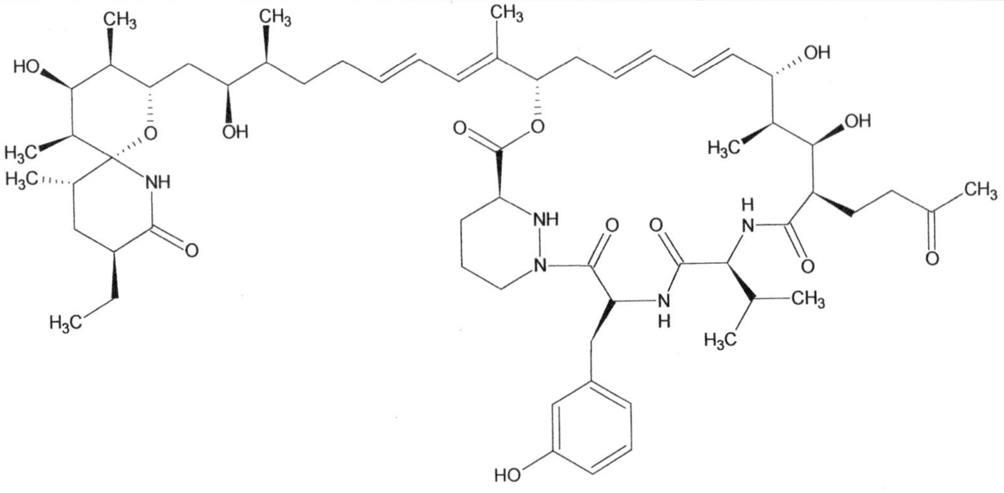

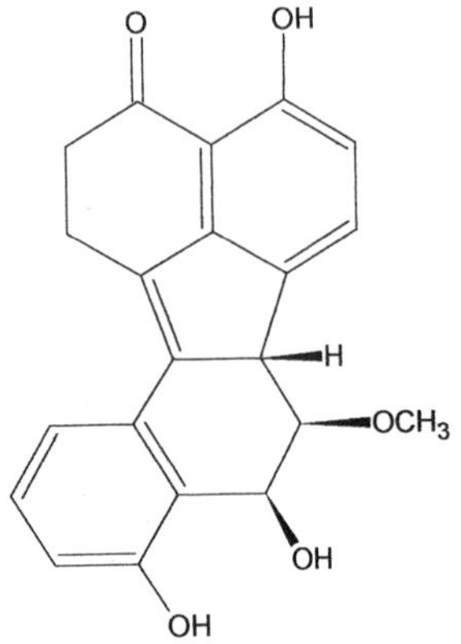

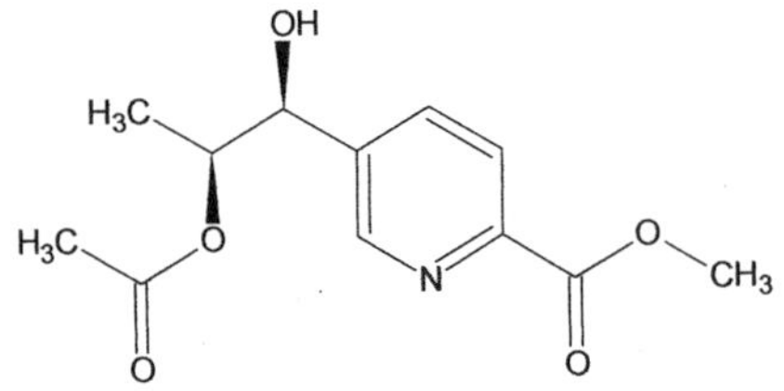

tion and washing, the bound cyclophilin was determined by adding streptavidin-linked alkaline phosphatase. Cymbimicin A bound to cyclophilin A with a slightly lower potency than cyclosporin A but was inactive in T-cell activation assays, an outcome predicted for many compounds identified in immunophilin-binding assays (Hertzberg, 1993). This lack of activity for cymbimicin A was ascribed to a lack of cellular penetration (Fehr et al., 1997). The

sanglifehrins, macrolides that are structurally distinct from cyclosporin A, FK-506, and rapamycin (Fig. 3), had higher cyclophilin A-binding affinities than cyclosporin A and did exhibit immunosuppressive activities on T cells, albeit with lower potency than cyclosporin A in the mixed lymphocyte reaction (Sanglier et al., 1999). Sanglifehrin A inhibits IL-2-induced T-cell proliferation at the GI phase of the cell cycle. Although its antiproliferative actions show some similarities to rapamycin, a recent study concluded that sanglifehrin A has a novel mechanism of action (Zhang and Liu, 2001).

Two other cytokines produced by activated macrophages and centrally involved in immune-inflammatory processes as mediators are tumor necrosis factor α (TNF-α) and IL-1. An extensive program conducted to identify small molecule inhibitors of cytokine-receptor binding and cytokine signal transduction illustrates some of the challenges involved in such microbial product discovery programs. Early discovery assays were based on IL-1 and TNF-α binding to membrane preparations and cell surfaces containing their respective receptors. These assays suffered from high false-positive rates due to membrane-disruptive agents such as polyene macrolides or compounds such as the actinomycete metabolites teleocidins, which activate protein kinase C and cause receptor shedding (MacAllan et al., 1997). A refined version of the TNF-α ligand-receptor assay was developed based on the binding of samarium chelate- and biotin-labeled TNF-α to europium chelate-labeled TNF-α p55 receptor in streptavidin-coated polystyrene plates (MacAllan, 1997). Monitoring the time-resolved fluorescence of the two lanthanide chelate labels allowed real inhibitors of the TNF-α–p55 receptor interaction to be distinguished from nonspecific binding inhibitors causing stripping of the TNF-α from the plates. This assay operated with an extremely low false-positive rate, but no significant inhibitors of this complex protein-protein interaction were found. Cell-based signal transduction assays were more successful in identifying new molecules of interest but were susceptible to interfering compounds such as those exemplified in "Microbial Product Diversity, Reactivity, and Toxicity" above. Once molecules of potential interest were identified, pinpointing their target and mechanism of action became a major challenge. An assay termed the "macrophage activation assay" designed to detect inhibitors of lipopolysaccharide (LPS)-induced TNF-α production by human histolytic lymphoma U937 cells activated by phorbol myristate acetate identified two classes of compounds of potential interest. Resorcyclic acid lactones such as 5-Z-7-oxo-zeaenol from the fungus Cochliobolus lunatus caused potent inhibition of TNF-α production. Extensive mechanism-of-action studies showed that this compound inhibited the phosphorylation and activation of mitogen-activated protein kinase (MAP kinase) (Rawlins et al., 1999). A closely related resorcylic acid lactone, L-783,277 from a Phoma sp., was separately shown to be a potent and specific inhibitor of MEK (MAP kinase kinase) (Zhao et al., 1999). The second compound of interest identified by the macrophage acti-

vation assay was a novel (6S)-4,6-dimethyldodeca-2E,4E-dienoyl ester of phomalactone isolated from a fermentation of a Phomopsis sp. This compound caused potent inhibition of TNF-α production in U937 cells but was found to have selective inhibitory effects on IL-1 production in a more physiologically relevant system comprising LPS-stimulated peripheral blood monocytes. The exact mechanism of action remained elusive but was posttranslational at the level of IL-1β secretion (Wrigley et al., 1999). A second cell-based signal transduction assay for inhibitors of CD28-induced IL-2 production in Jurkat E6-1 cells as potential immunosuppressants resulted in the identification of a novel benzo[j]fluoranthen-3-one, XR774 (Fig. 3), from Cladosporium cf. cladosporioides (Wrigley et al., 2001). This molecule was an effective inhibitor of IL-2 production that was found to be a competitive tyrosine kinase inhibitor with respect to ATP (Sadhegi et al., 2001). There have been a number of other recent reports of cytokine production inhibitors of microbial origin. Examples of potent inhibitors are the pyridine derivative CJ-14,897 (Fig. 3) from a Marasmiellus sp. (Ichikawa et al., 2001a) and a series of diterpenes from Oidiodendron griseum (Ichikawa et al., 2001b).

Other natural products with immunosuppressive properties are discussed in the review by Mann (2001). Although it is clear that microbial metabolites remain a major, if not the preeminent source of new immunosuppressants, it is also worth repeating the point stressed by Mann that "it is often difficult to differentiate this supposed [immunosuppressive] activity from the associated cytotoxicity of the compounds."

Anticancer Agents

Antitumor antibiotics isolated from various Streptomyces spp.—namely, the anthracyclines, daunomycin, doxorubicin, and their semisynthetic derivatives; the glycopeptide bleomycins; the peptolide actinomycins; the mitosanes such as mitomycin C; and the glycosylated anthracenone, mithramycin—have all proved to be clinically important cancer chemotherapeutic agents (Cragg et al., 2000). These compounds all have strong antibacterial as well as antitumor activity and were discovered in the era where the principal method of screening was for antimicrobial activity. Other microbial metabolites proving to be useful in the clinic, or reaching clinical evaluation, are listed by Cragg et al. (2000). Cell-based and mechanism-based antitumor discovery approaches continue to be productive in terms of new compounds identified, and a range of notable examples at various stages of development or evaluation is discussed here.

Epothilones A and B (Fig. 4) from the myxobacterium *Sorangium cellulosum* were originally detected by their narrow-spectrum antifungal activity. The purified compounds were found to have potent antitumor activity (Gerth et al., 1994). The level of interest in the activity of the epothilones increased significantly after the discovery that epothilones A and B acted via a mechanism similar to the successful antitumor agent of plant origin, taxol (Paclitaxel), by stabilizing polymerized microtubules (Bollag et al., 1995). The epothilones were also attractive as leads for anticancer agent development because they had a good activity profile against taxol-resistant tumor cell lines and were relatively straightforward to produce by fermentation. The in vivo activity of the epothilones was found only to be modest, however. A program of semisynthetic modification has now resulted in a lactam analogue of epothilone B, BMS-247550, with impressive activity in a battery of preclinical chemotherapy studies, and this compound is now being developed as an anticancer agent (Lee et al., 2001).

The angiogenesis-inhibitory properties of fumagillin (Fig. 4) were discovered following the serendipitous observation that a fungal contaminant affected the morphology of cultivated capillary endothelial cells in a fashion similar to previously characterized angiogenesis inhibitors. The fungal contaminant was isolated and identified as *Aspergillus fumigatus,* and the compound responsible for the observed activity was identified as the known metabolite fumagillin. Purified fumagillin inhibited endothelial cell proliferation in vitro and tumor-induced angiogenesis in vivo. Synthesis of fumagillin analogues yielded potent angiogenesis inhibitors with reduced side effects (Ingber et al., 1990). A synthetic fumagillin derivative, TNP-470, is now one of the leading angiostatic compounds in clinical development. Fumagillin and TNP-470 inhibit tumor neovascularization via endothelial cell cycle arrest in the late G1 phase. The mechanism of action of fumagillin involves covalent binding to and inhibition of the methionine aminopeptidase MetAP-2 (Sin et al., 1997). Fumagillin is an example of a reactive compound that has yielded synthetic derivatives with selective properties.

The calicheamicin enediyne-containing antitumor antibiotics (Fig. 4) from *Micromonospora echinospora* subsp. *calichensis* were discovered using the "biochemical induction assay," an *Escherichia coli*-based prophage induction assay using a genetically engineered *E. coli* strain that responds to DNA-damaging agents by overproducing β-galactosidase (Lee et al., 1993). They are among the most potent antitumor agents ever characterized. The calicheamicins are sequence-specific, minor groove DNA binding agents. Bioreductive activation of the enediyne

moiety results in the formation of a highly reactive diradical species that causes double-stranded DNA cleavage. This DNA cleaving capability probably forms the basis of the potent cytotoxicity of these compounds. The calicheamicins showed great potential as anticancer agents in vitro, but their therapeutic potential was restricted by dose-limiting lethality in animal models. A number of modified derivatives were conjugated to tumor-selective antibodies, and these monoclonal antibody conjugates displayed outstanding antitumor activity with reduced toxicity (Lee et al., 1993). One of these has recently become the first antibody-targeted chemotherapy agent to be approved for use in oncology. Mylotarg (gemtuzumab ozagamicin) is a conjugate of a calicheamicin derivative with an antibody specific to the CD33 antigen and was initially approved for use in cases of relapsed acute myeloid leukemia (Hamann and Berger, 2002).

Antitumor activity is relatively common among microbial metabolites, and new compounds found in screening programs based on cytotoxicity to tumor cell lines continue to be reported frequently. The rate of rediscovery in such programs is high. A more sophisticated approach involves primary or secondary screening against a panel of diverse tumor cell lines and looking for profiles of activity indicative of specific or new mechanisms of action. The National Cancer Institute operates a panel of 60 human cancer cell lines for this purpose (Cragg et al., 2000). A recent example of a novel compound found with such an approach is XR842 (Fig. 4). Microbial fermentation extracts active against a sensitive tumor cell line were subjected to a strenuous decision pathway designed to dereplicate known compounds and mechanisms. Extracts of a fermentation of *Streptomyces cyaneus* were selected for further characterization based on a unique pattern of activity against a tumor cell line panel. The active component, XR842, an unusual zinc complex, had potent in vitro cytotoxic properties and induced G1 arrest in a p53-independent manner. Further studies indicated that XR842 had a novel mechanism of action or mixed modes of action and were suggestive of the involvement of both iron chelation and DNA intercalation (A. T. Menendez et al., Discovery and characterization of XR842, an apoptotic natural product with novel structure and mechanism of action, Abstr. 93rd Annu. Meet. Am. Assoc. Cancer Res., abstr. 4656, 2002).

The current main approach to the discovery of compounds with promise as anticancer leads is via target-directed screening. An early target of interest was protein kinase C (PKC), a key signaling enzyme; overactivation of some of its isoenzymes was postulated to be important in cancer and some other disease states.

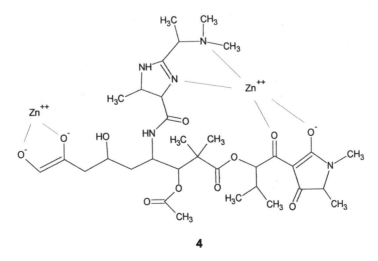

Figure 4. Structures of antitumor agents epothilone A (structure 1) from *Sorangium cellulosum*; fumagillin (structure 2) from *Aspergillus fumigatus*; calicheamicin γ_1^1 (structure 3) from *Micromonospora echinospora* subsp. *calichensis*; and XR842 (structure 4) from *Streptomyces cyaneus*.

The indole carbazole staurosporine, originally discovered as a *Saccharothrix* sp. metabolite in the course of a physicochemical screening program for microbial alkaloids, was subsequently found to be a potent inhibitor of PKC (Ōmura et al., 1995a). A screening program for more selective PKC inhibitors resulted in the discovery of a staurosporine derivative, UCN-01, 7-hydroxystaurosporine (Fig. 5) from a *Streptomyces* sp. (Takahashi et al., 1989). UCN-01 is currently in clinical trials as an antitumor agent (Newman et al., 2000). Another potent inhibitor discovered in a PKC screening program is balanol, a metabolite of the fungus *Verticillium balanoides* (Kulanthaivel et al., 1993). Balanol has attracted interest as a template for total chemical synthesis, and systematic structural modifications have resulted in improved enzyme inhibitory properties and cellular activity (Shu, 1998). Notable microbial inhibitors of other kinase enzymes include the tyrosine kinase inhibitor erbstatin, from a *Streptomyces* sp. (Umezawa, 1986), and the fungal terpenoid wortmannin, a potent inhibitor of phosphatidylinositol 3-kinase (Cardenas et al., 1998).

Some more recently identified targets illustrate the capability of microbial secondary metabolites to provide leads for the development of diverse mechanism-based anticancer therapeutics. Farnesyl protein transferase (FTPase) catalyzes the prenylation of Ras proteins, the products of *ras* genes, and this plays a major role in the proliferation of both normal and cancerous cells. *ras* genes with point mutations are associated with unregulated cell growth, are found in approximately 30% of all human tumors, and are the most common oncogenes associated with human carcinogenesis. Inhibitors of FTPase also inhibit the growth of tumor cells, are able to block *ras*-dependent tumorigenesis, and are potentially useful anticancer agents. An extensive microbial natural product screening program identified a range of FTPase inhibitors that fell into three different classes: compounds such as chaetomellic, oreganic, and actinoplanic acids that were competitive with farnesyl pyrophosphate and reversible FTPase inhibitors; clavaric acid (Fig. 5), which was competitive with Ras and was a reversible FTPase inhibitor; and other inhibitors that were noncompetitive or uncompetitive to Ras (Vilella et al., 2000). Clavaric acid, from the basidiomycete *Clavariadelphus truncatus,* was of particular interest not only as an FTPase inhibitor competitive with Ras but also as a nontoxic inhibitor of Ras processing in NIH3T3 *ras*-transformed cells. Clavaric acid was used as a template for chemical derivatization to obtain more therapeutically effective FTPase inhibitors (Lingham et al., 1998).

P53 is a phosphoprotein that acts as a powerful transcription factor and plays a key regulatory role in cell proliferation, controlling cell growth and the development of genetic abnormalities by inducing G1 arrest or apoptosis in response to DNA damage. Dysfunction of the *p53* gene is often associated with the development of cancers, and over 50% of all human tumors contain a mutated form. The cellular protein MDM2 is involved in the regulation of p53, binding to and concealing its DNA-binding domain and thereby preventing transcription of the p53-activated genes needed for G1 arrest or apoptosis. Compounds that disrupt the interaction between MDM2 and p53 could potentially treat tumors expressing abnormally high levels of MDM2, such as human sarcomas. Screening microbial extracts using an assay measuring the binding of a recombinant glutathione S-transferase–MDM2 fusion protein to a recombinant p53-myelin binding protein fusion protein with europium-based time-resolved fluorescence detection led to the identification of chlorofusin (Fig. 5), a novel metabolite of the fungus *Microdochium caespitosum* consisting of a cyclic peptide conjugated to an azaphilone-derived moiety (Duncan et al., 2001). Chlorofusin is a reversible MDM2 binder and inhibitor of this protein-protein interaction (Duncan et al., 2003). Other potential p53-related therapeutic approaches could involve small molecules that induce cell cycle arrest in a p53-independent manner or restore the active conformations of mutant forms of p53 that are transcriptionally inactive. Screening with transfected H1299/tsp53 cells, human non-small lung cancer cells that stably express a temperature-sensitive (Ala138 > Val) human p53, which adopts an inactive conformation at 37°C, identified microbial extracts that arrested cell cycle progression at this temperature. These included actinomycete fermentation extracts that contained the histone deacetylase inhibitor trichostatin A and novel derivatives, which caused arrest at both the G1 and G2 phases (Ueki et al., 2001), and a novel fungal metabolite, lucilactaene, which caused arrest at the G1 phase (Kakeya et al., 2001). In addition, these compounds significantly activated the p21[WAF1] promoter, a target of the *p53* gene, in H1299/tsp53 cells transfected with a luciferase reporter gene linked to the p21[WAF1] promoter. This latter cell line forms the basis of a robust screening assay to detect compounds with similar activities (Nie et al., 2001).

The proteasome is another target attracting much interest as a route to the discovery of potential cancer therapeutics. It is a complex comprising many different proteins that degrades proteins selected by labeling with ubiquitin. Rapid and irreversible proteasomal protein degradation is the key to the activation or repression of many cellular processes, including cell cycle progression and apoptosis. Two interesting natural product proteasome inhibitors were initially dis-

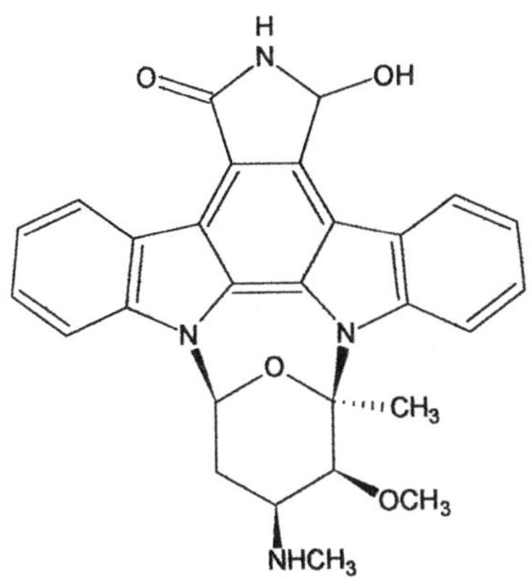

Figure 5. Structures of mechanism-based antitumor lead compounds UCN-01 (structure 1) from a *Streptomyces* sp.; clavaric acid (structure 2) from *Clavariadelphus truncatus*; chlorofusin (structure 3) from *Microdochium caespitosum*; lactacystin (structure 4) from a *Streptomyces* sp.; and telemostatin (structure 5) from *Streptomyces anulatus*.

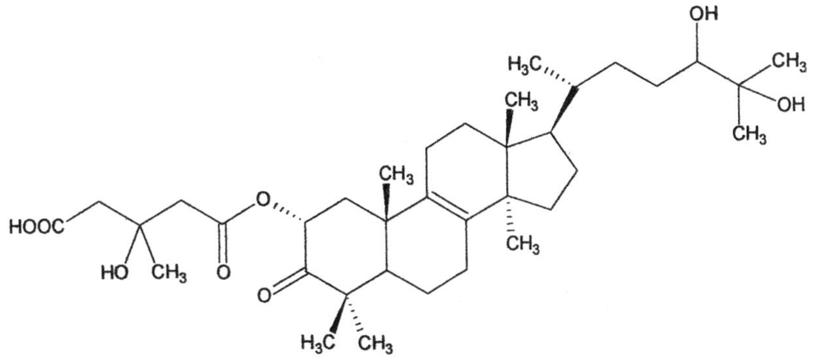

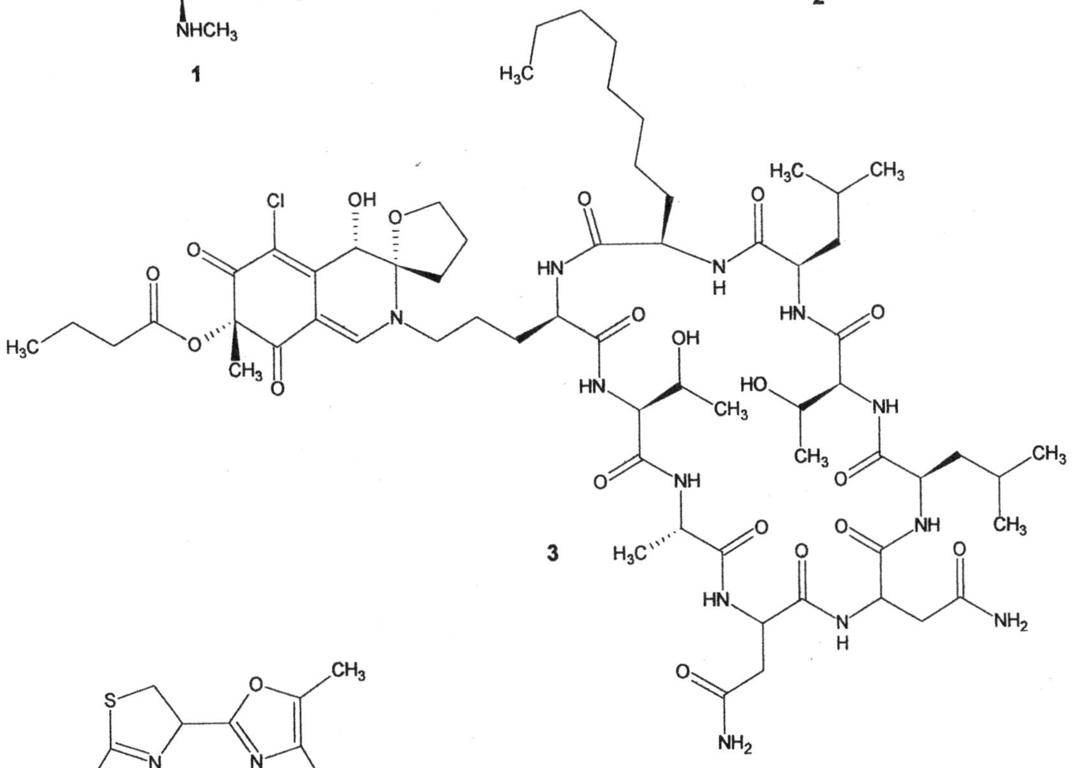

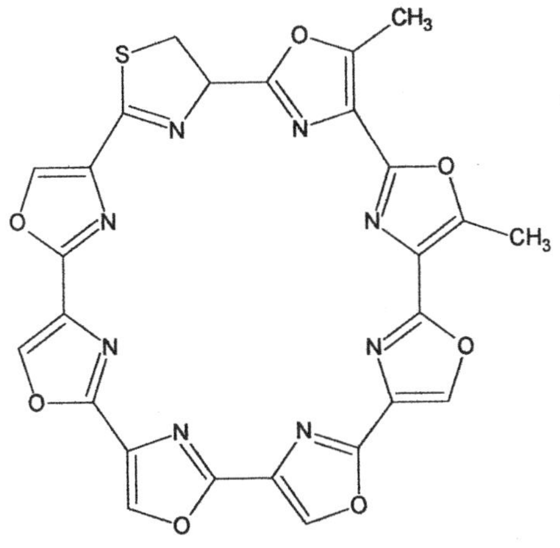

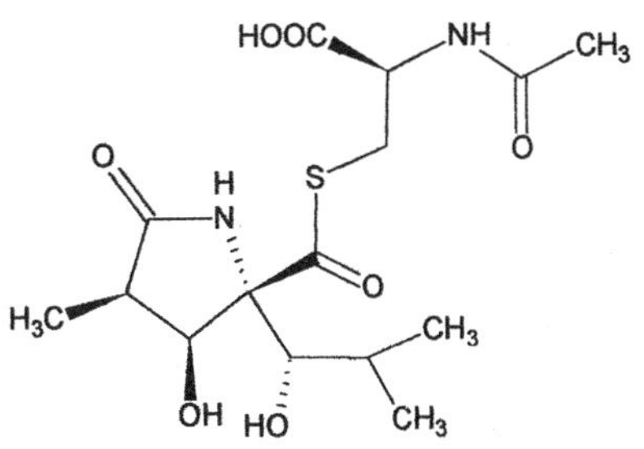

covered by other approaches. The *Streptomyces* sp. metabolite lactacystin (Fig. 5) was originally discovered as an inducer of neurite outgrowth and differentiation of neuroblastoma cells (Ōmura et al., 1991). It was subsequently one of the first proteasome inhibitors to be identified. In aqueous solution, lactacystin undergoes lactonization to form *clasto*-lactacystin β-lactone, which has better cell permeability than lactacystin and is the active proteasome inhibitory species, binding to and acylating catalytic proteasome subunits (Dick et al., 1997). The actinomycete metabolite epoxomicin, an α',β'-epoxyketone-containing tripeptide, was originally discovered as a result of its potent in vitro antitumor activity and in vivo efficacy against B16 melanoma (Hanada et al., 1992). Epoxomicin has also been found to act on the proteasome, binding to its catalytic β subunits and selectively inhibiting their major proteolytic activities. One of the key roles of proteasome-mediated degradation is in the activation of NF-κB-mediated proinflammatory signaling, and in addition to its antitumor properties, epoxomicin has in vivo antiinflammatory activity (Meng et al., 1999). Screening for new microbial metabolites with proteasome inhibitory properties is an active research area. Two recently reported microbial metabolite series discovered by this approach are tyropeptins A and B, peptidyl aldehydes produced by a *Kitasatosporia* sp. (Momose et al., 2001), and the cyclic hexapeptide phepropeptins from a *Streptomyces* sp. (Sekizawa et al., 2001).

Each time a cell divides, repeat guanine-rich DNA sequences at the end of its chromosomes called telomeres are eroded. Tumor cells acquire proliferative immortality by activating an enzyme, telomerase, that maintains telomere length. Telomerase is a large RNA-dependent DNA polymerase that uses its own associated RNA template to catalyze the addition of telomeric repeats to the 3' end of the single-stranded DNA telomere. Telomerase inhibitors therefore represent a new class of potentially useful cancer chemotherapeutic agents. A recent microbial metabolite telomerase screening program resulted in the identification of telomostatin from *Streptomyces anulatus*. Telomostatin (Fig. 5) is a macrocycle assembled from five oxazole, two methyloxazole, and one thiazoline ring. It is the most potent and selective telomerase inhibitor yet reported (Shin-ya et al., 2001).

Cardiovascular and Metabolic Disease Areas

One of the most successful classes of drugs over the past decade has been the antihypercholesterolemic agents termed the statins. The first statins were discovered by screening for inhibitors of hydroxymethylglutaryl coenzyme A (HMGCoA) reductase, the enzyme involved in the rate-limiting step in human cholesterol biosynthesis, the reduction of HMGCoA to yield mevalonic acid. The first statin HMGCoA reductase inhibitor to be discovered was mevastatin (Fig. 6) from *Penicillium citrinum* (Endo and Hasumi, 1993). This compound was independently isolated and identified as compactin, an antifungal agent from *Penicillium brevicompactum* (Brown et al., 1976). Shortly afterwards, lovastatin (or mevinolin, Fig. 6) was isolated and identified as a potent, competitive, and reversible HMGCoA reductase inhibitor and cholesterol-lowering agent from *Aspergillus terreus* (Alberts et al., 1980). Lovastatin was the first HMGCoA reductase inhibitor to be commercialized, followed by simvastatin, a synthetic analogue of lovastatin, and an exocyclic lactone ring-opened analogue, pravastatin, prepared by microbial biotransformation (Endo and Hasumi, 1993). Comparison of the ring-opened lactone structure common to all these compounds showed a resemblance to mevalonic acid. This recognition led to the design and synthesis of three totally synthetic statins, fluvastatin, cerivastatin, and atorvastatin (Newman et al., 2000). The crystal structure of the catalytic portion of HMGCoA reductase, when complexed with six different statins, including mevastatin, simvastatin, and atorvastatin, shows the statins occupying a portion of the binding site for HMGCoA, thus blocking access of this substrate to the enzyme's active site (Istvan and Deisenhofer, 2001). Recent reports suggest that the statins may have wider therapeutic applications. In addition to lowering cholesterol levels and reducing the risk of coronary heart disease and of stroke in post-heart attack patients, statin therapy appears to have a number of other potentially beneficial effects, including promoting nitric oxide-mediated angiogenesis (Kureishi et al., 2000), stimulation of bone formation (Mundy et al., 1999), and antioxidant and antiinflammatory properties (Simons, 2000).

An earlier enzyme in the cholesterol biosynthetic pathway that has been investigated as a site for pharmacological intervention to achieve cholesterol-lowering therapy is squalene synthase, which catalyzes the first pathway-specific step in sterol biosynthesis, the conversion of farnesyl pyrophosphate to squalene. Independent screening programs for squalene synthase inhibitors identified the same series of fungal metabolites, named either zaragozic acids or squalestatins, from different fungal sources. As zaragozic acids A, B, and C, these compounds were separately isolated from an unidentified fungal culture, *Sporormiella intermedia*, and *Leptodontium elatius*, respectively (Bergstrom et al., 1993). As the

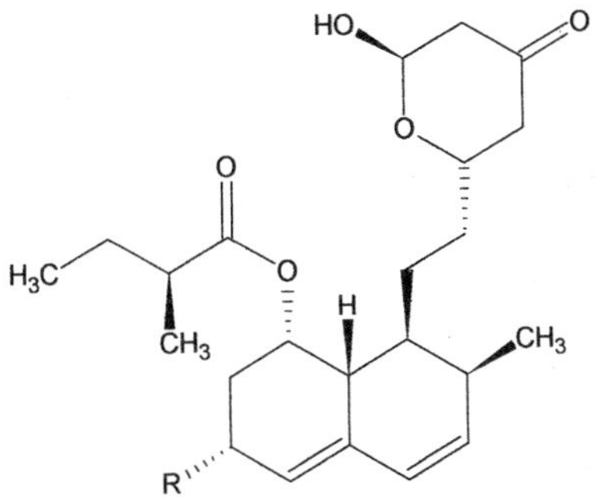

Figure 6. Structures of the cardiovascular drugs and lead compounds lovastatin (structure 1) from *Aspergillus terreus*; mevastatin (structure 2) from *Penicillium citrinum* and *Penicillium brevicompactum*; squalestatin S1 from a *Phoma* sp. or zaragozic acid A from an unidentified fungus (structure 3); and pyripyropene A (structure 4) from *Aspergillus fumigatus*.

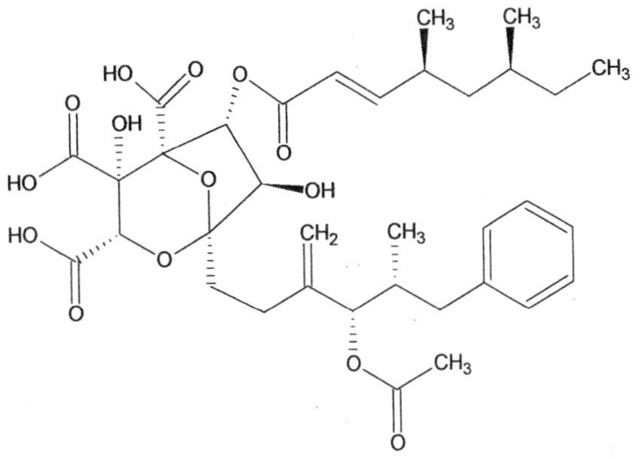

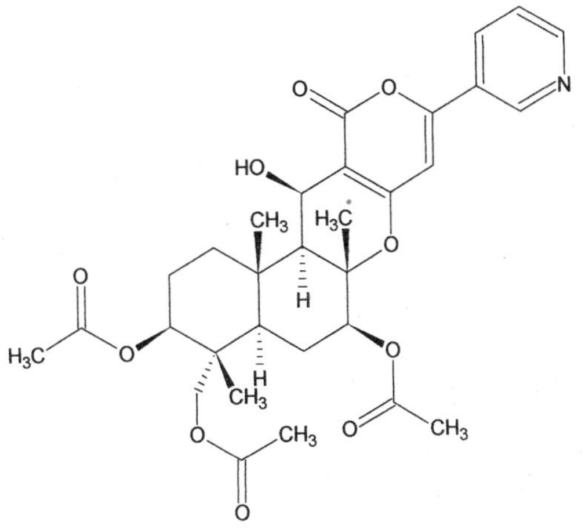

squalestatins, they were isolated from a *Phoma* sp. (Dawson et al., 1992). The zaragozic acids and squalestatins (Fig. 6) are extremely potent inhibitors of squalene synthase in in vitro assays and of cholesterol synthesis in vivo. The potential for development of these compounds was limited by hepatotoxicity (Shu, 1998). Novel squalene synthase inhibitors of microbial origin continue to be reported, such as the bisabosquals from a *Stachybotrys* sp. (Minagawa et al., 2001) and CJ-15,183 from *Aspergillus aculeatus*

(Watanabe et al., 2001), although there is now more emphasis on the antifungal potential of these compounds than on their potential to treat hypercholesterolemia.

Another enzyme involved in the human metabolism of cholesterol is acyl-CoA: cholesterol acyltransferase (ACAT), which catalyzes the formation of long-chain cholesterol esters. ACAT inhibitors could be used to treat hypercholesterolemia and atherosclerosis as they have the potential to inhibit the absorption of dietary cholesterol, inhibit lipoprotein production and stimulate cholesterol excretion as bile, and inhibit cholesteryl ester accumulation at the arterial wall. A screening program for microbial ACAT inhibitors was successful in identifying a number of novel ACAT inhibitors, the most potent of which was pyripyropene A (Fig. 6), isolated from *Aspergillus fumigatus*. ACAT inhibition by pyripyropene A was reversible, noncompetitive with the substrate oleoyl-CoA, and probably competitive with cholesterol. Pyripyropene A caused in vivo inhibition

of cholesterol absorption. Semisynthetic modification resulted in improved in vitro ACAT and in vivo cholesterol absorption inhibitory properties (Ōmura et al., 1995b).

Microbial metabolites have provided enzyme inhibitors with therapeutic utility in the treatment of obesity and diabetes as they reduce the gastrointestinal absorption of dietary factors important for these diseases. Lipstatin (Fig. 7) was isolated from *Streptomyces toxytricini* as a potent inhibitor of pancreatic lipase, a key enzyme involved in intestinal fat diges-

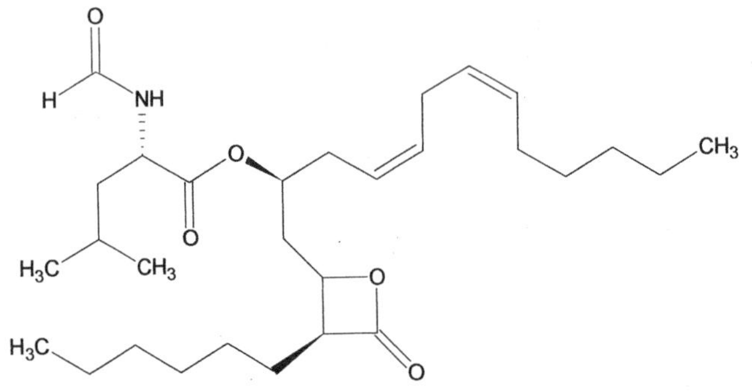

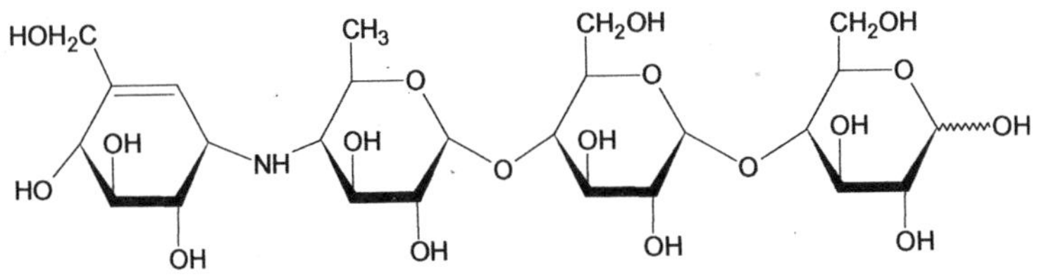

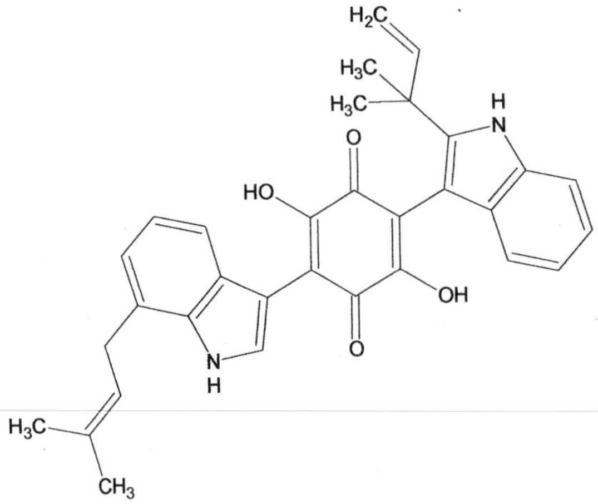

Figure 7. Structures of lipstatin (structure 1) from *Streptomyces toxytricini;* acarbose (structure 2) from an *Actinoplanes* sp.; and L-783,281 (structure 3) from a *Pseudomassaria* sp.

tion. Absorption of the dietary tryglyceride triolein, but not of oleic acid, was inhibited in mice, showing that the hydrolytic step of lipid digestion was specifically inhibited by lipstatin (Weibel et al., 1987). A semisynthetic derivative of lipstatin, tetrahydrolipstatin, has been developed as Orlistat, a potent and selective inhibitor of gastric and pancreatic lipases. Orlistat acts by binding covalently, through its β-lactone moiety, to the serine residue of the active sites of these lipases and has little or no inhibitory activity against amylase, trypsin, chymotrypsin, and phospholipases. It is used in the treatment of obesity. When administered with fat-containing foods, Orlistat partially inhibits the hydrolysis of triglycerides, reducing the subsequent absorption of monoglycerides and free fatty acids. At therapeutic doses, the reduction in dietary fat absorption achieved equates to approximately 30% of normal dietary fat absorption. Orlistat also has hypolipidemic properties due to its effects on triglyceride and cholesterol absorption and potentiates the hypolipidemic action of pravastatin (Guerciolini, 1997).

Acarbose (Fig. 7) is a pseudotetrasaccharide isolated from fermentations of an *Actinoplanes* sp. It binds reversibly and competitively to the oligosaccharide binding site of α-glucosidases and is an effective inhibitor of pancreatic α-amylase and intestinal enzymes such as glucoamylase, sucrase, and maltase. In vivo, acarbose not only delays the digestion of sucrose but is also a very potent inhibitor of starch degradation. This retarded carbohydrate digestion reduces postprandial increases in the levels of blood glucose and serum insulin (Muller et al., 1980). Oligosaccharide hydrolysis inhibition by acarbose lasts for 4 to 6 h, provided it is administered at the start of a meal to ensure that it is present at the site of enzymic action at the same time as the oligosaccharides. Acarbose is used to reduce postprandial hyperglycemia and hyperinsulinemia in cases of type 2 diabetes. It is the first-line treatment for newly diagnosed patients, for those who have high postprandial blood glucose, and for patients where dietary treatment alone provides inadequate glycemic control (Laube, 2002).

A more recent approach to find compounds with the potential to treat diabetes used a screen to detect activators of the human insulin receptor tyrosine kinase. This resulted in the identification of a non-peptidyl quinone-containing metabolite of a fungus, a *Pseudomassaria* sp., L-783,281 (Fig. 7), which acted as an insulin mimetic in biochemical and cellular assays. L-783,281 stimulated the phosphorylation of the insulin receptor in cellular assays with up to 100 times more potency than other very closely related natural products. The activation effect of L-783,281 was selective for the insulin receptor tyrosine kinase

over other receptor tyrosine kinases. Oral administration to mouse models of diabetes demonstrated significant lowering of blood glucose levels, showing the promise of insulin receptor activators for the development of new therapies for diabetes (Zhang et al., 1999).

CONCLUSIONS, TRENDS, AND PROSPECTS

The microbial products discussed in this chapter range from successful drugs to recently described lead compounds and provide ample testament to the ability of microbial secondary metabolism to provide compounds with real or potential therapeutic utility against a diverse and extensive range of pharmacological targets. Their structural diversity illustrates the versatility of microbial biosynthetic chemistry in providing multiple templates capable of binding to molecular targets and regulating enzyme activity and receptor signaling with potency and selectivity. At a time, however, when a significant increase in the number of pharmacological targets for drug discovery is expected, many pharmaceutical companies are turning away from natural products as a source of chemical diversity for screening. The reasons for this include the difficulties inherent in working with natural product samples with respect to the time needed to progress from screening hit to identified active compound in discovery operations aiming to complete dozens of drug discovery projects per annum; the challenges of scaling up natural product supply, semisynthetic modification, and/or total synthesis at the lead optimization stage; and faith that broad high-throughput screening of synthetic chemical diversity can provide good "druglike" leads for modification and optimization. There are opportunities here for smaller biotechnology companies and other organizations focusing on particular target or therapeutic areas to apply natural product diversity to pharmacological targets in less time-driven environments. The keys to future success in organizations of any size will lie in clever assay and screening strategy design and in the development and application of technologies to provide good hit prioritization and enable the rapid identification of active compounds.

The focus of many current microbial product screening operations is returning to antibiotic discovery. It is worth remembering that the discovery route of some very significant pharmacologically active agents of microbial origin has been indirect. Cyclosporin and rapamycin were initially discovered as antifungal agents, while mevastatin was isolated as an antifungal agent concurrently with its discovery as an HMGCoA reductase inhibitor. Novel natural products with interesting structures should be screened widely to establish their biological activity profiles; opportunities to do this are provided by organizations such as the National Cancer Institute (Cragg et al., 2000).

REFERENCES

Ainsworth, A. M., S. K. Wrigley, and U. Fauth. 2001. Molecular and taxonomic diversity in drug discovery: experience of chemical and biological screening approaches, p. 131–156. *In* S. B. Pointing and K. D. Hyde (ed.), *Bio-Exploitation of Filamentous Fungi*, Fungal Diversity Research Series 6. Fungal Diversity Press, Hong Kong.

Alberts, A. W., J. Chen, G. Kuron, V. Hunt, J. Huff, C. Hoffman, J. Rothrock, M. Lopez, H. Joshua, E. Harris, A. Patchett, R. Monaghan, S. Currie, E. Stapley, G. Albers-Schonberg, O. Hensens, J. Hirshfield, K. Hoogsteen, J. Liesch, and J. Springer. 1980. Mevinolin: a highly competitive inhibitor of hydroxymethylglutaryl-coenzyme A reductase and a potent cholesterol-lowering agent. *Proc. Natl. Acad. Sci. USA* 77:3957–3961.

Bergstrom, J. D., M. M. Kurtz, D. J. Rew, A. M. Amend, J. D. Karkas, R. G. Bostedor, V. S. Bansal, C. Dufresne, F. L. Van-Middlesworth, O. D. Hensens, J. M. Liesch, D. L. Zink, K. E. Wilson, J. Onishi, J. A. Milligan, G. Bills, L. Kaplan, M. Nallin Omstead, R. G. Jenkins, L. Huang, M. S. Meinz, L. Quinn, R. W. Burg, Y. L. Kong, S. Mochales, S. Mojena, I. Martin, F. Pelaez, M. T. Diez, and A. W. Alberts. 1993. Zaragozic acids: a family of fungal metabolites that are picomolar competitive inhibitors of squalene synthase. *Proc. Natl. Acad. Sci. USA* 90:80–84.

Bindseil, K. U., J. Jakupovic, D. Wolf, J. Lavayre, J. Leboul, and D. van der Pyl. 2001. Pure compound libraries; a new perspective for natural product based drug discovery. *Drug Discovery Today* 6:840–847.

Bollag, D. M., P. A. McQueney, J. Zhu, O. Hensens, L. Koupal, J. Liesch, E. Lazarides, and C. M. Woods. 1995. Epothilones, a new class of microtubule-stabilizing agents with a taxol-like mechanism of action. *Cancer Res.* 55:2325–2333.

Borel, J. F., C. Feurer, H. U. Gubler, and H. Stahelin. 1976. Biological effects of cyclosporin A: a new antilymphocytic agent. *Agents Actions* 6:468–475.

Brown, A. G., T. C. Smale, T. J. King, R. Hasenkamp, and R. H. Thompson. 1976. Crystal and molecular structure of compactin, a new antifungal metabolite from *Penicillium brevicompactum*. *J. Chem. Soc. Perkin Trans.* I:1165–1170.

Cardenas, M. E., A. Sanfridson, N. S. Cutler, and J. Heitman. 1998. Signal-transduction cascades as targets for therapeutic intervention by natural products. *Trends Biotechnol.* 16:427–433.

Chang, R. S. L., V. J. Lotti, R. L., Monaghan, J. Birnbaum, E. O. Stapley, M. A. Goetz, G. Albers-Schonberg, A. A. Patchett, J. M. Liesch, O. D. Hensens, and J. P. Springer. 1985. A potent nonpeptide cholecystokinin antagonist selective for peripheral tissues isolated from *Aspergillus alliaceus*. *Science* 230:177–179.

Chen, J. K., W. S. Lane, and S. L. Schreiber. 1999. The identification of myriocin-binding proteins. *Chem. Biol.* 6:221–235.

Cragg, G. M., M. R. Boyd, Y. F. Hallock, D. J. Newman, E. A. Sausville, and M. K. Wolpert. 2000. Natural products drug discovery at the National Cancer Institute. Past achievements and new directions for the new millennium, p. 22–44. *In* S. K. Wrigley, M. A. Hayes, R. Thomas, E. J. T. Chrystal, and N. Nicholson (ed.), *Biodiversity: New Leads for the Pharmaceutical and Agrochemical Industries*. Royal Society of Chemistry, Cambridge, United Kingdom.

Dawson, M. J., J. E. Farthing, P. S. Marshall, R. F. Middleton, M. J. O'Neill, A. Shuttleworth, C. Stylli, R. M. Tait, P. M. Taylor, H. G. Wildman, A. D. Buss, D. Langley, and M. V Hayes. 1992. The squalestatins, novel inhibitors of squalene synthase produced by a species of *Phoma*. I. Taxonomy, fermentation, isolation, physico-chemical properties and biological activity. *J. Antibiot.* 45:639–647.

Dick, L. R., A. A. Cruikshank, A. T. Destree, L. Grenier, T. A. McCormack, F. D. Melandri, S. I. Nunes, V. J. Palombella, L. A. Parent, L. Plamonden, and R. L. Stein. 1997. Mechanistic studies on the inactivation of the proteasome by lactacystin in cultured cells. *J. Biol. Chem.* 272:182–188.

Drews, J. 2000. Drug discovery: a historical perspective. *Science* 287:1960–1964.

Duncan, S. J., M. A. Cooper, and D. H. Williams. 2003. Binding of an inhibitor of the p53/MDM2 interaction to MDM2. *Chem. Commun.* 2003:316–317.

Duncan, S. J., S. Gruschow, D. H. Williams, C. McNicholas, R. Purewal, M. Hajek, M. Gerlitz, S. Martin, S. K. Wrigley, and M. Moore. 2001. Isolation and structure elucidation of chlorofusin, a novel p53-MDM2 antagonist from a *Fusarium* sp. *J. Amer. Chem. Soc.* 123:554–560.

Endo, A., and K. Hasumi. 1993. HMG-CoA reductase inhibitors. *Nat. Prod. Rep.* 10:541–550.

Fehr, T., V. F. J. Quesniaux, J. J. Sanglier, L. Oberer, L. Gschwind, M. Ponelle, W. Schilling, S. Wehrli, A. Enz, G. Zenke, and W. Schuler. 1997. Cymbimicin A and B, two novel cyclophilin-binding structures isolated from actinomycetes. *J. Antibiot.* 50:893–899.

Franco, C. M. M., and L. E. L. Coutinho. 1991. Detection of novel secondary metabolites. *Crit. Rev. Biotechnol.* 11:193–276.

Fujita, T., K. Inoue, S. Yamamoto, T. Ikumoto, S. Sasaki, R. Toyama, K. Chiba, Y. Hoshino, and T. Okumoto. 1994. Fungal metabolites. Part 11. A potent immunosuppressive activity found in *Isaria sinclairii* metabolite. *J. Antibiot.* 47:208–215.

Gerth, K., N. Bedorf, G. Hofle, H. Irschik, and H. Reichenbach. 1996. Epothilons A and B: antifungal and cytotoxic compounds from *Sorangium cellulosum* (Myxobacteria). Production, physico-chemical and biological properties. *J. Antibiot.* 49:560–563.

Grabley, S., and R. Thiericke. 1999. Bioactive agents from natural sources: trends in discovery and application. *Adv. Biochem. Eng. Biotechnol.* 64:101–154.

Guerciolini, R. 1997. Mode of action of orlistat. *Int. J. Obesity* 21:S12–S23.

Hamann, P. R., and M. S. Berger. 2002. Mylotarg: the first antibody-targeted chemotherapy agent, p. 239–254. *In* M. Page (ed.), *Cancer Drug Discovery and Development: Tumor Targeting in Cancer Therapy*. Humana Press Inc., Totawa, N.J.

Hanada, M., K. Sugawara, K. Kaneta, S. Toda, Y. Nishiyama, K. Tomita, H. Yamamoto, M. Konishi, and T. Oki. 1992. Epoxomicin, a new antitumour agent of microbial origin. *J. Antibiot.* 45:1746–1752.

Henkel, T., R. M. Brunne, H. Muller, and F. Reichel. 1999. Statistical investigation into the structural complementarity of natural products and synthetic compounds. *Angew. Chem. Int. Ed.* 38:643–647.

Hertzberg, R. P. 1993. Whole cell assays in screening for biologically active substances. *Curr. Opin. Biotechnol.* 4:80–84.

Hill, D. C., S. K. Wrigley, and L. J. Nisbet. 1998. Novel screen methodologies for identification of new microbial metabolites with pharmacological activity. *Adv. Biochem. Eng. Biotechnol.* 59:73–121.

Huang, S., and P. J. Houghten. 2002. Inhibitors of mammalian target of rapamycin as novel antitumour agents: from bench to clinic. *Curr. Opin. Invest. Drugs* 3:295–304.

Ichikawa, K., H. Hirai, M. Ishiguro, T. Kambara, Y. Kato, Y. I. Kim, Y. Kojima, V. Matsunaga, H. Nishida, Y. Shiomi, N. Yoshikawa, and N. Kojima. 2001a. Novel cytokine production inhibitors produced by a basidiomycete, *Marasmiellus* sp. *J. Antibiot.* 54:703–709.

Ichikawa, K., H. Hirai, M. Ishiguro, T. Kambara, Y. Kato, Y. J. Kim, Y. Kojima, Y, Matsunaga, H. Nishida, Y. Shiomi, N. Yoshikawa, L. H. Huang, and N. Kojima. 2001b. Cytokine production inhibitors produced by a fungus, *Oidiodendron griseum*. *J. Antibiot.* 54:697–702.

Ingber, D., T. Fujita, S. Kishimoto, K. Sudo, T. Kanamaru, H. Brem, and J. Folkman. 1990. Synthetic analogues of fumagillin that inhibit angiogenesis and suppress tumour growth. *Nature* 348:555–557.

Istvan, E. S., and J. Deisenhofer. 2001. Structural mechanism for statin inhibition of HMG-CoA reductase. *Science* 292:1160–1164.

Kakeya, H., S.-I. Kageyama, L. Nie, R. Onose, G. Okada, T. Beppu, C. J. Norbury, and H. Osada. 2001. Lucilactaene, a new cell cycle inhibitor in p53-transfected cancer cells, produced by a *Fusarium sp. J. Antibiot.* 54:850–854.

Kelland, L. R., S. Y. Sharp, P. M. Rogers, T. G. Myers, and P. Workman. 1999. DT-diaphorase expression and tumor cell sensitivity to 17-allylamino,17-demethoxygeldanamycin, an inhibitor of heat shock protein 90. *J. Natl. Cancer Inst.* 91:1940–1949.

Kino, T., H. Hatanaka, M. Hashimoto, M. Nishiyama, T. Goto, M. Okuhara, M. Kohsaka, H. Aoki, and H. Imanako. 1987a. FK-506, a novel immunosuppressant isolated from a *Streptomyces*. I. Fermentation, isolation, and physico-chemical and biological characteristics. *J. Antibiot.* 40:1249–1255.

Kino, T., H. Hatanuka, S. Miyata, N. Inamura, M. Nishiyama, T. Yajima, T. Goto, M. Okuhara, M. Kohsaka, H. Aoki, and T. Ochiai. 1987b. FK-506, a novel immunosuppressant isolated from a *Streptomyces*. II. Immunosuppressive effect of FK-506 *in vitro. J. Antibiot.* 40:1256–1265.

Kulanthaivel, P., Y. F. Hallock, C. Boros, S. M. Hamilton, W. P. Janzen, L. M. Ballas, C. R. Loomis, and J. B. Jiang. 1993. Balanol: a novel and potent inhibitor of protein kinase C from the fungus *Verticillium balanoides. J. Am. Chem. Soc.* 115:6452–6453.

Kureishi, Y., Z. Luo, I. Shiojima, A. Bialik, D. Fulton, D. J Lefer, W. C. Sessa, and K. Walsh. 2000. The HMG-CoA reductase inhibitor simvastatin activates the protein kinase Akt and promotes angiogenesis in normocholesterolemic animals. *Nat. Med.* 6:1004–1010.

Lam, Y. K. T., C. F. Wichmann, M. S. Meinz, L. Guariglia, R. A. Giacobbe, S. Mochales, I. Kong, S. S. Honeycutt, D. Zink, G. F. Bills, L. Huang, R. W. Burg, R. L. Monaghan, R. Jackson, G. Reid, J. J. Maguire, A. T. McKnight and C. I. Ragan. 1992. A novel inositol mono-phosphatase inhibitor from *Memnoniella echinata*. Producing organism, fermentation, isolation, physico-chemical and *in vitro* biological properties. *J. Antibiot.* 45:1397–1403.

Laube, H. 2002. Acarbose: an update of its therapeutic use in diabetes treatment. *Clin. Drug Invest.* 22:141–156.

Lee, F. Y. F., R. Borzilleri, C. R. Fairchild, S.-H. Kim, B. H. Long, C. Reventos-Suarez, G. D. Vite, W. C. Rose, and R. A. Kramer. 2001. BMS-247550: a novel epothilone analog with a mode of action similar to paclitaxel but possessing superior antitumor efficacy. *Clin. Cancer Res.* 7:1429–1437.

Lee, M. D., F. E. Durr, L. M. Hinman, P. R. Hamann, and G. A. Ellestad. 1993. The calicheamicins. *Adv. Med. Chem.* 2:31–66.

Lingham, R. B., K. C. Silverman, H. Jayasuriya, B. M. Kim, S. E. Amo, F. R. Wilson, D. J. Raw, M. D. Schaber, J. D. Bergstrom, K. S. Koblan, S. L. Graham, N. E. Kohl, J. B. Gibbs, and S. B.

Singh. 1998. Clavaric acid and steroidal analogues as Ras- and FPP-directed inhibitors of human farnesyl-protein transferase. *J. Med. Chem.* 41:4492–4501.

MacAllan, D. September 1997. Dual label solid phase binding assay. International patent application WO 9816833.

MacAllan, D., J. Sohal, C. Walker, D. Hill, and M. Moore. 1997. Development of a novel TNFα ligand-receptor binding assay for screening NatChem™ libraries. *J. Recept. Sig. Trans. Res.* 17:521–529.

Mann, J. 2001. Natural products as immunosuppressive agents. *Nat. Prod. Rep.* 18:417–430.

Meng, L., R. Mohan, B. H. B. Kwok, M. Eloffson, N. Sin, and C. M. Crews. 1999. Epoxomicin, a potent and selective proteasome inhibitor, exhibits *in vivo* anti-inflammatory activity. *Proc. Natl. Acad. Sci. USA* 96:10403–10408.

Minagawa, K., S. Kouzuki, K. Nomura, T. Yamaguchi, Y. Kawamura, K. Matsushima, H. Tani, K. Ishii, T. Tanimoto, and T. Kamigauchi. 2001. Bisabosquals, novel squalene synthase inhibitors. I. Taxonomy, fermentation, isolation and biological properties. *J. Antibiot.* 54:890–895.

Momose, I., R. Sekizawa, H. Hashizume, N. Kinoshita, Y. Homma, M. Hamada, H. Iinuma, and T. Takeuchi. 2001. Tyropeptins A and B, new proteasome inhibitors produced by *Kitasatosporia* sp. MK993-dF2. I. Taxonomy, isolation, physico-chemical properties and biological activities. *J. Antibiot.* 54:997–1003.

Morris, R. E., and B. M. Meiser. 1989. Identification of a new pharmacologic action for an old compound. *Med. Sci. Res.* 17:877–878.

Muller, L., B. Junge, W. Frommer, D. Schmidt, and E. Truscheit. 1980. Acarbose (BAY g 5421) and homologous α-glucosidase inhibitors from actinoplanaceae, p. 109–122. *In* V. Brodberk (ed.), *Enzyme Inhibitors. Proceedings of the Meeting*. Verlag Chemie, Weinheim, Germany.

Mundy, G., R. Garrett, S. Harris, J. Chan, D. Chen, G. Rossini, B. Boyce, M. Zhao, and G. Gutierrez. 1999. Stimulation of bone formation *in vitro* and in rodents by statins. *Science* 286:1946–1949.

Newman, D. J., G. M. Cragg, and K. M. Snader. 2000. The influence of natural products upon drug discovery. *Nat. Prod. Rep.* 17:215–234.

Nie, L., M. Ueki, H. Kakeya, and H. Osada. 2001. A facile and effective screening method for p21[WAF1] promoter activators from microbial metabolites. *J. Antibiot.* 54:783–788.

Nisbet, L. J., and Porter, N. 1989. The impact of pharmacology and molecular biology on the exploitation of microbial products, p. 309–342. *In* S. Banmberg, I. Hunter, and M. Rhodes (ed.), *Society for General Microbiology Symposium*, 44. Cambridge University Press, Cambridge, United Kingdom.

Ōmura, S., T. Fujimoto, K. Otoguro, K. Matsuzaki, R. Moriguchi, H. Tanaka, and Y. Sasaki. 1991. Lactacystin, a novel microbial metabolite, induces neuritogenesis of neuroblastoma cells. *J. Antibiot.* 44:113–116.

Ōmura, S., Y. Sasaki, Y. Iwai, and H. Takeshima. 1995a. Staurosporine, a potentially important gift from a microorganism. *J. Antibiot.* 48:535–548.

Ōmura, S., H. Tomoda, and T. Sunazuka. 1995b. ACAT inhibitors from microorganisms, p. 37–49. *In* W. Kuhn and H.-P. Fiedler (ed.), *Sekundarmetabolismus bei Mikroorganismen*. Attempto Verlag, Tubingen, Germany.

Rawlins, P., T. Mander, R. Sadhegi, S. Hill, G. Gammon, B. Foxwell, S. Wrigley, and M. Moore. 1999. Inhibition of endotoxin-induced TNFα production in macrophages by 5Z-7-oxo-zeaenol and other fungal resorcylic acid lactones. *Int. J. Immunopharmacol.* 21:799–814.

Rishton, G. M. 1997. Reactive compounds and *in vitro* false positives in HTS. *Drug Discovery Today* 2:382–384.

Roe, S. M., C. Prodromou, R. O'Brien, J. E. Ladbury, P. W. Piper, and L. H. Pearl. 1999. Structural basis for inhibition of the Hsp90 molecular chaperone by the antitumour antibiotics radicicol and geldanamycin. *J. Med. Chem.* **42**:260–266.

Sadhegi, R., P. Depledge, P. Rawlins, N. Dhanjal, A. Manic, S. Wrigley, B. Foxwell, and M. Moore. 2001. Differential regulation of CD3- and CD28-induced IL-2 and IFNγ production by a novel tyrosine kinase inhibitor XR774 from *Cladosporium* cf. *cladosporioides*. *Int. Immunopharmacol.* **1**:33–48.

Sanglier, J.-J., V. Quesniaux, T. Fehr, H. Hofmann, M. Mahnke, K. Memmert, W. Schuler, G. Zenke, L. Gschwind, C. Maurer, and W. Schilling. 1999. Sanglifehrins A, B, C and D, novel cyclophilin-binding compounds isolated from *Streptomyces* sp. A92-308110. I. Taxonomy, fermentation, isolation and biological activity. *J. Antibiot.* **52**:466–473.

Sekizawa, R., I. Momose, N. Kinoshita, H. Naganawa, M. Hamada, Y. Muraoka, H. Iinuma, and T. Takeuchi. 2001. Isolation and structure elucidation of phepropeptins A, B, and C, and D, new proteasome inhibitors, produced by *Streptomyces* sp. *J. Antibiot.* **54**:874–881.

Shin-ya, K., K. Wierzba, K. Matsuo, T. Ohtani, Y. Yamada, K. Furihata, Y. Hayakawa, and H. Seto. 2001. Telomestatin, a novel telomerase inhibitor from *Streptomyces anulatus*. *J. Am. Chem. Soc.* **123**:1262–1263.

Shu, Y.-Z. 1998. Recent natural products based drug development: a pharmaceutical industry perspective. *J. Nat. Prod.* **61**:1053–1071.

Sievers, T. M., S. J. Rossi, R. M. Ghobrial, E. Arriola, P. Nishimura, M. Kawano, and C. D. Holt. 1997. Mycophenolate mofetil. *Pharmacotherapy* **17**:1178–1197.

Simons, M. 2000. Molecular multitasking: statins lead to more arteries, less plaque. *Nat. Med.* **6**:965–966.

Sin, N., L. Meng, M. Q. W. Wang, J. J. Wen, W. G. Bornmann, and C. M. Crews. 1997. The anti-angiogenic agent fumagillin covalently binds and inhibits the methionine aminopeptidase, MetAP-2. *Proc. Natl. Acad. Sci. USA* **94**:6099–6103.

Stahelin, H. F. 1996. The history of cyclosporin A (Sandimmune®) revisited: another point of view. *Experientia* **52**:5–13.

Takahashi, I., Y. Saitoh, M. Yoshida, H. Sano, H. Nakano, M. Morimoto, and T. Tamaoki. 1989. UCN-01 and UCN-02, new selective inhibitors of protein kinase C. II. Purification, physicochemical properties, structural determination and biological activities. *J. Antibiot.* **42**:571–576.

Takeuchi, T., H. Iinuma, S. Kunimoto, T. Masuda, M. Ishizuka, M. Takeuchi, M. Hamada, H. Naganawa, S. Kondo and H. Umezawa. 1981. A new antitumor antibiotic, spergualin: isolation and antitumor activity. *J. Antibiot.* **34**: 1619–1621.

Ueki, M., T. Teruya, L. Nie, R. Usami, M. Yoshida, and H. Osada. 2001. A new trichostatin derivative, trichostatin RK, from *Streptomyces* sp. RK98-A74. *J. Antibiot.* **54**:1093–1095.

Umezawa, H. 1982. Low-molecular-weight enzyme inhibitors of microbial origin. *Ann. Rev. Microbiol.* **36**:75–99.

Umezawa, H., M. Imoto, T. Sawa, K. Isshiki, N. Matsuda, T. Uchida, H. Iinuma, M. Hamada, and T. Takeuchi. 1986. Studies on a new epidermal growth factor-receptor kinase inhibitor, erbstatin, produced by MH453-hF3. *J. Antibiot.* **39**:170–173.

Vertesy, L., H. Kogler, A. Markus, M. Schiell, M. Vogel, and J. Wink. 2001. Memnopeptide A, a novel terpene peptide from *Memnoniella* with an activating effect on SERCA2. *J. Antibiot.* **54**:771–782.

Vezina, C., A. Kudelski, and S. N. Sehgal. 1975. Rapamycin (AY-22,989), a new antifungal antibiotic. I. Taxonomy of the producing streptomycete and isolation of the active principle. *J. Antibiot.* **28**:721–726.

Vilella, D., M. Sanchez, G. Platas, O. Salazar, O. Genilloud, I. Royo, C. Cascales, I. Martin, T. Diez, K. C. Silverman, R. B. Lingham, S. B. Singh, H. Jayasuriya, and F. Pelaez. 2000. Inhibitors of farnesylation of Ras from a microbial natural products screening program. *J. Ind. Microbiol. Biotechnol.* **25**:315–327.

Watanabe, S., H. Hirai, M. Ishiguro, T. Kambara, Y. Kojima, T. Matsunaga, H. Nishida, Y. Suzuki, A. Sugiura, H. J Harwood, Jr., L. H. Huang, and N. Kojima. 2001. CJ-15,183, a new inhibitor of squalene synthase produced by a fungus, *Aspergillus aculeatus*. *J. Antibiot.* **54**:904–910.

Weibel, E. K., P. Hadvary, E. Hochuli, E. Kupfer, and H. Lengsfeld. 1987. Lipstatin, an inhibitor of pancreatic lipase produced by *Streptomyces toxytricini*. I. Producing organism, fermentation, isolation and biological activity. *J. Antibiot.* **40**:1081–1085.

Williams, D. H., M. J. Stone, P. R. Hauck, and S. K. Rahman. 1989. Why are secondary metabolites (natural products) biosynthesized? *J. Nat. Prod.* **52**:1189–1208.

Wood, K. A., D. A. Kau, S. K. Wrigley, R. Beneyto, D. V. Renno, A. M. Ainsworth, J. Penn, D. Hill, J. Killacky, and P. Depledge. 1996. Novel β-methoxyacrylates of the 9-methoxystrobilurin and oudemansin classes produced by the basidiomycete *Favolaschia pustulosa*. *J. Nat. Prod.* **59**:646–649.

Wrigley, S. K., R. Sadhegi, S. Bahl, A. J. Whiting, A. M. Ainsworth, S. M. Martin, W. Katzer, R. Ford, D. A. Kau, N. Robinson, M. A. Hayes, C. Elcock, T. Mander, and M. Moore. 1999. A novel (6*S*)-4,6-dimethyldodeca-2*E*,4*E*-dienoylester of phomalactone and related α-pyrone esters from a *Phomopsis* sp. with cytokine production inhibitory activity. *J. Antibiot.* **52**:862–872.

Wrigley, S. K., A. M. Ainsworth, D. A. Kau, S. M. Martin, S. Bahl, J. S. Tang, D. J. Hardick, P. Rawlins, R. Sadhegi, and M. Moore. 2001. Novel reduced benzo[j]fluoranthen-3-ones from *Cladosporium* cf. *cladosporioides* with cytokine production and tyrosine kinase inhibitory properties. *J. Antibiot.* **54**:479–488.

Zhang, B., G. Salituro, D. Szalkowski, Z. Li, Y. Zhang, I. Royo, D. Vilella, M. T. Diez, F. Pelaez, C. Ruby, R. L. Kendall, X. Mao, P. Griffin, J. Calaycay, J. R. Zierath, J. V. Heck, R. G. Smith, and D. E. Moller. 1999. Discovery of a small molecule insulin mimetic with antidiabetic activity in mice. *Science* **284**:974–977.

Zhang, L.-H., and J. O. Liu. 2001. Sanglifehrin a, a novel cyclophilin-binding immunosuppressant, inhibits IL-2-dependant T cell proliferation at the G$_1$ phase of the cell cycle. *J. Immunol.* **166**:5611–5618.

Zhao, A., S. H. Lee, M. Mojena, R. G. Jenkins, D. R. Patrick, H. E. Huber, M. A. Goetz, O. D. Hensens, D. L. Zink, D. Vilella, A. W. Dombrowski, R. B. Lingham, and L. Huang. 1999. Resorcyclic acid lactones: naturally occurring and potent inhibitors of MEK. *J. Antibiot.* **52**:1086–1094.

Microbial Diversity and Bioprospecting
Edited by Alan T. Bull
© 2004 ASM Press, Washington, D.C.

Chapter 33

Bioprospecting for Industrial Enzymes: Importance of Integrated Technology Platforms for Successful Biocatalyst Development

THOMAS SCHÄFER AND TORBEN VEDEL BORCHERT

Enzymes are established major contributors to clean industrial products and processes (Bull et al, 2000; van Beilen and Li, 2002; OECD, 1998). They show a variety of advantages over chemicals, for example, their specificity, their high efficiency, and their compatibility with the environment. Enzymes can be produced from renewable resources and are in turn degraded by microbes in nature. Various industries have substituted old processes using chemicals causing detrimental effects on the environment and equipment with new processes using biodegradable enzymes under conditions that are less corrosive.

The number of applications in which enzymes are used are many and diverse (Kirk et al., 2002). So far technical enzymes represent the largest part of the market with a value of approximately $1 billion in 1999 with enzymes for detergent being the largest single market for enzymes valued at around $0.5 billion (Schaefer et al., 2002). The other dominating markets are baking, beverage, and dairy, as well as feed and paper and pulp. All these industries are traditional users of enzymes. Overall, the estimated value of the worldwide use of industrial enzymes has grown from $1 billion (Godfrey and West, 1996) in 1995 to $1.5 billion in 2000 (McCoy, 2000).

Currently industrial enzymes are manufactured by three major suppliers, Novozymes A/S (headquartered in Denmark), Genencor International Inc. (headquartered in the United States), and DSM N.V. (headquartered in The Netherlands). Novozymes A/S is the largest supplier in each of the mentioned industries with an estimated market share between 41 and 44% of the industrial enzyme market in 1999. Genencor International Inc., which operates in the technical and feed segment, and DSM N.V., with focus on food and feed, had, according to Novozymes A/S estimates, market shares of around 21% and 8%, respectively, that year. The rest of the market is divided among a few smaller enzyme producers, some of which produce enzymes for their own use in the United States, Canada, Europe, and Japan, as well as a number of local producers in China.

The discovery and development of novel enzymes are based on several key parameters such as (a) the performance of the enzyme in the target application, (b) safe and economic production of the protein, and (c) patentability of the protein and/or the application to gain product protection (Gordon, 1999). Economic production of an enzyme candidate starts with the development of a production strain and involves many steps including fermentation development, product recovery, and product and process approval. These important steps of enzyme development are reviewed elsewhere (Cheetham, 1998; Liberman et al., 1999; Reisman, 1999). Here the focus is on the discovery of enzymes from natural environments and on optimization of enzymes by protein engineering and directed molecular evolution to develop products that show optimal performance in their target application.

THE NEEDLE IN THE HAYSTACK: FUNCTIONAL ASSAYS ARE APPLIED IN ITERATIVE SCREENING CYCLES FOR SELECTION OF THE BEST CANDIDATES

A screening program for industrial enzymes starts with a problem to be solved using enzymes and a business justification (Cheetham, 1998). To find the one ideal performing enzyme in the huge diversity of molecules is like finding a needle in a haystack. Once the target has been identified, two main questions have to be answered: "how to screen and where to screen" (Bull et al., 2000); what screening procedure should be used, and which group of organisms is the most promising source for the desired activity. How do industrial microbiologists tackle the tremendous

Thomas Schäfer and Torben Vedel Borchert • Novozymes A/S, Krogshøjvej 36, 2880 Bagsvaerd, Denmark.

diversity presented by nature? How exactly is the enzyme discovered that lives up to the specific conditions defined by the industrial application? Or should the enzyme derive from an approach based on optimization of an existing enzyme? Many possibilities exist, but the final approach will depend on the accumulated knowledge with respect to available diversity and the desired application.

In general, most searches follow a general pattern. First, screening criteria are defined that include the substrate to be converted and conditions such as temperature, pH, and presence or absence of ions. These are used to design functional screening assays that are able to select for the enzymes that show the best performance. A prerequisite for this step is an in-depth understanding of the industrial application including enzyme substrate interaction in the chemicophysical environment. Often a working hypothesis is implied in the assay, as for many applications it is impossible to totally understand and downscale the application. In these instances assays are developed that are as close to real conditions as possible.

Once the assay is established, the screening program is started, which generally constitutes an iterative process divided into several phases (Bull et al., 1992). In a broad primary screening all microbes, clones, or variants that are positive in the applied assay are selected. The primary screening assay is sometimes not very selective, in order to capture a wide variety of positive hits. In the next round these hits are subjected to a secondary screening, where a highly selective enzyme assay is applied that allows ranking and selection of the best-performing enzyme candidates. Accordingly, screening is seldom a simple linear and straightforward process but rather a process with learning loops running in several cycles. An important next step is the cloning, expression, purification, and characterization of the enzymes to make the top candidates available for large-scale testing. Only at this point can the hypothesis underlying the assay and thereby the whole screening be tested, i.e., do the selected candidates fail or pass the real application test. Both failure and success can be used to optimize the assay and thereby generate even more and better diversity. Finally, the remaining enzymes are compared with respect to performance, production feasibility, and economy as well as patentability to form a short list of top candidates for further development programs.

According to the above, the quality and nature of the screening assay has a central role during the whole selection and deselection program, as the quality of the assay determines the quality of the resulting candidate. The key to successful screening of industrial enzymes is to detect not a large diversity of proteins, but rather those few that can perfectly match the application conditions. Much effort is invested to develop screening assays and novel screening technologies, including high-throughput technology, that give enhanced freedom in assay design. Accordingly, a variety of publications and patent applications cover this field, and only selected examples can be referred to (Joo et al., 1999; Ruijssenaars and Hartmans, 2001; Meeuwsen et al., 2000; Preisig and Byng, 2001; Kongsbak et al., 1999; Short and Keller, 2001; Schellenberger, 1997).

In the overall discovery flow, the implementation of the screening assay is followed by the next concrete question, namely, "where to look for diversity" (Bull et al., 2000). There are various potential sources as input to screening programs, e.g., candidates from previous programs, known strains from in-house or external culture collections, cooperations, and known enzymes with or without protein structures. If the project is about new protein molecules, new diversity has to be made available from natural sources, and in contrast to this it is highly reasonable to start protein engineering or artificial evolution programs in cases where enzymes with similar functions and performance already exist.

INFINITE BIOLOGICAL DIVERSITY IS A NATURAL LIBRARY FOR BIOPROSPECTING OF INDUSTRIAL ENZYMES

The biological diversity of living organisms on the planet Earth is infinite. The real size of existing biodiversity can be estimated only roughly as many species have yet to be discovered. The most important basis for screening novel biocatalysts is the natural diversity of microorganisms, which during 3.5 billion years of evolution have adapted to thrive in all econiches (Woese, 1987). The general importance of microbial diversity for global ecology and discovery of natural compounds such as enzymes is underlined by the estimation that more than 90% of life's phylogenetic diversity, including its metabolic, molecular, and ecological diversity, lies in the microbial world (Olsen and Woese, 1997). However, our complete picture of biodiversity in terms of species richness and local as well as global distribution and function in the ecosystem still remains very incomplete (Bull et al., 2000).

Microorganisms can be found in all ecological niches on Earth, in both natural and man-made habitats such as hydrothermal vents in the deep sea, geothermal areas such as acidic solfataric fields, and soda and saline lakes, as well as in and on plants, and recently even growth in clouds has been reported

(Sattler et al., 2001). The amount of diversification and adaptation within this group is vast, which implies growth from below 0°C to above the boiling point of water, from pH values below 1 to above 10, from low to high salinity, and even at high pressure (Demirjian et al., 2001; Hallberg and Johnson, 2001; Psenner and Sattler, 1998; Rothschild and Mancinelli, 2001). This calls for a huge diversity of enzymes that act under a wide variety of conditions and on virtually all natural and many man-made substrates. The extent of bacterial diversity is unknown and cannot be extrapolated in a reasonable manner (Tiedje and Stein, 1999); for fungi, the number of known species is about 74,000 and the total number of fungal species is conservatively estimated to be 1.5 million (Hawksworth, 2001). In addition to defined species, Tiedje and Stein (1999) and Bull et al. (2000) draw attention to variants within species, for example, tolerances to high or low temperatures, modified growth rates, and modified yields of metabolic products, which both increase diversity and the chances of detecting special producers of secondary metabolites or enzymes.

Today, it is generally accepted that only minor parts of the microbial diversity have been cultured or might even be amenable to growth in the laboratory (Torsvik et al., 2002), thereby leaving not only a huge set of questions concerning our understanding of the role of microbes in their habitats, but also an enormous potential of yet undiscovered physiological and biochemical traits including enzyme-genes in the so-called metagenome (Rondon et al., 1999). Accordingly, this chapter considers the traditional screening approach, which has its roots in cultivation of the microbes, and the metagenomics approach, which directly exploits gene diversity without prior cultivation.

Traditional Enzyme Screening Is Based on Culturing a Broad Diversity of Microorganisms

The vast majority of industrial enzymes that are on the market today are extracellular enzymes that mainly derive from bacteria and fungi isolated from ecological niches and cultured in the laboratory. Screening large numbers of culturable microbes as described above is considered one of the most successful and efficient means of finding novel biological compounds such as biocatalysts, and this classical route of screening will continue to be applied with success in the future (Cheetham, 1998; Ogawa and Shimizu, 2002). Efforts are being made to isolate "new" microorganisms with novel features, including the previously described "unculturables" (Franco and McClure, 1998). Upcoming publications and indeed

our own unpublished results describe the successful isolation of formerly nonculturable, novel microbes (Takai et al., 2002; Watve et al., 2000). Accordingly, the classification of isolates as either novel or related to already known groups is of importance (Bull et al., 2000). Many methods for the classification of microbes are available and useful, but each has specific limitations. Some have proven to be preferred tools of industrial microbiologists, e.g., the sequence analysis of the 16S or 18S rDNA genes after the pioneering work of Carl Woese and his coworkers who, in addition to establishing this novel concept (Olsen et al., 1986), made a database publicly available (http://rdp.cme.msu.edu/html/). This enables researchers around the world to compare data and evaluate the uniqueness and relationships of their isolates uncultured organisms. As small ribosomal subunit sequencing is not suited for large numbers of isolates from screening campaigns, this method is often combined with high-throughput methods such as, e.g., Fourier-transform infrared (Nelson, 1991), that can be used for pregrouping.

Each screening program is unique, and it must specifically be decided which physiological or taxonomic groups most probably produce the desired activity (see chapter 28). Here the experience of taxonomists and physiologists comes into play, using their knowledge to recommend the taxa and ecological niches that are most likely to hold rich producers of the enzyme of interest. Often this approach is combined with screening of environmental samples for the isolation of new and unique strains, applying either enrichment or direct cultivation methods and highlighting the importance of sampling strategies (Felse and Panda, 2000; Tiedje and Stein, 1999). Physiologists culture natural strains under a variety of conditions to ensure induction of the desired enzymes. Induction is also a key prerequisite for generation and screening of cDNA libraries as described below.

By combining novel isolation technologies and microbial physiology with phylogenetic and taxonomic studies, duplicate isolates and thereby costly rediscovery of already known compounds are avoided while discovery of even better performing molecules in close neighbors of the best producer is made possible. Accordingly, investing in reliable taxonomies as part of the culture collection and isolation work still leads to higher speed and quality of the overall screening process (Bull et al., 1992). Screening efforts can vary from a few microbes known to produce the enzyme from previous experience to several thousands of bacteria, fungi, or samples. Ando et al. (1989) isolated 5,000 strains from soil samples to detect transglutaminase-producing *Streptoverticillium*, and Felse and Panda (2000) review extensive screening campaigns for chitinase-producing bacteria and fungi.

As a consequence of the above-described screening route, microbiologists have gathered microbial isolates for many years and have built up large and highly diverse culture collections that still continue to grow. The aim of the culture collections from an industrial point of view is to have representatives of as many taxonomic, physiological, ecological, and geographical groups of microorganisms as possible and to be optimally equipped when the screening of any type of enzyme is initiated. These microorganisms are viewed as a vast genetic library in which single species offer a variety of solutions for a range of challenges and applications, e.g., antibiotics, enzymes, compatible solutes, or structural components that can be continually screened for different functionalities.

Cloning as an important step to obtain monocomponent enzymes for application testing

Once the best performing candidates have been selected by primary and secondary screening, genes from the top candidates are cloned and expressed by a variety of different methods (Sambrock and Russel, 2001). In most cases a technique called expression cloning is used, which is an effective means of isolating a gene from a gene library based on its encoded activity. For fungal genes this technology is based on isolation of mRNA and reverse transcription to cDNA as exemplified for phytases from basidiomycete fungi (Lassen et al., 2001). Expression enables production of larger quantities of pure enzyme for thorough characterizations and application testing. Most microbes produce several enzymes simultaneously, and classical enzyme products are thus often mixtures of different enzymes in addition to the enzyme of interest. Although these mixtures can be used in many applications, monocomponent enzymes produced by recombinant DNA technology are preferred in small-scale applications to clearly refer measured effects to a given protein. Additionally, this procedure gives valuable indications as to the production potential of the enzyme, and those candidates that are poorly expressed are deselected early in the screening program. Cloning and expression represent essential steps in the production of industrial enzymes today. This approach allows the production of enzymes from microbes that are difficult to grow, thereby increasing the array of available biocatalysts for industrial applications. An additional advantage is that enzymes produced by gene technology show improved product quality, ecobalance, and cost-effectiveness (Cheetham, 1998, OECD, 2001). The first enzyme product produced by gene technology was marketed in 1984 by Novozymes, an amylase from a *Bacillus* isolate heterologously expressed in *Bacillus subtilis* (Diderichsen and Christiansen, 1988). The vast majority of enzymes today are produced by high-level expression in selected industrial host organisms, mainly *Bacillus* (Widner et al., 2000), *Aspergillus* (Dunn-Coleman and Prade, 1998; Archer, 1994), *Fusarium,* and *Trichoderma* (Pandey et al., 2001; Archer and Peberdy, 1997). These hosts are the result of many years of strain improvement work during the past 20 to 40 years prior to the advance of gene technology when industrial enzymes were produced in the original donor. This donor often showed economically unrealistic low yields, and yield improvement programs including classical mutagenesis and screening for high-yield mutants were performed to solve that problem.

Sequence-based approaches: molecular screening

Functional screening as described above is often supplemented with screening techniques, which are based on similarities between enzyme-encoding gene sequences (Dalbpge and Lange, 1998; Precigou et al., 2001). Sequence information from a set of related enzyme genes is used to screen for and to clone additional genes from other organisms. An alignment of either the nucleotide sequences or the corresponding amino acid sequences, followed by identification of evolutionarily conserved regions, allows the design of degenerate PCR primers. These primers can be used to amplify and sequence a fragment of a homologous gene from a sample of genomic DNA. Using this method, a number of genes homologous to the initial gene sequences can quickly be identified. The limitation of the method is that enzyme variants rather than totally novel enzymes are detected. The advantages, on the other hand, are that this method is not reliant on either expression of the gene in question (in both donor and host) or functional assays to detect the enzyme activity. As actively living cultures are not a prerequisite, this approach can also be used to screen environmental libraries (see below). Schülein et al. (1996) have described screening and cloning of 32 cellulase genes from cellulose family 45 (Henrissat and Romeu, 1995) representing all the major fungal taxa (namely, ascomycota, zygomycota, chytridiomycota, and basidiomycota) based on only four homologous gene sequences known at the start of the project. These cellulases can be used for detergents, laundering, textile, and paper and pulp applications. Dalbøge et al. (1997) have described a modification of the above method: the obtained PCR product is cloned, sequenced, and linked to both a 5' and 3' structural gene using splicing by overlap extension PCR. The result is a hybrid protein that still shows enzymatic activity after cloning of the full open reading frame as exemplified by xylanase and endoglucanase. The authors used this approach with uncharacterized DNA samples to fish genes of interest from complex microbial communities (see below).

Bioinformatics—a central discipline in enzyme discovery

It is apparent that the above-described screening approaches require tools to handle and compare sequence data obtained from cloning projects. Additionally, the expanding inventory of cloned enzyme genes in the public domain, rapidly increasing the available sequence information for a large selection of enzymes and enzyme families, has to be retrieved. The situation became more complex and challenging when whole bacterial, fungal, and archael genomes were sequenced, as exemplified by *Bacillus subtilis* (Kunst et al., 1997), *Aspergillus nidulans* (Dunn-Coleman and Prade, 1998), and *Methanococcus jannaschii* (Bult et al., 1996). The current status on established genomes and those under way can be obtained by visiting the home page of the Institute for Genomic Research (TIGR) (http://www.tigr.org/).

Figure 1 gives an overview of the development of the number of gene sequences submitted to public databases by scientists worldwide, which illustrates the need for bioinformatic tools. This large amount of microbial genome mapping and sequence information required tools for data retrieval, mining, analysis, comparison, and grouping of enzymes to families (Henrissat and Bork, 1996), as well as data management and interpretation—in short, a new research discipline, namely, bioinformatics (Rosteck et al., 1999; see Chapter 24, this book).

A basic demand in the discovery of new enzymes is the detection of protein encoding regions, their comparison with other gene sequences and with biochemically characterized proteins, and finally their grouping into protein families (Li and Godzik, 2002; Callebaut et al., 1997). This can often be used to deduce putative function or more concrete information on enzymatic activities encoded by those genes. Interesting hits found this way can subsequently be analyzed in more detail by cloning and expression of the gene, followed by purification and characterization of the corresponding enzyme. A primary challenge for bioinformatics is to establish a link between sequence and biochemical function, and also between sequence and structure and ultimately, performance of a given protein prior to expression and testing, including genes with an unknown function (Merlin et al., 2002). For modeling of enzymes where a rational design approach is preferred, bioinformatics is clearly a prerequisite (see below).

Genome analysis for discovery of novel genes

Whole-genome sequencing, bioinformatics, array studies, and proteomics are novel key technologies for the targeted improvement of production strains. This has been illustrated for lysine production in *Corynebacterium glutamicum* (Ohnishi et al., 2002). Whole-genome sequencing that completely maps all genes, however, is not ideal for discovery of *selected* genes, e.g., those encoding for enzymes and especially for those enzymes that match defined application criteria. Assuming an average genome size for a bacterium of ca. 4 Mb, for yeast of 13 Mb (Zagulski et al., 1998), and for filamentous fungi of the order of 30 to 40 Mb (Dunn-Coleman and Prade, 1998; Radford and Parish, 1997), the costs for sequencing programs of total genomes are unreasonably high for discovery purposes, especially considering the wide diversity of microbes that are interesting for enzyme screening. From the 4,100 open reading frames of the *B. subtilis* genome, only a fraction are possibly relevant for industrial enzymes. As described above, extracellular enzymes are of major importance for industrial applications, and it is estimated that *B. subtilis* produces 150 to 180 secreted proteins (Hirose et al., 2000). For filamentous fungi, the number of secreted enzymes might be of the order of 200 to 400, corresponding to their larger genome sizes. This indicates that only 2 to 5% of the open reading frames in

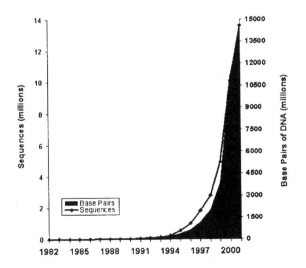

Figure 1. Number of gene sequences submitted to public databases from 1982 to 2000. Data were taken directly from the National Center for Biotechnology Information (NCBI) homepage (http://www.ncbi.nlm.nih.gov/Genbank/genbankstats.html).

a complete genome are of primary interest for enzyme discovery.

Accordingly, whole-genome sequencing can hardly be justified for enzyme discovery purpose due to high costs and binding of many resources; as a consequence, alternative approaches have been developed to selectively mine microbial genomes for secreted enzymes. One option is random sequencing of gene/cDNA libraries (Sanchez et al., 2001) with the possibility of identifying genes with signal sequences. Gene trapping approaches can be used, for example, to randomly disrupt functional genes and to generate large numbers of mutants with novel phenotypes that can be linked to genes by sequencing (Durick et al., 1999). One trapping approach suited to the screening of genes encoding secreted proteins is described as transposon-assisted signal trapping of gene libraries (Duffner et al., 2001). A genomic or cDNA library is treated with a transposon carrying a reporter gene that codes for a secreted protein with its own secretion signal sequence removed. A signal-less beta-lactamase has been used as a reporter that can, upon insertion in a gene with an active secretion signal, be transported out of the cell as a fusion protein. This results in ampicillin-resistant phenotypes that can be selected on agar plates. Genes encoding secreted proteins are subsequently sequenced and identified by comparison to databases using bioinformatic tools. In contrast to traditional screening of gene libraries with functional assays for selected activities, the entire genome as represented in the library is trapped for known and novel enzymes simultaneously.

Metagenomics or Cloning from Noncultivable Microorganisms

Metagenomics is an alternative approach in screening for a diversity of enzymes and is close to the idea of screening a "biodiversity library" (Handelsman et al., 1998; Rondon et al., 2000; Lorenz and Schleper, 2002; Handelsman, chapter 11, this book). The elegance of this strategy lies in the fact that it does not rely on cultivation of microorganisms, but instead on DNA or mRNA that is directly isolated from an environmental sample, purified, digested, and cloned into suitable cloning vectors to construct complex environmental libraries. These gene libraries need to be screened as described above using either sequence-based techniques or activity assays including some novel constraints due to the complexity of the library. Ideally, cultivation-independent approaches enable microbiologists to fully exploit the biological potential of a microbial community in its totality.

The fundamentals for this fascinating but also challenging approach were described by microbial ecologists that applied novel molecular techniques to study microbial diversity of communities as early as the beginning of the 1980s:

- A huge microbial diversity is present in soil and sediment samples. Bacterial communities in pristine soil and sediments may contain approximately 4,000 totally different genomes (Torsvik et al., 1990); for some soils, even more than 10,000 different bacterial types (Torsvik et al., 1998).
- Only a few of the microbes from a given sample can be cultivated, and the total microbial diversity comprises phylogenetic groups of organisms not related to any organism so far described (Schmidt et al., 1991; Amann et al., 1996; Hugenholtz and Pace, 1996). Totally new groups of bacteria and archaea have been identified in soil, some of which may be new phyla or even new kingdoms as they are deep branches in the phylogenetic tree of life (Liesack and Stackebrandt, 1992; Kuske et al., 1997).
- DNA can be isolated directly from these samples (Torsvik, 1980; Olsen et al., 1986; Steffan et al., 1988). PCRs can be carried out on environmental DNA templates (Stahl et al., 1985; Olsen et al., 1986, Steffan et al., 1988).
- DNA can be digested and cloned in suitable cloning vectors. Genes can be screened and cloned directly from environmental libraries, as shown in 1991 by Schmidt et al. for rRNA-containing clones and rDNA genes and proposed for "other genes of interest" with the methods being "generally applicable in microbial ecology" (Schmidt et al., 1991).

Metagenomics is a fascinating approach to study both the diversity of microorganisms and the diversity of genes in complex ecological habitats that was enabled by the development and application of molecular biology tools. It is thereby tempting to exploit this approach also for the discovery of industrial enzymes (Short, 1997) and other natural compounds (Brady et al., 2001), although there are also some important constraints and limitations. The most noteworthy are the quality of the environmental DNA, normalization of the library in order to avoid repetitive screening of highly abundant genes, and the size of metagenomics libraries, which for many screenings presupposes high-throughput screening equipment and consequent investment, and applicability of the approach for eukaryotic microbes undoubtedly present in the environment. Also, the production potential of the detected gene in a production organism, as well as product and production approval which in many countries might require a defined source, i.e., a bacterial or fungal species, are important criteria. An additional critical feature is that the goal in industrial screening is not to identify the theoretical total diversity of novel enzymes in the metagenome but rather to select for and detect the enzyme that is perfectly suited for a certain application, which is the actual art of screening.

Interestingly, published research lacks a metagenome strategy for eukaryotes, e.g., fungi. This is noteworthy, as approximately 50% of all industrial enzymes are of fungal origin. The only described approach is from a patent application (Sandal et al., 2000). In the described approach, mRNA was directly isolated from environmental samples, e.g., cow rumen or termite gut microflora, and used to construct cDNA libraries that were successfully screened for cellulases.

Published results to date show that environmental libraries can be screened for functional activities. Table 1 summarizes the screening of metagenome libraries as described in the literature so far. In general it can be stated that screening efficiency seems surprisingly low. The number of positive clones detected in the metagenome libraries is particularly low compared with the number of clones that were screened and compared with the expected diversity in the samples. One explanation of the low hit rates is that successful exploitation of metagenomics for discovery is hampered by heterologous gene expression, as not all genes will be expressed in *Escherichia coli* (Lorenz and Schleper, 2002). A more severe problem is high-yield expression of the detected genes in industrial production strains in downstream processing. Accordingly, no product derived from this approach has yet reached a position in the industrial enzyme market.

In conclusion, metagenomics has yielded evidence for new organisms and new genetic resources containing millions of uncharacterized genes. However, construction and screening of environmental libraries is still in its infancy, and screening efficiency is low compared with costs and technical complexity. Undoubtedly, the metagenomics approach offers a new tool to analyze and discover the as yet unexplored gene diversity (Lorenz and Schleper, 2002; Handelsman et al., 1998), and should be regarded as complementary to the conventional routes. The future will show whether this approach is valuable for ecological studies or will result in novel products for industrial applications. The initial results are promising, but methodological problems need to be solved before the virtually boundless diversity of genes hidden in the noncultivable world of microbes can be fully accessed (Rondon et al., 1999; Tiedje and Stein, 1999; Ogram, 2000).

ENZYME OPTIMIZATION

Industrial applications utilizing enzymes as biocatalysts often put extreme demands on the properties of the enzymes due to the very harsh conditions often applied under such processes. Examples of such application conditions include high temperatures, extreme pH, and the presence of chemicals known to induce protein unfolding or metal ion depletion. Enzyme properties other than these protein stability-related parameters are also important in a functional biocatalyst. High substrate turnover rates, correct substrate and product specificity, sufficiently low immunogenicity, and the possibility of achieving high expression levels in production hosts represent additional characteristics that are important in industrial biocatalysts. There is no guarantee that the enzymatic answer to an industrial process can necessarily be isolated from nature, as such a guarantee would demand that survival of an organism would depend on an enzyme functional under the conditions relevant for that particular application. In these cases where nature cannot directly provide the enzyme, the answer to the biocatalysis question has to be developed in the laboratory.

Two approaches for improving natural proteins in the laboratory have traditionally been taken, rational protein engineering and directed molecular evolution. Each of these two approaches is described separately below; however, the borderline between the technologies has become less distinct in recent years.

Rational Protein Engineering

The central dogma describes how a gene is transcribed into mRNA that subsequently is translated into a protein, thus establishing the close link between the DNA sequence and the resulting amino acid sequence in the expressed protein. This direct link allows the production of variant proteins in a host cell by altering the sequence of the encoding gene.

The development of molecular biology technology in the 1980s, especially techniques allowing for efficient site-directed mutagenesis, promoted the era of rational protein engineering (Smith, 1985). The ability to create protein variants with designed and deliberate amino acid alterations at any desired position provided the capability of precise probing of structure-function relationships in proteins.

Table 1. Screening of metagenomics libraries: summary of described results[a]

DNA source	Vector	Insert size	Enzyme activity	Substrate	No. of clones screened	No. of positive clones	Reference
Soil	Plasmid	5–8 kb	4-OH-butyrate dehydro-genase	Functional complemen-tation: growth on 4-OH-butyrate	930,000	5	Henne et al. (1999)
Soil	Plasmid	5–8 kb	Lipase	Triolein/ rhodamine agar	430,000[b] + 180,000[c] + 120,000[d] = 730,000	0[b] + 1[c] + 0[d] = 1	Henne et al. (2000)
			Esterase	Tributyrin agar	73,000[b] + 115,000[c] + 98,000[d] = 286,000	1[b] + 0[c] + 2[d] = 3	
Soil (1)	Plasmid	5–8	Na^+/H^+ antiporter	Functional complemen-tation: growth on 450 mM NaCl	1,480,000	2	Majernik et al. (2001)
Seawater (estuary water)	Lambda	2–10	Chitinase	MUF-di-NAG	75,000 (plaque assay)	9	Cottrell et al. (1999)
Seawater (coastal water)	Lambda	2–10	Chitinase	MUF-di-NAG	750,000 (plaque assay)	2 (identical clones)	Cottrell et al. (1999)
		2–10			230,000 (mt plate assay)	13	
Seawater (coastal water)	Lambda	2–10	Cellulase	MUF-cellobiose	750,000	0	Cottrell et al. (1999)
Soil	BAC	27 kb (SL 1) (2)	DNase	DNase methyl green agar	3,648	1	Rondon et al. (2000)
Soil	BAC	27 kb (SL 1)	Amylase	Starch agar plates	3,648	1	Rondon et al. (2000)
Soil	BAC	27 kb (SL 1)	Antibacterial	Nitrocefin	3,648	1	Rondon et al. (2000)
Soil	BAC	27 kb (SL 1)	Lipase	Lipid	3,648	2	Rondon et al. (2000)
Soil	BAC	27 kb (SL 1)	Cellulase chitinase, esterase, keratinase, proteinase hemolytic activity	Ostazin, chitin powder, Tween 20, keratin, dry milk, blood agar	3,648	0	Rondon et al. (2000)
Soil	Plasmid	1.5–5 kb 1.8–5 kb 5–13 kb	Oxygenase	Indole	479,000 3,870,000 304,000	2 2 1	Lorenz and Schleper (2002)
Enrichment from soil	Cosmid	30–40 kb	Biotin synthesis operon	Functional complemen-tation: growth in absence of biotin	ND	7	Entcheva et al. (2001)

[a] Host: *E. coli.* Abbreviations: MUF-di-NAG, 4-methylumbelliferyl-β-D-N,N'-diacetylchitobioside; mt plate, microtiter plate; ND, not described.
[b] The same libraries as in Henne et al. (1999, 2000) were used.
[c] SL 1, metagenomic library 1; insert size is an average value.
[d] SL 2, metagenomic library 2; insert size is an average value.

Naturally, the concept of designing and tailoring better enzymes from naturally occurring ones is a scientifically appealing idea. A trained protein engineer should be able to use the combination of a detailed knowledge of three-dimensional protein structure, the biochemical and biophysical properties of a native protein, and the specific requirements for efficient function in a particular application as a guide in choosing which residues should be altered in order to achieve a molecule with the exact desired properties. However, it must be recognized that our limited understanding of protein function presents a serious limitation for rational protein engineering. We are still far from the situation where the exact consequences of an amino acid substitution can be predicted. Years of trial and error in protein engineering, however, have significantly improved our knowledge of protein function and its dependency on sequence and structure. A situation has been reached where the protein engineer often is able to propose a limited number of alterations, of which a decent substantial fraction will turn out to be improved for the desired property.

The methods normally used for planning mutagenesis strategies include visual inspection of protein structures obtained from X-ray crystallography or nuclear magnetic resonance studies, and computational approaches such as studies of the dynamic properties of enzymes by molecular dynamics simulations (Hansson et al., 2002). Protein electrostatics calculations are used as guiding tools when investigating properties involving charges, i.e., the effect of pH on catalysis or inactivation of the protein (Honig and Nicholls, 1995). Docking studies are used to gain knowledge about protein interactions with ligands; i.e., enzyme interactions with substrates, inhibitors, and cofactors in the case where the primary interest is within catalysis.

Examples of successful rational engineering can be found in the literature for numerous classes of enzymes—the examples are especially abundant when properties are related to stability toward denaturation by high temperature or other means (e.g., see Van den Burg et al., 1998). The reason for the high success rate when addressing stability parameters probably lies in the fact that many stability-related issues can be addressed at many different positions in the protein and by various engineering concepts that we do understand. The measurement of the thermodynamic stability of the molecule when unfolding is a reversible process called ΔG_{fold} and is defined as the difference of the free energy between the folded state and the denatured state of the protein. The ΔG_{fold} of a protein can be changed by many different substitutions. One example of a well-established approach to

protein stabilization is the identification of positions in a protein that could potentially accommodate proline residues and the substitution of the naturally occurring amino acid for proline (Suzuki et al., 1989). Proline residues are known to stabilize proteins due to the entropic contribution to the ΔG of folding for the unfolded state of the molecules such that the molecule is stabilized if the proline can be accommodated in the folded state.

More complicated properties such as catalysis, substrate, and product specificities and so forth have been successfully addressed by rational means (e.g., see Cedrone et al., 2000, Beier et al., 1999). We refer to a number of recent reviews on applying protein engineering for the probing of structure function relationships and for improving the properties of a number of industrially relevant enzymes including subtilsin, alpha-amylase, cellulase, lipase, glucose isomerase, glucoamylase, and cyclodextrin glycosyl transferase (Bryan, 2000; Nielsen and Borchert, 2000; Schülein, 2000; Svendsen, 2000; Hartley, 2000; Sauer et al., 2000; Van den Veen et al., 2000).

Engineered versions of a number of hydrolytic enzymes such as proteases, amylases, lipases, and cellulases are currently commercially available. An example of rational engineering to reach commercial goals is presented by Bisgaard-Frantzen et al. (1999). They describe the successful development of new alpha-amylases for two very diverse applications, namely, as an additive to detergent formulations to aid in the removal of starch-containing stains in laundry and as the catalyst in the starch liquefaction step in the high-fructose corn syrup process. These applications are very different with respect to demands on the enzyme catalyst, as a wash is typically carried out at alkaline conditions and starch liquefactions are carried out at extreme temperature, initially above 100°C, and acidic pH.

One of the prerequisites of successful rational engineering work is the access to high-quality protein structures or a high-quality homology model, which again also demands access to high-quality and related protein structures. Within the field of protein structure determination we have also observed major improvements with respect to equipment, computing power, software, and techniques. These developments have resulted in the availability of many more protein structures of higher quality (Brunger and Laue, 2000).

Molecular Evolution

Humans have applied directional evolutionary principles since ancient times by selective breeding of the fastest horses, the highest-yielding crops, and milking cows, etc. If one wishes to breed for faster horses, starting with a fast stallion and mare improves the chances of obtaining even faster offspring. Darwin's evolutionary discoveries in the middle of the 18th century described how the same principles function in nature: variation between individuals is constantly generated, and the fittest variant will have a selective advantage and higher chance of passing its genetic variation to the next generation. Natural variation arises as errors are regularly incorporated during DNA replication due to the intrinsic error frequency of DNA polymerases. In addition to the natural variation generated by such errors, the possibility of recombination of traits in the fittest individuals is a basic element in evolution. There has been a natural desire to mimic these existing processes in the laboratory and even to speed up nature's methods of random mutagenesis, selection of the fittest individuals, and recombination. The basic principle is to carry out the more or less random introduction of mutations, thereby generating DNA libraries consisting of thousands or even millions of variant genes. The DNA variation is transcribed and translated into protein diversity in a variant library where the tight connection between the variant protein and its encoding gene has to be established. This library is subsequently exposed to a selection or screening procedure in which the individual variants are ranked according to the fitness with respect to the particular property or the set of properties that are screened for. The best candidates will then form the starting point for the next round of the repeated directed evolution cycle. The three components of directed molecular evolution are described in more detail below.

DNA variation

Complete gene random mutagenesis techniques include exposure of DNA to chemical or physical mutagens by passing the DNA through mutator strains or by error-prone PCR (Greener et al., 1996; Fabret et al., 2000). Although each method has its drawbacks, especially with respect to intrinsic bias in mutagenesis spectrum, the latter has turned out to be the preferred method due to ease of use and versatility.

The major drawback in complete gene random mutagenesis is the fact that none of the methods described so far results in all possible amino acid substitution in a specific position. In a variant library, one generally aims for only a few amino acid substitutions per protein in order to avoid the incorporation of too many detrimental mutations; i.e., one has to aim for a rather modest mutagenesis frequency. At such a low level of mutagenesis, it is extremely rare that two nucleotide substitutions fall within the same

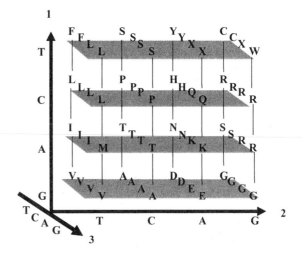

Figure 2. The codon cube shows the correlation between the amino acid and the encoding codons in the DNA represented by the three nucleotides along the *y, x,* and *z* axes.

codon, thus limiting the number of amino acid substitutions to typically 5 or 6 out of the theoretical 19 possible substitutions. The reason for this fact is exemplified in Fig. 2, the "codon cube" that shows the relationship between codons and the encoded amino acid. For example, starting from a tryptophan that is encoded only by one codon, TGG, the alteration of one nucleotide in this codon leads to new codons encoding the following amino acids: cysteine, arginine, glycine, serine, and leucine (or two stop codons). With a reasonable mutagenesis frequency, only five amino acids are reached from tryptophan in a variant library. If one has the desire to probe all 19 amino acid substitutions in a position-encoding tryptophan, obviously a different strategy has to be applied.

Alternatively, mutagenesis can be targeted to specific positions or a selected region of the protein by the use of doped mutagenizing oligonucleotides. In those cases, the design of the mutagenic oligonucleotide determines the mutation level and type of randomization of the designated region of the gene. For such a strategy, rational input is necessary. A consideration of protein structure can guide mutagenesis to specific regions of the protein such as the active site, the substrate binding regions, or the overall surface. Also, known engineering principles identical to those applied in rational protein engineering combined with the protein structure can guide mutagenesis to specific positions or regions.

Coupling of DNA mutants and protein variants

The exact connection between the protein displaying the properties and the DNA that encodes the protein is absolutely necessary for the experiment, as fitness evaluation is based on the functional properties of the protein, but the DNA is passed to the next generation.

The most frequently used method of coupling diversity is to express the encoded protein by the use of a cellular system, e.g., bacteria, yeast, fungus, or insect cells. The obtainable size of the library is limited by the transformation frequency of the selected host. If the most popular molecular biology host, *E. coli,* is utilized, libraries will typically be limited to around 10^7 individuals. Even smaller libraries can be expected if less-developed molecular biology hosts are utilized.

Variations on establishing the connection between genotype and phenotype exist. In some cases the protein can be generated by in vitro translation, thereby eliminating the transformation step and rendering the cells superfluous. Such a procedure is typically applied if a coupling between the genetic diversity and the translated protein can be made, e.g., by the elegant mRNA display technique (Hanes and Plückthun, 1997). If the subsequent enrichment procedure is efficient such that single-protein molecules are the basis for the isolation of the improved candidates, this method has the advantage of generating and screening extremely large libraries of the size of 10^{15}.

The coupling of phenotype and genotype can also be carried out by transformation into a cellular system that promotes the display of the protein (phenotype) on the surface of the cell or phage, thus establishing the connection to the genotype. The typical example is phage display (O'Neil and Hoess, 1995), but cells have also been utilized for this purpose (Stähl and Uhlen, 1997). The same demand for selection at the single-molecule level applies to these systems as that described above for the enrichment procedures used for mRNA display.

Determining which candidates should be the basis for the next generation

The method of ranking the individual according to particular properties and selecting the fittest individual is one of the key elements in directed evolution. It is a very demanding task to set up a meaningful assay that addresses all relevant properties for a particular industrial application. Such procedures can be based on established screening methods that typically are downscaled to small volumes in microtiter plates in order to enhance the throughput. Screening technology was described above, but it should be mentioned that directed evolution often puts more demands on the precision of the screening tools, as

the goal is to distinguish marginal improvements in a well-functioning wild-type enzyme rather than the "on/off answer" when dealing with screening of natural diversity. Various attempts to look at large libraries in screening procedures have been carried out (Wolcke and Ullmann, 2001).

If expression of a particular property can provide a growth advantage to the host, selection procedures can be applied. Naki et al. (1998) demonstrated that efficient production of a subtilisin resulted in a growth advantage to the host cells during growth in a defined medium with serum albumin as the sole nitrogen source.

The iterative aspect

The first directed evolution cases constituted repeated rounds of random mutagenesis with selection in each round for the single, best-performing clone to be used as the starting point for new rounds of mutagenesis, library generation, and isolation of the fittest candidate. Although a number of successful examples have been reported, it is well agreed that this method has disadvantages when compared with the newer recombination-based formats pioneered by Stemmer (1994a, 1994b). The disadvantages reside mainly in the fact that beneficial mutations as well as detrimental mutations tend to accumulate over a number of iterative cycles of mutagenesis. Another important disadvantage is that the selection of only one clone to form the starting point of the next diversification round means that beneficial mutations in all other clones are discarded.

The invention of DNA shuffling eliminated these problems (Stemmer, 1994a, 1994b). The genes encoding all the improved clones from the first round are randomly fragmented, and these fragments are used in PCR where they act both as primers and as templates of each other. After sufficient rounds of annealing and extension, the full-length gene is reassembled and typically reamplified in another PCR. A library is generated and screened, and the complete process is repeated until the desired properties have been obtained. This method provides clear advantages as improvements are not discarded from a round of evolution, and it allows for removal of detrimental diversity (for a review, see Tobin et al., 2000). Variations on this theme have been reported covering both alternative in vitro techniques (Zhao et al., 1998; Coco et al., 2001) and in vivo formats utilizing the recombination potential of living cells (Cherry et al., 1999).

Rather than using synthetic variants of a gene, the starting material can be a number of homologous genes from different species, resulting in a technology denoted "family shuffling" (Crameri et al., 1998).

The initial genes thus represent unique points in sequence space that all represent unique answers to a particular catalytic property. The shuffling reaction creates random hybrids between the starting genes, thus generating new spots in sequence space addressing the same question. This technology results in a surprisingly high fraction of active hybrids that might reflect the fact that the complete starting diversity has been preselected by nature. In other words, in each position, the amino acids found in the progeny have already fulfilled the requirements for function in one of the parents. The evolutionary steps are thus larger than observed in the case where the starting material is point mutations in one starting gene.

Ness et al. (1999) have carried out a family shuffling experiment with a very large family of 26 subtilisin fragments. The resulting library was screened for five different properties, and candidates that were improved for each of the screened properties could be identified from a library of less than 600 active clones. Even new subtilisins that improved simultaneously for two or more properties were identified from the screening. Alternative formats of artificial hybrid formation have also been reported (Ostermeier et al., 1999, Sieber et al., 2001).

Combining Evolution and Rational Protein Engineering

Recently, hypotheses for explaining and understanding the molecular evolutionary process have been presented. Voigt et al. (2001) have tried to explain the outcome of directed molecular evolution experiments from structural and computational considerations, and they propose the hypothesis that beneficial mutagenic or recombinogenic events cause the least disturbance of the structure. It will be interesting to see if such hypotheses result in the future design of better and more successful directed evolution techniques.

Is there a clear distinction between rational protein engineering and directed evolution-based protein optimization technologies? In our opinion the answer is a clear "no," as we see these technologies as complementary to or even supportive of each other.

The capacity of any screening procedure will depend on the properties being screened for. Most formats are able to handle thousands or up to a maximum of a million clones. Considering that the number of positions that could be probed for all possible 20 amino acids is only three to five (see Table 2), it is clear that any rational input that could enhance the quality of a diversification scheme is highly welcome. The continuously enhanced understanding of the mechanisms of protein function from rational

Table 2. Size of theoretical libraries encoding all combinations of the 20 amino acids in the specified number of positions

No. of positions	Theoretical library
1	20
2	400
3	8,000
4	160,000
5	3,200,000

protein engineering and protein structure determination can indeed aid this process.

It also makes sense to try to understand the reason for the observed improvements in variants originating from random libraries, thus enhancing our understanding of structure-function relationships of proteins. This enhanced knowledge can subsequently improve randomization strategies in future experiments.

The efficient development of future biocatalysts relies on combining our current technologies. We foresee a situation in which directed molecular evolution and rational engineering as complementing technologies will enable the future development of an ever-increasingly broad range of enzyme-based industrial catalysis.

CONCLUSIONS

Discovery of industrial enzymes is a multidisciplinary effort involving a wide array of different technologies. In this chapter we have outlined some major routes for the discovery of enzymes for industrial applications. Nature holds a wonderful diversity of organisms and the corresponding wealth of enzymes. Accordingly, conservation of this biodiversity, sustainable development of biological resources, and equitable sharing of benefits between users and owners of biodiversity resources are of utmost importance. This extremely interesting part of bioprospecting could not be discussed here, and the reader is referred to Bull et al. (2000) and to the home page of the Convention of Biological Diversity (http://www.biodiv.org/). For a variety of applications, even nature's assortment faces some limitations. It is the challenge for scientists to optimize the natural enzymes and to generate additional and "artificial" diversity to tailor enzymes for a given application. Finally, many more enzymes have already been made available as products for several applications. For all approaches it is important to stress that it is not the broadest possible diversity, but rather the highest possible quality of diversity that will lead to the ultimate goal, namely, a novel product. In this respect, selection/deselection via perfectly designed assays is of utmost importance, indicating the significance of linking process understanding to biochemistry. Natural diversity approaches and optimization strategies are complementing routes, and both are equally important to develop a high-quality diversity of enzymes. The importance of both approaches is reflected by the fact that Novozymes on average launched 5 to 10 novel enzyme products each of the past few years with around 50% deriving directly from nature and 50% being tailor-made proteins. As Cheetham (1998) points out, it needs to be highlighted that "people with many different skills are needed to create successful biocatalysts, and with the contribution of each specialism absolutely vital for the success." It will be the combination of skills, knowledge, new technologies, and novel enzymes that will ensure future benefits by the development of truly ecofriendly processes.

Acknowledgments. We are grateful to M.A. Stringer, Fiona Becker, and Corinne Squire for critically reading the manuscript.

REFERENCES

Amann, R., J. Snaidr, M. Wagner, W. Ludwig, and K. H. Schlerfeu. 1996. In situ visualization of high genetic diversity microbial community. *J. Bacteriol.* 178:3496–3500.

Ando, H., M. Adachi, K. Umeda, A. Matsuura, M. Nonaka, R. Uchio, H. Tanaka and M. Motoki. 1989. Purification and characteristics of a novel transglutaminase derived from microorganisms. *Agric. Biol. Chem.* 53:2613–2617.

Archer, D. B. 1994. Enzyme production by recombinant Aspergillus. *Bioprocess Technol.* 19:373–393.

Archer, D. B., and J. F. Peberdy. 1997. The molecular biology of secreted enzyme production by fungi. *Crit. Rev. Biotechnol.* 17:273–306.

Beier, L., A. Svendsen, C. Andersen, T. P. Frandsen, T. V. Borchert, J. R. Cherry. 2000. Conversion of the maltogenic a-amylase Novamyl into a CGTase. *Protein Eng.* 13:509–513.

Bisgaard-Frantzen, H., A. Svendsen, B. Norman, S. Pedersen, S. Kjærulff, H. Outtrup, and T. V. Borchert. 1999. Development of industrially important alpha-amylases. *J. Appl. Glycosci.* 46:199–206.

Brady, S. F., C. J. Chao, J. Handelsman, and J. Clardy. 2001. Cloning and heterologous expression of a natural product biosynthetic gene cluster from eDNA. *Org. Lett.* 3:1981–1984.

Brunger, A. T., and E. D. Laue. 2000. Biophysical methods. New approaches to study macromolecular structure and function. *Curr. Opin. Struct. Biol.* 10:557.

Bryan, P. 2000. Protein engineering of subtilisins. *Biochem. Biophys. Acta-Protein Struct. Mol. Enzymol.* 1553:203–222.

Bull, A. T., M. Goodfellow, and J. H. Slate. 1992. Biodiversity as a source of innovation in biotechnology. *Ann. Rev Microbiol* 46:219–252.

Bull, A. T., A. C. Ward, and M. Goodfellow. 2000. Search and discovery strategies for biotechnology: the paradigm shift. *Microbiol. Mol. Biol. Rev.* 64:573–606.

Bult, C. J., O. White, G. J. Olsen, L. X. Zhou, R. D. Fleischmann, G. G. Sutton, J. A. Blake, L. M. Fitzgerald, R. A. Clayton, J. D. Gocayne, A. R. Kerlavage, B. A. Dougherty, J. F. Tomb, M. D. Adams, C. I. Reich, R. Overbeek, E. F. Kirkness, K. G. Weinstock,

J. M. Merrick, A. Glodek, J. L. Scott, N. S. M. Geoghagen, J. F. Weidman, J. L. Fuhrmann, D. Nguyen, T. R. Utterback, J. M. Kelley, J. D. Peterson, P. W. Sadow, M. C. Hanna, M. D. Cotton, K. M. Roberts, M. A. Hurst, B. P. Kaine, M. Borodovsky, H. P. Klenk, C. M. Fraser, H. O. Smith, C. R. Woese, and J. C. Venter. 1996. Complete genome sequence of the methanogenic archaeon, methanococcus-jannaschii. *Science* 273:1058–1073.

Callebaut, I., G. Labesse, P. Durand, A. Poupon, L. Canard, J. Chomilier, B. Henrissat, and J. P. Mornor. 1997. Deciphering protein sequence information through hydrophobic cluster analysis (HCA): current status and perspectives. *Cell. Mol. Life Sci.* 53:621–645.

Cedrone, F., A. Menez and E. Quemeneur. 2000. Tailoring new enzyme functions by rational redesign. *Curr. Opin. Struct. Biol.* 10:405–410.

Cheetham, P. S. J. 1998. What makes a good biocatalyst? *J. Biotechnol.* 66:3–10.

Cherry, J. R., M. H. Lamsa, P. Schneider, J. Vind, A. Svendsen, A. Jones, and A. H. Pedersen. 1999. Directed evolution of a fungal peroxidase. *Nat. Biotechnol.* 17:379–384.

Coco, W. M., W. E. Levinson, M. J. Crist, H. J. Hektor, A. Darzins, P. T. Pienkos, C. H. Squires, and D. J. Monticello. 2001. DNA shuffling method for generating highly recombined genes and evolved enzymes. *Nat. Biotechnol.* 19:354–359.

Cottrell, M. T., J. A. Moore, and D. L. Kirchman. 1999. Chitinases from uncultivated marine microorganisms. *Appl. Environ. Microbiol.* 65:2553–2557.

Crameri, A., S. A. Raillard, E. Bermudez, and W. P. Stemmer. 1998. DNA shuffling of a family of genes from diverse species accelerates directed evolution. *Nature* 391:288–291.

Dalbøge, H., and L. Lange. 1998. Using molecular techniques to identify new microbial biocatalysts. *Trends Biotechnol.* 16:265–272.

Dalbøge, H., Th. Sandal, S. Kauppinen, and B. Diderichsen. August 2001. Method of providing a hybrid polypeptide exhibiting an activity of interest. U.S. Patent 6,270,968.

Demirjian, D. C., F. Moris-Varas, and C. S. Cassidy. 2001. Enzymes from extremophiles. *Curr. Opin. Chem. Biol.* 5:144–151.

Diderichsen, B., and L. Christiansen. 1988. Cloning of a maltogenic alpha-amylase from *Bacillus stearothermophilus*. *FEMS Microbiol. Lett.* 56:53–60.

Duffner, F., R. Wilting, and K. Schnorr. 2001. Signal sequence trapping. Patent application W00177315-A1.

Dunn-Coleman, N., and R. Prade. 1998. Toward a global filamentous fungus genome sequencing effort. *Nat. Biotechnol.* 16:5.

Durick, K., J. Mendlein, and K. C. Xanthopoulos. 1999. Hunting with traps: genome-wide strategies for gene discovery and functional analysis. *Genome Res.* 9:1019–1025.

Encheva, P., W. Liebl, A. Joham, T. Hartsch, and M. R. Strevt. 2001. Direct cloning from enrichment cultures, a reliable strategy for isolation of complete operons and genes from microbiol consortua. *Appl. Environ. Microbiol.* 67:89–99.

Fabret, C., S. Poncet, S. Danielsen, T. Borchert, S. D., Ehrlich, and L. Janniero. 2000. Efficient gene targeted random mutagenesis in genetically stable *Escherichia coli* strains. *Nucleic Acids Res.* 28:21 e95 1–5.

Felse, P. A., and T. Panda. 2000. Production of microbial chitinases: a revisit. *Bioprocess Eng.* 23:127–134.

Franco, C. M. M., and N. C. McClura. 1998. Isolation of microorganisms for biotechnological application. *J. Microbiol. Biotechnol.* 8:101–110.

Godfrey, T., and S. I. West. 1996. Introduction to industrial enzymes, p. 1–8. *In* T. Godfrey and S. I. West, (ed.), *Industrial Enzymology*, 2nd ed. Macmillan Press Inc., London, United Kingdom.

Gordon, J. 1999. Intellectual property, p. 309–320. In A. L. Demain and J. E. Davies (ed.), Manual of Industrial Microbiology and Biotechnology. ASM Press, Washington, D.C.

Greener, A., M. Callahan, and B. Jerpseth. 1996. An efficient random mutagenesis technique using an E. coli mutator strain. Methods Mol. Biol. 57:375–385.

Hallberg, K. B., and D. B. Johnson. 2001. Biodiversity of acidophilic prokaryotes. Adv. Appl. Microbiol. 49:37–84.

Handelsman, J., M. R. Rondon, S. F. Brady, J. Clardy, and R. M. Goodman. 1998. Molecular biological access to the chemistry of unknown soil microbes: a new frontier for natural products. Chem. Biol. (London) 5:R245–R249.

Hanes, J., and A. S. Plückthun. 1997. In vitro selection and evolution of functional proteins by using ribosome display. Proc. Natl. Acad. Sci. USA 94:4937–4942.

Hansson, T., C., Oostenbrink, and W. F. Gunsteren. 2002. Molecular dynamics simulations. Curr. Opin. Struct. Biol. 12:190–196.

Hartley, B. S., N., Hanlon, R. J. Jackson, and M. Rangarajan. 2000. Glucose isomerase: insights into protein engineering for increased thermostability. Biochem. Biophys. Acta Protein Struct. Mol. Enzymol. 1553:294–335.

Hawksworth, D. L. 2001. The magnitude of fungal diversity: the 1.5 million species estimate revisited. Mycolog. Res. 105:1422–1432.

Henne, A., R. Daniel, R. A. Schmitz, and G. Gottschall. 1999. Construction of environmental DNA libraries in Escherichia coli and screening for the presence of genes conferring utilization of 4- hydroxybutyrate. Appl. Environ. Microbiol. 65:3901–3907.

Henne, A., R. A. Schmitz, M. Bomeke, G. Gottschalk, and R. Daniel. 2000. Screening of environmental DNA libraries for the presence of genes conferring lipolytic activity on Escherichia coli. Appl. Environ. Microbiol. 66:3113–3116.

Henrissat, B., and P. Borla. 1996. On the classification of modular proteins. Protein Eng. 9:725–726.

Henrissat, B., and A. Romeu. 1995. Families, superfamilies and subfamilies of glycosyl hydrolases. Biochem. J. 311:350–351.

Hirose, I., K. Sano, I. Shioda, M. Kumano, K. Nakamura, and K. Yamano. 2000. Proteome analysis of Bacillus subtilis extracellular proteins: a two-dimensional protein electrophoretic study. Microbiology 146:65–75.

Honig, B., and A. Nicholls. 1995. Classical electrostatics in biology and chemistry. Science 268:1144–1149.

Hugenholtz, P., and N. R. Pace. 1996. Identifying microbial diversity in the natural environment: a molecular phylogenetic approach. Trends Biotechnol. 14:190–197.

Joo, H., A. Arisawa, Z. L., Lin, and F. H. Arnold. 1999. A high-throughput digital imaging screen for the discovery and directed evolution of oxygenases. Chem. Biol. 6:699–706.

Kirk, O., T. V. Borchert, and C. C. Fuglsang. 2002. Industrial enzyme applications. Curr. Opin. Biotechnol. 13:345–351.

Kongsbak, L., K. S. Jørgensen, C. T. Jørgensen, T. L. Husum, S. Ernst, and S. Møller. 1999. A fluorescence polarisation screening method. Novo Nordisk A/S, patent application WO9945143-A2.

Kunst, F., N. Ogasawara, I. Moszer, A. M. Albertini, G. Alloni, V. Azevedo, M. G. Bertero, P. Bessifres, A. Bolotin, S. Borchert, R. Borriss, L. Boursier, A. Brans, M. Braun, S. C. Brignell, S. Bron, S. Brouillet, C. V. Bruschi, B. Caldwell, V. Capuano, N. M. Carter, S. K. Choi, J. J. Codani, I. P. Connerton, and A. Danchir. 1997. The complete genome sequence of the gram-positive bacterium Bacillus subtilis. Nature 390:249–256.

Kuske, C. R., S. M., Barns, and J. D. Busch. 1997. Diverse uncultivated bacterial groups from soils of the arid southwestern United States that are present in many geographic regions. Appl. Environ. Microbiol. 63:3614–3621.

Lassen, S. F., J. Breinholt, P. R. Ostergaard, R. Brugger, A. Bischoff, M. Wyss, and C. C. Fuglsang. 2001. Expression, gene cloning, and characterization of five novel phytases from four basidiomycete fungi: Peniophora lycii, Agrocybe pediades, a Ceriporia sp., and Trametes pubescens. Appl. Environ. Microbiol. 67:4701–4707.

Li, W. Z., and A. Godzila. 2002. Discovering new genes with advanced homology detection. Trends Biotechnol. 20:315–316.

Liberman, D. F., R. Fink, and F. Schaefer. 1999. Biosafety and biotechnology, p. 300–308. In A. L. Demain and J. E. Davies (ed.), Manual of Industrial Microbiology and Biotechnology. ASM Press, Washington, D.C.

Liesack, W., and E. Stackebrandt. 1992. Occurrence of novel groups of the domain Bacteria as revealed by analysis of genetic material isolated from an Australian terrestrial environment. J. Bacteriol. 174:5072–5078.

Lorenz, P., and C. Schleper. 2002. Metagenome—a challenging source of enzyme discovery. J. Mol. Catal. 19–20C:13–19.

Majernik, A., G. Gottschalk, and R. Daniel. 2001. Screening of environmental DNA libraries for the presence of genes conferring Na+(L+)/H+ antiporter activity on Escherichia coli: characterization of the recovered genes and the corresponding gene products. Bacteriol. 183:6645–6653.

McCoy, M. 2000. Novozymes emerges. Chem. Eng. News 19:23–25.

Meeuwsen, P. J. A., J. P. Vincken, G. Beldman, and A. G. J. Voragen. 2000. A universal assay for screening expression libraries for carbohydrases. J. Biosci. Bioeng. 89:107–109.

Merlin, C., S. McAteer, and M. Masters. 2002. Tools for characterization of Escherichia coli genes of unknown function. J. Bacteriol. 184:4573–4581.

Naki, D., C. Paech, G. Ganshaw, and V. Schellenberger. 1998. Selection of a subtilisin-hyperproducing Bacillus in a highly structured environment. Appl. Microbiol. Biotechnol. 49:290–294.

Nelson, W. N. 1991. Modern Techniques for Rapid Microbiological Analysis. VCH Publishers Inc., Deerfield Beach, Fla.

Ness, J. E., M. Welch, L. Giver, M. Bueno, J. R. Cherry, T. V. Borchert, W. P. C. Stemmer, and J. Minshull. 1999. DNA shuffling of subgenomic sequences of subtilisin. Nat. Biotechnol. 17:893–896.

Nielsen, J. E., and T. V. Borchert. 2000. Protein engineering of bacterial alpha-amylases. Biochem. Biophys. Acta Protein Struct. Mol. Enzymol. 1553:253–274.

O'Neil, K. T., and R. H. Hoes. 1995. Phage display: protein engineering by directed evolution. Curr. Opin. Struct. Biol. 5:443–449.

OECD (Organisation for Economic Co-operation and Development). 1998. Biotechnology for Clean Industrial Products and Processes: Towards Industrial Sustainability. OECD, Paris, France.

OECD (Organisation for Economic Co-operation and Development). 2001. The Application of Biotechnology to Industrial Sustainability. OECD, Paris, France.

Ogawa, J., and S. Shimizs. 2002. Industrial microbial enzymes: their discovery by screening and use in large-scale production of useful chemicals in Japan. Curr. Opin. Biotechnol. 13:367–375.

Ogram, A. 2000. Soil molecular microbial ecology at age 20: methodological challenges for the future. Soil Biol. Biochem. 32:1499–1504.

Ohnishi, J., S. Mitsuhashi, M. Hayashi, S. Ando, H. Yokoi, K. Ochiai, and M. Ikeda. 2002. A novel methodology employing Corynebacterium glutamicum genome information to generate a new L-lysine-producing mutant. Appl. Microbiol. Biotechnol. 58:217–223.

Olsen, G. J., and C. R. Woese. 1997. Archaeal genomics: an overview. *Cell* 89:991–994.

Olsen, G. J., D. J. Lane, S. J. Giovannoni, N. R. Pace, and D. A. Stahl. 1986. Microbial ecology and evolution: a ribosomal RNA approach. *Annu. Rev. Microbiol.* 40:337–365.

Ostermeier, M., J. H. Shim, and S. J. Benkovic. 1999. A combinatorial approach to hybrid enzymes independent of DNA homology. *Nat. Biotechnol.* 17:1205–1209.

Pandey, A., G. Szakacs, C. R. Soccol, J. A. Rodriguez-Leon, and V. T. Soccol. 2001. Production, purification and properties of microbial phytases. *Bioresource Technol.* 77:203–214.

Precigou, S., P. Goulas, and R. Duran. 2001. Rapid and specific identification of nitrile hydratase (NHase)-encoding genes in soil samples by polymerase chain reaction. *FEMS Microbiol. Lett.* 204:155–161.

Preisig, C., and G. Byng. 2001. Applications of mass spectrometry in screening for new biocatalysts. *J. Mol. Catal. B* 11:733–741.

Psenner, R., and B. Sattler. 1998. Microbial communities—life at the freezing-point. *Science* 280:2073–2074.

Radford, A., and J. H. Parish. 1997. The genome and genes of Neurospora crassa. *Fungal Genet. Biol.* 21:258–266.

Reisman, H. B. 1999. Economics, p. 273–288. *In* A. L. Demain and J. E. Davis (ed), *Manual of Industrial Microbiology and Biotechnology.* ASM Press, Washington, D.C.

Rondon, M. R., R. M. Goodman, and J. Handelsman. 1999. The Earth's bounty: assessing and accessing soil microbial diversity. *Trends Biotechnol.* 17:403–409.

Rondon, M. R., P. R. August, A. D. Bettermann, S. F. Brady, T. H. Grossman, M. R. Liles, K. A. Loiacono, B. A. Lynch, I. A. MacNeil, C. Minor, C. L. Tiong, M. Gilman, M. S. Osburne, J. Clardy, J. Handelsman, and R. M. Goodman. 2000. Cloning the soil metagenome: a strategy for accessing the genetic and functional diversity of uncultured microorganisms. *Appl. Environ. Microbiol.* 66:2541–2547.

Rosteek, P. R., B. S. DeHoff, F. H. Norris, and P. K. Rockey. 1999. Bacterial genomics and genome informatics, p. 495–500. *In* A. L. Demain and J. E. Davies (ed.), *Manual of Industrial Microbiology and Biotechnology.* ASM Press, Washington, D.C.

Rothschild, L. J., and R. I. Mancinelli. 2001. Life in extreme environments. *Nature* 409:1092–1101.

Ruijssenaars, H. J., and S. Hartmans. 2001. Plate screening methods for the detection of polysaccharase-producing microorganisms. *Appl. Microbiol. Biotechnol.* 55:143–149.

Sambrock, J., and D. W. Russel. 2001. *Molecular Cloning*, 3rd ed. Cold Spring Harbor Laboratory Press, Cold Spring Harbor, N.Y.

Sanchez, D. O., R. O. Zandomeni, S. Cravero, R. E. Verdun, E. Pierrou, P. Faccio, G. Diaz, S. Lanzavecchia, F. Aguero, A. C. C. Frasch, S. G. E. Andersson, O. L. Rossetti, O. Grau, and R. A. Ugalda. 2001. Gene discovery through genomic sequencing of *Brucella abortus. Infect. Immun.* 69:865–868.

Sandal, Th., C. Sjoholm, T. Schaefer, L. Lange, and F. Duffner. 2000. Method for generating a gene library. Novo Nordisk A/S, (patent), W0200024882-A1.

Sattler, B., H. Puxbaum, and R. Psenner. 2001. Bacterial growth in supercooled cloud droplets. *Geophys. Res. Lett.* 28:239–242.

Sauer, J., B. W. Sigurdskjold, U. Christensen, T. P. Frandsen, E. Mirgorodskaya, M. Harrison, P. Roepstorff, and B. Svensson. 2000. Glucoamylase: structure/function relationships, and protein engineering. *Biochem. Biophys. Acta Protein Struct. Mol. Enzymol.* 1543:275–293.

Schaefer, T., O. Kirk, T. V. Borchert, C. C. Fuglsang, S. Pedersen, S. Salmon, H. S. Olsen, R. Deinhammer, and H. Lind. 2002. Enzymes for technical applications. *In* A. Steinbuechel (ed) *Biopolymers.* Wiley VCH, New York, N.Y.

Schellenberger, V. 1997. Compartmentalization method for screening microorganisms. Genencore International Inc., patent 87/37036.

Schmidt, T. M., E. F. DeLong, and N. R. Pace. 1991. Analysis of a marine picoplankton community by 16S rRNA gene cloning and sequencing. *J. Bacteriol.* 173:4371–4378.

Schülein, M. 2000. Protein engineering of cellulases. *Biochem. Biophys. Acta-Protein Struct. Mol. Enzymol.* 1553:239–252.

Schülein, M., L. N. Andersen, S. F. Lassen, L. Lange, S. Kauppinen, R. Nielsen, O. Ihara, and S. Takagi. 1996. Novel Endoglucanases. U.S. patent 6,001,639.

Short, J. M. 1997. Recombinant approaches for accessing biodiversity. *Nat. Biotechnol.* 15:1322–1323.

Short, J. M., and M. Keller. 2001. High-throughput screening for novel enzymes. U.S. patent 6,174, 673 B1.

Sieber, V., C. A. Martinez, and F. A. Arnold. 2001. Libraries of hybrid proteins from distantly related sequences. *Nat. Biotechnol.* 19:456–460.

Smith, M. 1985. In vitro mutagenesis. *Annu. Rev. Genet.* 19:423–462.

Stahl, D. A., D. J. Lane, G. J. Olsen, and N. R. Pace. 1985. Characterization of a Yellowstone USA hot spring microbial community by 5S ribosomal RNA sequences. *Appl. Environ. Microbiol.* 49:1379–1384.

Stähl, S., and M. Uhlep. 1997. Bacterial surface display. *Trend of Biotechnol.* 15:185–192.

Steffan, R. J., J. Goksoyr, A. K. Bej, and R. M. Atlas. 1988. Recovery of DNA from soils and sediments. *Appl. Environ. Microbiol.* 54:2908–2915.

Steipe, B. 1999. Evolutionary approaches to protein engineering. *Curr. Top. Microbiol. Immun.* 243:55–86.

Stemmer, W. P. 1994a. DNA shuffling by random fragmentation and reassembly: *in vitro* recombination for molecular evolution. *Proc. Natl. Acad. Sci. USA* 91:10747–10751.

Stemmer, W. P. 1994b. Rapid evolution of a protein by in vitro DNA shuffling. *Nature* 370:389–391.

Suzuki, Y. 1989. A general principle of increasing protein thermostability. *Proc. Jpn. Acad.* 65:146–148.

Svendsen, A. 2000. Lipase protein engineering. *Biochem. Biophys. Acta Protein Struct. Mol. Enzymol.* 1553:223–238.

Takai, K., H. Hirayama, Y. Sakihama, F. Inagaki, Y. Yamato, and K. Horikoshi. 2002. Isolation and metabolic characteristics of previously uncultured members of the order Aquificales in a subsurface gold mine. *Appl. Environ. Microbiol.* 68:3046–3054.

Tiedje, J. M., and J. L. Stein. 1999. Microbial diversity: strategies for its recovery, p. 682–692. *In* A. L. Demain and J. E. Davies (ed.), *Manual of Industrial Microbiology and Biotechnology.* ASM Press, Washington, D.C.

Tobin, M. B., C. Gustafsson, and G. W. Huisman. 2000. Directed evolution: the "rational" basis for "irrational" design. *Curr. Opin. Struct. Biol.* 10:421–427.

Torsvik, V. L. 1980. Isolation of bacterial DNA from soil. *Soil Biol. Biochem.* 12:15–22.

Torsvik, V., F. L. Daae, R. Sandaa, and L. Ovreas. 1998. Novel techniques for analysing microbial diversity in natural and perturbed environments. *J. Biotechnol.* 64:53–62.

Torsvik, V., J. Goksoyr, and F. L. Daao. 1990. High diversity in DNA of soil bacterial. *Appl. Environ. Microbiol.* 56:782–787.

Torsvik, V., L. Ovreas, and T. F. Thingstad. 2002. Prokaryotic diversity—magnitude, dynamics, and controlling factors. *Science* 296:1064–1066.

van Beilen, J. B., and Z. Li. 2002. Enzyme technology: an overview. *Curr. Opin. Biotechnol.* 13:338–344.

Van den Burg, B. G. Vriend, O. R. Veltman, G. Venema, and V. G. H. Eijsink. 1998. Engineering an enzyme to resist boiling. *Proc. Natl. Acad. Sci. USA* 95:2056–2060.

van der Veen, B. A., J. C. M. Uitdehaag, B. W. Dijkstra, L. Dijkhuizen. 2000. Engineering of cyclodextrin glycosyltransferase reaction and product specificity. *Biochem. Biophys. Acta* **1543**:336–360.

Voigt, C. A., S. Kauffman, and Z. G. Wang. 2001. Rational evolutionary design: the theory of in vitro protein evolution. *Adv. Protein Chem.* **55**:79–160.

Watve, M., V. Shejval, C. Sonawane, M. Rahalkar, A. Matapurkar, Y. Shouche, M. Patole, N. Phadnis, A. Champhenkar, K. Damie, S. Karandikar, V. Kshirsagar, and M. Jog. 2000. The "K" selected oligophilic bacteria: a key to uncultured diversity?. *Curr. Sci. (Bangalore)* **78**:1535–1542.

Widner, B., M. Thomas, D. Sternberg, D. Lammon, R. Behr, and A. Sloma. 2000. Development of marker-free strains of *Bacillus subtilis* capable of secreting high levels of industrial enzymes. *J. Ind. Microbiol. Biotechnol.* **25**:204–212.

Woese, C. R. 1987. Bacterial evolution. *Microbiol. Rev.* **51**: 221–271.

Wölke, J., and D. Ullmann. 2001. Miniaturized HTS technologies—uHTS. *Drug Discovery Today* **6**:637–646.

Zagulski, M., C. J. Herbert, and J. Rytka. 1998. Sequencing and functional analysis of the yeast genome. *Acta Biochim. Polon.* **45**:627–643.

Zhao, H., L. Giver, Z. Shao, J. A. Affholter, and F. H. Arnold. 1998. Molecular evolution by staggered extension process (StEP) in vitro recombination. *Nat. Biotechnol.* **16**:258–261.

Microbial Diversity and Bioprospecting
Edited by Alan T. Bull
© 2004 ASM Press, Washington, D.C.

Chapter 34

Plant Growth-Promoting Agents

JAMES M. LYNCH

The term "rhizosphere" was first used by Hiltner (1904) to describe the symbiotic interaction between bacteria and legume roots. However, the term was soon applied to the nonsymbiotic interactions involving a wide range of processes. Of particular fascination has always been the way in which bacteria and fungi associating with roots can stimulate plant growth. The direct stimulation of plant growth is mediated by the microbial production of plant growth-regulating metabolites. Improved plant nutrition can be provided from symbiotic associations, mainly the rhizobia that associate with legume roots to fix N_2 and the mycorrhizal fungi that associate with all roots, except those of the *Cruciferae,* to improve phosphorus uptake. Nonsymbiotic interactions of roots and microorganisms can also often provide such benefits to the plant. However, these symbiotic and nonsymbiotic benefits are not usually considered as plant growth promotion.

A seminal paper by Kloepper et al. (1980) described plant growth-promoting rhizobacteria or PGPRs, which acted by producing siderophores to chelate ferric iron, making it unavailable to plant pathogens and deleterious rhizobacteria and thereby indirectly stimulating plant growth by releasing pathogen stress. Such indirect action is generally considered as plant growth promotion and is so considered here. Similarly, there has been interest recently in the use of PGPRs in bioremediation of soils that are polluted with heavy metals. Here the mechanism is to promote rhizosphere catabolism around the plant exposed to negative growth regulator stress, thereby making it more capable of taking up toxic heavy metals from soils. Again, this indirect mechanism of plant growth promotion will be considered here.

Among the organisms that have received most attention in the field are the bacteria *Pseudomonas* spp., *Azospirillum* spp., and *Enterobacter* spp., together with the fungus *Trichoderma* spp.; the specific illustrations of their applications are therefore considered here.

PLANT GROWTH REGULATORS

The main groups of plant growth regulators are outlined in Fig. 1 (Lynch, 1990). Few investigators have shown that the formation of these substances by microorganisms exogenous to the root had any significance to the plant itself. The critical experiments of extracting and characterizing the metabolite and then reapplying it to the root at a natural dose range have not been done. Bacteria that penetrate the cortex and form products in the endorhizosphere could be more significant for plant growth than those formed in the ectorhizosphere that surrounds the roots. Whereas the observations of root morphology (increased extension, thickness, and root hair formation) following bacterial inoculation might be consistent with growth regulator action, definite evidence has to be offered in support of the presence of a specific regulator.

Ethylene is probably formed mainly at the sites of decomposing organic matter in soil, such as plant residues or straw. However, it can easily diffuse to the rhizosphere. It accumulates mainly in waterlogged soils where there is a sufficiently low redox potential but also where diffusion from the site of production is restricted because of the lack of continuity of air-filled porosity.

However, under these conditions, the root itself becomes stressed, and endogenous production of the gas can also become important (Drew and Lynch, 1980). One response of plants, especially rice, to ethylene is to develop aerenchyma (air spaces in the cortex), although it is not clear whether ethylene per se or anaerobiosis is the trigger for aerenchyma formation. The extra air spaces allow a more rapid movement of air from the shoots to the roots and thereby the amelioration of the anaerobic stress.

James M. Lynch • Forest Research, Alice Holt Lodge, Farnham, Surrey GU10 4LH, United Kingdom.

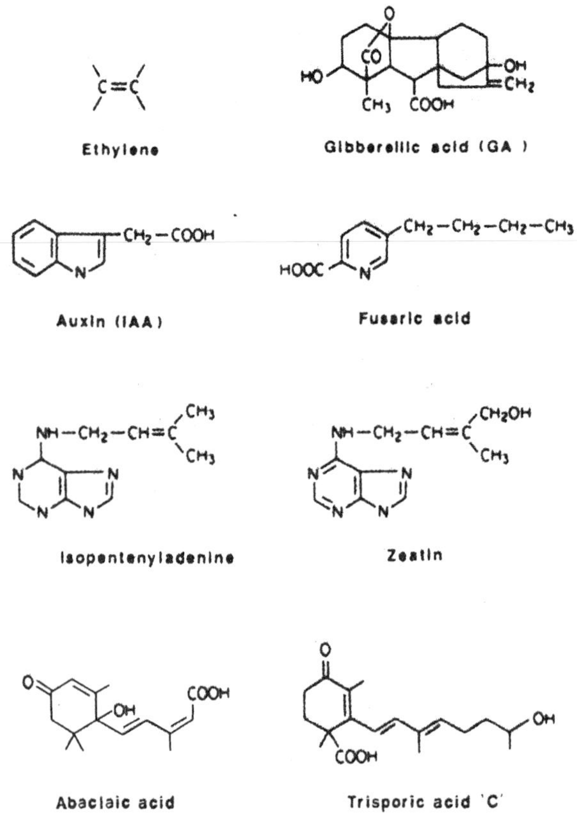

Figure 1. Plant growth regulators.

There has been considerably less study on the microbial formation of the nonvolatile plant growth regulators (gibberellins, auxins, cytokinins, and abscisic acid). However, it should always be remembered that gibberellic acid (GA₃) was the first of the gibberellins to be characterized, and that stemmed from the observation by Kurosawa (1926) in Taiwan that *Gibberella fujikuroi* (an imperfect form of *Fusarium moniliforme*) killed rice plants by causing foot rot after initially causing them to become abnormally tall. The list of gibberellins known to occur in plants now exceeds 50, but few of them have been searched for in root-infecting or free-living rhizosphere organisms. By contrast, auxin is a single structure and is formed readily by a wide range of microorganisms from tryptophan. Even bearing this in mind, however, the identities of auxins and gibberellins as microbial metabolites involved in the regulation of plant growth have been most commonly based on in vitro bioassays.

Cytokinins such as isopentyladenine and zeatin promote cell division in plant tissues. As aminopurines they can link to phosphate esters of sugars to form nucleosides and nucleotides. Whereas there is much evidence that bacteria degrade tRNA, there have been few studies on cytokinin formation per se.

An interesting factor in the pathogenicity of *Agrobacterium tumefaciens*, however, is that oncogene 4 of the wide-host-range plasmid pTiAch5 codes for the enzyme responsible for cytokinin biosynthesis *(cyt)*, and the lack of recognition of certain plants by a particular biotype is mainly due to the absence or inactivity of a *cyt* gene in the T region of the narrow-host-range Ti plasmid (Hoekema et al., 1984). Abscisic acid has been reported as a microbial metabolite of the plant pathogenic fungus *Cercospora rosicola* (Neill et al., 1982). Also, the structurally related trisporic acids are products formed during sexual mating of mucoraceous fungi. The instability of these compounds makes effects on roots difficult to determine (Lynch and White, 1977).

Accepting the principle that most substances can be toxic to plants if supplied at a sufficient dose rate, it follows that the size of accumulated metabolite pools is a critical factor in determining microbial effects in the root region. Microbial phytotoxic metabolites are represented mainly by aliphatic and aromatic organic acids, but also include hydrogen cyanide and hydrogen sulfide. Sublethal doses can sometimes stimulate plant growth, and therefore in this respect they have a growth regulator-like activity.

IONOPHORES

It has been recognized for a long time that bacteria and fungi produce a range of siderophores that chelate iron (Neilands, 1984). Yield increases by inoculation of *Pseudomonas fluorescens* and *Pseudomonas putida* onto potato seed pieces were subsequently attributed to the production of an extracellular siderophore, pseudobactin (Teintze and Leong, 1981). The action was supported by the observations of Kloepper et al. (1980), which showed that addition of ferric iron—Fe(III)—abolished the antagonistic action in vitro. Mutations induced by chemicals or UV light or transposon mutagenesis were deficient in siderophore production and were ineffective as antagonists. Pure pseudobactin could mimic the action of the bacteria.

The great excitement generated by these observations at the time was tempered by the failure to obtain consistent responses to inoculation in field trials. Why should iron chelation be so critical in soil-plant-microbial ecology? Iron is usually present in soil as Fe(OH)₃, the solubility constant of which is $10^{-38.7}$; thus the concentration of soluble iron in soil is usually very low. The concentration of Fe(III) can be greater than 10^{-6} M, for example, in anaerobic or low-pH soils; here the Fe transport system is considered to be sufficient, low-iron-affinity systems only being necessary for ferric ion uptake to occur. At lower concentrations the uptake is specific, and high-affinity uptake systems are necessary to dissolve and transport Fe to microbial cells. This can be achieved by complexing to siderophores. These are large molecules (1,000 to 1,500 Da) and cannot pass normal membrane pores without the special high affinity system. Siderophore production is usually switched on only when the Fe(III) concentration is less than 20 μM, and then its affinity is constant and its binding is heavily pH dependent. Hydroxymate siderophores are products of fungi and bacteria and are widespread in soils, but catecholate siderophores are products only of bacteria (Crowley et al., 1987). Without siderophores there is 1,000-fold decrease in Fe(III) availability per unit pH increase, making it essentially unavailable above a pH of 4. Fe(II) solubility decreases similarly but is also controlled by redox, with the result that, above pH4, Fe(II) is the most available form of soluble inorganic iron. Siderophores also from stable complexes with Cr(III) and the rare Ga(III) (Gascoiogne et al., 1991). Although Al(III) is abundant, the affinity for this and all divalent cations, including Fe(II), is many orders of magnitude below that of Fe(III). No system analogous to the siderophores has been found for any other metal ion, and Fe(III) seems to be unique in requiring such specific ligands. Since the initial studies by Schroth's group in California, several groups investigated the potential of siderophore producers as rhizosphere inoculants. Most study has been by Schippers' group in The Netherlands (Schippers et al., 1987). A range of *Pseudomonas* isolates has been shown to be effective in stimulating potato growth (presumably by controlling minor pathogens), and it seems that siderophore production is a prerequisite for the growth stimulation by *P. putida* WCS358 in the field (Bakker et al., 1986). However, responses are variable; for example, trials between 1981 and 1984 showed only a positive response in 1981, although the baseline yield was highly variable (Geels et al., 1986).

Bakker and Schippers (1987) suggested that PG-PRs were primarily active by outcompeting cyanide-producing rhizobacteria. They showed that 50% of potato rhizosphere pseudomonads produce HCN in vitro and that 5 μM HCN inhibits plant cytochrome oxidase respiration by 40% (again in vitro). Cyanide production by inhibitory rhizobacteria depends on the concentration of Fe(III) in growth media. Growth-promoting bacteria do not produce cyanide in vitro, and their capacity for siderophore production enables them to compete with the inhibitory bacteria. When cyanide inhibits the cytochrome pathway in roots that store energy as ATP, it stimulates the alternative cyanide-resistant pathway, which is less energy effective and can be inhibited by salicylhydroxamic acid. Operation of the latter pathway causes depressed root cell energy metabolism, which in turn could impair nutrient uptake.

Despite the many studies in recent years on siderophore-producing bacteria, no one has yet convincingly identified their role in plant growth. Certainly in crops as diverse as wheat and mushrooms, bacterial siderophore production is not always a prerequisite for disease biocontrol, but there is clearly a need for evaluation of its direct and indirect antibiotic roles. It should also be noted that the effectiveness of the PGPR siderophore-producing bacteria as inoculants has been highly variable. One of the major factors likely to be responsible for this inconsistency is that production and biological activity (Fe competition) of siderophores of the fluorescent pseudomonads are influenced by pH, which of course can vary greatly between soils. Action in the field may be limited to a few cases where the soil pH, or more precisely the rhozosphere pH, is alkaline, as Fe availability is inversely correlated with increased pH (Scher and Baker, 1982; Mishaghi et al., 1988).

Nearly all the effort on rhizosphere ionophore production has concerned siderophores. However, investigation of other ionosphores in the rhizosphere is likely to pay dividends. For example, a manganese-chelating agent was produced by bacteria growing on the roots of solution-grown barley plants, and this promoted Mn uptake by roots (Barber and Lee, 1974).

AZOSPIRILLUM

In studying the microbial associations of tropical grasses in the 1970, Joanna Dobereiner in Brazil identified that dinitrogen-fixing *Azospirillum* spp. could associate with roots to promote growth. Initially it was considered that this might be due to dinitrogen fixation, but studies by many groups showed that the effect was more likely due to hormonelike activity stimulating plant growth. Notable among these groups have been those led by Yacov Okon in Israel and Jos Vanderleyden in Belgium, who have particularly focused on the the molecular biology, and Yoav Bashan, who has published many research papers and reviews (e.g., Bashan and Holguin, 1997). Some commercial inoculants have been produced on a small scale, and they do not seem to have a very specific plant host range. However, the major problem in exploitation has been inconsistency and unpredictability of response.

Five bacterial species are recognized (*Azospirillum amazonense, A. brasilense, A. irakense, A.*

lipoferum, and *A. halopraeferens*). One of the outcomes of inconsistency has been to coinoculate *Azospirillum* with other microorganisms. For example, when *Azospirillum* sp. strain DNG4 was coinoculated with a mixture of cellulolytic fungi, its nitrogenase activity increased 22-fold (Halsall, 1993). Other coinoculants have included *Pseudomonas, Bacillus, Agrobacterium, Klebsiella, Azotobacter,* and *Enterobacter,* together with the symbiotic nodule-forming bacterium *Rhizobium* and the symbiotic vesicular-arbuscular mycorrhizal *Glomus.* As with all rhizosphere inoculants, secure attachment to host root surfaces is essential for a long-term association. The preferred attachment sites are on root hair zones by a two-phase attachment, often with the aid of the polar flagellum.

Under various stress conditions the bacteria are capable of cyst and floc formation, often associated with polyhydroxybutyrate synthesis, and this usually assists survial. The polar flagella can promote chemotaxis to the root, which is vital for the bacteria to encounter the emerging root.

Despite more than 20 years of research, the mode of action of *Azospirillum* is not agreed on. Dinitrogen fixation is probably involved to an extent, as is nitrite production and the production of some modified signal molecules that interfere with plant metabolism. However, most attention has focused on the plant hormones auxin or indole acetic acid (IAA), ethylene, gibberellins, and cytokinins. Probably most attention in recent years, however, has focused on IAA where, by understanding the genetics and biochemistry of the synthesis, it has been possible to construct genetically modified strains with known production levels (i.e., IAA minus, IAA attenuated, and IAA overproducers). The genetically modified bacteria have been tested for growth promotion and nitrogen uptake under field conditions, and certainly the drastic change in root morphology following inoculation is due to auxin production by the bacteria (Okon and Vanderleyden, 1991; Lambrecht et al., 1997).

TRICHODERMA

Of all microbial inoculants, probably the greatest interest has been shown in the fungus *Trichoderma.* This is largely due to its effective deployment as a biocontrol agent, which has been exploited commercially by several companies. Currently the largest producer of *Trichoderma harzianum* T22 is probably Bioworks, Geneva, N.Y. The biocontrol action involves the production of antibiotics and lytic enzymes, and the most successful strains are rhizosphere competent. Two companion volumes have been produced on the fungus (Harman and Kubicek, 1998).

One of the pioneers of *Trichoderma* research was Tex Baker in Colorado; early on he noted that the fungus had the capacity to stimulate plant growth over and above the relief of inhibition of growth generated by pathogen stress (Lindsey and Baker, 1967). It was subsequently shown that *Trichoderma* spp. increased plant growth in a range of bedding and crop plants, including alyssum (*Berteroa incana*), carnation (*Dendranthema grandiflora*), marigold (*Tagetes* sp.), periwinkle (*Vinca* sp.), moss rose (*Portulaca* sp.), petunia (*Petunia* x *hybrida*), snapdragon (*Antirrhinum majus*), cucumber (*Cucumis sativus*), eggplant (*Solanum melongena*), pea (*Pisum sativum*), pepper (*Piper nigrum*), radish (*Raphanus sativus*), tobacco (*Nicotiana* sp.), and tomato (*Lycopersicon esculentum*) (Baker et al., 1984; Chang et al., 1981; Paulitz et al., 1986; Baker, 1988). The increase in plant growth depends on the crop and the form of inoculum, a peat-bran inoculum often being best.

Subsequently, others have demonstrated increases in yield as a result of *Trichoderma* treatment (Harman et al., 1989; Lynch et al., 1991). Depending on the strain, inhibition of growth and germination can occur (Ousley et al., 1993), and the challenge is therefore as always to get consistency of plant growth stimulation. In a series of repeated trials, six *Trichoderma* spp., applied as a diesel powder from a liquid fermentation in molasses/yeast medium, proved to be consistent in promoting the growth of lettuce (*Latrica sativa* L.) seedlings grown in a peat-sand potting compost in the glasshouse (Ousley et al., 1994a) (Table 1). Subsequently it was shown that the best strains could promote the flowering and shoot growth of bedding plants (Ousley et al., 1994b)

Table 1. Effects of *Trichoderma* spp. strains applied at 1% dry (wt/wt) compost on shoot fresh weight[a] of glasshouse lettuce in potting compost (Ousley et al. 1994)

Trichoderma spp. strains	Mean % of control
92	156[b]
20	154[b]
84	152
WT	150[b]
T35	146[b]
93	146
47	143[b]
75	138[b]
38	107
65	91
T1+1	70

[a] Fresh weight of controls, 100%.
[b] Values significantly greater than control ($P = 0.05$).

(Color Plates 10 and 11). Interestingly, in relation to cyanide toxicity referred to in the siderophore strain, it has been shown that *Trichoderma* spp. can catabolize cyanide (Ezzi and Lynch, 2002).

ACC DEAMINASE

Ethylene has a range of effects on plants. Most notably it is the classical inhibitor of root growth in flooded soils, either from endogenous root production, causing arenchyma (air spaces) to form in roots, or from exogenous microbial sources. 1-Aminocyclopropane-1-carboxylate (ACC) is synthesized in roots and transported to plant shoots, where it is converted to ethylene by ACC oxidase. The synthesis of ethylene can be inhibited by the enzyme ACC deaminase. The ACC deaminase has been found in a range of strains of rhizosphere bacteria (*Enterobacter cloacae, Pseudomonas* spp., *Kluyvera ascorbata*) that appear to promote plant growth by inhibiting ethylene stress (Burd et al., 1998, 2000; Grichko and Glick, 2000, 2001a; Li and Glick, 2000; Ma et al., 2001; Shah et al., 1998; Wang et al., 2000). Not only do the plants become flood tolerant, but the destressing effect enables them to accumulate heavy metals and therefore become potential agents of bioremediation.

Transgenic tomato plants have been produced with the bacterial gene, under the transcriptional control of either two tandem 35S cauliflower mosaic virus promoters (constitutive expression), the *rolD* promoter from *Agrobacterium rhizogenes* (root-specific expression), or the pathogenesis related *PRB-1b* promoter from tobacco, to generate flooding tolerance and ability of the plants to accumulate Cd, Co, Cu, Ni, Pb, and Zn (Grichko et al., 2000; Grichko and Glick, 2001b). Thus microbial inoculants or transgenic plants with novel metabolic abilities to enhance bioremediation might become a very exciting new initiative to improve the soil environment.

CONCLUSIONS

Most of this discussion has focused on the use of the relatively small groups of plant growth-promoting agents that have been known for a long time. Many of the stimulatory activities of microorganisms depend on tests that have not been characterized in chemical terms; for example, the *Trichoderma* growth stimulation falls into this category. This is sometimes because there has been no attempt to characterize, but it might also be that the mode of action involves nonestablished growth regulators. An important recent example has been the use of *P. fluorescens* F113, which pro-

duces an antibiotic, 2,4-diacetylphloroglucinol (DAPG), that is effective in controlling plant diseases. We showed that the antibiotic-producing bacterium enhanced nodulation of pea roots by *Rhizobium leguminosarum* in comparison with a non-DAPG-producing deletion mutant (Andrade et al., 1998). Most recently we also showed that only the DAPG producer increased root mass production by more than 50% in all soil types providing that the soil conditions did not limit growth (De Leij et al., 2002). The presence of DAPG was associated with increased root length and root weight and transiently enhanced lateral root formation of the pea plants. It is suggested that DAPG can act as a plant hormone-like substance. It is likely that in the pool of soil microbial biodiversity there will be other microbial secondary metabolites capable of such activity which might be harvested to optimize crop productivity or microbially mediated phytoremediation processes. A new horizon for microbially mediated plant growth regulation has been heralded in the past few years.

REFERENCES

Andrade, G., F. A. A. M. De Leij, and J. M. Lynch. 1998. Plant mediated interactions between *Pseudomonas fluorescens, Rhizobium leguminosarum* and arbuscular mycorrhiza. *Lett. Appl. Microbiol.* 26:311–316.

Baker, A. W., and B. Schippers. 1987. Microbial cyanide production in the rhizosphere in relation to potato yield reduction and *Pseudomonas* spp.—mediated plant growth stimulation. *Soil Biol. Biochem.* 19:451–457.

Baker, R. 1988. *Trichoderma* spp. as plant growth stimulants. *CRC Crit. Rev. Biotechnol.* 7:97–106.

Baker, R., E. Elad, and I. Chet. 1984. The controlled experiment in scientific method with special emphasis in biological control. *Phytopathology* 74:1019–1021.

Bakker, P. A. H. M., J. G. Lamers, A. W. Bakker, J. D. Marugg, P. J. Weisbeek, and B. Schippers. 1986. The role of siderophores in potato tuber yield increase by *Pseudomonas putida* in a short rotation of potato. *Neth. J. Plant Path.* 92:249–256.

Barber, D. A., and R. B. Lee. 1974. The effect of micro-organisms on the absorption of manganese by plants. *New Phytol.* 73:47–106.

Basham, Y., and G. Holguin. 1997. *Azospirillum*-plant relationships environmental and physiological advances (1990–1996). *Can. J. Microbiol.* 43:103–121.

Burd, G. I., D. G. Dixon, and B. R. Glick. 1998. Plant growth-promoting bacteria that decrease heavy metal toxicity in plants. *Can. J. Microbiol.* 46:237–245.

Chang, C., Y. Chang, R. Baker, O. Kleifield, and I. Chet. 1986. Increased plant growth in the presence of *Trichoderma harziarium*. *Plant Dis.* 70:145–148.

Crowley, D. E., C. P. P. Reid, and P. J. Szaniszlo. 1987. Microbial siderophores as iron sources for plants, p. 375–386. *In* G. Winkelmann, R. van den Helm, and J. Neilands (eds.), *Iron Transport in Microbes, Plants and Animals* VCH Verlagsgesellschaft mbH, Weinheim, Germany.

De Leij, F. A. A. M., J. E. Dixon-Hardy, and J. M. Lynch. 2002. Effect of 2,4-diacetylphloroglucinol producing and non-producing strains of *Pseudomonas fluorescens* on root development of pea seedlings in three different soil types and its effect on nodulations by *Rhizobium. Biol. Fert. Soils* 35:114–121.

Drew, M. C., and J. M. Lynch. 1980. Soil anaerobiosis, micro-organisms and root function. *Annu. Rev. Phytopathol.* 18:37–67.

Ezzi, M. I., and J. M. Lynch. Cyanide catabolizing enzymes in *Trichoderma* spp. *Enzyme Microb. Technol.* 31:1042–1047.

Gascoigne, D., J. A. Connor, and A. T. Bull. 1991. Capacity of siderphore-producing alkalophilic bacteria foaccomulate iron, gallium and alluminium. *Appl. Microbiol. Biotechnol.* 36:136–141.

Geels, F. P., J. G. Lamers, O. Hoekstra, and B. Schippers. 1986. Potato plant responses to seed tuber bacterization in the field in various rotations. *Neth. J. Plant Pathol.* 92:257–272.

Grichko, V. P., and B. R. Glick. 2000. Identification of DNA sequences that regulate the expression of the *Enterobacter cloacae* UW4 1-aminocylopropane-1-carboxylic acid deaminase gene. *Can. J. Microbiol.* 46:1159–1165.

Grichko, V. P., and B. R. Glick. 2001a. Amelioration of flooding stress by ACC deaminase containing plant growth-promoting bacteria. *Plant Physiol. Biochem.* 39:11–17.

Grichko, V. P., and B. R. Glick. 2001b. Flooding tolerance of transgenic tomato plants expressing the bacterial enzyme ACC deaminase controlled by the *35S, rol*D or *PRB-lb* promoter. *Plant Physiol. Biochem.* 39:19–25.

Grichko, V. P., B. Filby, and B. R. Glick. 2000. Increased ability of transgenic plants expressing the bacterial enzyme ACC deaminase to accumulate Cd, Co, Cu, Ni, Pb and Zn. *J. Bacteriol.* 81:45–53.

Halsall, D. M. 1993. Inoculation of wheat straw to enhance lignocellulose breakdown and associated nitrogenase activity. *Soil Biol. Biochem.* 25:419–421.

Harman, G. E., and C. P. Kubiceck. 1998. *Trichoderma and Glioderdium,* vol. 1 and 2. Taylor & Francis, London, United Kingdom.

Harman, G. E., A. G. Taylor, and T. E. Stasz. 1989. Combining effective strains of *Trichoderma harziarium* and solid matrix priming to improve biological seed treatments. *Plant Dis.* 73:631–637.

Hiltner, L. 1904. Uber neuere Erfahrungen und Problem auf dem Gebiet der Bodenbakteriologie und unter besonder Berucksichtigung der Grundungung und Brache. *Arb. Dtsch, Landwirt. Ges.* 98:59–78.

Hoekema, A., B. S. de Pater, A. J. Fellinger, P. J. J. Hooykass, and R. A. Schilperoort. 1984. The limited host range of an *Agrobacterium tumefaciens* strain extended by a wide host range T-region. *EMBO J.* 3:3043–3047.

Kloepper, J. W., J. Leong, M. Teintze, and M. N. Schroth. 1980. Enhanced plant growth by siderophores produced by plant growth-promoting rhizobacteria. *Nature* 286:885–886.

Kurosawa, E. 1926. Experimental studies on the secretaion of *Fusarium heterosporum* on rice plants. *J. Nat. Hist. Soc. Formosa* 16:213–227.

Lambrecht, M., Y. Okon, A. Vande Broek, and J. Vanderleyden. 2000. Indole-3-acetic acid, a reciprocal signalling molecule in bacteria-plant interaction. *Trends Microbiol.* 8:298–300.

Li, J., and B. R. Glick. 2001. Transcriptional regulation of the *Enterobacter cloacae* UW4 1-aminocylopropane-1-carboxylate (ACC) deaminase gene (acds). *Can. J. Microbiol.* 47:359–367.

Lindsey, D. L., and R. Baker. 1967. Effect of certain fungi on dwarf tomatoes grown under gnotobiotic conditions. *Phytopathology* 78:1262–1263.

Lynch, J. M. (ed.). 1990. *The Rhizosphere.* Wiley, Chichester, United Kingdom.

Lynch, J. M., and N. White. 1997. Effect of some non-pathogenic microorganisms on the growth of gnotobiotic barley plants. *Plant Soil* 47:161–174.

Lynch, J. M., K. L. Wilson, M. A. Ousley, and J. M. Whipps. 1991. Response of lettuce to *Trichoderma* treatment. *Lett. Appl. Microbiol.* 12:59–61.

Ma, W., K. Zalec, and B. R. Glick. 2001. Biological activity and colonization patterns of the bioluminescence-labelled plant growth-promoting bacterium *Kluyvera ascorbata* SUD 165/26. *FEMS Microbiol. Ecol.* 35:137–144.

Mishaghi, I. J., M. W. Olsen, P. J. Cotty, and C. R. Donndelinger. 1988. Fluorescent siderophore-mediated iron deprivation—a contingent biological control mechanism. *Soil Biol. Biochem.* 20:573–574.

Neilands, J. B. 1984. Siderophores of bacteria and fungi. *Microbiol. Sci.* 1:9–14.

Neill, S. J., R. Horgan, D. C. Walton, and T. S. Lee. 1982. The biosynthesis of abscisic acid in *Cercospora rosicola. Phytochemistry* 21:61–65.

Okon, Y., and J. Vanderleyden. 1997. Root-associated *Azospirillum* species can stimulate plants. *ASM News* 63:366–370.

Ousley, M. A., J. M. Lynch, and J. M. Whipps. 1993. Effect of *Trichoderma* on plant growth: a balance between inhibition and growth promotion. *Microb. Ecol.* 26:277–285.

Ousley, M. A., J. M. Lynch, and J. M. Whipps. 1994a. Potential of *Trichoderma* app. as consistent plant growth simulator. *Biol. Fertil. Soils* 17:85–90.

Ousley, M. A., J. M. Lynch, and J. M. Whipps. 1994b. The effects of *Trichoderma* inocula on flowering shoot growth of bedding plants. *Sci. Hort.* 59:147–155.

Paulitz, T., M. Windham, and R. Baker. 1986. Effect of peat: vermiculite mixes containing *Trichoderma harziarium* on increased growth response of radish. *J. Am. Soc. Hort. Sci.* III:810–816.

Scher, F. M., and R. Baker. 1982. Effect of *Pseudomonas putida* and a synthetic iron chelator in induction of suppressiveness to *Fusarium* wilt pathogens. *Phytopathology* 72:1567–1573.

Schippers, B., B. Lugtenberg, and P. J. Weisbeek. 1987. Plant growth control by fluorescent pseudomonads, p. 19–39. *In* I. Chet (ed.), *Innovative Approaches to Plant Disease Control.* John Wiley & Sons, New York, N.Y.

Shah, S, J. Li., B. A. Moffat, and B. R. Glick. 1998. Isolation and characterisation of ACC deaminase genes from two different plant growth-promoting rhizobacteria. *Can. J. Microbiol.* 44:833–843.

Teintze, M., and J. Leong. 1981. Structure of pseudobactin A, a second siderophore from plant growth promoting *Pseudomonas* B10. *Biochemistry* 20:6446–6457.

Wang, C., E. Knill, B. R. Glick, and G. Defago. 2000. Effect of transferring 1-amiocylopropane-1-carboxylic (ACC) deaminase gene into *Pseudomonas fluorescens* strain CHAO and its gacA derivative CHA96 on their growth-promoting and disease-suppressive capacities. *Can. J. Microbiol.* 46:898–907.

Microbial Diversity and Bioprospecting
Edited by Alan T. Bull
© 2004 ASM Press, Washington, D.C.

Chapter 35

Biotreatment

LINDA LOUISE BLACKALL AND CHRISTINE YEATES

Biotreatment covers a broad field and it likely has different definitions in the view of different individuals. To set the scene for this chapter, we use the following definition for biotreatment: all biological (focusing on microbiological) treatment processes, including bioremediation, that detoxify or degrade wastes (including solids, liquids, and gases) before they enter the environment. Typically, naturally occurring whole microorganisms (Iranzo et al., 2001) are exploited, but on occasion macroscopic organisms including plants, fungi, and algae are included; in some cases appropriate enzymes to degrade specific compounds are used, and in specific cases genetically manipulated whole organisms are employed. Mixed communities of microorganisms rather than pure or single cultures are standard when the technology requires whole organisms (Hamer, 1997). The spectrum of wastes to be biotreated is extensive, ranging from seemingly innocuous compounds like nutrients (nitrogen, phosphorus, carbon) through recalcitrant paleobiochemical compounds (e.g., polychlorinated biphenyls) to anthropogenic chemicals or xenobiotics (e.g., tetrachloroethene), which are chemically synthesized compounds that do not occur naturally (Holliger, 1995). Because some xenobiotics are structurally similar to naturally occurring compounds, organisms capable of degradation of the structural analogue can also degrade the xenobiotic. Very often, there is a mixture of different compounds in the waste stream, making the choice of a treatment strategy complicated (e.g., Hamer, 1997; Kumar et al., 2000). There are several different types of professionals engaged in the biotreatment field. A nonexhaustive list would include biologists (several different types), biotechnologists, chemists, engineers (several different types), modelers, bioinformaticists, health professionals, environmental regulators, lawyers, patent attorneys, financial investors and entrepreneurs, and economists. This is clearly a field where one must be multidisciplinary. Collections of these specialists

should work collaboratively toward a common goal of biotreatment to achieve environmental sustainability. This latter term is a fairly contemporary one. It incorporates more than simply biotreating wastes by also encompassing cleaner production and reuse and recycling and, as such, integrates many individuals who are not biotreatment professionals. For example, national and global environmental programs called "Clean up Australia" and "Clean up the World," respectively, were instigated and now headed by Ian Kiernan, a builder and yachtsman by profession, but philanthropically, an environmental crusader. In this chapter we do not focus on the cleaner production and recycling topics, but it must be remembered that they will play a substantial role in the overall discipline (e.g., Bull, 2001, and Chapter 1 in this book; Hamer, 1997). The title of this book (*Microbial Diversity and Bioprospecting*) sets the focus for this chapter. We cover some aspects of biotreatment, then briefly discuss different biotreatment topics in relation to the book title, and conclude with a detailed case study.

BIOTREATMENT—SOME GENERALIZATIONS

The biotreatment of wastes can occur in many different configurations. The location of the waste and its transportability play major roles in dictating whether it will be treated in situ or ex situ, but the majority of controlled treatments are likely to occur ex situ and thus involve the transport of the waste to a treatment facility. Nevertheless, there are many examples of in situ biotreatment of contaminated or polluted groundwaters, marine waters, coastal environments, and soils, and commonly, bioremediation would be done in situ. The infrastructure for the transport of some wastes such as urban-generated wastes, including municipal solid waste or domestic wastewater, is a major cost consideration in the

Linda Louise Blackall and Christine Yeates • Advanced Wastewater Management Centre, The University of Queensland, St. Lucia, Queensland 4072, Australia.

biotreatment industry but a necessary component of modern communities in developed countries. Once a city is created with these infrastructure components, it is difficult to modify, and if, for example, a city lacks a sewerage system, the cost of its construction can be prohibitive. Ex situ treatment typically occurs in open or closed bioreactors of many different configurations and employs many different physicochemical influences. The phase in which the waste is located (solid, liquid, or gas) has an effect on the type of treatment and on the physical parameters such as mass transfer of pollutant and nutrients to the biodegrading organisms. Bioreactor design is a major field of biotreatment (Deshusses et al., 1997; Grady and Filipe, 2000). For example, in systems using microorganisms, they can be grown in complex communities in suspension as flocs (e.g., 50 to 500 μm in diameter) or granules (500 μm to several millimeters in diameter) or they can be grown as biofilms on inert substrata. Solid wastes treated in traditional landfills are special cases where the microorganisms mediating the degradation grow as biofilms but the biofilm substratum is actually biodegradable and not inert. In any case, the wastes for treatment by the microorganisms must be presented to the degraders in aqueous media so that the microorganisms can access them. Solid wastes can also be suspended in aqueous media and treated as slurries in either suspended or biofilm systems. Gaseous wastes (Kennes and Thalasso, 1998) can be collected and treated in biofilm systems (e.g., Attaway et al., 2001) or in bioscrubber liquid systems (e.g., Burgess et al., 2001). Wastes that are or can be dissolved in aqueous form are potentially treatable by many different treatment strategies.

The physicochemical conditions employed in bioreactors have a substantial influence on the types of microorganisms selected and on their expressed phenotype or physiology. The residence time of the organisms in the bioreactor (by flow rate control) and aeration status (aerobic, anoxic, or anaerobic) are two parameters that in general can be manipulated by the bioreactor operator, and both of these have a profound influence on the types of microorganisms that can prevail. Residence time affects the time the organisms have to degrade or detoxify the waste, and some bioprocesses can occur only under specific conditions of aeration. For example, microorganisms capable of aerobic nitrification (collectively ammonia oxidized to nitrate) are relatively slow growers (generation time in the order of several days), and, if the hydraulic residence time of a bioreactor is shorter than the generation time of the nitrifiers, they will not be able to grow and reproduce fast enough to survive in the system (Bernal et al., 2000). In another example, the anammox process is mediated by strict anaerobes (Jetten et al., 2001). If the redox potential of the water is too high, the activity of these organisms will be negligible or nonexistent.

Temperature is one parameter that only occasionally is capable of being modified. There are certainly technologies that are operated at elevated temperatures (e.g., Mason et al., 1992), but if the waste to be treated is massive in volume, it is unlikely that temperature can be modulated effectively.

BIOTREATMENT—PRACTICALITY IN MOTION

There are many different technologies used in biotreatment from classical to procedures that are still under development. We now focus on the approaches from the biotreatment industry and some of the limitations of biotreatment. The biotreatments covered in this chapter presume the employment of microorganisms. In many cases, complex communities of different microorganisms are present, and the biotreatment operator uses various strategies in attempts to select the correct types of microorganisms such that the treatment objectives are met. This strategy applies to both in situ and ex situ biotreatment operations. In general, this process of selection of suitable strategies precluded the involvement of microbiologists, but process operators could consult textbooks on microbial physiology and they did so to design the processes. Alternatively, processes were developed in a completely ad hoc manner and by serendipity, and on occasion, they actually worked.

A good example of the approach of the process operator designing a system for biotreatment was in the removal of nitrogen from domestic wastewaters by a combination of nitrification and denitrification. From chemical analyses, process engineers knew that the major form of nitrogen in this waste stream was ammonia, and the ammonia levels were far beyond that which could be accommodated in biomass of microorganisms as they grew since domestic wastewater is typically carbon limited. Thus, during the biotreatment process, the carbon would be removed by growth of microorganisms, but ammonia would still remain in solution in the effluent water. Nitrifiers are aerobic, autotrophic organisms comprised of ammonia-oxidizing bacteria (AOB), which produce nitrite, and nitrite-oxidizing bacteria (NOB), which produce nitrate. The treatment environment at the domestic wastewater treatment plants could easily facilitate nitrifier growth conditions—nitrifiers simply require aerobic conditions and the wastewater would provide the ammonia. However, if the wastewater was mixed with the biotreatment microbial communities and subjected to aerobic conditions, heterotrophic microorganisms from the microbial communities would also grow and aerobically utilize the carbon compounds in solution. Therefore, if aerobic-only conditions were employed, carbon would be removed by heterotrophic microbial metabolism (end products would be CO_2 and more-heterotrophic biomass) and ammonia would be oxidized to nitrate by a combination of AOB and NOB.

In practice, however, this does not seem to work. Likely it is related to the three-dimensional growth forms of wastewater microbial communities (called flocs, about 50 to 500 μm in diameter) creating substrate diffusional boundaries and heterotrophs outcompeting the nitrifiers for oxygen. Nitrate, like ammonia, is environmentally damaging in that it can promote eutrophic conditions in effluent receiving bodies. A mechanism was therefore required to remove the formed nitrate. This could be facilitated by passing this nitrate-rich wastewater through a gently mixed zone such that little or no oxygen would be brought into solution and denitrifiers (nitrate to N_2) could grow. Most denitrifiers require conditions devoid of oxygen, i.e., anaerobic. However, with nitrate or nitrite present as a terminal electron acceptor, and oxygen absent, the conditions are called "anoxic." The question was, what carbon source would the heterotrophic denitrifying bacteria use since the organic carbon from the wastewater was already used by aerobic heterotrophs. The challenge was to design a process that could exploit nitrification and denitrification for ammonia metabolism to N_2 but which could also use the carbon from the influent wastewater for the anoxic denitrifiers. A system was developed (Fig. 1) in which the carbon was capable of being used by the denitrifiers in the created anoxic conditions and in which the nitrifiers produced the required nitrate for denitrifiers. The process works quite well for nitrogen removal from domestic wastewaters, but the nitrogen is not capable of easy recycling. To recover the nitrogen from the atmosphere is an energetically demanding process.

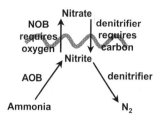

Figure 2. Diagram of nitrification (left arrows) and denitrification (right arrows). Shaded squiggle line shows pathways that could be eliminated and save energy and carbon for biotreatment processes.

One improvement in the process of nitrification to nitrate and then denitrification is to bypass two of the steps (Fig. 2). This shortcut will attain

- 25% reduction in aeration requirements,
- 40% reduction in carbon requirements, and
- 50% reduction in the size of the denitrification tank (Turk and Mavinic, 1989).

To do this reproducibly, the bioprocess operator needs to have a good understanding of the growth properties of the nitrifiers and the denitrifiers such as their growth rates, substrate affinities, and the effect of physicochemical parameters on their growth. As can be seen in Fig. 2, there will be competition for nitrite between NOB and denitrifiers capable of using nitrite as their terminal electron acceptor. The microbiological knowledge of the process is too elementary as yet but provides an interesting challenge for collaborative research.

BIOTREATMENT—SOME NOVEL BIODIVERSITY DISCOVERED

Several novel microorganisms have been discovered in the biotreatment industry, and many of them are as yet not in pure culture. Many of these novel organisms were found to be responsible for biochemical transformations that were unknown before the discovery of the organisms. A selection of some would include the following: *Dehalococcoides ethenogenes*, a tetrachloroethene (a xenobiotic compound contaminating many groundwaters of the developed world) dechlorinator (Maymó-Gatell et al., 1997); "*Candidatus* Brocadia anammoxidans" (Jetten et al., 2001) and "*Candidatus* Kuenenia stuttgartiensis" (Schmid et al., 2000), *Planctomycetales* bacteria responsible for anaerobic ammonium oxidation or the anammox process for combined removal of ammonia and nitrite, and *Nitrospira* sp. (Burrell et al., 1998; Juretschko et al., 1998), which was found to be a major NOB in wastewater treatment plants, contrary to the common belief that the organism responsible for ni-

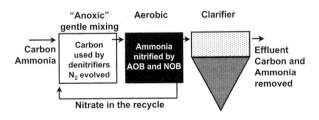

Figure 1. Schematic diagram of the configuration of wastewater treatment plants designed for complete nitrogen removal by pre-denitrification.

trite oxidation was *Nitrobacter* sp. Several filamentous organisms from the bacterial phylum *Chloroflexi* are abundant in activated sludge flocs but their role is as yet obscure (Björnsson et al., 2002; Juretschko et al., 2002).

A plethora of studies have explored the microbial community structures of different biotreatment systems where novel biodiversity is a common theme. Many different molecular inventorying approaches (detailed elsewhere in this book) have been applied. Typically, when several different studies of a specific type of biotreatment system are compared, the sorts of microorganisms, including novel ones, as gleaned from rRNA gene sequences, are quite similar. However, the different biotreatment processes collectively accommodate an extremely wide spectrum of the diversity of microorganisms.

BIOTREATMENT—A DETAILED CASE STUDY

Enhanced Biological Phosphorus Removal: The Process

Arguably one of the largest and most important biotreatment industries is that of wastewater treatment (Keller et al., 2002), yet the complete understanding of many parts of the processes involved remains elusive. The biological removal of phosphorus from wastewater is an example of one such process. The release of phosphorus (P) into the environment can stimulate the growth of cyanobacteria, resulting in eutrophication of global waterways. Wastewater effluent is one of the leading sources of nutrients such as phosphorus, resulting in high demands on treatment processes to remove nutrients before release. Enhanced biological phosphorus removal (EBPR) is a widely applied process to facilitate the removal of P from wastewater via microbial activity. The process involves the circulation of activated sludge through anaerobic and aerobic phases, with the introduction of influent wastewater during the anaerobic period. Although EBPR has widespread application, the process does regularly fail for unknown reasons. The fundamental microbial processes behind EBPR are not well understood, therefore remedial actions are based on conjecture.

To mimic the cycling seen in full-scale treatment processes, laboratory-scale sequencing batch reactors are used. Studies in sequencing batch reactors have led to the description of biochemical models describing the processes associated with EBPR. These models are based on the observed chemical transformations and data generated from the analyses of complex activated sludge biomass, assuming that the transformations occur within a certain type of bacterial cell. The supporting microbiological data for these models are currently lacking. The key to the success of EBPR was the serendipitous selection for organisms capable of utilizing carbon from the influent wastewater under anaerobic conditions. The organisms capable of this are instrumental in the process of EBPR and have been called polyphosphate-accumulating organisms (PAOs). Because there are no pure cultures of PAOs, their biochemical mechanisms have been modeled based on chemical analysis, and a summary of one well-endorsed biochemical model for PAOs is shown in Fig. 3A. During the anaerobic phase, PAOs hydrolyze their intracellular reserves of polyphosphate (polyP) to provide energy to uptake volatile fatty acids (VFAs) from the influent, which are then stored in the form of polyhydroxyalkanoates (PHA) and glycogen. During the aerobic phase, the PAOs replenish their polyP by utilizing their stores of PHA and glycogen, with the uptake of orthophosphate from the influent. Removal of biomass at this stage leads to the net removal of phosphorus from the system. Other organisms subsequently discovered in laboratory-scale EBPR sludges (Liu et al., 1994) can also sequester VFAs under anaerobic conditions, but they utilize intracellular glycogen as their sole energy source for anaerobic processes. These glycogen-accumulating organisms (GAOs) carry out similar carbon transformations as PAOs, but do not participate in the P transformations (Fig. 3B). Like PAOs, GAOs have not been obtained in pure culture despite several attempts. GAOs are thought to be detrimental to EBPR because they purloin some of the sparingly available VFAs required by PAOs, thus leading to reduced P removal performance The ability to stably remove P via EBPR is severely limited by the lack of microbiological evidence to confirm the biochemical models, such that it is impossible to predict success or failure of EBPR.

The Microbiology of EBPR

The source of the organisms that are key to the process of EBPR is activated sludge. Activated sludge can be described as a flocculent suspended growth culture of microorganisms (Grady et al., 1999), and as such is a complex microbial community. The inhabitants of activated sludge include bacteria, viruses, protozoa, fungi, algae, and metazoa (Seviour, 1999). The organisms responsible for the P-removal phenotype (Fig. 3A) in EBPR constitute a small part of this community in full-scale systems (Crocetti et al., 2000), yet play a significant role in the biotreatment process.

Although in no way a quantitative representation of the microbial community, the 16S rRNA gene clone library approach has been useful in confirming

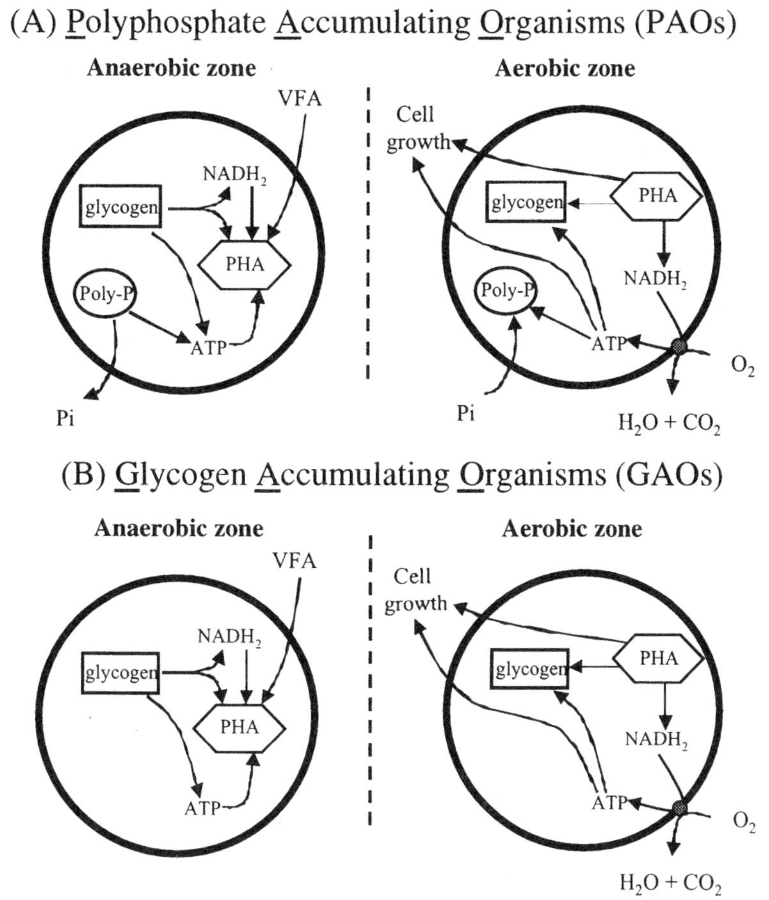

(A) Polyphosphate Accumulating Organisms (PAOs)

(B) Glycogen Accumulating Organisms (GAOs)

Figure 3. Diagrams showing the hypothesized transformations of major components of (A) PAOs and (B) GAOs in EBPR.

the diverse nature of the microbial community within EBPR biomass. In the first 16S rRNA gene clone libraries generated from laboratory-scale EBPR sludges, the community composition was dominated with clones from the *Betaproteobacteria*, followed by the *Alphaproteobacteria*, *Planctomycetales* phylum, *Cytophagales* phylum, and the *Gammaproteobacteria* (Bond et al., 1995).

The Key Microbial Players

The key microorganisms responsible for the process of EBPR are the PAOs. Culture-based studies dominated the first attempts to describe these organisms, which appeared in initial microscopic observations of EBPR biomass from the aerobic zone as clusters of cells that stained positively for polyphosphate by the Neisser staining protocol (e.g., Duncan et al., 1988). When culture methods were used, a prevalent isolate comprised coccobacillus clusters, which were identified as belonging to *Acinetobacter* species (Fuhs and Chen, 1975). However, cultures of *Acinetobacter* did not show all the characteristics of typical EBPR

sludges (van Loosdrecht et al., 1997). The application of culture-independent methods, such as 16S rDNA clone libraries and fluorescent in situ hybridization (FISH), revealed that *Acinetobacter* existed in low abundance in EBPR biomasses (Bond et al., 1995; Wagner et al., 1994). It is now widely accepted that *Acinetobacter* is not a PAO responsible for the phenotype associated with EBPR (Fig. 3A). The reliance on culture-based methods incorrectly represented microbial diversity within EBPR biomass, and this is true for many other biotreatment systems.

The construction of 16S rRNA gene libraries from laboratory-scale EBPR reactors led to the identification of PAOs as bacteria closely related to *Rhodocyclus* in the *Betaproteobacteria*, with the consequent design of rRNA-targeting FISH probes making visualization, quantitation, and determination of the spatial distribution of PAOs in laboratory-scale and full-scale EBPR biomasses possible (Crocetti et al., 2000; Hesselmann et al., 1999). The development of post-FISH chemical staining techniques for polyP and PHA enabled linking of the identified PAOs to aspects of their anaerobic and aerobic phenotype

(Crocetti et al., 2000). The PAOs have now been definitively identified as gram-negative bacteria, related to *Rhodocyclus* (Crocetti et al., 2000) and named "*Candidatus* Accumulibacter phosphatis" (Hesselmann et al., 1999), but their physiology remains unknown.

The unusual ability of PAOs to sequester VFAs and utilize polyP as their energy source under anaerobic conditions is responsible for their selection over other microorganisms. In full-scale EBPR processes, VFAs are sparingly available, and their supply for EBPR in full-scale processes generally needs to be promoted. The only identified GAO is a *Gammaproteobacteria* strain called "*Candidatus* Competibacter phosphatis" (Crocetti et al., 2002). Although GAOs are considered detrimental to EBPR, the triggering factors for their proliferation and why and how they can compete with PAOs like "*Candidatus* Accumulibacter phosphatis" are unknown. Although PAOs and GAOs are instrumental to EBPR, questions remain about their physiology and the means by which they carry out the transformations observed during EBPR and summarized in Fig. 3.

Toward Improved Bioprocess Control

Knowledge of the biochemistry of the PAOs and GAOs involved in EBPR would likely lead to process modifications to promote reliable P removal. Although the PAOs and GAOs have now been taxonomically identified, virtually nothing is known of the genetics or biochemistry of these organisms beyond their 16S rRNA gene sequence. The impediment to progress on this front has been the inability to culture these organisms in the laboratory. This is a common obstacle when examining environmental organisms in biotreatment processes. The importance of developing techniques to promote the culturing of such organisms should not be neglected. However, much research is now proceeding with examining environmental organisms at a molecular genetic level. The development of DNA extraction methods for environmental samples coupled with the power of the technique of PCR has enabled the detection and analysis of metabolic genes in environmental samples (Yeates et al., 2000). To investigate genes present within EBPR biomass, McMahon et al. (2002) designed PCR primers to target the gene-encoding polyP kinase (PPK), an enzyme responsible for the synthesis and degradation of polyP. The PPK genes detected in EBPR biomass were similar to those from *Rhodocyclus tenuis* (McMahon et al., 2002); however, direct linking of these genes to PAOs within the sludge is a more difficult task. A recent approach to obtain genetic information from environmental organisms involves the generation of large insert libraries in fosmids (Stein et al., 1996) or in bacterial artificial chromosomes (Béjà et al., 2000b). Large insert libraries have provided useful advances in microbial ecology, including the description of genes linked to identifiable rRNA genes from marine organisms (Stein et al., 1996), the expression of cloned genes from uncultured soil microbes in *E. coli* (Rondon et al., 2000), the detection of natural products (Brady et al., 2001), and the isolation of whole operons from microbial consortia (Entcheva et al., 2001). A large insert library generated from the marine environment enabled the detection and expression of a bacterial rhodopsin gene from an organism belonging to the SAR86 cluster of the *Gammaproteobacteria*, a group of organisms abundant in the marine environment with an unknown physiology until now (Béjà et al., 2000a). These results provide promise for the examination of microbial communities essential for biotreatment and in particular EBPR wastewater treatment systems, with a vision to understanding the biochemical pathways underlying this process.

With the DNA information being generated from PCR-based studies and the large insert library projects, the question arises as to the level of expression of these genes in the systems being studied. The development of microarrays containing rRNA genes (Guschin et al., 1997) or functional genes (Wu et al., 2001) will lead to better methods to describe communities and determine the physiological state of the organisms. The next step after examining gene transcription is to establish gene expression at the level of proteins. This has been tested in activated sludge by investigating the effect of exposure to toxic compounds (Bott and Love, 2001) and by the use of two-dimensional gel electrophoresis to determine the effect of substrate composition (Huber et al., 1998). Studies at the molecular level have provided significant advances in our understanding of microbial ecology, and now the area of wastewater treatment remains to be prospected using these techniques.

REFERENCES

Attaway, H., C. H. Gooding, and M. G. Schmidt. 2001. Biodegradation of BTEX vapors in a silicone membrane bioreactor system. *J. Indust. Microbiol. Biotechnol.* **26**:316–325.

Béjà, O., L. Aravind, E. V. Koonin, M. T. Suzuki, A. Hadd, L. P. Nguyen, S. Jovanovich, C. M. Gates, R. A. Feldman, J. L. Spudich, E. N. Spudich, and E. F. DeLong. 2000a. Bacterial rhodopsin: evidence for a new type of phototrophy in the sea. *Science* **289**:1902–1906.

Béjà, O., M. T. Suzuki, E. V. Koonin, L. Aravind, A. Hadd, L. P. Nguyen, R. Villacorta, M. Amjadi, C. Garrigues, S. B. Jovanovich, R. A. Feldman, and E. F. DeLong. 2000b. Construction and analysis of bacterial artificial chromosome libraries from a marine microbial assemblage. *Environ. Microbiol.* 2:516–529.

Bernal, M. A., B. O. Gonzalez, and M. S. Gonzalez. 2000. Nutrient removal and sludge age in a sequencing batch reactor. *Bioprocess. Eng.* 23:41–45.

Björnsson, L., P. Hugenholtz, G. W. Tyson, and L. L. Blackall. 2002. Filamentous *Chloroflexi* (green non-sulfur bacteria) are abundant in wastewater treatment processes with biological nutrient removal. *Microbiology* 148:2309–2318.

Bond, P. L., P. Hugenholtz, J. Keller, and L. L. Blackall. 1995. Bacterial community structures of phosphate-removing and non-phosphate-removing activated sludges from sequencing batch reactors. *Appl. Environ. Microbiol.* 61:1910–1916.

Bott, C. B., and N. G. Love. 2001. The immunochemical detection of stress proteins in activated sludge exposed to toxic chemicals. *Water Res.* 35:91–100.

Brady, S. F., C. J. Chao, J. Handelsman, and J. Clardy. 2001. Cloning and heterologous expression of a natural product biosynthetic gene cluster from eDNA. *Org. Lett.* 3:1981–1984.

Bull, A. T. 2001. Biotechnology for industrial sustainability. *Korean J. Chem. Eng.* 18:137–148.

Burgess, J. E., S. A. Parsons, and R. M. Stuetz. 2001. Developments in odour control and waste gas treatment biotechnology: a review. *Biotechnol. Adv.* 19:35–63.

Burrell, P. C., J. Keller, and L. L. Blackall. 1998. Microbiology of a nitrite-oxidizing bioreactor. *Appl. Environ. Microbiol.* 64:1878–1883.

Crocetti, G. R., P. Hugenholtz, P. L. Bond, A. Schuler, J. Keller, D. Jenkins, and L. L. Blackall. 2000. Identification of polyphosphate-accumulating organisms and design of 16S rRNA-directed probes for their detection and quantitation. *Appl. Environ. Microbiol.* 66:1175–1182.

Crocetti, G. R., J. F. Banfield, J. Keller, P. L. Bond, and L. L. Blackall. 2002. Glycogen accumulating organisms in laboratory-scale and full-scale activated sludge processes. *Microbiology* 148:3353–3364.

Deshusses, M. A., W. Chen, A. Mulchandani, and I. J. Dunn. 1997. Innovative bioreactors. *Curr. Opin. Biotechnol.* 8:165–168.

Duncan, A., G. E. Vasiliadis, R. C. Bayly, J. W. May, and W. G. C. Raper. 1988. Genospecies of *Acinetobacter* isolated from activated sludge showing enhanced removal of phosphate during pilot-scale treatment of sewage. *Biotech. Lett.* 10:831–836.

Entcheva, P., W. Liebl, A. Johann, T. Hartsch, and W. R. Streit. 2001. Direct cloning from enrichment cultures, a reliable strategy for isolation of complete operons and genes from microbial consortia. *Appl. Environ. Microbiol.* 67:89–99.

Fuhs, G. W., and M. Chen. 1975. Microbiological basis of phosphate removal in the activated sludge process for the treatment of wastewater. *Microbial. Ecol.* 2:119–138.

Grady, C. P. L., and C. D. M. Filipe. 2000. Ecological engineering of bioreactors for wastewater treatment. *Water Air Soil Poll.* 123:117–132.

Grady, C. P. L., G. T. Daigger, and H. C. Lim. 1999. *Biological Wastewater Treatment.* Marcel Dekker, Inc., New York, N.Y.

Guschin, D. U., B. K. Mobarry, E. Proudnikov, D. A. Stahl, B. E. Rittmann, and A. D. Mirzabekov. 1997. Oligonucleotide microchips as genosensors for determinative and environmental studies in microbiology. *Appl. Environ. Microbiol.* 63:2397–2402.

Hamer, G. 1997. Microbial consortia for multiple pollutant biodegradation. *Pure Appl. Chem.* 69:2343–2356.

Hesselmann, R. P. X., C. Werlen, D. Hahn, J. R. van der Meer, and A. J. B. Zehnder. 1999. Enrichment, phylogenetic analysis and detection of a bacterium that performs enhanced biological phosphate removal in activated sludge. *Syst. Appl. Microbiol.* 22:454–465.

Holliger, C. 1995. The anaerobic microbiology and biotreatment of chlorinated ethenes. *Curr. Opin. Biotechnol.* 6:347–351.

Huber, S., S. Minnebusch, S. Wuertz, P. A. Wilderer, and B. Helmreich. 1998. Impact of different substrates on biomass protein composition during wastewater treatment investigated by two-dimensional electrophoresis. *Water Sci. Technol.* 37:363–366.

Iranzo, M., I. Sainz-Pardo, R. Boluda, J. Sanchez, and S. Mormeneo. 2001. The use of microorganisms in environmental remediation. *Ann. Microbiol.* 51:135–143.

Jetten, M. S. M., M. Wagner, J. Fuerst, M. van Loosdrecht, G. Kuenen, and M. Strous. 2001. Microbiology and application of the anaerobic ammonium oxidation ("anammox") process. *Curr. Opin. Biotechnol.* 12:283–288.

Juretschko, S., G. Timmermann, M. Schmid, K.-H. Schleifer, A. Pommerening-Röser, H.-P. Koops, and M. Wagner. 1998. Combined molecular and conventional analyses of nitrifying bacterium diversity in activated sludge: *Nitrococcus mobilis* and *Nitrospira*-like bacteria as dominant populations. *Appl. Environ. Microbiol.* 64:3042–3051.

Juretschko, S., A. Loy, A. Lehner, and M. Wagner. 2002. The microbial community composition of a nitrifying-denitrifying activated sludge from an industrial sewage treatment plant analyzed by the full-cycle rRNA approach. *Syst. Appl. Microbiol.* 25:84–99.

Keller, J., Z. Yuan, and L. L. Blackall. 2002. Integrating process engineering and microbiology tools to advance activated sludge wastewater treatment research and development. *Rev. Environ. Sci. Biotechnol.* 1:83–97.

Kennes, C., and F. Thalasso. 1998. Waste gas biotreatment technology. *J. Chem. Technol. Biotechnol.* 72:303–319.

Kumar, M. S., A. N. Vaidya, N. Shivaraman, and A. S. Bal. 2000. Biotreatment of oil-bearing coke-oven wastewater in fixed-film reactor: a viable alternative to activated sludge process. *Environ. Eng. Sci.* 17:221–226.

Liu, W.-T., T. Mino, K. Nakamura, and T. Matsuo. 1994. Role of glycogen in acetate uptake and polyhydroxyalkanoate synthesis in anaerobic-aerobic activated sludge with a minimized polyphosphate content. *J. Ferment. Bioeng.* 5:535–540.

Mason, C. A., A. Haner, and G. Hamer. 1992. Aerobic thermophilic waste sludge treatment. *Water Sci. Technol.* 25:113–118.

Maymó-Gatell, X., Y. Chien, J. M. Gossett, and S. H. Zinder. 1997. Isolation of a bacterium that reductively dechlorinates tetrachloroethene to ethene. *Science* 276:1568–1571.

McMahon, K. D., D. Jenkins, and J. D. Keasling. 2002. Polyphosphate kinase genes from activated siduge carrying out enhanced biological phosphorus removal. *Water Sci. Technol.* 46:155–162.

Rondon, M. R., P. R. August, A. D. Bettermann, S. F. Brady, T. H. Grossman, M. R. Liles, K. A. Loiacono, B. A. Lynch, I. A. MacNeil, C. Minor, C. L. Tiong, M. Gilman, M. S. Osburne, J. Clardy, J. Handelsman, and R. M. Goodman. 2000. Cloning the soil metagenome: a strategy for accessing the genetic and functional diversity of uncultured microorganisms. *Appl. Environ. Microbiol.* 66:2541–2547.

Schmid, M., U. Twachtmann, M. Klein, M. Strous, S. Juretschko, M. S. M. Jetten, J. W. Metzger, K.-H. Schleifer, and M. Wagner. 2000. Molecular evidence for genus-level diversity of bacteria capable of catalyzing anaerobic ammonium oxidation. *Syst. Appl. Microbiol.* 23:93–106.

Seviour, R. J. 1999. The normal microbial communities of activated sludge plants, p. 76–98. *In* R. J. Seviour and L. L. Blackall (ed.), *The Microbiology of Activated Sludge.* Kluwer Academic Publishers, London, United Kingdom.

Stein, J. L., T. L. Marsh, K. J. Wu, H. Shizuya, and E. F. DeLong. 1996. Characterization of uncultivated prokaryotes: isolation and analysis of a 40-kilobase-pair genome fragment from a planktonic archaeon. *J. Bacteriol.* **178:**591–599.

Turk, O., and D. S. Mavinic. 1989. Stability of nitrite build-up in an activated sludge system. *J. Water Pollut. Control Fed.* **61:**1440–1448.

van Loosdrecht, M. C. M., G. J. Smolders, T. Kuba, and J. J. Heijnen. 1997. Metabolism of microorganisms responsible for enhanced biological phosphorus removal from wastewater. *Antonie Leeuwenhoek* **71:**109–116.

Wagner, M., R. Erhart, W. Manz, R. Amann, H. Lemmer, D. Wedi, and K.-H. Schleifer. 1994. Development of an rRNA-targeted oligonucleotide probe specific for the genus *Acinetobacter* and its application for in situ monitoring in activated sludge. *Appl. Environ. Microbiol.* **60:**792–800.

Wu, L., D. K. Thompson, G. Li, R. A. Hurt, J. M. Tiedje, and J. Zhou. 2001. Development and evaluation of functional gene arrays for detection of selected genes in the environment. *Appl. Environ. Microbiol.* **67:**5780–5790.

Yeates, C., A. J. Holmes, and M. R. Gillings. 2000. Novel forms of ring-hydroxylating dioxygenases are widespread in pristine and contaminated soils. *Environ. Microbiol.* **2:**644–653.

Microbial Diversity and Bioprospecting
Edited by Alan T. Bull
© 2004 ASM Press, Washington, D.C.

Chapter 36

Bioprospecting Novel Antifoulants and Anti-Biofilm Agents from Microbes

Carola Holmström, Peter Steinberg, and Staffan Kjelleberg

Surfaces immersed in aqueous environments are rapidly colonized by living organisms. This process is known as biofouling, and it is a major problem for marine and aquatic industries. Fouling on the surfaces of ships or other submerged structures results in corrosion, a decrease in hydrodynamic efficiency, transport of introduced pests, and many other problems worldwide. Fouling is estimated to cost the marine shipping industry >$5 billion per year, and high costs from biofouling also occur in other marine and aquatic industries, e.g., aquaculture, offshore oil production, and heat exchangers used for cooling coastal power plants (Clare, 1996). Fouling is also a major problem for other marine organisms, with fouling of sponges, seaweeds, and other living marine surfaces causing effects on the hosts ranging from decreased gas exchange to death (Kushmaro et al., 2001; Littler and Littler, 1995). Fouling generally occurs in a predictable sequence, with colonization by bacteria and biofilm formation happening in hours to days followed by colonization of higher organisms such as barnacles, tube worms, and algae (Henschel and Cook, 1990; Wahl, 1989). The number of propagules of all fouling organisms in a milliliter of seawater typically exceeds 10^6, so fouling pressure is intense in most waters.

Other than repeated cleaning of surfaces, by far the most common commercial approach to fouling control is to coat surfaces with antifouling paints that slowly release toxic compounds to the surface, deterring initial colonization of the surface by fouling organisms or killing newly settled foulers. The most common such paints use heavy metals (copper or tin) as the killing agents, although organic biocides (isothiazalones, pyrethiones) are also common, typically as cobiocides with copper or tin. The main problems

with such coatings are the nontarget environmental effects of the heavy metals released from the coatings. For this reason, tributyltin-based coatings are in the process of being banned by the International Maritime Organisation, with a complete restriction on the use of tributyltin-based paints scheduled for 2008 (Brady, 2000).

Two major alternatives to heavy metal-based paints have been proposed. The first, which are commercially available, are the so-called "nonstick" or foul-release coatings. These paints are most commonly based on silicone elastomers or fluoropolymers (Brady, 2000). Because they rely on water motion to remove weakly attached organisms from the surface, they are not appropriate for many uses. Also, because of boundary layer effects and the small vertical scale of biofilms, these paints may be less effective against bacterial colonization and biofilms than against higher organisms.

The second major class of alternatives to heavy metal-based paints are organic compounds, either synthetic or naturally derived. The later category includes the so-called natural antifoulants (see Clare [1996] and Rittschof [2001] for reviews), which are metabolites derived from marine organisms. The logic behind the use of natural antifoulants (de Nys and Steinberg, 2002; Rittschof, 2001) in antifouling coatings is based on several premises: (i) they have evolved for the specific function of antifouling: (ii) because producing organisms cannot in general coat themselves with toxic metabolites, the compounds may have specific, nontoxic modes of action; and (iii) as natural carbon sources, they should be biodegradable. We note that in the literature the term natural antifoulants has often been used to describe marine natural products ("secondary metabolites") of any

Carola Holmström and Staffan Kjelleberg • School of Biotechnology and Biomolecular Sciences and Centre for Marine Biofouling and Bio-Innovation, University of New South Wales, Sydney 2052, Australia. **Peter Steinberg** • School of Biological Science and Centre for Marine Biofouling and Bio-Innovation, University of New South Wales, Sydney 2052, Australia.

origin or function that happens to have some effect against a fouling organism in the laboratory, and thus premises (i) and (ii) do not necessarily apply to many putative natural antifoulants in the literature.

MODULATION OF SURFACE COLONIZATION BY BACTERIA

The focus on natural antifoulants to date has primarily been on marine invertebrates and algae (Clare, 1996; Rittschof, 2001). However, it is now clear that (i) microbial biofilms are ubiquitous on submerged natural surfaces and (ii) these communities release chemical signals that mediate colonization—either positively or negatively—of surfaces by other organisms (Holmström and Kjelleberg, 2000; Maki, 1999; Wieczorek and Todd, 1997). For example, the cosmopolitan marine fouling polychaete *Hydroides elegans* settles only in the presence of biofilms (Hadfield et al., 1994; Lau et al., 2002), and signals emanating from biofilms act as positive cues for settlement for a variety of other marine invertebrates or algae (Maki et al., 1989; Paul et al., 1997; Szewzyk et al., 1991). Of perhaps most interest to the topic of this review, biofilms can also inhibit settlement, either as a function of general properties of the biofilm such as age (Neal and Yule, 1994; Wieczorek and Todd, 1997), individual bacterial strains (Egan et al., 2001a; Holmström, et al., 1992; Maki et al., 1988; Mary et al., 1993) or, in a few examples, specific chemical inhibitors of settlement (Holmström and Kjelleberg, 2000). Below we highlight several case studies of inhibitory effects of bacteria on marine organisms. We also address methodological limitations in the research of marine microbial defenses and the progress that is being made in this research field.

Production of Antifouling Compounds by Members of the Genus *Pseudoalteromonas*

The genus *Pseudoalteromonas* is common in the marine environment and is often isolated from living surfaces, as demonstrated by both traditional culturing methods (Egan et al., 2000; Hentschel et al., 2001; Holmström et al., 1996; Ivanova et al., 1998a; Lemos et al., 1985) and culture-independent techniques such as denaturing gradient gel electrophoresis (Skovhus et al., 2001). In various screening studies performed to assess antifouling activities of marine bacterial isolates, *Pseudoalteromonas* species have repeatedly been found to express the strongest inhibitory activities among the tested strains (Egan et al., 2000; Hentschel et al., 2001; Holmström et al., 1996; Lemos et al., 1985). For example, Holmström

et al (1996) found that out of 24 dark pigmented strains the most inhibitory strains against invertebrate larvae were two species of the genus *Pseudoalteromonas* (Egan et al., 2000), *P. ulvae* (Egan et al., 2001a), and *P. tunicata* (Holmström et al., 1998).

Interestingly, there appears to be a strong correlation between pigment(s) production and antifouling activity by the bacteria (Holmström et al., 1996; Lemos et al., 1985). This has been further investigated by generating *P. tunicata* transposon mutants with different pigmentation (Egan et al., 2002a). A decrease in pigmentation correlated with a reduced production of antifouling activity and white mutants were found to have lost all their inhibitory activities. One of the white mutants contained a transposon insertion in a gene that encodes a ToxR-like protein (Egan et al., 2002b) suggesting that this gene can coordinately control the expression of pigment production and antifouling activity. ToxR and ToxR homologous proteins are often used by bacteria to sense changes in the environment and coordinately control the expression of several gene products in adaptive responses (Lin et al., 1993; Murley et al., 1999). We hypothesize that this might be a mechanism for *P. tunicata* to sense conditions in the biofilm on living surfaces and regulate the expression of antifouling compounds.

The antifouling activity of *P. tunicata* has been studied in detail. This bacterium has been isolated from various living surfaces such as a tunicate (Holmström et al., 1992), a green alga (Egan et al., 2000) and from a phytoplankton bloom (Skerrett, 2001). Later studies have demonstrated that *P. tunicata* produces specific compounds against invertebrate larvae, algal spores, fungi, bacteria, and diatoms (reviewed in Holmström and Kjelleberg, 2000). To determine whether this broad range of inhibition of fouling organisms is unique for *P. tunicata* or if other members of this genus also demonstrate a similar range of inhibitory activities, 10 *Pseudoalteromonas* species were tested in various biofouling assays (Holmström et al., 2002). As a general trend, pigmented rather than nonpigmented *Pseudoalteromonas* cells demonstrated a stronger activity against the tested fouling organisms. The dark green *P. tunicata* exhibited the strongest and also the broadest antifouling activity by inhibiting all tested organisms. The results showed that *P. tunicata* was active against different classes of organisms whereas some of the *Pseudoalteromonas* species employed a species-specific target activity (Holmström et al., 2002).

These studies indicate that members of the *Pseudoalteromonas* genus produce a broad range of inhibitory compounds that are active against both eukaryotic and prokaryotic organisms. The antilarval compound produced by *P. tunicata* is a polar compound less then 500 Da in size (Holmström et al., 1992), and the antialgal compound is a peptide approximately 3 kDa in size (Egan et al., 2001b). The mode of action of these compounds appears to be cidal rather than inhibitory (Holmström et al., 2002). The antibacterial compound is a novel protein, 190 kDa in size (James et al., 1996). This compound is the largest known antibacterial protein and inhibits the growth of most bacterial strains tested to date (Holmström et al., 2002; James, et al., 1996). Based on both molecular and chemical-based approaches, the antifungal compound has been identified to be a novel yellow pigment that inhibits the growth of both yeast and fungal strains at very low concentrations (Franks, 1998).

The expression of pigments in bacteria may be regulated by nutrient availability and composition (Gauthier, 1976, 1979; Gauthier and Flatau, 1976). For example, P. tunicata grown on a rich medium were nonpigmented and did not produce any antifouling activity. The finding that nutrients can control the expression of antifouling activity may speak to events in the interaction between host organisms and bacteria. We have suggested that marine host organisms that lack their own chemical defense mechanisms may be colonized by antifouling-producing bacteria, such as Pseudoalteromonas species (Kjelleberg and Steinberg, 2002). These bacteria may defend the host against further colonization through the production of bioactive compound(s). Such an in situ expression of antifouling activity may be triggered by nutrients available to the bacteria at the surface of the host organisms or by other bugs.

Self-control mechanisms also exist among bacteria. Interestingly, the three most active Pseudoalteromonas species against fouling organisms are also autoinhibitory (Holmström et al., 2002). Such autoinhibitory activity may maintain the bacterial diversity in various habitats, controlling the effect by which successful bacterial colonists form biofilm. The autotoxic compound in P. tunicata is the novel 190 kDa protein (James, 1996). The P. tunicata cells become less sensitive to the autoinhibitory compound at a later stage of growth. This has been demonstrated for both free-living cells and cells in biofilms.

Many of the Pseudoalteromonas species that display a broad range of antifouling activity have originally been isolated from different algae. These are P. tunicata (Egan et al., 2000), P. ulvae (Egan et al., 2001a), P. aurantia, P. rubra, P. citrea, P. luteoviolacea (Gauthier and Breittmayer, 1992), and isolates identified to the Pseudoalteromonas spp. level in various screenings studies. It is observed that marine plants and animals are often host to different microorganisms compared to the surrounding seawater

(Maximilien et al., 1998; Wahl et al., 1994). These findings indicate that colonization of bacteria on seaweeds is not a randomly occurring fouling process. As mentioned above, we suggest that some living surfaces may be colonized by antifoulant-producing bacteria that can help in protecting the host organisms from biofouling. For example, the green alga Ulva lactuca and the tunicate Ciona intestinalis often remain free of macrofouling in the field but are not known to produce deterrent secondary metabolites. Interestingly, both eukaryotes are hosts for P. tunicata. Additional members isolated from U. lactuca are P. ulvae and P. aurantia, which also produce antifouling compounds (Holmström et al., 2002).

In a study by Harder and Qian (2000) it was demonstrated that compounds from the green alga U. reticulata could inhibit the attachment and metamorphosis of the polychaete Hydroides elegans. They further cultured bacterial strains from this alga and tested them in antifouling bioassay. Inhibitory strains against the settlement of H. elegans were found, and the most inhibitory strain belonged to the genus Vibrio. Other genera represented on the surface of the seaweed were Altermononas and Pseudoalteromonas (Dobretsov and Qian, 2002). Importantly, these genera do not just colonize chemically undefended hosts. Their association with higher organisms may also reflect the occurrence of more specific chemical interactions between the bacteria and the host.

Other Case Studies of Marine Bacteria on Living Surfaces

Surface-dwelling bacteria isolated from living surfaces, and in particular bacteria isolated from various algal surfaces, were found to be highly active in inhibiting the settlement of invertebrate larvae (Holmström et al., 1996). Seventy percent of the bacteria isolated from algal surfaces inhibited the settlement of invertebrate larvae whereas 30% of the bacteria isolated from marine animal surfaces and only 10% of bacterial isolates from rock surfaces demonstrated inhibitory activity. Boyd et al. (1999) found that 21% of epiphytic marine bacteria exhibited antimicrobial activity. Many of these strains secreted repellents that influenced the colonization behavior of a fouling bacterium. It was speculated that both growth inhibition and repellents influencing the colonization behavior of competing bacteria might be a potential way for epiphytic bacteria to control the microbial population on the seaweed surfaces.

Burgess et al. (1999) also screened for the production of antimicrobials by using surface-associated bacteria isolated from seaweeds and invertebrates. They found that over 25% of the bacteria from a common starfish were able to produce antimicrobial compounds while isolates from sea slug displayed no activity. Enhanced antimicrobial activity was observed when cell-free supernatants from other strains were added to the different isolates. Interestingly a seaweed isolate that did not display any activity under normal test conditions demonstrated antimicrobial activity after exposure of supernatants from other marine epibiotic bacteria. This cross-species induction of activity was also demonstrated by Mearns-Spragg et al. (1998) and suggests that chemical interactions between different species might control the production of antimicrobial compounds. Armstrong et al. (2001) further demonstrated that bacteria isolated from seaweeds produce a stronger activity when grown on a surface compared with growth in suspension. A modification of surface-induced activity was also demonstrated by Ivanova et al. (1998b). They found that hydrophilic surfaces induced the highest antimicrobial activity despite the fact that *Pseudoalteromonas* cells were more abundant on hydrophobic surfaces.

These findings demonstrate that the interaction between bacteria and host organisms is very complex and suggest that there are various ways host organisms or cooccurring bacteria may influence the production of antifouling metabolites by associated bacteria. A change in the secretion of nutrients by the host may induce or switch off bacterial production of active compounds. This could also affect the hydrophobicity of the surface and as a consequence the production of antifouling compounds. A drastic change in the hydrophobicity might also change the species composition of bacteria on a surface given that the ability of bacteria to attach to a surface is fundamentally affected by its physical properties (Loosdrecht et al., 1989). This may thus affect the production of antifouling compounds given that cross-species induction of activity occurs (Burgess et al., 1999).

Technology Development

As with all natural antifouling technologies, the step from biology and compound identification to the development of an antifouling coating is a substantial one. Several approaches to this challenge have been made with bacteria. Armstrong et al. (2000) explored the incorporation of active bacterial compounds and culture supernatants from two different bacterial isolates previously shown to have growth inhibitory activity in antimicrobial assays. The bioactive compounds were incorporated into a water-based paint resin, Revacryl 380 Tm and tested for antimicrobial activity. The result demonstrated that both cell and supernatant extract from one of the two strains were significantly better than the tested controls after 4 days incubation.

Our laboratory tested a "living paint" in which the inhibitory bacterium *P. tunicata* was immobilized in a polyacrylamide gel (Gatenholm et al., 1995). This was the first report that demonstrated that bacteria could be immobilized into a hydrogel and be viable and produce antifouling compounds at concentrations high enough to kill barnacle larvae. In a further study we demonstrated an improved technique to prolong the life span of the immobilized hydrogel (Holmström et al., 2000). A 10% polyvinylalcohol gel was found to provide the optimal matrix. The immobilized bacteria maintained their viability for as long as 2 months and the life span of these gels was increased to more than 2 months by incorporation of small beads.

An additional technological approach, used in the aquaculture industry, is that of using bacteria as biological control agents—probiotics—against pathogens and fouling. For example, in hatchery systems of the Chilean scallop *Argopecten purpuratus*, mortalities of larvae can be as high as 90%. This has been attributed to the presence of biofilms of pathogenic marine bacteria such as the *Vibrio* species *V. anguillarum*, *V. parahaemolyticus*, and *V. splendidus*. Riquelme et al. (1997) isolated a *Vibro* sp. C33 from seawater used in the mass culture of scallops that produced bactericidal substances against the *Vibrio* pathogens as well as other bacteria. By adding *Pseudomonas* sp. 11 and *Bacillus* sp. B2 to scallop larvae, a completion of the pelagic larval phase without the use of antibiotics could be demonstrated (Riquelme et al., 2001). Additional bacterial strains associated with the scallop that demonstrated beneficial effects were identified as *Pseudoalteromonas haloplanktis* (Riquelme et al., 1996). Other studies have also have reported the beneficial effects obtained following the addition of *Pseudoalteromonas* (previously *Alteromonas*) species in aquaculture (Gil-Turnes et al., 1989; Ruiz et al., 1996; Tanasomwang et al., 1998; Uchida et al., 1997).

Many bacteria have species-specific effects against germination of algal spores and settlement of invertebrate larvae. This distinction of activity might be a particularly important criterion for the choice of developing effective antifoulants in aquaculture, where some organisms need to deter against settlement and other enhance colonization. The optimal choice may also be a mixed culture of different antifouling-producing bacterial strains. Also, the selection of bacterial strains from biofilms that are indigenous to the site of application avoids the introduction of exotic bacterial strains and agents to the system.

EXPANDED APPROACHES

Although research on marine microbial defenses is progressing, to date we have clearly explored only a fraction of the chemical diversity of marine microbes. This is because (i) the biomass of individual strains is generally too low to directly extract and characterize metabolites from field populations, (ii) the widely cited problems with culturing bacteria from the environment, (iii) the difficulty in getting bacteria to express metabolites even when culturable, and (iv) the difficulties of characterizing nonculturable strains. Progress to address these difficulties is being made along a number of fronts.

Several approaches are currently employed to enhance the culturability of cells. These include careful selection of media constituents (Bruns et al., 2002) and extinction and dilution procedures for removal of competitors from the culture medium (Connon and Giovannoni, 2002). This high-throughput procedure by dilution of natural microbial communities recently allowed for the successful isolation of representatives of the SAR11 clade, which for years were considered to be unculturable species (Rappe et al., 2002). Coculturing of strains also has a strong effect on the production of metabolites (Burgess et al., 1999).

Techniques that allow for a characterization of nonculturable species allow for a more targeted approach to mining specific taxa for metabolites. These techniques are also fundamental in understanding ecological interactions and may greatly facilitate both the identification of metabolites that mediate those interactions and the characterization of bacterial communities on surfaces that have been observed to deter settlement of fouling organisms. The increasing use of molecular-based tools for community analysis of prokaryotes (reviewed by Dahllöf, 2002) is rapidly generating information on the composition of bacterial assemblages on marine surfaces. Commonly used approaches include clone libraries that often are complemented with restriction fragment length polymorphism analyses, a fingerprinting method used to facilitate the selection of clones to sequence (Holmes et al., 2001). Other fingerprinting methods include terminal restriction fragment length polymorphism (Marsh, 1999) and denaturing gradient gel elecrophoresis (Dahllöf et al., 2000; Muyzer et al., 1993). These methods separate bacterial species based on size differences and differences in nucleotide sequences, respectively. They are both rapid methods that offer an overview of the microbial diversity in a community, and large numbers of samples can be processed in a short period of time. A more quantitative method for community analysis is to use fluorescence in situ hybridization based on RNA-targeted oligonucleotide probes that have been developed for identification of individual microbial cells (Wagner and Loy, 2002). Several studies have also used fluorescence in situ hybridization together with microautoradiography (Lee et al., 1999), which allows for in situ identification and provides information on substrate utilization in complex microbial communities (Daims et al., 2001; Nielsen and Nielsen, 2001). Other methods for community-wide analyses use DNA microarrays in which template genes are probed with DNA or PCR products from the environmental sample (Zhou and Thompson, 2002).

An important recent alternative to traditional chemical approaches to mining for active metabolites is the use of bacterial artificial chromosome (BAC) libraries (Beja et al., 2000). Here, large inserts of environmental DNA from all bacteria within an assemblage (culturable or not) are cloned into BAC vectors that can maintain very large DNA inserts stable in *Escherichia coli*. The *E. coli* BAC system can then be directly screened for expression of different biological activities (Ball and Trevors, 2002). Although not yet used for detection of genes involved in the production of antifouling metabolites, BAC libraries have been used to search for genes encoding nonribosomal peptides. Such libraries will most probably be the basis in the future to link phylogenetic and functional information from uncultured microorganisms in various environments (see Handelsman, Chapter 11, this book).

Many of the issues raised above for novel means of identifying new antifoulants from marine microorganisms are also relevant for identifying novel metabolites for other uses. In this respect, it is useful to understand some of the commercial realities of the antifouling industry versus that of, for example, the pharmaceutical industry. Total sales of marine paint are approximately $1 billion per year. This contrasts with the pharmaceutical industry, in which a single "blockbuster" drug would exceed this amount. In addition, current actives such as heavy metals on organic biocides range in price from several dollars per ton to ~$50/kg, and this has important effects on the pricing of coatings. In contrast, the actual price of the active metabolite in a drug is usually irrelevant to the price of the drug. These issues necessarily strongly influence the suitability of novel metabolites for the antifouling paint market and also limit the capacity of the industry to support research in this area. Yet new technologies are needed in combating biofouling on artificial surfaces, and only by a thorough understanding of the biological system can new products and management procedures be successfully developed.

Acknowledgments. The contributions made by members of the authors' laboratories are gratefully acknowledged. The work was founded by the Australian Research Council, the Centre for Marine Biofouling and Bio-Innovation, and the University of New South Wales.

REFERENCES

Armstrong, E., K. G. Boyd, A. Piscane, C. J. Peppiatt, and J. G. Burgess. 2000. Marine microbial natural products in antifouling coatings. *Biofouling* **16**:215–224.

Armstrong, E., L. Yan, K. G. Boyd, P. C. Wright, and J. G. Burgess. 2001. The symbiotic role of marine microbes on living surfaces. *Hydrobiologia* **461**:37–40.

Ball, K. D., and J. T. Trevors. 2002. Bacterial genomics: the use of DNA microarrays and bacterial artificial chromosomes. *J. Microbiol. Methods* **49**:275–284.

Beja, O., M. T. Suzuki, E. V. Koonin, L. Aravind, A. Hadd, L. P. Nguyen, R. Villacorta, M. Amjadi, C. Carrignes, S. B. Jovanovich, R. A. Feldman, and E. F. DeLong. 2000. Construction and analysis of bacterial artificial chromosome libraries from a marine microbial assemblage. *Environ. Microbiol.* **2**:516–529.

Boyd, K. G., D. R. Adams, J. G. Burgess. 1999. Antibacterial and repellent activities of marine bacteria associated with algal surfaces. *Biofouling* **14**:227–236.

Brady, R. F. 2000. Clean hulls without poisons: devising and testing nontoxic marine coatings. *Tech. Articles* **72**:45–56.

Bruns, A., H. Cypionka, and J. Overmann. 2002. Cyclic AMP and acyl homoserine lactones increase the cultivation efficiency of heterotrophic bacteria from Central baltic sea. *Appl. Environ. Microbiol.* **68**:3978–3987.

Burgess, J. G., E. M. Jordan, M. Bregu, A. Mearns-Spragg, and K. G. Boyd. 1999. Microbial antagonism: a neglected avenue of natural products research. *J. Biotechnol.* **70**:27–32.

Clare, A. S. 1996. Marine natural product antifoulants: status and potential. *Biofouling* **9**:211–229.

Connon, S. A., and S. J. Giovannoni. 2002. High-throughput methods for culturing microorganisms in very-low-nutrient media yield diverse new marine isolates. *Appl. Environ. Microbiol.* **68**:3878–3885.

Dahllöf, I. 2002. Molecular community analysis of microbial diversity. *Curr. Opin. Biotechnol.* **13**:213–217.

Dahllöf, I., H. Baillie, and S. Kjelleberg. 2000. rpoB-based microbial community analysis avoids limitations inherent in 16S rRNA gene intraspecies hetergeneity. *Appl. Environ. Microbiol.* **66**:3376–3380.

Daims, H., J. L. Nielsen, P. H. Nielsen, K. H. Schleifer, and M. Wagner. 2001. *In situ* characterisation of *Nitrospira*-like nitrite-oxidising bacteria active in wastewater treatment plants. *Appl. Environ. Microbiol.* **67**:5273–5284.

de Nys, R., and P. Steinberg. 2002. Linking marine biology and biotechnology. *Curr. Opin. Biotechnol.* **13**:244–248.

Dobretsov, S., and P. Y. Qian. 2002. Effect of bacteria associated with the green alga *Ulva reticulata* on marine micro- and macrofouling. *Biofouling* **18**:217–228.

Egan, S., T. Thomas, C. Holmström, and S. Kjelleberg. 2000. Phylogenetic relationship and antifouling activity of bacterial epiphytes from the marine alga *Ulva lactuca*. *Environ. Microbiol.* **2**:343–347.

Egan, S., C. Holmström, and S. Kjelleberg. 2001a. *Pseudoalteromomas ulvae* sp., nov a bacterium with antifouling activities isolated from the surface of a marine alga. *Int. J. Syst. Bacteriol.* **51**:1499–1504.

Egan, S., S. James, C. Holmström, and S. Kjelleberg. 2001b. Inhi-
bition of algal spore germination by the marine bacterium *Pseudoalteromonas tunicata*. *FEMS Microbiol. Ecol.* **35**:67–73.

Egan, S., S. James, C. Holmström, and S. Kjelleberg. 2002a. Correlation between pigmentation and antifouling compounds produced by *Pseudoalteromonas tunicata*. *Environ. Microbiol.* **4**:433–442.

Egan, S., S. James, and S. Kjelleberg. 2002b. Identification and characterization of a putative transcriptional regulator controlling the expression of fouling inhibitors in *Pseudoalteromonas tunicata*. *Appl. Environ. Microbiol.* **68**:372–378.

Franks, A. 1998. An investigation into the antifungal properties of *Pseudoalteromonas tunicata*. Honours thesis, University of New South Wales, Sydney, Australia.

Gatenholm, P., C., Holmström, J. S. Maki, and S. Kjelleberg. 1995. Toward biological antifouling surface coatings: marine bacteria immobilized in hydrogel coatings. *Biofouling* **8**:293–301.

Gauthier, M. J. 1976. Morphological, physiological, and biochemical characteristics of some violet-pigmented bacteria isolated from seawater. *Can. J. Microbiol.* **22**:138–149.

Gauthier, M. J. 1979. *Alteromonas rubra* sp.nov., a new marine antibiotic-producing bacterium. *Int. J. Syst. Bacteriol.* **26**:459–466.

Gauthier, M. J., and V. A. Breittmayer. 1992. The genera *Alteromonas* and *Marinomonas*, the prokaryotes, p. 3046–3070. *In* A. Balows (ed.), *A Handbook on the Biology of Bacteria: Ecophysiology, Isolation, Identification, Applications*, vol. III. Springer, Berlin, Germany.

Gauthier, M. J., and G. N. Flatau. 1976. Antibacterial activity of marine violet-pigmented *Alteromonas* with special reference to the production of brominated compounds. *Can. J. Microbiol.* **22**:1612–1619.

Gil-Turnes, M. S., M. E. Hay, and W. Fenical. 1989. Symbiotic marine bacteria defend crustacean embryos from a pathogenic fungus. *Science* **240**:116–118.

Hadfield, M. G., C. Unabia, C. M. Smith, and T. M. Michael. 1994. *Settlement Preferences of the Ubiquitous Fouler Hydraides elegans*. A. A. Balkema, Rotterdam, The Netherlands.

Harder, T., and P. Y. Qian. 2000. Waterborne compounds from the green seaweed *Ulva reticulata* as inhibitive ones for larval attachment and metamorphosis in the polychaete *Hydroides elegans*. *Biofouling* **16**:205–214.

Hentschel, J. R., and P. A. Cook. 1990. The development of a marine fouling community in relation to the primary film of microorganisms. *Biofouling* **2**:1–11.

Hentschel, U., M. Schmid, M. Wagner, L. Fieseler, C. Gernert, and J. Hacker. 2001. Isolation and phylogenetic analysis of bacteria with antimicrobial activities from the mediterranean sponges *Aplysina aerophoba* and *Aplysina cavernicola*. *FEMS Microbiol. Ecol.* **35**:305–312.

Holmes, A. J., N. A. Tujnia, M. Holley, A. Contos, J. M. James, B. Rogers, and M. R. Gillings. 2001. Phylogenetic structure of unusual aquatic microbial formations in Nullarbor caves, Australia. *Environ. Microbiol.* **3**:256–264.

Holmström, C., and S. Kjelleberg. 2000. Bacterial interactions with marine fouling organisms, p. 101–117. *In* L. V. Evans (ed.), *Biofilms: Recent Advances in Their Study and Control* Overseas Publishing Associates (UK) Ltd., Amsterdam, The Netherlands.

Holmström, C., D. Rittschof, and S. Kjelleberg. 1992. Inhibition of settlement by larvae of *Balanus amphitrite* and *Ciona intestinalis* by a surface-colonizing marine bacterium. *Appl. Environ. Microbiol.* **58**:2111–2115.

Holmström, C., S. James, S. Egan, and S. Kjelleberg. 1996. Inhibition of common fouling organisms by pigmented marine bacterial isolates. *Biofouling* **10**:251–259.

Holmström, C., S. James, B. A. Nellan, D. C. White, and S. Kjelleberg. 1998. *Pseudoalteromonas tunicata* sp. nov., a bacterium that produces antifouling agents. *Int. J. Syst. Bacteriol.* **48:**1205–1212.

Holmström, C., P. Steinberg, V. Christov, and S. Kjelleberg. 2000. Bacteria immobilized into hydrogels: a novel concept to prevent development of biofouling communities. *Biofouling* **15:**109–117.

Holmström, C., S. Egan, A. Franks, S. McCload, and S. Kjelleberg. 2002. Antifouling activity expressed by *Pseudoalteramonas* species. *FEMS Microbiol. Ecol.* **41:**47–58.

Ivanova, E. P., E. A. Kiprianova, V. V. Mikhailov, G. F. Levanova, A. D. Garagulya, N. M. Gorschkova, M. V. Vysotskii, D. V. Nicolau, N. Yumoto, T. Taguchi, and S. Yoshikawa. 1998a. Phenotypic diversity of *Pseudoaltermonas citrea* from different marine habitats and emendation of the description. *Int. J. Syst. Bacteriol.* **48:**247–256.

Ivanova, E. P., D. V. Nicolau, N. Yumoto, T. Taguchi, K. Okamoto, Y. Tatsu, and S. Yoshikawa. 1998b. Impact of conditions of cultivation and adsorption on antimicrobial activity of marine bacteria. *Mar. Biol.* **130:**545–551.

James, S., C. Holmström, and S. Kjelleberg. 1996. Purification and characterisation of a novel antibacterial protein from the marine bacterium D2. *Appl. Environ. Microbiol.* **62:**2783–2788.

Kjelleberg, S., and P. Steinberg. 2002. Defenses against bacterial colonisation of marine plants, p. 157–172. *In* S. E. Lindow, E. I. Hecht-Poinar, and V. J. Elliott (ed.), *Phyllosphere Microbiology.* APS Press, St Paul, Minn.

Kushmaro, A., E. Banin, Y. Loya, F. Stackebrandt, and E. Rosenberg. 2001. *Vibrio shilio* sp. nov., the causative agent of bleaching of the coral *Oculina patagonica. Int. J. Syst. Bacteriol.* **51:**1383–1388.

Lau, S. C. K., K. K. W. Mak, F. Chen, and P.-I. Qian. 2002. Bioactivity of bacterial strains isolated from marine biofilms in Hong Kong waters for the induction of larval settlement in the marine polychaete *Hydroides elegans. Mar. Ecol. Prog. Ser.* **226:**301–610.

Lee, N., P. H. Nielson, K. H. Andreasen, S. Juretschko, J. L. Nielsen, K. H. Schleifer, and M. Wagner. 1999. Combination of fluerscent *in situ* hybridization and microautoradiography—a new tool for structure-fiction analyses in microbial ecology. *Appl. Environ. Microbiol.* **65:**1289–1297.

Lemos, M. L., A. E. Toranzo, and J. L. Barja. 1985. Antibiotic activity of epiphytic bacteria isolated from intertidal seaweeds. *Microb. Ecol.* **11:**149–163.

Lin, Z., K. Kumagai, K. Baba, J. J. Mekalanos, and M. Nischibuchi. 1993. *Vibrio parahaemolyticus* has a homolog of the *Vibrio cholerae toxRS* operon that mediates environmentally induced regulation of the termostable direct hemolysis gene. *J. Bacteriol.* **175:**3844–3855.

Littler, M. M., and D. S. Littler. 1995. Impact of CLOD pathogen on Pacific coral reefs. *Science* **267:**1356–1360.

Loosdrecht, M. C. M. v., J. Lyklema, J. Norde, and A. J. B. Zehnder. 1989. Bacterial adhesion: a physicochemical approach. *Microb. Ecol.* **17:**1–15.

Maki, J. S. 1999. The influence of marine microbes on biofouling, p. 147–171. *In* M. Fingerman, R. Nagabhushanam, and M. F. Thompson (ed.), *Biofilms, Bioadhesion, Corrosion and Biofouling,* vol. 3. Science Publishers, Inc. New Delhi, India.

Maki, J. S., D. Rittschof, J. D. Costlow, and R. Mitchell. 1988. Inhibition of attachment of larval barnacles, *Balanus amphitrite,* by bacterial surface films. *Mar. Biol.* **97:**199–206.

Maki, J. S., D. Rittschof, A. S. Schmidt, and R. Mitchell. 1989. factors controlling attachment of bryozoan larvae. A comparison of bacterial films and unfilmed surfaces. *Biol. Bull.* **177:**295–302.

Marsh, T. L. 1999. Terminal restriction fragment length polymorphism (T-RFLP): an emerging method for characterizing diversity among homologous populations of amplification products. *Curr. Opin. Biotechnol.* **2:**323–327.

Mary, A. S., V. S. R. Mary, D. Rittschof, and R. Nagabhushanam. 1993. Bacterial-barnacle interaction: potential using juncellins and antibiotica to alter structure of bacterial communities. *J. Chem. Ecol.* **19:**2155–2167.

Maximilien, R., R. deNys, C. Holmström, L. Gram, M. Givskov, K. Crass, S. Kjelleberg, and P. Steinberg. 1998. Chemical mediation of bacterial surface colonisation by secondary metabolites from the red alga *Delisea pulchra. Aquat. Microb. Ecol.* **15:**233–246.

Mearns-Spragg, A., M. Bregu, K. G. Boyd, and J. G. Burgess. 1998. Cross-species induction and enhancement of antimicrobial activity produced by epibiotic bacteria from marine algae and invertebrates after exposure to terrestrial bacteria. *Lett. Appl. Microbiol.* **27:**142–146.

Murley, Y. M., P. A. Carroll, K. Skorupski, R. K. Taylor, and S. B. Calderwood. 1999. Differential transcription of the tcpPH operon confers biotype-specific control of the *Vibrio cholerae* ToxR virulence regulon. *Inf. Immunol.* **67:**5117–5123.

Muyzer, G., E. C. de Waal, and A. G. Uitterlinden. 1993. Profiling of complex microbial communities by denaturing gradient gel elecrophoresis analysis of polymerase chain reaction-amplified genes coding for 16S rRNA. *Appl. Environ. Microbiol.* **59:**695–700.

Neal, A. I., and A. B. Yule. 1994. The interactions between *Elminius modestus* Darwin cyprids and biofilms of *Deleya marina* NCMB 1877. *J. Exp. Mar. Biol. Ecol.* **176:**127–139.

Nielsen, J. L., and P. H. Nielsen. 2001. Enumeration of acetate-consuming bacteria by microautoradiography under oxygen and nitrite respiring conditions in activated sludge. *Water Res.* **36:**421–428.

Paul, V. J., C. Unabia, M. G. Hadfield, and P. J. Scheuer. 1997. Chemical cues from the marine bacterium *Bacillus* sp. that induce settlement of the tube-building worm *Hydroides elegans. In* R. F. Brady (ed.), *U.S.-Pacific Rim Workshop on Emerging Nonmetallic Materials for the Marine Environment.* Honolulu, Hawaii, section 3.16–3.20. U.S. Office of Naval Research, Washington, D.C.

Rappe, M. S., S. A. Connon, K. L. Vergin, and S. J. Giovannoni. 2002. Cultivation of the ubiquitous SAR11 marine bacterioplankton clade. *Nature* **418:**630–633.

Riquelme, C., G. Hayashida, R. Araya, A. Uchida, M. Satomi, and Y. Ishida. 1996. Isolation of a native bacterial strain from the scallop *Argopecten purpuratus* with inhibitory effects against pathogenic vibrios. *J. Shellfish Res.* **15:**369–374.

Riquelme, C., R. Araya, N. Vergara, R. Rojas, M. Guanita, and M. Candia. 1997. Potential of probiotic strains in the culture of the Chilean scallop *Agropecten purpuratus. Aquaculture* **154:**17–26.

Riquelme, C. E., M. A. Jorquera, A. I. Rojas, R. E. Avendano, and N. Reyes. 2001. Addition of inhibitor-producing bacteria to mass cultures of *Agropecten purpuratus* larvae. *Aquaculture* **192:**111–119.

Rittschof, D. 2001. Natural products antifoulants and coatings development, p. 543–566. *In* J. B. McClintock and B. J. Baker (ed.), *Marine Chemistry and Ecology.* Marine Science Series. CRC Press, London, United Kingdom.

Ruiz, C. M., G. Roman, and J. L. Sanchez. 1996. A marine bacterial strain effective in producing antagonisms of other bacteria. *Aquaculture Internat.* **4:**289–291.

Skerrett, J. 2001. Algal and bacterial interactions in a Tasmanian estuary. Ph.D. thesis, University of Hobart, Tasmania, Australia.

Skovhus, T. L., S. Kjelleberg, and N. B. Ramsing. 2001. Abstract 9th International Symposium on Microbial Ecology, 26–31 August, Amsterdam, The Netherlands.

Szewzyk, U., C. Holmström, M. Wrangstadh, M. O. Samuelsson, J. S. Maki, and S. Kjelleberg. 1991. Relevance of the exopolysaccharide of marine *Pseudomonas* sp. strain S9 for attachment of *Ciona intestinalis* larvae. *Mar. Ecol. Prog. Ser.* 75:259–265.

Tanasomwang, V., T. Nakai, Y. Nishimura, and K. Muroga. 1998. *Vibrio*-inhibiting marine bacteria isolated from tiger shrimp hatchery. *Fish Pathol.* 33:459–466.

Uchida, M., K. Nakata, and M. Maeda. 1997. Conversion of *Ulva* fronds to a hatchery diet for *Artemia* nauplii utilizing the degrading and attaching abilities of *Pseudoalteromonas espejiana*. *J. All. Phycol.* 9:541–549.

Wagner, M., and A. Loy. 2002. Bacterial community composition and function in sewage treatment systems. *Curr. Opin. Biotechnol.* 13:218–227.

Wahl, M. 1989. Marine epibiosis. 1. Fouling and antifouling: some basic aspects. *Mar. Ecol. Prog. Ser.* 58:175–189.

Wahl, M., P. R. Jensen, and W. Fenical. 1994. Chemical control of bacterial epibiosis on ascidians. *Mar. Ecol. Prog. Ser.* 110:45–57.

Wieczorek, S. K., and C. D. Todd. 1997. Inhibition and facilitation of bryozoan and ascidian settlement by natural multi-species biofilms: effects of film age and the roles of active and passive larval attachment. *Mar. Biol.* 128:463–473.

Zhou, J., and D. K. Thompson. 2002. Challenges in applying microarrays to environmental studies. *Curr. Opin. Biotechnol.* 13:204–207.

VII. CONSERVATION OF MICROBIAL GENE POOLS

PREAMBLE

In chapter 38 James Borneman examines the evidence pertaining to the loss of microbial diversity. There is no question that a wide range of anthropogenic activities have adversely affected *species richness* in soils and sediments, for example, but very few hard data point to the permanence of such loss. Indeed, what little information is available suggests that perturbed ecosystems often recover their microbial species richness following restorative actions. However, care is required when interpreting species richness, especially if it is based on broad-brush molecular methods such as DNA reassociation that will not reveal the loss of particular species. Moreover, anthropogenic effects on microbial diversity are not invariably adverse, and instances of neutral and positive effects have been observed. Borneman acknowledges that attempts to resolve the question of microbial gene pool loss are accompanied by considerable challenges; nevertheless, carefully designed projects, the use of new methods and approaches such as microarray technology, and autecological investigations "should facilitate a better understanding of the impacts of human activities on microbial diversity and ecosystem functioning."

Reference to the loss of macrobiological evolutionary history is relevant in this context. We have commented earlier in this book on the widespread and diverse types of associations that occur between micro- and macroorganisms. Consequently the extinction of or threats to animals and plants must be taken into account when reviewing potential loss of microbial gene pools. The American Academy of Microbiology (2002) report also touches on this matter and opines that the most readily observed microbial extinctions may occur with the symbionts of threatened animal and plant species. Two issues in particular warrant our attention: the question of whether macrobiological extinction is *random* or *nonrandom*, and what are the consequences of *environmental degradation* for microbial depauperation? These issues are discussed in chapter 37.

However, does it matter if, indeed, microbial diversity is being lost? A "yes" response is sustained by reference to the evolutionary, ecosystem functioning, and bioprospecting significance of microbial diversity, despite the difficulties we currently have of defining extinction in a microbiological context. The question of biodiversity redundancy (i.e., surplus to requirement) in ecosystems is one that has exercised

ecologists for many years, but now there is substantial agreement among macroecologists for the view expressed by Lawton and Brown (1993), "that all species are not equal, that the loss of some species is more important than that of others, and that species loss may be tolerated up to some critical threshold."

Researching this question at the microbial level is singularly difficult but what evidence there is has persuaded some microbiologists that *functional redundancy* exists within at least some groups of microorganisms. A good example of the experimental strategy that can be deployed to assess functional redundancy is found in the study by Yin et al. (2000) of the diversity of bacterial groups capable of in situ growth on selected substrates. Population responses were determined along a gradient comprising plant-denuded tin mine spoil to an adjacent preserved forest. Functional redundancy increased in relation to plant recolonization and was highest in the forest soil. On the other hand, strikingly different patterns of mycorrhizal plant and fungal diversity suggest that extinction of certain mycorrhizal fungi (notably arbuscular species) may result in significant shifts in how plants acquire resources (Allen et al., 1995) and hence compete in particular habitats. Probably the only certain conclusion that can be drawn at this stage is the need for greater research effort focused on microbial diversity loss and the implications for ecosystem maintenance and as a technological resource.

The exploitative value of microorganisms is a principal theme of this book and as such the *conservation* of microbial gene pools becomes a major issue. The final chapter in this section would have provided a comprehensive discussion of microbial conservation. However, in its absence, the article by Canhos and Manfio (2000) is recommended as a good overview of the subject; in particular this article is a valuable source of Internet sites that cover culture collections and complementary databases and networks and those that deal with regulatory and biosafety issues. We have contributed to the debate on culture collections on a previous occasion (Bull et al., 2000) and highlighted the problems of organism acquisition and distribution and funding and capacity building, especially in the context of developing countries. Traditionally, ex situ conservation has required skilled personnel and incurred high maintenance costs. Now services in addition to their custodial one are being asked of culture collections, among them being (i) as suppliers of large collections of strains for screening programs, (ii) as agencies for patent deposits, (iii) as repositories of authenticated strains for quality control, and (iv) for identification. Moreover, among the many problems facing culture collections is that of maintaining pure and authentic archival material, as recently highlighted by Johnson et al. (2001). To summarize, the traditional approach to microbial conservation through the establishment of culture collections is now recognized as a very inadequate strategem. Consequently there is a compelling case for developing complementary ex situ and in situ programs for conserving microorganisms (Bull et al., 2000).

The case for developing a major program of in situ preservation of microorganisms has been presented by Fuerst and Hugenholtz (2000). They emphasize the startling fact that of the 40 or more divisions of *Bacteria* (Smalla, chapter 9, this book), only 4 are even moderately well represented in culture collections, whereas many of them do not contain a single cultivated member. There are three options that need to be thoroughly investigated for in situ preservation purposes: (i) the conservation of prioritized ecosystems and habitats, (ii) the collection and preservation of environmental samples from different habitats, and (iii) the preservation of DNA obtained from environmental samples. The conservation of habitats raises a further set of issues including which biomes, ecosystems, or habitats should be protected; how they should be protected; how should access to protected localities be regulated and by whom and on what terms? These issues and related ones are considered by the contributors to section VIII.

Similarly, what criteria should guide the selection of environmental samples for preservation? Fuerst and Hugenholtz (2000) suggest two priorities—*endangered habitats* and potentially *ephemeral habitats* among which extreme environments, such as hydrogeothermal springs and vents, are likely to feature strongly. Personal experience of the latter situation occurred when, within months of when I collected a chronoseries of volcanic soils in central Java, Mt. Merapi erupted and completely obliterated the sampling locality. A similarly dramatic event overtook a collecting expedition on the island of Montserrat; within 6 months of when a number of sites had been surveyed on the island, the Soufrière Hills volcano erupted and devastated all but one of the hydrothermal sources (Atkinson et al., 2000). Consequently, the materials

collected 4 years earlier represent "the last remnants of the pre-existing thermophilic microbiology of these sites." Of course, various archives of environmental samples do exist and include terrestrial soil and marine sediment cores, ice cores, and herbarium materials. However, samples are usually collected and stored without particular regard for the needs of preventing contamination or preserving the associated microbiota or nucleic acids. Systematic studies of sample preservation for microbial conservation purposes are rarely made, but concern for such matters appears to be growing (see, for example, Harry et al., 2000; Sessitsch et al., 2002). Concerted efforts to establish protocols for prioritizing collecting and long-term storing of materials for in situ conservation of microbial gene pools would be timely.

REFERENCES

American Academy of Microbiology. 2002. *Microbial Ecology and Genomics: A Crossroads of Opportunity.* American Academy of Microbiology, American Society for Microbiology, Washington, D.C.

Allen, E. B., M. F. Allen, D. J. Helm, J. M. Trappe, R. Molina, and E. Rincon. 1995. Patterns and regulation of mycorrhizal plant and fungal diversity. *Plant Soil* 170:47–62.

Atkinson, T., S. Gairns, D. A. Cowan, M. J. Danson, D. W. Hough, D. B. Johnson, P. R. Norris, N. Raven, C. Robinson, and R. J. Sharp. 2000. A microbiological survey of Montserrat Island hydrothermal biotypes. *Extremophiles* 4:305–313.

Bull, A. T., A. C. Ward, and M. Goodfellow. 2000. Search and discovery strategies for biotechnology: the paradigm shift. *Microbiol. Mol. Biol. Rev.* 64:573–606.

Canhos, V. P., and G. P. Manfio. 2000. Microbial resource centres and ex-situ conservation, p. 421–446. *In* F. G. Priest and M. Goodfellow (ed.), *Applied Microbial Systematics.* Kluwer Academic Publishers, Dordrecht, The Netherlands.

Fuerst, J. A., and P. Hugenholtz. 2000. Microorganisms should be high on DNA preservation list. *Science* 290:1503.

Harry, M., B. Gambier, and E. Garnier-Sillam. 2000. Soil conservation for DNA preservation for bacterial molecular studies. *Eur. J. Soil Biol.* 36:51–55.

Johnson, J. R., P. Delavari, A. L. Stell, G. Prats, U. Carlino, and T. A. Russo. 2001. Integrity of archival strain collections: the ECOR collection. *ASM News* 67:288–289.

Lawton, J. H., and V. K. Brown. 1993. Redundancy in ecosystems, p. 255–270. *In* E.-D. Schulze and H. A. Mooney (ed.), *Biodiversity and Ecosystem Function.* Springer-Verlag, Berlin, Germany.

Sessitsch, A., S. Gyamfi, N. Stralis-Pavese, A. Wilharter, and H. Pfeifer. 2002. RNA isolation from soil for bacterial community and functional analysis: evaluation of different extraction and soil conservation protocols. *J. Microbiol. Methods* 51:171–179.

Yin, B., D. Crowley, G. Sparovek, W. J. de Melo, and J. Borneman. 2000. Bacterial functional redundancy along a soil reclamation gradient. *Appl. Environ. Microbiol.* 66:4361–4365.

Microbial Diversity and Bioprospecting
Edited by Alan T. Bull
© 2004 ASM Press, Washington, D.C.

Chapter 37

Extinction and the Loss of Evolutionary History

ALAN T. BULL

Biological extinction in the context of microorganisms may appear to be a totally fanciful or even erroneous subject for discussion; little wonder, therefore, that microbiologists have given it scant serious attention (Staley, 1997). The argument runs, since we have such a vague understanding of microbial diversity per se and so widely varying estimates of numbers of species, how is it possible to quantify the loss of microbial species or to identify those that might be under threat of extinction or of becoming threatened? Although agreed baselines for such a discussion may be difficult, even impossible, I believe that the loss of microbial evolutionary history is an issue that should engage the attention of microbiologists and the biotechnology community. Conservation of the microbiota can be celebrated on numerous grounds including conserving biodiversity as a matter of principle, as an underpinning element of ecosystem functioning and homeostasis at local and global scales, and in the context of this book especially, as a primary resource for sustainable biotechnology. The counterargument of species redundancy is often raised in such discussions and, although ecologists generally would agree that the world is unlikely to come to a halt if some organisms are lost (see below), that state of uncertainty over microbial diversity argues strongly in favor of the precautionary principle. I develop this discussion by briefly reviewing the condition of macroorganisms and attempting some prudent extrapolations from the available information.

THE LOSS OF BIODIVERSITY AND MASS EXTINCTIONS

That biological extinction has occurred in the past is beyond dispute and is well documented; its causes are both natural (catastrophic events such as volcanic eruptions and asteroid strikes and by natural selection) and anthropogenic in character. The enor-

mity of environmental degradation as a consequence of human intervention is known all too well through the effects of land, sea, and air pollution; deforestation; drainage of wetlands; desertification; urbanization; global warming; and so on (Bull, 1996). Thus many analysts have concluded that we are launched upon a period of mass extinction comparable to but more rapid than the major extinction spasms in geological history (Myers, 1979; Wilson, 1992; Ehrlich, 1995). Since the end of the Ordovician period (approximately 440×10^6 years [Gy] BP) there have been five mass extinctions, although "the thrust of biodiversity . . . . has generally been upward" (Wilson, 1992). However, whereas the extinction dynamics of different animal and plant taxa have been defined from the fossil record during these global extinctions (Benton, 1994), comparable information for the microbiota is lacking. The microfossil research of Schopf (1996) and the molecular chronometer studies of Doolittle et al. (1996) have established a framework for the divergent times of the major microbial groups: 3.5 Gy for the cyanobacterial-like lineage, 2.1 Gy for the last ancestral type common to all three domains, *Archaea-Eukarya* separation at 1.96 Gy, gram-positive and gram-negative bacterial separation at 1.5 Gy, and the divergence of protists at 1.23 Gy. But as Sogin et al. (1996) observed, if the bacterial lineages separated 2.1 Gy ago, "much of our evolutionary history was dominated by undescribed life-forms, the most primitive of which left cyanobacterial-like stromatolite fossils."

At this juncture it is apposite to refer to work at the Japan Marine Science and Technology Center (JAMSTEC), where research in Koki Horikoshi's laboratory promises to throw light on geological near-time changes in microbial community structures through the analysis of the deep-sea subsea floor biosphere (see Parkes and Wellsbury, chapter 12). Inagaki et al. (2001) recently defined archaeal phylotypes in a 1.4-m core taken from the West Philippine

Alan T. Bull • Research School of Biosciences, University of Kent, Canterbury, Kent CT2 7NJ, United Kingdom.

Basin at a depth of 5,719 m; the deepest section of the core was deposited between 2 and 2.25 × 10⁶ years (My) ago. Phylotypes of the ubiquitous marine crenarchaeota group I dominated the surface zone, whereas the deep sediment contained an unusual community of extremophilic archaea including extreme halophiles and hyperthermophiles. The latter may be microbial relicts of the Pleistocene period. Subsequently this group has started to construct depth profiles for bacterial communities in deep-sea sediments. Analysis of a short (0.16-m) core sampled from a cold seep in the Japan Trench (5,343 m) revealed that proteobacterial phylotypes were dominant throughout the core (Inagaki et al., 2002) and that with increasing depth the majority of phylotypes shifted from *Deltaproteobacteria* to *Epsilonproteobacteria*. Although these studies are at an early stage, the next phase of JAMSTEC's deep subsurface program will enable geologically much older sediments to be examined. The riser drilling vessel *Chikyu*, launched in 2001, will have the capability to recover, aseptically, 7-km cores from below the seafloor. Thus, major long-term shifts in microbial community composition should be revealed in such cores and then it may be possible to correlate some of these to microbial extinction events.

Returning now to the question of mass extinction, a term that describes a relatively short period of geological time during which large and diverse components of the Earth's biota become extinct. After each mass extinction, biodiversity has recovered, albeit over very long periods of time—nearly 90 My for the Mesozoic fauna to recover to the degree of Palaeozoic era diversity following the Permian mass extinction; ipso facto, major losses of biodiversity are unlikely to be recovered within the time intervals subject to human attention, and extinct biota represent permanent losses (Signor, 1994). Moreover, because many survivors of mass extinctions do not participate in postrecovery diversifications, e.g., "Dead Clade Walking" (Jablonski, 2002), the effects of such extinctions spread inexorably beyond the event itself.

One of the most intensive debates in biology over the past 25 years or so has centered on whether we have entered into the opening phase of a (sixth) mass extinction (Myers, 1993). In a stimulating essay, Myers raised a number of pertinent issues including intra- as well as interspecies extinction, the future of evolution, triage strategies for threatened biota, and the question of how long we have to avert disaster. However, such dire predictions of biodiversity loss have received strong criticism, among the most recent and well publicized being that from Lomborg (2001). Lomborg is especially critical of the high biodiversity loss estimates (Myers, 1979) and the way they have

been made. On the basis of data compiled by the World Conservation Monitoring Centre and analyzed by Smith et al. (1993), Lomborg estimates that the extinction rate for all animals will be of the order of 0.7% per 50 years. An extinction rate of this magnitude "is not trivial. It is a rate about 1,500 times higher than the natural background extinction. However, it is a much smaller figure than the typically advanced 1–100% over the next 50 y" (Lomborg, 2001). Smith et al. (1993) and others have emphasized that assessments of impending extinction can only be very rough, which again challenges the value of attempting to investigate microbial loss. One point of departure is to readdress the question posed by Staley (1997): "are free-living microbial species threatened with extinction?" The pertinent information is sparse and often anecdotal (Bull et al., 1992; Staley, 1997), and recent assessments are discussed in chapter 38. Nevertheless, two other points warrant our attention in this context: (i) is extinction random or nonrandom, and (ii) what is the state of habitat destruction?

Is Extinction Random?

When the extinction rates of geological spasms are analyzed in terms of taxonomic rank, the loss of species is greatest and that of higher taxa the least. Thus, it has been claimed that at the end of the Palaeozoic era up to 95% of the species of marine animals and foraminiferans became extinct, but only ca. 80% of the genera and 55% of the families disappeared. This random vulnerability according to taxonomic rank has been termed the "field of bullets" scenario (Raup, 1991; Wilson, 1992). Nee and May (1997) simulated mass extinctions by developing approximate equations for the fractional amount of evolutionary history that would be preserved if one assumed the field of bullets or algorithms that maximized the amount of evolutionary history preserved. They concluded that ca. 80% of the tree of life would survive even if 95% of species became extinct, irrespective of whether the survivors were selected on the basis of a maximizing algorithm or at random. However, nonrandom risks of extinction have been documented and "any phylogenetic clumping of factors that promote the risks would increase the chance of all species in polytypic taxa being lost" (Purvis et al., 2000). Using data from the World Conservation Union's 1996 Red List for mammals and birds, these authors calculated that 120 more genera were at risk compared with numbers predicted by random extinction, and the extra loss of mammalian evolutionary history would equate to the loss of a monotypic phylum. Other simulations suggest that loss of evolutionary history is determined, at least in part, by the correlation between speciation rate and extinction risk (Smith et al., 1993; Heard and Mooers, 2000) such that in the worse-case scenario a negative correlation can dramatically increase history loss. As far as microorganisms are concerned, a particularly interesting assessment of extinction and threat is that of McKinney (1999), who argues that attempts to estimate relative rates of extinction and threat among contemporary biota are impeded by differences in the quality of data for different taxa (cf. vertebrates and plants versus insects, nematodes, and microorganisms). McKinney tested the hypothesis that similar rates of extinction and threat exist among taxa at global and regional scales and concluded that the wide disparity in such rates among different groups is an artifactual distortion. Thus, the future threat for highly species-rich taxa including insects, marine invertebrates, and microorganisms "may be vastly underestimated." Consequently, if we want improved estimates, the focus should be on the use of well-studied proxy taxa, on comparisons of taxa only in well-studied regions, and most critically, on increased efforts to evaluate the threat to understudied taxa.

Habitat Destruction

Answers to the second question, on the state of habitat destruction from ecosystem to biome, are more readily found and, for the most part, agree. Degradation of natural environments undoubtedly poses threats to microorganisms. Staley (1997) points to threats to free-living hyperthermophiles and sea ice communities as a consequence of power generation schemes and climate change, respectively, and the enhancement of such effects if they are accompanied by biogeographic distribution and endemism. The threat comes further into focus when symbiotic microorganisms are considered, a back cloth to which was presented above. Various terrestrial locations that are "distinguished by their exceptional levels of biodiversity and endemism have been defined as 'hot spots' on the basis of their floral and faunal diversity. These hot spots . . . contain about 20% of the world flora in only 0.5% of its land area. Of the 18 hot spots, 14 are in tropical forests and the remainder in Mediterranean biomes; 5 of the hot spots already have lost 90% or more of their original integrity, and the rest are under considerable threat" (Bull et al., 2000). Among other strong contenders for hot-spot status are marine ecosystems, notably coral reefs. Roberts and his colleagues recently reported that, while the 10 richest multitaxon centers of marine endemism that they identified covered only ca. 16% of the world's coral reefs, they contained between 45 and 55% of restricted-range species (Roberts et al., 2002). The

diversity of microbial symbionts associated with corals and the predicted high degree of cryptic speciation of corals highlights these hot spots as very serious threats to microbial diversity loss.

One conclusion to be drawn from these facts is that microbiologists should be much more active in identifying and pressing the case for the protection of habitats that are or are likely to be hot spots of microbial diversity, in other words be prepared to adopt a triage strategy. As Myers (1993) challenges us, "Shall we attempt to apply triage? but, more to the point, How shall we apply triage to better effect?" Only a more systematic, science-based approach will provide useful answers.

REFERENCES

Benton, M. J. 1994. Finding the tree of life: matching phylogenetic trees to the fossil record through the 20th century. *Proc. R. Soc. London Ser. B* **268:**2123–2130.

Bull, A. T. 1996. Biotechnology for environmental quality: closing the circles. *Biodiver. Conserv.* **5:**1–25.

Bull, A. T., M. Goodfellow, and J. H. Slater. 1992. Biodiversity as a source of innovation in biotechnology. *Annu. Rev. Microbiol.* **42:**219–257.

Bull, A. T., A. C. Ward, and M. Goodfellow. 2000. Search and discovery strategies for biotechnology: the paradigm shift. *Microbiol. Mol. Biol. Rev.* **64:**573–606.

Doolittle, W. F., D. Feng, S. Tsang, G. Cho, and E. Little. 1996. Determining divergence times of the major kingdoms of living organisms with a protein clock. *Science* **271:**470–477.

Ehrlich, P. R. 1995. The scale of the human enterprise and biodiversity loss, p. 214–226. *In* J. H. Lawton and R. M. May (ed.), *Extinction Rates.* Oxford University Press, Oxford, United Kingdom.

Heard, S. B., and A. Ø. Mooers. 2000. Phylogenetically patterned speciation rates and extinction risks change the loss of evolutionary history during extinctions. *Proc. R. Soc. London Ser. B* **267:**613–620.

Inagaki, F., K. Takai, T. Komatsu, T. Kanamatsu, T. Fujioka, and K. Horikoshi. 2001. Archaeology of Archaea: geomicrobiological record of Pleistocene events in a deep-sea subseafloor environment. *Extremophiles* **5:**385–392.

Inagaki, F., Y. Sakihama, A. Inoue, C. Kato, and K. Horikoshi. 2002. Molecular phylogenetic analyses of reverse-transcribed bacterial rRNA obtained from deep-sea cold seep sediments. *Environ. Microbiol.* **4:**277–286.

Jablonski, D. 2002. Survival without recovery after mass extinctions. *Proc. Natl. Acad. Sci. USA* **99:**8139–8144.

Lomborg, B. 2001. *The Skeptical Environmentalist.* Cambridge University Press, Cambridge, United Kingdom.

McKinney, M. L. 1999. High rates of extinction and threat in poorly studied taxa. *Conserv. Biol.* **13:**1273–1281.

Myers, N. 1979. *The Sinking Ark.* Pergamon Press, Oxford, United Kingdom.

Myers, N. 1993. Questions of mass extinction. *Biodiver. Conserv.* **2:**2–17.

Nee, S., and R. M. May. 1997. Extinction and the loss of evolutionary history. *Science* **278:**692–694.

Purvis, A., J. L. Gittleman, G. Cowlishaw, and G. M. Mace. 2000. Predicting extinction risk in declining species. *Proc. R. Soc. London Ser. B* **267:**1947–1952.

Raup, D. M. 1991. *Extinction: Bad Genes or Bad Luck?* W. W. Norton, New York, N.Y.

Roberts, C. M., C. J. McClean, J. E. N. Veron, J. P. Hawkins, G. R. Allen et al. 2002. Marine biodiversity hotspots and conservation priorities for tropical reefs. *Science* 295:1280–1284.

Schopf, J. W. 1996. Are the oldest fossils cyanobacteria?, p. 23–61. *In* D. McL. Roberts, P. Sharp, G. Alderson, and M. A. Collins (ed.), *Evolution of Microbial Life*. Cambridge University Press, Cambridge, United Kingdom.

Signor, P. W. 1994. Biodiversity in geological time. *Am. Zool.* 34:23–32.

Smith, F. D. M., R. M. May, R. Pellew, T. H. Johnson, and K. S. Walter. 1993. Estimating extinction rates. *Nature* 364:494–496.

Sogin, M. L., J. D. Silberman, G. Hinkle, and H. G. Morrison. 1996. Problems with molecular diversity in the Eukarya, p. 167–184. *In* D. McL. Roberts, P. Sharp, G. Alderson, and M. A. Collins (ed.), *Evolution of Microbial Life*. Cambridge University Press, Cambridge, United Kingdom.

Staley, J. T. 1997. Biodiversity: are microbial species threatened? *Curr. Opin. Biotechnol.* 8:340–345.

Wilson, E. O. 1992. *The Diversity of Life*. Allen Lane, The Penguin Press, London, United Kingdom.

Microbial Diversity and Bioprospecting
Edited by Alan T. Bull
© 2004 ASM Press, Washington, D.C.

Chapter 38

What Is the Evidence for the Loss of Microbial Diversity?

JAMES BORNEMAN

This chapter examines the effects of anthropogenic activities on microbial species richness and diversity. This is a challenging area of study because current investigative methods do not allow accurate assessment of species composition and diversity (see Chapters 8 to 10 in this book for a discussion of the current methods). Although this limitation, among others, does not allow firm conclusions to be made, an examination of the literature showed that most reports identified a negative relationship between anthropogenic activities and microbial species richness and diversity. In this chapter, I review these studies, discuss the importance of the findings, and describe some of the challenges for future investigations.

ANTHROPOGENIC ACTIVITIES CAUSING REDUCTIONS OF MICROBIAL SPECIES RICHNESS AND DIVERSITY

A sampling of the literature showed that anthropogenic activities such as agricultural practices, mining, various pollutants, and the loss of plant species were frequently associated with reductions in microbial species richness and diversity.

Agricultural practices including deforestation, the use of herbicides, fish farming, sewage sludge amendments, and copper amendments have all been shown to lower microbial diversity. In Tai National Park (Ivory Coast), conversion from tropical rain forest to cultivation reduced the diversity of soil fungi; however, when cultivation was discontinued after one crop, which is the traditional practice in this region, fungal diversity quickly recovered (Persiani et al., 1998). The diversity of type I methanotrophs was decreased by an herbicide treatment in a long-term (18-year) field trial in Belgium (Seghers, 2001). A DNA reassociation analysis showed that fish farming practices reduced bacterial diversity in sediment by a factor of 200; 4 years after abandonment of the fish

farming operation, diversity increased but not to the levels of the pristine sediments (Torsvik et al., 1993, 1996). DNA reassociation was also used to show that soils amended with sewage sludge containing heavy metals decreased bacterial diversity in a dose-dependent manner (Sandaa et al., 1999). In a rice plot experiment, diatom diversity decreased after a one-time application of the insecticide chlorpyrofus (Nelson et al., 1976). Finally, an examination of soils from a long-term field experiment showed that bacterial diversity was decreased by copper contamination (Smit et al., 1997).

Declines in microbial species richness and diversity have also resulted from various mining practices. In a freshwater reservoir, sediments exposed to acid mine drainage exhibited lower heterotrophic bacterial diversity than those exposed to lower amounts of drainage or those from adjacent pristine regions (Wassel and Mills, 1983). Fungal species richness was lower in river samples containing coal mine effluent compared with uncontaminated samples that were collected upstream of the output site; the contaminated river samples also had lower rates of leaf litter decomposition, suggesting that the effluent from this coal mine altered the functioning of this ecosystem (Maltby and Booth, 1991). Finally, soils containing reconstituted spent oil shale had lower bacterial richness and diversity compared with uncontaminated control soils (Segal and Mancinelli, 1987). At the same site, increases in bacterial richness and diversity correlated positively with the amount of time under restoration, which was promoted by fertilization and revegetation.

Other anthropogenic activities that negatively affected microbial species and diversity included gasoline, air pollution, heavy metals, phenols, methane, and chloroform. Short-term gasoline contamination may represent a relatively common anthropogenic insult. A soil that received a single exposure to gasoline exhibited decreased bacterial diversity, which did not return to the original level 15 weeks after the contam-

James Borneman • Department of Plant Pathology, University of California, Riverside, CA 92521.

ination (Diltz et al., 1992). Records kept over the past century have shown that the number of ectomycorrhizal fungi collected per foray remained relatively constant from 1900 through the 1960s (Jaenike, 1991). However, in the 1970s, a considerable drop in species collected per foray started to occur. Explanations for these apparent species reductions include air pollution and acid precipitation (Arnolds, 1988). High levels of mercury decreased the microbial diversity in packed-bed bioreactors designed for mercury remediation; however, when mercury levels were returned to near-original levels in these bioreactors, the microbial diversity was restored to the original levels (von Canstein et al., 2002). In model sewage plants, phenol amendments decreased bacterial diversity (Eichner et al., 1999). A DNA reassociation analysis showed that methane and nitrogen exposure lowered bacterial diversity in soil (Ovreas et al., 1998). Finally, chloroform fumigation decreased soil bacterial diversity in a dose-dependent manner (Griffiths et al., 2000).

Microbial species are also lost because of their close relationships with other organisms such as plant and animals (Staley, 1997). Throughout the planet, numerous species of plants and animals have and will continue to go extinct as habitats for these organisms diminish. The number of microorganisms lost because of these extinctions could be relatively high, as, on average, six fungal species may be uniquely associated with one plant species (Hawksworth, 1991). In addition, reductions in tropical plant diversity will continue to cause considerable losses of microbial genetic diversity as tropical endophytic fungi represent a highly diverse assemblage of microorganisms (Arnold et al., 2000).

ANTHROPOGENIC ACTIVITIES THAT HAD A VARIABLE OR UNDETECTABLE EFFECT ON MICROBIAL SPECIES RICHNESS AND DIVERSITY

Several studies have shown that heavy metals and deforestation had a variable or undetectable impact on microbial species richness and diversity. The effects of heavy metals on plankton, sediment, and epilithic bacterial species diversity were examined in a river system in Fort Wayne, Indiana. An examination of six sites containing varying levels of heavy metals showed no relationship between diversity and metal contamination (Dean-Ross and Mills, 1989). These authors suggested that this result was due, in part, to a reduction in metal toxicity from high pH. In another study examining the relationship between bacterial diversity and cadmium contamination, no statistically significant relationship was found, al-

though the diversity levels were the lowest in the soils possessing the highest cadmium concentrations (Barkay et al., 1985). Arsenic is a pollutant originating from a variety of sources including mines, wastewater from geothermal power stations, and fungicides. When four soils with varying levels of arsenic were examined, one of the two soils with high levels of arsenic had significantly lower fungal diversity (Hiroki, 1993). When fungal diversity was examined between soils from clear-cut (7 to 9 years after the cutting) and unharvested regions, no significant relationships were identified (Houston et al., 1998).

ANTHROPOGENIC ACTIVITIES LEADING TO AN INCREASE IN MICROBIAL DIVERSITY

Several reports described anthropogenic activities that have increased microbial diversity. Water samples collected near a municipal wastewater effluent had higher diatom diversity than samples collected farther from the output site (Stevenson and Stoermer, 1982). Bacterial diversity was higher in sediments from a petroleum-contaminated salt marsh than from sediments in an adjacent pristine environment (Hood et al., 1975). A DNA reassociation analysis ($C_0t_{3/4}$) of a microcosm experiment showed total soil microbial genetic diversity increased due to the addition of either the herbicide 2,4,5-T, a bacterium capable of degrading 2,4,5-T, or both of these components (Bej et al., 1991). However, when these same samples were examined using a $C_0t_{1/2}$ analysis, the 2,4,5-T and the 2,4,5-T-degrading-bacterium treatments both led to a decrease in diversity.

DOES MICROBIAL SPECIES LOSS MATTER?

Even if anthropogenic activities are definitively shown to decrease microbial species richness and diversity, does it matter? One concept that should be explored when addressing this issue is Martinus Beijerinck's idea that "everything is everywhere." If everything were truly everywhere, then a local species loss would not lead to the extinction of that organism. To date, few studies have tested this concept. Evidence supporting this idea came from an intensive investigation of the ciliated protozoa inhabiting a few sediment samples, as this small study identified a large fraction of the species previously observed from the same well-studied sites (Fenchel et al., 1997). However, when 3-chlorobenzoate-degrading bacteria from five continents were examined, few genotypes from the different locations were similar (Fulthorpe et al., 1998), implying that everything may not be everywhere. Further studies assessing the validity of the concept that "everything is everywhere" are certainly needed. In addition, refinements to this concept are also warranted including the genotypic level that defines "everything" and the geographical scale that defines "everywhere" (Tiedje, 1995).

For investigations targeted at identifying useful compounds or products from microorganisms, species loss may be important. Numerous products and processes have been derived from microorganisms. Microbes are used in food production and as animal feed supplements to replace antibiotic and growth hormone treatments (Palm and Chapela, 1997; Porter and Fox, 1993; Tannock, 1999). They are used to produce genetically engineered products, catalyze chemical reactions, and serve as resources for useful compounds including pharmaceuticals (Palm and Chapela, 1997; Porter and Fox, 1993). Thus, a significant loss of microbial species may hinder the development of important industrial processes and therapeutic compounds.

Another area where microbial species loss may be important is ecosystem functioning. The relationship between diversity and ecosystem functioning has certainly been hotly debated. A recent article by Jennifer Hughes and Owen Petchey summarized the highlights from a conference titled "Biodiversity and Ecosystem Functioning: Synthesis and Perspectives" (Hughes and Petchey, 2001). Although this conference primarily dealt with macroecological issues, some of the principal concepts may be relevant for microbial ecology. One of the main conclusions from this article was that diversity matters when species numbers are low. Since microbial species numbers are very high in most ecosystems, is it likely that changes in microbial diversity will affect the functioning of ecosystems? Two other observations from this conference were that many studies showed no relationship between diversity and ecosystem functioning and that soil microbial diversity appears to have little effect on decomposition rate, an important parameter of ecosystem functioning. The next two paragraphs describe studies addressing the relationship between microbial diversity and ecosystem functioning.

Several reports have provided evidence for a positive relationship between microbial diversity and ecosystem functioning. To test this concept, Naeem and Li (1997) established microcosms with varying numbers of microbial species within several trophic groups. These experiments showed that increasing species diversity led to more consistent biomass and organism densities, suggesting that microbial diversity is positively correlated with ecosystem reliability. In another microcosm experiment where the number and types of organisms were carefully controlled, increased mycorrhizal fungal diversity led to an increase in plant productivity (van der Heijden et al., 1998).

Another approach for creating soils with varying levels of microbial diversity is to remove portions of the existing community by exposing the samples to fumigants for varying time periods. Griffiths and colleagues used this approach to show that soils with higher levels of microbial diversity were more resistant to copper stress, as measured by the rate of grass decomposition (Griffiths et al., 2000). Finally, in another microcosm experiment, combinations of two fungi exhibited increased respiration compared with individuals, suggesting that diversity enhanced ecosystem functioning (Robinson et al., 1993).

Several other reports have shown a variable or neutral relationship between microbial diversity and ecosystem functioning. A microcosm experiment showed that ectomycorrhizal species richness may be positive, negative, or neutral in regard to its effect on tree productivity and that the results depended on the tree species and fertility of growth substrate (Jonsson et al., 2001). In a study where straw decomposition was examined by burying litterbags in soil, no relationship between decomposition and the diversity of protozoa, nematodes, and microarthropods were found (Andren et al., 1995). Griffiths and colleagues showed that fumigation-induced reductions in microbial diversity led to increases in decomposition of plant materials over the short term; however, these differences did not persist over longer time periods (Griffiths et al., 2000).

CONCLUSIONS AND FUTURE RESEARCH

Microorganisms are essential for life as we know it (Colwell, 1997). They perform numerous functions essential for the biosphere including nutrient cycling and environmental detoxification. They are important in food production; are sources of antibiotics and other pharmaceuticals; and they provide enzymes for industrial catalysis, forensics, and products used in our daily lives. Because of these crucial roles in human civilization and the global ecosystem, it is important to determine if human activities are dwindling this resource or interfering with their functional roles in the biosphere.

Given the current knowledge base, discussed in part in this chapter, no firm conclusions can be made concerning the relationship between anthropogenic activities and microbial diversity. An examination of the literature supports the supposition that human activities can reduce microbial diversity. However, even though some progress has been made in defining these relationships, numerous questions remain unresolved. Which anthropogenic activities cause species loss? Do these factors always cause species losses? If not, what are the other factors that influence these processes? Does diversity recover after an anthropogenic stressor?

Is everything everywhere? Does local species loss result in extinction? Will it matter if some microbial diversity is lost? Is ecosystem functioning associated with microbial diversity? What is a bacterial species? Does gene transfer between microorganisms influence the amount of genetic diversity that is lost through species loss?

Numerous challenges hinder the accurate analysis of microbial diversity. One problem is the remarkable number of extant species. Estimates suggest that 4,000 to 40,000 bacterial species can inhabit a single gram of soil (Tiedje, 1995; Torsvik et al., 1990). Another problem has been the limitations of the experimental methods. Many methods exist, yet none provide a cost-effective means to thoroughly analyze community composition and diversity. As a result, no complex microbial community has been comprehensively described thus far. To overcome these problems, several approaches have been taken.

One approach is to develop methods allowing thorough and accurate analysis of microbial species composition and diversity. The recent development of array-based methods, which permit thousands of hybridization events to be examined in parallel, has brought great promise to the field of microbial ecology. In this approach, labeled rRNA or rDNA from environmental samples are analyzed by their hybridization to oligonucleotide probes attached to a substrate (Behr et al., 2000; Guschin et al., 1997; Rudi et al., 2000; Small et al., 2001; Urakawa et al., 2002). An alternative array-based approach is oligonucleotide fingerprinting of rRNA genes (Valinsky et al., 2002). With this approach, rDNA clone libraries are sorted into taxonomic groups by a series of hybridization experiments, each using a single DNA oligonucleotide. An additional existing method that should not be forgotten is DNA reassociation analysis, which provides a relatively inexpensive means to estimate total genetic diversity (Torsvik et al., 1990).

Another strategy for examining microbial ecosystems is to limit the number of organisms being studied. In this approach, the role of microbial diversity in ecosystem functioning can be examined by constructing model experiments where a limited number of species are added to a microcosm. Most of the experiments that used this approach showed a positive relationship between microbial diversity and ecosystem functioning (Naeem and Li, 1997; Robinson et al., 1993; van der Heijden et al., 1998). An alternative approach is to decrease natural levels of diversity through biocidal treatments (Griffiths et al., 2000). This strategy has several advantages over the constructive approach, including the fact that species richness will be more realistic and the organisms investigated will include those not readily cultured in the laboratory. Future work in this area should include investigations of microbial communities containing the typical complement of organisms found in the natural world.

Additional problems in examining the effects of anthropogenic activities include the numerous variables that may influence the factor being studied. To obtain a thorough understanding of when, why, and how anthropogenic activities can cause species reductions, these additional factors must be considered. For example, in a study by Dean-Ross and colleagues, it was proposed that high pH reduced the solubility of the heavy metals, resulting in a decrease in toxicity (Dean-Ross and Mills, 1989). It has also been suggested that much of the variation in toxicity studies may result from factors influencing bioavailability (Giller et al., 1998). Understanding these additional variables should contribute to the development of strategies to both predict species loss and manage these processes.

In sum, although the challenges are considerable, carefully designed studies coupled with further methodological advances should facilitate a better understanding of the impacts of human activities on microbial diversity and ecosystem functioning.

REFERENCES

Andren, O., J. Bengtsson, and M. Clarholm. 1995. Biodiversity and species redundancy among litter decomposers, p. 141–151. *In* H. P. Collins (ed.), The *Significance and Regulation of Soil Biodiversity.* Kluwer Academic Publisher, Dondrecht, The Netherlands.

Arnold, A. E., Z. Maynard, G. S. Gilbert, P. D. Coley, and T. A. Kursar. 2000. Are tropical fungal endophytes hyperdiverse? *Ecol. Lett.* 3:267–274.

Arnolds, E. 1988. The changing macromycete flora in the Netherlands. *Trans. Br. Mycol. Soc.* 90:391–406.

Barkay, T., S. C. Tripp, and B. H. Olson. 1985. Effect of metal-rich sewage sludge application on the bacterial communities of grasslands. *Appl. Environ. Microbiol.* 49:333–337.

Behr, T., C. Koob, M. Schedl, A. Mehlen, H. Meier, D. Knopp, E. Frahm, U. Obst, K. H. Schleifer, R. Niessner, and W. Ludwig. 2000. A nested array of rRNA targeted probes for the detection and identification of enterococci by reverse hybridization. *Syst. Appl. Microbiol.* 23:563–572.

Bej, A. K., M. Perlin, and R. M. Atlas. 1991. Effect of introducing genetically engineered microorganisms on soil microbial community diversity. *FEMS Microbiol. Ecol.* 86:169–175.

Colwell, R. R. 1997. Microbial diversity: the importance of exploration and conservation. *J. Ind. Microbiol. Biotechnol.* 18:302–307.

Dean-Ross, D., and A. L. Mills. 1989. Bacterial community structure and function along a heavy metal gradient. *Appl. Environ. Microbiol.* 55:2002–2009.

Diltz, M. S., C. E. Hepfer, E. Hartz, and K. H. Baker. 1992. Recovery of heterotrophic soil bacterial guilds from transient gasoline pollution. *Hazard. Waste Hazard. Mater.* 9:267–273.

Eichner, C. A., R. W. Erb, K. N. Timmis, and I. Wagner-Dobler. 1999. Thermal gradient gel electrophoresis analysis of bioprotection from pollutant shocks in the activated sludge microbial community. *Appl. Environ. Microbiol.* 65:102–109.

Fenchel, T., G. F. Esteban, and B. J. Finlay. 1997. Local versus global diversity of microorganisms: cryptic diversity of ciliated protozoa. *Oikos* 80:220–225.

Fulthorpe, R. R., A. N. Rhodes, and J. M. Tiedje. 1998. High levels of endemicity of 3-chlorobenzoate-degrading soil bacteria. *Appl. Environ. Microbiol.* **64:**1620–1627.

Giller, K. E., E. Witter, and S. P. McGrath. 1998. Toxicity of heavy metals to microorganisms and microbial processes in agricultural soils: a review. *Soil Biol. Biochem.* **30:**1389–1414.

Griffiths, B. S., K. Ritz, R. D. Bardgett, R. Cook, S. Christensen, F. Ekelund, S. J. Sorensen, E. Baath, J. Bloom, P. C. de Ruiter, J. Dolfing, and B. Nicolardot. 2000. Ecosystem response of pasture soil communities to fumigation-induced microbial diversity reductions: an examination of the biodiversity-ecosystem function relationship. *Oikos* **90:**279–294.

Guschin, D. Y., B. K. Mobarry, D. Proudnikov, D. A. Stahl, B. E. Rittmann, and A. D. Mirzaberkov. 1997. Oligonucleotide microchips as genosensors for determinative and environmental studies in microbiology. *Appl. Environ. Microbiol.* **63:**2397–2402.

Hawksworth, D. L. 1991. The fungal dimension of biodiversity: magnitude, significance, and conservation. *Mycol. Res.* **95:**641–655.

Hiroki, M. 1993. Effect of arsenic pollution on soil microbial population. *Soil Sci. Plant Nutr.* **39:**227–235.

Hood, M. A., W. S. Bishop, F. W. Bishop, S. P. Meyers, and T. Whelan. 1975. Microbial indicators of oil-rich salt marsh sediments. *Appl. Microbiol.* **30:**982–987.

Houston, A. P. C., S. Visser, and R. A. Lautenschlager. 1998. Microbial processes and fungal community structure in soils from clear-cut and unharvested areas of two mixedwood forests. *Can. J. Bot.* **76:**630–640.

Hughes, J. B., and O. L. Petchey. 2001. Merging perspectives on biodiversity and ecosystem functioning. *Trends Ecol. Evol.* **16:**222–223.

Jaenike, J. 1991. Mass extinction of European fungi. *Trends Ecol. Evol.* **6:**174–175.

Jonsson, L. M., M. C. Nilsson, D. A. Wardle, and O. Zackrisson. 2001. Context dependent effects of ectomycorrhizal species richness on tree seedling productivity. *Oikos* **93:**353–364.

Maltby, L., and R. Booth. 1991. The effect of coal-mine effluent on fungal assemblages and leaf breakdown. *Water Res.* **25:**247–250.

Naeem, S., and S. Li. 1997. Biodiversity enhances ecosystem reliability. *Nature* **390:**507–509.

Nelson, J. H., D. L. Stoneburner, E. S. Evans, N. E. Pennington, and M. V. Meisch. 1976. Diatom diversity as a function of insecticidal treatment with a controlled-release formulation of chlorpyrifos. *Bull. Environ. Contam. Toxicol.* **15:**630–634.

Ovreas, L., S. Jensen, F. L. Daae, and V. Torsvik. 1998. Microbial community changes in a perturbed agricultural soil investigated by molecular and physiological approaches. *Appl. Environ. Microbiol.* **64:**2739–2742.

Palm, M. E., and I. H. Chapela. 1997. Mycology in sustainable development: expanding concepts, vanishing borders. *In* M. E. Palm and I. H. Chapela (ed.), *Mycology in Sustainable Development: Expanding Concepts, Vanishing Borders.* Parkway Publishers, Boone, N.C.

Persiani, A. M., O. Maggi, M. A. Casado, and F. D. Pineda. 1998. Diversity and variability in soil fungi from a disturbed tropical rain forest. *Mycologia* **90:**206–214.

Porter, N., and F. M. Fox. 1993. Diversity of microbial products: discovery and application. *Pestic. Sci.* **39:**161–168.

Robinson, C. H., J. Dighton, J. C. Frankland, and P. A. Coward. 1993. Nutrient and carbon dioxide release by interacting species of straw-decomposing fungi. *Plant Soil* **151:**139–142.

Rudi, K., O. M. Skulberg, R. Skulberg, and K. S. Jakobsen. 2000. Application of sequence-specific labeled 16S rRNA gene oligonucleotide probes for genetic profiling of cyanobacterial abundance and diversity by array hybridization. *Appl. Environ. Microbiol.* **66:**4004–4011.

Sandaa, R. A., V. Torsvik, O. Enger, F. L. Daae, T. Castberg, and D. Hahn. 1999. Analysis of bacterial communities in heavy metal-contaminated soils at different levels of resolution. *FEMS Microbiol. Ecol.* **30:**237–251.

Segal, W., and R. L. Mancinelli. 1987. Extent of regeneration of the microbial community in reclaimed spent oil shale land. *J. Environ. Qual.* **16:**44–48.

Seghers, D. 2001. Do conventionally and biologically cultivated soils differ in bacterial diversity and community structure? *Meded. Fac. Landbouwkd. Toegep. Biol. Wet. Univ. Gent* **66:** 381–388.

Small, J., D. R. Call, F. J. Brockman, T. M. Straub, and D. P. Chandler. 2001. Direct detection of 16S rRNA in soil extracts by using oligonucleotide microassays. *Appl. Environ. Microbiol.* **67:**4708–4716.

Smit, E., P. Leeflang, and K. Wernars. 1997. Detection of shifts in microbial community structure and diversity in soil caused by copper contamination using amplified ribosomal DNA restriction analysis. *FEMS Microbiol. Ecol.* **23:**249–261.

Staley, J. T. 1997. Biodiversity: are microbial species threatened? *Curr. Opin. Biotechnol.* **8:**340–345.

Stevenson, R. J., and E. F. Stoermer. 1982. Abundance patterns of diatoms on *Cladophora* in Lake Huron with respect to a point source of wastewater treatment plant effluent. *J. Great Lakes Res.* **8:**184–195.

Tannock, G. W. 1999. *Probiotics: A Critical Review.* Horizon Scientific Press, Norfolk, England.

Tiedje, J. M. 1995. Approaches to the comprehensive evaluation of prokaryote diversity of a habitat, p. 73–87. *In* D. Allsopp, R. R. Colwell, and D. L. Hawksworth (ed.), *Microbial Diversity and Ecosystem Function.* CABI Publishing, Wallingford, United Kingdom.

Torsvik, V., J. Goksøyr, and F. L. Daae. 1990. High diversity in DNA of soil bacteria. *Appl. Environ. Microbiol.* **56:**782–787.

Torsvik, V., J. Goksøyr, F. L. Daae, R. Sorheim, J. Michaelsen, and K. Salte. 1993. Diversity of microbial communities determined by DNA reassociation technique, p. 375–378. *In* R. Guerrero and C. Pedros-Alio (ed.), *Trends in Microbial Ecology.* Spanish Society for Microbiology, Barcelona.

Torsvik, V., R. Sorheim, and J. Goksøyr. 1996. Total bacterial diversity in soil and sediment communities—a review. *J. Ind. Microbiol.* **17:**170–178.

Urakawa, H., P. A. Noble, S. El Fantroussi, J. J. Kelly, and D. A. Stahl. 2002. Single-base-pair discrimination of terminal mismatches by using oligonucleotide microarrays and neural network analyses. *Appl. Environ. Microbiol.* **68:**235–244.

Valinsky, L., G. D. Vedova, A. J. Scupham, S. Alvey, A. Figueroa, B. Yin, R. J. Hartin, M. Chrobak, D. E. Crowley, T. Jiang, and J. Borneman. 2002. Analysis of bacterial community composition by oligonucleotide fingerprinting of rRNA genes. *Appl. Environ. Microbiol.* **68:**3243–3250.

van der Heijden, M. G. A., J. N. Klironomos, M. Ursic, P. Moutoglis, R. Steriwolf-Engel, T. Boller, A. Wiemken, and I. R. Sanders. 1998. Mycorrhizal fungal diversity determines plant biodiversity, ecosystem variability and productivity. *Nature* **396:**69–72.

von Canstein, H., S. Kelly, Y. Li, and I. Wagner-Dobler. 2002. Species diversity improves the efficiency of mercury-reducing biofilms under changing environmental conditions. *Appl. Environ. Microbiol.* **68:**2829–2837.

Wassel, R. A., and A. L. Mills. 1983. Changes in water and sediment bacterial community structure in a lake receiving acid mine drainage. *Microb. Ecol.* **9:**155–169

VIII. CONVENTION ON BIOLOGICAL DIVERSITY: IMPLICATIONS FOR MICROBIAL PROSPECTING

PREAMBLE

This penultimate section of the book draws attention to particular aspects of microbiology and, most emphatically, microbial prospecting that are highlighted in the post-Convention on Biological Diversity (CBD) world. The CBD is a landmark treaty regarding the universality of its scope and approach; as Kerry ten Kate makes clear at the outset of her chapter, "Microbiologists today face a challenge not only to stay abreast of a fast-moving and interdisciplinary field of science, *but also to be conversant with the law and policy affecting them*" (my italics). The objectives of the CBD are threefold: conservation of biodiversity, sustainable use of biodiversity, and fair and equitable benefit sharing arising from the utilization of genetic resources. Issues relating to benefit sharing are raised by all of the contributors to this section. The scene is set by Hanne Svarstad (chapter 40), who reviews a number of bioprospecting case studies that include the discovery of the blockbuster immunosuppressive drug cyclosporin, a natural product from the fungus *Tolypodium inflatum*. Svarstad presents a sequence of case studies as examples of the laissez faire or *open-access* attitude to genetic resources that prevailed pre-CBD and that could obviate payments to the source country (or peoples). Post-CBD *access and benefit-sharing* arrangements, several of which are discussed in this section, offer the possibilities of direct monetary benefits, technical training and capacity building, and promoting conservation measures.

Several commentators have asserted that one consequence of the CBD has been to generate complex sets of conflicts between the biotechnology industries and biodiversity-rich countries. In other words, such countries may have been overoptimistic in their anticipation of the payback from their biological resources. Whether such conflicts, real or apparent, may have helped catalyze alternative, synthetic routes for developing new drugs is debatable (see MacIlwain, 1998). But in any case, as we have discussed earlier, combinatorial chemistry is increasingly reliant on novel natural chemical skeletons for elaborating novel bioactive compounds (see chapter 1), a fact not infrequently elided by proponents of combinatorial chemistry. Accordingly, it is valuable to look at some of the modalities that have been established between cooperating countries and clients and to evaluate the outcomes. The research collaborative agreement established in 1991 between the National Institute of Biodiversity (INBio, Costa Rica) and Merck & Co. is generally regarded as a landmark

event and has been used as a model for similar agreements with other companies and with academe. These agreements embody up-front payments, benefit sharing, technology transfer, and capacity building (including an element of locally conducted research) and, significantly, a requirement for nondestructive use of biodiversity. The initiative first made with Merck & Co. has led to a further 22 agreements being set up with INBio. Giselle Tomayo and colleagues review the achievements of INBio research agreements of the past decade (chapter 41) and summarize the biodiversity and applications targets of their partners. Many of these agreements started only in the late 1990s which makes it too early to expect commercial products; nevertheless, significant outputs are reported, and these authors also make the point that simpler products, resulting from Inter American Development Bank support, are likely to be commercialized before any blockbuster product reaches the international market. A somewhat different modality of cooperative natural product prospecting, but having many of the objectives and criteria found in INBio agreements, has been promoted by the U.S. National Institutes of Health. The International Cooperative Biodiversity Groups program of the National Institutes of Health currently comprises eight multifarious projects involving biodiversity hot-spot countries in Africa, South and Central America, Vietnam, and Laos. This program also has been in operation for 10 years, and progress toward its objectives is discussed in chapter 43 by Josh Rosenthal and Flora Katz.

The final bioprospecting case study in this section raises additional questions and examines the *contractual complexities* that have embroiled Yellowstone National Park (YNP) and, by extension, the whole National Park Service in the United States. The catalyst for this particular debate was the signing of a cooperative research and development agreement between YNP and Diversa Corporation in 1997. Holly Doremus (chapter 42) discusses this case in detail against the background that contains open access versus benefit-sharing agreements, the nonparticipation of the United States in the CBD, and the major commercial success of a YNP-derived product, namely, *Taq* polymerase. Further complications arose in 1998 when the YNP-Diversa agreement was legally challenged. Doremus' analysis of this case raises a host of issues that present a very broad fame of reference for bioprospecting, especially in the context of the developed world that "already enjoys strong legal protection of its natural resources."

This selection of case studies calls attention to a variety of socioeconomic and ethical matters associated with bioprospecting and biotechnology development (see ten Kate and Laird, 1999). Although these issues are not explored in this book, they are important, and, perhaps rather belatedly, microbiologists also will be required to heed them. On the one hand microbiologists have a history of working to codes of safety and ethics, as in the case of dangerous pathogens and experiments on animals, for example. However, guidelines or codes for collecting material may follow in the wake of precluded open access to genetic resources. For example, the International Society of Ethnobiology has drafted guidelines for research, collections, databases, and publications to be incorporated into its code of ethics (http://guallart. dac.uga.edu/ISE/).

Finally, we return to the *conservation* of biodiversity, an objective enshrined in the CBD, and the question of whether bioprospecting can help to fund conservation. Arguments continue to be made for and against this proposition. On the basis of numerical simulations, Rausser and Small (2000) concluded that, "under plausible conditions," the bioprospecting value of certain genetic resources could be sufficient to support "market based conservation of biodiversity." However, cautiously optimistically views such as these are challenged on a variety of grounds (see Barrett and Lybbert, 2000; Firn, in press, for example). It is fitting, therefore, that the final words in this book are left to the environmental economist David Pearce (chapter 44), who argues that biodiversity must be seen from the perspective of having *conferred value* and who explores the intricacies of valuing biodiversity.

REFERENCES

Barrett, C. B., and T. J. Lybbert. 2000. Is bioprospecting a viable strategy for conserving tropical ecosystems? *Ecol. Econ.* 34:293–300.

Firn, R. D. 2003. Bioprospecting—why is it so unrewarding? *Biodiv. Conserv.* 12:207–216.

MacIlwain, C. 1998. When the rhetoric hits reality in debate on bioprospecting. *Nature* 392:535–540.

Rausser, G. C., and A. A. Small. 2000. Valuing research leads: bioprospecting and the conservation of genetic resources. *J. Polit. Econ.* 108:173–206.

ten Kate, K., and S. A. Laird (ed.). 1999. *The Commercial Use of Biodiversity.* Earthscan Publications Ltd., London, United Kingdom.

Microbial Diversity and Bioprospecting
Edited by Alan T. Bull
© 2004 ASM Press, Washington, D.C.

Chapter 39

The Convention on Biological Diversity and Benefit Sharing

Kerry ten Kate

Microbiologists today face a challenge not only to stay abreast of a fast-moving and interdisciplinary field of science, but also to be conversant with the law and policy affecting them, ranging from treaties and national laws to contracts and intellectual property rights. The Convention on Biological Diversity (CBD) and the other policy instruments that have arisen in its wake are among the most significant of these.

The CBD is both an international treaty—and thus a source of international law—and an institutional framework for the continual development of legal, policy, and scientific initiatives on biological diversity. The CBD marks a milestone in international treaties. Where earlier treaties dealt with specific aspects of biodiversity, such as trade, particular ecosystems, geographic areas, or species, the CBD is comprehensive in its approach. Its scope is global, covering all components of biological diversity, from ecosystems and habitats, species and communities to genomes and genes, and it deals not only with the conservation of biological diversity in situ and ex situ, but with sustainable use and benefit sharing.

The objectives of the CBD are the conservation of biological diversity, the sustainable use of its components, and the fair and equitable sharing of the benefits arising from the use of genetic resources, including, through access to genetic resources, technology transfer and funding. The breadth of the issues it covers can be seen in Table 1, which sets out the titles of some of its key substantive articles.

The convention is constantly evolving. Since it was opened for signature at the 1992 Rio Earth Summit, its 187 parties have made various important commitments to implementation of its articles through the Subsidiary Body on Scientific, Technical, and Technological Advice (SBSTTA) and the primary political decision-making body, the biennial Conference of the Parties (COP). There have been over 180 COP decisions and SBSTTA recommendations, two further treaties (on biosafety and on agricultural plant genetic resources), six international programs on economics and on five different ecosystems, and 23 rosters of experts. The parties have adopted a Global Taxonomy Initiative, Guiding Principles on Invasive Alien Species, and a Global Strategy for Plant Conservation, among many other science-based tools. National biodiversity strategies required by the CBD have been undertaken in 150 countries, and the CBD's financial mechanism, the Global Environment Facility, has allocated some $4 billion to developing countries to implement the CBD (ten Kate, 2002).

The CBD's fundamental contribution to science will be to conserve the resource base for life sciences (and life itself): biological diversity. But it provides other opportunities as well as constraints. Treaties are agreements between states, but by and large it is organizations and not governments that are equipped to do the work. This offers a bargaining chip to universities, research institutes, companies, and communities. In return for helping governments achieve their commitments, microbiologists and their collaborators can participate in national policy-making, raise their profiles, derive a fresh mandate and renewed legitimacy for their work, and perhaps even use the CBD as a lever to help fund their work. As individual countries implement CBD work programs, apply COP guidelines, and execute national strategies, the influence of the CBD on science is likely to grow. One mechanism will be the allocation of public funding, another the advent of laws and policies that control the direction and methodologies of scientific research. One of the principal areas in which this is already happening is the regulation of access to genetic resources.

Kerry ten Kate • Insight Investment, 33 Old Broad Street, London EC2N 1HZ, United Kingdom.

Table 1. Key provisions of the CBD

Article	Provision
1	Objectives
6	General measures (including national strategies on conservation and sustainable use and mainstreaming biodiversity into other areas of policy)
7	Identification and monitoring
8	In situ conservation
9	Ex situ conservation
10	Sustainable use of components of biological diversity
12	Research and training
13	Public education and awareness
14	Impact assessment and minimizing adverse impacts
15	Access to genetic resources
16	Access to and transfer of technology
17	Exchange of information
18	Technical and scientific cooperation (and the clearing-house mechanism)
19	Handling of biotechnology and distribution of its benefits

ACCESS TO GENETIC RESOURCES AND BENEFIT SHARING IN THE CBD AND NATIONAL ACCESS LEGISLATION

The world's biological diversity is distributed largely in inverse proportion to scientific and technological capacity (Macilwain, 1998). The need for access to genetic resources by industry and the benefits sought by biologically diverse countries, asked by the international community to conserve biodiversity, set the scene for an exchange. The CBD reflects a commitment by governments to facilitate access to genetic resources in return for a fair and equitable sharing of benefits such as technology transfer (CBD Article 1), an exchange that has been described as a grand bargain (Gollin, 1993).

As defined in the CBD, genetic resources are any material of plant, animal, microbial, or other origin containing functional units of heredity of actual or potential value. (The exact definitions in Article 2 of the CBD define "genetic resources" as "genetic material of actual or potential value" and "genetic material" as "any material of plant, animal, microbial or other origin containing functional units of heredity.") Access to this significant chunk of life (humans are excluded) is vital for education and research in the life sciences, as well as for research on the conservation and sustainable use of biodiversity. Access also underpins commercial discovery and development. Global sales of products derived from genetic resources (pharmaceuticals, botanical medicines, major crops, horticulture, crop protection products, cosmetics and personal care products, and a broad range of biotechnologies) lie between $500 billion and $800 billion and are set to increase (ten Kate and Laird, 1999) (Table 2).

Table 2. Ballpark high and low estimates for annual markets for various categories of products derived from genetic resources[a]

Sector	Market (billions of U.S. dollars)		Notes
	Low	High	
Pharmaceuticals	75	150	Some products derived from genetic resources. Low estimate: natural products from 25% of global market. High estimate: 50%. (Chapter 3)
Botanical medicines	20	40	All products derived from genetic resources. (Chapter 4)
Agricultural produce (commercial sales of agricultural seed)	300+ (30)	450+ (30)	All products derived from genetic resources. Low estimate: final value of produce reaching consumer 10 times commerical sales of seed to farmers. High estimate: 15 times commercial sales of seed to farmers. (Chapter 5)
Ornamental horticultural products	16	19	All products derived from genetic resources. Low estimate: based on available data. High estimate: allows for unreported sales and products. (Chapter 6)
Crop protection products	0.6	3	Some products derived from genetic resources. High estimate includes wholly synthesized analogues, as well as semisynthesized products. (Chapter 7)
Biotechnologies in fields other than health care and agriculture	60	120	Some products derived from genetic resources. Low and high estimates based on assessments of environmental biotechnologies. (Chapter 8)
Cosmetics and personal care products	2.8	2.8	Some products derived from genetic resources, including cosmetics and "natural" products. (Chapter 9)
Rounded total	500	800	

[a] Source: ten Kate and Laird (1999). For more detail, please see ten Kate and Laird (1999). Corresponding chapters in that publication are indicated in parentheses.

The CBD seeks to balance the sovereignty and authority of national governments with the obligation for states to facilitate access to genetic resources for environmentally sound purposes. According to Article 15, access is to be subject to governments' prior informed consent on terms, mutually agreed by the provider and recipient, that promote the fair and equitable sharing of benefits (Table 3). Similarly, Article 8(j) provides that, subject to national law, access to the knowledge, innovations, and practices of indigenous and local communities requires the prior approval of the holders of that knowledge.

The CBD itself is a framework convention, setting out the provisions according to which states should regulate access to genetic resources. A great deal of discretion is left to national governments to determine how to regulate access to genetic resources. In response, some 100 countries—largely those that are home to the bulk of the world's biodiversity—have introduced, or are now considering, laws that regulate access by scientists to genetic resources, biochemicals, and associated traditional knowledge. These typically require national and foreign scientists alike to obtain permission for access and to work with partners from the countries providing the genetic resources, in the process, sharing benefits such as royalties, technology, joint research, and information.

(Regional groups, national governments, or state governments already regulating access to genetic resources to ensure prior informed consent and benefit sharing include the Andean Pact [Bolivia, Colombia, Ecuador, Peru, Venezuela]; Australia [the states of Western Australia and Queensland]; Brazil [at the federal level and the states of Acre and Amapa]; Cameroon; Costa Rica; the Republic of Korea; Malaysia [the state of Sarawak]; Mexico; Nicaragua; the United States of America [within Yellowstone and other national parks]; Thailand; and the Philippines. Those planning to regulate access to genetic resources to ensure prior informed consent and benefit sharing include the member countries of the Association of South-East Asian Nations [ASEAN—10 members]; Australia [the Commonwealth]; Ivory Coast; Cuba; Ethiopia; Eritrea; Fiji; the Gambia; Denmark [Greenland]; Guatemala; India; Indonesia; Kenya; Lao PDR; Lesotho; Malawi; Malaysia [at the national level and the state of Sabah]; Mozambique; Namibia; Nepal; Nigeria; the African Union [53 members]; Pakistan; Papua New Guinea; Samoa; the Seychelles; the Solomon Islands; South Africa; Sri Lanka; Tanzania; Thailand; Uganda; Vanuatu; Vietnam; and Yemen. Belize, China, El Salvador, Ghana, Guyana, Hungary, Iceland, Panama, the Russian Federation, and Zimbabwe may also be planning to regulate access to genetic resources in the near future. [Personal communication with delegates at the sixth meeting of the Conference of the Parties to the CBD and with Lyle Glowka, April 2002].)

Table 3. Summary of provisions in the CBD on access to genetic resources, on the knowledge, practices and innovations of local and indigenous communities, and on benefit sharing

Article	Provision
8 (j)	Promote the wider application of the knowledge, innovations and practices of indigenous and local communities with their approval and involvement and encourage the equitable sharing of the benefits arising from the utilization of the knowledge, innovations and practices of indigenous and local communities.
15.1	Sovereign rights of States over their natural resources; the authority of national governments to determine access to genetic resources.
15.2	Endeavor to create conditions to facilitate access to genetic resources for environmentally sound uses by other Contracting Parties and not to impose restrictions that run counter to the objectives of the CBD.
15.3	Articles 15, 16, and 19 only apply to genetic resources acquired "in accordance with this Convention," i.e., not to those obtained prior to its entry into force or from non-parties.
15.4	Access, where granted, to be on mutually agreed terms and subject to the provisions of Article 15.
15.5	Access to genetic resources to be subject to prior informed consent of the Contracting Party providing such resources, unless otherwise determined by that Party.
15.6	Endeavor to develop and carry out scientific research based on genetic resources provided by other Contracting Parties with the full participation of, and where possible in, such Contracting Parties.
15.7	Take legislative, administrative or policy measures, as appropriate, . . . with the aim of sharing in a fair and equitable way the results of research and development and the benefits arising from the commercial and other utilisation of genetic resources with the Contracting Party providing such resources. Such sharing to be upon mutually agreed terms.
16.3	Access to and transfer of technology using genetic resources to countries providing the genetic resources.
19.1	Effective participation by providers of genetic resources in biotechnological research on the genetic resources they provide.
19.2	Priority access on a fair and equitable basis by countries (especially developing countries) providing genetic resources to the results and benefits arising from biotechnologies based on them. Such access to be on mutually agreed terms.

THE MOVE TO FAIRER PARTNERSHIPS

Has this legal and policy framework at the international and national levels helped clarify the responsibilities of those involved in access arrangements and ensured fair partnerships? The practical implementation of these principles poses an enormous challenge for the governments in countries that have ratified the CBD, for the many sectors of industry that need access to genetic resources for product discovery and development, and for the communities that are custodians of resources. Together, they must find workable rules and procedures that reflect the rights of sovereign states, communities, research institutions, individuals, and companies but deliver partnerships that are fair and equitable in the context of the risks and rewards of product development. The rules and procedures need to be speedy, simple, and efficient. A number of factors conspire to make this difficult (ten Kate and Laird, 2000).

To begin with, despite its importance to mankind, the biological diversity at the heart of the exchange is being eroded. Conservative estimates place current extinction rates for well-documented groups of vertebrates and vascular plants at 50 to 100 times larger than the expected natural background (World Resources Institute, 1999, www.wri.com; Janetos, 1997). Second, the countries, institutions, communities, and companies involved in the exchange of genetic resources have extremely different perceptions about the relative value of the genetic resources, information, innovation, and research and development that are needed for product discovery and development. Their different expectations often block the successful conclusion of partnerships for scientific research and commercialization. For some, any commercial use of genetic resources is "biopiracy," because it is believed that the legal and policy environment does not adequately ensure prior informed consent and adequate benefit sharing (Shiva, 1998; RAFI, 1999, www.rafi.org; Genetic Resources Action International, 1999, www.grain.org). Others believe that countries have an unrealistic and overinflated estimation of the value to industry of access to their genetic resources. They fear that the "grand bargain" may be misconceived because there is insufficient commercial demand for access to genetic resources to generate benefits that will create the incentive to conserve biological diversity or to help countries develop. (For commentary on the range of perceptions discussed here, see "The Different Perspectives," ten Kate and Laird [1999, page 6].)

Furthermore, commercial demand for access is unreliable. Over the past 30 years, interest in accessing biodiversity for pharmaceutical development has been cyclical. In many sectors, research dollars are flowing out of natural products and into synthetic chemistry for rational drug design, combinatorial approaches, and genetics that focus largely on human material. A goal in many national biodiversity strategies is to help alleviate poverty, support sustainable livelihoods, and raise standards. Countries might do well to use the untapped potential for research on genetic resources to meet domestic needs, for example, through low-cost botanical medicines, rather than

seeking only to supply fickle international markets. They could also ensure that regulations distinguish between commercialization and the more steady demand for access for vital conservation research in fields such as ecology and systematics.

Another challenge is that benefits are not always forthcoming to countries facilitating access to genetic resources. Much genetic material used for research and development is obtained from collections made before the CBD entered into force, for which there are generally no benefit-sharing arrangements (see Svarstad, Chapter 40, this book). Any benefits that are negotiated rarely trickle down to local communities or to conservation. Scientific organizations tend to benefit most. Countries could require a certain proportion of benefits to be dedicated to conservation, as Costa Rica and Western Australia have done. Countries could also adapt growing experience with trust funds and other mechanisms to ensure that local people benefit and have an incentive to support conservation measures.

A third problem is that a number of features of the transfer of genetic resources and the discovery and development of products make the monitoring and enforcement of access and benefit-sharing agreements extremely difficult (Parry, 1999; ten Kate and Laird, 1999). Material often travels from countries of origin to the private sector through a complicated route, passing through many hands from collection to commercialization, with value being added at each stage. The product that is commercialized is frequently not physically linked to the original genetic resources collected, but may have been manufactured from scratch based on modifications of chemical structures originally found in nature. Consequently, it is difficult to track the exchange of genetic resources and link it to the sharing of benefits. The resulting lack of transparency, compounded by the common requirement for confidentiality in commercial partnerships, does not help to dispel the high levels of distrust prevalent between potential partner countries, companies, and institutions (ten Kate and Laird, 2000).

There is evidence that the anticipated bureaucracy, delay, and expense of compliance with the first wave of access laws have deterred foreign and domestic scientists and thus have unwittingly stifled not only commercial research but also essential conservation work. Confusion over which government bodies are authorized to grant access has not helped. Encouragingly, there is growing acknowledgment of the need for a more strategic and flexible approach and, following a COP decision, each government is also to nominate a single focal point to streamline access enquiries.

Three recent developments—the Bonn Guidelines, the International Treaty on Plant Genetic Resources for Food and Agriculture, and practical guidelines developed by researchers—should help to tackle these problems.

RECENT DEVELOPMENTS

In April 2002, the parties to the CBD adopted the Bonn Guidelines on Access and Benefit Sharing (see www.biodiv.org). These guidelines provide guidance to countries in the development of law and policy on access to genetic resources, associated traditional knowledge, and benefit sharing and to stakeholders such as university researchers, companies, and communities in the negotiation of access and benefit-sharing agreements. The guidelines encourage countries to take a strategic and flexible approach. They set out provisions on prior informed consent and mutually agreed terms and list key elements of the roles and responsibilities of countries and organizations as they provide and use genetic resources and associated traditional knowledge. They also contain suggestions on administrative aspects such as a national focal point for each country and potential functions of any competent national authorities established by governments to regulate access.

The guidelines encourage the participation of stakeholders, which, in the field of health care, might include health research organizations and pharmaceutical companies, but also traditional healers and their associations and indigenous and local communities providing knowledge associated with genetic resources that are accessed. The appendices set out suggested elements for inclusion in the Material Transfer Agreements commonly used to access and exchange genetic resources and typical monetary and nonmonetary benefits.

Another important recent development on access and benefit sharing, relevant to crop development but not to pharmaceuticals, is the International Treaty on Plant Genetic Resources for Food and Agriculture (IT). This treaty was finalized in Rome in November 2001, in harmony with the CBD. One of its important elements is a multilateral system for facilitated access to 35 crop genera and 29 forage species in the public domain worldwide for food and agriculture and associated benefit sharing through the exchange of information, access to and transfer of technology, capacity building, and a commercial benefit-sharing package. "Food and agriculture" explicitly excludes "chemical, pharmaceutical, and/or other nonfood and feed industrial uses." The IT will require seed companies to pay royalties on patented products derived from the genes accessed.

In addition, the International Agricultural Research Centres (IARCs) are invited to enter into agreements with the governing body of the IT concerning not only Annex I materials, but other materials in their collections. Benefit-sharing obligations—through the exchange of information, access to, and transfer of technology—capacity building, and sharing the benefits of commercialization are also described (see ftp://ext-ftp.fao.org/waicent/pub/cgrfa8/iu/ITPGRe.pdf or http://www.fao.org/ag/cgrfa/News.htm).

A range of individual companies, professional associations, and gene banks such as botanic gardens, culture collections, and indigenous communities' groups have developed institutional policies in line with the CBD that provide principles and practical guidance for their employees and associates (Laird, 2002). One example is the one-page set of Principles on Access and Benefit-Sharing for Participating Institutions. Twenty-eight botanic gardens and herbaria from 21 countries, led by the Royal Botanic Gardens, Kew, developed common standards on access to genetic resources and benefit sharing (Latorre Garcia et al., 2001; www.rbgkew.org/conservation; Table 2). Another example is the Micro-Organisms Sustainable Use and Access Regulation International Code of Conduct (MOSAICC), developed by a consortium of 16 organizations working with microorganisms (see next paragraph). Through initiatives such as these, practical tools such as material transfer agreements, steps for obtaining prior informed consent, guidelines for the preparation of strategies, and illustrative examples of benefit-sharing arrangements are now available (www.biodiv.org; ten Kate and Wells, 2001).

Finally, the prospect of further intergovernmental agreement on access and benefit sharing has arisen, on which the input of groups such as microbiologists has been invited. One paragraph of the Plan of Implementation of the World Summit on Sustainable Development adopted by heads of state in September 2002 called for governments to "negotiate within the framework of the Convention on Biological Diversity, bearing in mind the Bonn Guidelines, an international regime to promote and safeguard the fair and equitable sharing of benefits arising out of the utilization of genetic resources." At a CBD meeting in March 2003, governments, communities, and interested organizations were invited to submit views on the desirable nature of such a regime to the Secretariat, which will compile them so that the CBD's Working Group on Access and Benefit Sharing can advise the meeting of the COP in March 2004 on how it may wish to address this issue.

MOSAICC

The Micro-Organisms Sustainable Use and Access Regulation International Code of Conduct (MOSAICC) is an initiative of the Belgian Co-ordinated Collections of Micro-organisms (BCCM), together with 16 other organizations working with microorganisms from around the world. MOSAICC is a voluntary code of conduct (i) to facilitate access to microbial genetic resources in line with the CBD and other applicable national and international law (including the Budapest Treaty on the 1997 International Recognition of the Deposit of Micro-Organisms for the Purposes of Patent Procedure, and the 1994 WTO Agreement on Trade-Related aspects of Intellectual Property Rights [TRIPs Agreement]) and (ii) to ensure that the transfer of material takes place under appropriate agreements between partners and is monitored to secure benefit sharing. MOSAICC is a living document and, like the Bonn Guidelines on Access and Benefit Sharing, is open to further improvement.

MOSAICC seeks to assist microbiologists by enabling them to secure prior informed consent (PIC) for access, in line with CBD Article 15.5, as well as to negotiate a Material Transfer Agreement (MTA) for access to and transfer of genetic resources and associated technology, fair and equitable benefit sharing, and scientific and technical cooperation. It aims to assist countries providing microbial genetic resources by suggesting procedures to issue PIC, as well as to monitor the transfer of genetic resources, enabling fair and equitable sharing of the possible benefits arising from their utilization. MOSAICC also provides recommendations and model documents (MTAs, as well as access application forms and certificates) to be considered as guidelines for optimal implementation of the CBD. The code and its accompanying documents can be downloaded from ⟨www.belspo.be/bccm/mosaicc⟩. The European Biological Resource Centres Network is currently adapting MOSAICC to develop its own Material Transfer Agreement, equivalent to the Uniform Biological Material Transfer Agreement that has been introduced in the United States introduced in 1995 by the National Institutes of Health and the Public Health Service. (Source: Wells and ten Kate [in press], European Community Thematic Report to the CBD on Benefit-Sharing.)

BENEFIT SHARING IN PRACTICE

The benefits that are shared by those involved in a partnership involving access to genetic resources can be as varied as the imaginations of the partners allow. The type and magnitude of the benefits that arise from access to genetic resources and how they can fairly be shared will vary from case to case, not only in terms of whether use is for academic or commercial purposes, but also in terms of the specific use that is being made of the genetic resources. Experience is growing with Access and Benefit Sharing (ABS) partnerships between different users and providers of genetic resources, and benefit-sharing practices are now increasingly well-documented (see, for example, UNEP/CBD/COP/3/20, UNEP/CBD/COP/4/22, and the benefit-sharing case studies listed in ten Kate and Wells, 2001). Partnerships involving access to genetic resources can offer benefits of three main types: improved conservation, training and capacity building, and sustainable economic development. Table 4 provides examples of both monetary and nonmonetary benefit sharing (see ten Kate and Laird [1999] for a

Table 4. Common benefits shares in ABS partnerships

Benefit	Description
Monetary	Up-front fees, either for access to genetic resources or to cover the costs of any preparation of samples, research conducted on them, and handling and shipping costs
	Milestone payments when various stages in discovery and development are reached (either independent payments or set off against any royalties that may be incurred in the future
	Royalties—it is important to clarify the basis of royalty payments, for example, whether they are calculated on gross or net sales
Nonmonetary[a]	Participation of source-country scientists (who may be third parties) in research
	Transfer of equipment, software, and know-how
	Exchange of staff and training
	In-kind support for conservation (in situ and ex situ)
	Acknowledgment of providers in research publications patents, and other forms of intellectual property rights
	Joint patent rights or other intellectual property right
	Sharing of research results, including notification of discoveries and ensuring that copies of publications concerning research on the genetic resources are sent to the source country
	Voucher specimens to be left in national institutions
	Rights to license technologies developed from research on the genetic resources (at discounted rates)

[a] Nonmonetary benefits are an increasingly important component of ABS agreements. First, many uses of genetic resources are noncommerical and only give rise to nonmonetary benefits. Second, many forms of commercial discovery and development only to lead to a successful product (and thus monetary benefits) in a small proportion of cases, and often after many years. Finally, nonmonetary benefits often boost competitiveness by building long-term capacity in source countries, e.g., increased scientific capacity through joint research (Source: ten Kate and Wells [2001]).

more detailed discussion of benefit sharing in the pharmaceutical and other industries).

The division between commercial and noncommercial use of genetic resources is difficult to draw and can become blurred. However, benefits of a different nature and magnitude are involved in academic research and commercial development, and some scientific users such as students and botanic gardens cannot afford the transaction costs involved in complex ABS negotiations better suited to commercialization agreements. Also, the risk of strategic and commercially significant loss to a country is less in the case of access by a herbarium to prepare specimens than by a company to prepare extracts for screening. Although some activities are close to the boundary, the majority can be clearly categorized as commercial or noncommercial, so it is probably more pragmatic to draw this distinction and thus keep transaction costs and administration to a minimum. Any fast-track system of ABS for scientific use could rest on ensuring that scientific recipients are legally bound only to use the genetic resources for noncommercial purposes and not to pass them on to third parties for any other (commercial) purposes (ten Kate and Wells, 2001).

Benefits negotiated as part of access agreements are usually associated with the use of the genetic resources for which access is permitted, e.g., participation in biochemical research, or access to the technology used to conduct research on the genetic resources being accessed. Sometimes, however, benefits can be negotiated that are related more to the provider country's own priorities than to a user's activities. For example, a provider-country institution working with an access applicant can seek screens and training in therapeutic areas of priority to the country, e.g., malaria or river blindness, in exchange for samples that the applicant company will screen against its own therapeutic priorities, such as cancer and HIV. Some benefits, particularly for use at the community level, can be quite unrelated to the intended use of the genetic resources and are really in-kind equivalents of monetary benefits. Examples from earlier benefit-sharing partnerships include funds for communities to revitalize local health traditions, buy cows, or create an airstrip (ten Kate and Wells, 2001).

However, although accounts of ABS partnerships provide information on nonmonetary benefits such as joint research and on monetary benefits such as royalties, there are fewer examples of how ABS partnerships support conservation. National strategies on ABS can help ensure adequate consideration of sustainable harvesting as well as dedication of some share of benefits to conservation activities. Some partnerships already offer illustrations of how

this can be done. In December 1993, the Western Australian Department of Conservation and Land Management (CALM) entered into an agreement with an Australian pharmaceutical company, AM-RAD, to access and commercialize a species of smokebush (genus *Conospermum*) which had shown promising activity against cancer. In return, AMRAD agreed to provide CALM $1.15 million (Australian), a share in royalties, rights to conduct research on the active compound, and $500,000 (Australian) for further research on some eight smokebush patents lodged by CALM. CALM used funds generated by the agreement to support Western Australia's conservation infrastructure. $300,000 (Australian) was allocated for the conservation of rare and endangered flora and fauna, and a further $300,000 (Australian) was set aside for other conservation activities, including geographical information systems and other information technology.

The work of Costa Rica's National Biodiversity Institute (INBio) is based on a cooperative research agreement with the Ministry of Environment and Energy (MINAE). This sets the terms and conditions for INBio's biodiversity inventory and bioprospecting activities. According to this agreement, INBio will donate 10% of all bioprospecting budgets and 50% of all income from royalties to conservation efforts by MINAE (see Tamayo et al., Chapter 41, this book).

AGREEMENTS

Prior informed consent and the nature of the benefits outlined above to be shared are increasingly commonly committed to an agreement. A number of different contractual arrangements may contain provisions related to access to genetic resources, traditional knowledge, benefit sharing, and intellectual property rights. These include:

- intellectual property licenses
- material transfer agreements
- environmental permits
- real estate leases and land tenure
- shrinkwrap license
- option agreements
- letters of intent
- memorandum of understanding

Each of these is outlined in Gollin (2002). Whatever the nature of the agreement, it will typically involve terms and conditions on:

- Introductory provisions: preambular reference to the CBD, IT, and any other applicable international, regional, or national law; description of the legal status of the provider and user of genetic resources; mandate of the parties and their general objectives in establishing the agreement.

- Conduct of the collaboration: the roles, rights, and responsibilities of the different parties in the collaborative research process; confidentiality; duty to minimize environmental impacts of collecting activities.
- Access and benefit sharing: prior informed consent and legal acquisition of the genetic resources and/or associated traditional knowledge; description and permitted uses of genetic resources and associated traditional knowledge covered by the agreement (e.g., research, breeding, commercialization); conditions under which user may seek intellectual property rights; benefits to be shared and benefits with whom they are to be shared; clauses on whether the recipient of the resources and knowledge may pass them on to third parties, and, if so, on what terms.
- A range of legal provisions: definitions; length of agreement; notice to terminate the agreement; fact that the obligations in certain clauses (e.g., benefit sharing) survive the termination of the agreement; independent enforceability of individual clauses in the agreement; events limiting the liability of either party (such as Act of God, fire, flood, etc.); arbitration and alternative dispute settlement arrangements; assignment or transfer of rights; choice of law.

See ten Kate and Laird (2002); Bonn Guidelines, UNEP/CBD/COP/6/20, www.biodiv.org; and Latorre et al. (2001).

CONCLUSIONS

The principles of the CBD are finding their way into national laws and policies and into the working practices of scientists. A notable example of this is the regulation of access to genetic resources and benefit sharing, which will affect the manner in which scientists can obtain the basic materials for research from around the world.

Access and benefit-sharing partnerships can be a source of sustainable economic development, providing developing countries and their stakeholders with benefits such as improved capacity for conservation, new products, and income to meet basic needs such as health care, as well as support for value-added scientific research. However, these partnerships touch on a complex, varied, and unpredictable set of issues

linked to policy-making in many areas of government, as well as to domestic and global markets. The uses of genetic resources are diverse, and the stakeholders involved range from multinational companies to indigenous communities, each with different priorities. In addition, demand for access to genetic resources fluctuates significantly and can be difficult to predict in the medium to long term. Without an informed strategy to address this complexity, access and benefit-sharing law and policy can miss opportunities to contribute to conservation and sustainable development, constrain science, inadvertently stifle equitable and beneficial partnerships, and alienate stakeholders.

The most beneficial partnerships are likely to be achieved with the guidance and support of a mixture of policy measures, such as simple and flexible legislation, the Bonn Guidelines, indicators and model agreements tailored to the different uses and users of genetic resources, and case studies illustrating best practice. Among the most important tools is a national strategy on access and benefit sharing, closely linked to national strategies in related areas, such as health, science and technology, and industrial competitiveness (ten Kate and Wells, 2001). This can help define informed and realistic policy that meets the practical requirements of scientists and priority needs of other stakeholders, such as communities. It can also help a country to remain competitive in the face of uncertainty and rapid changes in the scientific, technological, economic, and legal context.

Governments will need to consider the kind of measures they can introduce to foster value-added partnerships to streamline administrative access and benefit-sharing requirements and make them cost effective and flexible. Countries with a stable policy framework and those whose institutions can demonstrate high scientific calibre, reliability of supply of materials, and provide highly diverse samples are most likely to be able to enter into and benefit from access and benefit-sharing partnerships. Competitiveness in access and benefit sharing rests on a country's capacity to contribute value-added services—research rather than raw materials. This, in turn, depends on the presence of institutions of high scientific and technological calibre. Capacity building through participation in research is thus a key benefit to seek from partnerships.

Over the past 10 years, there has been a growing experience and awareness of the best practices in benefit sharing, particularly in the pharmaceutical industry. Many of the demands of providers of genetic resources, such as royalties, technology transfer, and collaborative research, regarded as novel 10 years ago are now common practice. Scientists—sometimes reluctantly—are coming to terms with the fact that they must deal with law and policy if they are to be able to continue their research. Many regularly deal with the practicalities of prior informed consent and benefit sharing, and some have taken the initiative of developing codes of conduct that match their activities and circumstances.

Scientists can—and do—participate in the development of the guiding international and national law. But there is room for more input to the treaty from scientists, by lobbying or joining national government delegates, participating through representative scientific organizations or industry associations as nongovernmental organizations at U.N. meetings, and serving on expert panels. Scientific organizations can become accredited and attend meetings of the CBD's COP and SBSTTA. Such participation is vital to ensure that the treaty is based on sound science and promotes, rather than hinders, conservation. Scientific organizations can also participate in the implementation and evolution of the Bonn Guidelines on Access and Benefit Sharing, so that regulations on access to genetic resources worldwide facilitate science and support fair partnerships. Their voices need to be heard on the broad range of scientific and technical issues covered by the CBD. Finally, scientific organizations should also work with federal, state, and local governments to ensure a coordinated approach and should push for consistent decisions in the range of other multilateral environmental agreements under the auspices of the United Nations, as well as with trade issues in the World Trade Organization.

REFERENCES

Gollin, M. A. 1993. An intellectual property rights framework for biodiversity prospecting. *In* W. V. Reid, S. A. Laird, C. A. Meyer, R. Games, A. Sittenfeld, D. H. Janzen, M. A. Gollin, and C. Juma (ed.), *Biodiversity Prospecting: Using Genetic Resources for Sustainable Development.* World Resources Institute, Washington, D.C.

Janetos, A. C. 1997. Do we still need nature? The importance of biological diversity. *Consequences* 3:1.

Laird, S. A. (ed.) 2002. *Biodiversity and Traditional Knowledge: Equitable Partnerships in Practice.* People and Plants Conservation Series. Earthscan Publications Ltd., London, United Kingdom.

Latorre Garcia, F., C. Williams, K. ten Kate, and P. Cheyne. 2001. Results of the Pilot Project for Botanic Gardens: Principles on Access to Genetic Resources and Benefit-Sharing, Common Policy Guidelines to Assist with Their Implementation and Explanatory Text. Royal Botanic Gardens, Kew, Richmond, United Kingdom.

Macilwain, C. 1998. When rhetoric hits reality in debate on bioprospecting. *Nature* 392:535–541.

Parry, B. 1999. The fate of the collections: social justice and the annexation of plant genetic resources. *In* C. Zerner (ed.), *People, Plants and Justice: The Politics of Nature Conservation.* Columbia University Press, New York, N.Y.

Shiva, V. 1998. *Biopiracy: The Plunder of Nature and Knowledge.* Green Books, Devon, United Kingdom.

ten Kate, K. 2002. Science and the Convention on Biological Diversity. *Science* 295:2371–2372.

ten Kate, K., and S. Laird. 1999. *The Commercial Use of Biodiversity: Access to Genetic Resources and Benefit-Sharing.* Commission of the European Communities and Earthscan Publications Ltd., London, United Kingdom.

ten Kate, K., and S. Laird. 2000. *Biodiversity and Business: Coming to Terms with the "Grand Bargain." In International Affairs,* vol. 76. Royal Institute of International Affairs, London, United Kingdom.

ten Kate, K., and S. Laird. 2002. Bioprospecting agreements and benefit-sharing with local communities. *In* P. Schuler and M. Finger (ed.), *Intellectual Property and Communities.* World Bank, Washington D.C.

ten Kate, K., and A. Wells. 2001. Preparing a national strategy on access to genetic resources and benefit-sharing. A pilot study. Royal Botanic Gardens, Kew and UNDP/UNEP Biodiversity Planning Support Programme.

Wells, A., and K. ten Kate. European Community Thematic Report to the CBD on Benefit-Sharing, Environment Directorate, European Commission, in press.

Microbial Diversity and Bioprospecting
Edited by Alan T. Bull
© 2004 ASM Press, Washington, D.C.

Chapter 40

The Historical Context of Present Bioprospecting—Four Cases

HANNE SVARSTAD

During all times, human beings have explored nature in the search for useful resources. Today's bioprospecting, aided by advanced biotechnologies, can be seen as a new phase of this important part of our history. In the last decade of the 20th century, there was a growing concern for the rights of source country parties to control bioprospecting and to participate in benefit sharing. This has become a North-South issue because the tropics are constituted by countries rich in biodiversity but otherwise poor. In addition, benefits from bioprospecting have been linked to potentials for funding conservation. Two factors have changed the open-access situation for bioprospecting. First, patents related to biotechnology inventions and strengthened plant breeders rights imply private ownership and commodification of the genetic resources that have been subject to scientific alteration. Second, and to a large extent a reaction from developing countries on the first change, the Convention on Biological Diversity (CBD) provides an international legal instrument for source countries' national sovereignty over their genetic resources. The CBD came into force in the end of 1993. It states that any access to genetic resources, which, when granted, is to be subject to prior informed consent and mutually agreed terms. The CBD provides a weight to balance the interests of bioprospectors for intellectual property rights and the interests of source country parties by the right to control access and share benefits. During the 1990s a number of countries started the process of developing legislation to implement access regulations.

In the following, I look at some historical cases that illustrate the ways that bioprospecting took place before the CBD, and thus for most of the times in open-access situations. First, two cases are emphasized in the presentation of colonial bioprospecting, and thereafter I show two cases from the period after the Second World War. These cases provide an entrance to the context of the present situation in which new bioprospecting is to be controlled and subject to benefit sharing. Old cases of bioprospecting are often used in present debates related to conservation and bioprospecting. As we shall see, this is not always done with a basis in solid data.

BUSY LIZZIE AND *SAINTPAULIA*: TWO CASES OF COLONIAL BIOPROSPECTING

When David Livingstone explored Africa in the middle of the 19th century, he was accompanied on his second trip by John Kirk, who collected plants for the Royal Botanical Gardens in Kew, Great Britain. Later, Kirk became the British Political Resident and subsequently British General Consul in Zanzibar, and he continued to send plants from East Africa to Kew. Busy Lizzie (*Impatiens sultani*) was among these plants (Hooker, 1882; Juma, 1989). In Europe and the United States, Busy Lizzie is today a very popular ornamental. Commercial gardeners and flower shops sell millions of Busy Lizzie plants in pots every year.

Saintpaulia (also called the African violet) is another popular potted plant. In Norway, for instance, it can be found in almost every home. John Kirk collected a sample of the genus *Saintpaulia* in 1884, and he wrote that it was found at the coast opposite Zanzibar. The sample was sent to Kew, but it did not result in regeneration. In the 1880s, Tanganyika came under German control. The colonial administrator Baron Walter von Saint-Paul-Illaire was Regional Commissioner of Tanga. In 1893 he sent two samples of the plant to his father in Germany. These species were described by a botanical garden in Hannover, and von Saint-Paul, Sr., sold them to a firm, which by 1894 had the first seeds for this flower on the market. A hundred years later—in the 1990s—more than 40,000 different cultivares of *Saintpaulia* had been named. Most of

Hanne Svarstad • Norwegian Institute for Nature Research, P. O. Box 736, Sentrum, 0105 Oslo, Norway.

these have been derived from the two species that von Saint-Paul, Jr., sent to Europe (Baatvik, 1993).

As the two cases of the ornamental plants from East Africa illustrate, botanical gardens played key roles during colonial times for the efforts to collect, systematize, and make global transfers of biodiversity entities. Moreover, the movements of plants around the globe was one of the major features of the British colonial power in the 19th century. For instance, rubber (*Hevea brasiliensis*) was transferred from Brazil to Southeast Asia, and cinchona was brought from the Andes to India and other parts of the world. Cinchona yields quinine, which can be used as a cure for malaria, and its transfers facilitated military, political, and economic expansion. In the cases of both rubber and cinchona, collections were conducted in South American countries in abuse of national laws that were established in an effort to keep others from producing and competing in the trade with these products (Brockway, 1979). Thus, rubber and cinchona exemplify exemptions from situations that we usually characterized by open access for bioprospecting. Obviously, the export bans in these cases did not provide efficient protection for the source countries.

The Kenyan social scientist Calestous Juma argues that the introduction of genetic material and related technology into the economic system is among the most crucial sources of economic growth, together with accompanying institutional reorganization. Whereas the exchange of genetic resources helped Great Britain to expand by its colonies, the introduction of genetic resources was, Juma argues, a major factor in the development of the United States into a leading agricultural nation. In the beginning of the 19th century, American consuls and marine officers were ordered to collect seeds and plants that could be useful for agriculture in the United States. Thomas Jefferson (U.S. President 1801–1809) once said that the greatest service that could be rendered to any country was to add a useful plant to its culture (Juma, 1989).

A moderate estimate shows that in Norway alone, a country of only 4.5 million people, more than 1 million potted plants of Busy Lizzie and more than 2.5 million *Saintpaulia* were sold in 1996 for at least $13 million U.S. (Svarstad, 2002a). In the early 1990s, Baatvik (1993) assessed the total trade in *Saintpaulia* in Europe and North America to much more than $100 million annually. None of these revenues go back to Tanzania. Some think that this is unfair and propose that a percentage of such revenues should be sent back to the countries from where the plants originated.

In nature, *Saintpaulia* is very rare. Twenty-one species of the genus have been published, and recently they have all been considered to be in need of protection. These flowers grow in limited areas in Tanzania, and two species also grow in Kenya. The plant is found in low numbers in fragmented populations of the Eastern Arc Mountains, especially in the Usambara Mountains, and some also in the coastal lowlands (Mather, 1989; Baatvik, 1993; Simiyu et al., 1996). There are botanists and environmentalists who have specifically proposed that a part of the income from *Saintpaulia* should be given to Tanzania to finance conservation efforts (Larsen, 1989; Mather, 1989). However, historical cases of transfer of plants from the South to the North are today not subject to negotiations over payments, and this restriction is a basic principle of the CBD.

THE PERIWINKLE AND THE "HARDANGERVIDDA FUNGUS": TWO CASES OF POSTWAR BIOPROSPECTING

The Eli Lilly Company of the United States has been active in natural product collection since the 1950s. Research groups at Eli Lilly, as well as at a university laboratory in Ontario, Canada, independently examined the reputation of the periwinkle (*Catharanthus roseus*) as an oral hypoglycemic agent of traditional medicine. They ended up discovering profound properties for the treatment of various cancers, first of all by the two alkaloids vincristine and vinblastine (Johnson et al., 1963; Noble, 1990). Over 1,000 tons of periwinkle are used in the United States every year for the manufacture of vinblastine and vinchristine (Laird, 1993).

The Swiss life science giant Novartis was created by the 1996 merger of Sandoz and Ciba-Geigy. In 1958, Sandoz began a screening program for antifungal antibiotics. Since then, employees on business trips and on vacation routinely carried plastic bags for collection of soil samples for the screening of microorganisms (Tribe, 1998). In 1969, a Sandoz biologist on holiday crossed the Norwegian mountain plateau Hardangervidda in a hired car with his wife. They often stopped to take pictures, and each time the biologist used the opportunity to collect a handful of soil. From one of the samples, the fungus *Tolypocladium inflatum* was found to produce the biochemical cyclosporin. In 1972 Sandoz researchers discovered the strong immunosuppressive property of this strain of cyclosporin. In 1983 the first cyclosporin medicine was ready for sale on the market, and in the late 1990s, this constituted Novartis's main drug, with sales of $1.2 billion in 1997. The medicine is important for patients who have had organ transplantations (Svarstad et al., 2000).

STRONG CLAIMS—BUT OFTEN WEAK DATA

In the current debates related to conservation and bioprospecting, old cases are often used to prove specific points. However, a close examination of the cases presented above shows that they are not always used in valid ways. In the following, I look at two major ways discursive icons are constructed concerning bioprospecting. The term discursive icon is used for features that constitute symbols within an established discourse about a topic. Discursive icons are seldom subject to critical examination.

First, species of old successful cases of bioprospecting are often used to argue the importance of conservation. On mistaken premises, the periwinkle has become an icon for the conservation of tropical rainforests. Catherine Caufield, for instance, is a science journalist who early published an engaging book about tropical rainforests, and there the periwinkle is highlighted as a "major anticancer drug from tropical rainforests" (Caufield, 1984, p. 221). Similarly, Chivian (1997) uses the periwinkle as one of a few examples of today's most useful medicines that came from tropical rainforests. Likewise, in an educational book for children called *Rainforest,* a chapter on rainforest destruction uses the periwinkle as the sole example of medicines "originally made from rainforest plants" (Parker, 1997). In advertisements for environmental nongovernmental organizations (NGOs), the periwinkle is often focused to show the importance of their work with conserving tropical rainforests. Two examples of this from 2001 include a television commercial of Worldwide Fund for Nature (WWF) Norway and a pamphlet from the Rainforest Foundation Norway.

However, the periwinkle is not at all a rainforest plant. In the *Flora of Southern Africa,* the plant is characterized as a "weed of waste places in warm, dry areas" (Dyer et al., 1963). Spreading as a weed and widely disseminated in tropical and subtropical areas of the world, the actual character of the periwinkle does not substantiate the symbolic role that this plant has been given in struggles against erosion of biodiversity.

Similarly, the "Hardangervidda fungus" has often been used by Norwegian biologists and conservationists as a discursive icon for conservation (e.g., Høyland and Ryvarden, 1990; WWIW Norway, 1998). The message of this narrative is that important medicines for humankind can be found in biodiversity, even in barren Norwegian mountains with relatively few species. Thus, conservation is important. Like the periwinkle, however, *T. inflatum* is a widespread species (Svarstad et al., 2000), and it is not likely to face extinction, no matter how the Norwegian mountains are managed.

Looking at the two examples presented above of colonial bioprospecting in East Africa, Busy Lizzie is a widespread plant in Africa, and it has not been used in arguments for conservation. *Saintpaulia,* on the other hand, is the only one of the examples presented here of a truly rare species in need of protection. Thus, on a legitimate basis, environmental organizations in East Africa often use *Saintpaulia* as a conservation icon.

Patenting constitutes the other main type of discursive icons related to bioprospecting. Patenting provides the assignee with the control of the use of the invention for a period of usually about 20 years. Participants in the "biopiracy discourse" dismiss bioprospecting as exploitation, and patenting is focused on as a major cause (Svarstad, 2000, 2002b). Old cases of bioprospecting are sometimes used to prove these points. In the case study I conducted with colleagues on the "Hardangervidda fungus," we discussed possible benefits and provided the specificities of the case from the point of the present norms for access and benefit sharing. We found that Norway, in a situation of regulated access, could have had a reasonable claim of 2% annual royalties of Novartis's sale of the cyclosporin medicines. In 1997, 2% of the income from sales of this blockbuster drug amounted to $24.3 million U.S. Our study showed that in this case the lack of regulations and agreement is the cause of the income loss to the source country of biodiversity, while the patents related to cyclosporin are not relevant for this loss (Svarstad et al., 2000).

Data from this study have been reproduced several times in Norwegian media. However, the conclusions presented by journalists have tended to be remarkably different from those we made in our study. Based exclusively on our research, the Norwegian NGO magazine *Folkevett* wrote about the Norway-Novartis case under the headline "EU patented" on the front page with a picture from Hardangervidda. The following quote is from that article.

> The patent wave is seriously hitting us. Norway loses millions annually because a foreign company has patented a medicine made of fungus from Norwegian nature. The developing countries lose even more, when pharmaceutical producers grab a piece of nature and patent so-called biological innovations without providing any benefits to source country (Munkejord, 1998 [my translation]).

With *Folkevett* as the source, other media have repeated this narrative, such as a news bulletin on the major Norwegian TV channel (Dagsrevyen, NRK1) and in Norway's largest subscription newspaper, *Aftenposten.* In this case, patenting is claimed in the media to be the source of exploitation, even though the research from which it was generated provides the opposite conclusion (Svarstad, 2002b).

The periwinkle has also become subject to a biopiracy narrative. For instance, Rural Advancement Foundation International (RAFI) (an NGO based in North America; in 2001 it changed its name to ETC Group) writes that the "patented biopiracy of Africa's innovations and resources is becoming endless," with the following as one of the examples:

> Vital compounds from a medicinal plant in Madagascar, the Rosy Periwinkle, were patented by a U.S. pharmaceutical company and are now making it hundreds of millions of dollars each year in sales as cancer drugs (RAFI, 1999).

Do patenting and sales of cancer medicines made of the periwinkle imply an exploitation of Madagascar? The plant is believed to have its origins in Madagascar, and it is often called the Madagascar periwinkle. However, during the past few hundred years it has spread out extensively, and today it is widely naturalized in the tropics and subtropics of both hemispheres (Dyer et al., 1963). Moreover, the researchers discovering its medicinal effects for cancer treatment first noticed traditional usage in Jamaica and the Philippines. The researchers at Eli Lilly selected and collected the plant in the Philippines, while the Canadian research group at the same time obtained the plant for screening from Jamaica (Noble, 1990). Thus, if this had taken place after the CBD came into force, the access to the biodiversity in question could have been made subject to a generally accepted claim of benefit sharing for the Philippines and Jamaica. In this case, the patents related to the medicines would have protected the interests not only of Eli Lilly and the Canadian institute, but also of the two source countries against pirate production without benefit sharing by other companies.

However, it is not possible to substantiate the claims of exploitation and losses for Madagascar in relation to the sales and patents of the periwinkle medicines. On the contrary, the medicine production in the 1970s and 1980s led to considerable export earnings for Madagascar as well as for some other countries such as India, Mozambique, and the Philippines. Thus, in 1983, Madagascar exported approximately 1,100 tons of periwinkle (Rasoanivo, 1990). However, the revenues from this export have diminished over time because the company has gradually shifted its raw material source to a plantation in Texas (Sheldon et al., 1997).

Old cases of bioprospecting can be useful to examine in order to shed light on debates on how to handle the issue today. However, when studying the use of old cases in the present debates, one often finds unsubstantiated claims in which discursive icons play an important role.

All the presented cases show that bioprospecting sometimes result in products of high benefits. For source countries to control bioprospecting activity and gain any benefits from it, it is imperative for them to implement the CBD by establishing adequate legislation and institutional competence and capacity.

REFERENCES

Baatvik, S. T. 1993. The genus *Saintpaulia* (Gesneriaceae) 100 years: history, taxonomy, ecology, distribution and conservation. *Fragm. Flor. Geobot.* (Suppl.) 2:97–112.

Brockway, L. 1979. *Science and Colonial Expansion: The Role of the British Royal Botanic Gardens.* Academic Press, New York, N.Y.

Chivian, E. 1997. Global environmental degradation and biodiversity loss: implications for human health. *In* F. Grifo and J. Rosenthal (ed.), *Biodiversity and Human Health.* Island Press, Washington, D.C.

Caufield, C. 1984. *In the Rainforest.* The University of Chicago Press, Chicago, Ill.

Dyer, R. A., L. E. Codd, and H. B. Rycroft. 1963. *Flora of Southern Africa. The Republic of South Africa, Basutoland, Swaziland and South West Africa.* The Republic of South Africa, Pretoria.

Hooker, Sir J. D. 1882. Impatiens Sultani. *Curtis's Botanical Magazine,* vol. XXXVIII of the Third Series.

Høyland, K. and L. Ryvarden 1990. *Er det liv er det sopp. Sopp i miljoe og kulturhistorie* [If there is life, there is fungus (mushroom). Fungus in the environment and cultural history]. Fungiflora, Oslo, Norway.

Johnson, I. S., J. G. Armstrong, M. Gorman, and J. P. Burnett, Jr. 1963. The vinca alkaloids: a new class of oncolytic agents. *Cancer Res.* 23:1390–1397.

Juma, C. 1989. *The Gene Hunters. Biotechnology and the Scramble for Seeds.* Zed Books Ltd., London, United Kingdom.

Laird, S. A. 1993. Contracts for biodiversity prospecting. *In* W. V. Reid, S. A. Laird, C. A. Meyer, R. Gámez, A. Sittenfeld, D. H. Janzen, M. A. Gollin, and C. Juma (ed.), *Biodiversity Prospecting: Using Genetic Resources for Sustainable Development.* World Resources Institute, Washington, D.C.

Larsen, A. F. 1989. *Usambara. Afrikas groenne magnet* [Usambara. Africa's green magnet]. Gonzo/Mellemfolkeligt Samvirke, Copenhagen, Denmark.

Mather, S. 1989. Saintpaulia. *In* A. C. Hamilton and R. Bensted-Smith (ed.), *Forest Conservation in the East Usambara Mountains, Tanzania.* Gland and IUCN, Cambridge, United Kingdom.

Munkejord, M. 1998. Fra hoeyfjells-sopp til menneskegener. *Folkevett* 20:5.

Noble, R. L. 1990. The discovery of the vinca alkaloids—chemotherapeutic agents against cancer. *Biochem. Cell Biol.* 68:1344–1351.

Parker, J. 1997. *Rainforests.* Quarto Children's Books Ltd., London, United Kingdom.

RAFI (Rural Advancement Foundation International). 1999. TRIPS traps for small farmers. *RAFI News,* 19 May.

Rasoanaivo, P. 1990. Rain forests of Madagascar: sources of industrial and medicinal plants. *Ambio* 19:421–424.

Sheldon, J. W., M. J. Balick, and S. A. Laird. 1997. Medicinal plants: can utilization and conservation coexist? The New York Botanical Garden: *Adv. Econ. Bot.* 12.

Simiyu, S. W., P. Muthoka, J. Jefwa, B. Bytebier, and T. R. Pearce. 1996. The conservation status of the genus *Saintpaulia* in Kenya, p. 341–344. *In* L. J. G. Van der Maesen et al. (ed.), Dordrecht, *The Biodiversity of African Plants.* Kluwer Academic Publishers, Dordrecht, The Netherlands.

Svarstad, H. 2000. Reciprocity, biopiracy, heroes, villains and victims. *In* H. Svarstad and S. S. Dhillion (ed.), *Responding to Bioprospecting. From Biodiversity in the South to Medicines in the North.* Spartacus, Oslo, Norway.

Svarstad, H. 2002a. Biologisk mangfold: Ressurser i Soer og interesser i Nord [Biological diversity: Resources in the South and interests in the North]. *In* T. A. Benjaminsen and H. Svarstad (ed.), *Samfunnsperspektiver på Miljoe og Utvikling* [Social Science Perspectives on Environment and Development], 2nd rev. ed. Universitetsforlaget, Oslo, Norway.

Svarstad, H. 2002b. Analysing conservation-environment discourses: the story of a biopiracy narrative. *Forum Dev. Studies* 29(1):31–61.

Svarstad, H., H. C. Bugge, and S. S. Dhillion 2000. From Norway to Novartis: cyclosporin from tolypocladium inflatum in an open access regime. *Biodivers. Conserv.* 9:1521–1541.

Tribe, H. T. 1998. The discovery and development of Cyclosporin. *Mycologist* 12:20–22.

WWIF Norway. 1998. Soppen fra Hardangervidda [The fungus from Hardangervidda]. *Verdens Nat.* 2.

Microbial Diversity and Bioprospecting
Edited by Alan T. Bull
© 2004 ASM Press, Washington, D.C.

Chapter 41

Biodiversity Prospecting: The INBio Experience

GISELLE TAMAYO, LORENA GUEVARA, AND RODRIGO GÁMEZ

Bioprospecting, although defined and shaped in the past century, is associated with the survival, adaptation, and evolution of the human species. Our ancestors widely used elements from biodiversity. To cite a few examples, plant and animal fibers were used for clothing, oils and waxes as fuels and ointments, and the fruiting body of *Piptoporus betulinus* was used by the "Ice Man" to treat an endoparasite infection (Mateo et al., 2001).

Bioprospecting is now defined as the systematic search for genes, compounds, designs, and organisms that might have a potential economic use and might lead to a product development. Although this broad definition apparently describes a mere scientific challenge with the final goal of economic development, it harmonizes with other important social aspects such as the conservation of wild lands through sustainable use of biodiversity and the technical and scientific capacity for building local human resources (Sittenfeld et al., 1999; Tamayo et al., 1997). This new interpretation of the above definition is consistent with the mission of Costa Rica's National Institute of Biodiversity (INBio): "to promote a new awareness of the value of biodiversity, and thereby achieve its conservation and use to improve the quality of life" (INBio, 2000; Gámez, 2000).

RESEARCH COLLABORATIVE AGREEMENTS

In 1991 INBio set a landmark when it signed the first commercial bioprospecting research collaborative agreement (RCA) between a pharmaceutical industry and a biodiversity-rich country organization: Merck & Co. and INBio. Although there have been plenty of such agreements with other companies all over the world, this particular one has been taken as a model for different case studies (ten Kate and Laird, 1999; Reid et al., 1994; Sittenfeld and Villers, 1994)

and surprisingly enough, it is still of interest. Several issues, such as the legal framework and conservation strategy that synergistically shaped the RCA with Merck, have been discussed in detail in a previous publication (Reid et al., 1994). The standing RCAs that INBio has signed during the past few years conserve the same structure and principles described therein. More recently, the Diversa-INBio RCA was highlighted during the meeting of the Conference of Parties (COP6) in 2002 in The Hague as a "well-known example of an access and benefit-sharing" agreement (CBD, 2002). This distinction validates INBio's efforts toward fair and equitable benefit-sharing mechanisms.

The Merck & Co. and INBio RCA incorporated all necessary elements to ensure the fulfilment of INBio's mission and set strategic milestones for future negotiations. Thereby, the role of bioprospecting has been to promote and support the valorization of the Costa Rican biological diversity. This is achieved in the negotiation process and is based on a set of ethical criteria, which will ensure the access to and the integrity of national genetic resources under a fair and equitable benefit-sharing perspective. These basic criteria are as follows.

- Access to a limited amount of samples from natural resources is facilitated for a limited period of time (exclusivity terms are limited) through a framework legal agreement with the Ministry of the Environment and Energy (MINAE).
- A significant part of the research should be carried out locally, and associated research costs should be entirely covered by the industrial partner (defined as research budget).
- Up-front payments should be made for conservation (a minimum of 10% of the research budget is transferred to MINAE for conservation purposes).

Giselle Tamayo, Lorena Guevara, and Rodrigo Gámez • National Institute of Biodiversity (INBio), 3100 Santo Domingo, Heredia, Costa Rica.

- Benefit-sharing mechanisms should be negotiated beforehand and should include among others
  - milestone payments for the discovery and development phases of a potential product to be shared 50:50 with MINAE (as a nonprofit organization, INBio would invest its corresponding part in compliance with its biodiversity conservation mission)
  - a percentage of royalties on net sales of the final product (covering also derivatives from the original natural scaffold and/or any technology derived thereof), also to be shared 50:50 with MINAE
  - intellectual property rights to include participation of INBio's scientists if applicable (joint patents and publications)
- Technology transfer and capacity building of local scientists should be significant and should include state-of-the-art technologies.
- The discovery and development of a product must engage nondestructive uses of natural resources and be consistent with the national legislation regarding access of genetic resources and development thereof.

INBio, as a nonprofit organization created in 1989, has been able to build internal capacity on capturing information on natural resources and to process and transfer that information to society in different formats. Such framework rests on strategic alliances with the government (INBio-MINAE agreements 1992, 1997, and renewal in 2002) and with the academic sector at the national and international levels.

Costa Rica has a tradition of establishing legal frameworks to deal with conservation of genetic resources and its sustainable use. The Biodiversity Law, approved in 1998 by the Costa Rican Congress in full compliance with the CBD, redefines the structure in which bioprospecting activities should be formulated in Costa Rica. On this issue, INBio's experience was very important for defining the benefit-sharing and intellectual property rights mechanisms for future RCAs.

Throughout the years, this approach has proven to be successful. To date, INBio has signed more than 20 agreements with industry (see Table 1) that repre-

Table 1. Most significant research collaborative agreements with industry and academia, 1991–2002

Industry or academic partner	Natural resources accessed or main goal	Application fields	Research activities in Costa Rica
Cornell University	INBio's capacity building	Chemical prospecting	1990–1992
Merck & Co.	Plants, insects, microorganisms	Human health and veterinary	1991–1999
British Technology Group	DMDP, compound with nematocidal activity[a]	Agriculture	1992–present
Empresas ECOS S. A. (ECOS)	*Lonchocarpus felipei*, source of DMDP[a]	Agriculture	1993–present
Cornell University and National Institute of Health	Insects	Human health	1993–1999
Bristol Myers and Squibb	Insects	Human health	1994–1998
Givaudan Roure	Plants	Fragrances and essences	1995–1998
University of Massachusetts	Plants and insects	Insecticidal components	1995–1998
Diversa	DNA from bacteria	Industrial enzymes	1995–present
INDENA SPA	Plants[a]	Human health	1996–present
Phytera Inc.	Plants	Human health	1998–2000
Strathclyde University	Plants	Human health	1997–2000
Eli Lilly	Plants	Human health and agriculture	1999–2000
Akkadix Corporation	Bacteria	Nematocidal proteins	1999–2001
Follajes Ticos	Plants	Ornamentals	2000–present
La Gavilana S.A.	*Trichoderma* spp.[a]	Ecological control of pathogens of *Vanilla*	2000–present
Laboratorios Lisan S.A.	None[a]	Production of standardized phytopharmaceuticals	2000–present
Bouganvillea S.A.	None[a]	Production of standardized biopesticide	2000–present
Agrobiot S.A.	Plants[a]	Ornamentals	2000–present
Guelph University	Plants[a]	Agriculture and conservation purposes	2000–present
Florida Ice and Farm	None[a]	Technical and scientific support	2001–present
Chagas Space Program	Plants, fungi[a]	Chagas disease	2001–present
SACRO[b]	Plants[a]	Ornamentals	2002–present

[a] These agreements involve a significant component of technical and scientific support from INBio.
[b] Saving Costa Rican Orchids Foundation.

sent $0.5 million (U.S.) per year for bioprospecting activities and $0.5 million per year for capacity building, technology transfer, and empowerment. The latter are of transcendental relevance for INBio's negotiations, which are focused on agreements where the scientific participation of INBio could be maximized. It is vital for INBio's positioning in a very competitive technological sector to increase the local capacity by means of training and technology transfer. This is essential to ensure INBio's participation in the overall process of discovery and development of final products.

This general strategy has provided two outcomes of importance to the future development of bioprospecting. As shown in Table 1, most of the standing agreements encompass a significant technological and scientific participation of INBio in the development of a final product. Second, it involves sharing not only benefits (monetary and nonmonetary), but also risks with the industrial partner. This has been essential in building long-lasting partnerships.

MICROBIAL BIOPROSPECTING AGREEMENTS AT INBIO–A SHORT REVIEW

The relevance of microorganisms on bioprospecting is undisputable. From the 30 best-selling ethical pharmaceuticals in 1997 (Wood Mackenzie's PharmaQuant, Edinburgh, United Kingdom, 1999), 13 were either derived or developed from natural products; and from them, a significant number were derived from microorganisms (O'Neill and Lewis, 1993).

Costa Rica's biodiversity is extremely rich in microorganisms (Obando, 2002). It is expected that more than 25,000 species of bacteria and microalgae exist in Costa Rica, but only some 5% of cultured specimens have been collected for bioprospecting. Taxonomical work on this particular group has not been emphasized at INBio.

On the other hand, the inventory of fungi was initiated in 1998, and from the expected 65,000 species for Costa Rica only 3% have been taxonomically described (Obando, 2002). Some experiments conducted on samples from Costa Rican rainforests (Bills and Polishook, 1994a, 1994b) illustrate the vast biodiversity on microfungi found in the Costa Rican rainforest. Adapting a particle filtration technique to maximize the number of isolates obtained from a substrate, they found 80 to 145 species of fungi per sample from leaf litter samples, whereas four replicates of the same substrate yielded 300 to 400 different species.

INBio's portfolio of research collaborative agreements includes some in the biotechnological area, which involve microorganisms. By 1995 INBio

set up a small microbiology laboratory for in-house antibiotic testing and began the long-lasting partnership with Diversa. A brief description of such collaborations follows.

Merck & Co-INBio RCA

In 1996 and with a strong support from Merck & Co., INBio initiated a special program on collecting and culturing representative specimens of ascomycete, basidiomycete, coelomycete, hyphomycete, and sapromycete fungi, as well as some endophytic fungi for bioprospecting purposes. The collaboration emphasized training for bioprospectors on collecting and culturing techniques and set up two laboratories, one at the Guanacaste Conservation Area and the other at the INBio headquarters. Collection was initiated on entomopathogen microfungi and was enlarged to other groups towards the end of the research activities in Costa Rica. The work initiated under this RCA has continued at INBio, but is focused on collection, culturing and curation of ascomycete and basidiomycete fungi and has been supported by Global Environmental Facility (GEF)-World Bank grants and INBio itself.

Diversa-INBio RCA

It is well known that >85% of bacterial organisms cannot be cultured (Liu et al., 1997; Staley, 1998; Finlay and Clarke, 1999). This experimental limitation has been overcome by a different technological approach in which environmental DNA is used for building large gene libraries that, when expressed, could produce enzymes with potential industrial applications. The collaboration with Diversa, signed originally in 1995 and renewed twice, involves training for collecting techniques and DNA extraction and processing and it initiated the molecular biology laboratory at INBio in 1998.

Akkadix-INBio RCA

From 1999 to 2001, INBio collected environmental samples and isolated and cultured bacterial strains. These isolates were further tested for their nematocidal activity on a *Caenorhabditis elegans* test. As a result, a significant collection of bacterial isolates was obtained, and a 10% hit rate on the preliminary and secondary screenings was achieved. This agreement terminated abruptly due to the bankruptcy of the company. INBio recovered the custody of all genetic material and is evaluating different course of actions for the development of the promising protein activity found.

La Gavilana S.A.

In 1999 INBio initiated an exploration of the *Hypocrea/Trichoderma* potential for agricultural uses and in 2000 signed a RCA with a local producer of *Vanilla* spp.—La Gavilana S.A.—thanks to partial nonrefundable support from the Inter American Development Bank (IADB). The goal of the project is to produce *Vanilla* spp. in an eco-friendly fashion, by controlling fungal and bacterial pathogens using the native species of *Trichoderma*.

Chagas Space Program

In 1997 INBio agreed to be part of a Pan American initiative that seeks a cure for Chagas disease. This initiative that was led by the Escuela de Agricultura de la Región del Trópico Húmedo (EARTH) and sponsored by the U.S. National Space Agency received financial support from the Costa Rican Congress and a substantial contribution from the Congress of the United States, in 2001. INBio proposed to evaluate plants extracts and fermentations products from microfungi. Current evaluation indicates a hit rate of 1% in the fungal extractions using a single enzymatic assay. These results are encouraging, and the probability of finding inhibitors with potential for the treatment of Chagas disease will increase when the number of enzymes to be tested increases; this is expected to happen by 2004.

ACHIEVEMENTS OF BIOPROSPECTING— THE EXPERIENCE OF INBio

It is popular knowledge that the development of a product might take 5 to 20 years of research depending on the field (agricultural, biotechnological, or pharmaceutical applications) and that it might take hundreds of millions of U.S. dollars until it reaches the market. A new estimation made at Tufts University indicates that the investment needed on an 11-year period of research is over $800 million (Watkins, 2002). Taking into consideration the investment needed—the actual expenditures on bioprospecting and the odds (1:10,000 to make it to market)—it is still too early to expect products from Costa Rican biodiversity to be launched into the commercial market. It is most likely that simpler products from the IADB-funded projects would be commercialized before any blockbuster in the United States or Europe. This is not only the case for INBio but for other bioprospecting initiatives around the world (Moran et al., 2001; ten Kate and Laird, 1999).

Table 2 summarizes the tangible benefits arising from bioprospecting activities at INBio. A special mention is made here on the significant contribution

Table 2. Monetary and nonmonetary benefits of bioprospecting

Category	Benefit
Monetary	100% of research budgets, technology transfer and infrastructure, up-front payments for conservation, significant contribution for the Guanacaste Conservation Area and universities, milestone and royalty payments to be shared with MINAE
Nonmonetary	Trained human resources, empowerment of human resources, negotiations expertise developed, market information, improvement of local legislation on conservation issues

of more than $2 million that has been transferred to MINAE's Guanacaste Conservation Area, to the University of Costa Rica, and to National University. These organizations have been part of strategic alliances for the execution of projects. The Guanacaste Conservation Area contributes with a strong scientific advisory board, infrastructure, and other resources that facilitate such collaborative agreements. On the other hand, the main public universities, the University of Costa Rica and the National University, support the technical and scientific development of selected research activities.

Table 3 presents concrete outputs of projects since 1992. Although there are no products on the

Table 3. Outputs generated since 1992 as a result of RCA with INBio

Project	Year initiated	Output
Merck & Co.	1992	27 patents
British Technology Group and ECOS	1992	DMDP on its way to commercialization
National Cancer Institute	1999	Secondary screening for anticancer compounds
Givaudan Roure	1995	None yet
INDENA	1996	Two compounds with significant antibacterial activity
Diversa	1998	Two potential products at initial stages; publication under way
Phytera Inc.	1998	None yet
Eli Lilly & Co.	1999	None yet
Akkadix	1999	52 bacterial strains with nematocidal activity
Costa Rica-USA Foundation	1999	One compound with significant antimalarial activity
Lisan	2000	Two phytopharmaceuticals in process
Caraito	2000	None yet
Follajes Ticos	2000	None yet
Bougainvillea	2001	None yet
La Gavilana	2001	None yet
Agrobiot	2001	None yet
SACRO	2002	None yet

market yet, the RCA with Merck has provided 27 patents to date. Most of these patents are related to microorganisms and describe compounds with antibacterial, antiprotozoal, antifungal, and HIV integrase inhibitory activities. The development of a natural nematocide compound, $2R,5R$-dihydroxymethyl-$3R,4R$-dihydroxypyrrolidine (DMDP), isolated from a legume tree, constitutes the most advanced project in the pipeline. Greenhouse experiments and field trials were conducted on tropical crops, and INBio is evaluating with the British Technology Group and ECOS, a local company, the scaling up and development of commercial products. Finally, the strategic alliance with a Costa Rican pharmaceutical company, Laboratorios Lisan S.A., with the aim of developing standardized selected extracts from medicinal plants and the manufacturing of phytopharmaceuticals, will materialize with the launching of two products next year (Guevara, 2002).

THE FUTURE

INBio's initiatives have provided important benefits for Costa Rica. It is clear that INBio will continue to promote the development of biotechnological activities, with emphasis on microbial prospecting, isolation, and identification of genes with potential application in the industry and agriculture.

In the field of chemical prospecting, INBio foresees agreements with academia and biotech international enterprises, thanks to the acquisition through donations from several preparative automated fractionators, which will allow the isolation of significant amounts of compounds in a high-throughput fashion. This will enable the screening of a large number of natural products in an efficient way, with the additional advantage of securing resupply of compounds.

Bioprospecting has provided Costa Rica and INBio with a vast and complex experience on access, legislation, and uses of genetic resources. Most important, the acquisition of state-of-the-art technologies and training of human resources has allowed INBio to offer innovative products and services with significant added value. The projects financed partially by the IADB demonstrated that agreements with local endeavors are not only possible, but will result in developing final marketable products in a shorter period of time, with the subsequent promotion of local economic development. This will have a significant impact on public awareness and will contribute to increased valorization of biodiversity and its conservation.

REFERENCES

Bills, G. F., and J. D. Polishook. 1994a. Abundance and diversity of micofungi in leaf litter of a lowland rain forest in Costa Rica. *Mycologia* 86:187–198.

Bills, G. F., and J. D. Polishook. 1994b. Microfungi from decaying leaves of *Heliconia mariae*. *Brenesia* 41/42:27–43.

CBD (Convention on National Biological Diversity). 2002. First-ever global guidelines adopted on genetic resources Biodiversity conference also acts on forests and invasive species. Press Release. In media only. The Hague, The Netherlands.

Finlay, B. J., and K. J. Clarke. 1999. Ubiquitous dispersal of microbial species. *Nature* 400:828.

Gámez, R. 2000. *Perspectivas del Desarrollo Científico Costarricense: La Vision Desde una Institución Privada*, Vol. III. p. 115–130. Academia Nacional de Ciencias, San José, Costa Rica.

Guevara, A. L. 2002. Los Aportes de La Bioprospección realizada por el INBio. *Rev. Ambientico* 100:7.

Liu, W., T. L. Marsh, H. Cheng, and L. J. Forney. 1997. Characterization of microbial diversity by determining terminal restriction fragment length polymorphisms of genes encoding 16S rRNA. *Appl. Environ. Microbiol.* 63:4516–4522.

Matamoros, A., and R. Garcia (ed.). 2000. *Institutional Sustainability Strategy*. Internal Documents Series no. 1. Editorial INBio, Instituto Nacional de Biodiversidad, Heredia, Costa Rica.

Mateo, N., W. Nader, and G. Tamayo. 2001. Bioprospecting, p. 471–478. In S. A. Levin (ed.), *Encyclopedia of Biodiversity*, vol. 1. Academic Press, New York, N.Y.

Moran, K., S. King, and C. Carlson. 2001. Biodiversity prospecting: lessons and prospects. *Annu. Rev. Anthropol.* 30:505–526.

Obando, V. 2002. *Biodiversidad en Costa Rica: Estado del Conocimiento y Gestión*, D. Avila, (ed.). Editorial INBio, Heredia, Costa Rica.

O'Neill, M. J., and J. A. Lewis. 1993. The renaissance of plant research in the pharmaceutical industry, p. 48–78. In D. Kinghorn and M. Balandrin (ed.), *Human Medicinal Agents from Plants*. American Chemical Society, Washington, D.C.

Reid, W., et al. 1994. *Prospección de la Biodiversidad: El uso de los Recursos Genéticos para el Desarrollo Sostenible*. Traducción Centro Universitario de Traducción (CUTRA). Instituto Nacional de Biodiversidad, San José, Costa Rica.

Sittenfeld, A., and R. Villers. 1994. Costa Rica's INBio: Collaborative biodiversity research agreements with the pharmaceutical industry. Case study 2, p. 500–504. In G. K. Meffe and C. R. Carrol (ed.), *Principles of Conservation Biology*. Sinauer Associates, Sunderland, Mass.

Sittenfeld, A., et al. 1999. Costa Rican International Cooperative Biodiversity Group: using insects and other arthropods in biodiversity prospecting. *Pharm. Biol.* 37(Suppl.):1–14.

Staley, J. 1998. Economic value of microbial diversity. In O. Kirsop and V. Perez (ed.), *Workshop on The Economic Value of Microbial Genetic Resources*. Eighth International Symposium on Microbial Ecology. WFCC (World Federation for Culture Collections), Halifax, Canada.

Tamayo, G., W. F. Nader, and A. Sittenfeld. 1997. Biodiversity for bioindustries, p. 255–280. In J. A. Callow, B. V. Ford-Lloyd, and H. J. Newbury (ed.), *Biotechnology and Plant Genetic Resources*. CAB International, Wallingford, United Kingdom.

ten Kate, K., and S. Laird. 1999. *The Commercial Use of Biodiversity: Access to Genetic Resources and Benefit-Sharing*. European Communities, Brussels.

Watkins, K. 2002. Fighting the clock: pharmaceutical and biotechnology companies seek ways to reduce the time required to discover and development medicines. *Chem. Eng. News* Jan., p. 34.

Microbial Diversity and Bioprospecting
Edited by Alan T. Bull
© 2004 ASM Press, Washington, D.C.

Chapter 42

Contracts for Bioprospecting: the Yellowstone National Park Experience

HOLLY DOREMUS

In 1997, the National Park Service (NPS), Yellowstone National Park (YNP), and Diversa Corporation, a California-based biotechnology company, signed the first benefit-sharing agreement for commercial bioprospecting on federally owned lands in the United States. The agreement, which followed a period of intense controversy over bioprospecting in Yellowstone, remains suspended pending resolution of litigation. It appears, however, that the litigation will soon be resolved and the Yellowstone-Diversa agreement will become a model for benefit-sharing agreements in Yellowstone, other U.S. national parks, and potentially other federal lands.

Globally, bioprospecting has been touted as a way to fulfill the three goals of the Convention on Biological Diversity (CBD): conservation of biological diversity, sustainable use of biological resources, and equitable sharing of the benefits of using those resources (e.g., Reid et al. 1993; tenKate, Chapter 39, this book). Because the United States is not a party to it, the CBD does not apply to Yellowstone bioprospecting. Nonetheless, the idea that the CBD's three goals can be fruitfully combined underlies the Yellowstone agreement. The experience of Yellowstone, the world's first national park, with bioprospecting provides an important case study because Yellowstone, by sharp contrast with the earliest targets of bioprospecting, is in the developed world and already enjoys strong legal protection of its natural resources.

YELLOWSTONE BIOPROSPECTING AND Taq POLYMERASE

Yellowstone National Park, the only essentially undisturbed geyser basin remaining in the world, is a unique geological and biological location. It contains some 10,000 geothermal features, including geysers, hot springs, mudpots, and fumaroles. According to NPS, the park contains the greatest diversity of thermal features in the world. Those features exhibit a wide range of temperature, pH, and mineral content (Yellowstone National Park, Division of Interpretation [YNP], 2001; see Stoner et al., chapter 23, this book). They are inhabited by a rich and extraordinarily diverse microbiota (Pace, 1997; Hugenholtz et al., 1998), only a fraction of which has been characterized.

The study of life in Yellowstone's thermal features began in 1898 with the issuance of the first known scientific research permit, authorizing a researcher at the University of California to collect microorganisms (Varley and Scott, 1998). Much later, in the 1960s, Thomas Brock, then at the University of Indiana, began sampling in the hot springs. He and his students identified microbes living at higher temperatures than had previously been thought possible, learned to culture some of those microbes, and deposited them with the American Type Culture Collection (ATCC), making them available to other interested researchers (Brock, 1995, 1997).

The molecular biology revolution brought commercial attention to Brock's thermophile work. In the 1980s, Kary Mullis at Cetus Corporation developed PCR (the polymerase chain reaction), a novel technique for rapidly amplifying DNA that brought Mullis the Nobel prize and quickly became a staple technique of DNA labs, commercial and academic, around the world. PCR uses alternating cycles of high and low temperatures. Heating separates the DNA strands. Cooling allows primers to bind to the separated strands and DNA polymerase to replicate those strands (Eeles and Stamps, 1993). Conventional DNA polymerases could not survive the heating step and therefore had to be added at the outset of each

Holly Doremus • School of Law, 400 Mrak Hall Dr., University of California at Davis, Davis, CA 95616.

replication cycle. Automation, which has made PCR inexpensive and easy to use, was not possible until Mullis proposed using a heat-stable polymerase that could survive the heating cycles. David Gelfand found one in *Thermus aquaticus*, one of the thermophilic bacteria deposited in the ATCC by Brock and his colleagues, who had isolated it from a YNP hot spring (Rainbow, 1996). Cetus obtained a series of patents on the PCR technique and the purified *Taq* polymerase, and PCR became an enormous commercial success. Cetus reportedly sold the rights to the process to Hoffman-LaRoche for $300 million (Rainbow, 1996). The market for *Taq* polymerase itself was estimated in 1996 at $80 to $85 million annually (Barinaga, 1996).

The economic success of *Taq* polymerase drew attention to the potential economic value of Yellowstone's unique biotic resources, setting off a small-scale bioprospecting gold rush. In Yellowstone's extreme habitats, bioprospectors hope to find more enzymes like *Taq* polymerase, whose ability to maintain activity at high temperatures, under extremely acidic or basic conditions, or in the presence of high concentrations of metals will make them useful in industrial and other processes. By late 1996, YNP management believed that 13 thermophilic microbes from Yellowstone had demonstrated potential commercial value for biotechnology applications (Smith, 1997).

THE DIVERSA COOPERATIVE RESEARCH AND DEVELOPMENT AGREEMENT

Before negotiation of the Yellowstone-Diversa agreement, research in the park was conducted on an essentially open-access basis. Permits were required, but were routinely issued if the planned research would not damage park resources. Permit holders were required to make collected specimens and data available to the public, to file annual reports, and to provide copies of publications to the park. Release of proprietary information was not required.

In the early 1990s, YNP officials began searching for a mechanism to share in any profits from the commercialization of products developed from Yellowstone extremophiles. At the time some 39 scientists, including 5 from private firms, had permits to collect microbes in Yellowstone (Chester, 1996). John Varley, Yellowstone's chief scientist, hit on the idea of a bioprospecting benefit-sharing agreement analogous to those developed under the CBD in Costa Rica and other locations outside the United States.

Some biotechnology companies reacted with outrage, arguing that the government already benefited from commercial bioprospecting through income taxes on corporate profits. But others were more receptive. Diversa Corporation, which had experience with bioprospecting contracts in Costa Rica and Iceland, volunteered to be a partner in benefit sharing. Assisted by a nonprofit organization, the World Foundation for Environment and Development, officials from YNP, NPS, and the U.S. Department of Interior negotiated a bioprospecting agreement with Diversa. The contract was announced with great fanfare on August 17, 1997, at the gala celebration of the 125th anniversary of Yellowstone's designation as a national park.

The Yellowstone-Diversa agreement is couched as a Cooperative Research and Development Agreement (CRADA), a type of agreement specifically authorized by the U.S. Federal Technology Transfer Act (FTTA). That format was chosen to allow YNP to retain revenues from the agreement, rather than remitting them to the federal treasury (Doremus, 1999). Like other national parks, Yellowstone must send all revenues other than those from entrance fees to the U.S. Treasury. Congress controls what is done with those funds; there is no guarantee that they will be reallocated to the individual park, or even to NPS. The CRADA format offered a way around that requirement. To encourage government laboratories to transfer their knowledge to the private sector for commercialization, the FTTA allows the federal partner to keep all CRADA payments up to a ceiling of 5% of its total annual budget. At current YNP funding levels, that ceiling would be roughly $1.4 million per year. YNP would keep 25% of any revenues above that amount, with the remainder going to the U.S. Treasury. Although CRADAs do offer the park hope of substantial revenues, those revenues would not come entirely without strings. The FTTA limits use of CRADA revenues to enumerated purposes, including rewarding laboratory employees for inventions, supporting scientific research, and supporting education and training of employees consistent with the laboratory's research mission.

The Yellowstone-Diversa CRADA provides for systematic sampling of park habitats, including but not limited to thermal features, over a period of 5 years. Diversa intends to assemble a collection of samples encompassing the microbial genetic diversity of the park. Gene libraries generated from the samples will be used to identify enzymes, genes, and other compounds with commercial value. Essentially all of Diversa's work with Yellowstone samples is destructive. Samples may be cultured briefly, but eventually all are destroyed to harvest their nucleic acids.

Sampling techniques are selected jointly by Diversa and YNP. Collecting must be done out of public view, entry to restricted areas must be separately authorized, and an NPS liaison must be present during all collecting. Sampling requires a scientific collection permit and is subject to park regulations, which prohibit significant impacts to park resources and other park uses. It is plausible, as both Diversa and the NPS contend, that direct physical impacts will be minimal. Although the CRADA does not expressly limit the size of samples to be removed, Diversa has no reason to remove large samples.

The CRADA calls for Diversa to pay YNP a flat fee of $20,000 per year, creditable against any royalties due in that year, donate equipment and scientific training worth about $75,000 per year, and prepare phylogenetic trees illustrating the likely evolutionary relationships among Yellowstone's microbes. The details of the royalty arrangement have not been made public, but reportedly include a sliding scale ranging from 0.5 to 10%. Those royalties would be payable on gross revenues from inventions conceived or first actually reduced to practice in the course of work under the CRADA. Diversa must disclose inventions that may be patentable to the park and must maintain records, subject to NPS audit up to once a year. The federal government will be offered a no-cost license for any inventions the CRADA produces.

The Yellowstone-Diversa CRADA is nonexclusive. It does not preclude similar contracts with other bioprospectors, and indeed NPS anticipates entering into similar agreements with a variety of researchers. It does not give Diversa exclusive rights to collection sites, park information or access to park personnel, or product development in any particular field.

Diversa has donated additional scientific work to the park. At the request of YNP officials, Diversa prepared a phylogenetic tree illustrating the genetic relationships among the park's wolves and analyzed an unusual microbial sample found deep in Yellowstone Lake. Park officials regard these donations as tremendously valuable. They believe that similar work could have been done by university researchers, but only at high cost and not nearly so quickly (John Varley, personal communication).

LEGAL CHALLENGES TO THE CRADA

The Yellowstone-Diversa CRADA was announced in a high-profile setting because both parties thought it would be universally hailed as a boon to the park, an important industry, and the nation. They soon learned, however, that not everyone was pleased. In March 1998 three small nonprofit groups filed a lawsuit challenging the CRADA. The three were Alliance for the Wild Rockies, which describes itself as working to secure the ecological integrity of

the northern Rockies bioregion; the Edmonds Institute, primarily interested in the regulation and implications of biotechnology; and the International Center for Technology Assessment, which concentrates on the social, environmental, and other impacts of new technologies, including biotechnology.

The plaintiffs raised both procedural and substantive claims. Procedurally, they alleged that the NPS was required by the National Environmental Policy Act (NEPA) to study the environmental impacts before executing the CRADA. Substantively, they claimed that the agreement violated laws governing technology transfer and management of national parks. Plaintiffs prevailed on the procedural issue, but lost on the substantive one.

Environmental Study Is Required

NEPA requires federal agencies to study the potential environmental impacts of proposed actions before making irrevocable commitments. Environmental studies under NEPA serve two purposes. They provide the agencies with detailed information about environmental consequences to factor into their decisions. And they make that information public, facilitating public participation in the decision-making process.

NEPA compliance can take one of three forms. Agencies must prepare an environmental impact statement (EIS) for any action that will significantly affect the quality of the human environment. Actions that will not individually or cumulatively have a significant environmental impact are "categorically excluded" and do not require environmental study. For actions between these two extremes, agencies must prepare an environmental assessment to determine if the environmental impacts will be significant, requiring a full EIS (Bear, 1989).

YNP and NPS claimed that a categorical exclusion for day-to-day resource management and research activities applied to the CRADA. In *Edmonds Institute* v. *Babbitt*, 42 F. Supp. 2d 1 (D.D.C. 1999), a federal district court rejected that contention on two grounds. First, the categorical exclusion had not been invoked until after the agreement was finalized; the court refused to accept *post hoc* invocation of the exclusion during litigation as an excuse for failing to undertake environmental analysis before the decision was made. Second, U.S. Department of the Interior guidelines limited the categorical exclusion defendants cited to actions that would not establish a precedent or represent a decision in principle about future actions with potentially significant environmental effects. The court ruled that, because the Yellowstone-Diversa CRADA was clearly intended as a model for future benefit-sharing agreements, the categorical exclusion did not apply. Implementation of the CRADA was ordered suspended, pending compliance with NEPA. YNP and NPS decided not to appeal the decision.

The Agreement Is Not Unlawful

In a subsequent decision [*Edmonds Institute* v. *Babbitt*, 93 F. Supp. 2d 63 (D.D.C. 2000)], the same judge rejected the plaintiffs' substantive challenge to the Yellowstone-Diversa CRADA. Plaintiffs contested the validity of the agreement first under the FTTA, which authorizes federal "laboratories" to enter into CRADAs. Plaintiffs claimed that YNP was not a "laboratory" within the meaning of the statute and therefore could not be a party to a CRADA. The court disagreed. The term "laboratory" is broadly defined by the FTTA to include any facility "a substantial purpose of which is the performance of research" [15 U.S.C. 3710a(d)(2)]. The legislative history describes this definition as encompassing "the widest possible range of research institutions operated by the Federal Government" (U.S. Senate, Committee on Commerce, Science, and Transportation, 1986). The court ruled that the broad statutory definition easily encompassed YNP, which employs more than 40 scientists, conducts and oversees a broad range of research related to the park's resources, and maintains its own laboratories and other scientific research facilities within the park.

Plaintiffs also alleged that the CRADA violated the law governing national parks by allowing consumptive use and sale of park resources. The NPS Organic Act mandates management of the parks in accordance with their fundamental purposes, which are to conserve and provide for the enjoyment of their scenery, natural and historic objects, and wildlife in a manner that will leave them unimpaired for future generations. Commercial and consumptive uses of park resources are generally considered inconsistent with that mandate unless authorized by park-specific legislation. The court found, however, that the CRADA was consistent with the Organic Act because it would contribute both funds and scientific knowledge to support YNP's conservation mission. The court also concluded that the CRADA itself did not authorize impermissible commercial activity because it did not confer the right to collect any research specimens and plaintiffs had not challenged the scientific collection permit that did. Challenges to bioprospecting permits remain possible (Doremus, 1999), although this court does not seem inclined to view them favorably.

Plaintiffs also contended that the CRADA violated an NPS regulation prohibiting the sale or commercial use of natural materials from the park. Noting that an agency's own interpretation of its regulations is entitled to substantial judicial deference, the court upheld the defendants' conclusion that the CRADA did not involve sale or commercial use of park resources because only scientific information, not park resources, would be commercialized. Plaintiffs filed an appeal of this decision, but subsequently dropped it.

THE CURRENT STATE OF BIOPROSPECTING IN THE U.S. NATIONAL PARKS

To comply with the court's NEPA decision, the NPS has decided to prepare an environmental impact statement on benefits-sharing agreements throughout the national park system. The EIS is projected to be released in the fall of 2002. It will consider whether and in what form the parks should enter into benefit-sharing agreements with bioprospectors, but will not consider revision of the general rules for scientific research in the parks (National Park Service, 2001). Because the NPS is currently conditioning all research permits on agreement to enter into a CRADA if commercial products result, it seems highly likely that the preferred alternative proposed in the EIS will call for benefit-sharing agreements like the Yellowstone-Diversa CRADA. The EIS will immunize the NPS against future NEPA challenges to benefit-sharing agreements.

In the meantime, Diversa has had to halt specimen collection under the CRADA, but is continuing to work on commercialization of products based on its early collections. A search of the U.S. Patent and Trademark Office database revealed two patents issued to Diversa, one for thermostable glycosidases and the other for thermostable esterases, that reference Yellowstone microbes as a source. Because initial applications for both were filed prior to the CRADA, it is not clear whether they are subject to that agreement. Diversa also has patented and applied for patents on the methods it uses to screen genetic libraries such as those created from YNP samples. Again, it is unclear whether the CRADA would cover these patents. Diversa reportedly has at least one commercial product for which it is paying royalties under the CRADA into an escrow account pending resumption of the agreement (John Varley, personal communication).

Other researchers are not subject to the judicial moratorium and are continuing to sample microbes in Yellowstone and other national parks. The NPS is no longer allowing entirely open access, however. Since 1995, the NPS has asserted continuing ownership of specimens taken out of the park, as well as cultures grown from those specimens (Milstein, 1995). Because the NPS has challenged the right of the ATCC to distribute samples of YNP organisms already in its collection, those organisms are currently unavailable (John Varley, personal communication). The standard terms for scientific research and collecting permits throughout the system have also been modified. They now include the following paragraph:

> Any specimens collected under this permit, any components of any specimens (including but not limited to natural organisms, enzymes, or other bioactive molecules, genetic materials, or seeds), and research results derived from collected specimens are to be used for scientific or educational purposes only, and may not be used for commercial or other revenue-generating purposes unless the permittee has entered into a Cooperative Research and Development Agreement (CRADA) or other approved benefit-sharing agreement with the NPS. The sale of collected research specimens or other unauthorized transfers to third parties is prohibited. Furthermore, if the permittee sells or otherwise transfers collected specimens, any components thereof, or any products or research results developed from such specimens or their components without a CRADA or other approved benefit-sharing agreement with NPS, permittee will pay the NPS a royalty rate of twenty percent of gross revenue from such sales or other revenues. In addition to such royalty, the NPS may seek other damages to which the NPS may be entitled including but not limited to injunctive relief against the permittee. (National Park Service, 2001b)

YNP's permit conditions, borrowing nearly verbatim from the Yellowstone-Diversa CRADA, add the requirement that the permittee notify the park of any patent application and any invention that may be patentable.

These new requirements do not seem to have discouraged microbiological research in YNP. The park reported 35 microbiology research projects in 1999 (Yellowstone Center for Resources, 2000), compared with 23 in 1996 (Yellowstone Center for Resources, 1997). Only two of the 1999 studies were undertaken directly by biotechnology companies, but others may have commercial implications. Other national park locations also appear to be drawing some bioprospecting interest, particularly those with their own extreme environments, such as Carlsbad Caverns and Mammoth Cave.

AN EARLY EVALUATION OF BIOPROSPECTING IN U.S. NATIONAL PARKS

Because the Yellowstone-Diversa agreement is only 5 years old, and its implementation has been blocked for most of that time by litigation, it is too early to say with confidence what effects it will ultimately have. Because the agreement is soon likely to become a model for bioprospecting in the national parks, however, it is worth attempting a preliminary evaluation. The three goals of the CBD, conservation, sustainable use, and equitable sharing of benefits, offer reasonable metrics for that evaluation.

Conservation

Discovery of another *Taq* polymerase under the CRADA could bring YNP a substantial return. Sales of *Taq* polymerase in the late 1990s were estimated at roughly $80 million a year. Given the reported range of royalty rates and the ceiling on revenues, an equivalent product would add between $400,000 and $3 million to the current YNP annual budget of roughly $28 million. If YNP shared in process patents for a process that becomes as popular as PCR, those numbers could go much higher. The odds that the park would be embroiled in costly litigation, of course, would increase with the value of the patents. Nonetheless, it is no surprise that park officials, who face chronic budget shortages, find bioprospecting CRADAs appealing.

There are, however, substantial limits to the ability of benefit sharing to support conservation at U.S. national parks generally. Even at Yellowstone, the park most appealing to bioprospectors, the costs of operating the park and conserving its resources far exceed the maximum conceivable income from benefit sharing. Few other parks can expect any income from bioprospecting. Nor can prospecting for genetic information pay for conservation indefinitely. Once the park's microbial resources have been thoroughly sampled, a process that park officials believe will take a long time but that surely has an end point, the biotechnology industry will have no further incentive to pay for protection of the park.

Furthermore, the choice of the CRADA format significantly restricts the uses to which revenues can be put. The FTTA limits the use of CRADA revenues to management and performance of the CRADA itself, intellectual property management, scientific research, and education or training of employees. CRADA funds cannot be used for many day-to-day park activities essential to conservation, such as law enforcement and maintenance of facilities. It is difficult to imagine a permissible use of those funds that could reduce the greatest threat to Yellowstone's geothermal resources, the potential for geothermal development outside the park.

In light of these limitations, it is distressing that park officials have joined the company and free-market advocates in touting the potential economic returns of bioprospecting as demonstrating the value of conservation. No U.S. national park is likely to produce a net income stream, with or without bioprospecting. National parks are protected because their noneconomic values, the inspiration and wonder they provide, outweigh any economic return exploitation of their resources might bring.

Enthusiasm for bioprospecting threatens to confuse the public and policy makers about the balance of noneconomic and economic values in parks, with potentially serious negative consequences for conservation. Congress, as chronically strapped for funds as the parks, could easily decide to reduce appropriations to the parks by the amount of income from bioprospecting agreements. Much worse, encouraged by the bioprospecting rhetoric, Congress or the American people could decide that even in the parks conservation should be tested by a strict economic cost-benefit test. Undoubtedly many parks, although perhaps not Yellowstone, would fail such a test.

Park officials contend that the CRADA will also further conservation by increasing the park's knowledge of its microbial resources. That knowledge should help the park effectively manage its resources, as well as provide new opportunities to expose the public to the wonders of nature. Carefully structured CRADAs could maximize that kind of scientific benefit to the parks. But to this point there has not been a public discussion of whether noncommercial academic research could produce the same sorts of information for the parks, how long that process would take, and how it might be funded. Until those questions are aired, the knowledge of park resources that might be developed under benefit-sharing arrangements should not drive aggressive pursuit of those arrangements.

Sustainable Use of Biotic Resources

The Yellowstone-Diversa CRADA is expected to produce scientific and economic benefits without threatening YNP's unique resources. Certainly that is true of the screening step, which requires removal of only very small samples of material. So long as the information gained from those small samples suffices to produce commercial products, as it should for many genes and the enzymes they encode, later steps will also be environmentally benign. Should bioprospectors discover a valuable chemical compound other than a gene or enzyme, the resources could come under more pressure. Unless and until synthetic production methods are developed, isolation of useful quantities of natural compounds often requires large quantities of starting material (Farrier and Tucker, 2001). The CRADA does not, however, authorize large-scale extraction, nor could YNP or NPS allow it

without modification of their governing legislation. Such changes currently seem unlikely, although too much emphasis on economic returns from parks, together with sampling that is sufficient to give a glimpse of the riches that Yellowstone's biota might contain, could conceivably shift the balance toward greater exploitation.

A more realistic concern about benefit sharing as the NPS is seeking to practice it is that it could actually retard the progress of scientific knowledge. NPS's contention that it continues to own samples removed from parks, as well as subsequent cultures from such samples, is essential to the validity of the commercial park bioprospecting agreements under current law because it allows NPS to assert that no park resources are being "sold." But that position may limit research. *Thermus aquaticus* was available as the source of *Taq* polymerase for PCR because Thomas Brock had deposited it years earlier in the ATCC without any demand for royalties should it prove commercially useful. Brock also freely sent samples of his cultures to other researchers. That sort of sharing is now impermissible.

What effect that will have on the availability of research materials, and ultimately the progress of academic and commercial science, is difficult to predict. Perhaps NPS will share samples or cultures with any researcher willing to commit to profit sharing, but the inevitable paperwork and recordkeeping requirements could discourage academic researchers. Alternatively, researchers themselves might be inspired by the park's attitude to demand that they too share in any eventual profits, no matter how distantly related to their contribution. If so, that could exacerbate the problem of excessive intellectual property claims on biological materials, which is already complicating the development of commercial biotechnology products (Heller and Eisenberg, 1998). This issue, too, deserves greater discussion before the Yellowstone-Diversa CRADA is adopted as a general model.

Benefit Distribution

Assuming that profits are to be made from park resources, it is difficult to argue with some return going to the United States, the ultimate steward of the park lands. Conceding that, two tough questions remain about the distribution of benefits.

The first is whether the precise rate of return should be public. The refusal to disclose the Yellowstone-Diversa royalty schedule has been a major source of contention. Park officials are convinced that publishing royalty rates would greatly discourage bioprospectors. It is unclear, however, either how strong the industry's desire for secrecy is or whether it is supported by reasons that the public should regard as legitimate.

In the context of public contracts, secrecy has two potential costs. The first is that the government party, typically less experienced than the private party in the market, might not negotiate the best possible deal. Park officials did receive advice from a nonprofit agency and believe they are sufficiently aware of royalties in other bioprospecting contracts to negotiate from a position of strength. But secrecy prevents the public from evaluating that claim. Secrecy also offers the potential for multiple benefit-sharing deals with different terms. Standard, public royalty rate schedules would remove the possibility of corrupt influence and ensure that any differences in royalty terms reflect public policy rather than the whims of the individual contract negotiator.

The second question is precisely how any revenues should be distributed. It could be argued that revenues should go to the U.S. Treasury, to be distributed by the ordinary democratic process. But if the point is to tie together the goals of conservation and sustainable use, revenues should be dedicated to conservation. Payments also should be distributed to the entire system of conserved lands, not just to the lands from which valuable products are obtained, in order to minimize the effect they could have of linking conservation too closely to economic productivity. The Yellowstone-Diversa CRADA offers neither of these advantages.

The CRADA format does funnel revenues directly to agency use, avoiding the political barriers to appropriation by the Congress that have made the Land and Water Conservation Fund ineffective. But CRADAs, which were developed for a very different context, do not dedicate revenues to conservation. Instead, as explained above, they may preclude spending on important conservation activities. Nor do they provide a formula for distributing revenues. If individual parks are permitted to enter into CRADAS, as YNP has been, those parks will keep the vast majority of any resulting revenues. The NPS might adjust its use of discretionary funds to compensate, but the CRADAs themselves will not require such an adjustment. The NPS should develop a model benefit-sharing agreement specifically for bioprospecting and seek legislation that would allow it to retain revenues obtained from such agreements. Congress, which in 1998 specifically authorized the NPS to negotiate benefit-sharing arrangements, would probably heed such a request.

Taking a Broader View of the Issues

The most disappointing aspect of the Yellowstone bioprospecting experience to date is that the NPS and YNP have taken an exceedingly narrow view of the issues warranting consideration and public discussion. Although plaintiffs in the CRADA litigation challenged only the benefit-sharing agreement, public anxiety about the Yellowstone-Diversa deal is not so limited. It also encompasses doubts about the extent to which genetic resources should be subject to privatization generally and questions about whether park resources specifically should be open to such privatization. The NPS and YNP have refused to factor these concerns into their bioprospecting decisions.

That narrowing of the issues is understandable. The NPS and YNP do not control intellectual property law or policy. They cannot force the U.S. Patent and Trademark Office or the federal courts to reconsider the doctrines that allow patenting of a gene or protein simply because it has been isolated from its natural source. So long as that remains the law, it is probably impossible to prevent private profit from park genetic resources without also squelching all scientific research in the parks. Clearly that result would be bad for the parks as well as for science.

But perhaps there are ways to limit the privatization of information originating in the parks without discouraging noncommercial science. The NPS should at least consider requiring that researchers make the genetic information they gain from the parks, and any cultures they develop, freely available to the public. Perhaps collection permits could forbid application for a patent on any gene, protein, or natural substance derived from a park specimen. Such a condition would not discourage scientists like Thomas Brock. The parks, the public, and the biotechnology industry would all benefit from the knowledge that scientists produce. Patents would still be possible on processes like PCR that employ those natural substances. In the long run, that approach might prove more consistent with conservation, sustainable use, and equitable benefit distribution than aggressive pursuit of commercial benefit-sharing agreements.

Acknowledgment. Mona Badie, U.C. Davis School of Law class of 2004, provided invaluable research assistance for this chapter.

REFERENCES

Barinaga, M. 1996. Promega wins round in fight over Taq. *Science* 273:1039.

Bear, D. 1989. NEPA at 19: a primer on an "old" law with solutions to new problems. *Environ. Law Rep.* 19:10060–10069.

Brock, T. D. 1995. The road to Yellowstone—and beyond. *Annu. Rev. Microbiol.* 49:1–28.

Brock, T. D. 1997. The value of basic research: discovery of *Thermus aquaticus* and other extreme thermophiles. *Genetics* 146: 1207–1210.

Chester, C. C. 1996. Controversy over Yellowstone's biological resources: people, property and bioprospecting. *Environment* 38:10–36.

Doremus, H. 1999. Nature, knowledge and profit: the Yellowstone bioprospecting controversy and the core purposes of America's national parks. *Ecol. Law Q.* **26**:401–488.

Eeles, R. A., and A. C. Stamps. 1993. *Polymerase Chain Reaction (PCR): The Technique and Its Applications.* R. G. Landes Co., Austin, Tex.

Farrier, D., and L. Tucker. 2001. Access to marine bioresources: hitching the conservation cart to the bioprospecting horse. *Ocean Dev. Int. Law* **32**:213–239.

Heller, M. A., and R. S. Eisenberg. 1998. Can patents deter innovation? The anticommons in biomedical research. *Science* **280**:698–701.

Hugenholtz, P., C. Pitulle, K. L. Hershberger, and N. R. Pace. 1998. Novel division level bacterial diversity in a Yellowstone hot spring. *J. Bacteriol.* **180**:366–376.

Milstein, M. 1995. Yellowstone managers stake a claim on a hot-springs microbe. *Science* **270**:226.

National Park Service. 2001a. Benefits-sharing for conservation? Environmental Assessment Scoping Newsletter. National Park Service, Washington, D.C.

National Park Service. 2001b. *General Conditions for Scientific Research and Collecting Permit.* National Park Service, Washington, D.C.

Pace, N. R. 1997. A molecular view of microbial diversity and the biosphere. *Science* **276**:734–740.

Rabinow, P. 1996. *Making PCR: A Story of Biotechnology.* University of Chicago Press, Chicago, Ill.

Reid, W. V., S. A. Laird, R. Gámez, A. Sittenfeld, D. H. Janzen, M. A. Collin, and C. Juma. 1993. A new lease on life, p. 1–52. *In* W. V. Reid, S. A. Laird, C. A. Meyer, R. Gámez, A. Sittenfeld, D. H. Janzen, M. A. Gollin, and C. Juma (ed.), *Biodiversity Prospecting: Using Genetic Resources for Sustainable Development.* World Resources Institute, Washington, D.C.

Smith, C. 1997. Yellowstone Park's deal: some call it "biopiracy." The Salt Lake Tribune, Nov. 9, 1997, p. A1.

U.S. Senate, Committee on Commerce, Science, and Transportation. 1986. Federal Technology Transfer Act of 1986. Committee Report No. 99–283. U.S. Government Printing Office, Washington, D.C.

Varley, J. D., and P. T. Scott. 1998. Conservation of microbial diversity a Yellowstone priority. *ASM News* **64**:147–151.

Yellowstone Center for Resources. 1997. *Investigators' Annual Reports, Yellowstone National Park, 1996.* Yellowstone Center for Resources, Yellowstone National Park, Wyo.

Yellowstone Center for Resources. 2000. *1999 Investigators' Annual Reports, Yellowstone National Park.* Yellowstone Center for Resources, Yellowstone National Park, Wyo.

Yellowstone National Park, Division of Interpretation. 2001. Yellowstone Resources and Issues. Yellowstone National Park, Wyo.

Microbial Diversity and Bioprospecting
Edited by Alan T. Bull
© 2004 ASM Press, Washington, D.C.

Chapter 43

Natural Products Research Partnerships with Multiple Objectives in Global Biodiversity Hot Spots: Nine Years of the International Cooperative Biodiversity Groups Program

JOSHUA P. ROSENTHAL AND FLORA N. KATZ

OVERVIEW OF THE ICBGs

The International Cooperative Biodiversity Groups (ICBGs) have been among the most ambitious bioprospecting endeavors because of the explicit intent to use the drug discovery and bioinventory research process to generate enhanced research capacity, opportunities for sustainable economic activity, and incentives for conservation at each host site. As outlined in previous chapters discussing the Convention on Biological Diversity (CBD), pharmaceutical, agricultural, and other biotechnological development from natural sources can no longer be carried out in complete isolation from the economic and conservation context of the source country of those materials. Today, any collection-based research program needs to be aware of this context and should at least indirectly address the public's expectation that more is at stake than one's specific scientific objectives. The ICBGs represent an attempt to address this broader context directly and provide working models for approaches that can be broadly useful to scientists working with biodiversity in developing countries.

The ICBG program is administered by the Fogarty International Center of the National Institutes of Health (NIH), with the participation of other NIH institutes including the National Cancer Institute, the National Institute of Allergy and Infectious Diseases; the National Institute of Mental Health; the National Institute on Drug Abuse; and the National Heart, Lung, and Blood Institute, together with the National Science Foundation and the U.S. Department of Agriculture's Foreign Agricultural Service. At the time of this writing a third Request for Applications (RFA; http://grants1.nih.gov/grants/guide/rfa-files/RFA-TW-03-004.html) was recently issued by the above funding agencies, with additional commitments for participation by the National Center for Complementary and Alternative Medicine, the Office of Dietary Supplements, the National Institute on Child Health, and the U.S. Forest Service.

In the first 10 years of the program, eight diverse projects have been supported for work in 12 developing countries. Eleven of these countries are included within Conservation International's 25 "biodiversity hot spots," regions gravely endangered by human activity that together contain 44% of all plant species and 35% of all terrestrial vertebrate species in only 1.4% of the planet's land area (Myers et al., 2000; http://www.biodiversityhotspots.org/xp/Hotspots/hotspots_by_region). These areas are in urgent need of bioinventory and protection. The eight programs that have been or are currently funded are outlined in Table 1, and the structure, objectives, and methods of each are described in detail elsewhere (Rosenthal et al. 1999). Following the requirements in the RFAs, each ICBG addresses the general goals of drug discovery, scientific and economic development, and biodiversity conservation. Specifically, they must include substantial and novel efforts in natural products drug discovery, biological inventory, research capacity building, and benefit sharing. Each ICBG is led by a U.S. Principal investigator and is organized into several associate programs, each of which addresses one or more of the objectives above. At least one of the associate programs must be based in and led by a developing country organization. In addition to these requirements, several other commonalities among the groups have emerged. All have done at least some work with terrestrial plants (mostly in and from tropical forests), all conduct research in multiple disease areas simultaneously, most have had some ethnomedical component

Joshua P. Rosenthal and Flora N. Katz • Fogarty International Center, National Institutes of Health, 31 Center Drive, Bethesda, MD 20892-2220.

Table 1. ICBG program summaries[a]

Years active	Project title	Group leader and AP leaders	Institution of the group/AP leader
1993–2003	Biodiversity utilization in Madagascar and Suriname	David G. I. Kingston	Virginia Polytechnic Institute and State University
	AP-1 Botany and systematics	James Miller	Missouri Botanical Garden
	AP-2 Ethnobotany, conservation, and development	Russell A. Mittermeier	Conservation International
	AP-3 Ethnobotany, sample processing, and phytomedicine development	Rabodo Adriantsiferana	Centre National d'Application et des Recherches Pharmaceutiques (Madagascar)
	AP-4 Sample processing and antimicrobial drug discovery	Jan Wisse	Bedrijf Geneesmiddelen Voorziening Suriname
	AP-5 Drug discovery from Surinamese and Madagascan plants	J. J. Kim Wright	Bristol-Meyers Squibb Pharmaceutical Research Institute
	AP-6 Natural products as agrochemical agents	B. Cliff Gerwick	DowElanco Agrosciences
	AP-7 Rain forest natural products as anticancer and other agents	David G. I. Kingston	Virginia Tech
1994–2000	Peruvian medicinal plant sources of new pharmaceuticals	Walter H. Lewis	Washington University (St. Louis)
	AP-1 Plant ethnomedicine	Walter H. Lewis	Washington University
	AP-2 Biotic inventories and conservation	Gerardo Lamas	San Marcos University (Lima)
	AP-3 Medically significant plants in the tropics	Abraham Vaisberg	Universidad Peruana Cayetano Heredia (Lima)
	AP-4 Phytochemistry	Dave Corley and Margaret Wideman	Searle-Monsanto Co.
	Collaboration with indigenous peoples	Cesar Sarasara	Confederation of Amazonian Nationalities of Peru
1993–1998	Chemical prospecting in a Costa Rican conservation area	Jerrold Meinwald	Cornell University
	AP-1 Ecology, systematics, bioprospecting, and training	Ana Sittenfeld Giselle Tamayo	INBio (Costa Rica) University of Costa Rica
	AP-2 Chemistry and chemical ecology	Jerrold Meinwald	Cornell University
	AP-3 Drug discovery and development	Dinesh Vyas	Bristol-Myers Squibb
1994–2003	Drug development and conservation of biodiversity in West and Central Africa	Brian G. Schuster	Walter Reed Army Institute of Research (WRAIR)
	AP-1 Biodiversity inventory and monitoring, conservation, and training	Elizabeth Losos	Center for Tropical Forest Science, Smithsonian
	AP-2 Phytochemistry and Africa-based bioassays and phytomedicine development	Johnson Ayafor	University of Dschang (Cameroon)
	AP-3 Antimalaria drug discovery and development	Wilbur Milhous	WRAIR
	AP-4 Antiparasitic drug discovery and development	Joan Jackson	WRAIR
	AP-5 Ethnobiology, socioeconomic value assessment and community-based conservation projects	Maurice Iwu	Bioresources Development and Conservation Programme
	AP-6 Nonparasitic drug discovery and development	Brian Schuster	WRAIR
1993–2003	Bioactive agents from dryland biodiversity of Latin America	Barbara Timmermann	University of Arizona
	AP-1 Inventory, Ethnobotany, and conservation	Enrique Suarez	Instituto Nacional de Tecnologia Agropec. (Argentina)
		Gloria Montenegro	Pontifica Universidad Catolica de Chile (Chile)
		Robert Bye	Universidad Nacional Autonoma de Mexico (Mexico)

(continued)

Table 1. Continued

Years active		Project title	Group leader and AP leaders	Institution of the group/AP leader
1993–2003 (con't)	AP-2	Drug discovery	Barbara Timmermann	University of Arizona
	AP-3	Information management, dissemination, and related training	Barbara Hutchinson	University of Arizona
1998–2002		Drug discovery and biodiversity among the Maya of Mexico	Brent Berlin	University of Georgia
	AP-1	Drug discovery and pharmaceutical development	David Puett Robert Nash	University of Georgia Molecular Nature Ltd. (UK)
	AP-2	Medical ethnobiology and biodiversity inventory	Brent Berlin Eloise Ann Berlin	University of Georgia
	AP-3	Conservation, sustained harvest, and economic growth	Luis Garcia-Barrios Jose Carlos Fernandez	El Colegio de la Frontera Sur (ECOSUR)
1998–2003		Ecologically guided bioprospecting in Panama	Phyllis Coley	Smithsonian Tropical Research Institute
	AP-1	Collections, coordination, and data management	Phyllis Coley Todd Capson Thomas Kursar	Smithsonian Tropical Research Institute
	AP-2	Panama-based screening, isolation, and characterization of biologically active natural products	Mahabir Gupta	University of Panama, School of Pharmacy
	AP-3	Screening biological materials for activity against tropical disease agents	Eduardo Ortega-Barria	Florida State University-Panama Gorgas Memorial Institute of Health Research
	AP-4	Biological inventories	Don Windsor	Smithsonian Tropical Research Institute
	AP-5	Pharmaceutical and agricultural discovery and development	Leslie Harrison	Monsanto
	AP-6	Conservation and Ethnobotany	Manuel Ramirez	Conservation International
1998–2003		Biodiversity of Vietnam and Laos	Djaja D. Soejarto	University of Illinois at Chicago
	AP-1	Inventory and conservation of Cuc Phuong National Park	Djaja D. Soejarto	University of Illinois at Chicago
	AP-2	Lao medicinal plants as potential source of new medicines	Boun Hoong Southavong	Research Institute for Medicinal Plants (Laos)
	AP-3	Drug discovery from plants of Vietnam and Laos for AIDS and malaria therapies	John Pezzuto Harry Fong	University of Illinois at Chicago
	AP-4	Biomass production and economic development	Le Thi Xuan	National Center for Natural Sciences and Technology (Vietnam)
	AP-5	Drug discovery and development	Melanie O'Neill	Glaxo Wellcome (UK)

[a]Over the course of the projects, there have been several changes in personnel and some institutions. This table identifies the partners as they were established at the beginning of their most recent grant cycle. Only the institution of the Associate Program (AP) leader is given in the rightmost column. For a full list of the 59 participating institutions and additional information on each ICBG, see Rosenthal, 1999, 37S.

to their field efforts, and most include collaboration with at least one industrial partner that finances its own research and development activities.

Beyond these similarities, a diversity of approaches, focuses, and strengths are seen among the eight cooperative groups. Together they encompass researchers from over 59 organizations in 12 countries on five different continents (see Table 1). One project is working entirely in arid and semiarid landscapes (Latin America ICBG). One has been focused primarily on insects and other arthropods both for in-ventory and as a source of novel compounds (Costa Rica ICBG). Two of the current projects have also included collections from microbiological sources (Latin America and Panama ICBGs). The field efforts of two groups (Panama and Costa Rica ICBGs) have been driven substantially by ecological cues regarding chemistry. One project has never had an industrial partner (West Africa ICBG), in part reflecting its principal focus on parasitic diseases for which there is little industrial interest. Another began work with a relatively small drug discovery company rather than a major pharmaceutical corporation (Maya ICBG). Although most have knowledge and interests that overlap, the group leaders include three chemists (Kingston, Meinwald, Timmermann), a physician (Schuster), an ecologist (Coley), an anthropologist (Berlin), a plant taxonomist (Soejarto), and an ethnobotanist (Lewis). These specialties reflect only a fraction of the diverse expertise and approaches represented in the projects.

Bioinventory Results

The major focus of bioinventory activity to date has been terrestrial plants. The ICBGs have recorded over 360,000 collections representing over 8,000 taxa. Although this is a small number of species, these collections have sampled over half of the 386 angiosperm families (Cronquist, 1988). Furthermore, five gymnosperm families, 11 fern families, the occasional alga or moss, and arthropods from 21 orders have also been examined.

A major goal of these efforts is to return a catalogue of biodiversity to the countries in which they are collected as a national heritage, for education and recreation, as a scientific contribution to world surveys of biodiversity, and to guide conservation policy in the country. At least 36 electronic databases for biodiversity collections have been compiled by the ICBGs, some of which are now publicly available. The database of voucher specimens collected in Madagascar and Suriname, for example, is available in the TROPICOS database maintained by the Missouri Botanical Gardens. The Latin America ICBG has compiled an illustrated Flora of Chile (http://misdb.bpa.arizona.edu/~guoxiang/chile/biod/webdesc.html), which provides botanical information on flowering plants, and the Vietnam-Laos ICBG has introduced an Atlas of Seed Plants of Cuc Phuong National Park (http://uic-icbg.pharm.uic.edu) to share the results of their inventory efforts in Vietnam, which will be continuously updated as the project progresses.

Many of the groups have made ethnobotanically based collections, with associated data on usage, and some have produced major reference works on ethnomedicine (see, for example, Lewis et al., 1998). In addition, two of the ICBGs (in Suriname and Panama) have supported Shaman apprentice programs to promote the transfer of centuries-old ethnomedical knowledge to a new generation, which for the most part has abandoned this rich traditional heritage.

Insect collections from the Costa Rica and the Panama ICBGs total 128,026 records, although only 771 taxa have as yet been identified. Insects have proven a particularly difficult source for drug discovery efforts, as contaminants from ingested plants and obtaining sufficient biomass remain significant obstacles (Sittenfeld et al., 1999).

In the past couple of years, two projects have begun collecting microorganisms for drug discovery screens. From 1,250 single cultures, the Panama ICBG has isolated 367 morphospecies of tropical endophytic fungi, a largely unexplored group of ubiquitous fungi that live inside the tissues of vascular plants and whose further description may contribute significantly to assessments of fungal biodiversity (Arnold et al., 2000). Endophytic fungi have been shown to contain bioactive compounds and are promising targets for drug discovery screening (Strobel, 2002). Unfortunately, the group set aside their microbial work after these initial efforts due to competing demands on their resources. The Latin American ICBG has collected soil microorganisms from arid lands and extreme environments of northern and central Chile. Focusing on actinomycetes, a family of bacteria that has historically been a rich source of antibiotics and bioactive compounds, they have recovered over 500 isolates to date, and preliminary data indicates that numerous previously undescribed taxa and a significant number of bioactive compounds are likely to emerge. Transfer of the technology for collection, growth, and isolation of these organisms from the industrial partner, Wyeth, to the Chilean University partner, Universidad Catolica, is a major feature of the project.

Microbes remain a rich source for drug discovery and biodiversity inventory in developing countries. Although the technologies of microbial isolation may be transferable with relatively small expense, the downstream aspects of large-scale fermentation, extraction, and taxonomy require additional equipment and expertise, including nucleic acid sequencing for species identification and access to large databases, some of which remain proprietary. At the same time, as the majority of microorganisms are unculturable by standard methods (Pace, 1997), there is a need to develop new technologies to capture the chemical diversity likely to be found in these populations. For the purposes of bioprospecting partnerships such as the ICBGs, these techniques must be simple, inexpensive, and scalable. An approach with these characteristics has recently been reported (Zengler et al., 2002), and more are needed. Overcoming all of these obstacles will be important in creating self-sustaining and independent microbial drug discovery efforts in the participating countries.

Bioprospecting Results

The primary effort of most groups is the collection, cataloging, and screening of diverse biota for activity against a range of diseases, followed by bioassay-guided chemical identification and modification of active agents. Together the groups have assayed over 275,000 samples from more than 8,000 species of plants, 642 species of arthropods (mostly insects), and 500 species of microorganisms. Each group carries out assays in multiple therapeutic areas. Almost all of the groups have some effort in cancer and malaria. Cancer is generally a research area for the industrial partners and some academic labs, whereas malaria research is at present entirely the domain of academic and government labs. Most groups also target a variety of infectious diseases, including parasitic and respiratory diseases that pose a high disease burden in the partner developing countries. Four groups have done some work in agricultural areas, including veterinary medicines and insect, weed, nematode, and fungal pest control, predominantly through the industrial partners. With eight groups running assays in multiple labs and multiple therapeutic areas, it is estimated that over 270 types of primary assays in 26 therapeutic and agricultural areas have been run over the life span of the project. The largest portion of these represent mechanism-based assays in high-throughput systems carried out by the industrial partners.

At least 490 bioactive compounds of interest have been isolated over the past 9 years, of which 233 are new to science. Although compounds have been studied in animal models in at least a half-dozen therapeutic areas, none to date has reached clinical trials. Approximately 25 compounds are considered active leads. These are mostly compounds with anti-infective, antiparasitic, and neuroactive properties. Currently, the most promising of these leads are in analog development programs for therapeutic potential against malaria, leishmaniasis, tuberculosis, and central nervous system disorders. Although numerous provisional patent applications have been filed, only one to date has been carried all the way through to patent issuance. This was recently issued to the West Africa ICBG for a novel approach to antiparasitic and antifungal disease therapy using lipid synthesis, metabolism, and/or excretion inhibitors (Jackson et al., 2002).

Interactions with Industrial Partners

Numerous changes in the pharmaceutical sector and in individual companies have made interactions with companies challenging and somewhat unpredictable for many of the scientists involved. In a number of cases, industrial partners have withdrawn from the projects unexpectedly in midterm. Moreover, while sharing research results among these nontraditional partners is crucial to successful work,

it has been a greater challenge than many have predicted. Companies are typically concerned that their competitive edge will be compromised if proprietary bioassays and related methodology, as well as the nature of any specific leads or the financial terms of an agreement, are shared readily with parties peripheral to this work. Furthermore, the unfamiliar objectives and conduct of conservationists, indigenous groups, and others raises concerns among industrial partners that their needs for secrecy will not be respected, and vice versa. However, we have found that, once committed to working in an ICBG, most companies have been very willing to negotiate novel research and benefit-sharing agreements, provide training opportunities, and transfer technology to their partners.

Access, Intellectual Property Rights, and Benefit Sharing

Careful attention by the ICBGs to access, intellectual property, and benefit-sharing (ABS) issues have yielded approximately 125 contracts, including research and benefit sharing, material transfer, confidentiality, know-how licenses, license option agreements, and trust funds. The groups have encountered very significant differences among host countries and within them regarding access regulation for plants, which are the major focus of the current ICBGs. However, seeking access to work with microbes will likely raise additional issues, including the anxiety of host country partners over samples that do not require recollection for development. Many governments view the need to return to the source for more material as an important control point for access, although the ability to do synthetic chemistry based on isolated compounds could arguably raise similar concerns in the arena of plant collections.

Unfortunately, despite the profound conceptual shift in treatment of genetic resources that the CBD signifies, the treaty provides little guidance to governments or private organizations on how to implement this new paradigm. Elaboration of a model system to implement access and benefit-sharing policies has been elusive, even 10 years after the treaty entered into force. In part, this is due to the complexity of the scientific, legal, and commercial elements of the model. To make matters worse, suspicion, resentment, and misunderstanding, fueled by colonial history and the politics of trade and intellectual property rights, have frequently brought discussion of the issues to a standoff in both multilateral and project-specific fora.

The ICBG program has had both the privilege and challenge of being one of the first large-scale and coordinated efforts to implement the access and benefit-sharing objectives of the CBD in specific projects. The groups have generally had to develop their access and benefit-sharing policies and agreements in the complete absence of any regulatory guidance beyond the general framework of the CBD (Macilwain, 1998) and the principles of the ICBG program (originally outlined by Schweitzer et al. [1991] and further developed in the 1997 and 2002 RFAs [http://www.nih.gov/fic/programs/rfa.html and above]).

Sometimes, regardless of how thoughtfully, transparently, or collaboratively a collection-based project and its approach to ABS are formulated, the political context in which it operates may ultimately make certain partnerships controversial. This is particularly the case when working with indigenous peoples that present relatively undefined or Western-incompatible systems of land tenure, governance, and intellectual property rights (Alexiades and Laird, 2002). In the policy vacuum that characterizes the current ABS situation in most countries, it is easy for anxiety and suspicion to proliferate about arrangements with indigenous peoples. As this aspect of the ICBG program is experimental, it is not surprising that some projects would need to adapt or even terminate. There have been two examples in the history of this program where this was the case, both involving the issue of prior informed consent and who is capable of giving consent for indigenous communities. In Peru (Peru ICBG 1995–1996), working with local representatives of communities organized into legal entities, barriers to the project could be resolved through further discussions and negotiations (Lewis et al., 1999). In the volatile political climate of Chiapas, Mexico (Maya ICBG), where similar representative organizations do not exist, the project was not able to continue. Analysis and debate about these experiences continue to unfold over time in the anthropology, development, and science policy communities (see, for example, Dalton, 2000, 2001; Nigh, 2002, and responses therein; Tobin, 2002). We view these experiences as difficult but valuable lessons in both the ethics and the politics of research and capacity-building partnerships.

Many countries, including Mexico, Argentina, Chile, Peru, and Suriname, have treated the ICBGs as testing grounds for their developing policies on access and benefit sharing for genetic resources. Although this has occasionally produced frustration for the investigators and has been a significant rate-limiting factor in some projects, overall we consider it to be a positive role for the program. We believe that the projects have offered concrete experiences for governments and a variety of resulting lessons. Some examples of these lessons, as we interpret them, include (i) access to biodiversity has both research and commercial goals, and it is best to recognize these as separate goals and treat the corresponding activities different from the outset, even in the context of the same project; (ii) policy measures that provide clarity and some security to providers and users of genetic resources are needed, but "one size fits all" approaches are impractical, and elaborate and inflexible access regulations in this diverse and changing field may hurt the interests of both; and (iii) a diversity of benefits may be available through such collaborations, and although biodiversity is of global value, monetary benefits from any specific project may be unpredictable.

Capacity and Capability Building

Training and the transfer of technology, including equipment, protocols, and assays, represent a significant investment of ICBG efforts. Since 1993, more than 2,800 individuals from 12 countries have received formal training through the program. Over 90% of these trainees represent developing country participants. They include Bachelor's, Master's, and doctoral students and postdoctoral fellows, as well as technicians, nonscientific community residents, and others. Approximately 70 of the trainees (44 from partner countries) have been enrolled in long-term degree-earning programs. The vast majority of others have participated in short-term training efforts such as workshops (1 to 7 days) and limited-duration visits (3 weeks to 6 months) to participating laboratories.

Training topics include almost every aspect of ICBG drug discovery work, including plant collection and drying in the field, extraction, testing, compound isolation, identification and modification, database development and maintenance, use of geographic information systems, contract development, and understanding of intellectual property rights. Numerous training events have also focused on other elements of ICBG work, including conservation and restoration, pollination biology, cell and tissue culture, taxonomy, ethnobotany, plant anatomy, agroecology, curatorial methods, grassroots community organization, and commercial production of medicinal plants.

Associated with training and research efforts, a substantial amount of equipment and infrastructure enhancement for both U.S. and developing country institutions is carried out by the groups. Through the ICBGs and their industrial collaborators, developing country scientific organizations have received items such as updated computers and software, geographic positioning systems, high-pressure liquid-chromatography equipment, rotovaporators, −70°C freezers, plate readers, water purification systems, fume and sterile hoods, vehicles, microscopes, bar code readers, cameras, herbarium cabinets, greenhouse supplies, laboratory glassware, and safety equipment. The ICBG program also partially supported purchase of a nuclear magnetic resonance imaging device in Panama. Participating community organizations also receive other simple but important contributions such as water tanks, fencing for gardens, shade cloth, boats, refrigerators, building enhancements, or travel funds.

Examples of site capacity milestones include independent chemical isolation capacity in Panama and Cameroon, in vitro screening capacity for malaria and establishment of a malaria research center in Madagascar, microorganism culture in Chile, geographic imaging system capacity in Suriname, establishment and development of the herbarium of Cuc Phuong National Park in Vietnam, and herbarium storage and data management capacity in all countries.

Conservation Outcomes

Several types of ICBG activities promote conservation and development. They include training personnel and research capacity enhancement at host country institutions, scientific research in support of biodiversity management, in situ and ex situ conservation projects, environmental education, and policy analysis. These are described below.

Capacity-building efforts

The capacity-building efforts have been characterized above. With better trained and equipped staff, as well as experience with nondestructive uses of biodiversity, developing country institutions involved in natural resource management will be better prepared to make informed decisions on important and pressing concerns such as logging or mining concessions or agricultural development projects. For example, the Suriname ICBG, through the Missouri Botanical Garden and Conservation International, has enhanced the facilities for preserving plant specimens and managing data at the National Herbarium as well as the associated technical skills of both herbarium staff and the National Forest Service. Increased capacity to identify and monitor regions of high biodiversity will help these and other Surinamese institutions assess the advisability and nature of commercial logging concessions that are under consideration by the government (Kingston et al., 1999b).

Science to support management of natural resources

Conservation is also advanced through scientific investigation that directly provides taxonomic, eco-logical, and economic data that are useful in managing natural resources. Here we describe a few illustrative examples among the many supported by the ICBG program. The forest dynamics and inventory plots of the West Africa ICBG in Cameroon (Schuster et al., 1999) provide information that will be useful in assessing long-term trends of reproduction and survivorship in tropical forests under threat. Studies that yield understanding of the patterns of feeding, distribution, and migration of butterflies and other insects in Costa Rica help identify areas and periods of high diversity for collection and can be used to project the impacts of development programs in a given site (Janzen and Gauld, 1997). Studies of morphological and anatomical predictors of plant tolerance to tissue harvesting (see, for example, Montenegro et al., 2001) are useful to promote sustainable collections of medicinal plants. Socioeconomic assessment of the value of biodiversity for local medicines and other nontimber forest products (West Africa ICBG, Suriname-Madagascar ICBG) can provide important information for local and national decisions regarding natural resource use.

Ex situ botanical conservation

Ex situ conservation in botanical gardens and seed banks will be an increasingly important resource for conservation as natural habitats are destroyed by development and other processes. Most of the ICBGs have made significant contributions to these efforts in established host country institutions, such as the Botanical Garden of the National University of Mexico, as well as at U.S. institutions, and several have started smaller medicinal plant gardens in communities, parks, and at universities.

Integrated conservation and development

The ICBG is, in part, an integrated conservation and development program (Brown and Wyckoff-Baird, 1992). That is, conservation of biodiversity is an expected outcome of development efforts that create an opportunity, means, and incentive to change patterns of resource use. In this framework, conservation-promoting activities are those that build scientific, commercial, and legal capacity, those that educate resource users and regulators about the alternatives to unsustainable practices; and finally, those that provide financial or other benefits to stakeholders in ways that may influence relevant behavior. Financial benefits that are relevant are all those that are a result of the project, including near-term compensation, milestones or royalties that may come from commercial partners in relation to research activities, and any commercial products that emerge. Because of the research focus of the ICBG program, local commercial use of biodiversity that generates significant near-term income for local populations has been a more limited feature of the program to date than classic integrated conservation and development programs usually promote. However, most of the groups are supporting projects such as development of traditional woodcraft enterprises (Suriname), propagation of ornamental plants (Mayan Mexico), and propagation of plants for widely sold herbal remedies (West Africa, Vietnam, Mexico, Peru). Almost all of the projects have provided income to community members as compensation for their time and skill expended as participants in the research and training efforts of the projects.

As the expected outcome of the integrated conservation and development approach is a shift in attitude and behavior by landowners, policy makers, and others who affect natural resource use, it is often difficult to tie specific events that occur on a national or regional level to the efforts of any individual project. However, one important example of this incentive and example effect on conservation is the role of the Suriname ICBG in the recent establishment of the Central Suriname Reserve, a four-million acre preserve of interior rainforest (Kingston et al., 1999b).

Outreach

All of the above-described contributions to conservation require dissemination of findings and outreach to other scientists, governments, and communities to be effective. All of the groups have made numerous presentations and publications on the process elements of their programs including partnership structures (see, for example, Kingston et al., 1999a; Timmermann, 1997), contractual arrangements (see, for example, Iwu, 1996; Rosenthal et al., 1999; others described in Laird, 2002), and the potential for economic use of preserved landscapes (Janzen, 1999). Outreach efforts such as environmental education are in themselves important means of advancing conservation. Several of the projects as well as the government funding agencies have held international conferences related to sustainable use of biodiversity and bioprospecting (see, for example, Timmermann and Montenegro, 1997; Iwu et al., 1997; Grifo and Rosenthal, 1997). Perhaps the most important long-term investment is education of grade school children upon whom future decisions regarding resource use will depend. The Costa Rica, Latin America, Panama and Mayan Mexico ICBGs have been very active with children and youth groups in this regard.

CONCLUSION

In summary, in the first 9 years of the program, the ICBGs have (i) discovered numerous bioactive compounds, some of which are leads of significant continuing interest; (ii) enhanced the technical capacity of over 2,000 developing country participants and their associated institutions; (iii) contributed to the scientific and policy process of conservation; and (iv) provided important models for governments and other organizations for collaborative research that supports multiple objectives, including those of the CBD.

This last accomplishment is perhaps the single most significant contribution of the program to date. The ICBG has pioneered the development of models for nontraditional international partnerships of universities, companies, and government and community organizations. It has shown that such organizations can work collaboratively to achieve their own objectives and contribute to larger goals. Over the past several years the funding agencies have received hundreds of queries and requests for guidance in this area from governments, universities, companies, multilateral banks, foundations, conservation organizations, and others. The investigators have undoubtedly received many as well. From the standpoint of the funding agencies, it has also provided a model for collaboration on complementary goals that could not be supported by one agency alone. The demand for ways to achieve the integration of goals that the ICBG represents is huge, whereas the examples of such integration are very few.

Even in the relatively narrow context of biomedical research needs, it is critical that we do not underestimate the value of this product of the program—models for collaboration. Drug discovery from natural products, as well as a wide range of biomedical research topics on physiology, genetics, and behavior, depend on access to tens of thousands of different organisms that may occur in very isolated places around the world. Beyond plants, access to diverse populations of biomedical study organisms, including sulfur bacteria, endophytic fungi, nematodes, grasshoppers, coral reef sponges, sea cucumbers, dolphins, and chimpanzees, to name a few, is increasingly threatened not only by the rapid disappearance of these organisms, but also by changing attitudes and suspicions about their use. The research community needs to demonstrate that this work can be done in a flexible and accommodating manner that recognizes the environmental and socioeconomic context in which these organisms exist, or we will lose access to them in the near term through politics, and eventually to extinction. It is in recognition of this imperative that some commentators have pointed to the ICBG program as one of the best opportunities to find a constructive path toward resolution of the apparently opposing points of view associated with prospecting for drugs and crop protection agents in biodiversity-rich developing countries (Nature, 1998).

REFERENCES

Alexiades, M. N., and S. A. Laird. 2002. Laying the foundation: equitable biodiversity research relationships, p. 1–15. *In* S. A. Laird (ed.), *Biodiversity and Traditional Knowledge: Equitable Partnerships in Practice.* Earthscan, London, United Kingdom.

Arnold, A. E., Z. Maynard, G. S. Gilbert, P. D. Coley, and T. A. Kursar. 2000. Are tropical fungal endophytes hyperdiverse? *Ecol. Lett.* 3:267–274.

Brown, M., and B. Wyckoff-Baird. 1992. *Designing Integrated Conservation and Development Projects.* The Biodiversity Support Program, World Wildlife Fund, The Nature Conservancy, World Resources Institute. Corporate Press, Inc., Landover, Md.

Cronquist, A. 1988. *The Evolution and Classification of Flowering Plants,* 2nd ed. The New York Botanical Garden, Bronx, N.Y.

Dalton, R. 2000. Political uncertainty halts bioprospecting in Mexico. *Nature* 408:278.

Dalton, R. 2001. The curtain falls. *Nature* 414:685.

Grifo, F., and J. Rosenthal (ed.). 1997. *Biodiversity and Human Health.* Island Press, Washington, D.C.

Iwu, M. 1996. Biodiversity prospecting in Nigeria: seeking equity and reciprocity in intellectual property rights through partnership arrangements and capacity building. *J. Ethnopharm.* 51:209–219.

Iwu, M. M., E. N. Sokomba, C. O. Okunji, C. Obijiofor, and I. P. Akubue. 1997. *Commercial Production of Indigenous Plants as Phytomedicines and Cosmetics.* BDCP Press, Silver Spring, Md.

Jackson, J. E., M. Iwu, C. Okunji, C. Bacchi, J. Talley, and J. Ayafor. 2002. Antifungal and antiparasitic compounds. U.S. patent 6,403,576.

Janzen, D. 1999. Gardenification of tropical conserved wildlands: multitasking, multicropping, and multiusers. *Proc. Natl. Acad. Sci. USA* 96:5987–5994.

Janzen, D. H., and I. D. Gauld. 1997. Patterns of use of large moth caterpillars (Lepidoptera: Saturniidae and Sphingidae) by ichneumonid parasitoids (Hymenoptera) in Costa Rica dry forest, p. 251–271. *In* A. D. Watt, N. E. Stork, and M. D. Hunter (ed.), *Forests and Insects.* Chapman & Hall, London, United Kingdom.

Kingston, D. G. I., H. van der Werff, R. Evans, R. Mittermeier, L. Famolare, M. Guerin-McManus, S. Malone, J. H. Wisse, D. Vyas, and J. J. K. Wright. 1999a. Biodiversity conservation, economic development, and drug discovery in Suriname, p. 39–59. *In* H. Cutler and S. Cutler (ed.), *Biologically Active Natural Products: Pharmaceuticals.* CRC Press, Boca Raton, Fla.

Kingston, D. G. I., M. Abdel-Kader, B.-N. Zhou, S.-W. Yang, J. M. Berger, H. Van Der Werff, J. S. Miller, R. Evans, R. Mittermeier, L. Famolare, M. Guerin-McManus, S. Malone, R. Nelson, E. Moniz, J. H. Wisse, D. M. Vyas, J. J. K. Wright, and G. S. Aboikonie. 1999b. The Suriname International Cooperative Biodiversity Group Program: lessons from the first five years. *Pharm. Biol.* 37S:22–34.

Laird, S. A. (ed.). 2002. *Biodiversity and Traditional Knowledge: Equitable Partnerships in Practice.* Earthscan, London, United Kingdom.

Lewis, W., D. Mutchier, N. Castro, M. Elvin-Lewis, and N. Farnsworth. 1998. *Ethnomedicine, Chemistry, and Biological Activity of South American Plants,* vol. 1–3. Chapman & Hall, London, United Kingdom.

Lewis, W. H., G. Lamas, A. Vaisberg, D. G. Corley, and C. Sarasara. 1999. Peruvian medicinal plant sources of new pharmaceuticals (International Cooperative Biodiversity Group—Peru). *Pharm. Biol.* 37S:69–83.

Macilwain, C. 1998. When rhetoric hits reality in debate on bioprospecting. *Nature* 392:535–540.

Montenegro, G., G. Patrick, P. Echenique, M. Gomez, and B. Timmermann. 2001. Mechanisms towards a sustainable use of *Chorizanthe vaginata* Benth. Var. maritime Remy: a medicinal plant from Chile. *Phyton* 2001:91–106.

Myers, N., R. A. Mittermeier, C. G. Mittermeier, G. A. B. da Fonseca, and J. Kent. 2000. Biodiversity hotspots for conservation priorities. *Nature* 403:853–858.

Nature. 1998. The complex realities of sharing genetic assets. *Nature* 392:525.

Nigh, R. 2002. Maya medicine in the biological gaze. *Curr. Anthropol.* 43:451–478.

Pace, N. 1997. A molecular view of microbial diversity and the biosphere. *Science* 276:734–740.

Rosenthal, J. P. (ed.). 1999. Drug discovery, economic development and conservation: the international cooperative biodiversity groups. *Pharm. Biol.* 375:6–144.

Rosenthal, J. P., D. Beck, A. Bhat, J. Biswas, L. Brady, K. Bridbord, S. Collins, G. Cragg, J. Edwards, A. Fairfield, M. Gottlieb, L. A. Gschwind, Y. Hallock, R. Hawks, R. Hegyeli, G. Johnson, G. T. Keusch, E. E. Lyons, R. Miller, J. Rodman, J. Roskoski, and D. Siegel-Causey. 1999. Combining high risk science with ambitious social and economic goals. *Pharm. Biol.* 37:6–21.

Schuster, B. G., J. E. Jackson, C. N. Obijiofor, C. O. Okunji, W. Milhous, E. Losos, J. F. Ayafor, and M. M. Iwu. 1999. Drug development and conservation of biodiversity in West and Central Africa: a new standard of collaboration with indigenous people. *Pharm. Biol.* 37S:84–99.

Schweitzer, J., G. Handley, J. Edwards, F. Harris, M. Grever, S. Schepartz, G. Cragg, K. Snader, and A. Bhat. 1991. Summary of the Workshop on Drug Development, Biological Diversity and Economic Growth. *J. Natl. Cancer Inst.* 83:1294–1298.

Sittenfeld, A., G. Tamayo, V. Nielsen, A. Jimenez, P. Hurtado, M. Chinchilla, O. Guerrero, M. A. Mora, M. Rojas, R. Blanco, E. Alvarado, J. M. Guttierez, and D. H. Janzen. 1999. Costa Rican International Cooperative Biodiversity Group: using insects and other arthropods in biodiversity prospecting. *Pharm. Biol.* 37S:55–68.

Strobel, G. A. 2002. Microbial gifts from rain forests. *Can. J. Plant Pathol.* 24:14–20.

Timmermann, B. N. 1997. Biodiversity prospecting models for collection of genetic resources: the NIH/NIF/USAID model, p. 233–249. *In* K. E. Hoagland and A. Y. Rossman (ed.), Proceedings of the Symposium on Global Genetic Resources: Access, Ownership, and Intellectual Property Rights, Washington, D.C. Association of Systematics Collections, Washington, D.C.

Timmermann, B., and G. Montenegro. 1997. Taller Internacional: Aspectos Ambientales, Eticos. Ideologicos y Politicos en el Debate sobre Bioprospeccion y Uso de Recursos Geneticos en Chile. *Noticiero de Biologia—Organo Oficial de la Sociedad de Biologia de Chile* 5:119.

Tobin, B. 2002. Biodiversity prospecting contracts: the search for equitable agreements, p. 287–309. *In* S. A. Laird (ed.), *Biodiversity and Traditional Knowledge: Equitable Partnerships in Practice.* Earthscan, London, United Kingdom.

Zengler, K., G. Toledo, M. Rappe, J. Elkins, E. J. Mathur, J. M. Short, and M. Keller. 2002. Cultivating the uncultured. *Proc. Natl. Acad. Sci. USA* 99:15681–15686.

IX. CONCLUSION

Microbial Diversity and Bioprospecting
Edited by Alan T. Bull
© 2004 ASM Press, Washington, D.C.

Chapter 44

The Value of Biodiversity

DAVID W. PEARCE

THE INEVITABILITY OF VALUING BIODIVERSITY

A voluminous literature exists on the nature of the value that resides in or is conferred on biological diversity (see, for example, Ehrenfield [1988] and Randall [1988]). For many, value is an intrinsic property of biodiversity and not something that can be conferred on it by human perception. For most economists, value is necessarily of and by people, and the notion of an intrinsic value independent of a valuer is meaningless. The distinction between intrinsic and conferred value is fundamental and occupies an even larger literature in moral philosophy (for a good discussion, see Beckerman and Pasek, 2001). But a crucial question is what we can do with competing notions of value in a world where the population will grow by a further 50% by 2100 independent of human action and where biodiversity-rich land is converted daily to far more homogeneous uses such as agriculture. For anyone who is pragmatic and action oriented and who is concerned about biodiversity loss, the notion of intrinsic value is unhelpful. If all the constituent parts of biodiversity have intrinsic value, and if intrinsic value is not measurable—as all would agree—then it is not possible to trade off biodiversity against anything else. Indeed, this nontradability is exactly what advocates of intrinsic value want. For if it is tradable, then we can find justifications for losing at least some biodiversity, and perhaps a justification for losing rather a lot of it. If value is intrinsic, there are no trades, no justifications for biodiversity loss. But trades there have to be. First, it is not remotely feasible to conserve all biodiversity—any real world resident would acknowledge that proposition. Second, conservation is not costless. All policy actions involve sacrifices, what the economist would call opportunity cost—the value of the thing we go without by selecting any given option. Land uses for conservation are not compatible with inten-

sive, high-yielding agriculture to feed people, for example. So, even if there is no money cost to a conservation action, there is always, and unavoidably, an opportunity cost. In an uncrowded world of a few million people and a large stock of biodiversity, those costs could be extremely low. In a crowded world they are likely to be high. So, if biodiversity is to be conserved, its value must be shown to be even higher. That is the logic of the economic approach to conservation. As an historical reflection, it might also be pointed out that those who do not acknowledge tradability have had ample opportunity to make a significant contribution to saving biodiversity. By all accounts, they have failed.

Whether the proximate destroyer of biodiversity is a polluter, a frontier agriculturist, or a logging company, the real causes of biodiversity loss are financial. Stripped of the many complexities that surround decisions to destroy biodiversity, the root cause is that it is more profitable to destroy biodiversity than to conserve it. What economists do is show that these financial equations are incomplete and shortsighted because economics is concerned with the well-being of individuals, whoever they are. Finance and economics are categorically not the same thing, despite the continuing misperception to the contrary. If financial equations drive biodiversity loss, then there is some hope that ensuring those equations are correctly specified in terms of human well-being will contribute to less biodiversity loss. There is no real likelihood of showing that no loss at all should occur: the opportunity costs must eventually become very high. But we can be equally sure that the current rate of diversity loss is far too high.

The only scope for avoiding wholesale trade-offs is either (i) to deny that satisfying human wants and needs is relevant to a conservation decision or (ii) to argue that there is a set of limits or boundaries to diversity loss and that trade-offs can take place only within this meta constraint. The former view charac-

David W. Pearce • Department of Economics, University College London, Gower Street, London WC1E 6BT, United Kingdom.

terizes some people's interpretation of the Gaian ethic in which humans are portrayed as ecologically irritating and rather unimportant in the evolutionary scale of things. The latter approach characterizes the safe minimum standards (SMS) approach in which the presumption is in favor of conservation unless the social cost of conservation (the opportunity cost) is, in some sense, very high (Page, 1977; Randall, 1988). This SMS approach is also consistent with the notion of a precautionary principle. Although the Gaian approach is attractive, we proceed on the assumption that human preferences do matter, but that they may need to be constrained within a wider envelope of constraints. The challenge then is to find valuation procedures that account for these constraints. One other way of minimizing trade-offs is to seek land uses and technologies that combine biodiversity with the demand for developmental uses of land, e.g., for forest lands via agro-forestry rather than conventional agriculture, or sustainable timber production rather than conventional logging. Unfortunately, many of these sustainable use options still fail the financial test, shifting the focus to a proper comparison of economic benefits and costs (Pearce et al., 2002).

THE EQUATION OF WELL-BEING

The above discussion can be encapsulated in a few simple equations. Define the benefit to human beings of biodiversity conservation as B_C. Define the money costs of conservation (e.g., park rangers, etc.) as C_C. Then let B_D and C_D be the respective benefits and costs of development, e.g., conversion of forest land to agriculture. Ignoring the time dimension, the economic (as opposed to the financial) equation that would justify conservation is

$$(B_C - C_C) - (B_D - C_D) > 0 \qquad (1)$$

One reason this is not a financial equation is that anyone considering the development of the land would consider only the magnitude $(B_D - C_D)$, i.e., whether the development makes a profit or not. They would not consider the loss in economic value due to the sacrifice of the net conservation benefits $(B_C - C_C)$. So even this simple modification of the basic financial formula has potentially huge consequences. Mathematically, it is obvious that development is far harder to justify if $(B_C - C_C)$ is a positive sum. Note also that, if human needs are pressing, i.e., if conversion to agriculture is really that economically valuable, B_D will be high and conservation will be less justifiable. This is the SMS principle coming to the fore: presume that equation 1 holds in general, but observe that

high development values will justify land conversion and the corresponding biodiversity loss.

B_D, however, can be unjustifiably high. For an economist, prices in freely functioning markets are reasonably good reflectors of scarcity and hence economic value. This explains the diamonds and water paradox: diamonds are scarce and have high prices; water is usually plentiful and has low prices. But water in a dry land or a drought is scarce and has a high price. Many prices, however, are not the outcome of market forces. They are distorted by government interference in markets, and most notably by subsidies. Current estimates of subsidies to industry, transport, and especially agriculture suggest that they amount to over $1 trillion (U.S.) annually (van Beers and de Moor, 2001). That is roughly 4% of the entire world's economic product. Most of those subsidies do environmental harm. In terms of equation 1 it is easy to see why. The financial condition for development ceases to be $(B_D - C_D) > 0$ and becomes $(B_D + S - C_D] > 0$, where S is the subsidy. Subsidies justify more development on financial grounds, but, from the economic point of view, a subsidy is a financial transfer from one sector of society (taxpayers) to another sector (e.g., farmers) and has no counterpart economic output. Hence equation 1 would remain valid: subsidies would be netted out in any economic analysis.

B and C in equation 1 are defined in terms of human well-being. In economics, well-being (or welfare, or, an old-fashioned term but still used by economists, utility) reflects human preferences. My well-being is higher in situation X than in Y if I prefer X to Y. Preferences show up in many activities, e.g., voting, but one pervasive form of revealing preferences is in market places. If any good in the supermarket has a price P, I vote for it if I am willing to pay P and against it if I am not willing to pay for it. Hence, willingness to pay (WTP) becomes a means of observing preferences. As a final step, economists have long shown that the actual magnitude of WTP is a measure of the intensity or amount of gross well-being, and the excess of WTP over price (WTP $- P$) is a measure of net well-being (or consumer's surplus, as it is known). The importance of this approach to economic value is critical for equation 1. B_C measures the economic value of conservation, but that value may well not be directly observable. Pursuing the forest land example, forest users may collect nontimber products which they may sell in local markets. The products have a market and hence a set of observable prices. But the carbon embodied in the biomass is an ecological asset that is also an economic asset. If it is released as carbon dioxide, it adds to the greenhouse effect and causes global warming damage. That damage is a loss of human well-being. If biodiversity is lost, then various losses occur in economic well-being, an issue explored below. But biodiversity and carbon may not have markets, in which case there are no observable prices. But people are willing to pay to avoid global warming damage, and they are willing to pay to reduce biodiversity loss. So the absence of observed prices does not mean there are no prices. The science of environmental economics is, among other things, concerned with eliciting the WTP for nonmarket ecological services such as carbon sequestration and biodiversity provision. As a general rule, all ecological services are also economic goods and all have an economic value. The challenge is to find it. In terms of equation 1 we seek to measure B_C through the use of nonmarket valuation techniques.

There is still more to say about equation 1. If B_C is measured in terms of WTP (and hence in money units), we can ask whether its value is likely to stay constant over time. If biodiversity is reducing each year, as is the case, then, like any economic asset, its relative value will tend to rise. Relative here means relative to the value of all other goods. The intuition is clear. The economic value attached to the last few thousand hectares left of forest land is likely to be much higher than the economic value of the existing area of forest land. We would expect there to be a gradient of economic value, rising as the quantity remaining declines. But there is another reason for supposing that B_C will rise through time: people tend to get richer over time. While economists have done surprisingly little work on the link between rising incomes and rising economic values of environmental assets (Pearce, in press), what we know is that WTP rises with income and that the elasticity of WTP with respect to income (the percentage rise in WTP divided by the percentage rise in income) is probably around 0.3 to 0.5. So B_C in equation 1 needs to be modified to read $(B_{C,0} [1 + p]^t)$ where p is now the rate at which B_C rises through time, and t is time. Arguably, B_D could rise through time as well, but the evidence suggests that B_C is likely to rise faster.

Notice that equation 1 now has to have time subscripts: the 0 signifying the value of B_C in an initial period. More generally, all of the components in equation 1 have time subscripts, so it should be rewritten as

$$[B_{C,0} (1 + p)^t - C_{C,t}] - (B_{D,t} - C_{D,t}) > 0 \quad (2)$$

Equation 2 is beginning to look a little more complex, but a moment's reflection shows that conservation is more likely to be justified in this formulation than in equation 1.

Putting time into the formula enables the final adjustment to be made to equation 1. Economic well-being is based on the notion of preference, the so-called consumer sovereignty ethical principle that underlies normative economics. But people are not indifferent to when benefits and costs occur. There are at least three reasons for this. First, most of us are fairly impatient. Even if we had a guarantee that a good would be available to us in one year's time and that we would be around to receive it, most people would prefer to have it now. Impatience means that we discount time: $1 of future benefit is not worth the same as $1 now. It is worth less. The second reason for discounting lies with the risk that we will not be around to collect the benefit—the risk of death. It is widely argued that this is not relevant to a social decision rule because, although individuals are mortal, society is immortal. Whereas perhaps once tenable, in a world of global risks to humankind, social immortality hardly seems tenable any more. Even though the risk may be small, it is not zero. The third reason for discounting is that we are likely to be richer in the future. The richer we are, the lower the well-being we tend to attach to an extra $1 of income. Intuition suggests that rich people do not care as much about an extra $1 compared with poor people. Economists say there is a diminishing marginal utility of income. The rate at which this importance of an extra $1 declines with income is observable in savings behavior (although some economists dispute this). Expressed as an elasticity, the rate would appear to be in the region of 0.8 to 1.0 in rich countries (Pearce and Ulph, 1999).

Combining the impatience, social risk, and utility of income arguments produces a social rate of discount. Economists argue endlessly over what this is, but there is a fair consensus that it would currently be around 3 to 4% per annum. Thus, $1 in one year's time would be equivalent to $1/(1 + r)$ where r is the social rate of discount. If $r = 4\%$, $1 in one year's time would be worth 96 cents ($1/1.04). But discounting is cumulative, so $1 in 50 year's time would be worth $1/(1.04)^{50} = \$1/7.1 =$ only 14 cents. The logic of discounting—it simply reflects human preferences—can immediately be seen to pose a potentially serious problem. If the main benefits of biodiversity conservation lie in the long term, then discounting will reduce those benefits in a potentially dramatic way. For example, at 4%, the $1 of benefit in 100 years is reduced to just 2 cents. Arguments have long gone on about choosing the right discount rate: 4% is clearly better for biodiversity than 10%. Only recently has it emerged that the behavior of the discount rate with respect to time also needs to be questioned, an issue addressed below.

The final equation can now be set out. It looks more complicated because it sums across time t. But the essence of it remains the same (note the plus before development costs, as we have removed the parentheses that appear in equation 1. If we are to justify conservation, net conservation benefits summed over time and adjusted for rising relative values must exceed net development benefits also summed over time.

$$\left[\sum_t B_{C,0} (1 + p)^t - \sum_t C_{C,t} - \sum_t B_{D,t} + \sum_t C_{D,t}\right](1 + r)^{-t} > 0 \qquad (3)$$

THE ECONOMIC BENEFITS OF BIODIVERSITY

The economic benefits of biodiversity comprise three general components: the contribution to ecosystem functions, nonuse values, and the contribution to ecosystem resilience (Pearce et al., 2002). Pursuing the forest example, ecosystem functions include the provision of direct services such as timber nontimber products, the provision of indirect services such as watershed regulation, the provision of information about ecosystem evolution (encompassing scientific information), the provision of global services such as climate regulation, and nonuse value (Pearce, 2001). Nonuse value reflects individuals' willingness to pay for biodiversity conservation regardless of the direct and indirect uses made of biodiversity. Imagine someone who has never seen a tropical forest and, for some reason, can never visit one. Independently of any indirect value they obtain (e.g., from a less-warmed climate), they may nonetheless be willing to pay to conserve the forest. Motivations for nonuse value vary because of some notion of stewardship, perhaps some notion of Nature's right to exist, a concern to leave an asset for children and grandchildren to use, and so on. It would be hard to explain donations to conservation societies if those donations were not partly based on nonuse value.

Some use values of ecosystem functions are derived from component parts of biodiversity, say a specific species. Other use values are very much dependent on the whole ecosystem. Nonuse values may similarly be heavily influenced by a single component (again, a single species, for example) or by an appreciation of the interconnected whole asset. Surprisingly, what it is that people value is seriously underresearched. There are reasonably good studies of individuals' valuations of wetlands or forests or rivers, but remarkably little on how they perceive these assets, i.e., what precisely it is that determines their willingness to pay. In contrast to the mixed part-whole determinants of use and nonuse values, the value of ecosystem resilience derives from the diversity of the components. Thus, whereas the economic value derived from ecological functions and nonuse motives could be the value placed on biological resources, the economic value of resilience is very much a value of biological diversity. Resources in a low-diversity system may still attract high-use and nonuse values—it is a matter of perception. But low-diversity systems are unlikely to contain high resilience values, although the precise links between diversity and resilience are debated and occupy a substantial ecological literature.

Views differ as to whether the value of resilience is local or global, or, as seems more likely, both. Pearce and Perrings (1995) are of the view that the main consequences of diversity show up mainly in local losses rather than global losses. Resilience shifts the focus to the way ecosystems may change in the presence of stresses and shocks. These processes of change might not be linear. For example, a modest change may result in some dramatic effect rather than a modest one. The process of change is marked by discontinuities and potential irreversibilities. Equally, some major changes may have little effect on the system. Resilience measures the degree of shock or stress than the system can absorb before moving from one state to another very different one. Diversity, it is argued, stimulates resilience perhaps because individual species that are threatened or affected by change can have their roles taken over by other species in the same system. The smaller the array of species, the less chance there is of this substitution process taking place.

From an economic standpoint, the issue is one of identifying and measuring this insurance value. Unfortunately, neither is easy. Identifying how close a system might be to collapse of some or all functions is itself extremely difficult, yet one would expect a willingness to pay to avoid that collapse to be related in some way to the chances that the collapse will occur. If the chances are known, the value sought is then the premium that would be paid to conserve resilience. Suggestions include the entire cost of managing nonresilient systems because these costs could be avoided if more diverse and therefore more resilient systems are adopted. In the agricultural context, for example, this would make the premium equal to the entire costs of ensuring that intensive agriculture is maintained, including such things as fertilizer and pesticide costs. Inverting the process, it could be argued that the premium is approximated by the cost of all the losses incurred by maintaining a resilient system. If, as is widely suggested, diverse and resilient systems are lower-productivity systems, then the loss of productivity from maintaining a resilient system might be thought of as the economic value of resilience, i.e., as the resources that have to be sacrificed to maintain diversity.

VALUATION TECHNIQUES

How might B_C in equations 1 to 3 be estimated? Valuation techniques occupy a vast literature in environmental economics, and only a flavor is provided here. For detailed analyses see Bateman et al. (2002) and Garrod and Willis (1999). Revealed preference techniques seek to find the WTP for a nonmarketed good by looking at other markets influenced by the good. For example, it is well known that pollution affects property markets, so that the WTP to avoid pollution can be found by looking at the differentials in housing prices between polluted and nonpolluted areas. Multivariate techniques are used to isolate the various effects. Thus, although pollution appears to have no market—pollution is not directly bought and sold—the value of pollution of avoidance is revealed in a separate market. More relevant to biodiversity, values may be revealed by travel behavior. The distance to a recreational site can be translated into a cost of travel, and that cost can in turn be shown to reflect WTP. If the recreational site is desired because of its biodiversity, then an implicit price of biodiversity emerges. More relevant still to biodiversity are stated preference techniques. Akin to market research, these approaches use questionnaires that elicit attitudes to the good in question, socioeconomic features of the respondents, and answers to questions about willingness to pay. Substantial efforts are made to ensure that the answers are as truthful as possible (incentive compatibility) by making scenarios realistic, by varying questionnaire formats, and using different value-elicitation questions. Although still controversial, these approaches have the obvious merit of creating a hypothetical market. So long as the necessarily hypothetical answers can be shown to be calibrated on real WTP, e.g., by conducting both real money transactions and hypothetical questionnaires at the same time, any degree of exaggeration can be corrected. As far as biodiversity is concerned, there are many questionnaire studies of single species or habitats, but very little is known about how people value the diversity of ecosystems. This remains a research priority.

DISCOUNTING AGAIN

It was observed above that discounting could have a dramatic effect on the costs and benefits of any policy decision that affects biodiversity. The discount rate is an economic value, but one that reflects the preferences for individuals with respect to the timing of gains and losses. Recent work has questioned the need to have a constant discount rate (3 or 4%, for example) with respect to time. Rather, it is argued

that the discount rate should decline the further into the future we look (Weitzman, 1997). The lower the discount rate, the less reduction the practice of discounting has on future benefits and costs. Even if near-term benefits and costs are discounted at rates like 3 and 4%, long-term impacts would be discounted at low and very low rates. The discount rate r needs to be distinguished from the discount factor $1/(1 + r)^t$. It is the discount factor that gives the preference weight applied to each time period. Suppose the discount rate and hence the discount factor is not known with certainty and is a random variable. This seems more than reasonable the longer we look into the future. No one can tell what will happen over long periods of time. Suppose it takes the values 1 to 6% each with a probability of 0.167. Table 1 shows the relevant values. Although the weighted average (expected value) of the discount rate stays the same in all periods (3.5%), the discount factor obviously varies with time. The value of the implicit discount rate, r^*, is given by the equation

$$\frac{1}{(1 + r^*)^t} = \frac{\sum DF_{t,i}}{n}, i = n \qquad (4)$$

where n is the number of possible discount rates, DF is the discount factor, and t is time.

Table 1 shows that the implicit discount rate goes down over time even though the average discount rate stays the same for each period. The implication for equation 4 is that r needs to be replaced by r^* and r^* will vary with time so that instead of r, the equation would contain $r^*(t)$ where t is time. Note again that the effect on the decision about conservation could be substantial. The benefits of biodiversity conservation are likely to be long-lived, whereas development benefits could be limited to a few decades. Where previously the long-run benefits of conservation were ignored because discounting effectively reduces them to zero, the new argument provides a rationale for allowing for those long-term benefits.

Table 1. Values of the discount factor[a]

r	DF_{10}	DF_{50}	DF_{100}	DF_{200}
1	0.9053	0.6080	0.3697	0.1376
2	0.8203	0.3715	0.1380	0.0191
3	0.7441	0.2281	0.0520	0.0027
4	0.6756	0.1407	0.0198	0.0004
5	0.6139	0.0872	0.0076	0.0000
6	0.5584	0.0543	0.0029	0.0000
Sum	4.1376	1.4898	0.5900	0.1589
Sum/6	0.7196	0.2483	0.0983	0.0265
r^*	3.34%	2.82%	2.34%	1.83%

[a] DF_{10} is the discount factor for year 10, etc. r^* is the value of r that solves equation 4.

BIODIVERSITY, INFORMATION, AND IRREVERSIBILITY

A final consideration in the valuation of biodiversity concerns the value of biodiversity as information. As is well known, the vast majority of the world's species remain uninvestigated in any detail. Some entire habitats remain largely unexplored—the deep oceans and forest canopies being two major ones. The context for decision making is therefore one of uncertainty—we cannot know what benefits reside in biodiversity. The process of learning is one of generating information about those benefits. But in the biodiversity context there is a second feature of decision-making: irreversibility. If biodiversity continues to be lost at the current rates, then there is every prospect that whatever information resides in that biodiversity will itself be irretrievably lost. Irreversibility and uncertainty combine to provide a powerful rationale for slowing the rate of loss of bioidversity. What is missing is the economic value of delaying the losses until more and better information is gathered about the information content of biodiversity. This missing value is variously described as the quasi-option value in the environmental economics literature and option value in financial investment theory (somewhat confusingly because option value refers to something else in environmental economics).

Given that we do not have full information about the functions and services of ecosystems, that this uncertainty is greatest for the biodiversity, especially in unexplored systems, and that ecosystems and species are being lost at a rapid rate, we can ask what the benefits and costs would be of slowing down the rate of biodiversity loss. Quasi-option value refers to the value attached to delaying the loss of forest canopies so that information from the canopies can be generated: it is the value attached to the information gained from the delay. In a correctly specified model of decision making about biodiversity, these potential information gains should be identified and accounted for. Hence, separately estimating quasi-option value should not be required. Rather, the concept is there to remind us to specify the model correctly. But the reminder is often essential because the majority of economic studies do not in fact make allowance for learning over time. The learning may occur naturally and independently of the decision to conserve biodiversity (exogenous learning) or, probably more relevant, it will occur precisely because the delay enables targeted research to take place (endogenous learning). As learning occurs, the probabilities implicitly or explicitly attached to the damage that accrues by not conserving biodiversity (Bayesian learning) emerge.

ECONOMIC VALUATION OF BIODIVERSITY IN PRACTICE

Although it would be desirable to be able to illustrate all of the previous components of a rigorous valuation decision model in the context of biodiversity, it is not possible to do so. The reason is that, even though there are many studies of the economic value of wildlife and habitats such as forests and wetlands, none incorporates all of the elements identified, especially irreversibility, time-varying discount rates, and rising relative values. Above all, little or nothing is known about the value of diversity per se (as opposed to biological resources), especially the value of ecosystem resilience. Consider a further example. It is known that, over long periods of time, the output of agricultural systems oscillates—there are cycles of expanded and reduced activity. Less well known is the fact that the amplitude of these cycles itself increases over time (Anderson and Hazell, 1989). The implication is that the downswings will themselves increase in size, perhaps risking collapse of the relevant system. The consequences of collapsed agricultural systems would, of course, be immense. Although there is a complex of factors contributing to the variability in yields, one factor appears to be the increasing homogenization of agriculture. This refers not just to homogeneous seed varieties, but also to homogeneous farming methods, especially those introduced in the Green Revolution. Diverse systems (e.g., organically based systems) are known to be less productive, but less variable. The relevant trade-off then is the damage that a collapsed system would do, multiplied by some probability that it will occur, versus the lost output by adopting a low productivity but diverse system. Although it would seem imperative that such a calculation be carried out, it has not been done.

Economists have come late to the biodiversity issue and this may have advantages in that, in the meantime, they have developed valuation and analytical techniques that are relevant to the biodiversity problem. But it has also had a cost because the chances of real insights into the rationale for biodiversity conservation have not been exploited. This remains one of the most important items on the biodiversity research agenda.

REFERENCES

Anderson, I., and P. Hazell. 1989. *Variability in Grain Yields: Implications for Agricultural Research and Policy in Developing Countries.* Johns Hopkins University Press, Baltimore, Md.

Bateman, I., R. Carson, B. Day, M. Hanemann, N. Hanley, S. Mourato, E. Özdemiroglu, D. W. Pearce, and R. Sugden. 2002. *Economic Valuation with Stated Preference Techniques: A Manual.* Edward Elgar, Cheltenham, United Kingdom.

Beckerman, W., and J. Pasek. 2001. *Justice, Posterity and the Environment.* Oxford University Press, Oxford, United Kingdom.

Ehrenfield, D. 1988. Why put a value of biodiversity? p. 212–216. *In* E. O. Wilson (ed.), *Biodiversity.* National Academy Press, Washington, D.C.

Garrod, G., and K. Willis. 1999. *Economic Valuation of the Environment: Methods and Case Studies.* Edward Elgar, Cheltenham, United Kingdom.

Page, T. 1977. *Conservation and Economic Efficiency.* Johns Hopkins University Press, Baltimore, Md.

Pearce, D. W. 2001. The economic value of forest ecosystems. *Ecosyst. Health* 7(4):1–11.

Pearce, D. W. *The Distribution of Benefits and Costs of Environmental Policies: Conceptual Framework and Literature Survey.* Organisation for Economic Co-operation and Development, Paris, France, in press.

Pearce, D. W., and D. Ulph. 1999. A social discount rate for the United Kingdom, p. 268–285. *In* D. W. Pearce, *Economics and the Environment: Essays in Ecological Economics and Sustainable Development.* Edward Elgar, Cheltenham, United Kingdom.

Pearce, D. W., D. Moran, and D. Biller. 2002. *Handbook of Biodiversity Valuation: A Guide for Policy Makers.* Organisation for Economic Cooperation and Development, Paris, France.

Pearce, D. W., and C. Perrings. 1995. Biodiversity conservation and economic development: local and global dimensions, p. 23–40. *In* C. Perrings, K.-G. Mäler, C. Folke, C. Holling, and B.-O. Jansson (ed.), *Biodiversity Conservation.* Kluwer, Dordrecht, The Netherlands.

Randall, A. 1988. What mainstream economists have to say about the value of biodiversity, p. 217–223. *In* E. O. Wilson (ed.), *Biodiversity.* National Academy Press, Washington, D.C.

van Beers, C., and A. de Moor. 2001. *Public Subsidies and Policy Failures: How Subsidies Distort the Natural Environment, Equity and Trade, and How to Reform Them.* Edward Elgar, Cheltenham, United Kingdom.

Weitzman, M. 1998. Why the far distant future should be discounted at its lowest possible rate. *J. Environ. Econ. Manage.* 36:201–208.

SUBJECT INDEX

DNA shuffling, 247, 385
DNA-based typing methods, 35
DNA-DNA hybridization
    delineation of bacterial species, 33–35
    microbial identification, 50–51, 55
    species concept based on, 41–42, 47
DNase, 382
Docetaxel, 349–350
Doratomyces, 208
Dormancy, 100–105
Dot blot hybridization, quantitative, 92–93
Doxorubicin, 349, 356, 363
Drosophila, 195
Drug resistance, 274–275, 338–339, 347–348
Dunaliella, 150
Dust storms, 221
Dysidea, 182
Dysidea arenaria, 184
Dysidea avara, 183
Dysidea herbacea, 177

E
E-CELL project, 247
Echinocandins, 328–329, 336, 340, 346, 349
Ecome, 243
Economics, environmental, 469–475
Ecosystem functioning, microbial diversity and, 423–424
Ecosystem functions, 472
Ecosystem resilience, 472, 474
Ecosystem restoration, 205, 207
Ecteinascidia turbinata, 186–187
Ecteinascidins, 186–187
Ectoines, 151
Ectomycorrhizal fungi, 206–207
Edmonds Institute v. *Babbitt*, 452–453
Elaiophylin, 300, 302–303, 306
Elementary mode analysis, 282
Emiliana huxleyi, 217–218
Encephalitozoon cuniculi, 62
Endangered habitats, 416
Endemism, *see* Microbial endemism
Endophytes, 204
    bacterial, 206
    fungal, 18
    of leaves, 208
Endosymbiont theory, 57
Energy yield, genome size and, 47
Enfumafungins, 346, 349
Enhanced biological phosphorus removal, 400–402
Enrichment methods, 83–84
Entamoeba histolytica, 58
Enterobacter, 391, 394
Enterobacter cloacae, 395
Enterococci, vancomycin-resistant, 338
Environmental DNA, *see* DNA, environmental
Environmental economics, 469–475
Environmental genomics, *see* Metagenomic libraries
Environmental impact statement, 452–453
Environmental permits, 437

Enzymes
    chemotaxonomy, 291
    enzyme inhibitory assays, 330
    extremophiles, 4, 151–152
    industrial, 4, 151–152, 317–320, 375–390
        bioinformatics in enzyme discovery, 379
        bioprospecting, 376–381
        cloning from nonculturable microbes, 380–382
        cloning to obtain enzymes for testing, 378
        directed evolution, 383–386
        genome analysis for novel genes, 379–380
        manufacturers, 375
        metagenomics approach, 380–382
        molecular screening, 378–379
        rational protein engineering, 381–383, 386
        screening based on culturing of microbes, 377–378
        screening programs, 375–376
        worldwide use, 375
Ephemeral habitats, 416
Epigallocatechin gallate, 348
Epothilones, 349–350, 364–365
Epoxomicin, 368
Epulopiscium fishelonii, 256
Equation of well-being, 470–472
Erbstatin, 366
Ergokinin A, 349
Ergot, 356
Ergotamine, 345
Ericoid mycorrhiza, 207
Erwinia, 148
Erythromycin, 326, 336–337, 341
Erythromycin resistance, 274
Escherichia coli, 225, 296, 299, 305
    comparison with primate host species, 42
    environmental, 110
    genome sequence, 20, 256, 298
    genome size, 46
    in silico strains, 247
    phenomics, 283–286
    proteome database, 263–264
    specific affinity, 166
Escherichia coli O157:H7, 255
Escovopsis, 198
Esterases, 382, 453
Ethericins, 328
Ethnomedicine, 461
Ethylene, 391–392, 394
Euglenids, 62
Euglenozoa, 61–62
Eukaryotic microbes
    alveolates, 59–61
    biogeography, 226–227
    definition, 57–59
    diplomonads, 63–64
    diversity, 21–22
    early, 58
    euglenozoa, 61–62
    evolutionary relationships, 57–58
    foraminiferans, 63
    heterokonts, 61
    lineages, 59–64